# Shear Waves in Marine Sediments

# Shear Waves in Marine Sediments

edited by

Jens M. Hovem
SACLANT Undersea Research Centre,
La Spezia, Italy

Michael D. Richardson
Naval Oceanographic and Atmospheric Research Laboratory,
Stennis Space Center, NY, U.S.A.

and

Robert D. Stoll
Lamont-Doherty Geological Observatory,
Columbia University, Palisades, NY, U.S.A.



KLUWER ACADEMIC PUBLISHERS
DORDRECHT / BOSTON / LONDON

Proceedings of the Conference on
Shear Waves in Marine Sediments
La Spezia, Italy
October 15–19, 1990

**Library of Congress Cataloging-in-Publication Data**

Shear waves in marine sediments / edited by Jens M. Hovem, Michael D.
Richardson, Robert D. Stoll.
p. cm.
Papers from a conference held in La Spezia, Italy, Oct. 15-19, 1990.
Includes index.
ISBN 0-7923-1357-7 (alk. paper)
1. Marine sediments--Congresses. 2. Shear waves--Congresses.
I. Hovem, Jens M., 1937- . II. Richardson, M. D. (Michael D.)
III. Stoll, Robert D., 1931- .
GC377.S54 1992
551.46'083--dc20 91-22213
CIP

ISBN 0-7923-1357-7

Published by Kluwer Academic Publishers,
P.O. Box 17, 3300 AA Dordrecht, The Netherlands.

Kluwer Academic Publishers incorporates the publishing programmes of
D. Reidel, Martinus Nijhoff, Dr W. Junk and MTP Press.

Sold and distributed in the U.S.A. and Canada
by Kluwer Academic Publishers,
101 Philip Drive, Norwell, MA 02061, U.S.A.

In all other countries, sold and distributed
by Kluwer Academic Publishers Group,
P.O. Box 322, 3300 AH Dordrecht, The Netherlands.

*Printed on acid-free paper*



Printed in the Netherlands

## TABLE OF CONTENTS

## *PREFACE*

Shear waves and closely related interface waves (Rayleigh, Stoneley and Scholte) play an important role in many areas of engineering, geophysics and underwater acoustics. In some cases interest is focused on large-amplitude waves of low frequency such as those associated with earthquakes and nuclear explosions; in other cases low amplitude waves, which have often travelled great distances through the sediment, are of interest. Both low and high frequency shear and interface waves are often used for seafloor probing and sediment characterization. As a result of the wide spectrum of different interests, different disciplines have developed lines of research and a literature particularly suited to their own problems. For example water-column acousticians view the seafloor sediment as the lower boundary of their domain and are interested in shear and interface waves in the near bottom sediments mainly from the standpoint of how they influence absorption and reflection at this boundary. On the other hand, geophysicists seeking deep oil deposits are interested in the maximum penetration into the sediments and the tell-tale characteristics of the seismic waves that have encountered potential oil or gas bearing strata. In another area, geotechnical engineers use shear and interface waves to study soil properties necessary for the design and the siting of seafloor structures.

The groups mentioned above, as well as many others, have developed unique theoretical models, data sets, and methods of data analysis that are of mutual interest and yet the amount of interaction and communication between groups is often very limited. For these reasons, we organized the conference 'Shear Waves in Marine Sediments' which was held in La Spezia, Italy on 15–19 October 1990. One of the prime objectives of the conference was to establish a dialogue between these different groups and, hopefully, stimulate more interaction in future research. Based on the lively discussions that occured at the conference and the wide variety of papers that were presented, it is clear that each group has much to offer the others.

Because of the diversity of disciplines and topics, we divided the conference papers into three broad sessions. We have maintained that organization in this book:

I. Physical Properties and Laboratory Measurements.

II. Field Experiments and Interpretation.

III. Modelling and Inversion Techniques.

Many of the papers are broad enough to fit into more than one of these categories so that the final ordering in this volume is somewhat arbitrary; nevertheless, we have attempted to arrange the papers into groups which have similar emphasis.

Conference attendees were asked to submit papers suitable for publication at the end of the conference. Papers were reviewed by the editors, session chairpersons and other peers. All of the papers in this volume have been altered, some extensively, by the authors to reflect not only presentations and discussions during the conference but the reviewers comments. We hope the peer-review process will increase the value of this volume.

We would like to thank the following session chairpersons and outside reviewers for their contributions to this volume: H.B. Ali, D.L. Bibee, P.J. Bucca, S.A. Chin-Bing, J.F. Gettrust, D.L. Lavoie, M.M. Rowe (NOARL), J.F. Andrews (ONR Europe), G. Leger (Ocean Routes Canada), D.M.F. Chapman, F. Desharnais (DREA), A.M. Davis, D. Taylor Smith

(University College of North Wales), R.R. Goodman (University of Southern Mississippi), R.M. Holt (IKU), F.B. Jensen, M. Porter (SACLANTCEN), G. Mavko, A.M. Nur (Stanford University), R. Meissner (Institut für Geophysik), T.G. Muir (Applied Research Laboratories), D. Rauch (USEA), H. Schmidt (MIT), P.J. Schultheiss (SCHULTHEISS Geotek), G. Sills (Oxford University), M. Snoek (MEER), T. Yamamoto (RSMAS), and other reviewers who wish to remain anonymous.

We would like to acknowledge the financial support of the Naval Oceanographic and Atmospheric Research Laboratory (NOARL), the Office of Naval Research (ONR) in Washington and Europe and the SACLANTCEN.

We wish to thank SACLANTCEN staff members for their help in organizing the conference and managing the myriad details that were so important to its success. In particular, thanks are due to the conference secretaries, Ornella Toso and Rosanna Giglione.

Jens M. Hovem
Michael D. Richardson
Robert D. Stoll

*shear waves in marine sediments*
*15 - 19 oct. 1990*

*shear waves in marine sediments*
*15 - 19 oct. 1990*

shear waves in marine sediments
15 - 19 oct. 1990

# Section 1

## *Physical properties and laboratory measurements*

# SHEAR WAVES IN MARINE SEDIMENTS - BRIDGING THE GAP FROM THEORY TO FIELD APPLICATIONS

ROBERT D. STOLL
Lamont-Doherty Geological Observatory of Columbia University
Palisades, New York 10964

ABSTRACT. The relationship between primitive variables such as porosity and mean grain size and the velocity and attenuation of shear waves propagating in a marine sediment can be established on the basis of correlations with many field measurements or, more basically, may be derived from the physics governing the response of particulate material in a water environment. In this paper we discuss some of the progress that has been made along the latter path wherein porosity, permeability and other variables that may be measured in the lab or estimated from geological descriptions may be used to estimate seismo-acoustic properties via a mathematical model that reflects all the significant physics of the problem. This model, which may be termed "poro-viscoelastic", is a complex research tool that may be used to derive much simpler "fleet-oriented" models for making estimates of properties in selected environments. We first briefly discuss the fundamental model and then compare the results of predictions made using the model with field measurements at several well-documented sites.

## 1. Introduction

Water-column acousticians view the sea floor as a boundary to be reckoned with in any manner which will produce a reasonable acoustic field in their domain. Thus, various fluid models, elastic models and anelastic models have been proposed in an effort to emulate the bottom response over a wide frequency range for a variety of different geological settings. Since many of these models are too simplified to represent the true physical processes which govern the seismo-acoustic behavior, it has often been necessary to choose unrealistic material parameters (i. e., excessively high shear moduli or thin "boiler plate" layers) in order to obtain satisfactory agreement with the measured acoustic field in the water. While this expedient is sometimes necessary in the short term, a practical worldwide characterization of acoustic response requires that bottom models be chosen to reflect the mechanics of deformation of the sediments in such a manner that simple primitive parameters related to the the geological history of an area may be used to estimate the seismo-acoustic response of the bottom.



*J. M. Hovem et al. (eds.), Shear Waves in Marine Sediments*, 3–12.

## 2. Basic Model

The physical model which we have used to describe unconsolidated marine sediments is based on the classical theory of the mechanics of porous media developed by Maurice Biot. This theory considers a two component medium composed of a compressible fluid which fills the interstices of a compressible porous frame that has shear stiffness and interconnected void spaces. Thus, the average overall motion of the fluid may be partially out of phase with the motion of the skeletal frame and part of the total energy loss will be caused by relative motion between the fluid and frame. Additional energy losses result from "local" fluid motion associated with distortion of the void spaces during straining (i. e., squeeze film motion in the annuli near each intergranular contact). The effects of local motion are included in the model by modeling the skeletal frame as viscoelastic while the overall relative motion of the components is controlled by a viscodynamic operator which reflects the influence of permeability and tortuosity.

In addition to the energy losses resulting from fluid motion there is dissipation resulting from intergranular friction. During each cycle of strain associated with the propagation of a disturbance of small amplitude, there is energy dissipated at each grain-to-grain contact owing to a small amount of "slip" occurring <u>within</u> the tiny areas of contact. Even though this is an inherently nonlinear process, the loss per cycle of loading is small enough that the response may be considered as "slightly" nonlinear and the effect may be included approximately in the linear viscoelastic model used to model the skeletal frame.

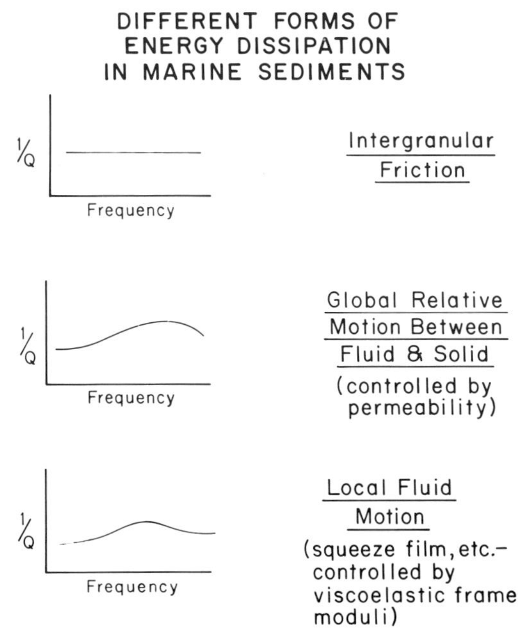


Fig. 1.
Effect of different forms of energy dissipation on the quality factor.

The isolated effect of each of the three loss mechanisms just described is shown graphically in Fig. 1, and the overall response of each different sediment depends on their relative dominance. For example in coarse, uniform sands of high permeability the overall motion of the fluid relative to the frame dominates at all but the lowest frequencies, whereas in fine-grained sediments local fluid losses and intergranular friction control the overall response at least in the low frequency-range. Fig. 2 shows the overall response predicted by the model for a marine silt including the effects of all three mechanisms of energy loss.

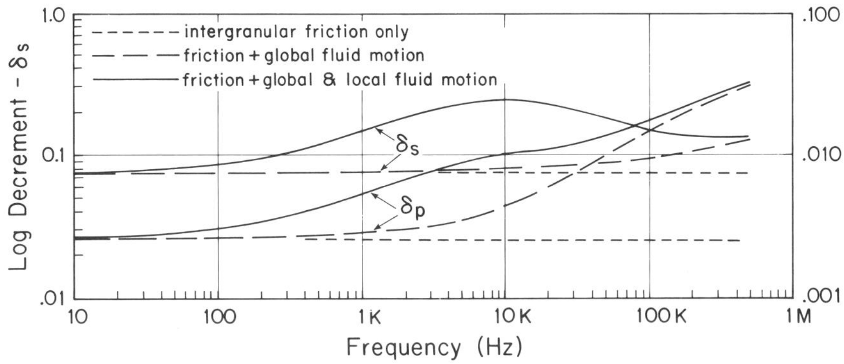


Fig. 2 Overall response of a marine silt based on the Biot theory. $\delta_s$ is log decrement for shear waves and $\delta_p$ is log decrement for dilatational waves of the "first kind". A description of the viscoelastic model used to model the effects of local fluid motion is given in Stoll (1989).

In the general case the Biot theory results in two pairs of coupled equations of motion, one pair governing the response of dilatational waves and the other the response of shear waves. For dilatational waves the equations are

$$\nabla^2(He - C\zeta) = \frac{\partial^2}{\partial t^2}(\rho e - \rho_f \zeta)$$

$$\nabla^2(Ce - M\zeta) = \frac{\partial^2}{\partial t^2}(\rho_f e - m\zeta) - \frac{\eta}{k}\frac{\partial \zeta}{\partial t}$$

where $e$ is the volumetric strain of an element of the skeletal frame and $\zeta$ is the increment of fluid that flows into or out of the element. $\rho$ is the total mass density of the sediment, $\rho_f$ is the mass density of the fluid and $m$ is an apparent mass, greater than $\rho_f/\beta$, that corresponds to the increment of fluid flow. $\beta$ is porosity, $\eta$ is viscosity and $k$ is permeability. The parameters $H$, $C$ and $M$ are complex moduli that depend on the compressibility of the skeletal frame, the fluid and the solid particles as well as the shear modulus of the skeletal frame. For a complete discussion of these equations and parameters see Stoll (1985) or Stoll (1989).

The above coupled equations of motion allow two kinds of dilatational wave to propagate. In waves of the "first kind" the fluid and skeletal frame move almost in phase, whereas in waves of the "second kind" they are nearly 180 degrees out of phase in the nature of a diffusion process. In a normal marine environment (without large amounts of gas in the water) the p-waves that are usually excited and observed are waves of the first kind.

In the original Biot theory, a second pair of coupled equations governs the propagation of shear waves allowing for only one kind of relative motion between fluid and frame (nearly in phase).

$$\mu \nabla^2 \vec{\omega} = \frac{\partial^2}{\partial t^2}(\rho \vec{\omega} - \rho_f \vec{\theta})$$

$$\eta / k \frac{\partial \vec{\theta}}{\partial t} = \frac{\partial^2}{\partial t^2}(\rho_f \vec{\omega} - m \vec{\theta})$$

where $\vec{\omega}$ is the rotation, $\vec{\theta}$ is the rotation of an increment of fluid and $\mu$ is the complex shearing modulus. In some recent work (Liu and Katsube, 1990), it has been suggested that a rotational wave of the "second kind", analogous to the dilatational wave of the second kind, can exist in a fluid-filled porous medium; however this kind of wave has yet to be observed.

In many cases the permeability of the sediment and the frequencies of interest are low enough that the increment of fluid content is negligibly small in the above equations. The equations of motion may then be simplified to the following forms which are equivalent to Gassmann's equations (Gassmann, 1951)

$$H \nabla^2 e = \frac{\partial^2}{\partial t^2} \rho e$$

$$\mu \nabla^2 \vec{\omega} = \frac{\partial^2}{\partial t^2} \rho \vec{\omega}$$

where $H - 4\mu/3$ is the "effective " bulk modulus of a saturated porous material in which the fluid is restricted against movement relative to the skeletal frame. In saturated sediments where the porosity remains constant during unjacketed loading, which is not a bad approximation in many cases, the modulus $H$ is given by the expression

$$H = \frac{(K_r - K_b)^2}{K_r(1 + \beta(K_r/K_f - 1)) - K_b} + K_b + 4\mu/3 \qquad (1)$$

where $K_b$, $K_r$ and $K_f$ are the bulk moduli of the skeletal frame, the individual rock particles and the fluid respectively. The bulk modulus of the skeletal frame is the inverse of the drained compressibility which is commonly measured in geotechnical testing. Eq. 1 is the keystone which we use to relate the the geoacoustic properties of the sediment to primitive geological variables such as porosity, grain-size distribution and geostatic stress since the bulk and shear moduli of the skeletal frame are directly related to these parameters.

In very soft sediments or suspensions when $K_b$ and $\mu$ become vanishingly small, we may recover Wood's equation for the compressibility of a water sediment mixture by letting $K_b = \mu = 0$ in the inverse of the expression for effective bulk modulus $H - 4\mu/3$ yielding

$$c_m = \frac{1}{K_r}(1 - \beta) + \frac{1}{K_f}\beta$$

where $c_m$ is the compressibility of the mixture (Wood, 1912).

From the above it is clear that the Biot theory is a natural extension of the earlier work of Wood and Gassmann which has been used with considerable success in geoacoustic modeling. Unfortunately, as the theory has become more comprehensive the number of parameters that must be evaluated has increased significantly (to as many as 16 for the full implementation of the theory). Hence there is a heavy burden on the user to supply meaningful parameters in an environment where nonhomogeneity is the rule rather than the exception. Thus, in many cases the main benefit from the general theory may be its help in choosing which parameters are significant and which may be neglected in designing simplified models.

## 3. Model Parameters

There are three groups of basic physical parameters that must be evaluated in order to use the general theory. The first is a group of frequency-independent variables that are fairly easy to obtain from standard tables and other compilations. These include the mass density and the compressibility of the individual sediment grains and the pore fluid.

The second group includes those variables that affect the global fluid motion such as permeability, viscosity of the pore fluid, pore size parameter, and a structure factor which determines the magnitude of the apparent mass density, $m$. These parameters were used by Biot to define a "viscodynamic " operator based on a simple model composed of interstitial channels of one size, corrected for the the fact that not all of the flow is in the direction of the average pressure gradient. More recently other investigators have proposed alternate formulations of "dynamic permeability" that account for the complicated pore geometry in real sediments. Johnson, Koplik and Dashen (1987) defined a new parameter related to pore geometry which depends on the ratio of surface area to pore volume. Yamamoto and Turgut (1988) modelled the pore space using a bundle of capillaries with a distribution of radii related to the pore-size distribution. While both of these variants are appealing in that they attempt to model more realistically the extremely complicated pore geometry in real sediments, they both require additional quantitative information that may be difficult to obtain or even estimate in many practical applications.

Finally, the third group includes the porosity and those variables controlling the frequency dependent response of the skeletal frame. We consider the skeletal frame to be anelastic and therefore two independent complex moduli are necessary to describe the response. For example, in the equations of motion given above, the moduli $H$ and $\mu$ are required. Since we are concentrating on unconsolidated sediments occurring near the seafloor where there is little or no lithification, it is appropriate to visualize the skeletal frame of the sediments as a particulate material. Hence, based on theoretical studies of the mechanics of granular media (e.g. see Stoll, 1989, Chap. 3), the moduli of the skeletal frame depend to a large extent on the number of intergranular contacts per particle and the area of each of these contacts. The primitive variable most important in determining both of these quantities is the porosity as a function of depth in the seafloor since it determines, for a given grain size distribution, the permissible packing geometries and the geostatic stress level at a given depth.

There have been many laboratory studies designed to relate the dynamic shear modulus, $\mu$, to mean effective stress, voids ratio or porosity, cyclic strain level, time since deposition, and other factors. These tests have shown that, for very small cyclic shear strain levels, the real part of the shear modulus depends primarily on the voids ratio, $e$ and the mean effective stress, $\sigma_0'$. Bryan and Stoll (1988) recently proposed an empirical equation based on a regression analysis of nearly 500 test results on a variety of different sediments ranging from sands to clay. Their equation is

$$\mu / p_a = a \cdot \exp(b \cdot e) \cdot (\sigma_0' / p_a)^n \qquad (2)$$

where $p_a$, the atmospheric pressure, is introduced to nondimensionalize the variables. The regression analysis yielded the following values for the three parameters $a$, b and n

$a = 2526$ b= -.15 n = .45

with a correlation coefficient of 0.971. Other similar formulas have been given by Hardin and Richart (1963) and Hardin and Black (1968). All of these formulas must be used within the range of voids ratio and pressure consistent with the data set from which they were derived. In the data used to derive Eq. 2, $e$ varied from about 0.4 to 1.5 and $\sigma_0'$ ranged from 14 to about 700 kPa. Moreover, it is important to remember that the above equation represents the results of laboratory experiments and it is well known that other factors such as sample disturbance and thixotropic effects due to aging must be taken into account before direct predictions of *in situ* response can be made. For example, field measurements (e. g. see Arango, Moriwoki and Brown (1978), Anderson, Espana and McLamore (1978), and Stoll (1989)) suggest that while b and n may not change significantly, the parameter $a$ may have to be increased by a factor in the range of 1.3 to 2.5 in order for the formula to be valid for *in situ* conditions. Much of our work over the past six years has been aimed at measuring the *in situ* dynamic response of sediments at sites where the geology was well known from prior borings and other geophysical studies so that the usefulness of the above equation and the most appropriate values of the parameter $a$ could be determined. In the remainder of this paper we describe some of the field experiments and the results obtained to date.

## 4. Field Measurements

Our field experiments have utilized an impulsive source and a geophone array, both located on the seafloor (Stoll, 1991). This configuration is very efficient in generating both interface waves and wide-angle refractions of both shear and p-waves. A new signal source powered by electrically detonated seismic shells (eight-gauge blank shotgun shells) has been used to focus energy into the bottom and eliminate much of the water-borne noise that accompanies the usual explosive sources. In most of the experiments a string of 24 gimballed geophones was placed on the bottom and moved along a line toward the source in order to produce stacked traveltime data of up to 72 equally spaced ranges. Examples of typical data are shown in Fig. 3

To date we have obtained data at over two dozen different sites in sediments ranging from soft organic clays where the shear-wave speed is as low as 30 m/s to stiff overconsolidated glacial sediments where the shear-wave velocity is several hundred m/s near the seafloor. In choosing sites we have not used a regional basis (i. e., Atlantic coast, Pacific coast, etc.) but rather have searched for regions where the "ground truth" is available and where certain typical geological histories and near-bottom sediment types are known to exist. We have used this approach because one of our objectives is to accumulate a library of sediment models that represent most of the "typical" conditions that exist *in situ*. In each case the goal has been to derive a bottom model by inverting the field data and then comparing the result with the predictions of the theoretical model described above, with particular emphasis on determining an appropriate value for the parameter $a$ in Eq. 1.

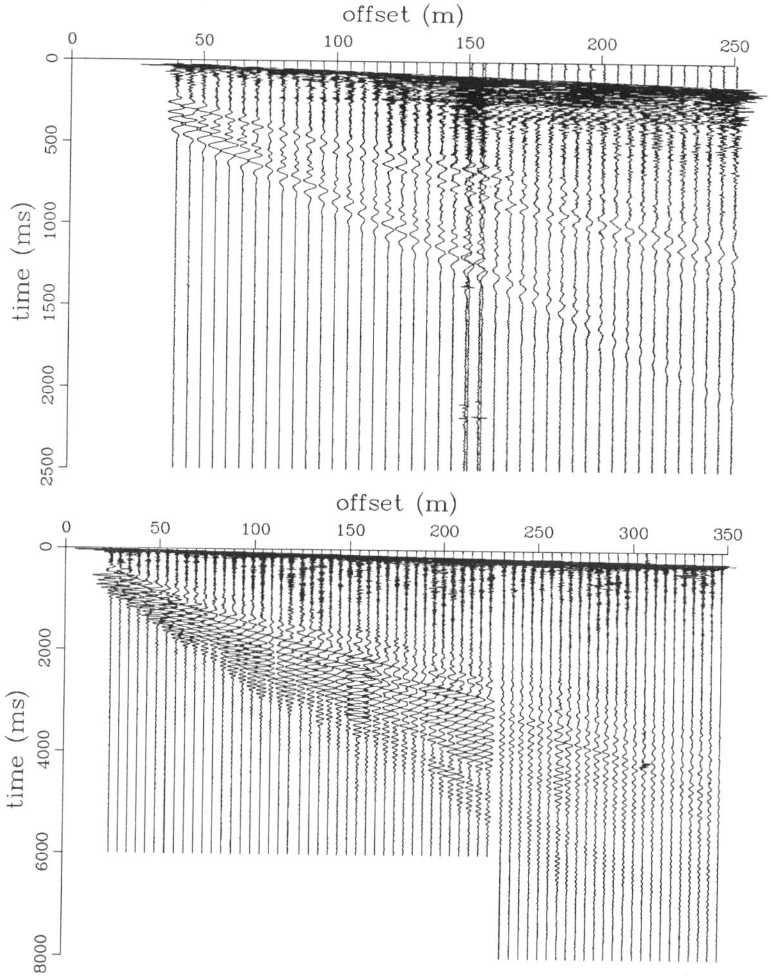


Fig. 3. Examples of stacked traveltime data for impulsive source and string of gimballed geophones on the seafloor. The top set, which shows a clearly defined interface wave and a refracted shear wave, is from a layered site off the New Jersey shore (Atlantic Generator site) and the bottom set is from the Gulf of La Spezia near Mariperman buoy.

Data of the kind shown in Fig. 3 may be analyzed in a number of different ways. In most cases the portion of the data corresponding to the Scholte wave will be dispersive due to the rapid increase in velocity with depth that results from the changing overburden stress or the layering that is generally present in terrigenous, shelf sediments. As a result, the multiple filter method of Dziewonski, Block and Landisman (1969) is often very effective in inverting the data to study the structure very near the bottom. This method produces a group velocity dispersion curve which must be matched by any model that is assigned to the bottom. We have developed a semiautomatic, iterative method of deriving a layered bottom model which uses partial derivatives of the interface-wave velocity with respect to the shear-wave velocity of the various layers (Stoll, 1991 and Ciati, Akal and Stoll, 1991). Given reasonable choices of layer thickness and starting velocities, this method results in a model that will replicate the group velocity dispersion curve; however the modeler is still faced with the question of uniqueness and whether a local minimum has been reached. It is therefore good practice to test the resulting model by calculating synthetic seismograms using a full-wave program such as SAFARI (e. g., see Schmidt and Tango, 1986).

In addition to using the interface waves to study the structure very near the bottom, the first arrival of refracted or diving shear and dilatational waves may be used to study the acoustic properties to greater depths. Fig. 4 is a schematic diagram that depicts the various techniques that we have used to derive geoacoustic information from data such as that shown in Fig. 3.

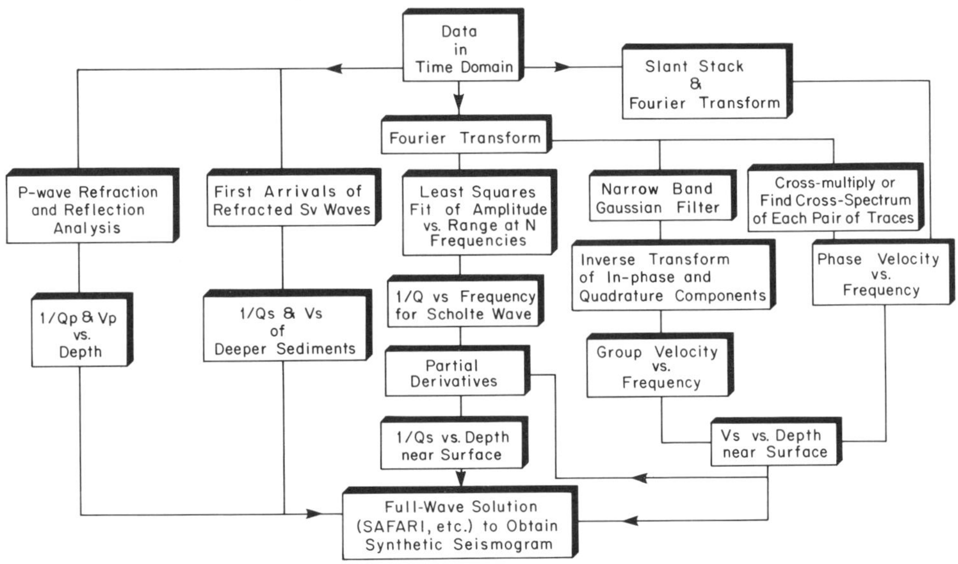


Fig. 4. Schematic showing various methods of analysis that have been adapted to study shear and interface waves. (Stoll, 1991)

## 5. Summary of Results

We are still in the process of analyzing the more than two dozen data sets in our library; however, the work completed so far suggests that the porosity, as determined by the type of sediment and the geologic history of a region, is the primative variable most significant in predicting the *in situ* geoacoustic response. Given the porosity as a function of depth and a reasonable estimate of the density of sediment particles, it is possible to calculate the overburden pressure from the equation

$$p_0(z) = \int_0^z \{(1-\beta)\cdot(G_s - G_w)\cdot\gamma_w\}\,dz$$

where $G_s$ and $G_w$ are the specific gravity of the sediment grains and the water respectively and $\gamma_w$ is the unit weight of water. The mean effective stress, $\sigma_0'$, which is necessary to evaluate Eq. 1, may then be estimated using

$$\sigma_0' = (1 + 2K_0)/3 \cdot p_0$$

where $K_0$ is the coefficient of earth pressure "at rest" which is normally in the range of .4 to .7 unless the sediment is highly overconsolidated or very soft.

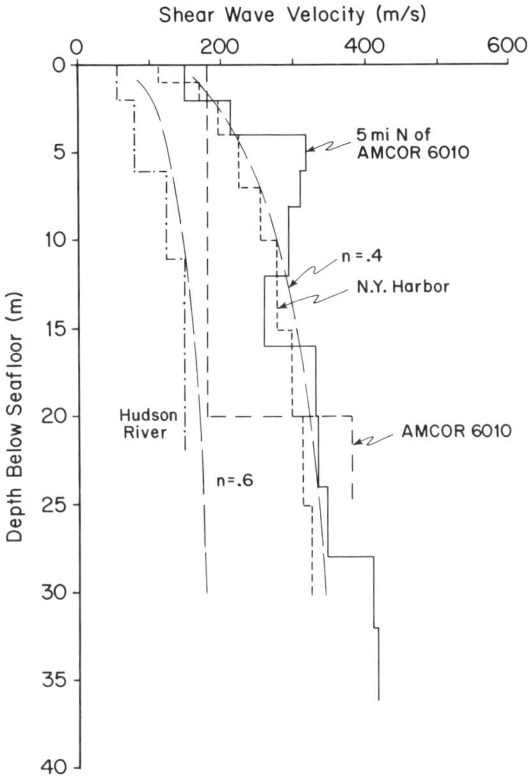


Fig. 5.
Shear wave velocity versus depth for several different sites. Smooth dashed lines are velocities predicted by Eq. 2 for two different constant porosities.

Fig. 5 shows a number of models derived from data of the sort shown in Fig. 3. In addition two curves calculated using Eq. 2 are shown for reference; one curve is for a constant porosity of 0.4 and the other for a porosity of 0.6. The curve labelled "Hudson River" is based on data obtained near the Tappan Zee bridge where there is a deep layer of clayey silt which decreases in porosity rather quickly with depth. Near the bottom the porosity averages about 65% (see Stoll, 1989). Data that resulted in the curve labeled "New York Harbor" was taken at a site near Coney Island, New York where there is a thick layer of sand. The other two models were obtained in deeper water (65-70m) on the continental shelf off the the east coast of the U. S. near the location of a boring made during the Atlantic Margin Coring program in 1969. As can be seen from this figure, velocities in the upper 30 m can vary from 50 or so m/s to well over 400 m/s depending on the geological origin. Moreover, by comparing with the computed curves, it can be seen that there is a general tendency for the velocity to increase with depth in a manner similar to what is predicted by Eq. 1 once the effects of distinct layering are averaged out.

Our current objective is to complete the analysis of all of the field data that is currently in the library and compare the predictions of the theory with the resulting models on a case-by-case basis. Only then can the applicability and full usefulness of the model be judged in a meaningful way.

## 5. Acknowledgements

The experiments and analysis described in this paper were sponsored by the Office of Naval Research, Code 1125OA under contracts N00014-87-K-0204 and N00014-89-J-1152. Lamont-Doherty Geological Observatory Contribution No. 4692.

## 6. References

Anderson, D. G., Espana, C. and McLamore, V. R. (1978) "Estimating in situ shear velocity and moduli at competent sites," Proc. Specialty Conf. on Earthquake Engrg. and Soil Dynamics, ASCE, New York, 181-197.

Aranga, I., Moriwoki, Y. and Brown, F. (1978) "In situ shear velocity and modulus," Proc. Specialty Conf. on Earthquake Engrg. and Soil Dynamics, ASCE, New York, 198-212.

Bryan, G. M. and Stoll, R. D. (1988) "The Dynamic Shear Modulus of Marine Sediments, "J. Acoust. Soc. Am. 83, 2159-2164.

Caiti, A., Akal, T. and Stoll, R. D. (1991) "Determination of shear velocity profiles by inversion of interface wave data," in this volume.

Dziewonski, A., Bloch, S. and Landisman, M. (1969) "A technique for analysis of transient seismic signals, "Bull. Seism. Soc. Am. 59, 427-444.

Gassmann, F. (1951) "Uber die elastizitat poroser medien," Bierteljahrsch. Naturforsch. Ges. 96, 1-23

Hardin, B. O. and Richart, F. E. Jr. (1963) "Elastic wave velocities in granular soils," J. Soil Mech. Found. Div., ASCE, 89, 33-65.

Hardin, B. O. and Black, W. L. (1968) "Vibration modulus of normally consolidated clay," J. Soil Mech. Found. Div., ASCE, 94, 353-369.

Johnson, D. L., Koplik, J., and Dashen, R. (1987) "Theory of dynamic permeability and tortuosity in fluid-saturated porous media," J. Fluid Mech. 176, 379.

Liu, Q. and Noriko, K. (1990) "The discovery of a second kind of rotational wave in a fluid-filled porous material," J. Acoust. Soc. Am. 88, 1045-1053.

Schimdt, H. and Tango, G. (1986) "Efficient global matrix approach to the computation of synthetic seismograms," Geophys. J. R. Astr. Soc. 84, 331-359.

Stoll, R. D. (1986) "Acoustic waves in marine sediments," in T. Akal and M. Berkson (eds), Ocean Seismo-Acoustics, Plenum Press, New York, pp. 417-434.

Stoll, R. D. (1989) Sediment Acoustics, Springer-Verlag, New York.

Stoll, R. D., Bryan, G. M., Mithal, R. and Flood, R. (1991) "Field Experiments to study seafloor seismo-acoustic response," J. Acoust. Soc. Am. 89, in press.

Wood, A. B. (1911) "A textbook of sound," G. Bell and Sons, London.

Yamamoto, T. and Turgut, A. (1988) "Acoustic wave propagation through porous media with arbitrary pore size distribution," J. Acoust. Soc. Am. 83, 1744-1751.

# SURFACE WAVES IN PORO-VISCOELASTIC MARINE SEDIMENTS

CHARLES W. HOLLAND
Planning Systems Incorporated
7925 Westpark Drive
McLean, Virginia 22102
U.S.A.

**ABSTRACT.** Marine sediments can be treated as a two-phase porous medium, consisting of a solid phase (the sediment grains) and a fluid phase (the interstitial pore fluid). The Biot theory of acoustic wave propagation in such a medium is used to derive the dispersion relation for surface waves at a fluid--poro-viscoelastic boundary. The frequency dependence of the surface wave speeds and attenuations, are calculated numerically for a sand, silty clay, and sandstone. These results are compared with the same quantities calculated with the usual approximation of a marine sediment as a single-phase, viscoelastic solid. The viscoelastic model yields a reasonable approximation to the surface wave velocity but a poor approximation to the attenuation, with respect to the poro-viscoelastic model.

## 1. Introduction

Various researchers have investigated related surface waves, however, it appears that the general case of surface waves at a fluid-poro-viscoelastic boundary (with complex wave speeds) has not been addressed. Deresewicz (1964) derived the secular equation without accounting for losses due to friction at the grain-to-grain contacts and Feng and Johnson (1983) considered the high frequency limit to the Biot (1962) theory where the viscous attenuation is zero.

In this study, the upper halfspace is considered to be an ideal fluid and the lower halfspace is considered to be a poro-viscoelastic medium using the Stoll (1974) form of Biot theory that accounts for viscoelastic effects in the sediment frame. The boundary conditions are stated for the "open pore" condition such that the sediment interstitial fluid and the fluid in the upper halfspace are in direct contact.

Biot theory requires the specification of 13 geophysical parameters. These parameters and their effect on the Type I, Type II, and shear wave velocity has been covered extensively in the literature (e.g., Stoll (1974) and Ogushwitz (1985)). Representative parameters used in this study for the three sediment types are given in Table 1.

The comparison of surface wave velocity and attenuation between the viscoelastic model and the poro-viscoelastic model is performed for three sediment types: a silty clay, a carbonate sand, and a sandstone. These sediment types were chosen as representative of common sediment types and because they represent distinct relationships between the ratio of their velocities to that of the overlying bottom water. For the silty clay, the Type I and shear wave velocity are both less than that of the upper halfspace; for the carbonate sand, the Type I wave velocity is greater and the shear velocity less than that of the upper halfspace; and for the sandstone, the Type I wave and shear velocity are both greater than that of the upper halfspace. The Type II velocity is less than that of the upper halfspace for all sediment types.



*J. M. Hovem et al. (eds.), Shear Waves in Marine Sediments*, 13–20.

**Table 1. Material Properties for Three Marine Sediments**

| Property | Units | Sand | Silty-Clay | Sandstone |
|---|---|---|---|---|
| Fluid density | $g/cm^3$ | 1.024 | 1.024 | 1.024 |
| Fluid bulk modulus | $dyne/cm^2$ | $2.38\ 10^{10}$ | $2.38\ 10^{10}$ | $2.38\ 10^{10}$ |
| Fluid viscosity | poise | .0101 | .0101 | .0101 |
| Grain density | $g/cm^3$ | 2.85 | 2.68 | 2.65 |
| Grain bulk modulus | $dyne/cm^2$ | $4.0\ 10^{11}$ | $3.5\ 10^{11}$ | $3.7\ 10^{11}$ |
| Porosity | --- | .487 | .68 | .203 |
| Permeability | $cm^2$ | $4.26\ 10^{-7}$ | $5.2\ 10^{-10}$ | $1.07\ 10^{-7}$ |
| Structure factor | --- | 1.75 | 3.0 | 2.0 |
| Pore size | cm | $4.18\ 10^{-3}$ | $1.25\ 10^{-4}$ | $2.5\ 10^{-4}$ |
| Frame bulk modulus | $dyne/cm^2$ | $3.33\ 10^{8}+ i6.67\ 10^{6}$ | $3.67\ 10^{7}+ i1.2\ 10^{6}$ | $8.32\ 10^{10}+ i4.15\ 10^{8}$ |
| Frame shear modulus | $dyne/cm^2$ | $2.5\ 10^{8}+ i3.75\ 10^{6}$ | $7.86\ 10^{6}+ i2.5\ 10^{5}$ | $8.53\ 10^{10}+ i4.25\ 10^{8}$ |

## 2. Theory

When solving the field integral for the response of a source above the seafloor, the poles of the plane wave (reflection and transmission) coefficients give rise to contributions from surface waves. The secular equation for the surface waves can be obtained by setting the denominator of the plane wave coefficients to zero and solving for the horizontal wavenumber ($k_x$).

The relation between the plane wave reflection coefficient and the three transmission coefficients (for the Type I , Type II and shear wave ) as a function of the fluid and sediment properties can be expressed as a set of four linear complex equations. The secular equation from Cramer's rule is:

$$\mathrm{Det}\begin{bmatrix} \alpha_o & \alpha_p(1-G_p) & \alpha_2(1-G_2) & -k_x(1-G_s) \\ \rho_o & \dfrac{CG_p - H}{V_p^2}+\dfrac{2\mu k_x^2}{\omega^2} & \dfrac{CG_2 - H}{V_2^2}+\dfrac{2\mu k_x^2}{\omega^2} & \dfrac{2k_x\alpha_s\mu}{\omega^2} \\ 0 & \alpha_p & \alpha_2 & \dfrac{\omega^2}{2V_s^2 k_x}-k_x \\ -\rho_o & \dfrac{C-MG_p}{V_p^2} & \dfrac{C-MG_2}{V_2^2} & 0 \end{bmatrix} = 0 \tag{1}$$

where the surface wave velocity, $V_{surf} = \omega/k_x$. The complex bulk wave speeds are denoted by $V_i$, the vertical wavenumbers by $\alpha_i$, and the ratios of the interstitial fluid-to-solid frame displacement potential coefficients by $G_i$ (i=p, 2, s). The subscripts "p", "2", and "s" correspond to the sediment Type I, Type II and shear wave, respectively. The subscript "o" refers to parameters in the upper halfspace. The complex moduli H, C, and

M account for the viscoelastic losses in the sediment frame and are a function of the frame shear modulus ($\mu$), and the fluid, grain and frame bulk moduli (see Stoll (1974)).

The secular equation for a single-phase viscoelastic solid is given by:

$$\alpha_o \frac{\rho_1}{\rho_o}\left[4k_x{}^2\,\alpha_p\,\alpha_s + \left(\frac{\omega^2}{V_s{}^2} - 2k_x{}^2\right)\right] + \alpha_p \frac{\omega^4}{V_s{}^4} = 0 \qquad (2)$$

which can be obtained from Eq 1 in the limiting case of the porosity, permeability and interstitial fluid properties approaching zero. The sediment bulk density is $\rho_1$ and the density of the fluid is $\rho_o$.

## 3. Numerical Results

The solution of Eqs 1 and 2 requires finding complex roots of a transcendental equation. While many methods exist for the real-zero problem only a handful have been investigated for complex zeros. The most robust method for this particular problem seemed to be the secant method extended to the complex plane. Even using this relatively simple method, the evaluation of the zeros in Eq 1 required a very high degree of numerical precision. This is primarily due to the disparate values of the material parameters that comprise a marine sediment. The difference between the smallest and largest parameter typically spans 15 to 20 orders of magnitude (e.g., see Table 1).

Three kinds of surface waves can occur in the solution: a true surface wave that travels slower than all the wave speeds, a pseudo-Scholte wave that travels between the shear wave velocity and the Type II wave velocity (that leaks energy to the slower Type II wave), and a pseudo-Rayleigh wave that occurs when the surface wave velocity exceeds the velocity of the overlying water column.

Mathematically, the pseudo-Rayleigh wave arises from a pole on the lower Reimann sheet where $\mathrm{Im}\{\alpha_i\} < 0$ (i=p,s) and $\mathrm{Im}\{\alpha_2\} > 0$. The true surface wave is located on the top Reimann sheet and the amplitude of the surface wave decays exponentially away from the boundary.

### 3.1 CARBONATE SAND $V_p > V_o$, $V_s < V_o$

The surface wave velocity as a function of frequency is plotted in Figure 1 along with the phase velocity of the Type II and shear wave. Note that the surface wave changes continuously from a pseudo-Scholte wave to a true surface wave at about 700 Hz. Also plotted in Figure 1 is the surface wave velocity in the single-phase viscoelastic solid approximation, where the complex phase velocities used to calculate the surface wave velocity were taken from Biot theory at 400 kHz (a common frequency for measuring sediment geoacoustic properties). The single phase approximation to the surface wave velocity is roughly 10% higher than the poro-viscoelastic predictions over this frequency band. The attenuation of the poro-viscoelastic surface wave shows a similar frequency dependence to the shear wave (Figure 2) and the single-phase viscoelastic solid approximation to the surface wave attenuation is in general, too conservative.

### 3.2 SILTY CLAY $V_p < V_o$, $V_s < V_o$

The surface wave velocity for the silty clay changes continuously from a pseudo-Scholte wave to a true surface wave at 170 kHz (see Figure 3). The velocity dispersion of the surface wave is small such that the surface wave velocity under the single-phase viscoelastic solid approximation is quite good for this case. The single-phase viscoelastic approximation to the attenuation (Figure 4), however, is quite poor.

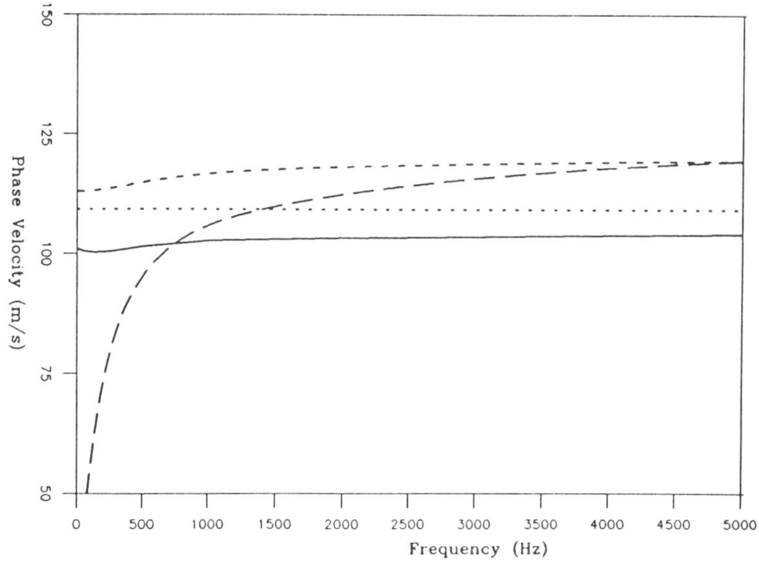


Figure 1. Phase velocity of various wave types in carbonate sand: pseudo-Scholte wave at a fluid-poro-viscoelastic boundary (———); pseudo-Scholte wave at a fluid-viscoelastic boundary (·······); shear wave in a poro-viscoelastic sediment (------); and Type II wave in a poro-viscoelastic sediment (— — —).

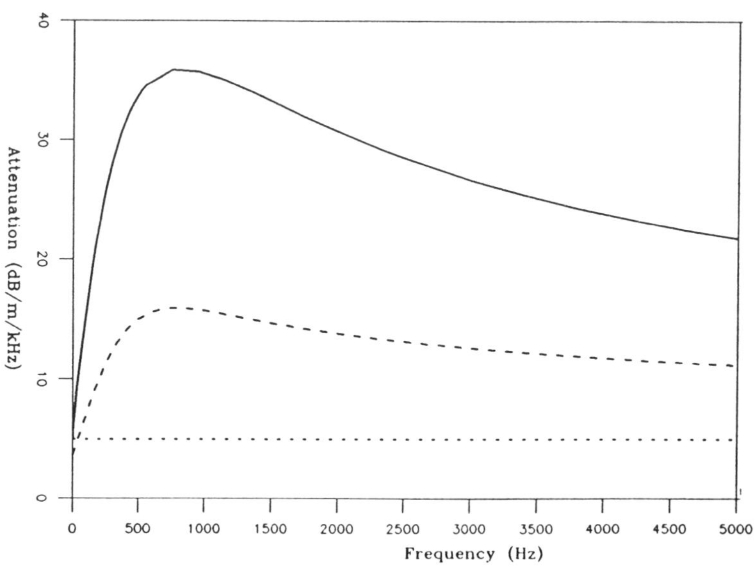


Figure 2. Attenuation of various wave types in carbonate sand: pseudo-Scholte wave at a fluid-poro-viscoelastic boundary (——); pseudo-Scholte wave at a fluid-viscoelastic boundary (·······); and shear wave in a poro-viscoelastic sediment (--------).

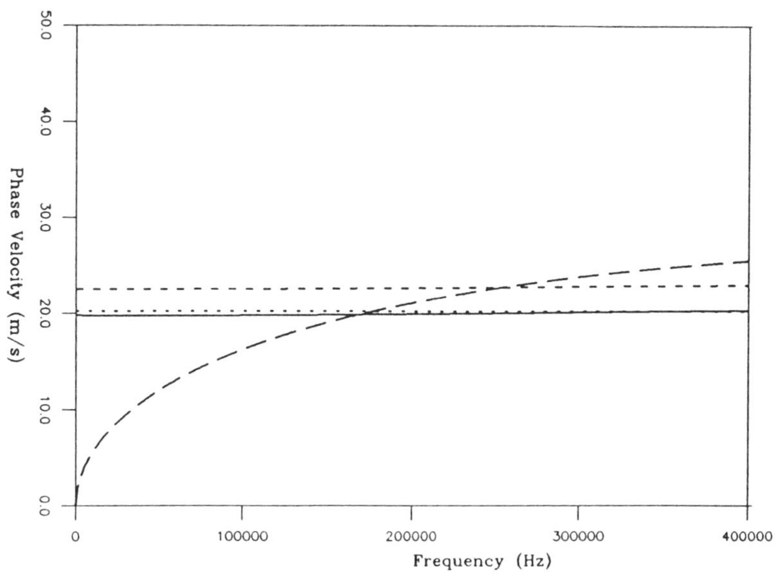


Figure 3. Phase velocity of various wave types in silty clay: pseudo-Scholte wave at a fluid-poro-viscoelastic boundary (———); pseudo-Scholte wave at a fluid-viscoelastic boundary (·······); shear wave in a poro-viscoelastic sediment (------); and Type II wave in a poro-viscoelastic sediment (— — —).

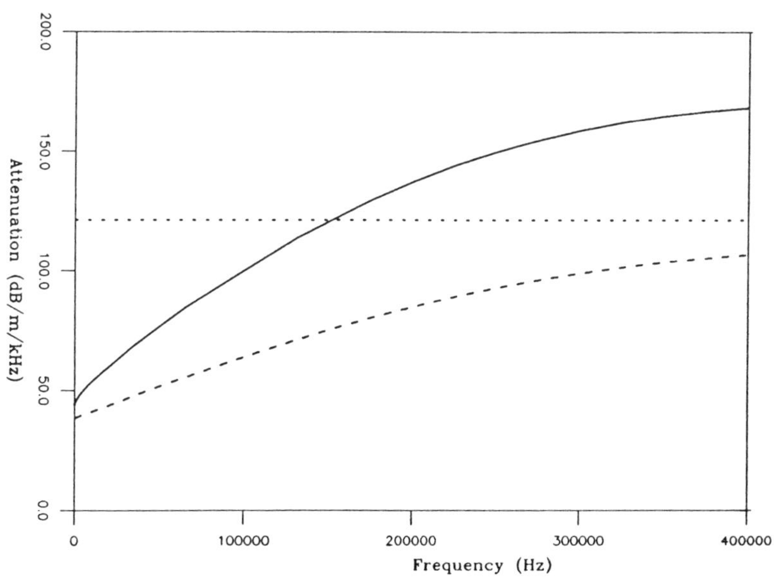


Figure 4. Attenuation of various wave types in silty clay: pseudo-Scholte wave at a fluid-poro-viscoelastic boundary (——); pseudo-Scholte wave at a fluid viscoelastic boundary (·········); and shear wave in a poro-viscoelastic sediment (------).

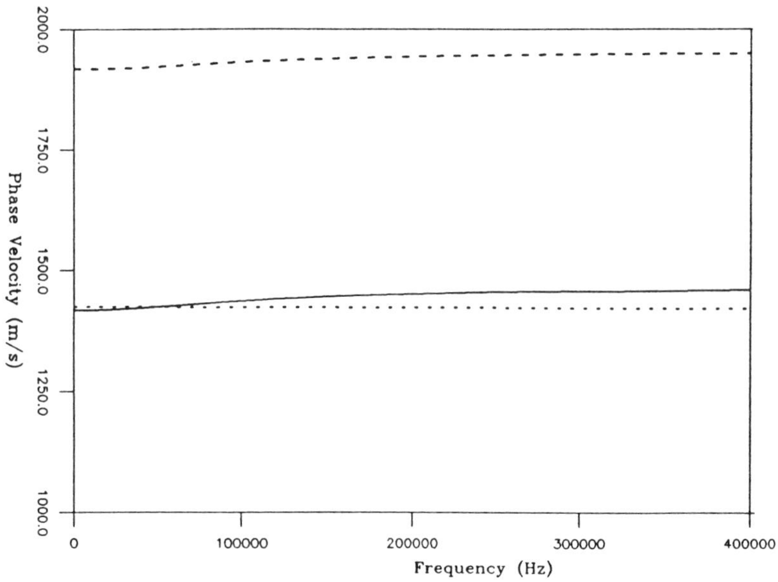


Figure 5. Phase velocity of various wave types in sandstone: pseudo-Scholte wave at a fluid-poro-viscoelastic boundary (———); pseudo-Scholte wave at a fluid-viscoelastic boundary (··········); and shear wave in a poro-viscoelastic sediment (---------).

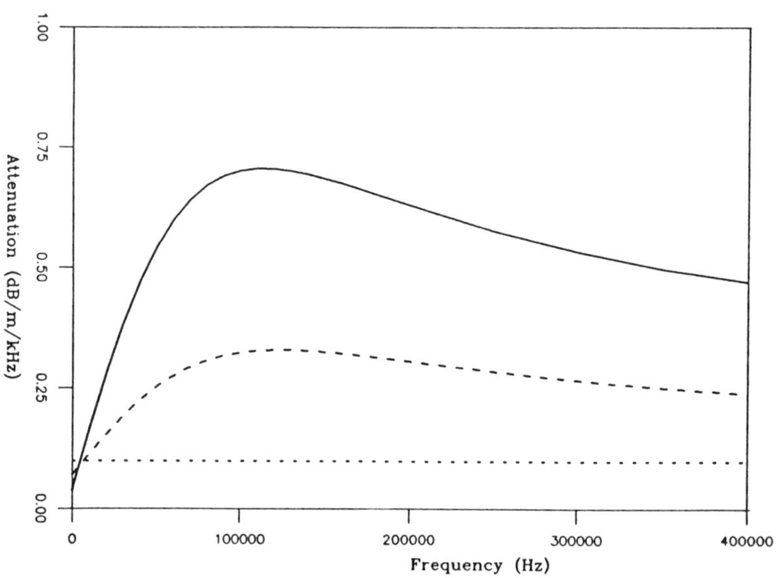


Figure 6. Attenuation of various wave types in sandstone: pseudo-Scholte wave at a fluid-poro-viscoelastic boundary (———); pseudo-Scholte wave at a fluid viscoelastic boundary (··········); and shear wave in a poro-viscoelastic sediment (---------).

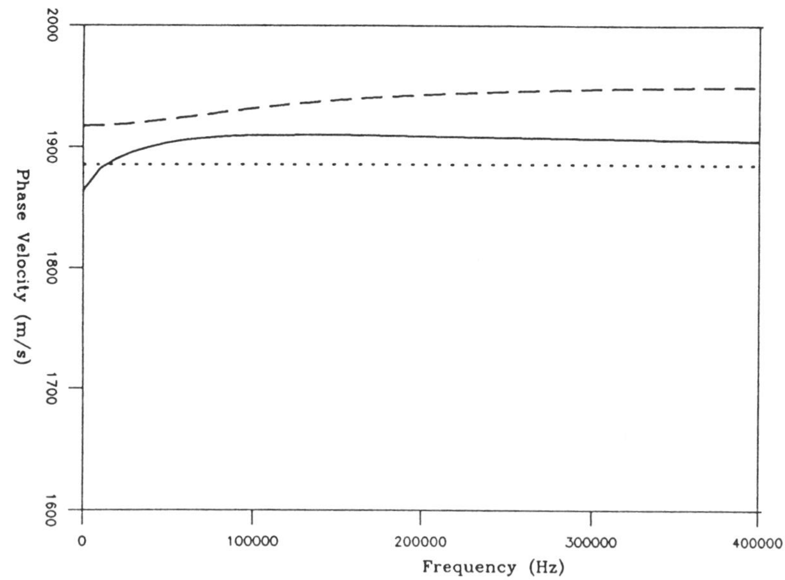


Figure 7. Phase velocity of various wave types in sandstone: pseudo-Rayleigh wave at a fluid-poro-viscoelastic boundary (———); pseudo-Rayleigh wave at a fluid-viscoelastic boundary (............); and shear wave in a poro-viscoelastic sediment (— — —).

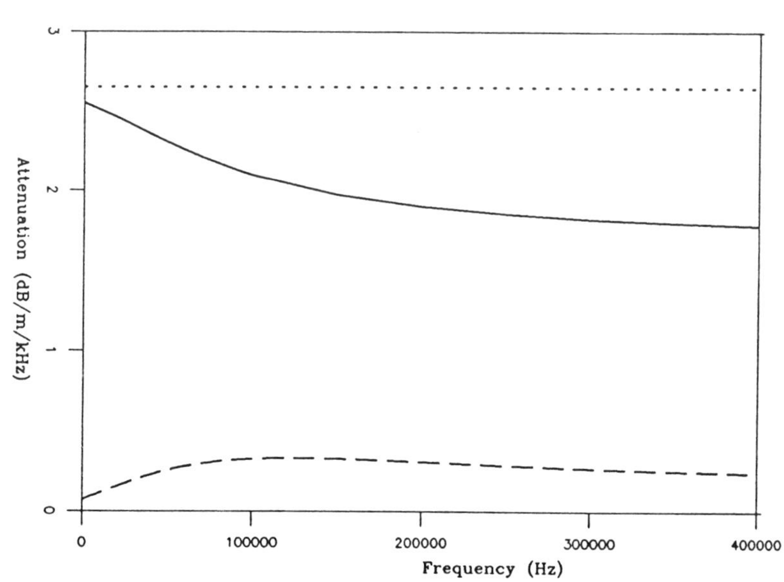


Figure 8. Attenuation of various wave types in sandstone: pseudo-Rayleigh wave at a fluid-poro-viscoelastic boundary (———); pseudo-Rayleigh wave at a fluid viscoelastic boundary (............); and shear wave in a poro-viscoelastic sediment (— — —).

### 3.3 SANDSTONE $V_p > V_o, V_s > V_o$

For this sediment, both the pseudo-Scholte wave and the pseudo-Rayleigh wave exist. For both the pseudo-Scholte (Figures 5 and 6) and the pseudo-Rayleigh (Figures 7 and 8) surface wave types, the single-phase viscoelastic approximation is reasonable for the velocity but quite poor for the attenuation.

For a pulse, Roever et al. (1959) demonstrate experimentally and theoretically that the zero-crossing of a pseudo-Rayleigh wave travels along the boundary at precisely the Rayleigh wave velocity; that is independent of the properties of the upper halfspace. The zero crossing of a pseudo-Rayleigh wave excited by a time harmonic source, however, does travel at $V_{surf}$ (i.e., is influenced by the properties of the upper halfspace). Since this study considers the response of the boundary of a time harmonic source, Figures 7 and 8 represent solutions to the secular equation accounting for the properties of the upper halfspace.

## 4. Summary and Conclusions

Surface wave velocities and attenuations predicted by Biot theory for porous sediments were compared with results which assume (as is commonly done for marine sediments) that the sediments are a single-phase viscoelastic solid. The Biot theory predicts slightly dispersive surface wave velocities and attenuations that are highly non-linear functions of frequency at a fluid-sediment interface. The viscoelastic model yields a reasonable approximation to the surface wave velocity but a poor approximation to the attenuation, with respect to the poro-viscoelastic model. Thus, the full poro-viscoelastic model may be required in predictive modeling of surface wave propagation and inversion of surface wave data.

## 5. References

Biot, M.A. (1962) 'Generalized Theory of Acoustic Propagation in Porous Dissipative Media,' J. Acoust. Soc. Am. 34, 1254-1264.

Deresiewicz, H. (1964) 'The Effect of Boundaries on Wave Propagation in a Liquid-Filled Porous Solid: VII. Surface Waves in the Presence of a Liquid Layer,' Bull. Siesm. Soc. Am. 54, 425-430.

Feng, S. and Johnson, D.C. (1983) 'High-Frequency Acoustic Properties of a Fluid/Porous Solid Interface. I. New Surface Mode,' J. Acoust. So. Am. 74, 906-914.

Ogushwitz, P.R. (1985) 'Applicability of the Biot Theory. I. Low-porosity Materials,' J. Acoust. Soc. Am. 77, 429-440.

Roever, W.L., Vining, T.F., and E. Strick (1959) 'Propagation of Elastic Wave Motion from an Impulsive Source Along a Fluid/Solid Interface,' Philos. Trans. R. Soc. London, Ser A 251, 455-523.

Stoll, R.D. (1974) 'Acoustic Waves in Saturated Sediments,' in Physics of Sound in Marine Sediments, ed. by L. Hampton, Plenum Press.

# AN INVESTIGATION OF CAUSALITY FOR BIOT MODELS BY USING KRAMERS-KRÖNIG RELATIONS

A. TURGUT
*University of Miami, RSMAS*
*4600 Rickenbacker Cswy.*
*Miami, Florida 33149, U.S.A.*

ABSTRACT. Kramers-Krönig integrals for the index of refraction are evaluated numerically to establish causality requirements between phase velocity and attenuation. Accuracy of numerical integrations is checked using a causal Biot (poro-elastic) model. A poro-viscoelastic Biot model is developed for acoustic wave propagation in porous media, and it is demonstrated that the new model satisfies the causality principle. Non-causality of a Biot model with a slightly inelastic frame assumption is analyzed for different types of marine sediments. Acoustic pulse propagation in marine sediments is also formulated for poro-viscoelastic and slightly nonlinear Biot models, and resulting pulse shapes are compared.

## 1. Introduction

Attenuation of acoustic/seismic waves in earth materials is a complex phenomenon and is still not well understood. Several linear and nonlinear attenuation mechanisms have been suggested to model acoustic/seismic wave propagation. Observation of constant energy loss per unit volume per cycle over a large frequency range suggests a nonlinear mechanism. However, agreement between pulse measurements and single frequency measurements implies that the principle of superposition is applicable and the attenuation mechanism must be linear. In a pulse experiment, linearity of the attenuation mechanisms can be investigated by using Kramers-Krönig relations between phase velocity and attenuation. In this paper, causality of three Biot models each having the same type of viscous losses but different types of frame losses (no frame loss, viscoelastic frame loss, and nonlinear friction loss) are examined by using these fundamental relations. Applicability and validity of using a 'slightly nonlinear' attenuation mechanism with a linear theory of wave propagation is also investigated.

## 2. Poro-viscoelastic Biot model

A complete theory for acoustic wave propagation in porous media was developed by Biot and presented in a series of papers. In his early papers [1,2], Biot treated the skeletal frame as an elastic medium and introduced viscous damping of acoustic waves due to motion of the pore fluid relative to the skeletal frame. Stoll [3] and Yamamoto [4] assumed a constant complex-shear and bulk moduli of the skeletal frame to treat the inelasticity of the frame.



*J. M. Hovem et al. (eds.), Shear Waves in Marine Sediments*, 21–28.

In this section, a poro-viscoelastic Biot model is developed by following Kjartansson's [5] derivation for the treatment of the inelasticity of the frame.

Biot's equations of motion with frequency dependent elastic moduli can be expressed as [2]

$$\mu\nabla^2\mathbf{u} + (H-\mu)\nabla\, e - C\nabla\,\zeta = \frac{\partial^2}{\partial t^2}\left(\rho\mathbf{u} + \rho_f\mathbf{w}\right), \tag{1}$$

$$C\nabla\, e - M\nabla\zeta = \frac{\partial^2}{\partial t^2}\left(\rho_f\mathbf{u} + m\mathbf{w}\right) + \rho_f\frac{\nu_{os}}{k_s}\frac{\partial\mathbf{w}}{\partial t}. \tag{2}$$

where $\mathbf{u}$ is the frame displacement vector, $\mathbf{w}$ is the seepage displacement vector, $e = \operatorname{div}\mathbf{u}$, $\zeta = \operatorname{div}\mathbf{w}$, $\rho$ is the bulk density, $[\rho = (1-\beta)\rho_r + \beta\rho_f]$, $\rho_r$ is the density of the grain, $\rho_f$ is the density of pore fluid, $\nu_{os}$ is the frequency-dependent kinematic viscosity, $k_s$ is the hydraulic coefficient of permeability, and $\mu$ is the shear modulus. Biot's elastic moduli $H, C$, and $M$ can be related to the bulk modulus of the grain $K_r$, the apparent bulk modulus of the pore fluid $K_f$, the bulk modulus of the skeletal frame $K_b$, and the shear modulus $\mu$ as [3]

$$H = (K_r - K_b)^2/(D_r - K_b) + K_b + \frac{4}{3}\mu, \tag{3}$$

$$C = K_r(K_r - K_b)/(D_r - K_b), \tag{4}$$

$$M = K_r^2/(D_r - K_b), \tag{5}$$

where $D_r = K_r[1 + \beta(K_r/K_f - 1)]$. For a poro-elastic Biot model, $\mu$, $K_b$, $K_r$ and $K_f$ are real constants. For a slightly inelastic Biot model, due to the assumption of slightly imperfect elasticity, $\mu$ and $K_b$ are complex constants. For a poro-viscoelastic Biot model, $\mu$ and $K_b$ are complex and frequency dependent. Frequency-dependent shear modulus $\mu(\omega)$ can be written as [5]

$$\mu(\omega) = \mu_0\left(i\omega/\omega_0\right)^{\frac{2}{\pi}\tan^{-1}(1/Q)}, \tag{6}$$

where $Q^{-1}$ is the specific attenuation and $\mu_0$ is the shear modulus at the arbitrary frequency $\omega_0$. Similarly, the frequency-dependent bulk modulus of the skeletal frame can be defined as

$$K_b(\omega) = K_{b_0}\left(i\omega/\omega_0\right)^{\frac{2}{\pi}\tan^{-1}(1/Q)}. \tag{7}$$

Then, frequency-dependent Biot elastic moduli $H(\omega)$, $C(\omega)$, and $M(\omega)$ can be calculated from Eqs.3, 4, and 5.

## 3. Kramers-Krönig relations for index of refraction

The causality principle can be formulated using a pair of integral equations (Kramers-Krönig relations) which connect phase velocity and attenuation. Kramers-Krönig relations for index of refraction $n(\omega)$ are used to verify if these three models satisfy the causality principle. These relations are given as [6]

$$n_r(\omega_1) = 1 + \frac{\omega_1}{\pi}P\int_{-\infty}^{\infty}\frac{d\omega\, n_i(\omega)}{(\omega-\omega_1)\omega}, \tag{8}$$

and

$$n_i(\omega_1) = -\frac{1}{\pi}P\int_{-\infty}^{\infty}\frac{d\omega\, n_r(\omega)}{\omega-\omega_1}, \tag{9}$$

where $P$ denotes principal value and $n_r$ and $n_i$ are the real and imaginary parts of the index of refraction $n(\omega)$. The index of refraction is defined in terms of phase velocity and attenuation as

$$n(\omega) = \frac{c_0}{\omega}\left(\frac{1}{c(\omega)} + i\alpha(\omega)\right), \tag{10}$$

where $c_0$ is the nondispersive low frequency limit of phase velocity and $\alpha$ is the attenuation coefficient. The interval of integrations in Eqs.8 and 9 can be reduced to a physical range of positive frequencies by using the crossing symmetry relation, $n_i(\omega) = -n_i(-\omega)$,

$$n_r(\omega_1) = 1 + \frac{2\omega_1^2}{\pi} P \int_0^\infty \frac{d\omega n_i(\omega)}{\omega(\omega - \omega_1)(\omega + \omega_1)}, \tag{11}$$

$$n_i(\omega_1) = -\frac{2\omega_1}{\pi} P \int_0^\infty \frac{d\omega n_r(\omega)}{(\omega - \omega_1)(\omega + \omega_1)}. \tag{12}$$

The above integrals are Cauchy principal value integrals and several numerical schemes are available to treat their singularities (e.g. Kutt, 1975 [7]). An IMSL subroutine is available to evaluate the integrals of the form

$$I(y) = \int_a^b \frac{f(x)}{x - y} dx, \tag{13}$$

which utilizes Clenshaw-Curtis and Gauss-Kronrod formulas with an extrapolation procedure [8].

Also, defining $f(\omega) = n(\omega) - n(\infty)$, Eqs.8 and 9 can be written as

$$f_r(\omega_1) = \frac{1}{\pi} P \int_{-\infty}^{+\infty} \frac{d\omega f_i(\omega)}{\omega - \omega_1}, \text{ and } f_i(\omega_1) = -\frac{1}{\pi} P \int_{-\infty}^{+\infty} \frac{d\omega f_r(\omega)}{\omega - \omega_1}, \tag{14}$$

which is the Hilbert transform pair. Three Fast Fourier Transform (FFT) methods have been given in Liu and Kosloff [9] for computing the Hilbert transform pair numerically. All three methods produce some numerical errors which can be reduced as sampling interval $\Delta\omega \to 0$ and number of sample points $N \to \infty$. In this paper, the simplest of these three methods is used to numerically calculate $f_r(\omega)$ from $f_i(\omega)$ by considering

$$F_r(\Omega) = \begin{cases} -iF_i(\Omega) & \text{for } \Omega > 0 \\ 0 & \text{for } \Omega = 0 \\ +iF_i(\Omega) & \text{for } \Omega < 0, \end{cases} \tag{15}$$

where $F_r(\Omega)$ and $F_i(\Omega)$ are the Fourier transforms of $f_r(\Omega)$ and $f_i(\Omega)$, respectively.

Bode [10] has used Kramers-Krönig relations for the response of an electrical amplifier and demonstrated that the phase shift of an electrical amplifier at any frequency is approximately related to the local rate of change of gain with frequency. The approximation is sufficiently accurate if both the gain and the phase shift are slowly varying over a limited frequency range of interest. O'Donnell *et al.* [11] derived a local approximate relation between real and imaginary parts of compressibility $K(\omega)$ for ultrasonic applications. Following O'Donnell *et al.* and making use of Eq.11, a similar approximate relation can be derived between real and imaginary parts of the index of refraction as

$$n_i(\omega) \approx -\frac{\pi\omega}{2} \frac{dn_r(\omega)}{d\omega}. \tag{16}$$

Introducing the intrinsic attenuation as

$$Q^{-1}(\omega) \approx \frac{2\alpha(\omega)c(\omega)}{\omega}, \tag{17}$$

the index of refraction in Eq.10 can be rewritten as

$$n(\omega) \approx \frac{c_0}{c(\omega)} \left(1 + i\frac{Q^{-1}(\omega)}{2}\right). \tag{18}$$

Then, Eq.16 becomes

$$\frac{Q^{-1}(\omega)}{c(\omega)} = -\pi\omega \frac{d}{d\omega}\left(\frac{1}{c(\omega)}\right) = \frac{\pi\omega}{c^2(\omega)} \frac{dc(\omega)}{d\omega}, \tag{19}$$

or

$$\frac{dc(\omega)}{c(\omega)} = \frac{Q^{-1}(\omega)}{\pi\omega} d\omega. \tag{20}$$

The solution of the above equation is

$$c(\omega) = c_0 \exp\left(\frac{1}{\pi}\int_{\omega_0}^{\omega} \frac{Q^{-1}(\omega)}{\omega} d\omega\right). \tag{21}$$

## 4. Numerical examples

In Fig.1, compressional and shear velocities for different types of sediments (see Table I) are compared with those calculated from attenuation using the direct integration scheme (Eq.11) and the approximate relation (Eq.21). Biot's poro-elastic model (no frame loss) is used to test these two equations. Biot's treatment of viscous losses is a causal relaxation phenomenon and velocities are expected to be identical to those calculated from attenuation. For given a low frequency limit of phase velocity $c_0$, the approximate relation (Eq.21) accurately predicts the phase velocity $c_\infty$ at the nondispersive high frequency limit. However, there is small discrepancy in the dispersion region where the attenuation and velocity variations with frequency are pronounced. In Fig.2, compressional and shear velocities are compared with those calculated from attenuation using the direct integration scheme for Biot slightly inelastic and Biot poro-viscoelastic models. In the slightly inelastic Biot model, compressional wave velocities calculated from attenuation introduce more discrepancy for sediments with high shear rigidity (i.e. consolidated sediments and porous rocks). There is 15% discrepancy at 1 kHz between shear waves and those calculated from attenuation for all types of sediments. This discrepancy increases with frequency, having the same velocity $c_0$ at the low frequency limit. Fig.2a,b shows that proposed poro-viscoelastic model with a constant frame inelasticity satisfies the causality principle for compressional and shear waves.

In Fig.3, the Hilbert transforms are used to calculate compressional wave speed from the attenuation (imaginary part of the index of refraction) for coarse sand. During the calculation, Eqs. 14 and 15 are numerically evaluated by using 65,536 discrete point FFT. There is still some discrepancy between actual and calculated speeds even for the causal

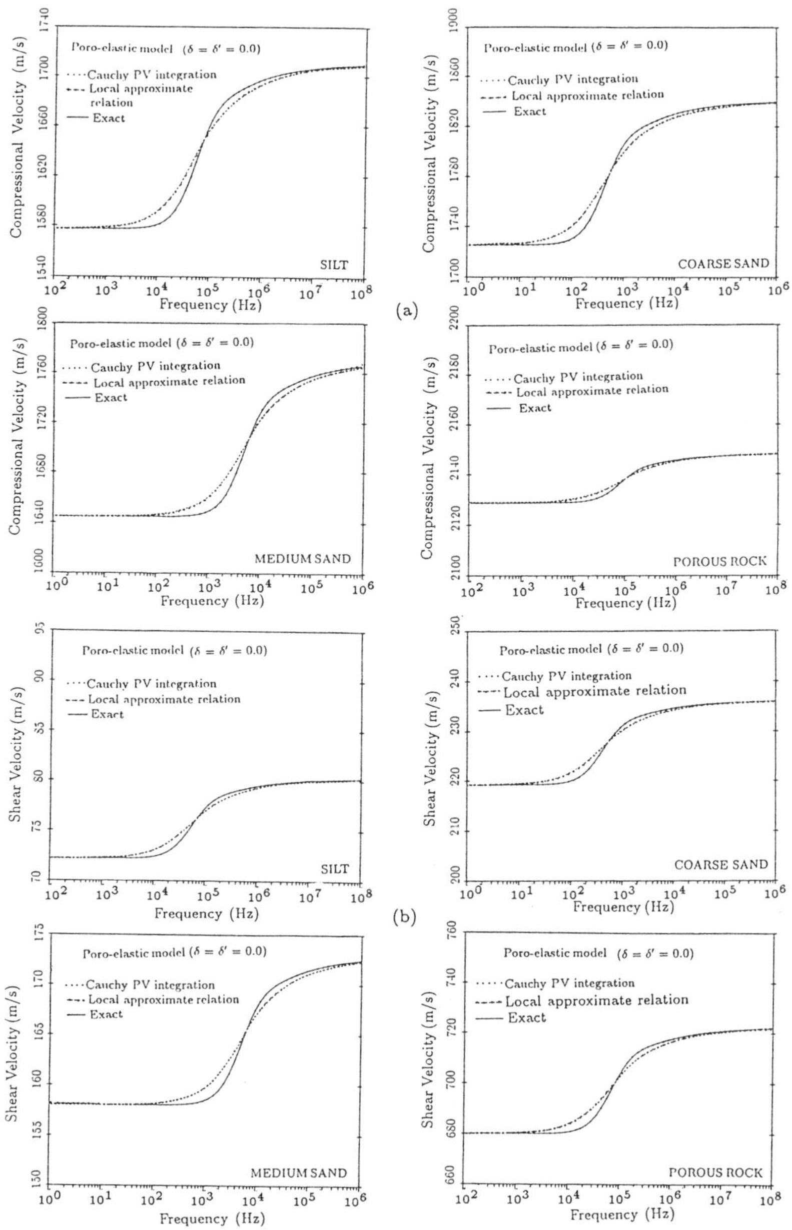


Figure 1: Comparison of a)fast compressional wave and b)shear wave velocities with those calculated from exact Kramers-Krönig relations and local approximate relation for poro-elastic Biot model.

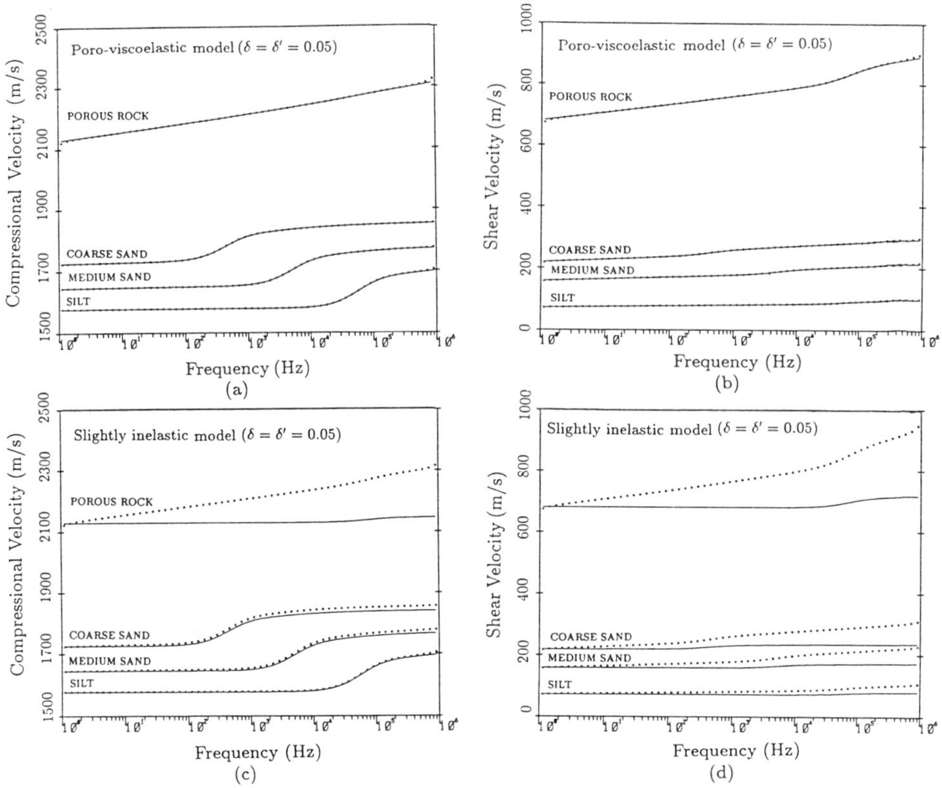


Figure 2: Comparison of fast compressional wave velocities and shear wave velocities (solid lines) with those calculated from attenuation (dots) using exact Kramers-Krönig relations for poro-viscoelastic and slightly inelastic Biot models.

models (Fig.3a,b). This is due to the band limitation of calculations and the error in defining the wave speed at infinity.

Finally, a 400 Hz Ricker wavelet propagating in porous rock and silt is shown in Fig.4 for compressional and shear wave modes. The spectrum of this wavelet is defined as

$$N(\omega) = \omega^2 e^{(-\omega^2/\omega_0^2)}, \tag{22}$$

where $\omega_0$ is the peak angular frequency. The difference between a causal and a non-causal pulse propagation is minor for silt, which supports our conclusion stating that sediments with less rigidity introduce less discrepancy in compressional wave propagation.

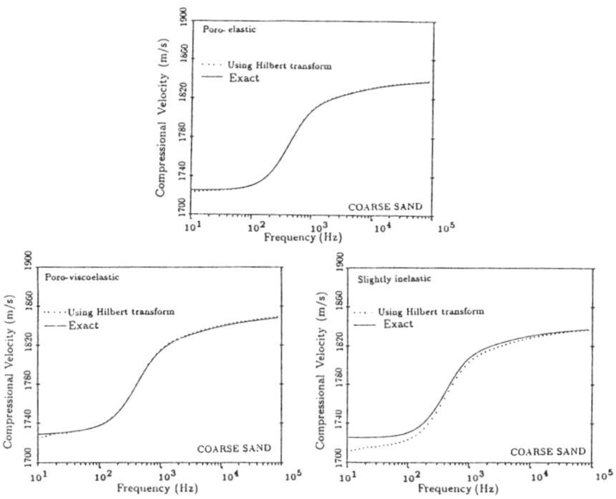


Figure 3: Comparison of fast compressional-wave velocities (solid lines) with those calculated from attenuation (dots) using the Hilbert transform for a)poro-elastic b)poro-viscoelastic, and c)slightly inelastic Biot models.

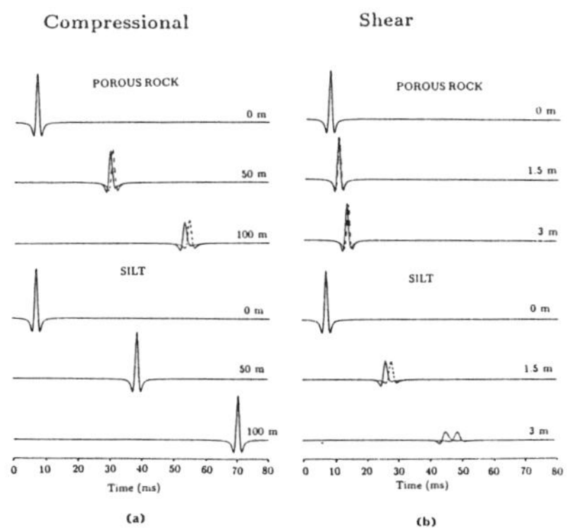


Figure 4: Causal (solid line) and non-causal (dashed line) propagation of Ricker wavelet ($f_0$ = 400 Hz) in porous rock and silt with compressional and shear modes.

Table I: Values of physical properties of sediments used in the calculations.

| Physical properties | Symbols (units) | Silt | Medium sand | Coarse sand | Porous rock |
|---|---|---|---|---|---|
| Dynamic viscosity of pore fluid | $\eta$ (Pa.s) | 1.0x10$^{-3}$ | 1.0x10$^{-3}$ | 1.0x10$^{-3}$ | 1.0x10$^{-3}$ |
| Permeability | $k_s$ (m$^2$) | 1.0x10$^{-12}$ | 1.0x10$^{-11}$ | 1.0x10$^{-10}$ | 5.0x10$^{-13}$ |
| Porosity | $\beta$ | .45 | 0.4 | 0.35 | 0.3 |
| Added mass coefficient of skeletal frame | $\alpha$ | 0.25 | 0.25 | 0.25 | 0.25 |
| Dynamic shear modulus | $\mu_0$ (N/m$^2$) | 1.0x10$^7$ | 5.0x10$^7$ | 1.0x10$^8$ | 1.0x10$^9$ |
| Bulk modulus of skeletal frame | $K_{b_0}$ (N/m$^2$) | 2.0x10$^7$ | 1.0x10$^8$ | 2.0x10$^8$ | 2.0x10$^9$ |
| Specific loss in the frame | $\delta, \delta'$ | 0.05 | 0.05 | 0.05 | 0.05 |
| Bulk modulus of fluid | $K_f$ (N/m$^2$) | 2.3x10$^9$ | 2.3x10$^9$ | 2.3x10$^9$ | 2.3x10$^9$ |
| Bulk modulus of grain | $K_r$ (N/m$^2$) | 3.6x10$^{10}$ | 3.6x10$^{10}$ | 3.6x10$^{10}$ | 3.6x10$^{10}$ |
| Density of fluid | $\rho_f$ (kg/m$^3$) | 1.0x10$^3$ | 1.0x10$^3$ | 1.0x10$^3$ | 1.0x10$^3$ |
| Density of grain | $\rho_r$ (kg/m$^3$) | 2.65x10$^3$ | 2.65x10$^3$ | 2.65x10$^3$ | 2.65x10$^3$ |

## 5. Conclusions

Several numerical methods are introduced to evaluate Kramers-Krönig integrals. Direct evaluation of these principal value integrals seems to be more accurate and efficient than Hilbert transform method. A local approximate relation is also given for the calculation of the phase velocity from the specific attenuation.

Using the Kramers-Krönig relations, it is shown that a newly developed poro-viscoelastic Biot model satisfies the causality principle and that it is physically consistent. It is also shown that a Biot model with slightly nonlinear frame loss should be used in the linear theory of wave propagation with some limitations. Non-causality of such a model is more apparent for the shear wave propagation than for the compressional wave propagation in marine sediments. However, non-causality is equally apparent for both shear and compressional wave propagations in sediments with high rigidity.

### References:

1. Biot, M. A. (1956a) 'Theory of propagation of elastic waves in a fluid saturated porous solid, I. Low-frequency range', J. Acoust. Soc. Am. 28, 168-178.
2. Biot, M. A. (1956b) 'Theory of propagation of elastic waves in a fluid saturated porous solid. II. Higher frequency range', J. Acoust. Soc. Am., 28, 179-191.
3. Stoll, R. D. (1974) 'Acoustic waves in saturated sediments', in Physics of Sound in Marine Sediments, edited by L. D. Hampton, Plenum, New York.
4. Yamamoto, T. (1983) 'Acoustic propagation in the ocean with a poroelastic bottom', J. Acoust. Soc. Am. 73, 1587-1596.
5. Kjartansson, E. (1979) 'Constant Q wave propagation and attenuation', J. Geophy. Res., 84, 4737-4748.
6. Futterman, W. I., (1962) 'Dispersive body waves', J. Geophys. Res., 67, 5279-5291.
7. Kutt, H. R. (1975) 'The numerical evaluation of principle value integrals by finite-part integration', Numer. Math., 24, 205-211.
8. IMSL user's manual, April 1987, Version 1.0, 593-595.
9. Liu H. P., Kosloff D. D. (1981) 'Numerical evaluation of the Hilbert transform by the Fast Fourier Transform (FFT) technique', Geophys. J. R. astr. Soc., 67, 791-799.
10. Bode, H. W. (1945) 'Network Analysis and Feedback Amplifier Design', Van Nostrand, Princeton, N.J.
11. O'Donnell, M., Jaynes, E. T., and Miller, J. G. (1981) 'Kramers-Krönig relationships between ultrasonic attenuation and phase velocity', J. Acoust. Soc. Am. 69, 696-701.

# SOIL PROPERTIES FOR SHEAR WAVE PROPAGATION

R.D. WOODS
University of Michigan
Ann Arbor, Michigan 48109-2125
USA

ABSTRACT. All methods of analyzing the dynamic response of soil/rock, including seafloor materials, depend ultimately on the parameters used to characterize the soil/rock. In the last 25 years much has been learned about the properties which affect dynamic response and wave propagation and ways to measure those properties. For analysis and design of structures and for interpretation of wave propagation phenomena, the key soil/rock properties are shear modulus, material damping and unit weight. This paper describes nine important parameters which affect shear modulus and damping of soil and rock.

## 1. BACKGROUND

The experience, knowledge, and techniques described here have come from land-based experience in soil dynamics and earthquake engineering. Some of that background does not translate directly to seafloor soils, however, a majority does. Furthermore, the organizers of this volume have specifically set out to include cross-disciplinary approaches to wave propagation phenomena, and the experience presented here clearly meets that goal.

The properties of earth materials (soil and rock) which are of most importance in dynamic design considerations are: shear modulus, damping, and unit weight. Some consider Poisson's Ratio also to be a major factor, however, it is so often the case that the soil/rock is not isotropic that Poisson's ratio does not have a clear meaning. All three properties are important in wave propagation phenomena, but shear modulus and damping are dominant. For machine and instrument foundations, soil damping is less important because geometric or radiation damping far overshadows material damping, Richart, Hall and Woods, (1970).

There has developed a habit in the fields of soil dynamics and earthquake engineering to refer to "dynamic soil properties." This was helpful in the beginning to alert engineers to certain features of soil properties which influenced the dynamic design and analysis process. It is clear now that some properties which have been called "dynamic" are simply basic properties specified for a defined strain range. Figure 1 shows a typical shearing stress vs shearing strain curve for a soil. That relationship has curvature from the very beginning, so the definition of a single modulus is problematical. There are several useful choices like initial tangent modulus, tangent modulus at a given strain, or secant modulus at a given strain.

For some time "dynamic modulus" was considered to be the initial tangent modulus at very low shearing stain. However, there have developed analytical techniques which require the secant modulus for a wide span of shearing strains as for example in amplification of ground



*J. M. Hovem et al. (eds.), Shear Waves in Marine Sediments*, 29–39.

motion during earthquakes. Therefore the "dynamic" designation should be dropped and the strain amplitude range should be substituted to describe shear modulus of earth materials. There already exists one widely accepted notation supporting this approach, namely $G_o$ or $G_{max}$ representing the initial tangent modulus. Other designations could be defined in terms of shearing strain range, for example $G_{-3}$ for shear modulus at $10^{-3}$ % shearing strain.

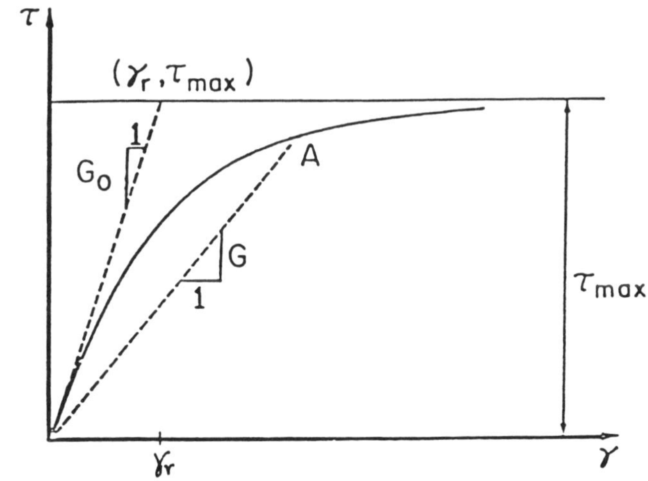


FIGURE 1 - Typical Shearing Stress vs. Shearing Strain Relationship for a soil.

## 2. PARAMETERS AFFECTING SHEAR MODULUS & DAMPING OF SOIL/ROCK

There are a number of important factors which affect the shear modulus and damping of earth materials. As a result of 25 years of evaluation, the most important factors have been identified for onshore applications. It is believed that the same factors are also important for seafloor wave propagation, however, additional factors may be identified for the special environment of the seafloor.

The following nine basic factors will be discussed: Void Ratio (porosity), Confining Pressure, Stress History, Geologic Aging, Shearing Strain Amplitude, Reference Strain, Degree of Saturation, Temperature and Overconsolidation Ratio (OCR).

2.1 Void Ratio (porosity) - From the earliest studies it became clear that shear modulus and damping were dependent on void ratio (e). Of course void ratio (e) and porosity (n) represent the state of soil density and are related through:

$$e = [n/(1-n)] \text{ or } n = [e/(1+e)] \qquad (1)$$

Some attempts have been made to relate shear modulus to relative density, but research continues to show that it is void ratio which is basic. Burmister (1948) in his classic paper on relative density showed that soils with the same relative density can have greatly differing dry unit weights, i.e. void ratios. It is strongly recommended that void ratio be used as the soil packing parameter for characterizing shear wave velocity and shear modulus of soils.

The empirical equations suggested by Hardin and Richart (1963) demonstrate the relationship between shear modulus and the basic parameters, void ratio, confining pressure and grain shape: ($G_o$ and $\sigma_o$ in psi)

$$G_o = 2630\,(2.17 - e)^2 / (1 + e)\,(\sigma_o)^{0.5} \tag{2}$$

for rounded grain soils, and

$$G_o = 1230\,(2.97 - e)^2 / (1 + e)\,(\sigma_o)^{0.5} \tag{3}$$

for angular grain soils,

in which e = void ratio, and $\sigma_o$ = average effective confining pressure. Clearly there must be a transition between extremes in grain shape and specific relationships could be defined. It is essential that the void ratio be included in specifying all soils for which shear modulus is reported.

2.2 Confining pressure - In the 1960's it was thought that shear modulus and damping were influenced only by the average effective confining pressure, $[(\sigma_1 + \sigma_2 + \sigma_3)/3]^{0.5}$, but Hardin and Black (1966) considered the effects of the initial state of shearing stress in the ground on modulus and damping. Those authors concluded that the average confining stress was adequate to describe the stress condition for estimating shear modulus. The question arose again in the mid-1970's and the writer along with Schmertmann (1978) again concluded that average confining pressure was sufficient to characterize confinement. However, Roesler (1979) and Yu and Richart (1984) found that there was a minor influence of the state of stress if the principal stress ratio was greater than 2.0. Yu and Richart presented an empirical equation for shear modulus of the following form:

$$G_o = A / F(e)\,(P_a)^{0.5}\,[(\sigma_a + \sigma_p)/2]^{0.5}\,[1 - 0.3\,(K_n)^{1.5}] \tag{4}$$

where $\sigma_a$ = the principal stress in the direction of wave propagation, $\sigma_p$ = the principal stress in the direction of particle motion, A = a dimensionless factor accounting for grain shape, F(e) is a function of void ratio, $P_a$ = atmospheric pressure, and $K_n$ is a stress history factor given by:

$$K_n = [(\sigma_1 / \sigma_3) - 1] / [(\sigma_1 / \sigma_3)_{max} - 1] \tag{5}$$

in which $\sigma_1$ is the maximum principal stress and $\sigma_3$ is the minimum principal stress, and $(\sigma_1 / \sigma_3)_{max}$ is the ratio at failure. Knox et al (1982) also studied the influence of the state of stress using wave propagation tests in a large triaxial bin. Their conclusions were that shear modulus depends about equally on the stresses in the direction of wave propagation and in the direction of particle motion, Fig. 2, which is a significant conclusion with respect to interpretation of field measurements with seismic methods.

2.3 Stress History - Stress history is a complex parameter because it has greatly different consequences depending how often and at what intervals the stress is changed and how large the stress changes have been. Equations 4 & 5 include stress history in terms of the factor $K_n$ which relates the current state of stress to failure stress conditions.

Figure 3 shows the low strain behavior of a sand subjected to large strain excursions at intervals far enough apart such that the modulus is fully recovered to "virgin" values between high strain applications. It can be seen that the modulus recovers completely in these tests, but the time required to total recovery depends on the number of cycles of high strain and the

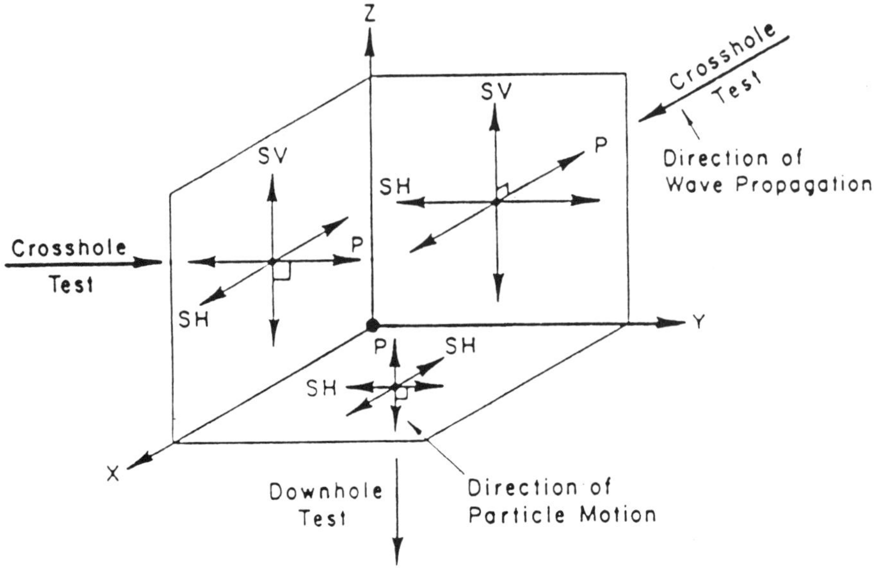


Figure 2 - Triaxial State of Stress For Seismic Wave Propagation (from Knox et al, 1982)

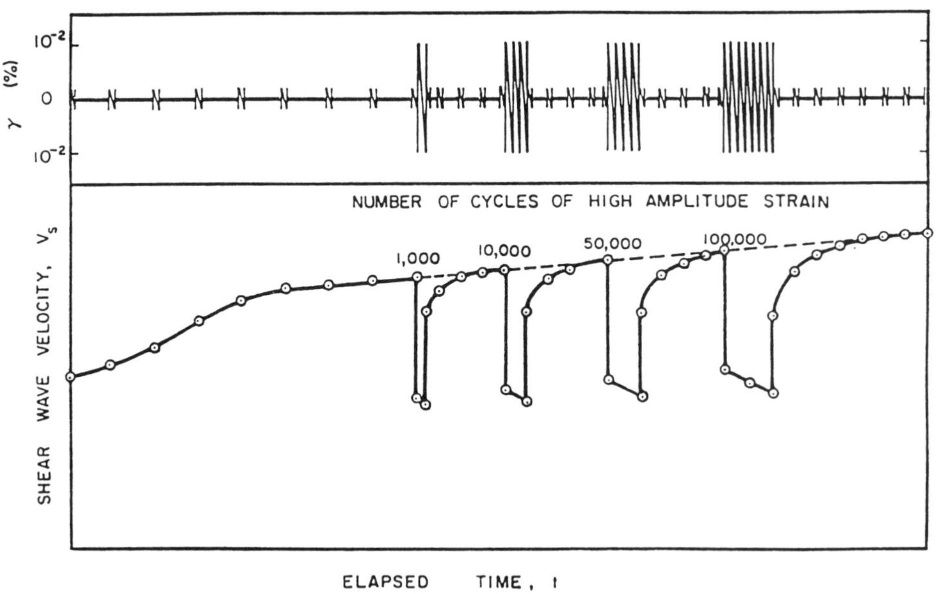


Figure 3 - Shear Modulus Loss and Regain from High Shearing Strain Excursions (from Anderson, 1974)

amplitude of the high strain excursions.

Stress history effects caused by very large amplitude phenomena as might occur because of blasting or earthquakes may cause a permanent reduction in shear modulus. Figure 4 from Thomann (1990) shows variation of shear modulus after shear strain excursions of 0.05% and 1.27%. After the 1.27% excursion, Fig. 4b, the shear modulus did not return to the original trend.

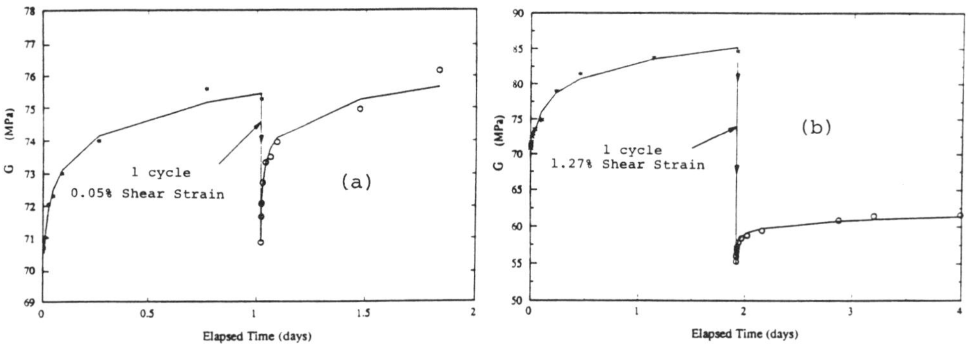


Figure 4 - Change in Shear Modulus of Agsco 100-200 Sand after High Strain Excursions (from Thomann, 1990)

2.4 Geologic Aging - An increase in stiffness and strength for soils as a function of aging has been recognized for many years. Figure 5 shows the increase in shear wave velocity as a function of time in a resonant column test. Laboratory tests out to about three years show a continuation of the trend in Fig. 5. The rate of increase in shear wave velocity or shear modulus with time is a function of particle size with smaller particles showing greater increase with time, Anderson (1974).

Schmertmann in his Terzaghi Lecture (1989) describes the aging effects on many soil properties. Included among them is the increase in shear modulus with time, Fig. 6 for a recently constructed embankment. Schmertmann suggests one explanation for the aging effect is an increase in the number and area of particle contacts per unit volume as a function of time. These changes occur with no measurable change in volume.

2.5 Shearing Strain Amplitude - Because of the continuous curvature of the shearing stress vs shearing strain curve for soils, Fig. 1, it is clear that the shear modulus must be a strain dependent property. Research has shown that modulus degradation with strain amplitude has a common shape for all soils. A typical set of degradation curves for $G/G_o$ vs. strain amplitude at several confining pressures are shown on Fig. 7a. The modulus ratio starts at a value of 1.0 for very small strains (on the order of $10^{-4}$%) and decreases to a value as low as 0.1 for shearing strains of 0.5% or higher. Damping in soil as represented by damping ratio has the opposite trend, i.e. damping increases as shearing strain amplitude increases. However, damping ratio is not normalized with respect to a low amplitude value, but is plotted as a arithmetic value vs. shearing strain amplitude, Fig. 8.

2.6 Reference Strain - Figure 1 also shows the definition of reference strain as that strain at which the initial tangent modulus intersects the maximum shear stress line. If the modulus degradation curves for a granular soil at different confining pressures, Fig. 7a, are normalized for reference strain, a single shear modulus vs. normalized strain curve is obtained, Fig. 7b.

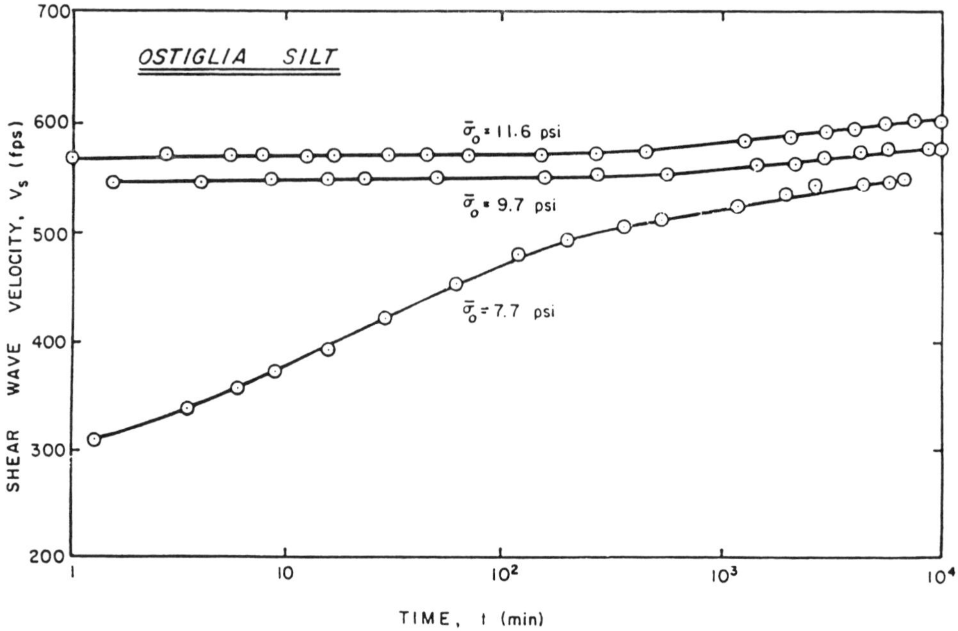


Figure 5 - Geologic Aging in Specimen in Resonant Column Test on Ostiglia Silt (note change in slope between 100 & 1000 minutes)

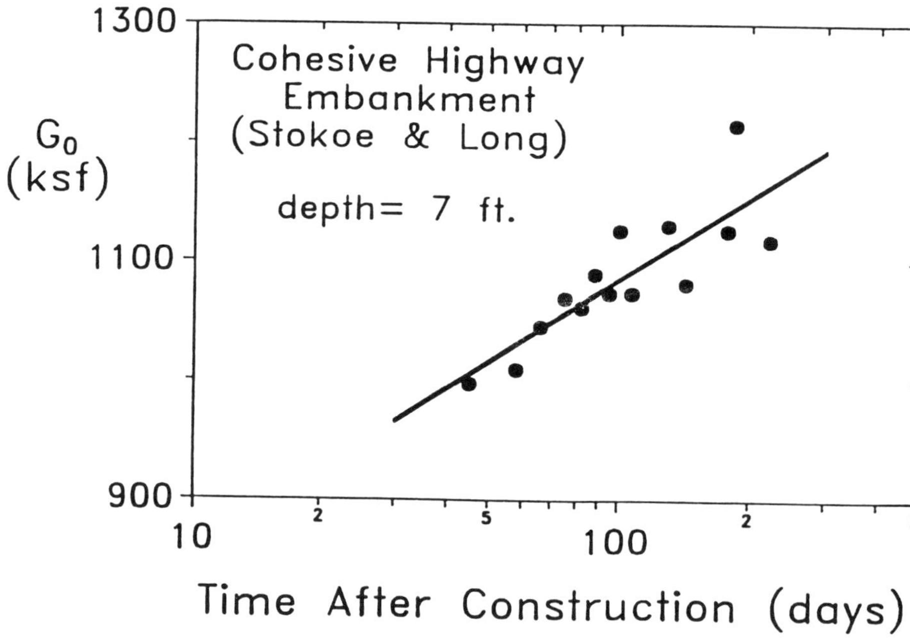


Figure 6 - Shear Modulus Increase vs. Time, Field Study (from Schmertmann, 1989)

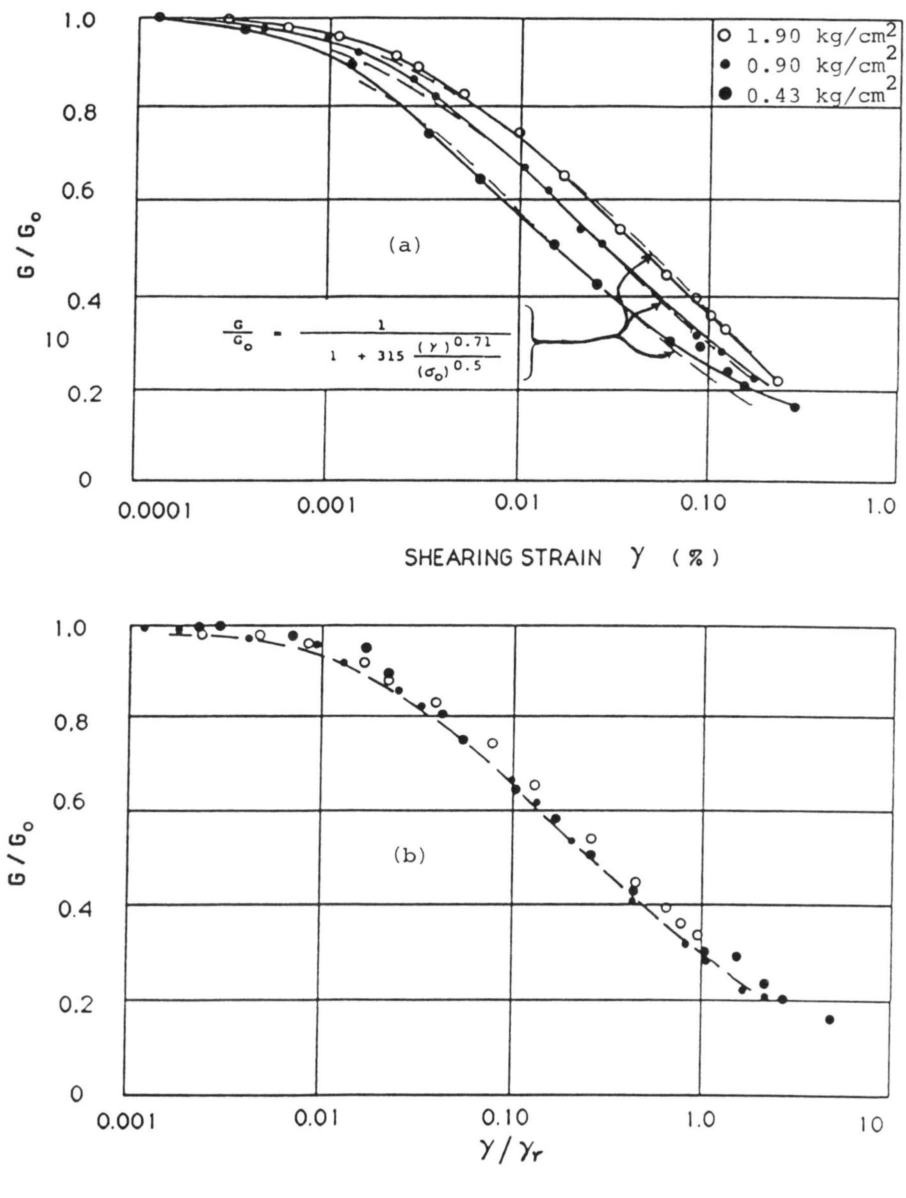


Figure 7 - Shear Modulus Ratio vs Shear Strain
(from Richart, 1987)

The concept of reference strain is particularly useful in evaluating liquefaction potential of a soil.

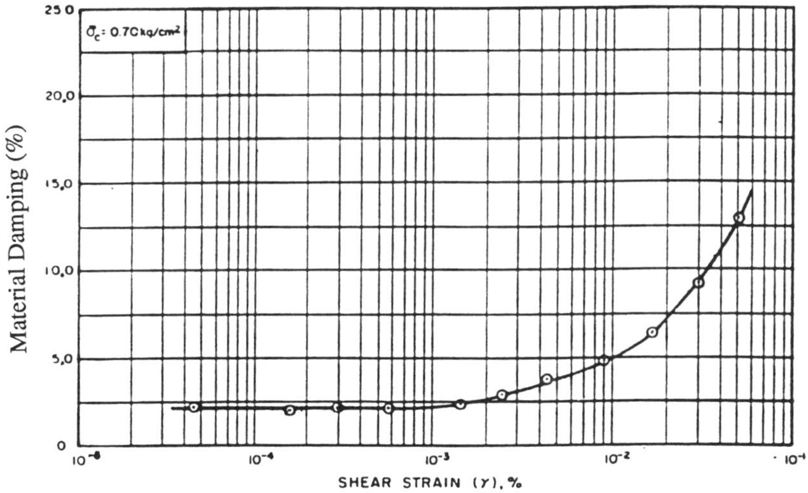


Figure 8 - Damping Ratio vs Shear Strain for Ottawa Sand

2.7 Degree of Saturation - Figure 9 shows shear wave velocity vs. confining pressure for the same soil in the dry, drained, and saturated conditions. The difference in the curves can be accounted for adequately by considering the difference in mass density. The mineral of this sand is not sensitive to moisture, i.e. is not a clay mineral. As long as the mineral constituting the soil does not react strongly with water, the above conclusion is satisfactory for the extreme conditions of dry and fully saturated.

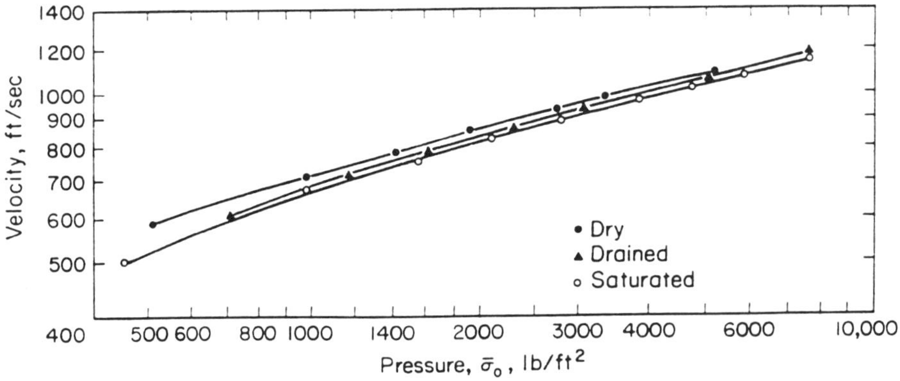


Figure 9 - Shear Wave Velocity vs Confining Pressure for Ottawa Sand At Three Saturations (from Richart et al, 1970)

For partially saturated sands and silts, however, there are responses to saturation which must be considered. Under conditions of low confining pressure and partial saturation, capillarity at particle to particle contacts can create an increase in confinement accompanied by a proportional increase in shear wave velocity or shear modulus. This artificial confinement can be destroyed either by complete drying or complete saturation of the soil.

Wu, Richart, and Gray (1984) have shown that for each soil there is an optimum degree of saturation which causes the maximum additional effective confinement, Fig. 10, and have estimated the optimum water content and the amount of increase in modulus. Qian, Woods, and Gray (1990) have gone further to predict the optimum degree of saturation and maximum increase in modulus for any sand and/or silt based on the particle size distribution.

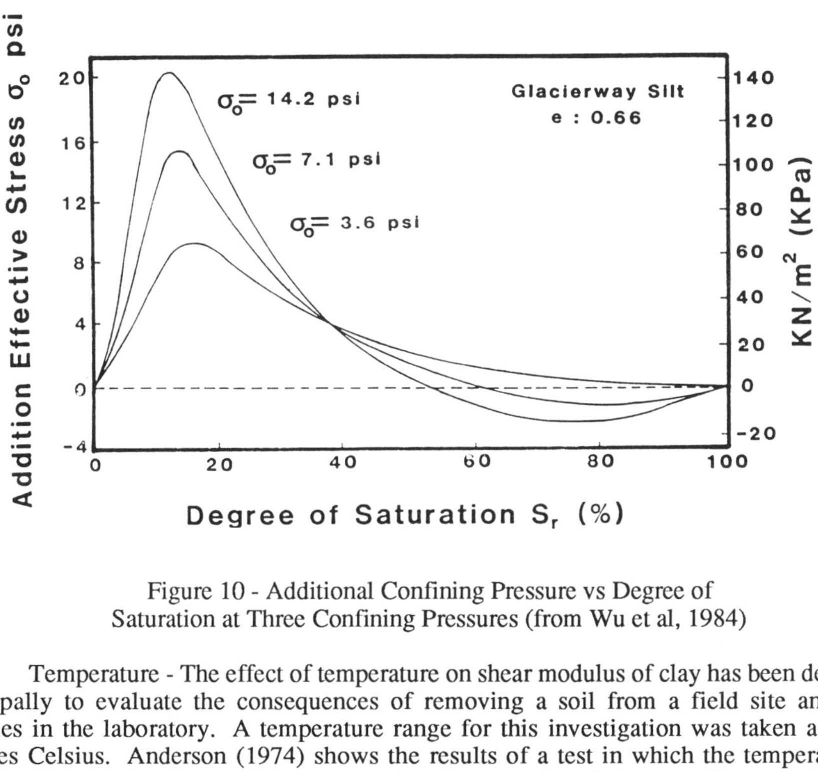


Figure 10 - Additional Confining Pressure vs Degree of Saturation at Three Confining Pressures (from Wu et al, 1984)

2.8 Temperature - The effect of temperature on shear modulus of clay has been determined principally to evaluate the consequences of removing a soil from a field site and testing samples in the laboratory. A temperature range for this investigation was taken as 4 to 22 degrees Celsius. Anderson (1974) shows the results of a test in which the temperature was increased from 4 degrees to 22 degrees, and then returned to 4 degrees. The sample demonstrated strong geologic aging as well as some temperature effects, however, the temperature effects were relatively minor and were totally recoverable.

2.9 Overconsolidation Ratio (OCR) - Although consolidation effects are usually small for granular soils, fine grained sand and silty sands may experience overconsolidation effects. (OCR is defined as the ratio of the maximum confining pressure felt by the soil to the current confining pressure.) In general the effects of OCR can be taken into account through an empirical relationship from Hardin and Black (1968) as:

$$G_{max} = 1230 \, [ \, (2.97 - e)^2 / (1 + e) \, ] \, (OCR)^K \, ( \sigma_o)^{0.5} \qquad (6)$$

where the factor K ranges from 0 to 0.5 as the plasticity index ranges from 0 to 100. It can be seen that the effect is incorporated through a power to the OCR itself where that power is based on the plasticity of the soil.

A factor which as not been evaluated thoroughly but which may well be important in sediments like those with void ratios greater than 4-6 as reported by Richards (1962) is underconsolidation. Land based considerations usually do not require investigation of underconsolidated soils.

## 3. SUMMARY

In this paper nine parameters which affect shear modulus and damping of soil/rock have been presented, and their effects demonstrated. There may be more parameters identified as research continues, but these basic parameters should be considered in all investigations.

## 4. REFERENCES

Anderson, D. G. (1974) 'Dynamic Modulus of Cohesive Soils', Ph.D. Dissertation, University of Michigan.

Burmister, D. M. (1948) 'The Importance and Practical Use of Relative Density in Soil Mechanics', Proc. of ASTM, Vol. 48, pp. 1249-1268.

Hardin, B.O. and Black, W.L. (1966) 'Sand Stiffness Under Various Triaxial Stresses', Journal of the Soil Mechanics and Foundations Division, ASCE, Vol. 92, No. SM2, Mar., pp 27-42.

Hardin, B.O. and Black, W.L. (1968), 'Vibration Modulus of Normally Consolidated Clay', Journ. Soil Mech. and Found. Div., Proc. ASCE, Vol. 94, No. SM2, March, pp. 353-369.

Hardin, B.O. and Richart, F.E., Jr. (1963) 'Elastic Velocities in Granular Soils', Journal of the Soil Mechanics and Foundations Division, ASCE, Vol. 89, No. SM1, Feb., pp. 33-65.

Knox, D.P, Stokoe, K.H.,II, and Kopperman, S.E. (1982) 'Effect of State of Stress on Velocity of Low-Amplitude Shear Waves Propagating along Principal stress Directions in Dry Sand', Geotechnical Engineering Report GR 82-23, University of Texas at Austin.

Qian, X, Woods, R.D., and Gray, D.H. (1990) 'A Procedure for Estimating the Dynamic Properties of Partially Saturated Sands', Submitted to Journal of Geotechnical Engineering, ASCE, October 1990.

Richards, A.F. (1962) 'Investigations of Deep-Sea Sediment Cores, II Mass Physical Properties', TR-106, U.S. Navy Hydrographic Office. Oct.

Richart, F.E., Jr. 'Foundation Dynamics - 1987', The Art and Science of Geotechnical Engineering, A Volume Honoring Ralph B. Peck, Prentice-Hall, Englewood Cliffs, N.J., 1989.

Richart, F.E., Jr., Hall, J.R., Jr. and Woods, R.D. (1970) Vibrations of Soils and Foundations, Prentice-Hall, New York, 412 pp.

Roesler, S.K. (1979) 'Anisotropic Shear Modulus due to Stress Anisotropy', Journal of the Geotechnical Engineering Division, ASCE, Vol. 105, No. GT 7, July, pp. 871-880.

Schmertmann, J.H. (1978) 'Effect of Shear Stress on the Dynamic Bulk Modulus of Sand', Technical Report S-78-16, Waterways Experiment Station, U.S. Army Corps of Engineers, Oct.

Schmertmann, J.H. (1989) 'The Mechanical Aging of Soils', The 25th Terzaghi Lecture, New Orleans, ASCE Annual Convention, October 1989.

Thomann, T.G. (1990) 'Stiffness and Strength Changes in Cohesionless Soils Due to Stress History and Dynamic Disturbance', Doctoral Dissertation, University of Michigan, Ann Arbor, Michigan.

Wu, S., Gray, D.H., and Richart, F.E., Jr. (1984) 'Capillary Effects on Dynamic Modulus of Sands and Silts', Journal of Geotechnical Engineering, ASCE, Vol. 110, No. GT9, pp. 1188-1203.

Yu, P. and Richart, F.E., Jr. (1984) 'Stress Ratio Effects on Shear Modulus of Dry Sands', Journal of Geotechnical Engineering, ASCE, Vol. 110, No. 3, March, pp. 331-345.

# THE RELEVANCE OF SHEAR WAVES FOR STRUCTURAL SUBSURFACE INVESTIGATIONS

R. MEISSNER, W. RABBEL, F. THEILEN,
*Institut für Geophysik,*
*Olshausenstr. 40*
*23 Kiel*
*Germany*

ABSTRACT. Combined P- and S-wave studies generally yield information on subsurface structures, on anisotropy effects, on the type of sediments and on physical properties. The mapping of gas-containing layers requires the application of shear waves, although the field techniques, especially in the marine environment, is rather complex. The great increase in petrophysical and structural knowledge obtained by additional S-wave studies in connection with a joint interpretation of all seismic parameters justifies its difficult and time consuming application. Results from ample laboratory studies and field measurements on land may be transferred to marine environments.

## 1. Introduction

Shear (S-) wave observations at sea will always be a very special, time-consuming and problematic way of gathering seismic information. Because of the overlying water, S-wave sources and receivers have to be employed at the sea bottom, a technique which never will compete with the rapid and effective way of using the routine, compressional (P-) wave techniques with sources (mostly airgun arrays) and receivers (large streamers) towed behind a research vessel at shallow depth. Excellent examples have recently been obtained by a kind of "mixed" technique, using airgun arrays in the water and 3-component ocean bottom geophones (OBS) at the sea bottom (Mjelde et al., 1990). Strong P-S conversions at the sea bottom from the near-vertical to the wide-angle range are recorded, and their use for sediment refraction and crustal studies seems to be generally possible.

The additional lithological information to be gained by S-wave studies is immense and may justify even more complex techniques. In this paper we want to assess the various advantages, pecularities and problems which are involved in S-wave studies.



*J. M. Hovem et al. (eds.), Shear Waves in Marine Sediments*, 41–49.

## 2. Recent Achievements of Shear (S) Wave Studies

Simple formulas on the S-wave velocity state that they propagate much slower than P-waves.

(1) $$V_S = \sqrt{\frac{\mu}{\rho}}$$

$$V_P = \sqrt{\frac{k + 4/3\,\mu}{\rho}} = \sqrt{\frac{\lambda + 2\mu}{\rho}}$$

k = compressional modulus
λ = Lame parameter
μ = shear modulus
ρ = density

Assuming μ is roughly equal to λ we get by assessing $V_S/V_P$:

(2) $$V_S/V_P = 1/\sqrt{3}$$

which is the same as assuming a Poisson's ratio σ of 0.25 as postulated in classical papers for the earth's crust and mantle.

In water-saturated sediments λ and k are much higher than μ; also σ becomes much higher than 0.25, and in recent years unexpected high σ-values have been found in bore-hole surveys, lab. and field studies of sediments (Kähler and Meissner, 1983), ranging from σ = 0.4 to 0.49. Their great influence on the radiation and receiver pattern has been studied by using classical formulas (Gutenberg, 1959). Figure 1 shows such a pattern for a horizontal point source. The deterioration of the P-wave for high σ is surprisingly favourable, the smoothing influence of high σ for SV wave becomes clear, and the independence of SH-waves against σ is evident. Vertical point sources do not produce any SH-waves.

Another advantage of SH-waves is their behaviour at horizontal boundaries, as often found in marine sediments. In contrast to P- or SV-waves they are not converted to other wave types, which helps considerably in the interpretation. Dipping boundaries, on the other hand, as well as anisotropies lead to a certain polarization of S-waves which is often difficult to predict. While P-waves, for all practical purposes, are polarized in the direction of the ray, S-waves can be polarized in any direction in the plane perpendicular to the direction of the ray (Douma and Helbig, 1987).

Polarization of S-waves in anisotropic media leads to different S-velocities in different directions and to a splitting of S-wave traveltimes. Crack-induced anisotropy (Nur, 1971; Nur and Simmons, 1970; Crampin, 1984) or "layered" anisotropy in sediments (Hudson, 1981, Krey and Helbig, 1956) produce the much debated "shear wave splitting". S-waves polarized parallel to the crack alignment and waves polarized parallel to sediment boundaries (mostly SH) travel faster than do S-waves perpendicular to them. In a crack-dominated medium S-waves parallel to the cracks travel faster than S-waves perpendicularly polarized. SH-waves in layered sediments travel with roughly the same velocity as SV-waves up to angles of incidence of 40°-50°, but after that they overtake SV-waves rapidly (Krey and Helbig, 1956).

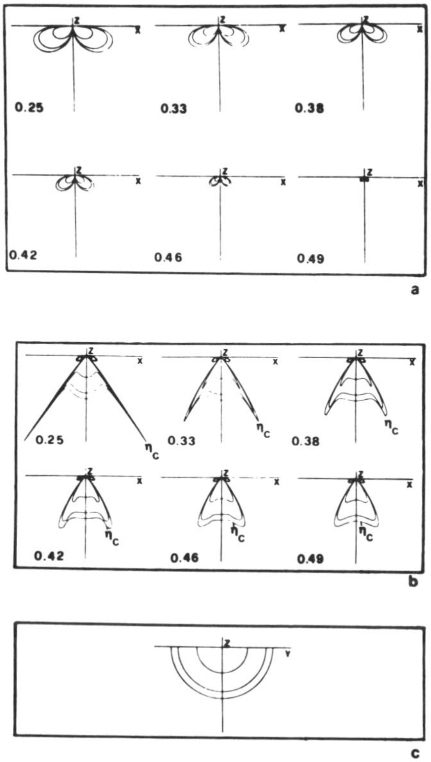


Fig.1 Radiation pattern for a horizontal surface point source and six different Poisson's ratios. P-wave pattern in $H_{\parallel}$-direction (a), SV-wave pattern in $H_{\parallel}$-direction (b), SH-wave pattern in $H_{\perp}$-direction (c). The contour lines are for $\Phi$=90°, 60°,30° for P-and SV-waves,and $\Phi$=0°,30°,60° for SH-waves.

Using 3-component geophones S-wave splitting can easily be recognized in crystalline rocks as well as in sedimentary layers. It provides an important information on the S-wave anisotropy, which often is related not only to cracks, layering, but also to present and past stress systems. Figure 2 demonstrates 3 examples of S-wave splitting.

Another peculiar property of S-waves is their absorption, preferably expressed by their specific attenuation $Q_S^{-1}$. Theoretical studies and laboratory evidence (Meissner and Theilen, 1983, Prasad, 1988) show that $Q_S^{-1}$ is higher than $Q_P^{-1}$ in most sedimentary layers, and especially strong in clay layers. On the other hand, $Q_S^{-1}$ is smaller than $Q_P^{-1}$ in partially saturated sands and in peat and other organogenic sediments. Complex methods to determine $Q^{-1}$ values have been developed recently (Tonn, 1988, Engelhard, 1986). The $Q_S^{-1}$ values of shear waves are favourable for a penetration through gas-containing layers, be it from hydrocarbon or other organic matter e.g., $CH_4$.

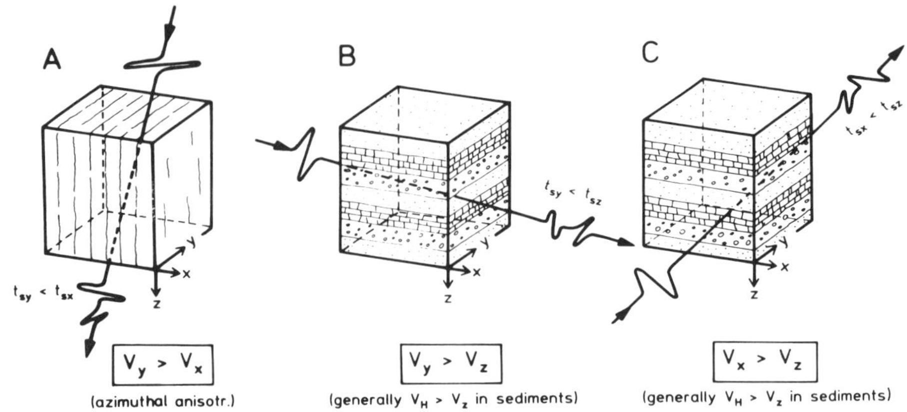


Fig. 2 **SHEAR WAVE SPLITTING**

**Components parallel to cracks or layering travel faster**

In such layers P-waves suffer from a strong decrease in their velocities and an extremely high absorption, which often makes them totally disappear, while S-wave velocities and absorption remain more or less the same as in totally saturated media. Examples from marine sediments (Theilen and Pecher, this volume) seem to confirm the theoretical framework. There are some areas in the Baltic Sea where organic gas prevents P-waves from penetration, producing only reverberations or peg-leg multiples when studied by high frequency, near-vertical surveys (Gimpel, 1987).

An increasing number of S-wave applications has recently been obtained by specific laboratory studies. These studies were performed in shore based laboratories (Prasad, 1988) as well as on board ship on freshly drilled box cores (Phillip, 1989). Results are summarized as follows. (1) While $V_S$ of sediments is uneffected by grain size, $V_P$ as well as $Q_S^{-1}$ and $Q_P^{-1}$ increase with the coarseness of the grains (Prasad, 1988). (2) Dry grains show higher $Q_S^{-1}$ than saturated grains. (3) $V_S$ and $V_P$ decrease with increasing porosity (Schön, 1983). (4) $Q_S^{-1}$ decreases sharply in coarse-grained material upon an increase of pressure while $Q_P^{-1}$ remains rather constant (Prasad, 1988). (5) $Q_S^{-1}$ and $Q_P^{-1}$ decrease moderately in fine-grained material upon an increase in pressure (Prasad, 1988). (6) Fluid flow predicted losses ($Q_S^{-1}$ and $Q_P^{-1}$) are increasing with frequency (Stoll, 1977). (7) Attenuation peaks depend on permeability and viscosity (Meissner and Theilen, 1983).

Extensive S-wave studies, often carried out together with P-wave measurements, were performed in sediments on land (Stümpel et al. 1984, Milkereit et al. 1986), and some of these data can be easily transferred to marine sediments. These studies include structural and hydrological objectives of the upper few 100m, often carried out together with geoelectrical measurements. But also S-wave studies for oil prospection with a penetration depth of a few 1000 m have been performed in research programmes of contractor companies. Differences in the reflection amplitude of P- and S-waves have been used as a direct hydrocarbon gas indicator (Tatham and Krug, 1985). Some tens of kilometers have been penetrated by crustal studies with S-waves, and SS reflections from the Moho ($S_MS$) as well as converted waves have been frequently recorded. In some cases indications on the petrological composition and anisotropy of the lower crust were obtained by a joint interpretation of corresponding P- and S-reflectivity patterns and velocities (Lüschen et al. 1990, Hollbrook et al. 1988).

Figure 3 gives an example of a joint P- and S-wave study in shallow structures. It shows 6 seismogram records from a peat area, obtained by sources acting in 3 directions (X , Y , Z) and 3 component geophones, recording also all 3 directions (Meyer, 1988). Here only the even patterns, i.e. X -->X ; Y -->Y, and Z -->Z directions, are plotted. During these studies in the highly porous peat a mysterious wave with a velocity between those of P and S was determined. It was polarized preferably in the direction of the ray, and its oscillation is prograde . While speculations on its nature are under way it is not known whether it exists in marine environments. But similar studies as those of Fig. 3 are definitely desirable in marine environments.

## 3. Generalized Interpretation

Many physical, as well as lithological parameters, can be assessed by using a generalized interpretation diagram, modified after Winkler and Nur, 1982. It is shown in Figure 4 and permits to assess certain areas in the $Q_P/Q_S$ and $(V_P/V_S)^2$ range.

Straight lines in the diagram represent relationships between the imaginary parts of compressional ($K_i$) and the shear modulus ($\mu_i$). They determine areas of comparable ($K_i = \mu_i$) and dominating absorption in compression ($K_i = 10\,\mu_i$).

There are data from laboratory, borehole, and field investigations incorporated into the diagram. The lab data show the large difference between dry and saturated samples. Dry samples show similar absorption for K and $\mu$ and a small range of $V_P/V_S$ values. Water-saturated samples show the dominating absorption in K and extremely high values of $V_P/V_S$. But also the data from boreholes and surface (field) measurements show the large influence of the compression-dominated absorption ($K_i$), while $V_P/V_S$ is moderately large. These data were all from saturated or partially saturated conditions and are in reasonable harmony with the laboratory data from saturated samples. All these data show comparable values of P- and S-absorption; sometimes $Q_P/Q_S$ is $<1$, but effects of differential scattering cannot be ruled out.

We consider this kind of diagram very helpful for a generalized interpretation, also for a future presentation of marine (saturated) samples and field data. These samples are presently obtained on board ship by box cores and studied by P- and S transmission techniques.

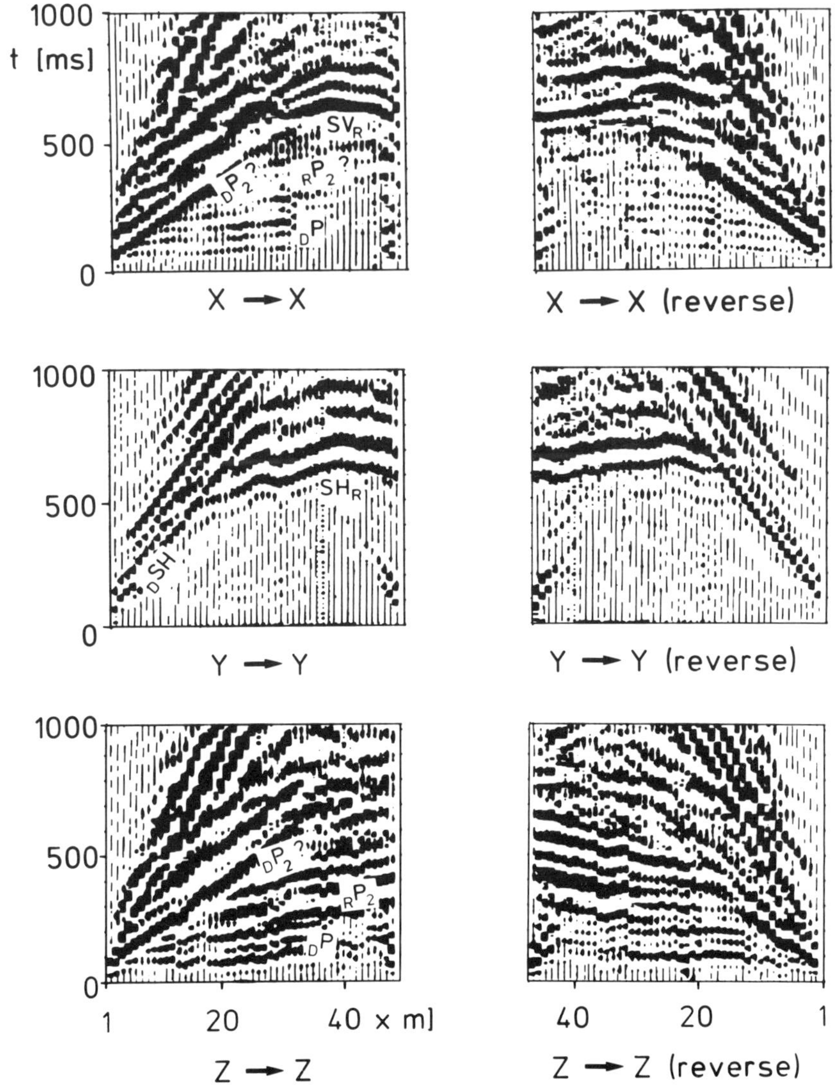


Fig.3 Seismogram Sections for 3 Component Observations in a Peat Area on Land

Note the strong SV and SH waves on the X → X and Y → Y arrangements; note also the mysterious secondary directed P wave $_DP_2$ on X → X and Z → Z configuration. See text.

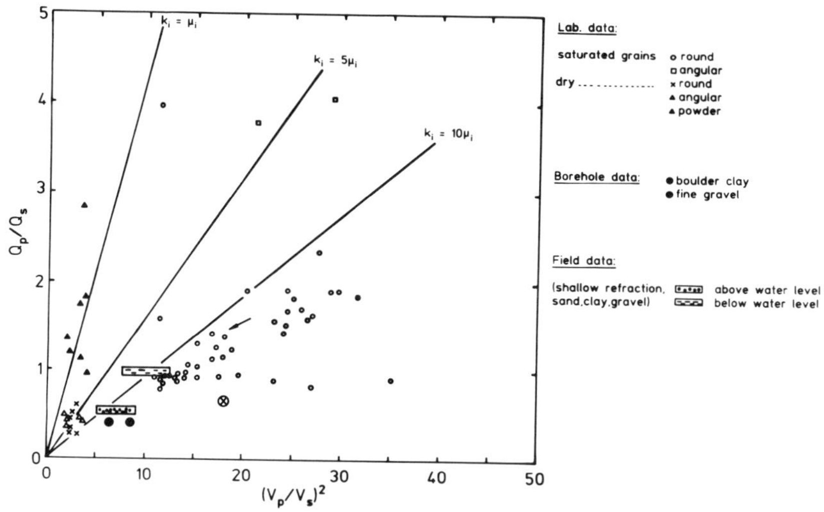


Fig. 4 THE GENERALIZED SEISMIC EVALUATION DIAGRAM (modified after Winkler and Nur, 1982)

$$\frac{Q_P}{Q_S} = \frac{\mu_i}{k_i + \frac{4}{3}\mu_i} \cdot \left(\frac{V_P}{V_S}\right)^2$$

$Q_P$ = quality factor for P-waves, $\mu_i$ = imaginary part of shear modulus, $V_P$ = velocity of P-waves
$Q_S$ = quality factor for S-waves, $k_i$ = imaginary part of compr. modulus $V_S$ = velocity of S-waves

## 4. Discussion and Conclusion

There is no doubt that S-waves are needed for a complete petrophysical description, preferably together with P-wave data. Not only the absolute values of P- and S velocities are essential when assessing petrological boundary conditions, also the ratios $V_P/V_S$ and $Q_P/Q_S$ are requested for a complete description of the medium to be studied.

The application of S-waves is especially useful for: (1) the investigation of gas-containing layers in high resolution near surface studies, (2) the determination of anisotropies either from crack orientation, layering or stress patterns, (3) the study of sediment types in the subsurface with respect to their seismic properties, (4) the indirect in-situ assessments of shear strength values, especially at depths below the penetration of core- and penetrometer devices, (5) the recognition of waste reservoirs.

Shear wave techniques on land and in shallow marine areas can nowadays be performed with high efficiency. However, in deep sea areas the methods are rather time consuming. Shear wave investigations in such areas can more easily be carried out with converted P-S-waves, as the load capacity at the sea floor does not permit the use of heavy generators of SH-waves. New methods for generating and recording S-waves on the sea floor have to be developed.

## 5. References

Crampin, S. (1984) 'Effective anisotropic elastic constants for wave propagation through cracked solids', Geophys. J.R. astron. Soc. 76, 135-145.

Douma, J. and Helbig, K. (1987) 'What can polarization of shear waves tell us?', First Break, 5, 95-105.

Engelhard, L. (1986) 'Determination of attenuation of seismic waves from actual field data', in DGMK-report 254 'Absorption of seismic waves', final report, 83-118.

Gimpel, P. (1987) 'Marine flachseismische Untersuchungen in der Kieler Bucht unter besonderer Berücksichtigung von Scherwellen', Diss. Math.-Nat. Fak. Univ. of Kiel, 1-123.

Gutenberg, B. (1959) 'Physics of the Earth's Interior', Acad. Press, New York, 1-359.

Hollbrook, W.S., Gajewski, D., Krammer, A., Prodehl, C. (1988) 'An interpretation of wide-angle compressional and shear wave data in South-West Germany; Poisson's ratio and petrological implications', J. Geophys. Res. 93, 12,081-12,106.

Hudson, D.A. (1981) 'Wave speed and attenuation of elastic waves in material containing cracks', Geophys. J.R. astr. Soc. 64, 133-150.

Lüschen, E., Nolte, B., Fuchs, K. (1990) 'Shear wave evidence for an anisotropic lower crust beneath the Black Forest', SW-Germany; Tectonophysics 173, 483-493.

Kähler, S. and Meissner, R. (1983) 'Radiation and receiver pattern of shear and compressional waves as a function of Poisson's ratio, Geophys. Prosp. 31, 421-435.

Krey, T. and Helbig, K. (1956) 'A theorem concerning anisotropy of stratified media and its significance for reflection seismics', Geophysical Prospecting 4, 294 pp.

Meissner, R. and Theilen, F (1983) 'Attenuation of seismic waves in sediments', Proc. 11th World Petr. Congr. 2, SP3, 363-379.

Meyer, J.H. (1989) 'Darstellung und Verarbeitung vektorieller seismischer Wellenfelder', Diss. Math.-Nat. Fak., Univ. Kiel, 1-115.

Milkereit, B., Stümpel, H., Rabbel, W. (1986) 'Shear wave reflection profiling for new surface ligmite exploration', Geophysical Prospecting 34, 845-855.

Mjelde, R., Sellevoll, M.A., Shimamura, H., Iwasaki, T., Kamozawa, T. (1990) 'A study of crustal structures off Lofoten, Norway, using 3-component ocean bottom seismographs', Terra abstracts 2, 28.

Nur, A. (1971) 'Viscous phase in rocks and the low velocity zone'. Journal Geophys. Research, 76, 1270-1277.

Nur, A., Simmons, G. (1970) 'The effect of viscosity of a fluid phase on velocity in low porosity rocks, Earth and Planetary Science Letters, 7, 99-108.

Phillip, J. (1989) 'Bestimmung der Kompressions- und Scherwellengeschwindigkeit mariner Sedimente an Kastenlotkernen', Institut für Geophysik der Universität Kiel, SFB 313, 1-95.

Prasad, M. (1988) 'Experimental and theoretical considerations of attenuation and velocity interactions with physical parameters in sands', Diss. Math.-Nat. Fak. Univ. of Kiel, 1-122.

Schön, J. (1983) 'Petrophysik', Akademic Verlag, Berlin 1983, 1-405.

Stoll, R.D. (1977) 'Acoustic waves in ocean sediments', Geopyhsics, 42, 715-725.

Stümpel, H., Kähler, S., Meissner, R., Milkereit, B. (1984) 'The use of seismic shear waves and compressional waves for lithological problems of shallow sediments', Geoph. Prosp. 32, 662-675.

Tatham, R.H. and Krug, E.H. (1985) '$V_P/V_S$ Interpretation' in A.A. Fitch (ed.), Developments in geophysical exploration methods, Vol. 6, Elsevier Applies Science Publishers, London, New York.

Theilen, F. and Pecher, K. (1990) 'Rheological properties of marine sediments from compressional and shear waves with respect of stability problems', this volume.

Tonn, R. (1989) 'Comparison of seven methods for the computation of Q', Physics Earth Plan. Inter., 55, 259-268.

Winkler, K.W. and Nur, A. (1982) 'Seismic attenuation effects of pore fluid and frictional sliding', Geophysics 47, 1-15.

# EXPERIMENTAL INVESTIGATION OF SEISMIC SURFACE WAVES IN THE SEAFLOOR

K.H. STOKOE, II, R.C. GAUER, AND J.A. BAY
*Department of Civil Engineering*
*ECJ 9.227*
*The University of Texas*
*Austin, Texas 78712-1076 USA*

ABSTRACT. The Spectral-Analysis-of-Surface-Waves (SASW) method is a nonintrusive seismic method which has been used on land to determine shear modulus profiles of geotechnical sites. The method involves field measurement of surface wave dispersion. Application of the method offshore is being investigated both analytically and experimentally. The experimental work has involved measurements in a small-scale facility in which bottom materials and water depths were varied. A "stiff" seafloor was modelled with a cured concrete while a "soft" seafloor was modelled with curing concrete, soft limestone, and lightly cemented sand. Wave velocities in the "soft" bottom were those of Scholte waves which agrees with analytical results presented in a companion paper. For the "stiff" bottom, wave velocities for both Scholte and pseudo-Rayleigh waves were measured, depending on the source/receiver arrangement.

## 1. Introduction

The Spectral-Analysis-of-Surface-Waves (SASW) method is a nondestructive seismic method which utilizes surface waves of the Rayleigh type to measure the shear stiffness of geotechnical systems on land. Some benefits of the method are: 1. all equipment is placed on the ground surface, 2. soft layers sandwiched between stiffer materials can be measured and 3. measurement of the soil skeleton stiffness is made for both unsaturated and saturated materials. These benefits also make the method attractive for offshore geotechnical applications. However, the presence of water needs to be considered when adapting the onshore SASW method to seafloor exploration. Therefore, the studies presented herein were undertaken.

## 2. Use of the SASW Method on Land

A brief discussion of the approach used to apply the SASW method on land is first presented to develop an understanding of the method. The surface wave of interest in this work which propagates along an air/solid interface is called a Rayleigh wave. Other terminology is used for surface waves propagating along a water/solid interface as noted later.

### 2.1 SURFACE WAVE DISPERSION

The dispersive nature of Rayleigh waves provides the basis for the SASW method. Dispersion occurs because Rayleigh wave particle motion decays exponentially with depth which results in



*J. M. Hovem et al. (eds.), Shear Waves in Marine Sediments*, 51–58.

different wavelengths stressing the material at different depths. The objective of field testing is to develop a representative set of dispersion measurements for the site. The measurement set is expressed as a dispersion curve, a plot of Rayleigh wave phase velocity ($V_R$) versus wavelength ($\lambda_R$) or frequency (f). Figure 1a illustrates a dispersion curve for plane Rayleigh waves propagating through a uniform half-space with a shear wave velocity ($V_S$) equal to one. In this case, phase velocity is independent of wavelength because all waves propagate through the same material. The dispersion curve for a system composed of a softer layer ($V_S = 1$) over a stiffer half-space ($V_S = 2$) is shown in Figure 1b. In this case, phase velocity varies with frequency. At high frequencies (wavelengths which are shorter than the top layer thickness), $V_R$ is equal to the Rayleigh wave velocity in the top layer. At low frequencies (wavelengths which are more than about 10 times the top layer thickness), $V_R$ is essentially equal to the Rayleigh wave velocity of the half-space. A transitional stage occurs in which $V_R$ is influenced by the stiffnesses of both layers.

## 2.2 EQUIPMENT AND FIELD TESTING

The general testing configuration of source, receivers, and recording equipment used on land is shown in Figure 2. Both the source and receivers are placed on the ground surface. A variety of vertical dynamic sources may be used to generate energy over the desired frequency range. Rayleigh wave propagation is monitored with a linear array of two receivers placed at distances $d_1$ and $d_2$ from the source, with $d_2$ usually selected to be twice $d_1$. Several different source-to-receiver spacings are used during testing to avoid measurement complications due to attenuation and near-field effects. A dual-channel Fast Fourier Transform (FFT) signal analyzer is used to record and analyze receiver outputs. This analyzer permits real time evaluation of frequency domain data in the field. A microcomputer can be used to develop the dispersion curve in the field. Interpretation of dispersion results is normally accurate for wavelengths less than or equal to the distance between the source and first receiver ($d_1$).

## 2.3 DISPERSION CALCULATIONS

SASW measurements obtained from various receiver spacings are used to determine the composite dispersion curve for a site. For each source/receiver spacing, the time histories recorded by the two receivers are transformed to the frequency domain. The cross power spectrum and the coherence function are then calculated. Figure 3 shows these data for one receiver spacing at the experimental facility used in this study. The coherence function represents a signal-to-noise ratio and should be nearly one in the range of acceptable data. Rayleigh wave velocity is determined at each frequency using the phase of the cross power spectrum and receiver spacing. The corresponding wavelength is determined from the velocity and frequency as illustrated for one frequency by the insert in Figure 3. The results of all velocity and wavelength calculations for one receiver spacing are a dispersion curve ($V_R$ versus $\lambda_R$). Individual dispersion curves from all receiver spacings are assembled together to form the composite dispersion curve for the site.

A 20-cm receiver spacing was used for the record shown in Figure 3. For this particular record, the frequency range used for dispersion calculations corresponds to wavelengths between 5 and 20 cm. The subsequent dispersion curve is shown in Figure 4. Phase velocity fluctuates but has an average value of approximately 2340 m/s. The fluctuations are primarily due to reflections encountered in the 2.4 m by 4 m concrete slab upon which testing was performed. Measurement wavelengths are less than or equal to the 20-cm receiver spacing. Since the thickness of the concrete slab is approximately 37 cm, all wavelengths are confined to the concrete slab which can be considered the top layer in this system. The surface wave shows essentially no dispersion because all wave energy propagates within the same medium, the top layer.

### 2.4 INVERSION

Individual dispersion curves from all receiver spacings are normally assembled together to form a composite dispersion curve for the site. The shear wave velocity profile is derived from the composite dispersion curve using an inversion process. The process is an iterative procedure based on forward modeling and is discussed in the companion paper by Roesset et al.

## 3. Experimental Facilities

A small-scale facility was constructed at The University of Texas at Austin to investigate the influence of water on the propagation velocity of interface waves of the Rayleigh type. This facility, shown in Figure 5, was constructed by excavating a pit approximately 2.4 m by 4 m in plan and 2.4 m deep. The geotechnical profile at the site is 1 m of clay over 1 m of weathered limestone over competent limestone ($V_R \cong 1310$ m/s). Wood shoring was placed along the inner walls to prevent deterioration of the walls from repeated filling and emptying with water. Three steel casings were placed in the bedrock limestone and extended about 1 m above the surface of the rock. The casings were arranged in a linear array for crosshole seismic testing.

A concrete slab was cast on the limestone. The slab measured approximately 37 cm in thickness for approximately half of the facility floor area. This area was the location where both SASW and crosshole seismic tests were performed. The remainder of the floor area gently sloped downward towards a sump used for water collection.

### 3.1 TESTING WITH A "STIFF" SEAFLOOR

SASW measurements were performed on the concrete slab as shown in Figure 5. In this case, the concrete represented a "stiff" seafloor as defined by Roever, et al (1959) and Rauch (1986) because the shear wave velocity of the concrete was greater than the compression wave velocity of the overlying water ($C_W$). Underwater accelerometers were used as receivers, a piezoelectric vibrator was used as an underwater seismic source, and a dynamic signal analyzer was used as the recording device. The vibrator was placed in a water-tight cannister with a protruding stainless-steel foot resting on the concrete slab. To couple the receivers to the concrete, flat-head nails were fixed in the concrete with hydrostone cement immediately after the concrete slab was cast. The nail heads, exposed on the slab surface, were arranged in a linear array and effectively produced two receiver spacings of nominally 10 cm and 20 cm. The underwater accelerometers were attached to the nail heads with magnets.

### 3.2 TESTING WITH A "SOFT" SEAFLOOR

Upon completion of the SASW tests on the "stiff" seafloor, a 46-cm thick layer of lightly cemented sand was placed over the concrete slab. The sand had approximately 3.5 percent by weight of cement added to it. Immediately after placement, flat-head nails were pushed into the cemented sand to create a linear SASW testing array with two receiver spacings of nominally 15 cm and 30 cm. The cemented sand was allowed to cure for five days before testing commenced. At that time, the cemented sand had a shear wave velocity of 760 m/s which modelled a "soft" seafloor as defined by Roever et al (1959) and Rauch (1986) because $V_{S\ seafloor}$ was less than $C_W$. The degree of saturation of the sand layer when it was submerged was estimated to be about 40 percent based on laboratory tests. Crosshole seismic tests performed in the sand during SASW testing gave a compression wave velocity of 1330 m/s.

## 4. Experimental Results

When water overlies a solid, the propagation velocity of surface waves may be affected by the water. If the wave velocity is affected by the water, the surface wave is termed a Scholte wave. If the wave velocity is unaffected, the surface wave has been termed a pseudo-Rayleigh wave (Roever et al, 1959). Both wave types were measured in this work. However, the easiest wave to measure was a Scholte wave propagating along a "soft" seafloor. The most comprehensive set of these data was generated by testing the lightly cemented sand layer. Dispersion curves measured on the sand with a 30-cm receiver spacing and water depths ranging from 18 to 71 cm are presented in Figure 6. The curves are expressed in terms of normalized velocity (Scholte wave velocity, $V_{SC}$, divided by $V_S$ of the cemented sand) versus normalized wavelength (Scholte wavelength, $\lambda_{SC}$, divided by the water depth, h). Selected data points are shown for the normalized dispersion curves to allow identification of measurements at individual water depths because the curves plot close together. Analytical results from the companion paper are also shown.

Measured dispersion data shown in Figure 6 have normalized Scholte wave velocities ranging from 0.83 to 0.90, with an average value of about 0.86. By comparing with the analytical results, it can be seen that the data fall in the "deep" water range, where water has a constant effect on propagation velocity. Scholte wave velocity is reported to be of the order of 0.88 $V_S$ for soft seabed conditions (Rauch, 1986), depending upon the properties of the sediment. The analytical model predicts a ratio $V_{SC}/V_S = 0.84$ for the range of normalized wavelengths. Although the experimental data fluctuates somewhat, the average value ($V_{SC}/V_S \approx 0.86$) compares favorably with the analytical model.

Additional tests were also performed on other "soft" seafloors. Measurements were made on the concrete slab while it was curing and also on the limestone bedrock before the concrete slab was placed. These results are shown in Figure 7 in terms of "deep" water measurements, where "deep" water was conservatively selected to be conditions when wavelengths were less than one-half the water depth. Good agreement exists between the experimental and analytical results. The good agreement also holds for the case of "stiff" seafloors ($V_S/C_W > 1$) which was determined on the cured concrete slab.

Dispersion curves measured on the cured concrete slab have been presented by Stokoe et al, 1990. Good agreement between experimental and analytical Scholte wave velocities was found for "deep" water tests with a 10-cm accelerometer spacing. However, for the 20-cm accelerometer spacing, interface wave velocities were measured which were not influenced by the overlying water layer but were essentially the same as $V_R$ measured when the conrete slab was tested dry. These velocities are referred to as pseudo-Rayleigh wave velocities by Roever et al (1959) and Rauch (1986). Work is continuing in this area to understand better the parameters which control measurement of Scholte versus pseudo-Rayleigh waves.

## 5. Summary and Conclusions

A small-scale facility was constructed to evaluate application of the SASW method offshore. Initial experimental studies were conducted to measure the velocity of surface waves propagating along various water/solid interfaces. Solid materials with differing stiffnesses were used to represent the seafloor. "Soft" seafloor conditions ($V_S < C_W$) were modelled with a lightly cemented sand, limestone bedrock and curing concrete. For these conditions, Scholte wave velocities were measured which are in good agreement with analytical predictions. "Stiff" seafloor conditions were modelled with a cured concrete slab. For this condition, both Scholte and pseudo-Rayleigh waves were measured, depending on the source/receiver arrangement, and work is continuing to understand better the controlling parameters.

## 6. Acknowledgements

This work was initiated as an experimental study with support from ISMES, Bergamo, Italy. Support for all phases was continued with funding from the National Science Foundation through the Offshore Technology Research Center, Texas A&M University and The University of Texas at Austin. Support from these organizations is gratefully acknowledged.

## 7. References

Rauch, D. (1986) "On the Role of Bottom Interface Waves in Ocean Seismo-Acoustics: A Review," Ocean Seismo-Acoustics: Low-Frequency Underwater Acsoutics Conference, La Spezia, Italy, June 1985, Marine Science, Vol. 16, Plenum Press, New York, pp. 623-641.

Roever, W.L., Vining, T.F. and Strick, E. (1959) "Propagation of Elastic Wave Motion from an Impulsive Source Along a Fluid/Solid Interface: I. Experimental Pressure Response, II. Theoretical Pressure Response, III. The Pseudo-Rayleigh Wave," Publication No. 173, Shell Development Company, Exploration and Production Research Division, Philosophical Transactions of the Royal Society of London, Vol. 251A, pp. 455-523.

Stokoe, K.H., II, Wright, S.G., Roesset, J.M., Gauer, R. and Sedighi-Manesh, M. (1990) "In Situ Measurement of Stiffness Profiles in Ocean Bottom Materials Using the SASW Method," Offshore Technology Conference, Houston, Texas, OTC Paper No. 6234, pp. 299-305.

Roesset, J.M, Wright, S.G. and Sedighi-Manesh, M. (1990) "Analytical Investigation of Seismic Surface Waves in the Seafloor," in J.M. Hovem, M.D. Richardson and R.D. Stoll (eds.), Shear Waves in Marine Sediments, Kluwer Academic Publishers, Dordrecht, (in press).

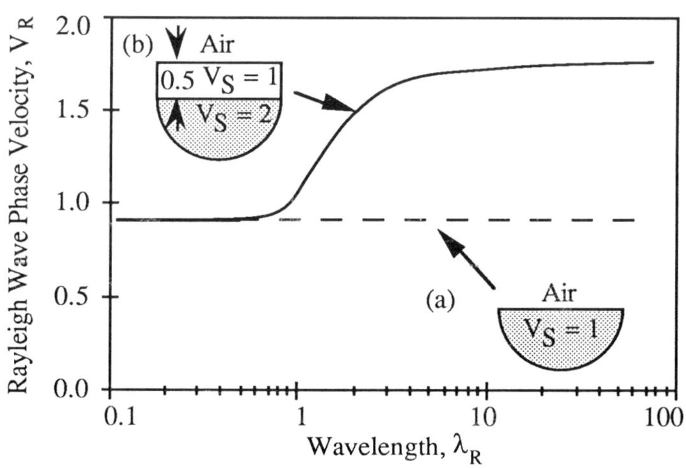


Figure 1 Dispersion curves for plane Rayleigh waves propagating in: (a) uniform half-space and (b) softer layer over stiffer half-space.

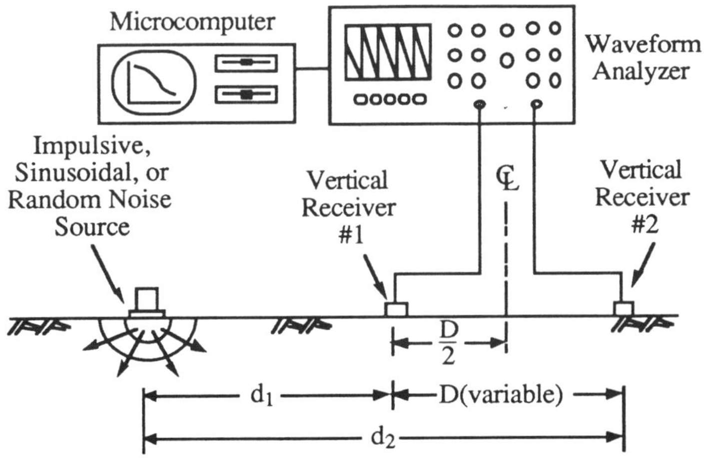


Figure 2 Typical equipment configuration for SASW testing on land.

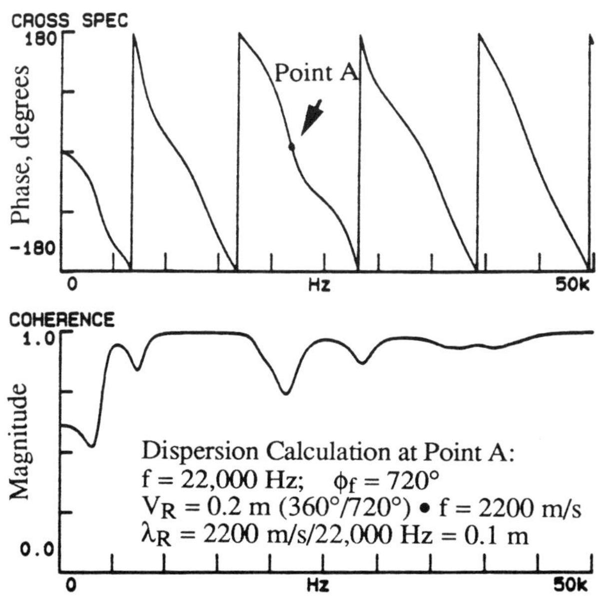


Figure 3 Cross-power spectrum and coherence measurements made on dry concrete slab in small-scale facility.

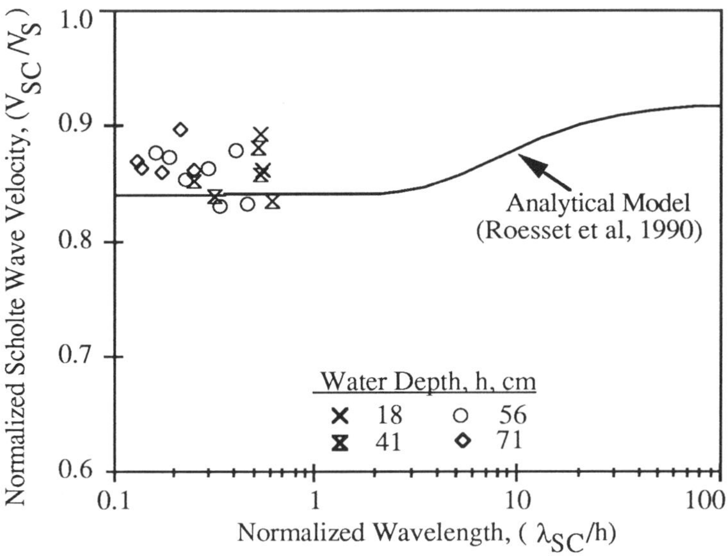


Figure 6 Comparison of experimental and analytical dispersion curves for SASW measurements on the lightly cemented sand layer.

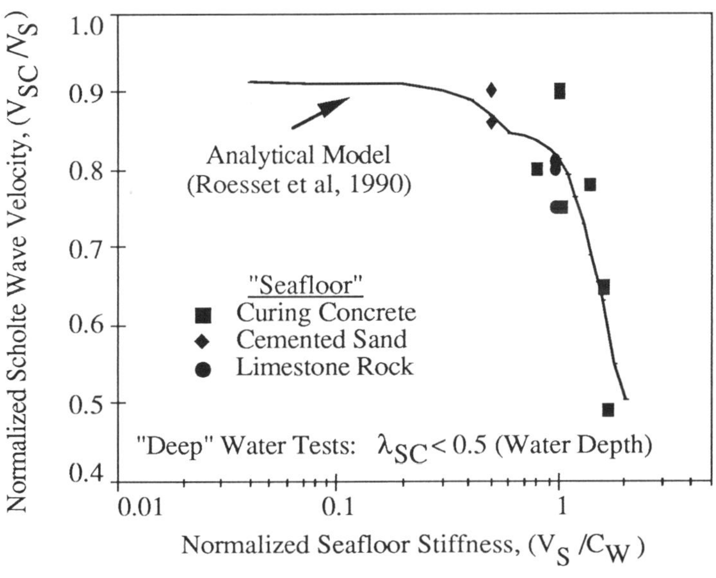


Figure 7 Comparison of experimental and analytical values of normalized Scholte wave velocities for seafloors of different stiffnesses tested in deep water.

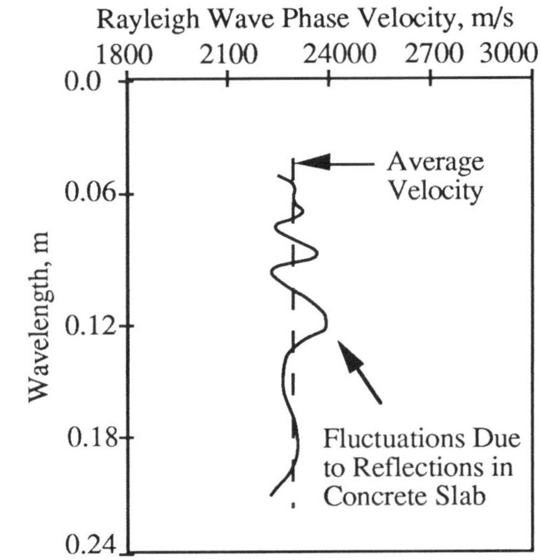


Figure 4 Dispersion curve resulting from phase record shown in Figure 3.

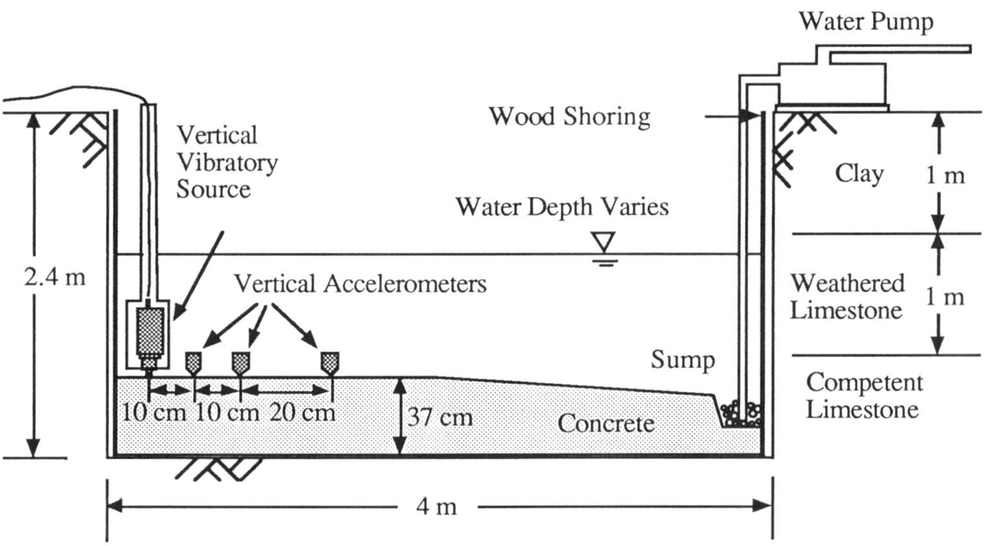


Figure 5 Small-scale facility used in experimental testing.

# LABORATORY STUDIES ON PULSED LEAKY RAYLEIGH WAVE COMPONENTS IN A WATER LAYER OVER A SOLID BOTTOM

Jacques R. Chamuel
Sonoquest Advanced Ultrasonics Research
P.O. Box 153, Wellesley Hills, MA 02181-5339
U.S.A.

ABSTRACT. Laboratory experimental results are presented demonstrating the existence of multiple pulsed leaky Rayleigh wave components in a liquid layer over a solid bottom. These broadband pulsed leaky Rayleigh waves are not revealed from normal mode phase and group velocity dispersion curves. Waveforms from a free liquid layer model are compared with waveforms from a liquid layer over a solid half-space model. The time separation between different leaky Rayleigh wave pulses is independent of range, however, the time separation between waveguide acoustic pulses in a free liquid layer increases with range. The results indicate that the amplitude of subsequent pulsed leaky Rayleigh wave components increase with range over finite regions creating late-arriving high-amplitude seismo-acoustic waves. The laboratory waveforms are related to underwater Canadian Arctic field data.

## 1. INTRODUCTION

At the interface of a liquid half-space in contact with a high-shear-speed solid half-space, a leaky Rayleigh wave exists [1-2]. The leaky Rayleigh wave is represented by a complex pole from the characteristic equation:

$$4[1-(V/\alpha)^2]^{\frac{1}{2}}[1-(V/\beta)^2]^{\frac{1}{2}}-[2-(V/\beta)^2]^2 = \frac{i\rho V^4[1-(V/\beta)^2]^{\frac{1}{2}}}{\rho_s \alpha^4[(V/C)^2-1]^{\frac{1}{2}}} ; \qquad (1)$$

where V, $\alpha$, $\beta$, C, $\rho$, and $\rho_s$ are the interface wave velocity, solid

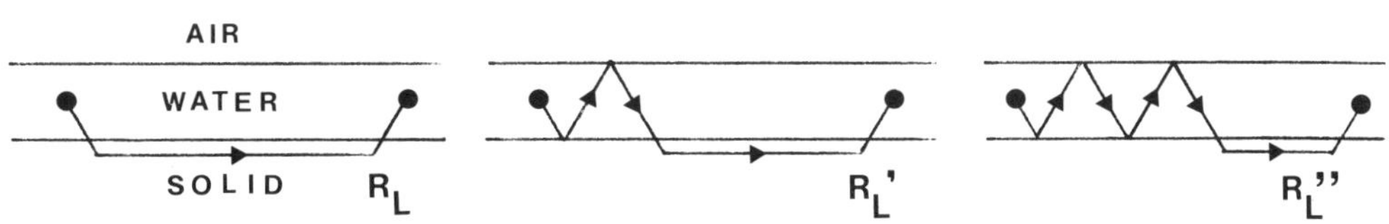


Fig. 1. Multiple pulsed leaky Rayleigh wave components.



J. M. Hovem et al. (eds.), Shear Waves in Marine Sediments, 59–66.

compressional wave velocity, shear wave velocity, fluid compressional wave velocity, fluid density, and solid density, respectively. The leaky Rayleigh wave decays with range and radiates its energy to the liquid with the angle $\Theta_{RL} = \sin^{-1} C_w/C_{RL}$, where $C_w$ is the water compressional wave velocity and $C_{RL}$ the leaky Rayleigh wave velocity.

When the liquid half-space is replaced with a liquid layer, the right-hand term of equation (1) is multiplied by D, where $D = \tan[(2\pi fH/V)((V/C)^2-1)^{\frac{1}{2}}]$. The lowest propagating mode is dispersive [1-3] and has its phase velocity approaches: a) the Scholte wave velocity at high kH (where k is the wavenumber and H is the layer thickness), and b) the free Rayleigh wave velocity at very low kH. The coexistence of a leaky Rayleigh wave at large kH is not indicated. The liquid-filled borehole problem in the geophysics field is somewhat analogous and there seem to be unresolved issues regarding the leaky Rayleigh wave [4-6]. These highly dispersive normal modes have been called "shear modes" by Phinney [7], "reflected conical waves" by Biot [8], "reflected waves" by Peterson [10], "pseudo-Rayleigh waves" by Tsang et al. [5], and "pseudo-Rayleigh (reflected) waves" by Cheng et al. [6]. It is important to differentiate between waves, modes, and rays.

In the liquid layer model, the radiated pulsed leaky Rayleigh wave propagates in the liquid to the top free surface and reflects back towards the solid. The reflected wave interacts with the solid at a later time generating a second leaky Rayleigh wave pulse $R_L'$ (Fig.1). The first leaky Rayleigh wave pulse travels in the solid along the interface with supersonic speed and does not interact with the late-arriving reflected acoustic wave.

Laboratory experimental investigations were conducted to assess the potential role of low-frequency leaky Rayleigh waves associated with the ocean bottom. Underwater acoustic waves propagating in a "perfect waveguide" air/water/air were compared with waves in a water layer over a hard solid half-space.

## 2. EXPERIMENTAL SET UP

The basic experimental set up was previously described [11-12]. The ultrasonic model consisted of a large aluminum block representing a hard solid half-space. The source and receiver were identical broadband (50-800 kHz) piezoelectric transducers (Sonoquest 3.5 mm). The source was excited with a single 3 μs pulse. The measured bulk elastic properties of the aluminum block were $C_p$ = 6328 m/s, $C_s$ = 3126 m/s, and density of 2.7 kg/m$^3$. The measured Rayleigh wave velocity, however, was $C_R$ = 2952 m/s indicating that the shear wave velocity was slightly higher near the free surface of the aluminum block ($C_s$ = 3168 m/s). The water compressional wave velocity was $C_w$ = 1480 m/s.

A 0.09 mm thick plastic sheet was placed on the top surface of the aluminum block. The edges of the plastic sheet were taped to the aluminum surface leaving a thin air layer trapped between the plastic sheet and the block. The block was then immersed in water with a water layer of thickness H above it. Source and receiver transducers were located near the top surface of the water layer. The source-receiver

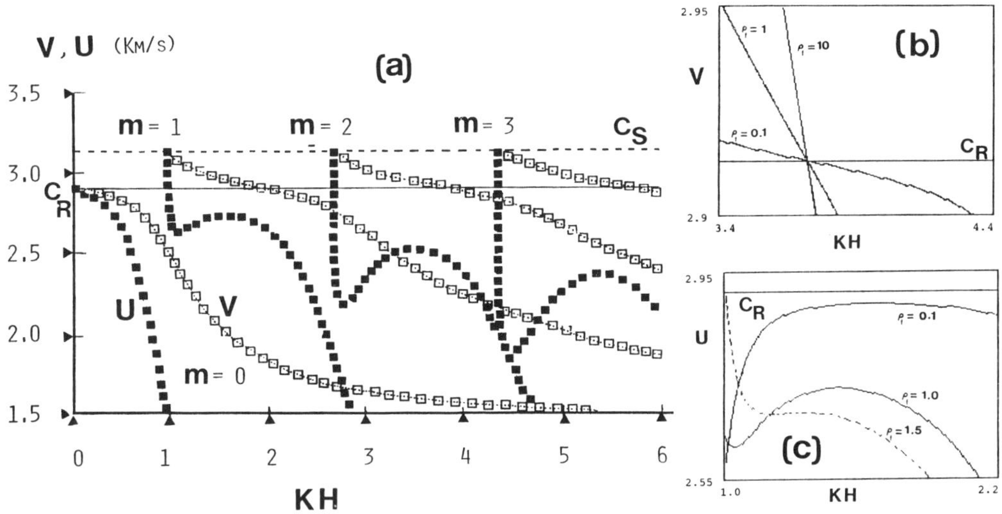


Fig. 2. Theoretical normal mode phase and group velocity dispersion curves for a water layer over an aluminum half-space.

range X was selected in each experiment. With the plastic sheet in place, "perfect waveguide" data were obtained and recorded from the air/water/air model. The plastic sheet was then removed without disturbing the set up to allow the water layer to become in contact with the aluminum block. The data from the air/water/aluminum model were obtained, recorded, and compared with the corresponding perfect waveguide results. The procedure was repeated for different X and H.

## 3. RESULTS

Theoretical phase V and group velocity U normal mode dispersion curves were calculated using slightly different water/aluminum model parameters $C_p$ = 6300 m/s, $C_s$ = 3122 m/s, $C_w$ = 1485 m/s, and a solid/liquid density ratio of 2.7 (Fig.2(a)). Each of the phase velocity curves had one single point at the free Rayleigh wave velocity when D = 0. Fig. 2(b) shows a segment of the third normal mode phase velocity expanded near the Rayleigh wave and calculated at three different fluid densities (0.1, 1.0, and 10 kg/m$^3$). All three phase velocity curves intersected precisely at $C_R$ similar to Paillet's results [4] for a liquid layer between two solids. The corresponding half-space leaky Rayleigh wave velocities, however, were greater as shown in Table 1. The laboratory aluminum model had a Rayleigh wave velocity $C_R$ = 2952 m/s and a measured leaky Rayleigh wave velocity $C_{RL}$ = 2964 m/s. Fig.2(c) shows the group velocity of the first mode expanded near $C_R$ again at the same three fluid densities. Notice that the group velocity curves of Fig. 2(c) did not intersect each other at $C_R$. The normal mode dispersion curves do

TABLE 1. Calculated leaky Rayleigh wave root

| Fluid Density ($x10^3$ $kg/m^3$) | Leaky Rayleigh Wave Root (km/s) |
|---|---|
| 0 | 2.91318 (Rayleigh wave velocity) |
| 0.1 | 2.9132 - 0.08592i |
| 1.0 | 2.9235 - 0.08584i |
| 5.0 | 3.2122 - 0.42269i |
| 10.0 | 5.1814 - 2.31610i |

not reveal the presence of nondispersive leaky Rayleigh waves. Strick has shown that the point of zero pressure at a liquid/solid interface travels with precisely the free Rayleigh wave velocity independent of the fluid. From Snell's law, the leaky Rayleigh wave angle $\Theta_{RL}$ was equal to 29.95°.

Waveforms from a free water layer model (H = 5 cm, and X = 9.5 cm) (Fig. 3(a)) were compared with waveforms from an air/water/aluminum model (Fig. 3(b)). Pulses 1 and 2 in Fig.3(a) corresponded respectively to one bottom water/air reflection, and to bottom/top/bottom water/air reflections as shown. When the water layer was in contact with the aluminum block, major leaky Rayleigh wave components were generated in addition to the specularly reflected acoustic waves. Ray number 4 in Fig.3(b) was incident at the Rayleigh angle, traveled along a portion of the interface with the leaky Rayleigh wave velocity, and radiated into the water with the same angle. At the generating point, the radiated leaky Rayleigh wave 3 was superimposed on the specularly reflected water wave. Ray 3 traveled upward, was reflected by the water/air boundary, and interacted again with the aluminum regenerating a second leaky Rayleigh wave component having its own radiated acoustic field in the water. The geometry of Fig.3 allowed near-overlapping of the specularly reflected acoustic wave with the radiated leaky Rayleigh wave. The calculated times of arrival of rays 1, 2, 3, and 4 were 93.2 μs, 149.6 μs, 151.2 μs, and 90.6 μs, respectively. The phase of each component depended on the number of reflections from the air or aluminum interfaces causing constructive or destructive interference.

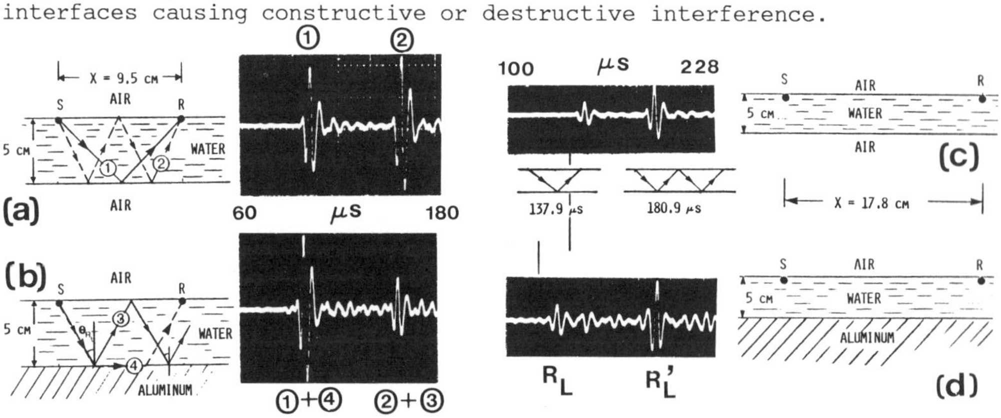


Fig. 3. Waveforms from a free water layer are compared with waveforms from a water layer over an aluminum bottom.

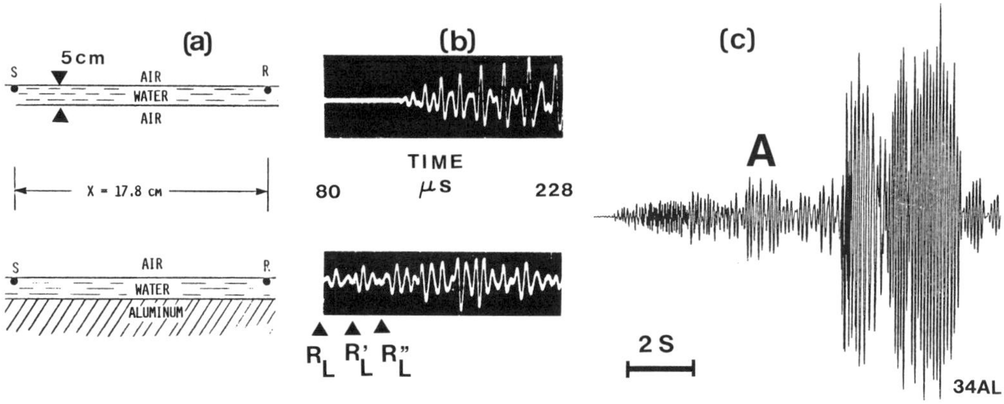


Fig. 4. Multiple pulsed leaky Rayleigh wave components observed in the laboratory (Fig. 4(b)) may explain the arrival of supersonic waves (Fig. 4(c) region A) obtained in the Canadian Arctic by G. H. Brooke/DREP.

In Fig.3(c-d), H remained equal to 5 cm, and X was increased to 17.8 cm to clearly separate the first acoustic bottom reflected wave (137.9 μs) from the leaky Rayleigh wave (118.6 μs). Notice that the leaky Rayleigh wave arrived before the bottom reflected acoustic wave.

A shallow-water example is represented in Fig. 4 (H = 1.5 cm, and X = 17.8 cm). The acoustic waves in the "perfect waveguide" consisted of a series of pulses starting at $t = X/C_W$ (Fig. 4(a)). The air/water/aluminum model had multiple low-frequency leaky Rayleigh wave components (Fig. 4(b)) arriving at 77.6 μs, 95.1 μs, and 112.7 μs before the "perfect waveguide" acoustic waves. These leaky Rayleigh wave components may explain the early arrival of low-frequency waves observed by Brooke in 1977 in the Canadian Arctic indicated by region "A" in Fig.4(c). The water depth was 150 m, the range 13 km, the source depth 20 m, the receiver was on the ocean bottom, and the shear wave velocity was roughly close to that of aluminum. The leaky Rayleigh waves described in this section exist in addition to the lateral waves mentioned by Brekhovskikh [9] and Brooke [10].

The previous results were obtained with the source and receiver placed near the top surface of the water layer. Acoustic waves in the water interfered with the leaky Rayleigh wave components. In order to isolate the leaky Rayleigh wave components, the transducers were mechanically isolated from the water and were allowed to couple only to the surface of the solid in the following experiments.

Fig.5 demonstrates the effect of increasing the water layer thickness on the leaky Rayleigh wave pulses. The effect of increasing the range is demonstrated in Fig.6(a). One observes that the time spacing between the leaky Rayleigh wave components is constant and does not depend on range unlike the "perfect waveguide." Also we observe that the amplitude of the direct leaky Rayleigh wave $R_L$ drops with range, however, the amplitude of $R_L'$ increases with range over a particular region. The relative amplitudes of $R_L$ and $R_L'$ as function of range are plotted in Fig.6(b). The star markers at the bottom of the graph indicate the location of the reflection regions of the water acoustic waves, which were generated by the leaky Rayleigh wave components after returning from the top surface.

The experiment of Fig.7 used a water layer over granite and a machined shallow crack to demonstrate that a surface crack could stop $R_L$'. In Fig.8, the source and receiver transducers were placed on the aluminum block and were acoustically decoupled from the water to respond only to elastic waves in the solid. The received waveform had a small $R_L$ and a large $R_L$' components (Fig. 8(a)). The fluid was then acoustically decoupled from the solid over a region to prevent the

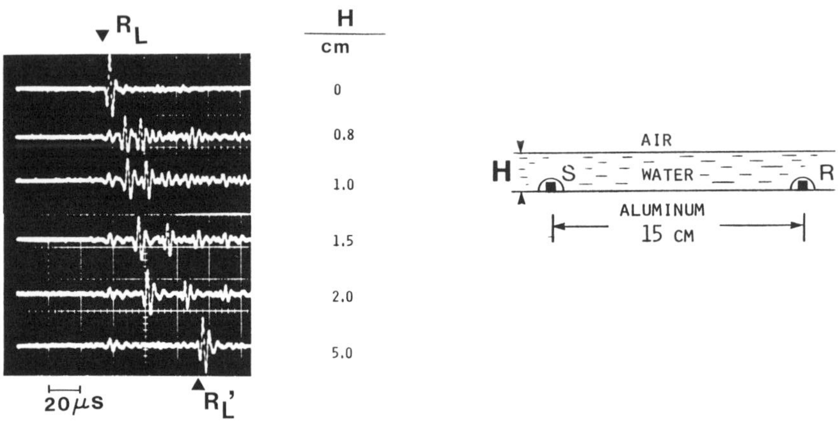


Fig. 5. Effect of water layer thickness on pulsed leaky Rayleigh waves.

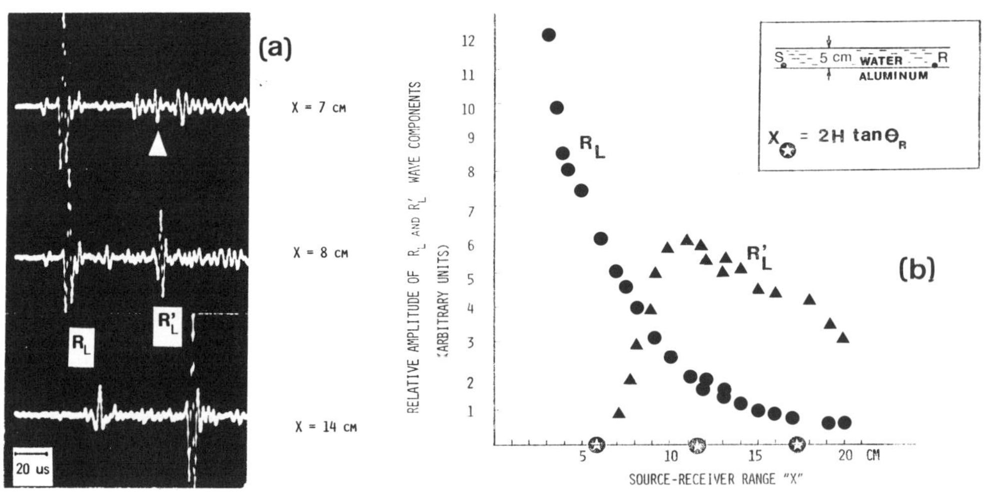


Fig. 6. Effect of range on leaky Rayleigh wave pulses. Notice increase of amplitude of $R_L$' with range over region.

interaction of reflected acoustic waves from the top surface of the water. The amplitude of $R_L$ increased, and $R_L'$ almost vanished. This experiment demonstrated that $R_L'$ was indeed generated by the reflected acoustic waves.

## 4. CONCLUSIONS

Laboratory experimental results were presented demonstrating the existence of multiple pulsed leaky Rayleigh wave components propagating in a liquid layer over a high-shear-speed bottom. The presence of leaky Rayleigh wave components is not revealed from normal mode phase velocity dispersion curves (although there exists one discrete point on each curve with the free Rayleigh wave velocity independent of the liquid density). The measured velocity of a pulsed leaky Rayleigh wave peak is greater than the free Rayleigh wave velocity. The amplitude of pulsed leaky Rayleigh wave components increase with range over finite regions creating late-arriving high-amplitude seismo-acoustic waves. Leaky Rayleigh waves affect the energy partitioning between underwater acoustic and seismic waves.

The findings indicate that transient leaky Rayleigh wave phenomena may affect the determination of ocean bottom reflection and attenuation coefficients, and inversion of acoustic field data to determine bottom parameters. The laboratory results provide physical insight into low-frequency shallow-water acoustic propagation and borehole geophysics.

## 5. ACKNOWLEDGEMENTS

This work was supported by the U.S. Office of Naval Research Code 1125AR and by the Defence Research Establishment Pacific, Canada.

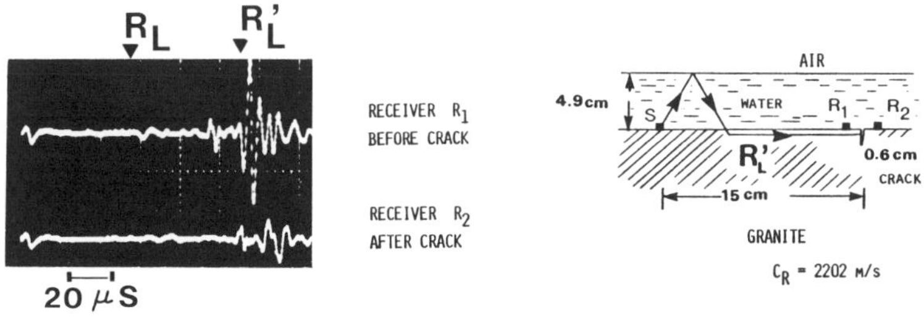


Fig. 7. Experiment demonstrating that $R_L'$ component can be interrupted by a surface shallow crack in granite.

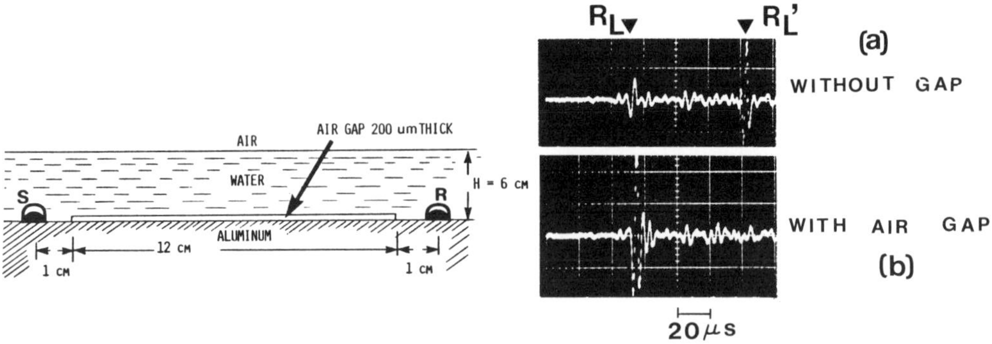


Fig. 8. Acoustic waves radiated from $R_L$ wave are responsible for the generation of the seismic $R_L'$ wave.

## 6. REFERENCES


1. Ewing, W. M., Jardetsky, W. S., and Press, F. (1957) Elastic Waves in Layered Media, McGraw-Hill Book Co., New York.
2. Brekhovskikh, L. M. (1960) Waves in Layered Media, Academic Press, New York.
3. Tolstoy, I. A. and Clay, C. S. (1987) 'Ocean Acoustics/ Theory and Experiment in Underwater Sound', 1987 edition, page 136, Amer. Inst. of Phys., New York.
4. Paillet, F. L. and White, J. (1982) 'Acoustic Modes of propagation in the borehole and their relationship to rock properties', Geophys. 47(8), 1215-1228.
5. Tsang, L. and Rader, D. (1979) 'Numerical evaluation of transient acoustic waveform due to a point source in a fluid-filled borehole', Geophysics 44(10), 1706-1720.
6. Cheng, C. H., Toksoz, M. N. and Willis, M. E. (1982) Determination of in situ attenuation from full waveform acoustic logs', J. Geophys. Res. 87, 5477-84.
7. Phinney, R. A. (1961) 'Leaking modes in the crustal waveguide: The oceanic PL wave' J. Geophys. Res. 66(5), 1445-1469.
8. Biot, M. A.(1952) 'Propagation of elastic waves in a cylindrical bore containing a fluid', J. Appl. Phys. 23, 997-1005.
9. Brooke, G. H., Thomson, D. J. and MacKinnon, R. F. (1986) 'Some characteristics of virtual modes in shallow water with high speed bottom', in Ocean Seismo-Acoustics Low Frequency Acoustics, edited by T. Akal and J. M. Berkson, NATO Conference Series IV:16, 233-242.
10. Peterson, E. W. (1974) 'Acoustic wave propagation along a fluid filled cylinder' J. Appl. Phys. 45, 3340-3350.
11. Chamuel, J.R. (1990) 'Arctic Acoustics Ultrasonic Modeling Studies' Sonoquest Advanced Ultrasonics Research, Massachusetts, Report No. JRC-10-90. Available from National Technical Information Center document No. AD-A224 165/1/GAR.
12. Chamuel, J. R., and Brooke, G. H. (1988) 'Transient Scholte wave transmission along rough liquid-solid interfaces', J. Acoust. Soc. Am. 83(4), 1336-1344.
13. Strick, E. (1959) 'Propagation of elastic wave motion from an impulsive source along a fluid/solid interface: III. The pseudo-Rayleigh wave', Phil. Trans. Roy. Soc. London, ser. A (251), 488-523.

ASSESSMENT OF SHEAR STRENGTH OF THE SEA BOTTOM FROM SHEAR WAVE VELOCITY MEASUREMENTS ON BOX CORES AND IN-SITU

FR. THEILEN and I. A. PECHER
*Institut für Geophysik der Christian Albrechts Universität zu Kiel*
*Olshausenstr. 40*
*2300 Kiel*
*Fed. Rep. of Germany*

**ABSTRACT.** The velocities of compressional (P-) and shear (S-) waves are important parameters used for the identification of sediment types and for geotechnical applications. Pulse transmission techniques have been developed for both types of waves and are applied directly to fresh box core samples. Various sediment cores from the Barents Sea were investigated with respect to seismic velocities, shear strength, density, porosity and other physical properties in the upper nine meters of sediment. Compressional wave velocities did not change significantly within this depth range; whereas, the shear wave velocities increased rapidly from 10 to 40 m/sec. Dynamic shear modulus and shear wave velocity were correlated with shear strength. This correlation allows in-situ estimations of shear stress in the subsurface from shear wave velocity data.

## 1. Introduction

The combined evaluation of compressional and shear waves together with wet bulk density yields compressional and shear moduli of the sediment. Both are dynamic moduli, which can depend on frequency. They are not identical to the static moduli which are used for engineering purposes. Nevertheless, these data are necessary for the investigation of sites for offshore engineering and installations.

P- and S-wave investigations have been performed by Muckelmann (1985) and Meissner and Theilen (1983) with respect to seismic velocity and attenuation on artificial samples of well sorted quartz sands. It was shown that: (i) velocity and seismic quality factor Q increased with the effective pressure; (ii) velocity and quality factor Q of P-waves were strongly dependent on small amounts of gas in the pore fluid; whereas, shear waves were not affected; and (iii) that S-waves had a broader spectrum of velocity due to changes of frame properties than P-waves. Ultrasonic P- and S-wave investigations by Matthews (1982) showed that velocities of S-waves can vary between values of 30 and 100 m/sec, whereas P-waves were nearly constant within a range of 1490 and 1510 m/sec. Extremely low S-wave velocities were also reported by Shirley and Hampton (1978) with values ranging between 2 and 40 m/sec. These authors investigated the relationships between velocity, attenuation and consolidation. Schultheiss (1985) investigated marine sediments of the DSDP-leg 86 in the Northwest Pacific with respect to bulk density, water content, gamma ray attenuation, porosity,



*J. M. Hovem et al. (eds.), Shear Waves in Marine Sediments*, 67–74.

P- and S-wave velocities, shear strength, consolidation and permeability. He found an increased shear wave velocity with shear strength at four different sites. All the investigations of the uppermost layers of the sea floor showed that S-waves have a broader range of velocity from a few m/sec up to 400 m/sec, compared to the relatively small range of P-wave velocity from 1480 up to 2250 m/sec. In general, S-waves seem to be more sensitive to changes in the frame structure of sediments than P-waves and are therefore very useful for the investigation of physical properties of sediments.

The dynamic elastic moduli characterize the elastic properties of sediments and rocks within the small linear stress-strain regime. The strain is always less than $10^{-6}$ for normal seismic waves. Shear strength, on the other hand, is defined for an extreme non-linear stress-strain regime. It characterizes the breaking point of the material. There is no physical model available which theoretically combines both types of data. Therefore shear strength cannot be easily predicted from the dynamic shear modulus. On the other hand, it seems plausible that a relation between shear rigidity and dynamic shear modulus exists. The relationships between shear moduli and shear strength is of importance for the solution of geotechnical problems. If relationships between both kinds of data are known for different sediment types, shear strength could be determined indirectly by in-situ measurements of P- and S-wave velocities at the seafloor. However, S-wave velocities and shear strength values have still to be correlated for different types of sediments.

## 2. Instrumentation and Methods

Different methods are available for the measurement of seismic velocities of marine sediments. Previous investigations for the determination of velocities, which are based on samples taken from a core, have been performed using a Hamilton frame (Boyce, 1971). Matthews (1982) and Schultheiss (1985) used a modified Hamilton frame for the investigation of S-waves. Selected samples within the core, however, are sometimes not representative of the structure of the whole core. Layers with extreme values due to compaction or diagenetic processes may be omitted by chance. This may lead to erroneous velocity profiles along the core that are difficult to correlate with high resolution reflection seismic profiles. A new pulse transmission method has been developed for box cores. Pulses are transmitted parallel to the axis of the core so that velocity-depth profiles can be obtained.

P- and S-wave measurements carried out on large square box cores are preferable to measurements on samples with smaller cross-sections for two reasons:

(1) A nearly plane wave can be observed due to the high cross-section of the core, and
(2) Core shortening is significantly smaller than e.g. for gravity cores.

The technical concept of the system for determination of P- and S-wave velocities is shown in Fig. 1. Box core samples have lengths of six meters and cross-sections of 30 x 30 cm. P-waves were generated by an echosounder transducer which was mounted at the front side of the core. The transducer has a resonance frequency of 33 kHz. The energy of the transducer was sufficient for the penetration of the signal through the entire core length. Two piezoceramic bender elements were used as receivers which were mounted on a frame that can be moved to any selected position along the core. At each position they were lowered to the surface of the core without penetrating the sediment. In this way the core is not destroyed for further geological investigations. Due to very low S-wave velocities in unconsolidated sediments and small signal penetration,

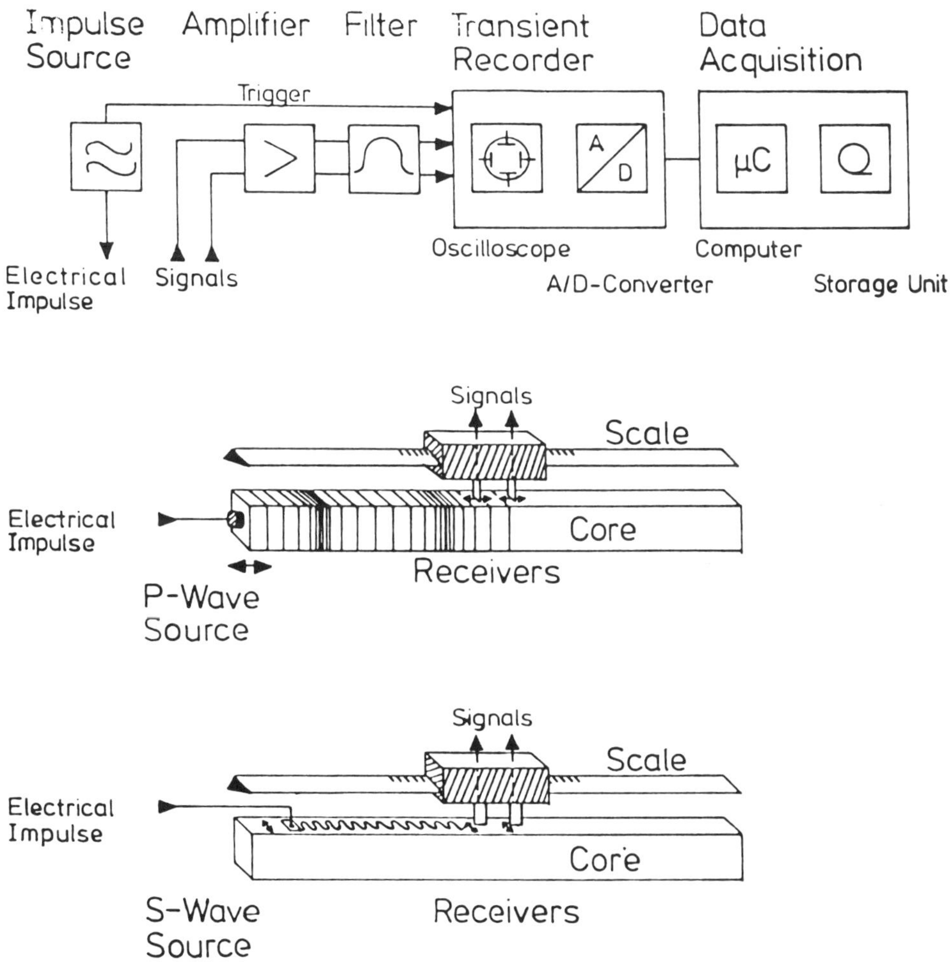


Fig. 1 Concept for pulse transmission measurements on sediment cores.

some modifications had to be introduced for S-wave measurements. A special electromagnetic transducer was developed that could be mounted on the side of the core generating horizontally polarized S-waves with a particle motion perpendicular to the core axis. The signal frequency is about 300 Hz. Due to strong attenuation of S-waves penetration was only about 50 cm. In order to get an S-wave profile along the whole core the transducer has to be moved stepwise along the core. The receivers were the same as for P-waves, but they had to be mounted in a way that the bending direction was the same as the particle motion. The signals were recorded on a two-channel transient recorder and stored on a Personal Computer. The processing and evaluation of the data can be performed directly on board ship.

### 3. Results

Velocities of P- and S-waves, sand content, and wet bulk density have been investigated for various cores from the Barents Sea. Sand content was generally low in the uppermost 3 meters with a high variability in the lower part of the core. Wet bulk density increased with depth from 1.3 to 1.8 gr/cm$^3$.

P-wave velocity of 1400 m/sec was very low in the uppermost 3 meters with the highest deviations in the lower part having a mean value of 1480 m/sec. The relative changes of velocities were small. Shear waves, on the other hand, showed a strong increase with depth from 10 to 40 m/sec at a depth of 8 meters below seafloor, as shown in Fig. 2. Correspondingly, the dynamic shear moduli also increased with depth.

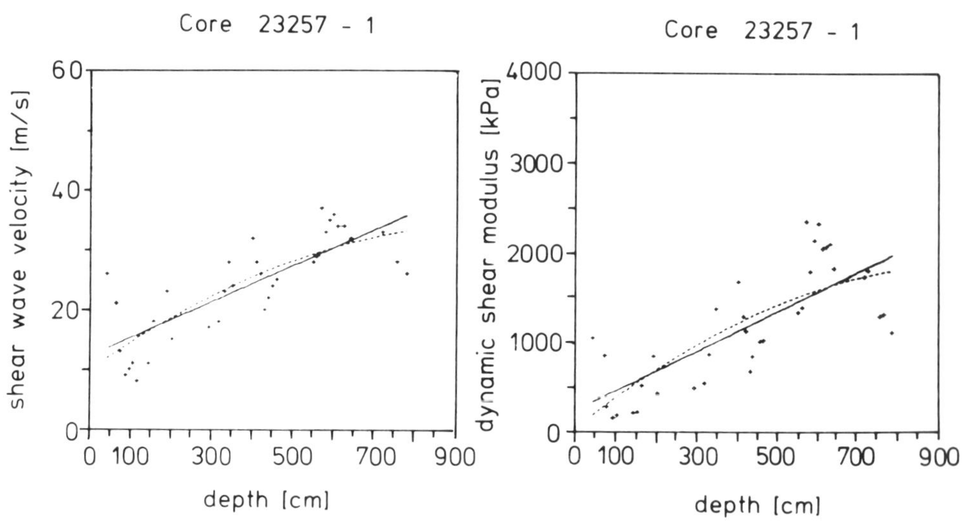


Fig. 2 Shear wave velocities and dynamic shear modulus as a function of depth.

A similar behaviour was observed for shear strength which was measured at small distances along the cores (Holler and Kassens 1989). These values, too, increased with depth. The correlations of shear wave velocity and dynamic shear modulus with shear strength is shown in Fig. 3. There was nearly a linear relationship between both parameters indicating that the dynamic shear modulus is about 200 times higher than the shear strength.

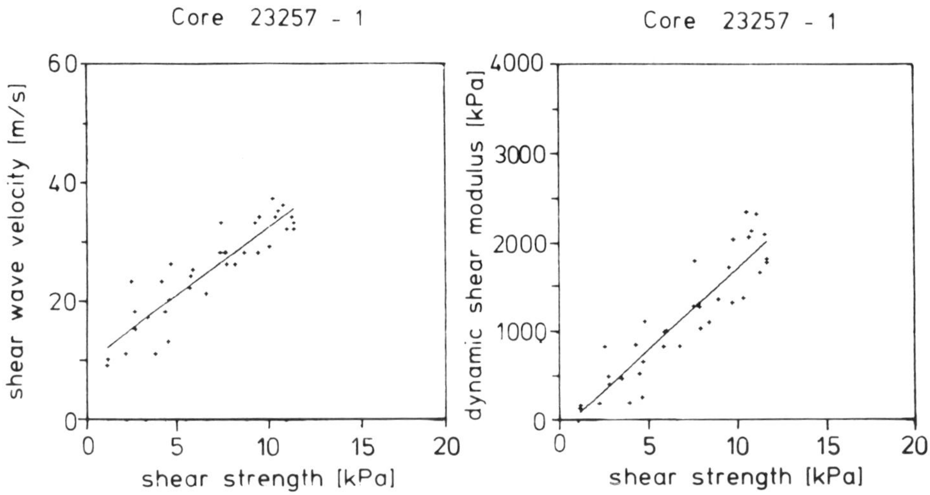


Fig. 3 Relationship between shear wave velocities, dynamic shear modulus and shear strength.

In all cases which have been investigated P-wave velocity was nearly independent of depth with the exception of peaks which can be attributed to changes in material or early diagenetic processes (Kassens, 1990). Density, S-wave velocity, and shear strength considerably increased with depth. This is probably due to compaction of the sediments with overburden pressure. For P- and S-waves an increase of velocities with increasing effective pressure (difference between overburden pressure and pore pressure) has also been observed from laboratory measurements (Meissner and Theilen, 1983; Muckelmann, 1985). The relative increase in S-waves velocity, however, is much stronger than for P-waves. The S-wave velocities depend mainly on grain structure and moduli of the sediment frame; whereas P-wave velocities are a function of elastic moduli of the grains, sediment frame, and pore fluid. As marine sediments near the sea floor have a high porosity, the effect of consolidation is much higher on S-wave than on P-wave velocities.

These effects can principally be described by the theory of Digby (1981) who calculated the effective shear- and compressional moduli for a package of grains under confining pressure. Adhesive forces between the grains are taken into consideration. They can explain an existing shear modulus even without any confining pressure. The adhesive forces and the effective pressure increase the contact radii between the grains resulting in a higher friction and stiffness of the sediment frame. Prasad (1988) showed that such a contact radius model can well describe the changes of velocity in a sediment sample under load. It is questionable whether the contact radius model can also describe the change of shear strength with confining pressure. Calculation of effective elastic moduli according to Digby (1981) is based on a linear theory in a small stress-strain regime. Shear strength is determined under extreme non-linear conditions. Shear strength of a sediment, however, is obviously also effected by increase in stiffness due to consolidation.

## 4. Methods for Shear Wave Measurements at the Sea Floor

The relationships found between shear velocities and shear strength values as a function of depth yield an access to in-situ data of strengths using shear wave velocity measurements at the sea floor.

A method for direct shear wave measurements at the sea floor has been developed by Gimpel (1987) and Gehrmann et al. (1984). This concept is applicable to shallow marine environments with depth up to 40 meters (Fig. 4).

The instrumentation consists of a shear wave source generating SH-waves at the sea floor. The pulses are generated with an airgun with horizontally oriented exhaust ports. Pulses are transferred to the sea floor by the edges of the sledge which carries the airgun. The source is towed at low speed by the ship.

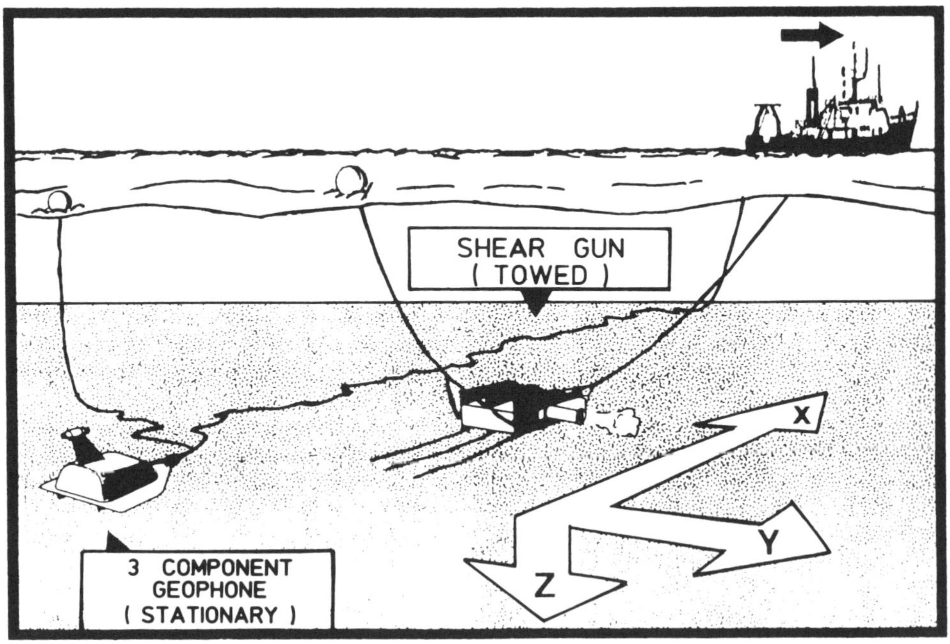


Fig. 4 Concept for shear wave measurements at the sea floor.

Signals are recorded by a three component geophone together with a hydrophone. In order to avoid noise, it is placed at a fixed position during the measurements. It is linked to the ship by a combined towing and data transmission cable with a length up to 1000 meters, which is layed out as the ship moves away. At the maximum length of the cable, measurements are stopped and the geophone is towed back to the ship and it is again lowered to the sea floor for another profile. By this method, expanded spread profiles can be gathered, yielding structure and velocities of the subsurface down to a depth of 50 meters.

Results from the Baltic Sea obtained by in-situ shear wave measurements showed a strong increase of shear wave velocities from 170 m/sec at the sea bottom up to 380 m/sec at a depth of about 20 meters below the sea floor.

## 5. Conclusions

The experiments have shown a good correlation between dynamic shear modulus and shear strength. Both depend on effective pressure and compaction. For this reason, both kinds of data increase with burial depth. However, the relationship between the shear strength and the shear modulus depends on sediment type and sedimentation history. This could be shown by comparison of data from the Baltic Sea and the Barents Sea. Knowledge of the relationship of both kinds of data by empirical means opens a new way for the estimations of in-situ shear strength data below the sea floor. Correlation of shear wave velocity with shear strength from laboratory measurements can be transferred to S-wave velocity data from in-situ experiments at the sea floor. This yields a new method for in-situ estimation of shear strength as a function of depth along profiles and allows lateral variation of these data to be estimated. These investigations have to be combined with high resolution P-wave reflection seismic measurements in order to get information on the sedimentary structures in the subsurface (Theilen and Pecher, 1990). Application of such an in-situ experiment can contribute to the investigation of stability problems at the sea floor with respect to offshore engineering activities. A weak stability of marine soft sediments may cause damage to any offshore installation at the sea floor. In-situ prediction of the shear strength is a contribution to a more precise presite survey.

## 6. References


Boyce, R.E. (1971) 'Appendix I: Physical properties methods', in: N. T. Edgar and J.B. Saunders et al.: Init. Repts. DSDP, 15, Washington (U.S. Govt. Printing Office), pp 1115-1128.

Digby, P.J. (1981) 'The effective elastic moduli of porous granular rocks', J. Appl. Mech., ASME, Vol. 48.

Gehrmann, T., Gimpel, P. and Theilen, F. (1984) 'Marine Shear Wave Profiling', in: Expanded Abstracts and Biographies, 1984 Technical Praogram, 54th Annual International SEG Meeting, Atlanta, USA.

Gimpel, P., (1987) 'Marine flachseismische Untersuchungen in der Kieler Bucht unter besonderer Berücksichtigung von Scherwellenmessungen', Ph. D. Ths., Univ. Kiel.

Holler, P. and Kassens, H. (1989) 'Sedimentphysikalische Eigenschaften aus dem Europäischen Nordmeer' (Datenreport F.S. METEOR Reise 7), Ber. Sonderforschungsbereich 313, Univ. Kiel, No. 15.

Kassens, H. (1990) 'Verfestigte Sedimentlagen und seismische Reflektoren: Frühdiagenese und Paläoozeanographie in der Norwegischen See', Ph. D. Ths., Univ. Kiel.

Matthews, J.E. (1982) 'Shear wave velocity measurements in marine sediments', Geo-Marine Letters Vol. 2, pp 215-217.

Meissner, R. and Theilen, F. (1983) 'Attenuation of seismic waves in sediments', in Proceedings of the 11th World Petroleum Congress, 2, John Wiley, London.

Muckelmann, R. (1985) 'Theoretische und experimentielle Untersuchungen von P- und S-Wellen in Sanden unter besonderer Berücksichtigung ihrer Dämpfungseigenschaften', Ph. D. Ths., Univ. Kiel.

Prasad, M. (1988) 'Experimental and theoretical considerations of attenuation and velocity interactions with physical parameters in sands', Ph. D. Ths., Univ. Kiel.

Schultheiss, P. J. (1985) 'Physical and geotechnical properties of sediments from the Northwest Pacific', in Hearth, G. R. and Burckle, L. H., et al., Init. Repts. DSDP, 86, Washington (U.S. Govt. Printing Office).

Shirley, D. J. and Hampton, L. D. (1978) 'Shear-wave measurements in laboratory sediments', J. Acoust. Soc. Am., Vol. 63, No. 2, pp 607-613.

Theilen, F. and Pecher, I. A. (1990) 'Shear-wave investigations of the sea floor', in: 'Estimation and Practical Use of Seismic Velocities'. EAEG/SEG Research Workshop. Expanded abstract, EAEG, Zeist, pp 364-365.

## 7. Acknowledgements

We thank the officers and crews of RV "Poseidon" and RV "Meteor" for their assistance during the surveys. This study was supported by the German Research Foundation. It was performed within the Joint Research Project SFB 313 at Kiel University.

# NUMERICAL MODELLING AND LABORATORY EXPERIMENTS ON UNDERWATER SOUND PROPAGATION OVER A SHEAR SUPPORTING BOTTOM

ALLEN HUNDLEY and STEWART A.L. GLEGG
*Center for Acoustics and Vibration*
*Department of Ocean Engineering*
*Florida Atlantic University*
*Boca Raton, FL 33431*

JUAN I. ARVELO
*Code U25, Naval Surface Warfare Center*
*Silver Spring, Maryland 20903-5000*

HERBERT UBERALL
*Department of Physics*
*The Catholic University of America*
*Washington, D.C . 20064*

ABSTRACT. This paper describes a detailed comparison between a numerical model and an experiment of underwater sound propagation over a bottom which can support shear waves. The experiment was carried out at model scale in a large water tank, and measured the transmission loss, as a function of depth and range, in a water column above a geoacoustic model of the sediment and sub-bottom. The material used to model the sediment was an epoxy resin ($c_p$ = 2450 m/s, $c_s$ = 1120 m/s) overlying a concrete basement ($c_p$ = 3300 m/s, $c_s$ = 2100 m/s). The numerical model is based on normal mode theory, which incorporates the effect of shear in the sediment. Good agreement between the measurements and predictions were found at high frequencies, where the acoustic wavelength was much smaller than the water depth. However, at low frequencies the transmission loss was overpredicted. This error was attributed to the interface wave which was not included in the numerical prediction code.

## 1. Introduction

Sound propagation in shallow water is strongly affected by the boundary conditions imposed by the bottom and sub-bottom, as well as temperature gradients in the water column itself. Although these properties can be measured at specific points there still remains a certain degree of uncertainty concerning the accuracy with which the environment has been sampled. Consequently, when full scale measurements are compared with theoretical prediction schemes there will always be doubt concerning the accuracy of the parameters which have been used as a basis for the calculations. For this reason theoretical models need to be checked against accurate laboratory scale measurements, where all the experimental parameters are known, before they can be applied with confidence to full scale environments. In underwater acoustics there appears to be very little laboratory scale measurement data available for this purpose. Previous studies are limited to those of Eby et al [1] and Chamuel and Brooke [2], but neither of these considered a multilayered sub-bottom which can have a important impact on the



*J. M. Hovem et al. (eds.), Shear Waves in Marine Sediments*, 75–81.

acoustic field in the water column. Consequently, to verify the existing numerical codes which predict long range propagation over multilayered shear supporting bottoms, detailed measurements are required in a range independent laboratory scale environment.

There are several numerical models which can be used to predict sound propagation in a range independent environment with a shear supporting bottom. Ellis and Chapman [3] included shear wave effects in the bottom of a simple Pekeris waveguide model and achieved reasonable agreement with full scale measurements. However, they recognized the need for a multilayered model. Collins [4] extended the PE method to accommodate bottom elasticity in a range dependent waveguide, and Schmidt [5] included shear waves in the Fast Field program. Arvelo and Uberall [6] have developed a normal mode solution, which includes shear waves and can be extended to two dimensional range dependent environments.

The purpose of this study is to present the results of a laboratory experiment of sound propagation over a multilayer, range independent bottom which can support shear waves. The experiments were carried out in a large water tank over a range of frequencies and source/receiver displacements, giving a comprehensive set of measurements of the transmission loss versus depth in the water column. These experiments are described in Section 3 of this paper. The results of these measurements have been compared with the normal mode prediction method given by Arvelo and Uberall[6], and a summary of this method is described in Section 2. The theoretical model and the experimental results are compared in Section 4.

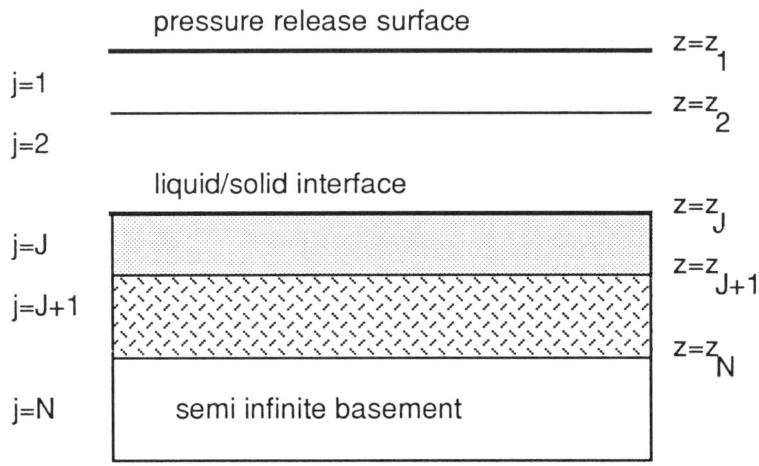


Figure 1: The multi-layered model of a range independent environment

## 2. The Numerical Model

The numerical model is based on normal mode theory and is capable of calculating acoustic propagation in range dependent ocean environments with sediments and substrates which can support shear waves. The model considers a multi-layered medium bounded at the top by a liquid layer with a pressure-release upper surface and at the bottom by a semi infinite solid (Figure l). At the depth $z = z_J$ there is a liquid solid interface, and the model is sufficiently general that the bulk modulus of the material in each layer can vary linearly with depth.The normal modes (eigenfunctions) and phase speeds (eigenvalues) are calculated using an up layer matching technique which couples the waves in each layer using the boundary conditions at the interfaces .This procedure allows the velocity potential in any layer to be defined in terms of the velocity potentials of the compressional wave and the shear wave in the semi-infinite basement. By invoking the pressure release boundary condition at the upper surface the complete solution for the wave field can be obtained.

Propagation only occurs at specific phase speeds, which define the normal modes. To evaluate the phase speed a characteristic equation is specified using the boundary condition that there must be zero tangential stress at the solid /liquid interface. Solutions to the characteristic equation are found by searching a limited region of the complex k-plane using a Levenberg-Marquardt minimization algorithm. The transmission loss for a point source in the water column is then obtained from the sum of all the propagating modes.

This method provides the complete solution for a range independent environment in terms of the densities and wave speeds in each layer. The attenuation coefficients of the waves can also be included by specifying an imaginary part for the wavenumbers in each layer.

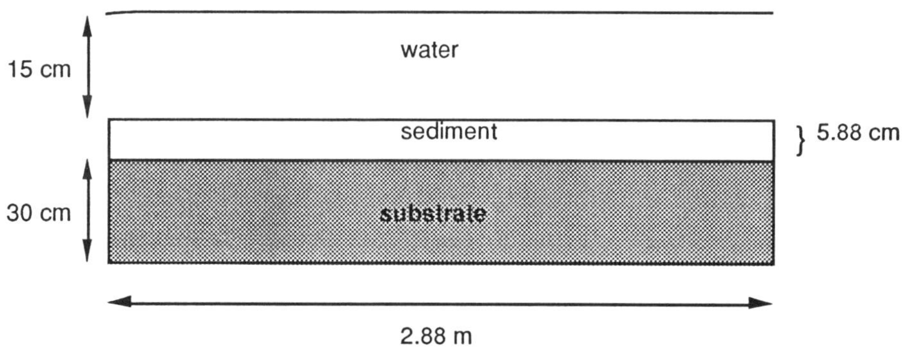


Figure 2: Illustration of the three layered model used in the experiment

## 3. The Experimental Model

A three layer model of the ocean was constructed, as illustrated in Figure 2. This is typical of many shallow water environments where a surficial sediment overlies a harder substrate rock such as basalt or limestone. The acoustic properties of substrate materials are specified in terms of their density, compressional and shear wave velocities, and absorption coefficients. It was found that concrete has similar properties to limestone and so it was used as the base for the model. The material used had a compressional wave speed of 3300 m/s , a shear wave speed of 2100 m/s and a density of 2300 $Kg/m^3$ .

The sediment layer was more difficult to model realistically with commonly available materials. Using a fine sand as a model sediment at laboratory scale can result in problems caused by air entrapment and inhomogeneities in the material over large areas. As an alternative, resin type materials were investigated. The properties of polyurethane resin and commercial two part laminating epoxy were compared with the values found in marine sediments [7]. The compressional and shear wave velocities in the resins are higher than those of most marine sediments, but the densities are lower. Calculations of the acoustic impedance for these materials show that they all lie in the same range. The optimum material was found to be an epoxy resin with a mixing ratio between the hardener and the base of 2:3 giving a compressional wave speed of 2450 m/s , a shear wave speed of 1120 m/s and a density of 1030 $Kg/m^3$ .

The wave speeds in the epoxy resin were measured on samples of the material using a pulse technique. However, it was not possible to measure the absorption coefficients. These were estimated from measurements reported by Hartman [8], at frequencies of over 1 MHz and were linearly extrapolated down to the frequencies used in the experiment. Consequently this may be a cause of error in the numerical calculations , but parametric studies have not indicated that this is significant.

The seafloor model was mounted at the bottom of a large test pool . Care was taken to level the surfaces to an accuracy of 2 mm so that all the surfaces were parallel (with a maximum error of $0.1^{o}$). Acoustic tests were carried out to check the effect of reflections from the pool side walls and the edges of the model. No reflections from the side walls were observed, and measurements using pulsed CW excitation or pure CW excitation were found to be the same. The homogeneity of the acoustic field in the water above the model was checked by moving a source and receiver with a fixed separation to different locations on the model. CW excitation was used at frequencies of 8 KHz, 15 KHz and 25 KHz and the results showed variations of less than 2 dB at ten different locations.

## 4. Experimental Results.

Measurements of transmission loss were made at ranges of 1 m through 2.2 m at 0.2 m intervals for CW excitation at frequencies of 8 KHz, 15 KHz, 20 KHz, 25 KHz and 30 KHz. For all measurements the source depth was 50 mm (equivalent to 0.33 water depths). Typical transmission loss versus depth profiles are shown in Figures 3 and 4 for frequencies of 8 kHz and 25 kHz at a fixed range of 1.4 m. The data are the average values of transmission loss from up to four repeats of the experiment over a two month period. The scatter in the measurements was always less than 2dB over the complete data set.

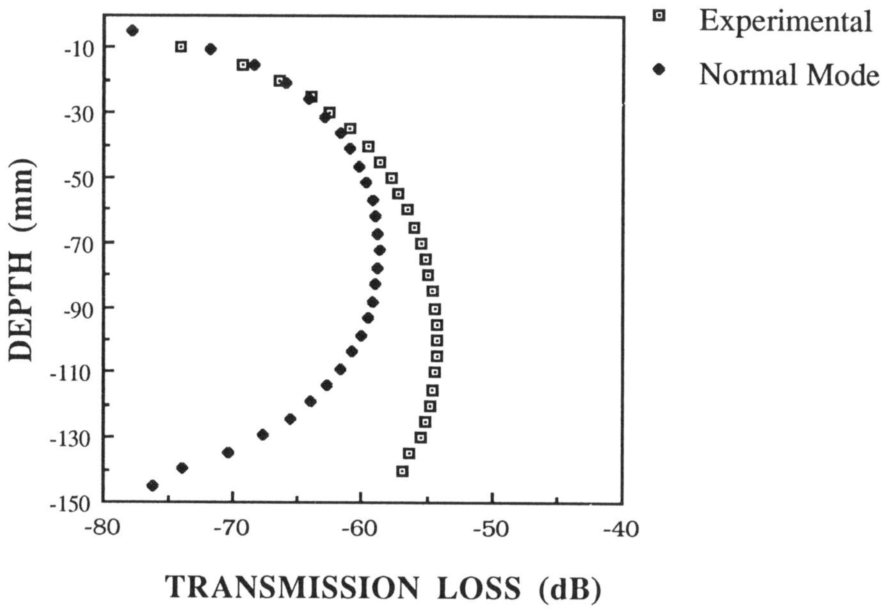


Figure 3: Comparison of the transmission loss versus depth curves ay 8kHz. and a range of 1.4 m. Note 60 dB has been added to the transmission loss values.

The results presented in Figures 3 and 4 are compared with the levels calculated using the numerical model and show the trends which were observed. The agreement between the calculated and measured values at frequencies above 20 kHz are within the bounds of experimental error. However, at the lower frequencies (8 kHz and 15 kHz) there is a significant overprediction of the transmission loss especially close to the bottom of the water column. The numerical prediction method does not include the interface wave and this is the most probable cause of the error. The interface wave propagates efficiently at low frequencies and its influence extends approximately a wavelength above the water sediment interface. At the lowest frequency considered here (8 kHz) the wavelength is 187 mm which is greater than the water depth. Consequently the interface wave can be expected to make a contribution to the sound field in the water column even when the source is located relatively close to the surface (50 mm). At higher frequencies the interface wave will not be excited as strongly and so the numerical prediction method is expected to be more accurate as shown by Figure 4 .

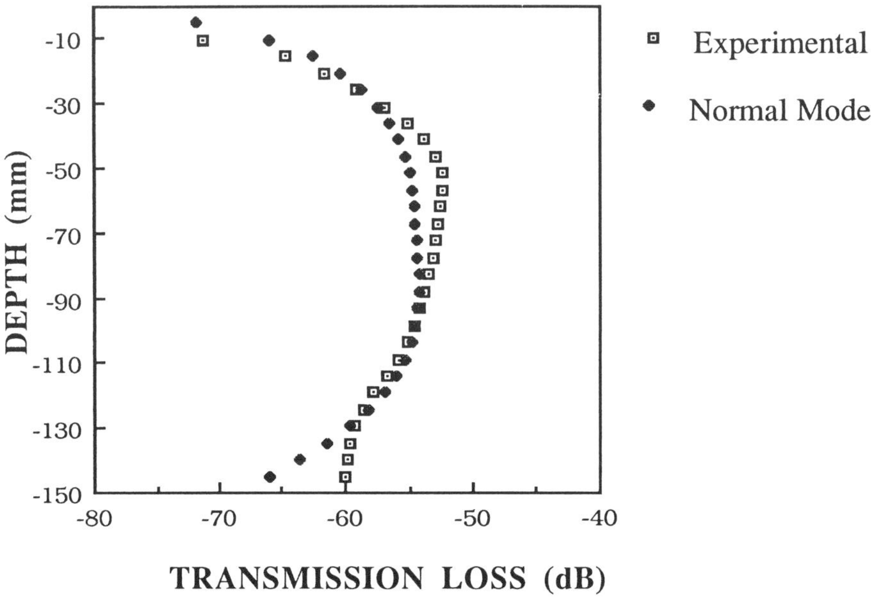


Figure 4: Comparison of the transmission loss versus depth curves ay 25kHz. and a range of 1.4 m. Note 60 dB has been added to the transmission loss values.

## 5. Conclusion.

This paper describes the results of a model scale experiment which measured the transmission loss in a range independent environment over a shear supporting bottom, and the comparison of the measurements with a numerical prediction method based on normal mode theory. The results validate the prediction method at high frequencies but show significant errors at low frequencies. When the acoustic wavelength is of the same order as the water depth, the transmission loss is overpredicted close to the water/sediment interface. This error is attributed to the interface wave which is not included in the numerical prediction method.

## Acknowledgements

Thanks are extended to Mr. Stephan Williams for performing wave velocity tests at the Allied Geophysical Laboratory, University of Houston. This work was supported by the Office of Naval Research.

**References**

[1] Eby, R.K., Williams, A.O., Ryan, R.P., and Tamarkin, P. (1960) "Study of Acoustic Propagation in a Two Layered Media", Journal of the Acoustical Society of America, Vol. 32 #1, 88-99, 1960.
[2] Chamuel, J.R., and Brooke, G.H. (1988) "Shallow Water Acoustic Studies Using an Air-Suspended Water Waveguide Model", Journal of the Acoustical Society of America, Vol. 84 , 1777-1786
[3]Ellis, D., and Chapman, D. (1985) "A Simple Shallow Water Propagation Model Including Shear Wave Effects", Journal of the Acoustical Society of America Vol 78(6), 2087- 2095
[4]Collins, M. (1989) "A Higher-Order Parabolic Equation for Wave Propagation in an Ocean Overlying an Elastic Bottom", Journal of the Acoustical Society of America Vol 86(4),1459-1464
[5]H. Schmidt (1988) "SAFARI User's Guide", SACLANTCEN SR-113, SACLANT Undersea Research Center, La Spezia, Italy
[6]Arvelo, J.I. and Uberall, H. (1990) "Adiabatic Normal Mode Theory of Sound Propagation Including Shear Waves in a Range Dependent Ocean Floor", Journal of the Acoustical Society of America, Vol 88(5), 2316-2325
[7]Hamilton, N.G. (1982) "Sound Velocity and Related Properties of Marine Sediments", Journal of the Acoustical Society of America, Vol. 72 #6, p. 1891-1903
[8]Hartman, B. (1984) Encyclopedia of Polymer Science and Engineering, Vol. 1 (reprint for U.S. Naval Surface Weapons Center), 131-160, Wiley and Sons, New York

# A REVIEW OF LABORATORY SHEAR WAVE TECHNIQUES AND ATTENUATION MEASUREMENTS WITH PARTICULAR REFERENCE TO THE RESONANT COLUMN

J. D. Bennell and D. Taylor Smith
School of Ocean Sciences
University of Wales, Bangor
Menai Bridge
Gwynedd, UK.

ABSTRACT. Developments in laboratory techniques used to measure the shear wave properties of marine sediments will be discussed in this paper, highlighting the particular problems associated with the different methods employed. Unconfined marine sediments pose particular problems when trying to make measurements without disturbing the sample and altering its properties. On the other hand confined sediments, simulating depth of burial, are easier to test for shear wave propagation in a resonant column device, although this too is shown to have its limitations. A modified resonant column system, will be described, which allows the measurement of shear wave velocity, shear modulus and shear damping in marine sediments utilising several different techniques simultaneously. This allows measurements over a wide range of frequencies and strains to be performed, providing a means of determining the non linear aspects of shear wave propagation.

## 1. Introduction.

The determination of dynamic shear velocity, modulus and damping in marine sediments has become increasing important over the last few years in fields as diverse as earthquake engineering, civil engineering, marine acoustics and marine geophysics. On land this can be achieved by the in situ measurement of shear wave velocity to assess the undisturbed characteristics of a large volume of material. In the marine environment the in situ measurement of shear wave velocity is only just becoming a viable procedure. A second best alternative has been to measure shear wave propagation characteristics on sediment samples in the laboratory. The effects of sample disturbance and the unrepresentative nature of any particular sample are often hard to evaluate or overcome. Some success has been achieved in predicting in situ values from laboratory tests by allowing for secondary consolidation effects which lead to a stiffening of the sample with time of confinement (Anderson and Stokoe 1978). Measurements in the carefully controlled environment of the laboratory do, however, offer the best means of evaluating the relationships between shear wave propagation and the physical and geotechnical properties of sediment samples under various loading conditions. Laboratory measurements have already allowed the establishment of many functional relationships between, for example shear modulus and effective stress, void ratio and shear strain amplitude, as well as allowing the evaluation of various theoretical shear wave propagation models. Laboratory results from appropriate samples have also been used for correcting in situ values of geophysically derived shear moduli to values more appropriate to engineering type problems which involve higher shear strain amplitudes (Bennell et al 1984).

Broadly speaking the measurement of shear wave velocity and damping in natural materials



*J. M. Hovem et al. (eds.), Shear Waves in Marine Sediments*, 83–93.

can be achieved by two main laboratory techniques

i) Pulse excitation, and

ii) continuous vibration torsional resonance

A third possibility might be included as a quasi-static method of determining shear modulus and damping from low frequency torsional cyclic loading of a confined sample.

## 2. Pulse Measurements.

This technique utilises piezoelectric transducers to generate and detect low strain amplitude, shear wave pulses which have travelled through a sample of material, rather like the propagation of ultrasonic compressional waves. Thus the pulse velocity can be determined over a known path length or lengths by measuring the transit times through the material. Several methods have been used to generate pulsed shear waves in solids some of which are applicable to marine sediments.

### 2.1. SHEAR PLATES AND RADIAL EXPANDER TRANSDUCERS.

An idealised shear plate transducer is shown in fig. 1. It utilises either a Y-cut quartz crystal or a piezoelectric ceramic that is polarised perpendicular to the applied field. As an electrical signal is applied across the crystal the material is deformed as shown. If the face which experiences transverse motion is in contact with a medium, a shear wave is propagated in a direction perpendicular to the transverse motion. Because such a transducer exhibits reciprocity it can be used both to generate and detect shear waves. A radial expander transducer is similar, being designed to impart maximum shear deformation by expanding laterally. Usually the radial mode resonant frequency is less than the thickness mode, but is still likely to be of the order of 100kHz for a reasonable sized transducer. The effectiveness of these transducers depends on how they are coupled to the material to be tested. For solids they can either be cemented or clamped on to the faces of the sample. For a sediment too much pressure may result in the sample being damaged. Laughton (1957) utilised Y-cut quartz crystals in a sediment consolidation cell to determine shear wave velocity. Only after applying a significant load, however, was he able to identify a shear wave arrival. Stephenson (1978) used radial expanding crystals of 90kHz resonant frequency to measure the shear wave velocity of reconstituted samples of silty clay at various degrees of saturation. The transducers were coupled to the sample using a compressive strength testing frame in order to maintain a constant pressure between the probes and the specimen. He found that he could easily identify the shear wave arrival fig.1, but the identification may have been aided by the fact that

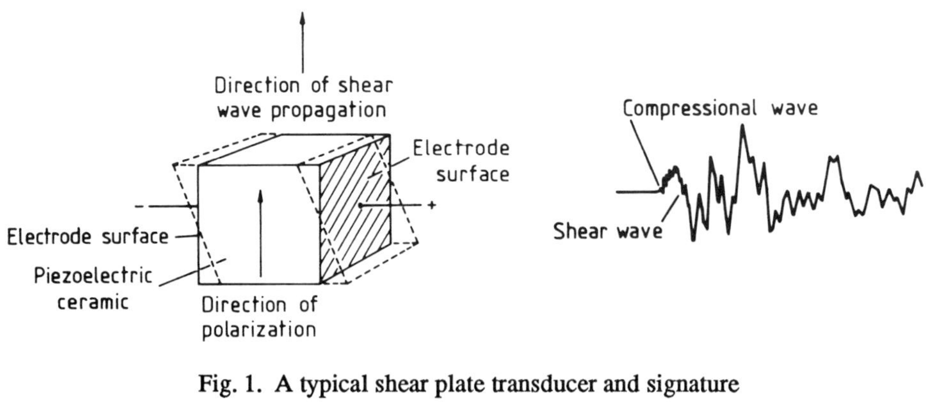


Fig. 1. A typical shear plate transducer and signature

the samples were not 100% saturated so that the compressional wave was highly attenuated. Even so the signature displayed is highly complicated and shear wave identification as a later arrival is somewhat subjective.

The main problems in using these transducers to measure shear waves in marine sediments are

i) Identification of the shear wave arrival itself and the need to apply sufficient pressure between the transducer and sample,

ii) without making the transducer too large only high resonant frequency crystals can be used where attenuation in the sediment will absorb most of the shear wave energy, and

iii) the amount of mechanical motion that can be transferred to the sample is small due to the high stiffness of the transducer and the low stiffness of a sediment sample. ie there is a large mismatch between the characteristic impedance of the transducer and that of the sediment.

## 2.2 BENDER TRANSDUCERS.

An entirely different type of piezoelectric element was developed by Shirley (1978) and Shirley & Hampton (1978), to overcome these problems. They fabricated a transducer that used ceramic bimorph or bender elements to generate and detect shear waves in sediment samples. Each bender element consists of two lead zirconate titanate, (PZT4), transverse expander plates bonded to a centre electrode. The polarisation of the ceramic material in each plate, and the electrical connections, are such that when a driving voltage is applied to the element one plate elongates and the other contracts. The net result is a bending displacement which is greater in magnitude than the length change in either plate. If this element is coupled to a sediment sample, a shear wave is propagated perpendicular to the soil particle motion.(fig. 2). The resonant frequency of such an element is inversely proportional to its length squared compared to a shear plate which is inversely proportional to its length. Thus a bender element has a much lower resonant frequency than a shear plate. Shirley and his co-workers used a stack of six bender elements (fig.2), separated by a layer of high compliance material, to increase the area of the radiating surface. The whole transducer is potted in epoxy resin to exclude moisture. This type of arrangement relies on the contact of the end of the transducer onto the sample surface. Even so they were able to measure shear wave velocities, as low as 2m/sec, in kaolinite clay. They were also able to make attenuation measurements by recording amplitudes of the received pulses at several separations.

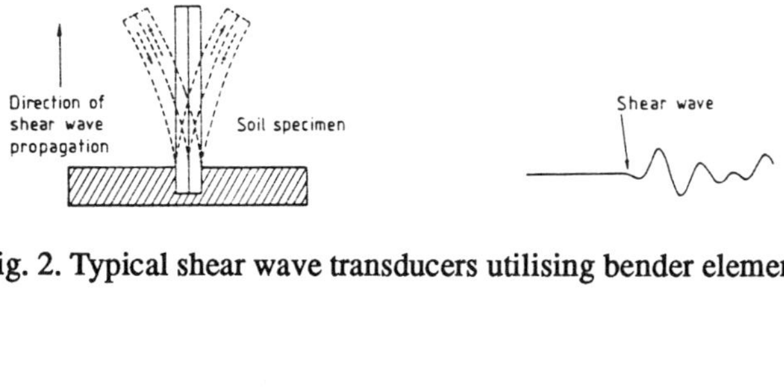


Fig. 2. Typical shear wave transducers utilising bender elements

A single bender element can also be used, Shirley(1978), which although not as robust as the stack of elements can be inserted into a sediment sample with minimum disturbance. This increases the radiating area of the transducer and hence the coupling of the shear waves into the sediment. The motion of the element is damped by the surrounding sediment, and thus the shear wave length in the sediment becomes the controlling factor in the frequency response of the element. The transducers can be mounted in different configurations and with various geometries. In the cantilever mode one end of the crystal is clamped whilst the other is free to vibrate. Alternatively the crystal need not be clamped at all, in which case it will tend to curl up like a leaf when subjected to an applied voltage. Two types of bender elements are commercially available, series connected and parallel connected. Schultheiss(1983) showed that a parallel connected element is twice as effective as a series element when used to generate shear waves; whereas a series connected element is more effective as a receiver. The bender element may be energised by either a high voltage spike, the leading edge of a low voltage square wave or a gated sine wave.

The great advantage of these bender transducers is that there characteristic impedance matches the sediment impedance far better than any shear plate transducer. They can be used just as easily to measure shear waves in unconsolidated sediments, as in sediments subjected to some form of loading or confining pressure. A number of investigators have used bender elements to measure shear wave velocity and attenuation in unconsolidated sediments by measuring arrival times and signal amplitudes at various transducer separations (Brunson and Johnson 1980). Attenuation estimates are difficult to perform because the radiation pattern of the transducers is often not known, so the correction for geometrical losses may be inappropriate. Measurements over a single separation can be made but unlike compressional wave transducers there is no "standard" medium available for reference measurements. In practice it is possible to calibrate a pair of transducers, by measurements over several separations, through a homogeneous clay sample.

Bender elements have also proved an excellent means of determining the anisotropic shear wave characteristics of overconsolidated clays (Bates 1990 and Davis 1982). The shear wave motion from the bender element is essentially polarised, so measurements can be made over the same travel path, both parallel and perpendicular to the axis of symmetry, by simply rotating the transducers by 90 degrees. In some cases the velocities for vertical and horizontally polarised shear waves can differ by as much as 100%.

Another technique used for unconsolidated, non-cohesive sediments has been to mount the bender elements into the walls of a compaction cell (Schulthesis 1983, Horn 1980), in order to measure the variation in shear wave propagation with porosity or void ratio. Starting at a maximum porosity the sample is vibrated in stages whilst measuring transit times, until a minimum porosity is achieved. The main difficultly with this equipment is to identify the shear wave arrival which travels through the sample rather than the wave that travels around the walls of the cell. Fortunately the shear wave is lower in frequency than the faster, interfering wave which also tends to be reduced in amplitude when the bender element becomes embedded in the sediment sample.

Bender transducers have also been successfully installed in various compression and triaxial cells where the effects of different types of load on shear wave propagation can be studied (Schultheiss 1981, Bates 1989, and Dyvik and Madshus 1985). The transducers are mounted in the top and bottom platens of the soil testing apparatus so they penetrate into the sample. The resulting transit times through the sample, as it consolidates, can be used to give the shear stiffness of the material under simulated depths of burial. Unlike shear plates, these bender elements work at very low or even zero load. As a sample is loaded, not only does the transit time decrease but the frequency response of the system also increases from around 1 to 50kHz., as the motion of the bender element becomes more restricted. One of the major problems with this sort of testing, however, is deciding on the correct arrival time. Although the signature of the arrival is much simpler than the signature from a shear plate, the identification of the correct onset is critical as there is only one travel path over which to calculate the velocity. A recommended identification criterion is to measure the time difference between the start of the driving signal and the first significant jump in the received signal.

## 2.3. THE PULSE SHEAROMETER.

This commercially available device (Rank Bros. England) is designed to propagate small strain shear waves through viscoelastic materials (James et al 1986). It works best with gels having a shear modulus in the range 50Pa to 0.5MPa. The apparatus consists of a pair of parallel discs mounted in a glass cylinder containing the test material. Each disc is in contact with a piezoelectric crystal, which induces small amplitude torsional displacements of one disc relative to the other to initiate a shear wave pulse. The measurement is made at the resonant frequency of the cell/sample system (normally about 200 Hz.). Transit time measurements can be made by either varying the disc separation to determine shear wave velocity, or keeping the discs at a constant separation, over a period of time, to investigate the thixotropic properties of the sample.

# 3 Torsional Resonance Using Continuous Vibrations

A number of early measurements on rock samples were made using the resonant bar technique in various vibration modes (longitudinal, torsional or flextural). The same principles have been employed in the field of Soil Dynamics for over 30 years in various forms of resonant column devices. In the USA, where these devices are widely available, the resonant column technique has become the standard ASTM method for determining the shear modulus and damping capacity of soils. More recently the resonant column has been used to determine the shear wave properties of samples of marine sediments (Stoll 1979,1985, Bennell et al 1984 and Davis and Bennell 1986).

The resonant column test consists of vibrating a cylindrical specimen in its first torsional mode to determine its dynamic shear modulus and in its first longitudinal mode to determine its dynamic compression modulus. By monitoring the specimen's end displacements as a function of frequency the resonant frequency for which the sinusoidal excitation force is in phase with the velocity of the vibration at the end of the specimen can be determined. This, together with the samples length, allows the computation of the wave velocities through the specimen (fig.3). The damping ratio, which can be related to Q or attenuation, can be determined from the steady state magnification factor at resonance, the width of the resonant response curve or the logarithmic decay of free vibrations.

There are a number of different designs of resonant column devices, the most popular employing fixed-free boundary conditions. That is one end of the sample is held rigidly fixed by a large inertia whilst the other end is forced to vibrate using an electrodynamic vibration system. The sample is confined in a similar manner to a triaxial sample and can be subjected to isotropic or anisotropic loading conditions. This allows an indication of a soils behaviour in terms of shear velocity, modulus and damping at any particular depth of burial. It should be noted, however, that most cohesive samples behave thixotropically and that even after the end of primary consolidation the samples stiffness and damping characteristics will vary with time (Anderson and Stokoe 1978).

Whereas pulse transducers operate at only relatively low strain amplitudes, the resonant column can test samples over a wide range of amplitudes (0.0001% to 1%), ie low amplitudes where the sample behaves in a linearly elastic fashion to higher amplitudes representative of earthquake strains where the sample behaves in an elastic-plastic fashion and permanent deformations can occur.

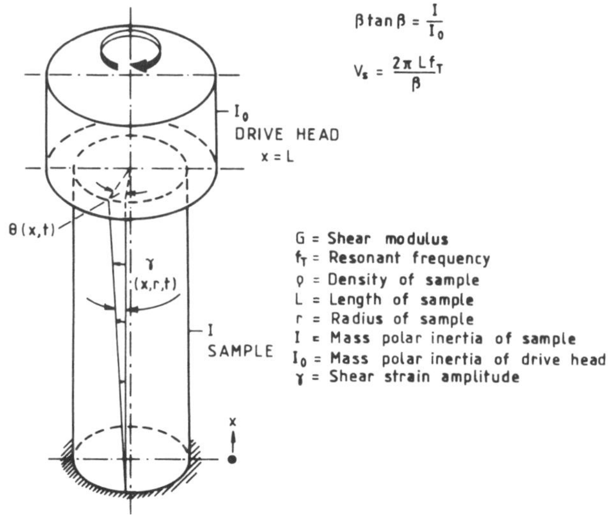


Fig. 3. Principle of the resonant column test

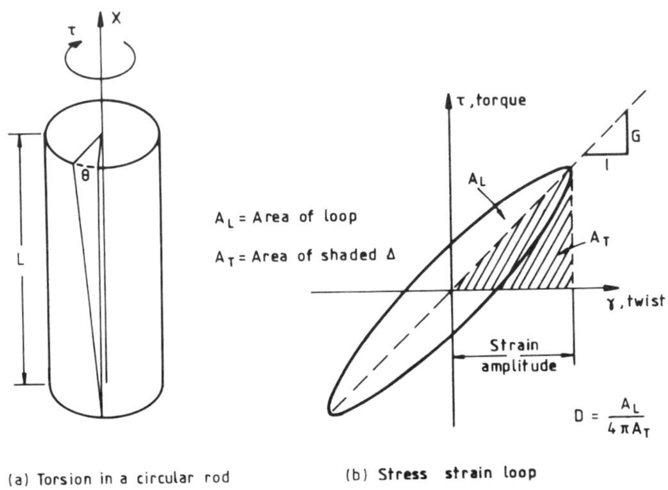


Fig. 4. Principle of torsional cyclic loading.

All soils show non-linear stress-strain behaviour even at small strains. The initial tangent to the shear stress-strain curve is often referred to as G*max.* and it is this modulus that is usually assessed by pulse measurements. Shear modulus can decrease significantly when strains as small as 0.001% are exceeded. Numerous resonant column tests have shown that at shear strains of 0.001%, even for sensitive clays, the testing procedure does not affect the sample, no excess pore pressures are developed, and the measured shear modulus approximates to $G_{max}$ (Massarsch and Drnevich 1978).

One disadvantage with pulse testing is that the strain amplitude of the vibrations is difficult to assess. In most cases it can be assumed that $G_{max}$ is being assessed. In some instances, especially with unconfined sediments, large strains are induced close to the transmitting transducer and a localised build up of excess pore pressure can cause liquefaction to occur. At the receiver, the shear strains are still very small, but often the quality of the received signal can be dramatically improved by reducing the amplitude of the driving signal into the transmitter. With resonant testing the shear strain amplitude is well known, although again it varies considerably through the sample. For a solid cylinder the strain is zero along the torsion axis and maximum at the circumference. Usually the average shear strain is defined as the strain developed at 0.707 of the sample radius. If hollow cylindrical samples are tested, then the strain field is far more uniform, but the sample is likely to be subjected to far more disturbance during preparation.

One of the greatest problems with the resonant column test is to ensure complete coupling between the top of the sample and the end driving platen, so that both undergo the same motion. If this condition is not satisfied then only the frictional stiffness at the top platen/sample boundary is being tested. For this reason, tests with effective stresses below 50kPa are not recommended, especially if the axial load is less than the confining pressure when slippage is most likely to occur. Also, when testing very stiff samples, complete coupling may be difficult to achieve without the aid of vanes protruding into the sample (Drnevich 1978). If exceedingly stiff samples are tested then the components of the apparatus itself may suffer some deformation and erroneous results would occur. A further problem that occurs with very sensitive marine sediments is that the top of the sample can twist and distort sideways due to uneven consolidation, causing increasing friction between the drive coils and magnets. This can often be recognized by increased damping values and possibly unexpectedly high resonant frequencies. In this situation it is necessary to interrupt the testing process to re-align the drive system. In order to establish the resonance condition and carry out the required measurements a large number of continuous vibration cycles, (typically 1000) are applied. Whilst this is not a problem at low strain amplitudes, at high strains large permanent deformations can occur, the sample can be weakened and in some cases will fail altogether.

## 4 Torsional Cyclic Loading

Torsional cyclic loading at very low frequencies, (0.01 - 0.1Hz), allows a more controlled application of shear stress reversals to a sample confined in a manner similar to a resonant column sample (Bolton and Wilson 1989, Drnevich 1972 and Davis and Bennell 1986). With this test single torsional shear stress reversals can be applied to the top of a sample to derive the hysteretic shear stress-strain loop shown in fig.4. The shear modulus, (and hence shear wave velocity), can be directly derived from the gradient of the line joining the end points of the loop whilst the damping ratio is proportional to the area of the loop. Again a range of strain amplitudes can be applied and at low strains $G_{max}$ is determined. By varying the frequency of loading it is also possible to determine the effect of strain rate as well as the number of loading cycles on shear modulus, velocity and damping. Ultimately torsional cyclic loading has a greater strain amplitude range than the resonant column, and a number of published shear modulus-shear strain curves are a compilation of results using both techniques. Of course torsional cyclic loading suffers from the same deficiencies as the resonant column when it comes to the coupling of the sample to the end platen. Pulse testing on the other hand is more likely to reflect the average properties of the sample as a whole.

## 5. Combinations of Pulse and Torsional Testing

The installation of bender elements in resonant column devices (Bennell et al 1984, Davis and Bennell 1986 and Dyvik and Madshus 1985) has allowed a direct comparison between both low strain amplitude methods, carried out on the same sample, under precisely the same conditions. This has also provided an opportunity to validate both techniques for determining laboratory shear wave properties.

Dyvik and Madshus (1985) have installed bender transducers in a Dnevich resonant column device, (Hardin Oscillator). Measurements of G*max* were made using both pulse and resonance techniques on three offshore clays, a Drammen clay and a sample of Haga clay. The results from both measurement methods were shown to be in excellent agreement over a wide range of soil stiffnesses (up to 140MPa), although there were some slight divergencies at the stiffness extremes. The shear wave travel times were assumed to apply to the travel path between the tips of the bender elements, and were taken to the first significant rise in the receiver signal.

Davis and Bennell (1986) have also described a modified Drnevich resonant column with bender elements installed. Again they showed that there was close agreement between bender shear wave velocities and low amplitude resonant column velocities for marine clays and sands. For very stiff clay samples, however, the bender shear wave velocity can be as much as 30% higher than the resonant column value. In some instances the difference can be attributed to poor coupling between the top platen and top of the sample thus reducing the resonant shear wave velocity. Fig.5 shows the variation of shear wave velocity with effective stress for a stiff clay sample in a resonant column capable of applying anisotropic stress conditions. In this particular case the axial load was less than the all around confining pressure so that there was insufficient coupling between the top of the sample and the drive platen. This resulted in a large deviation in pulse and resonance velocities as the effective stress increased. Bender elements can thus serve as a means of monitoring resonant tests for any problems that might arise as well as providing a simple means of determining G*max*.

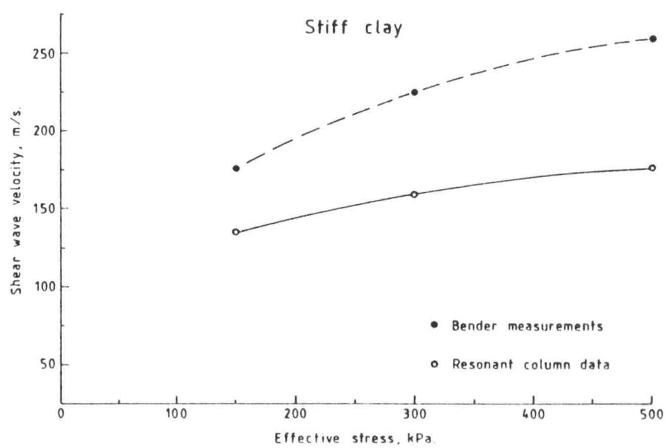


Fig. 5. Comparison of bender & pulse velocities through a stiff clay

Davis and Bennell(1986) also describe an additional modification to a Drnevich resonant column, which allows low frequency torsional cyclic loading to be performed in conjunction with pulse and resonance testing. By varying the frequency of torsional loading they showed that sediment samples behaved in a viscoelastic manner, in that the measured shear modulus decreased as the frequency of loading decreased, (ie in response to a torsional load the shear strain increased with time resulting in a strain rate effect). Fig. 6 shows how the shear wave velocity of a saturated sample of sand, measured using bender elements, the resonance

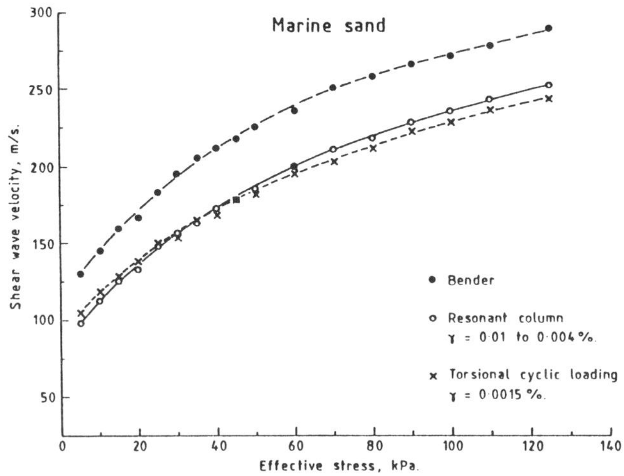


Fig. 6. Comparison of bender, resonance and cyclic loading velocities in a sand sample.

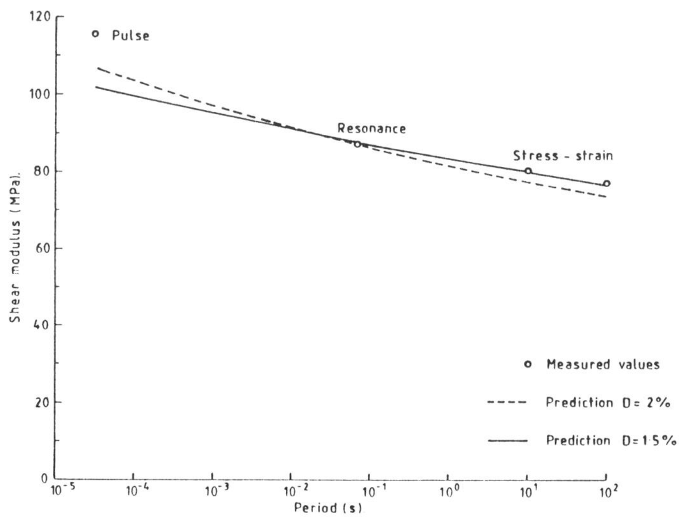


Fig. 7. Viscoelastic behaviour of sand at an effective stress 100 kPa.

technique and torsional cyclic loading, compare over a range of effective stresses. Note that the pulse measurements are significantly higher than either of the other techniques. The torsional cyclic loading data were obtained at a constant strain amplitude of 0.0015%, whilst the strain amplitude of the resonance tests had to be increased at the low effective stresses, to about 0.01%, in order to actually make the measurement at all. Thus neither the torsional testing methods provide data that can be assumed to be equivalent to $G_{max}$, although the difference is only likely to be small.

Fig.7 illustrates the variation of shear modulus with frequency, or period of loading, at an effective stress of 100 kPa and a shear strain amplitude of 0.005%, for the same sample as the data shown in fig.6. Also shown is a viscoelastic prediction of shear modulus at various periods assuming shear damping ratios of 1.5 and 2% (2% was the measured value from the resonance test). This seems to account reasonably well for the differences in resonance and the cyclic loading moduli. The discrepancy in the prediction for the high frequency pulse measurement is probably due to its shear strain being somewhat less than 0.001%. Of course the viscoelastic behaviour for clays will be somewhat more pronounced than the results for the sand sample shown here. Thus the differences in shear moduli, determined by the different testing procedures, can be accounted for in terms of the non-linear aspects of the shear stress-strain behaviour of marine sediments.

## 6. Conclusions

Currently there are a number of well established, laboratory techniques for determining the shear wave characteristics of marine sediments under a wide range, of confined or unconfined conditions. The use of a variety of techniques on the same sample, subjected to the same conditions and thus having identical physical properties, has shown the non-linear aspects of shear wave propagation in terms of shear strain amplitude and shear strain rate. The combination of techniques is also capable of determining any problems that might occur in any particular testing method, such as a lack of coupling in the resonance testing.

## Acknowledgements

Laboratory shear wave research has been supported over a number of years at the School of Ocean Sciences by several research grants from NERC and SERC. This continued support is gratefully acknowledged by the authors.

## References

Anderson, D.G. and Stokoe, K.H. (1978) 'Shear modulus: A time-dependent material property', ASTM STP 654, 66-90.

Bates, C.R. (1989) 'Dynamic soil property measurements during triaxial testing', Geotechnique 39, No.4, 721-726.

Bates, C.R. (1990) 'Seismic anisotropy as an indicator of marine sediment stability', PhD. Thesis, University of Wales.

Bennell, J.D., Taylor Smith, D. and Davis, A.M. (1984) 'Resonant column testing of marine soils', Proc. Oceanology Internat. Conf., Brighton: Soc. of Underwater Technol., OI 1.9.

Bolton, M.D. and Wilson, J.M.R. (1989) 'An experimental and theoretical comparison between static and dynamic torsional soil tests', Geotechnique 39, No.4, 585-599.

Brunson, B.A. and Johnson, R.K. (1980) 'Laboratory measurements of shear wave attenuation in saturated sand', J. Acoust. Soc. Am. 68(5), 1371.

Davis, A.M. (1982) 'Static-dynamic properties of sedimentary materials' PhD. Thesis, University of Wales.

Davis, A.M. and Bennell, J.D. (1986) 'Dynamic properties of marine sediments', in Ocean

Seismo-acoustics (ed. Akal and Berkson), Plenum Press, New York.

Drnevich, V.P. (1972) 'Undrained cyclic shear of saturated sand', J. Soil Mech. Found. Div., Proc. ASCE, 98, No. SM8, 807-825.

Drnevich, V.P. (1978) 'Resonant column testing-problems and solutions', Dynamic Geotechnical Testing, ASTM STP 654, 384-398.

Dyvik, R. and Madshus, C. (1985) 'Laboratory measurement of $Gmax$ using bender elements', in Advances in the art of soil testing under cyclic loads. Am. Soc. Civ. Engrs., New York.

Horn, I.W. (1980) 'Some laboratory experiments on shear-wave propagation in unconsolidated sands', Marine Geotechnology Vol. 4, 31-54.

James, A.E., Williams, D.J.A. and Williams, P.R. (1986) 'Small strain, low shear rate rheometry of cohesive sediments', in J Dronker and W. van Leussen(eds.), Physical Processes in Estuaries. Berlin: Springer -Verlag, 488-500.

Laughton, A.S. (1957) 'Sound propagation in compacted ocean sediments', Geophysics 22, 233-260.

Massarsch, K.R. and Drnevich V.P. (1979) 'Deformation properties of normally consolidated clays', Design parameters in geotechnical engineering, BGS, London 251-255.

Schultheiss, P.J. (1981) 'Simultaneous measurements of P and S wave velocities during conventional laboratory soil testing procedures', Marine Geotechnology, No. 4, 343.

Schultheiss, P.J. (1983) 'The influence of packing structure on seismic wave velocities in sediments. University College of North Wales, Marine Geological Report 83/1.

Shirley, D.J. (1978) 'An improved shear wave transducer', J. Acoust. Soc. Am. 63, 1643-1645.

Shirley, D.J. and Hampton, L.D. (1978) 'Shear wave measurements in laboratory sediments', J. Acoust. Soc. Am. 63, 607-613.

Stephenson, R.W. (1978) 'Ultrasonic testing for determining dynamic soil moduli', Dynamic Geotechnical Testing, ASTM STP 654, 179-195.

Stoll, R.D. (1979) 'Experimental studies of attenuation in sediments', J. Acoust. Soc. Am. 66, 1152-1160.

Stoll, R.D. (1985) 'Computer-aided studies of complex soil moduli', in Measurement and Use of Shear Wave Velocity for Evaluating Dynamic Soil Properties, ed. R.Woods, ASCE, New York, 18-33.

RELATIONSHIP BETWEEN ACOUSTIC AND MECHANICAL PROPERTIES OF TWO MARINE CLAYS

K.C. Baldwin, P.A. de Alba and A.N. Jones
Ocean Engineering Center
University of New Hampshire
Durham, NH USA 03824

ABSTRACT. Marine clay samples from two geographic locations were tested in a triaxial-acoustic testing chamber. This chamber was equipped with piezoelectric benders and thickness expanding disks in the end caps for the generation and reception of shear and compressional waves respectively. The chamber design facilitated both triaxial strength and acoustic measurements to be made on a particular sample without disturbing the test set-up. The acoustic data was digitized and processed for time delay assessment for velocity determination. Staged triaxial tests were used to maximize information from each sample. Measurements were made both under increasing consolidation pressure and during rebound. Preliminary analysis indicate that the relationship between undrained shear strength and shear wave velocity is both material and stress history dependent.

## 1. Introduction

This paper presents the results of a series of acoustic and geotechnical measurements performed on two marine clays. The acoustic parameter presented is shear wave velocity. These data are presented as a function of consolidation history and undrained shear strength as determined from staged triaxial tests. A modified triaxial chamber, which has acoustic transducers in the end caps, was used to generate the data presented here.

This work is part of ongoing research concerning possible relationships between large-strain mechanical behavior of sediments and low-strain acoustic properties. The efforts in shear wave velocity measurements in marine clays are an outgrowth of previous work in sands described by de Alba et al, 1984 and Baldwin et al, 1991.



*J. M. Hovem et al. (eds.), Shear Waves in Marine Sediments*, 95–102.

## 2. Measurement System

There are two distinct functions performed by the measurement system. The first is to determine the undrained shear strength via staged triaxial testing. The second is to measure the S and P-wave velocities at various stages of mechanical testing. A schematic of the complete system is shown in Figure 1.

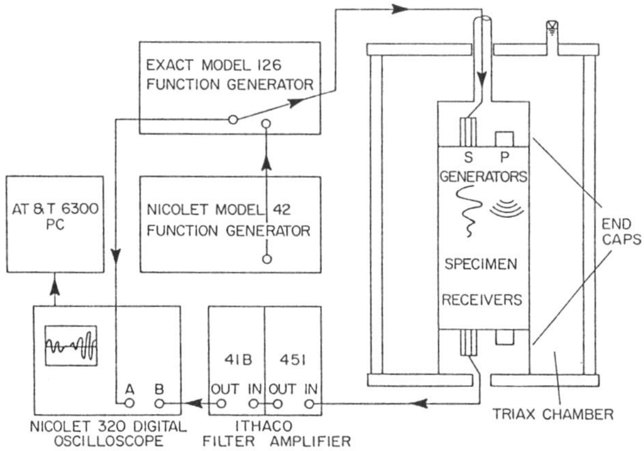


Figure 1. Schematic of triaxial acoustic chamber and electronic pulse-time delay measurement system.

### 2.1 Triaxial-Acoustic Chamber

The chamber is made from cast acrylic and can withstand pressures up to 862 kPa (125 psi), and is set up for salt water backpressure saturation. The chamber pressure and pore pressure are externally monitored with a single pressure transducer to avoid any differential calibration problems. The axial strain is measured with an LVDT. The externally applied axial load is transmitted to the clay specimen in the chamber via the loading ram which passes through an air bushing in the chamber top. It is monitored with a strain gage load cell. These transducers allow the determination of the mechanical behavior of the specimen during the staged triaxial test.

### 2.2 Acoustic Measurements

The acoustic pulse-time delay measurement system is necessary for determining acoustic wave velocities. The acoustic measurement system has two components in the triaxial chamber, the end cap acoustic transducers. The remaining electronic equipment is outside of the chamber (Figure 1). The electronic pulse-time delay measurement system is capable of making measurements for both S-and P- wave velocities, but this paper focuses on S-wave velocity measurements.

The S-wave transducers are four element bender arrays, similar to those described by Shirley et al. 1978, with one end mounted rigidly and the opposite end encapsulated in a compliant urethane window, which is in contact with the specimen. This window is watertight and pressure-compensated to maintain the integrity of the piezoelectric elements. The compliant window allows proper coupling and transmission of the S-wave to the clay specimens.

The pulse-time delay measurement system was configured to generate gated sine wave pulses of either positive or negative polarity. The pulses were applied to the top cap S-wave transducer (Figure 1) where they were transmitted through the clay specimen and received at the bottom end cap S-wave transducer. The signal was then amplified and filtered before being displayed and digitized with a digital storage oscilloscope.

The oscilloscope served many functions. The time base was triggered by the transmit pulse establishing time zero on the sweep. The received signals were stacked using internal software to capture transients and add them together. The resultant signal was stored and transferred to a PC using WAVEFORM Basic and saved for future analysis. The onset of the received signals indicated the travel time for the S-wave across the clay specimen

## 3. Experimental program

### 3.1 Sediments

The clay sediments tested were from two geographic locations: the Canadian Beaufort Sea and Portsmouth, NH. The Beaufort Sea specimens were supplied by the Atlantic Geoscience Centre of the Geological Survey of Canada. The Portsmouth, NH clays were obtained locally from an exposed marine clay deposit. Index properties of the sediments are summarized in Table 1.

TABLE 1. Sediment index properties

| Specimen | Depth (m) | OCR | Initial Void Ratio | Liquid Limit | Plasticity Index | USCS Symbol |
|---|---|---|---|---|---|---|
| A1 | 12.5 | 1.8 | 0.691 | 35 | 15 | CL |
| C3 | 3.05 | 2.6 | 2.001 | 64 | 35 | CH |
| D4 | 6.0 | 2.0 | 1.574 | 51 | 27 | CH |
| P1 | 6.7 | 1.6 | 0.988 | 34 | 15 | CL |
| P2 | 6.7 | 1.6 | 0.972 | 34 | 15 | CL |

The Beaufort Sea clays (A1,C3,D4) came from two different offshore borings in approximately 50 meters of water. The MacKenzie River is the main source of sediments in this area (Jefferies et al 1987). Two specimens (C3, D4) tested from one boring were described as gray marine clays of high plasticity and soft consistency with bioperturbations and lignite inclusions. Mineralogic analysis indicated that the clay was essentially composed of illites and chlorites. Specimen (A1), extracted from the other boring, was described as inorganic stiff to very stiff marine clay. The clay from the Portsmouth, NH (P1,P2) site was described as a dark gray soft silty marine illitic clay with silt and fine sand lenses, which was leached with fresh water. All the clays tested had an overconsolidation ratio (OCR) greater than one.

### 3.2 Staged Triaxial Tests

Since only one specimen was available from each depth, a conventional triaxial test would not have permitted any conclusions to be reached as to the variation of S-wave velocity with shear strength in the material. Consequently, we performed staged triaxial tests on each specimen following the procedure suggested by Nambiar et al. (1985). Since these materials were overconsolidated to a degree which was only approximately known, it was further decided to combine the staged test with the SHANSEP procedure described by Ladd and Foott (1974), in which the specimens were first consolidated to pressure levels in excess of the maximum past pressure and then rebounded to obtain known degrees of overconsolidation. This procedure had the additional advantage of minimizing the effects of sample disturbance.

Each specimen was first consolidated by stages to a value in excess of the estimated maximum past pressure. It was then sheared under undrained conditions, with pore pressure measurements, until a strain level of 2% to 4% was reached, sufficient to extrapolate the complete stress-strain curve to failure. The specimen was then reconsolidated to a higher effective stress, and again sheared, to the maximum effective stress value permitted by the equipment. It was then rebounded to a lower stress, producing a known OCR, and sheared again, as long as a satisfactory degree of saturation could be maintained; the final test on each specimen was taken to failure in the usual way.

It should be noted that all specimens were consolidated isotropically (CIU-tests). Jeffries et al. (1987) measured lateral stress values approaching, and in several cases exceeding, the vertical stress during pressuremeter tests in the Beaufort Sea, and also noted their apparent independence from geostatic stresses; it was therefore concluded that isotropic consolidation was appropriate in this case.

### 3.3 Shear wave velocity

Shear wave velocity measurements were made following the specimen mounting, saturation and each consolidation stage.

Gated sine waves of two cycles at 1500 Hz were used. These waves were generated with both a positive and a negative initial phase. Simultaneous display of the two signals allowed the phase reversal characteristic of shear waves to be used in a zero-crossing scheme to determine the acoustic time delay. A series of 35 signals were stacked to produce a single resultant waveform. The shear wave velocity was determined by dividing the specimen height by the acoustic time delay. Accurate values of the specimen height (aprox. 0.04%) were available from the axial strain data.

## 4. Discussion of Results

Staged tests were performed on three Beaufort Sea specimens, A1, C3 and D4, as well as on two Portsmouth specimens, P1 and P2. The last four specimens were all consolidated beyond their maximum past pressure into the normally-consolidated range, as previously described. Specimen A1 could only be tested in its original overconsolidated range; the high backpressures necessary to saturate this material limited the effective pressure that could be applied with our equipment.

It is interesting to remark that speciments C3, D4, P1 and P2 all exhibited normalized behavior as defined by Ladd and Foott (1974), with the undrained strength (Su) following a relationship for overconsolidated materials of the form:

$$S_u/\sigma' = S*(OCR)^m$$

where $\sigma'$ is the consolidation pressure and S is the Su/$\sigma'$ ratio for normally-consolidated material. This suggests that the combined stage testing- SHANSEP procedure gives reasonable results. This is further confirmed by the effective-stress friction angle of 21° obtained for the normally-consolidated Portsmouth clay, which compares well with that reported by Ladd (1972) for this material.

Correlation techniques were considered for time-delay determination. Reference signals were obtained and it was assumed that these signals would be known replicas of the received signals through the sediment. Problems occurred at elevated pressures due to air-bushing-induced noise received by the S-wave transducer and an apparent impedance change as the specimens were consolidated and strained. Air rushing past the loading ram produces a frictionless condition allowing axial load measurements external to the chamber. This air flow causes structural excitation which is received by the highly sensitive S-wave transducers. The apparent impedance change as specimens were consolidated and became stiffer and denser caused a shift in the resonant frequency. This shift made it extremely difficult to perform the required correlation with the reference signal, acquired at the beginning of the tests at atmospheric pressure, to extract the time delay necessary for velocity determination. These two difficulties lead to the use of the zero-crossing technique to extract time delay data.

The results most relevant to this symposium are summarized in Figures 2 and 3. Figure 2 shows the variation in Vs with consolidation pressure, for all consolidation stages including the shear tests. Two trends can clearly be discerned; one defined by specimens A1, P1 and P2, all of which have a plasticity index (PI) of 15, and the other by the more plastic materials C3 and D4, which have PI- values of 35 and 27 respectively (Table I). Curves fitting the general trends are shown in the figure; in both cases, the Vs-increase with consolidation pressure follows an exponential law of the form:

$$V_s=K(\sigma')^n$$

where n= 0.21 for the A1, P1 and P2 set; and 0.19 for the C3 and D4 data. These values compare reasonably well with n= 0.25 which may be inferred for clays from the resonant-column results of Hardin and Drnevich (1972).

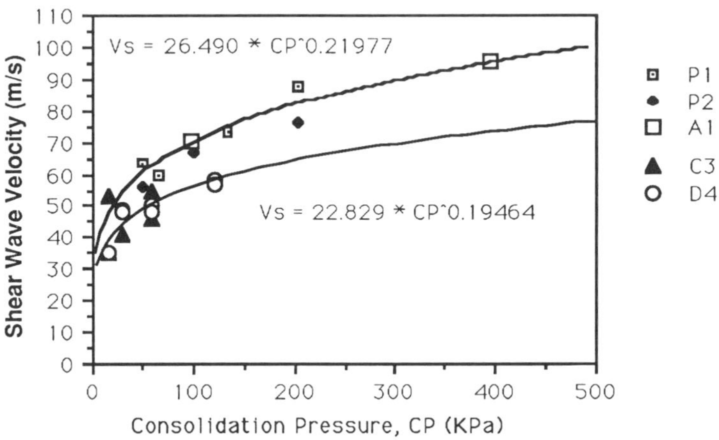


Figure 2. Variation of shear wave velocity (Vs) with effective consolidation pressure (CP).

The values of Vs, obtained immediately prior to shear testing, are plotted against the undrained shear strength (Su) in Figure 3. Two distinct relationships are defined, one for the Portsmouth materials P1 and P2, and another for the more plastic C3 and D4 materials.

Two additional points are defined by the much stiffer A1 material. Both the C3-D4 and P1-P2 data sets clearly show that distinct trends are defined by the clays in their

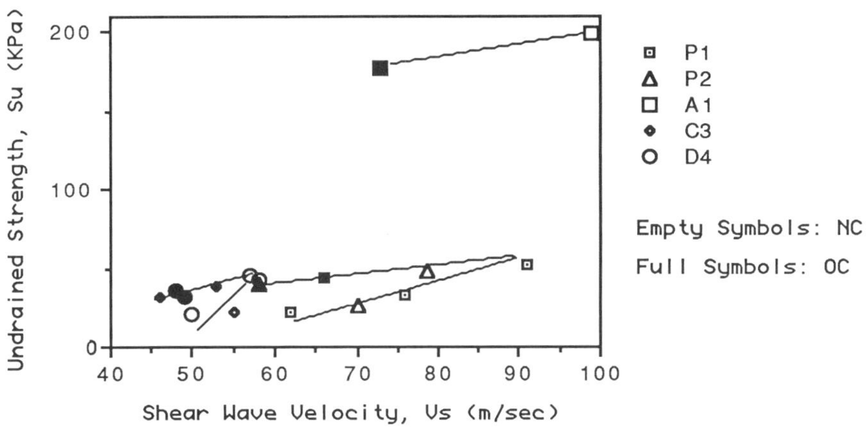


**Figure 3.** Shear wave velocity (Vs) vs. undrained shear strength (Su) relationships showing the material and stress history effects.

normally-consolidated (NC) versus their overconsolidated (OC) states. The plots also show that Vs is a very sensitive indicator of undrained strength change; e.g., for the P1-P2 data, the slope of the normally-consolidated (NC) branch is about 1.4 KPa/(m/s), and for C3-D4 it is approximately 4.7 KPa/(m/s).

## 5. Conclusions

While it is dangerous to make general statements based on this limited data set, the results obtained so far suggest the following conclusions:

1) Shear wave velocity is seen to be very sensitive to changes in undrained strength, Su, and is material-dependent for these marine clays.

2) Stress history is very important; the same clay in an overconsolidated state will define a different Su-Vs relationship than in a normally-consolidated state.

3) To use these relationships to determine sediment strength in situ through Vs-measurements, a testing program similar to that described in this paper must be carried out for each material, and the stress history of the deposit must be determined, i.e., through oedometer tests. Results suggest that Vs measurements could then be used for first-order estimates of undrained strength at other locations in the soil deposit.

4) Axial loading must be measured inside the chamber to eliminate the air bushing as a noise source.

## 6. Acknowledgements

This study was funded by the Atlantic Geoscience Center of the Geologic Survey of Canada. Their support is gratefully acknowledged.

## 7. References

Shirley,D.J.,(1978) "An Improved Shear Wave Velocity Transducer", J. Acoust. Soc. Am. 63 (5) p. 1643-1645.

de Alba, P., Baldwin, K., Janoo, V., Roe, G., and Celikkol, B.(1984) "Elastic-Wave Velocities and Liquefaction Potential", Geotechnical Testing Journal, GTJODJ, Vol. 7, No. 2, pp. 77-87.

Baldwin, K., de Alba, P., Menguc, I. and Jones, A. "Measurement of Shear and Compression Waves During Triaxial Testing", 70 th Meeting of the Transportation Research Board, 14 - 18 January 1991.

Jefferies, M.G., Crooks, J.H.A., Becker, D.E., and Hill, P.R.(1987) "Independence of Geostatic Stress from Overconsolidation in some Beaufort Sea Clays", Canadian Geotechnical Journal, 24, pp. 342-356.

Nambiar, M. R. M., Rao, G. V., and Gulhati, S. K.(1985) "Multistage Triaxial Testing: A Rational Procedure", Strength Testing of Marine Sediments: Laboratory and In-Situ Measurements, ASTM STP 883, R. C. Chaney and K. R. Demars, Eds., American Society for Testing and Materials, Philadelphia, pp. 274-293.

Ladd, C. C., and Foott, R. (1974) "New Design Procedure for Stability of Soft Clays", Journal of the Geotechnical Engineering Division, ASCE, Vol. 100, No. GT7, pp. 763-786.

Ladd,C.C. (1972) "Test Embankment on Sensitive Clay", ASCE, SMFD Speciality Conference on Performance of Earth and Earth Supported Structures, Purdue Univ., p.101-129.

Hardin, Bobby O., and Drnevich, V.P. (1972) "Shear Modulus and Damping in Soils: Measurement and Parameter Effects", Journal of the Soil Mechanics and Foundation Division, ASCE, Vol. 98, No. SM6, Proc. Paper 8977, pp. 603-624.

# A LABORATORY METHOD TO INVESTIGATE SHEAR WAVES IN A SOFT SOIL CONSOLIDATING UNDER SELF WEIGHT.

I. R. McDERMOTT
*School of Ocean Sciences*
*Menai Bridge*
*LL57 5EY*
*United Kingdom.*

ABSTRACT. A laboratory technique is described, incorporating an instrumented settling column, to monitor the changing shear wave velocity profile with depth and time during the self weight consolidation process of an Irish Sea clayey silt, initial density 1160 $kg/m^3$, initial voids ratio 11.5. The density of the sediment was measured using a non destructive X-ray technique. Such clays exhibit calculated shear moduli in a range of 1.9 ± 8.5% kPa to 122.5 ± 2.5% kPa with shear wave speeds from 1.2 ± 4.2% m/s to 9.1 m/s ± 1.5% for a range of voids ratio from 5.5 to 3.

## 1. Introduction.

A sedimentary soil develops by deposition through water, passing through a series of stages; from a fluid supported suspension, to a loosely structured soil becoming a soft soil with the passage of time. As the soil consolidates under its self weight, settlement is accompanied by large strains and increasing stiffness. The physical behaviour of such a soft soil can be described by one dimensional consolidation theories such as that of Gibson et. al. (1967). Such a theory does not impose the limitations of small strains and includes the stresses of self weight. Development of this theory, together with experimental observations led researchers Lee (1979), Been & Sills (1981), and Bowden (1989) to make considerable advances in the study of soft soils.

Shirley and Hampton (1978) described a laboratory experiment in which shear wave measurements were made in low concentrations of kaolinite clay. The experimental conclusions recommended modification to the shear wave transducer design, a means of improving signal to noise ratio, and an alternative technique of measuring the bulk density of the laboratory sediment. It is upon this background that the present research is based.

As part of the continuing programme at University of Wales (Bangor), relating geophysical measurements to geotechnical properties of sediments [Hamdi & Taylor Smith (1981), Schultheiss (1983), and Davis & Bennell (1986)], a laboratory technique was developed, incorporating an instrumented settling column, to monitor the changing shear wave



*J. M. Hovem et al. (eds.), Shear Waves in Marine Sediments*, 103–110.

velocity profile of an Irish sea soil consolidating under its self weight.

## 2. Materials and Methods.

An instrumented acrylic settling column, 1 m high and of 10 cm inner diameter, was set up. The column contained 8 pairs of shear wave transducers. A mud slurry of known density was introduced into the settling column and shear wave velocities and bulk densities were monitored over a 2000 hour period. A schematic of the experimental arrangement is shown in figure 1.

### 2.1. SOIL PREPARATION.

The aim of the experiment was to measure the behaviour of a real soil settling in idealized controlled conditions. An Irish Sea clayey silt sample was taken and mechanically stirred to a uniform density. As such a soil would naturally be deposited through salt water the settling column was partially filled with sea water. The mud slurry was then introduced into the column giving an initial density of 1160 $kg/m^3$.

### 2.2. SHEAR WAVE MEASUREMENTS.

The shear wave transducer construction is shown in figure 2. The transducer consists of a bender element of dimensions 20.4 x 5.5 x 0.5 mm set in epoxy resin and mounted in an acrylic housing. The transducer construction improves the robustness and electrical insulation of the bender element but does, however, result in a decrease in compliance, an increase in resonant frequency and a greatly diminished amplitude of vibration, as shown in figure 3. The transducers are able to detect shear waves travelling at speeds as low as 1.2 m/s through a soft soil of density 1244 $kg/m^3$, and void ratio 5.5.

When in position the transducer remains isolated from the sediment by a strong flexible rubber membrane. It protrudes approximately 1.5 cm into the sediment and vibrates in the horizontal shear mode. The supporting shear wave velocity measuring apparatus is shown in figure 1. The dc step from a 20 V square wave was used to drive the transducer. Transmission of the shear wave through a loosely structured soil severely attenuates the received signal. The received waveform therefore requires substantial amplification and signal averaging to improve the signal to noise ratio before the final waveform can be satisfactorily logged by a microcomputer.

The transducer separation was measured using the travel time of the compressional wave generated by the tip of the bender element. Utilizing pairs of compressional wave transducers, of known separation, located at the same height, but perpendicular to the shear wave transducers, the compressional wave velocity was determined at each level. This was then used to calculate the bender element separation to an average accuracy of ±1%.

The vertical spatial separation of the S-wave transducers was 5 cm. The overall accuracy of the shear wave velocity measurements depends

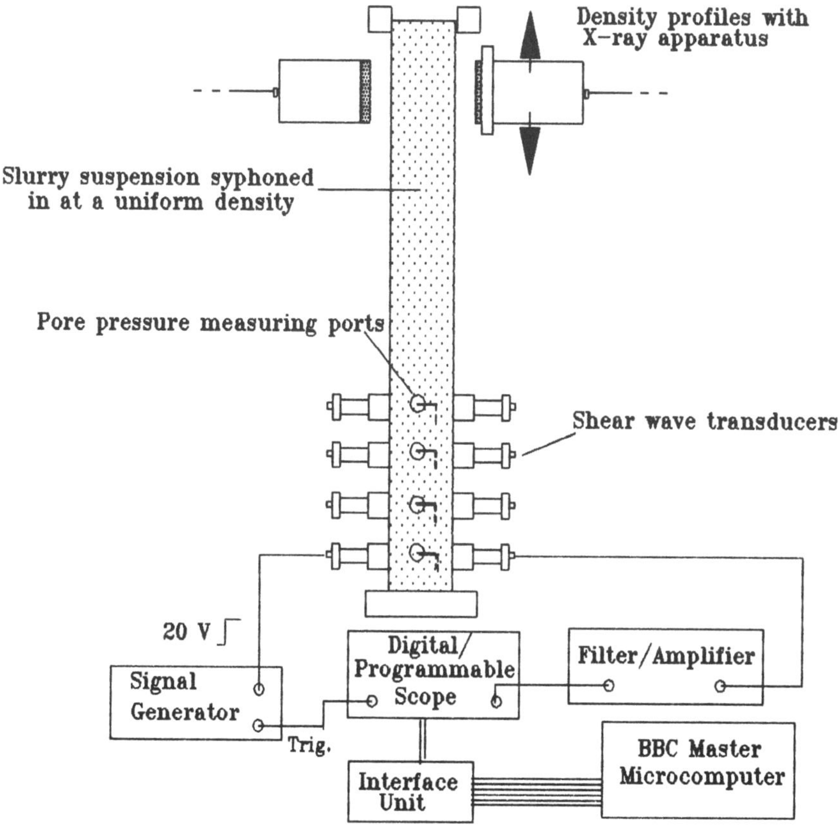


Fig. 1 Schematic of Consolidation Experiment.

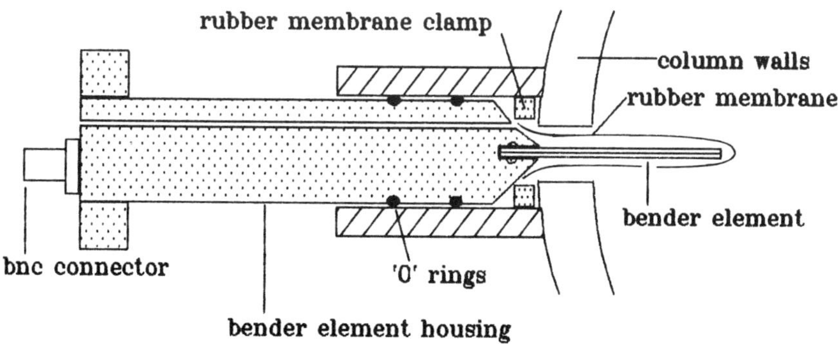


Fig. 2 Shear wave transducer construction.

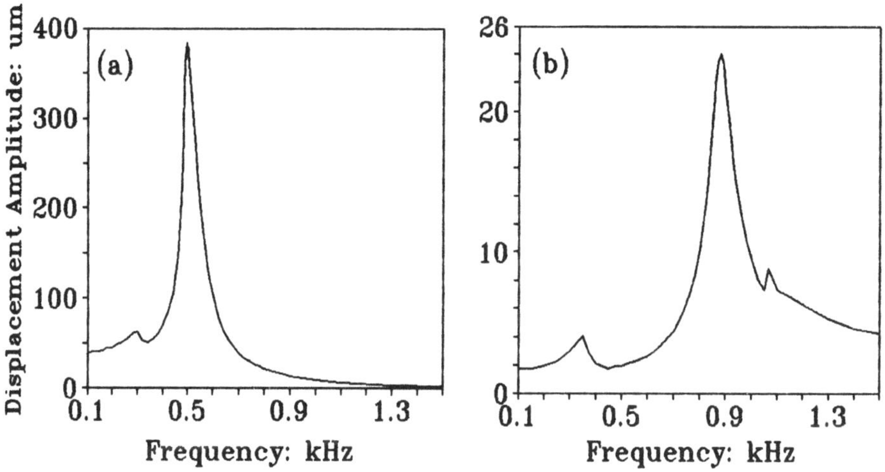


Figure 3. Frequency response of (a) unpotted bender element, and (b) potted bender element resonating in cantilever mode.

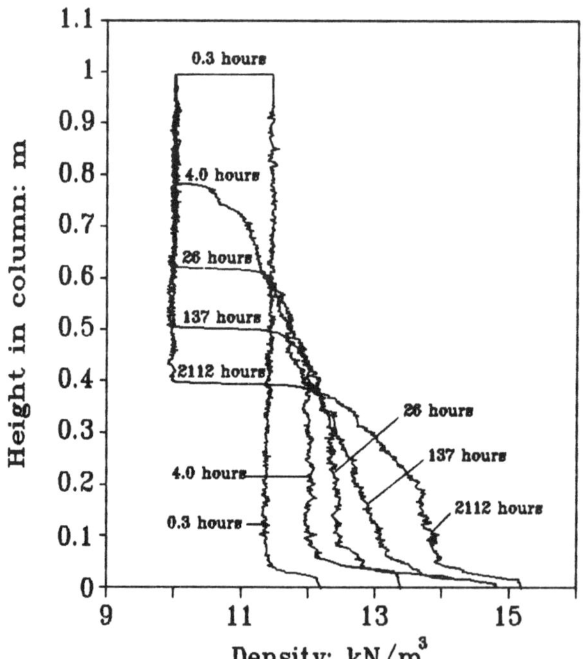


Figure 4. Density profiles, initial density 11.38 $kN/m^3$

greatly upon the degree of formation of the soft soils structural frame. At worst the accuracy fell to ±4.2%. However, with time, and the development of the soil structure, the shear wave velocity could be calculated to an accuracy of ±1.5%.

2.3. DENSITY MEASUREMENTS.

Bulk soil density was determined non-destructively using X-ray attenuation. The measurement technique is described by Been(1980). The energy level of the x-rays were 160 keV. The density measuring system has an overall accuracy of ±0.03 $kN/m^3$ and vertical resolution of the order of 1mm.

## 3. Results.

The results are shown in figures 4 to 7.

## 4. Discussion.

4.1. DENSITY PROFILES.

Figure 4 shows successive density profiles taken during the experiment. Initially the density is almost uniform except for a narrow region of high density material, consisting of the coarser particles, at the base of the column (profile 0.3 hours). A single layer exists above the dense region, consisting of the finer particles. It is this region which is under study in this experiment. The density of this layer gradually increases with time as indicated by a broadening of successive profiles. The void ratio is calculated from the measured density.

4.3 SHEAR WAVE VELOCITY RESULTS.

Shear waves were not observed in the soil until 50 hours had elapsed. The shear waves evolved from the base of the column upwards. Initially the waves were of low amplitude with velocities between 1-2 m/s. It was noted that once established, the shear wave amplitude increased rapidly. Shear wave velocity is shown to increase with decreasing voids ratio, figure 5. Figure 6 shows how shear wave velocity changes with height in the column. The data clearly demonstrate the changing velocity gradient, from a gradient of 4.8 (m/s)/m at 91.5 hours to one of 16.1 (m/s)/m at 1609 hours. It is thought that the increasing density, decreasing voids ratio of the soil, which gives rise to greater soil particle contact, leads to the increased strengthening of physical-chemical bonds (Anderson & Woods 1975) giving the soil rigidity and the ability to support shear wave propagation.

The shear modulus of the soil was calculated from bulk density and shear wave velocity data. A maximum value of 122.5 kPa for the shear modulus was recorded at the base of the column. The time-dependent behaviour of the soft soil shear modulus is demonstrated in figure 7. Two distinct phases may be identified. The initial phase, when the modulus increases rapidly with time, followed by a second phase when the

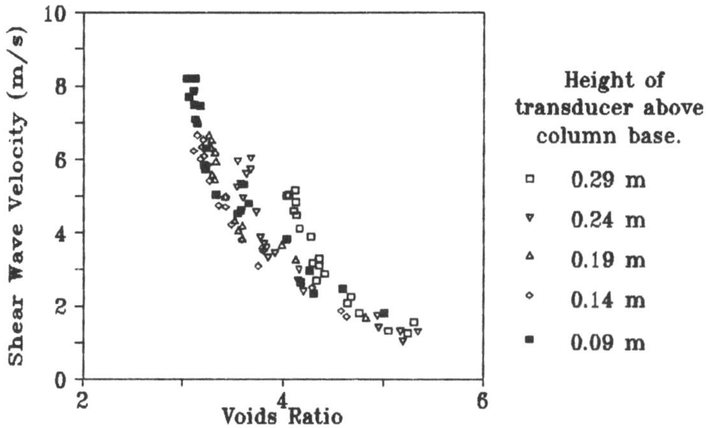


Figure 5. Shear Wave Velocity versus Voids Ratio.

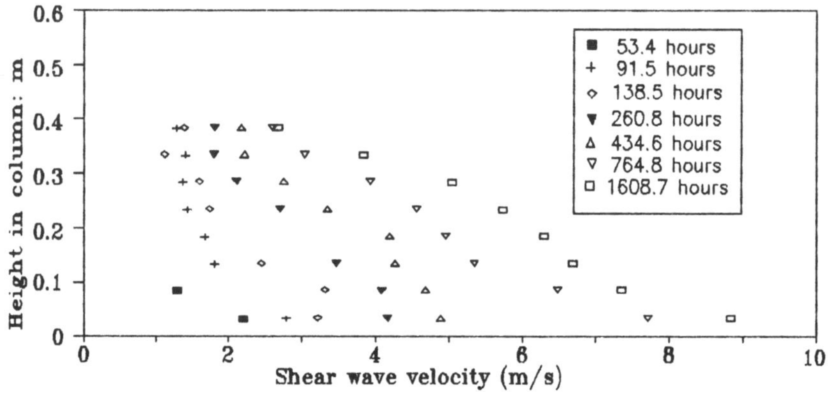


Figure 6. Shear Wave Velocity versus Height in Column.

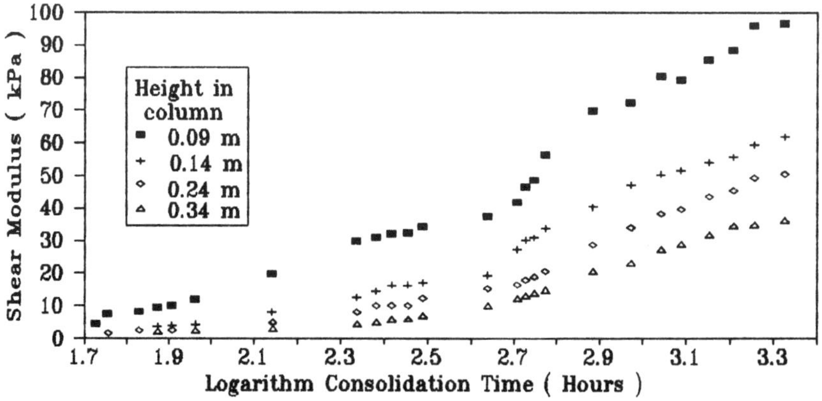


Figure 7. Shear Modulus versus Consolidation Time

modulus increases linearly with the logarithm of time. The point of transition between the first and second phase was found not to be related to the termination of primary consolidation. The range of coefficient of shear modulus increase with time $I_G$, as defined by Anderson and Stokoe 1978, was found to be 0.65 kPa for near surface soil and 60.0 kPa at the base of the column.

## 4.4 CONCLUSION.

A laboratory experiment has been described which enables the study of shear wave propagation in a soft soil consolidating under self weight. Shear wave transducers have been developed capable of measuring shear waves in laboratory sediments which have a bulk density as low as 1244 $kg/m^3$, a shear modulus of 1.91 kPa and a shear wave speed of 1.2 m/s. It is believed that some of the problems described by Shirley and Hampton (1978) have been overcome by using the above experimental apparatus and techniques.

## 5. Acknowledgements.

The instrumented sedimentation column was developed as part of a S.E.R.C. funded research project. The use of the X-ray equipment and computing facilities at the Soil Mechanics Laboratory, Department of Engineering Science, Oxford University is gratefully acknowledged. The author also wishes to acknowledge the assistance of Dr Sills and the technical staff at Oxford University, in particular Mrs G. Jenkins.

## 6. References.

Anderson, D.G. & Woods, R.D. (1975) Proceedings, In situ measurement of soil properties, American Society of Civil Engineers, Rayleigh, N.C., Vol. 1, 1975, 66-92.

Anderson, D.G. & Stokoe, K.H. (1978) 'Shear modulus: a time-dependent soil property'. Dynamic Geotechnical Testing, ASTM STP 654, American Society for Testing Materials, 66-90

Been, K. (1980). 'Stress strain behaviour of a cohesive soil deposited under water', DPhil thesis, University of Oxford.

Been, K. & Sills G.C. (1981) 'Self weight consolidation of soft soils: an experimental and theoretical study'. Geotechnique 31,No., 519-535.

Bowden, R.K. (1989) 'Compression behaviour and shear strength characteristics of a natural silty clay sedimented in the laboratory', DPhil thesis, University of Oxford.

Davis, A.M. & Bennell, J.D. (1986) 'Dynamic properties of marine sediments', in Ocean Seismo-acoustics (ed. Akal & Berkson) Plenum Press, New York.

Gibson, R.E.,England, G.L., & Hussey, M.J.L. (1967) 'The theory of one-dimensional consolidation of saturated clays', Geotechnique, 17, 261-273.

Hamdi, F.A.I. & Taylor Smith, D. (1981) 'Soil consolidation behaviour assessed by seismic velocity measurements',

Geophysical Prospecting, 29, 715-729.
Lee, K. (1979) 'An analytical and experimental study of large strain soil consolidation', DPhil thesis, University of Oxford.
Schultheiss, P.J. (1983) 'The influence of packing structure on seismic wave velocities in sediments', Marine Geological Report No. 83/1, University College of North Wales.
Shirley, D.J. & Hampton, L.D. (1978) 'Shear wave measurements in laboratory sediments', J.Acoust. Soc. Am. 63(2), 607-613.

# LABORATORY MEASUREMENTS OF ACOUSTIC PROPERTIES OF PERIPLATFORM CARBONATE SEDIMENTS

Dawn Lavoie
*Code 361, Naval Oceanographic and Atmospheric Research Laboratory*
*Stennis Space Center, MS 39529*
*USA*

Aubrey Anderson
*Department of Oceanography*
*Texas A&M University*
*College Station, TX 77843*
*USA*

ABSTRACT. Geoacoustic models of the seafloor are essential components of underwater and seafloor acoustic studies. Propagation velocities for compressional and shear waves in the seafloor material are important geoacoustic parameters. Previously reported values for shear wave velocities in seafloor carbonate oozes have shown wide disparity. New laboratory measurements of compressional and shear wave geoacoustic parameters have been made as a function of effective stress. Periplatform carbonate samples from the Little Bahama Bank and Exuma Sound in the western North Atlantic Ocean were used for these measurements. Conventional pulse timing compressional wave velocities were measured. Shear wave parameter measurements were made with a new technique using duomorph sensors embedded in the seafloor sample. The resulting compressional wave velocity values are consistent with previously reported measurements for similar material (e.g. Hamilton). The shear wave dynamic property values obtained are more consistent with the lower shear wave velocity values reported, for example by Schultheiss. These lower shear wave velocities also agree with values obtained by pulse timing measurements with bender bimorph transducers in samples from the same seafloor region.

## Introduction

Geoacoustic models of the seafloor are basic information required to understand acoustic propagation in bottom interacting situations. A geoacoustic model as defined by E. L. Hamilton (1980) is a model of the real sea floor with emphasis on measured, extrapolated and predicted values of those properties important in underwater acoustics and aspects of geophysics involving sound transmission. Implicit in this definition is the understanding that measured data are frequently not available for modeling and must be estimated and predicted. Valid predictions can only be made when the seafloor properties within the range of interest (i.e. the water-sediment interface to a depth of several hundred meters, depending upon frequency) are well understood. Among the sediment properties required for modeling acoustic propagation are compressional and shear wave velocity as a function of depth or effective stress. The effective stress is the load carried by the sediment particles minus the load-carrying contribution of the pore fluids.

Ideally, velocity measurements should be made *in situ*. However, due to difficulties obtaining *in situ* measurements in deep water, for a number of years *in situ* velocities have been computed from seismic studies (Houtz, 1974, Dobrin, 1976, Gettrust and others, 1988) or converted from laboratory measurements using temperature and pressure corrections. Another method is to apply laboratory consolidation techniques whereby vertical pressure is added to the sample under controlled conditions to simulate *in situ* pressures, and the properties of interest are measured during the process of consolidation. As part of a larger study, compressional and shear wave



*J. M. Hovem et al. (eds.), Shear Waves in Marine Sediments*, 111–120.

velocity and shear modulus of periplatform sediments were measured in a laboratory consolidation study. Sediment samples for these measurements were recovered from Exuma Sound and north of Little Bahama Bank (Figure 1). The measurement results provide an empirical data base for these acoustic properties of periplatform sediments as a function of depth or effective stress. Regression relations fitted to these data can be applied to sediments in similar environments but in different geographic locations.

## Background

Predicting the compressional or shear wave velocity for various subseafloor depths within shallow and mid-water depth carbonates sediments is not straightforward. This is illustrated by the information in Figures 2 and 3. Compressional wave velocity values shown in Figure 2 were measured on board the RV JOIDES RESOLUTION on samples recovered from Ocean Drilling Program (ODP) Sites 627 and 628. The plotted velocity versus depth data can be compared with the solid line which represents Hamilton's regression equation (1980, Figure 5 and Table IV) predicting expected velocity in carbonate sediments as a function of depth. The ODP measured compressional wave velocity values exhibit almost no depth dependences, diverging increasingly from Hamilton's predicted values with increasing depth. Hamilton's regression relation is based on *in situ* velocity values from sonobuoy measurements while the ODP data are uncorrected values measured under shipboard laboratory conditions of temperature and pressure.

Shown in Figure 3 are shear wave velocity values which were measured, using bender element transducers, in samples recovered from ODP Site 630, Northern Little Bahama Bank. As was true for the laboratory measured compressional wave data, the shear wave velocities in Figure 3 exhibit essentially no dependence on depth below the seafloor from which the sample was

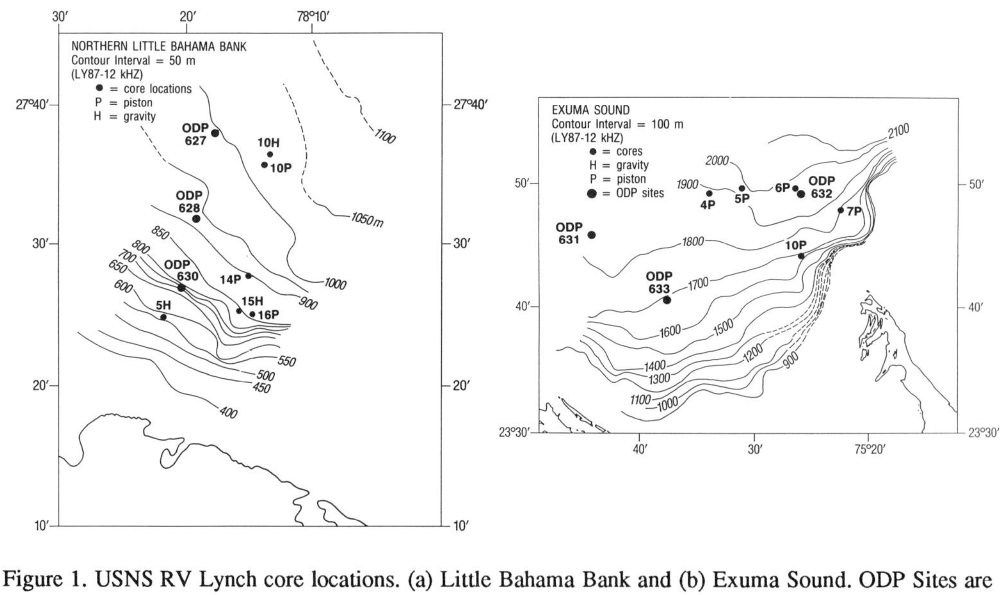


Figure 1. USNS RV Lynch core locations. (a) Little Bahama Bank and (b) Exuma Sound. ODP Sites are also noted.

recovered. For comparison, predicted *in situ* shear wave velocities are also shown versus sub-seafloor depth in Figure 3. These predicted values are from Hamilton (1980) and Richart (1970). As was true for compressional velocity values, the shear wave laboratory measurements increasingly deviate from predicted *in situ* values as sample depth increases. Such predicted behavior is not unexpected because it is known that both compressional and shear wave velocities typically increase as either pressure or temperature of a sample increase, and both overburden pressure and temperature increase with depth into the seafloor. However, the relevant *in situ* pressure or stress is the effective stress. Thus, correction of laboratory measurements to *in situ* values requires accurate knowledge of *in situ* effective stress as well as knowledge of the relationship between effective stress and velocity for the sampled seafloor material. Thus a laboratory experimental program was planned to investigate shear wave velocity in periplatform carbonate samples versus pressure.

The bender element shear wave transducers used at NOARL are designed to operate only under ambient pressures and the configuration of these devices makes them unsuitable for incorporation into a consolidometer. A search for a suitable technology led to the duomorph sensor. This device was developed in the early 1970's to measure the dynamic modulus of solid propellant *in situ* (Briar and others, 1976). The technology has been adapted in this study to measure the dynamic shear modulus of sediment during loading. From the shear modulus and density, the shear wave velocity of the sediment can be determined as a function of effective stress.

The duomorph is a bending plate device designed to operate in vibration. The device consists of a stainless steel plate sandwiched between a pair of piezoceramic crystals with a metallic strain gauge glued to the center of each crystal (Fig. 4). The piezoceramic crystals are low power, electromechanical transducers capable of converting electrical energy to mechanical energy and vice versa. When stimulated by an alternating current, the duomorph vibrates in a parabolic (drumhead) mode. To measure the modulus of a sediment, the duomorph is embedded in the sediment. The difference between the unconstrained bending of the duomorph in air and the bending

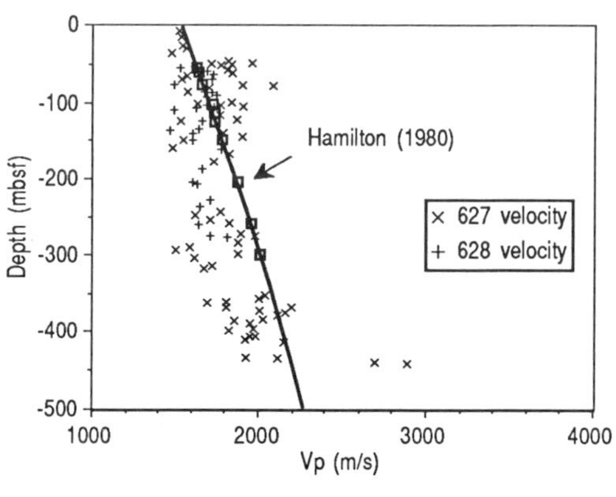


Figure 2. Compressional wave velocity measured from ODP Sites 627 and 628 are plotted as a function of depth. Predicted values by Hamilton (1980) are plotted for comparison.

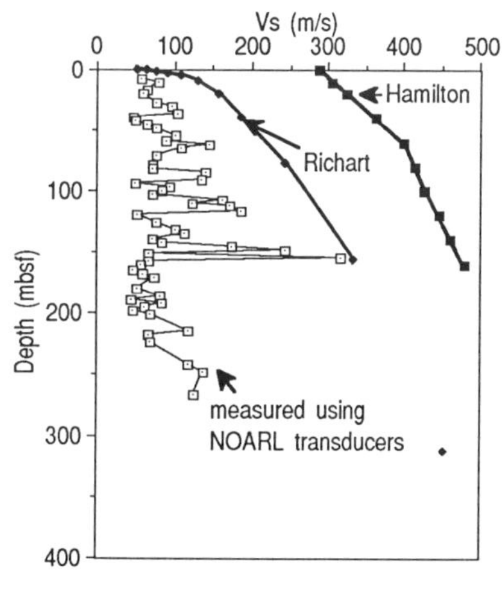


Figure 3. Shear wave velocity measured on periplatform samples from ODP Site 630 are compared with predicted values by Hamilton (1980) and Richart (1970).

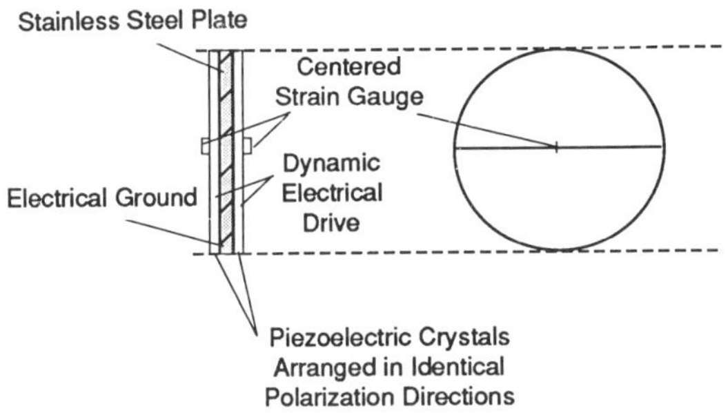


Figure 4. Schematic diagram of the duomorph sensor. The duomorph sandwich is excited with an alternating current; the resulting deflections are measured with strain gauge technology.

allowed by the sediment while embedded is a function of the sediment dynamic modulus. Duomorph deformation due to this motion is measured by the strain gages. Much of the theoretical basis for the duomorph is taken from Timoshenko's discussion of classical plate theory (Timoshenko and Woinowsky-Krieger 1959) and is further discussed by Briar and others (1976). This device is especially suited for the present studies because the sediment sample in which shear modulus is being measured can be confined in a consolidometer and subjected to external pressure. Compressional wave velocity was also measured on duplicate samples subjected to consolidation pressures. It was necessary to use duplicate samples for this measurement because the presence of the duomorph in the small sediment sample precluded compressional wave measurements. An alternate calculation of the dynamic stress modulus G which is used as a check of the duomorph results:

$$G = (\rho Vp^2 - k)(3/4) \text{ (Hamilton, 1970)}, \tag{1}$$

where $\rho$ is the bulk density measured using weight and volume techniques before and after testing. Intermediate values are interpolated. In using equation 1, values of k, the bulk modulus, are obtained from:

$$k = Av\left(\frac{1}{1 + e_0}\right) \tag{2}$$

where $e_0$ is the initial void ratio and Av, the coefficient of compressibility, is the slope from the consolidation curve at each loading

$$Av = \Delta e/\Delta p \ . \tag{3}$$

In equation 3, $\Delta e$ is the change in void ratio between two load increments, and $\Delta p$ is the change in pressure between the same two load increments. Compressional velocity results were temperature corrected using Medwin's Equation (Clay and Medwin, 1977) for surficial sediments and assuming a temperature gradient of 35°C per 1000 m of subbottom depth (Wright and Lauden, 1989).

Samples were examined with the Scanning Electron Microscope (SEM) to identify fabric relationships and constituent particles. The samples were classified as either matrix-supported (the grains are floating within the matrix as in Figure 5), grain supported (the grains are in contact with each other as in Figure 6), or in several cases where the classification was difficult using SEM, as matrix-supported with numerous grains. The distinction between matrix and grain supported sediments is important because it has been shown that matrix supported and grain supported samples will consolidate differently (Fruth and others, 1966, Rezak and Lavoie, 1990). The matrix material, generally a periplatform ooze, was of two main types; either predominantly a coccolith ooze or composed of a large proportion of aragonite needles. When grains and particles, because of breakage or other reasons, could not be positively identified as to origin, they were called unidentifiable grains or particles. Often grains and particles would be identified as shell fragments because they had an aragonite needle structure or because they were planes of calcite. Other descriptive properties including grain size, porosity, and density were measured as well. The Bahamas were chosen as a source of carbonate sediments for this research because they serve as a modern example for interpreting carbonate platforms and slope environments in the geological record.

## Results

Compressional wave velocity measured for effective stress values between zero and about 4500 kPa and corrected for temperature varied between 1426 m/s and 1617 m/s with an average velocity of 1548 m/s. A composite plot of compressional wave velocity as a function of effective stress/ depth for all these samples is illustrated in Figure 7a. Plotted for comparison are predicted values for a calcareous sediment computed using a regression equation developed for deep water calcareous sediments by Hamilton (1980). Figure 7b and 7c illustrate compressional wave velocity as a function of depth for matrix-supported samples and grain-supported samples respectively. The same Hamilton (1980) predicted curve is shown again in Figures 7b and 7c. Regression fits of second order equations to the data provided the following expressions:

$$Vp = 1552 + 2.0589 \times 10^{-2}(\sigma') + 1.274 \times 10^{-6}(\sigma'^2) \text{ for all samples,} \tag{4}$$

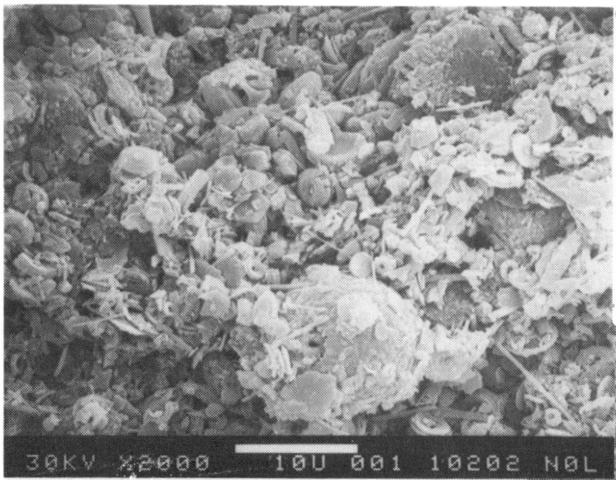


Figure 5. ES 10P-2 is matrix supported. Few grains could be found during SEM analysis. Matrix particles are predominantly coccoliths. Magnification is 2000X.

Figure 6. ES 7P-5 is grain supported. Grains are rounded and often unidentifiable. Note contacts between grains. Magnification is 200X.

$$Vp = 1588 + 5.1929 \times 10^{-2}(\sigma') - 6.5874 \times 10^{-6}(\sigma'^2) \text{ for grain supported samples,} \quad (5)$$

$$Vp = 1502 + 1.1894 \times 10^{-2}(\sigma') + 3.4344 \times 10^{-6}(\sigma'^2) \text{ for matrix supported samples,} \quad (6)$$

where $\sigma'$ represents effective stress in kPa and the velocities are in m/s. Similarly, the following regression equations represent compressional wave velocity in m/s as a function of depth (z) in meters.

$$Vp = 1552 + 0.13944\ (z) + 1.6276 \times 10^{-4}\ (z^2) \text{ for all samples,} \quad (7)$$

$$Vp = 1587 + 0.40797\ (z) - 4.1148 \times 10^{-4}\ (z^2) \text{ for grain supported samples,} \quad (8)$$

$$Vp = 1502 + 5.5272\ (z) + 3.535 \times 10^{-4}\ (z^2) \text{ for matrix supported samples.} \quad (9)$$

The individual compressional wave velocity data points measured during the consolidation testing were variable. Time delay was measured using a Vernier scale rather than reading from a digital time-delay readout, which resulted in some variability (±20 m/s) in results. In addition, during the consolidation process, especially immediately after a load was added, it was observed that the height of the sample changed quite rapidly, but the recorded time-delay did not. There is a time lag that occurs before accurate velocity measurements can be made. Measurements made during the reload cycle were accordingly eliminated. The measurements presented are compressional wave velocity values on the virgin curve. These and subsequent values were not temperature corrected in order to concentrate on the dependence on effective stress. To smooth the numbers, a regression equation rather than the actual data points, was used to calculate the shear modulus (Figure 8). Shear moduli measured using the duomorph and calculated from the compressional wave velocity and compressibility (derived from consolidation testing) are presented in Figure 8b. For the data of Figure 8b, and other measurements not shown, the measured shear modulus shows lower initial values, which possibly reflect seating problems in

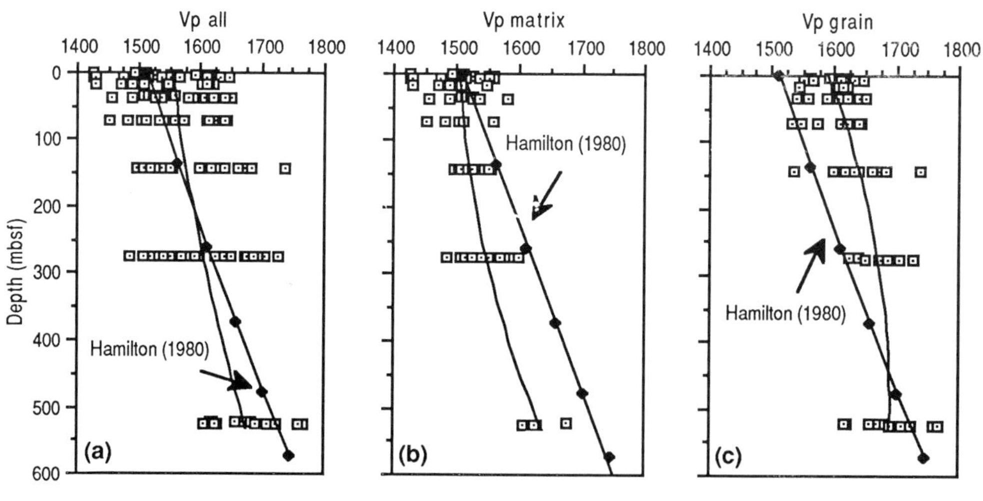


Figure 7. Compressional wave velocity as a function of effective stress and estimated subbottom depth. (a) All samples, (b) grain supported samples, and (c) matrix supported samples.

the consolidometer. Shear wave velocity values derived from both the calculated and measured shear moduli are shown in Figure 8c. Because the wet bulk density used at each effective stress was the same for shear wave velocity calculated from either the measured or calculated shear modulus, the shear modulus trends are mirrored in the shear wave velocity trends.

Shear velocity values exhibited little or no dependence on the effective stress for many of the samples (as illustrated for one sample in Figure 8). The shear modulus of most of the samples ranged between $1.69 \times 10^5$ N/m$^2$ and $4.26 \times 10^6$ N/m$^2$ at ambient laboratory pressures, which resulted in shear velocities that ranged between ~10 m/s and 49 m/s. Shear modulus at 120 kPa ranged between $2.03 \times 10^6$ N/m$^2$ and $4.96 \times 10^6$ N/m$^2$, which resulted in shear wave velocities between 35 and 55 m/s. Shear wave velocities based on compressional wave measurements at 4500 kPa ranged between 44 and 48 m/s, close to the values based on measured shear moduli (duomorph) at both 120 kPa and 4500 kPa.

Shear modulus and shear wave velocity as a function of effective stress for these samples can be described by the regression equations:

$$G = 2.95 \times 10^6 + 1.597 \times 10^3 (\sigma') - 0.264 (\sigma'^2) , \tag{10}$$

$$Vs = 37.67 + 1.007 \times 10^{-2} (\sigma') - 1.659 \times 10^{-6} (\sigma^2) , \tag{11}$$

where $\sigma'$ represents effective stress in kPa, G is the shear modulus in N/m$^2$, and Vs is shear wave velocity in m/s.

## Discussion

Accurate sediment compressional wave velocity measurements are important in acoustic prediction because at low grazing angles, low-frequency sound energy can be returned to the water when

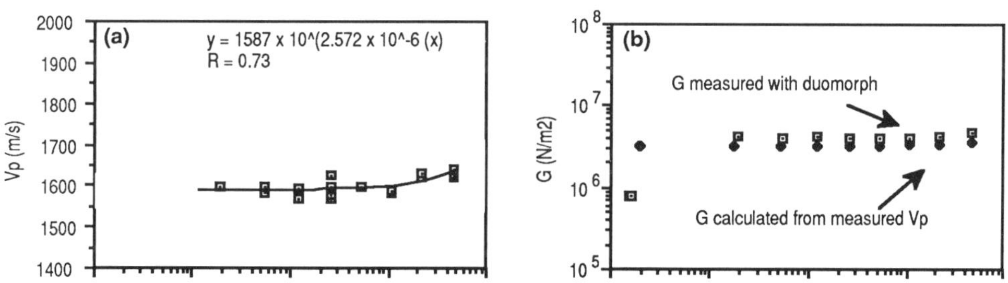


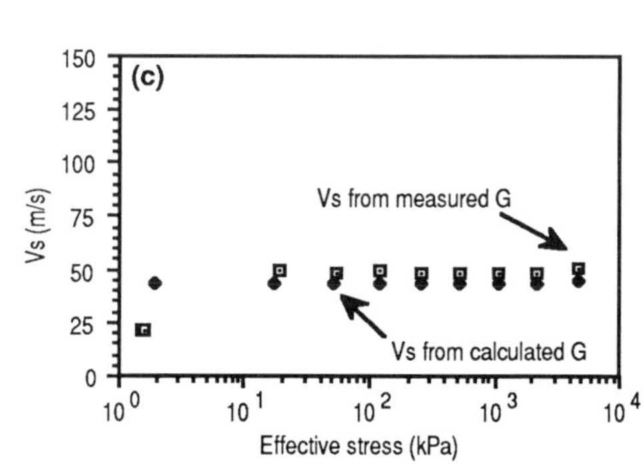


Figure 8. Laboratory measured acoustic properties of ES 7P-1. (a) Compressional wave velocity (Vp) values are taken from the virgin curve only, (b) comparison of measured shear modulus using duomorphs and calculated shear modulus, and (c) a comparison of shear wave velocity (Vs) derived using measured shear modulus and calculated shear modulus values.

positive velocity gradients are present (Hamilton, 1980). Velocity gradients can be used to compute layer thickness and to determine the depth of lithification in periplatform oozes. The compressional wave velocity gradient for the total sample set measured here is similar to that predicted by Hamilton (1980). It is possible, with a rudimentary estimate of fabric type, that is, grain- or matrix-supported, to estimate velocity more precisely using the appropriate regression equations derived for these samples. Such an estimate of fabric can often be made when factors in the environment of deposition, such as slope inclination, are known. For example, on the steep Exuma Sound slope, sediment was predominantly matrix-supported because much of the material composed of larger (and heavier) grains by-passed the slope and was deposited on the gentler toe of the slope. In contrast, turbidite deposition on Little Bahama Bank resulted in the majority of the samples recovered being grain supported. Similarly, a sediment that has a high proportion of sand sized grains is likely to be grain supported in contrast to a sediment that is predominantly clay sized.

One of the goals of this study was to characterize the shear wave velocity of periplatform carbonate sediments. As previously discussed, an early hypothesis was that shear wave velocity measurements made in the laboratory under ambient pressure may not be representative of *in situ* values and that the divergence between laboratory measurements and *in situ* values would continue to increase with increasing *in situ* subbottom depth. Our bender-element shear velocity transducers are useful only at ambient laboratory pressures. The duomorph technique developed by Briar and others (1972, 1976) offered an alternate and independent method for measuring the shear modulus and, therefore, the shear wave velocity if the techniques could be adapted for use in the consolidometer.

This study indicates that the duomorph technique is a feasible alternative to the bender-element shear wave transducers. The data show that the initial shear modulus and the shear wave velocity values are low, that they increase with increasing effective stress up to 100 to 150 kPa, and then level off. This pattern is consistent for all the samples tested. Two possible explanations for the constant shear wave property values at high effective stress are: 1) the duomorphs do not provide accurate values above 150 kPa, or 2) the shear modulus and shear wave velocity of the carbonate sediments simply do not increase significantly with effective stress above 150 kPa. In fact, shear modulus and shear wave velocity values calculated from measured compressional wave velocity at 120 kPa and higher were close to those measured by the duomorphs (Figure 8). The fully corrected measured compressional wave velocities (Fig. 7) agree well with the predicted values of Hamilton (1980). Thus the shear modulus and shear wave velocity values calculated from measured compressional wave velocity are probably good estimates and these results indicate the probable correct values obtained with the duomorphs even at high effective stress values.

It is probable that the initial (low effective stress) low shear wave velocities measured in the laboratory are also real. They approximate values reported by Schultheiss (1980, 1985) and Richardson and others (1988). Schultheiss reported values, under laboratory temperatures and pressures, between 35 and 39 m/s on nanofossil oozes from Deep Sea Drilling Project Site 577 for samples recovered from between 5 and 10 mbsf. For samples recovered from 93 mbsf, he reported a shear wave velocity of 39 m/s. This value translates to a shear modulus of $2.54 \times 10^6$ $N/m^2$ using his reported wet bulk density of 1.67 $Mg/m^3$. In a different type of sediment—beach sands of mixed grain size—Schultheiss (1980) reported only a slightly higher shear modulus of $6.6 \times 10^6$ $N/m^2$. For glass beads he reports $1.3 \times 10^6$ $N/m^2$ which compares to ~2 or $3 \times 10^6$ $N/m^2$ measured by the duomorphs in a glass bead test. Richardson and others (1988) report shear wave velocities of 20 m/s, in flocculent marine muds, to 80 m/s in hard packed sands. These values translate to $5.6 \times 10^5$ $N/m^2$ and $1.06 \times 10^7$ $N/m^2$ respectively. Although these reported values of shear velocity confirm the occurrence of low values they do not demonstrate the lack of effective stress dependence.

Hamilton's regression equations (1980), however, indicate that shear modulus and shear wave velocity should be significantly higher at effective stress levels comparable to those in this study. Measurements made by Turgut and others (1987) support Hamilton's predictions. They report shear modulus values an order of magnitude higher in marine sands (Trevorrow and others 1988) and oolitic Bahamian carbonates than those reported by Schultheiss (1980, 1985) and here. It is not clear why there should be such a discrepancy among reports of shear modulus and shear wave velocity in carbonate sediments. Hamilton's regression equations for shear wave velocity are based partly on seismic data and partly on surficial compressional wave velocity measurements. Differences between seismically derived and laboratory-measured parameters are common. Turgut and others' values are measured on ooids that may be very similar to hard-packed, equally sized sands, which would result in higher values of shear modulus. Indeed, this seems to be the case in the Schultheiss (1980) experimental study in which he reports that sands with a narrow range of grain size have higher shear moduli (on the order of $1.9 \times 10^7$ N/m$^2$) than mixed sands (on the order of $10^6$ N/m$^2$), which are similar to the grain distribution of the periplatform oozes.

In addition, it has been shown that carbonate sediments consolidate differently than marine clays. They consolidate less overall and do not reach as low a void ratio as do marine clays under mechanical consolidation (Rezak, 1974, Lavoie, 1989). Matthews (1980) noted that a marked increase in compressional wave velocity, for instance, occurs mainly at lithological boundaries as a result of cementation rather than consolidation. It is possible that for shallow water carbonates one should not expect a continuous increase in properties such as shear modulus and shear wave velocity in the absence of cementation. This would suggest that lithologic changes or the fabric of a periplatform sediment may be as important as overburden pressure in predicting gradients of acoustic properties in the upper sedimentary section.

## Conclusions

Periplatform carbonate sediments consolidate less than marine clays and do not reach as low a void ratio. Geoacoustic properties depend on sediment fabric as well as effective overburden stress. Average values of compressional wave velocities of periplatform oozes measured here, when corrected to *in situ* conditions, are similar to those predicted by Hamilton (1980). However, a more precise estimate of the compressional wave velocity/depth gradient can be made when the fabric of the periplatform sediment is taken into account.

Shear modulus and shear wave velocity values measured in these periplatform sediments are significantly lower than predicted by Hamilton but essentially the same as those measured by Schultheiss (1985) on nanofossil oozes under laboratory conditions. The independent measurements of shear wave velocity using duomorph produced values are similar to those resulting from bender element transducer measurements of samples from the same region when the values are compared at laboratory pressures. This increases the confidence in both measurements and suggests that duomorphs are a feasible alternative to shear wave bender-element transducers. The most thought-provoking suggestion from these measurements is the small dependence of shear wave properties on effective stress for these samples. The behavior exhibited as an example in Figure 8b and 8c should certainly be investigated further.

## Acknowledgments

The authors would like to thank Dr. Richard Rezak of Texas A&M University and Dennis Lavoie of NOARL for their help with the scanning electron microscopy. Special appreciation is offered to Dave Young of NOARL and Samantha Breeding of Planning Systems, Inc., for their support with the fabrication of laboratory equipment and the many sample analyses. This work was supported by NOARL's 6.1 research program, directed by Halcyon Morris, NOARL Program Element 61153N. NOARL contribution PR91:002:361.

## References

Briar, H., Bills, K., and Schapery, R. A. (1976) 'Design and test of the operational in-situ gage for solid propellant surveillance', Air Force Rocket Propulsion Laboratory-TR-76-36. Edwards, CA, 93523, 131 p.

Clay, C. S., and Medwin, H. (1977) Acoustical Oceanography, John Wiley and Sons, New York, N.Y., 544 p.

Dobrin, M. B. (1976) Introduction to Geophysical Prospecting, McGraw-Hill Co., NY, 630 p.

Fruth, L. S., Jr., Omre, G. R., Donath, F. A. (1966) 'Experimental compaction effects in carbonate sediments', JSP, 36(3):747-754.

Gettrust, J. F., Grimm, M. Madosik, S., and Row, M. (1988) 'Results of a deep-tow multichannel survey on the Bermuda Rise', Geophys. Res. Letters, 15(12):1413-1416.

Hamilton, E. L. (1971) 'Elastic properties of marine sediments', JGR 76(2): 579-603.

Hamilton, E. L. (1980) 'Geoacoustic modeling of the sea floor', JASA 68(5):1313-1340.

Houtz, R. (1974) 'Preliminary study of global sediment sound velocities from sonobuoy data', in Hampton L. (ed.), Physics of Sound in Marine Sediments, NY, Plenum Press, pp. 519-535.

Matthews, J. (1980) 'Heuristic physical property model for marine sediments', JASA 68(5):1361-1370p.

Rezak, R., and Lavoie, D. L. (1990) 'Consolidation-related fabric changes of periplatform sediments', Geo-Marine Letters 10(2):101-110.

Richardson, M. D., Muzi, E., Troiano, L. (1988) 'Shear waves in sediments: a comparison of in situ and laboratory measurements', Abstract 38, The microstructure of fine-grained terrigenous marine sediments-form muds to shale, NORDA, Stennis Space Center, MS.

Richart, F. A. (1970) Vibrations of Soils and Foundations, Prentice-Hall, Inc., NJ, 414 p.

Schultheiss, P. J. (1980) 'Simultaneous measurement of P & S wave velocities during conventional laboratory soil testing procedures', Marine Geotechnology, Wormley, Godalming, Surrey, U.K. 4(4):343-367.

Schultheiss, P. J. (1985) 'Physical and geotechnical properties of sediments from the Northwest Pacific: Deep Sea Drilling Project Leg 86', in Heath, G. R., Bruckle, L. H., and others, Initial Reports, Deep Sea Drilling Project, 86, US Gov. Printing Office, Wash., pp. 701-722.

Timoshenko, S., Woinowsky-Krieger, S. (1959) Theory of Plates and Shells. McGraw-Hill Co., NY, pp. 51-78.

Trevorrow, M., Yamamoto, T., Trugut, A., Abbott, C., Badiey, M., Goodman, D., and Ando, K. (1988) 'High resolution bottom shear modulus profiler experiments on the New Jersey Shelf, Summer, 1987', RSMAS TR-88-002, Miami, FL, 233 p.

Turgut, A., Yamamoto, T., Abbott, C., Badiey, M., Trevorrow, M., and Goodman, D. (1987) 'Higher resolution bottom shear modulus profiler: a shallow water real-time OBS array', RSMAS TR-87-007, Miami, FL, 48 p.

Wright, J. A., Lauden, K. E. (eds.) (1989) CRC Handbook of Seafloor Heatflow, CRC Press Inc., Boca Raton, FL, 498 p.

# COMPARISON OF MEASURED COMPRESSIONAL AND SHEAR WAVE VELOCITY VALUES WITH PREDICTIONS FROM BIOT THEORY

K. B. BRIGGS
*Naval Oceanographic and Atmospheric Research Laboratory*
*Code 361*
*Stennis Space Center, MS 39529-5004*
*USA*

ABSTRACT. Sediment samples were collected from homogeneous shallow-water areas off the coast of northern Florida for the purpose of characterizing the sediment surface as a part of high-frequency acoustic experiments. Relatively undisturbed cores up to 50 cm long were collected by divers from medium sand, fine sand, and mud. Values of permeability, porosity, and grain size in addition to compressional and shear wave velocities were determined in the laboratory from cores to make comparisons between predicted and measured acoustic velocities. Shear wave velocity measurements were made with a modified Hamilton frame and were used to calculate shear moduli for use as model inputs. Predictions of acoustic velocities are made from the Biot model, elastic model, Gassmann theory (close pack sphere model), and Wood formula. The ability of a theoretical model to predict velocities near measured values improves as more inputs describing the sediment composite system are incorporated into the model. Through manipulation of the pore size parameter in the Biot model, predicted and measured velocity values may be made identical. Of the discussed models, only Biot theory demonstrates the dispersion of velocity with frequency.

## 1. Introduction

Prediction of compressional and shear wave velocity in poro-elastic media has been accomplished since the seminal work of Biot (1956a & b; 1962). Further work in the laboratory has confirmed the theory of Biot and brought new applications. Although many workers have utilized Biot's theory in modeling acoustic propagation in the seabed, little if any data exists from the field to examine the theory's mechanisms. This lack of data is due to difficulties involved in collecting large numbers of detailed and reliable measurements. Variability or heterogeneity in the real world usually invalidates or at least reduces the applicability of environmental data. Fortuitously, high quality environmental data has been collected by the Navy from a variety of homogeneous sites on the seabed.

Environmental data were gathered and analyzed by the Naval Oceanographic and Atmospheric Research Laboratory (NOARL) for two high-frequency acoustic experiments in shallow water (10 to 22 m water depth). The multi-laboratory high-frequency shallow-water experiment (HFSW) was conducted northeast of Jacksonville, Florida in 1985 (Briggs et al., 1989) and the Acoustic Test Bed (ATB) experiment was conducted in waters surrounding Panama City, Florida in 1986. Compressional wave velocity and attenuation, shear wave velocity, porosity, permeability, and



*J. M. Hovem et al. (eds.), Shear Waves in Marine Sediments*, 121–130.

grain size were determined as a function of depth in the sediment. These data make possible a test of compressional and shear wave velocities calculated from Biot theory with those velocities measured in relatively undisturbed field samples. The Biot model employed in this investigation uses porosity, permeability, and shear modulus as free parameters to predict acoustic velocities within the sediment. That is to say, assumptions are made for values of such parameters as specific frame loss, added mass coefficient of skeletal frame, and bulk moduli of grains and fluid. Discrepancies between calculated and measured velocities may be reconciled by adjustment of such "fixed" parameters.

## 2. Materials and Methods

### 2.1. SITE DESCRIPTIONS

The site for the HFSW experiment was located about 50 km northeast of Jacksonville, Florida. The site was located at a water depth of 21 m in a loose, rippled medium sand. Two areas were investigated in the ATB experiment near Panama City, Florida: within the ship channel of St. Andrew Bay at a water depth of 10 m in mobile fine sand and in the bay at a water depth of 10 m in a lagoonal mud. The areas of the sea bottom under investigation were necessarily chosen to be uniform and homogeneous in order to interpret the experimental data.

### 2.2. SAMPLE COLLECTION AND ANALYSIS

Sediment samples for assessing geoacoustic properties were collected with the aid of scuba divers. Sediment samples were collected with 45-cm-long, 6.1-cm-ID (inside diameter), polycarbonate plastic cylindrical cores that were bevelled at one end to facilitate penetration into the sediment. Cores for the measurement of sediment permeability and shear wave velocity were similar but were 13 cm in length. Cores were capped at both ends immediately upon collection to retain the water overlying the sediment and kept in an upright position during transport to the laboratory for analysis.

Individual cores penetrated about 20-25 cm into the sediment at the experimental area. Sediment samples deeper than 20-25 cm were obtained by inserting cores adjacent to the initial core at successive, overlapping depths in the sediment as a diver-operated airlift was employed to excavate around the inserted cores. The 13-cm-long cores were inserted at every point at which long cores were inserted. Depth profiles of geoacoustic properties were ascertained from overlapping data and used to reconstruct the sediment depth for data from the 13-cm-long cores.

Measurement of sediment compressional wave velocity was made within 24 hours of collection. Selected cores were sectioned at 2-cm increments for analysis of porosity upon completion of velocity measurements. The core sections and unopened cores were then transported on ice to the laboratory at Stennis Space Center, Mississippi for the remaining analyses of geoacoustic properties.

Values of sediment compressional wave velocity were determined at 1-cm intervals in core samples as described in Briggs et al. (1989). Essentially, the velocity was measured with a 400 kHz signal transmitted via oil-filled silicone rubber transducer-receiver heads through the core liner and the intact core. Compilation of previous data reveal the accuracy of the compressional wave velocity measurements to be $\pm 5$ m/s. Sediment porosity was determined from weight loss of sediment when dried at 105°C for 24 hours. Average grain density was determined on dried samples with an air-comparison pycnometer. Sediment bulk density was calculated from measured values of porosity and average grain density. Sediment grain size distribution was determined by dry sieving samples on a set of sieves of quarter-phi increments on a CE-Tyler sieve shaker for gravel- and sand-sized particles and by the pipette method for percent silt and percent clay. Mud samples were analyzed with a Micromeritics Sedigraph. Detailed description of sample disaggregation and pipette analysis are given in Richardson et al. (1986). Measurements of sediment permeability were made on 13-cm cores in their entirety. Permeability was measured with Soiltest K-605 combination permeameter modified to accommodate the $6.1 \times 13$-cm core. The inside diameter of the permeameter cell and a groove for inserting a neoprene O-ring was milled out to allow insertion of the core into the permeameter. In this way, *in situ* measurements of permeability were approximated. A minimum of six determinations were made on each core by measuring the flow rate of fresh water with a constant head (falling head method was used on mud samples).

Shear wave velocities were measured on a Hamilton frame modified with shear wave transducers (Richardson, 1983) similar to those of Brunson and Johnson (1980) and a load cell to calibrate the pressure exerted on the samples. Measurements were made with the transducer incident upon the sample at a consistent pressure of 5 kPa. Shear wave velocities were calculated from time delay of shear wave transmission through the measured thickness of the extruded sediment sample at a frequency of 2 kHz (samples collected from muddy sediment were measured at 300 Hz). All samples were confined in a 6.1-cm diameter core liner cut to accommodate the sample extruded from the identical diameter core in which it was collected. A minimum of five determinations were made on each extruded sample by consistently introducing the designated pressure of the transducer and immediately noting the time delay before relaxation of the pressure due to the slight compaction of the sediment.

## 3. Results and Discussion

### 3.1. HFSW EXPERIMENT

The vertical distributions of sediment compressional wave velocity at 400 kHz porosity, and mean grain size ($\phi$) for cores collected in the HFSW experiment are displayed in Figure 1. Values of compressional wave velocity at *in situ* conditions range from 1721 to 1801 m/s and average 1761 m/s. Values of sediment porosity in the area where cores for permeability were collected average 35.9%. Porosity varies as much as ±1.4% from the mean value. Mean grain size for the sediments at the depths where permeability was measured is $1.18\phi$ (0.44 mm). Sediment permeability in core 126A is $3.7 \times 10^{-11}$ and $2.7 \times 10^{-11}$ m$^2$ for the 0-13 and 14-26 cm depth intervals, respectively. Permeability in core 126B is $2.9 \times 10^{-11}$ and $3.0 \times 10^{-11}$ m$^2$ for the 0-13 and 14-26 cm depth intervals, respectively. Permeability in core 130B is $2.7 \times 10^{-11}$, $3.5 \times 10^{-11}$, and $3.8 \times 10^{-11}$ m$^2$ for the 0-13, 13-25, and 24-36 cm depth intervals, respectively. There does not appear to be a discernible gradient with depth, thus, average permeability values for cores 126A, 126B, and 130B may be used. Values of shear wave velocity made on the same cores from which the permeability was measured are displayed in Figure 2. Shear wave velocity values from core 126A average 97 m/s; from core 126B average 94.5 m/s; from core 130B average 91.3 m/s.

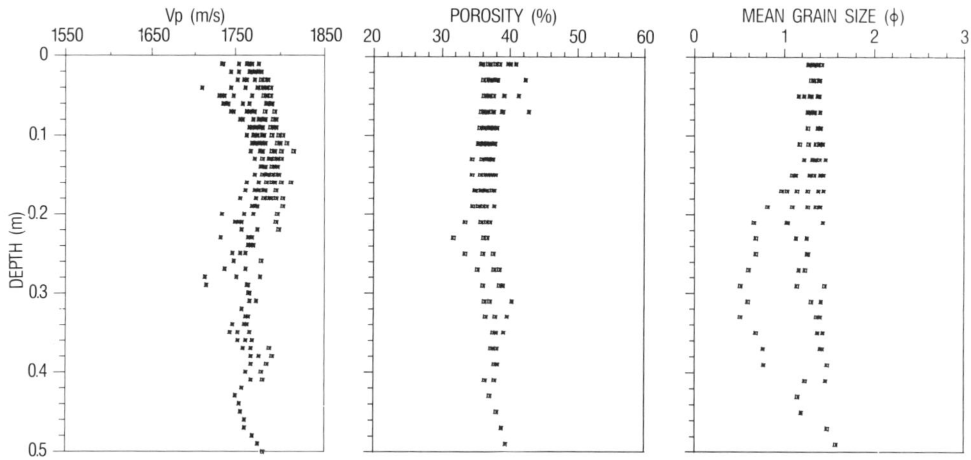


*Figure 1. Vertical distribution of sediment compressional wave velocity at 400 kHz, porosity, and mean grain size for all 45-cm cores collected for the HFSW experiment.*

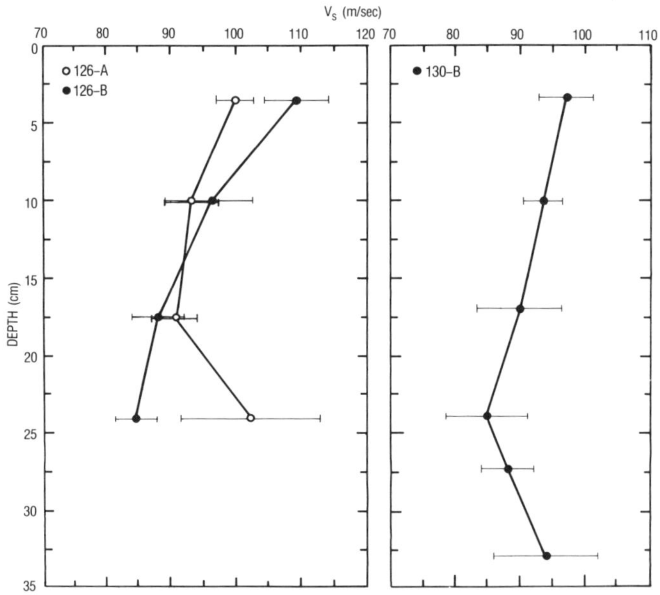


*Figure 2. Depth profiles of shear wave velocity ($V_s$) in m/s from 13-cm cores collected at two dives (126 and 130) at the site of the HFSW experiment. Bars depict ±1 standard deviation around mean of 5-6 determinations.*

Sediment shear modulus may be calculated from measured values of shear wave velocity and calculated values of bulk density after Hamilton (1971) and Young and Southard (1978). Using the relationship,

$$\mu = V_s^2 \rho \ , \tag{1}$$

the shear modulus, $\mu$, can be calculated directly. The range of values for the shear modulus as a function of depth in the sediment is small, especially compared with the variability of the shear wave measurements used to calculate the modulus. Hence, it is reasonable to use an average value of shear modulus for each core. The calculated shear moduli for cores 126A, 126B, and 130B are $1.95 \times 10^7$, $1.87 \times 10^7$, and $1.71 \times 10^7$ Pa, respectively.

### 3.2. ATB EXPERIMENT

Mean values of sediment compressional wave velocity at 400 kHz, porosity, and mean grain size ($\phi$) for cores collected in the ATB experiment are displayed in Table 1. Values of compressional wave velocity are quite uniform with depth in the sediment and average 1520 ($\pm$3) m/s in lagoonal mud (S6) and 1743 ($\pm$8) m/s in channel sands (B14). Values of sediment porosity are very high in the lagoonal mud (average: 88%) but vary little from the mean ($\pm$0.7%). The top 20 cm of sediment from cores 166 and 167 are easily differentiated from sediment below it by color and texture. The sediment-water interface at site S6 is repeatedly resuspended and, upon settling, forms a loosely compacted layer on top of more cohesive sediment layers. The low values of average grain density ($\rho_r$) are indicative of the high organic content of the sediment. Porosity values of the channel sands are typical for well-sorted fine sands and vary only $\pm$0.8% from the mean. Slight differences between the B14 cores reflect differences in the characteristics of sediment collecting in the trough (168-3) and depositing on the crest (168-4) of migrating sand waves.

The sediment is quite uniform in its physical properties as a function of depth in the sediment. Additional cores collected for x-ray analysis and sediment volume scatterer counts at site S6 show practically no inhomogeneities within the sediment. Sieving of cores from site B14 produce only a few coarse shell fragments in the relatively homogeneous top 12 cm of sediment.

Sediment permeability (k), shear wave velocity ($V_s$), and shear modulus ($\mu$) as a function of depth in the sediment are given in Table 1. Velocity of shear waves were measured at a frequency

TABLE 1. Geoacoustic data from 1986 ATB experiment. Mean grain size (MGS) is expressed in phi units, $\phi$ where = $-\log_2$ (dia. in mm).

| Core Nos. | Depth (cm) | $V_p$ (m/s) | Porosity | MGS ($\phi$) | $\rho_r$ (kg/m$^3$) | k (m$^2$) | $V_s$ (m/s) | $\mu$ (Pa) |
|---|---|---|---|---|---|---|---|---|
| S6-166/167 | 0 - 12 | 1521 | 0.89 | 11.0 | 2350 | $3.4 \times 10^{-15}$ | 6 | $4 \times 10^4$ |
| S6-166/167 | 10 - 22 | 1520 | 0.89 | 11.1 | 2350 | $2.4 \times 10^{-15}$ | 8 | $6 \times 10^4$ |
| S6-166/167 | 27 - 39 | 1520 | 0.86 | 10.8 | 2380 | $1.4 \times 10^{-15}$ | 10 | $1.1 \times 10^5$ |
| S6-166/167 | 41 - 53 | 1518 | 0.86 | 10.9 | 2400 | $1.0 \times 10^{-15}$ | 9 | $1.0 \times 10^5$ |
| B14-168-3 | 0 - 12 | 1741 | 0.39 | 2.3 | 2650 | $5.2 \times 10^{-12}$ | 83 | $1.4 \times 10^7$ |
| B14-168-4 | 0 - 12 | 1746 | 0.38 | 2.1 | 2650 | $4.8 \times 10^{-12}$ | 78 | $1.2 \times 10^7$ |

of 300 Hz in the mud (S6) cores and 2 kHz in the sand (B14) cores. The extremely low values for shear modulus (rigidity), however, cast doubt on the ability of the mud to support shear waves.

### 3.3. MODEL PREDICTIONS

Values of permeability, porosity and shear modulus are tabulated in Table 2 for use in the Biot model derived in Yamamoto (1983) and Yamamoto and Turgut (1988). These parameters control the model output to the greatest extent. On the other hand, variation in the Poisson's ratio of the skeletal frame in the range of 0.32 to 0.35 is responsible for a change in compressional wave velocity of only 1-2 m/s in the high-frequency realm considered in this comparison. The value for specific frame loss ($\delta$) in the model is chosen to be 0.02 after Yamamoto (1983). Three sediment types are chosen from the data collected from the two experiments. The medium sand represents the average properties of sediments of the HFSW experiment; the fine sand represents the average of the mobile channel sand (B14) in the ATB experiment; and the mud represents the average of the highly organic clay (S6) in the ATB experiment. Measured acoustic velocities are included for comparison with the predicted velocities depicted in Figures 3 and 4.

Figure 3 reveals the values of compressional wave velocity in medium sand to vary from 1677 m/s at 1 Hz to 1807 m/s at 1 MHz. The response of velocity to the change in frequency

TABLE 2. Parameters for Biot model and average velocities (m/s) measured in cores collected from each sediment texture.

| Sediment Type | k ($m^2$) | Porosity | $\mu$ (Pa) | $V_p$ | $V_s$ |
|---|---|---|---|---|---|
| medium sand | $3.1 \times 10^{-11}$ | 0.36 | $1.84 \times 10^7$ | 1761* | 94† |
| fine sand | $5.0 \times 10^{-12}$ | 0.39 | $1.32 \times 10^7$ | 1743* | 81† |
| mud | $2.1 \times 10^{-15}$ | 0.88 | $0.78 \times 10^7$ | 1520* | 8§ |

*400 kHz; †2 kHz; §300 Hz

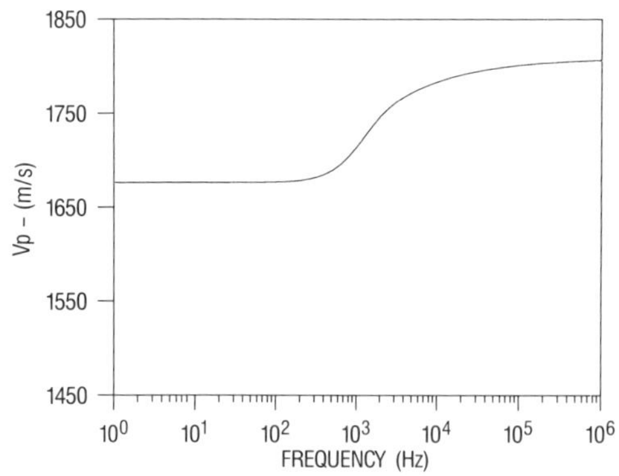


*Figure 3. Compressional wave velocity predicted for medium sand from Biot Theory.*

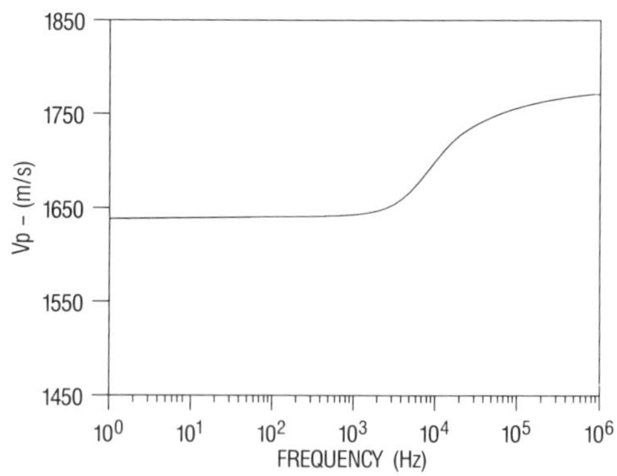


*Figure 4. Compressional wave velocity predicted for fine sand from Biot Theory.*

at the high end near 400 kHz is nearly flat. The predicted values of shear wave velocity in medium sand to vary from 94 m/s at 10 Hz to 102 m/s at 1 MHz. Measurements of shear wave velocity made at 2 kHz in medium sand are apparently at a steep gradient of velocity response to changing frequency. The region of maximum change in velocity with changing frequency is also the region of maximum attenuation (Yamamoto and Turgut, 1988). Dispersion in velocity is due to the changing oscillatory boundary layer thickness in the idealized pore spaces as the frequency increases (wavelength decreases).

This general pattern of frequency dependence by acoustic velocity holds for fine sand, as shown in Figure 4, although the maximum slope of the dispersion curve is shifted toward higher frequencies. The result of this shift in the dispersion curve for shear wave velocity in fine sand is that velocities measured at 2 kHz are not near the maximum slope and thus should exhibit less variation with frequency and less attenuation than in coarser sand. Values of compressional wave velocity in fine sand vary from 1642 m/s at 1 Hz to 1769 m/s at 1 MHz; values of shear wave velocity vary from 81 to 88 m/s over the same range of frequency.

The Biot model predicts little velocity dispersion in the frequency range 1 Hz to 1 MHz for the mud. The compressional wave velocity is predicted to be about 1479 m/s throughout the frequency spectrum. Such small values of measured permeability are to be expected in the mud samples (Table 1) and dictate a minuscule characteristic pore size for cohesive sediments. If $k$ is permeability and $\beta$ is porosity, then the characteristic pore size, $a_0$, is calculated as

$$a_0 = \sqrt{8\frac{k}{\beta}} \; . \tag{2}$$

The value for the relaxation frequency, where maximum velocity dispersion occurs, should be over 42 MHz according to Stoll (1977). The shear wave velocity shows a flat frequency response. The predicted shear wave attenuation shows an almost imperceptible increase throughout the frequency range used. The lack of predicted response of velocity and attenuation to frequency may be related to the lack of rigidity of the mud. With such a low shear modulus, the mud in St. Andrew Bay is a poor medium for the propagation of shear waves.

Due to the fluid-like nature of the mud at site S6, the Wood formula for speed of compressional waves in composite media may be applicable (White, 1983). This theory contrasts markedly with Biot theory in that all constituents of the sediment move together. From measured values of porosity and grain density and estimated values of bulk moduli of fluid and solid constituents (after Yamamoto, 1983), a range of 1463 to 1467 m/s is calculated. Table 3 displays the additional predicted values for medium and fine sand. These predictions underestimate measured values to a greater extent than those from the Biot model. The Wood formula predicts sound velocity in

TABLE 3. Comparison of predicted values of compressional wave velocities at 400 kHz from various models with measured values (m/s).

| Sediment Type | Meas. | Wood Formula | Close Pack | Elastic | Biot |
|---|---|---|---|---|---|
| medium sand | 1761 | 1655 | 1717 | 1664 | 1775 |
| fine sand | 1743 | 1622 | 1687 | 1629 | 1762 |
| mud | 1520 | 1465 | 1586 | 1488 | 1479 |

suspensions where rigidity of the sediment skeletal frame is negligible. Because of this failure to predict the sound velocity accurately, the mud, as well as the medium and fine sand, must indeed have appreciable rigidity.

The Gassmann theory derives the sound velocity from the characteristics of the sediment component grains and the porosity. The close pack sphere model of Duffy and Mindlin (1957) may be appropriate because a face-centered array of spheres more closely approximates the natural packing of sediments. From values of elastic constants for quartz (Poisson's ratio = 0.15, Young's modulus = $1 \times 10^{11}$ Pa) suggested in White (1983) and measured values of porosity and grain density, values of compressional wave velocity in a suspension of solid particles in a fluid is calculated for the three sediment types. The predictions attributed to the close pack model reported in Table 3 underestimate velocities in sands by 40 to 50 m/s and overestimate the velocity in mud by more than 60 m/s. Clearly, restricting model inputs to those parameters describing the component solid (and fluid) as Gassmann theory does, ignores the effects of the sediment as a composite system.

A simple elastic model for compressional wave velocity takes into account the bulk modulus of the sediment skeletal frame as well as the bulk modulus of the fluid/solid composite. Bulk moduli for the three sediment types are calculated from the frame bulk modulus used in the Biot model of Yamamoto (1983). The bulk modulus of the frame $K_s$ is related to the shear modulus $\mu$ and the Poisson's ratio of the skeletal frame $\sigma_s$ by the relationship

$$K_s = \mu\left(\frac{2\sigma_s}{1-2\sigma_s} + \frac{2}{3}\right). \tag{3}$$

Assuming Poisson's ratio for the skeletal frame is 0.333, values for bulk modulus for medium sand, fine sand, and mud are $5.79 \times 10^9$, $5.31 \times 10^9$, and $2.62 \times 10^9$ Pa, respectively. Values of compressional wave velocity are calculated from the bulk modulus $K$, the shear modulus $\mu$, and the bulk density $\rho$, by the Hookean relationship $V_p = (K + 4/3\ \mu/\rho)^{1/2}$.

The comparison in Table 3 shows that the elastic model underestimates sound velocities in sands and in mud. The velocities predicted from the elastic model, however, are very close to the low frequency limit of the Biot model predictions. Although the elastic model considers the sediment skeletal frame, the model makes two assumptions not made in the Biot model: that the pore water and solids move together and that there is no dispersion of velocity with frequency.

Any discrepancies between measured velocities and predictions from Biot theory are probably due to variations in the characteristic pore size, $a_0$. The choice of constant within the radical in Eq. (2) is determined by the tortuosity or shape factor of the pores in the media. Choice of a different tortuosity or shape factor may close the gap between measured and predicted values.

Reliability of measurements of permeability and porosity is good considering the relatively close agreement between measured velocities and velocities predicted by Biot theory in sands (Table 3 and Figures 3 and 4). The lack of agreement in the mud may be due to difficulties in measurement of permeability in clays. Clay particles typically are charged and react to fluid seeping through the particle matrix according to the electrolytic charge of the fluid. Lack of similarity in charge between clay and fluid may deform the pore structure by flocculation and yield lower values of hydraulic velocity.

Moreover, use of measured values of permeability rather than values calculated by Stokes' sphere model yields closer agreement between and measured and predicted acoustic velocities. The predicted Stokes' permeability, $k_s$, is given by

$$k_s = \frac{\beta^2 d^2}{18(1 - \beta)}, \tag{4}$$

where $\beta$ is porosity and $d$ is the mean grain size diameter. Values of $k_s$ for the medium sand, fine sand, and mud in Table 2 are $2.2 \times 10^{-7}$, $4.9 \times 10^{-8}$, and $8.3 \times 10^{-16}$ m$^2$, respectively. Adjustment of the grain size parameter to weight the finest particle sizes does not change the Stokes' permeability more than one order of magnitude. Grain size histograms from the HFSW experiment (medium sand) are used by Briggs et al. (1989) to produce cumulative percent curves. The lowest ten percent grain size diameter ($D_{10}$) is determined to be 2.25$\phi$ (0.21 mm) *vis-a-vis* 1.24$\phi$ (0.42 mm). The change in the Stokes' permeability is only one order of magnitude smaller and remains three orders away from agreement. Hence, laboratory measurement of permeability is more reliable than use of the Stokes' sphere model.

Comparisons between measured and predicted velocities, however, are inappropriate due to failure to sweep the frequency when determining sediment acoustic velocities from cores. Yet, measurement of sound velocity at one frequency from a variety of sediment grain sizes will vary the characteristic pore size. The frequency and characteristic pore size together affect oscillatory flow and thus propagation of sound through sediment according to Biot theory (Yamamoto, 1983). Measurement of velocity dispersion can be a difficult proposition because (1) most transducer-receiver assemblies designed for measuring sediment acoustic velocities are not broad band and (2) rarely can frequencies below 20 kHz be measured in cores with any certainty. Data for higher frequencies are scarce due to scattering of acoustic energy of small wavelength by individual sediment particles. The best efforts in measuring the velocity dispersion in sediments will come from *in situ* measurements with probes which are capable of transmitting energy at high amplitudes.

Constant frequency notwithstanding, compressional wave velocity has been measured accurately ($\pm$5 m/s) in undisturbed cores from regions of remarkable homogeneity. As the variable character of the media changed from a medium sand with some shell hash to a well-sorted fine channel sand to a uniformly fine-grained mud, the variability of the velocity measurement decreased from $\pm$40 to $\pm$8 to $\pm$3 m/s, respectively. One would expect the ability of theoretical models to predict the measured velocities to improve in accordance with increasing uniformity of substrate. However, this expected trend is not evident in the comparisons discussed in this study. Uniformity of particle size does not warrant an assumption of uniform particle arrangement or pore size. Myriad possibilities of particle stacking and intercalation exist depending on the sedimentological and biological processes controlling deposition, reworking, and diagenesis. Consequently, one would expect the ability of theoretical models to predict measured velocities to improve as more variables describing substrate characteristics are incorporated into the model. For instance, the Wood formula and Gassmann theory are insufficient because they fail to consider the sediment skeletal frame; the elastic theory is insufficient because it does not account for differential movement of the solid and fluid components of the system. The Biot model incorporates data on the sediment frame, component grains, fluid, and viscous interaction between solid and fluid components and consequently is the most satisfactory of the models considered for comparison. The gap between values measured in real sediments and values predicted by the Biot model may be eliminated by consideration of the pore size distribution.

## 4. Acknowledgments

The author would like to thank D. Young, M. Richardson, R. Ray, and S. Stanic for their expert assistance in collecting sediment samples. Special thanks to R. Ray and T. Orsi for assistance with laboratory and data analysis and to D. Lavoie and M. Richardson for help in measuring shear waves. This work was supported by NOARL Program Element 605857N. This is NOARL contribution number PR90:101:361.

## 5. References

Biot, M. A. (1956a) 'Theory of propagation of elastic waves in a fluid-saturated porous solid. I. Low frequency range', J. Acoust. Soc. Amer. 28, 168-178.

Biot, M. A. (1956b) 'Theory of propagation of elastic waves in a fluid-saturated porous solid. II. Higher frequency range', J. Acoust. Soc. Amer. 28, 179-191.

Biot, M. A. (1962) 'Mechanics of deformation and acoustic propagation in porous media', J. Appl. Phys. 33, 1482-1498.

Briggs, K. B., Fleischer, P., Lambert, D. N., Ray, R. I., and Sawyer W. B. (1989) 'Environmental support for high-frequency acoustic experiments conducted aboard the USNS Bartlett 2 August to 3 September', Naval Ocean Research and Development Activity, Stennis Space Center, MS, Tech. Note 458, 241pp.

Brunson, B. A. and Johnson, R. K. (1980) 'Laboratory measurement of shear wave attenuation in saturated sand', J. Acoust. Soc. Amer. 68, 1371-1375.

Duffy, J. and Mindlin, R. D. (1957) 'Stress-strain relations and vibrations of a granular medium', J. Appl. Mech. 24, 585-593.

Hamilton, E. L. (1971) 'Elastic properties of marine sediments', J. Geophys. Res. 76, 579-603.

Richardson, M. D. (1983) 'The effects of bioturbation on sediment elastic properties', Bull. Soc. Géol. France 25, 505-513.

Richardson, M. D., Briggs, K. B., Ray, R. I., and Jahn, W. H. (1986) 'Environmental support for high-frequency acoustic experiments conducted at the Quinault Range, April-May 1983', Naval Ocean Research and Development Activity, Stennis Space Center, MS, Report 132, 76p.

Stoll, R. D. (1977) 'Acoustic waves in ocean sediments', Geophysics 42, 715-725.

White, J. E. (1983) Underground Sound. Application of Seismic Waves, Elsevier Science Publishers, New York.

Yamamoto, T. (1983) 'Acoustic propagation in the ocean with a poro-elastic bottom', J. Acoust. Soc. Amer. 73, 1587-1596.

Yamamoto, T. and Turgut, A. (1988) 'Acoustic wave propagation through porous media with arbitrary pore size distributions', J. Acoust. Soc. Amer. 83, 1744-1751.

Young, R. N. and Southard, J. B. (1978) 'Erosion of fine-grained marine sediments: seafloor and laboratory experiments', Geol. Soc. Amer. Bull. 89, 663-672.

# WAVE VELOCITIES IN SEDIMENTS

AMOS NUR, DOMINIQUE MARION*, AND HEZHU YIN
*Rockphysics Laboratory*
*Department of Geophysics*
*Stanford University*
*Stanford, California, 94305*
*USA*
* *Now at Elf Aquitaine, Pau, France*

ABSTRACT. Systematic relations between porosity and compressional and shear velocities $V_p$ and $V_s$ in the three component sand, grains, clay and brine systems of (1) porous sandstone, (2) sands, and (3) suspensions, were obtained using experimental data and models. In cemented shaley sandstones *Vp and Vs* were found to correlate linearly with porosity and clay content. The velocities in clean sandstones are about 7% higher than those predicted by the linear fit, indicating that a small amount of clay significantly reduces the elastic moduli of sandstones.
For uncemented shaley sand, a model for the dependence of velocity and porosity on clay content and compaction was developed for sand with clay dispersed in the pore space and for shale with suspended sand grains. The model closely mimics the experimentally observed minimum for porosity and the peak in velocity versus clay content. The results explain much of the scatter in velocity data in-situ. Velocity in suspensions at $\phi$ = 39% of grains in brine is close to values predicted by the Reuss (Isostress) average. Our results highlight the crucial role of the critical porosity, $\phi_c$ at about 39%, and the transition between cemented to uncemented sands. Using a modified Voigt average, we obtain excellent agreement between $V_p$ and $V_s$ versus porosity for porosities below $\phi_c$ This average involves pure quartz on the one hand, and a critical suspension of sand grains on the other.

## 1. Introduction

Most existing models for the relation between velocity and porosity in sands and sandstones have generally failed to account for some of the main features observed in-situ or in the laboratory. Recent results by us (Han et al., 1986) and others have revealed the reasons for this: (1) the shapes of pores; (2) the presence of clay particles between grains and in pores, and (3) the contacts between grains. Using extensive experimental data, we derived general, complete, and rigorous relations between velocities and porosity, which are divided into three distinct domains: (1) Cemented sands, (2) Uncemented sands,and (3) Suspensions. The key to obtaining this relationship is the critical porosity, $\phi_c$, below which the solid is stress-bearing and above which the pore fluid is stress-bearing.



*J. M. Hovem et al. (eds.), Shear Waves in Marine Sediments*, 131–140.

## 2. Velocity Porosity Relation in Cemented Rocks

We have carried out a laboratory study of the effect of clay content and porosity on acoustic velocities on 75 sandstone samples with porosity from 2% to 30%, and clay content by volume from 0 to 50% (Han et. al, 1986). Velocities were measured ultrasonically (Birch, 1960) as functions of pressure, water saturation and pore pressure. The densities of the dry and saturated samples were determined. Porosity $\phi$ at room pressure of each sample was measured with a helium porosimeter, and clay content was obtained from point counts on thin sections.

Despite the scatter, the data clearly show that $V_p$ and $V_s$ decrease with increasing porosity (fig. 1). A simple empirical equation including clay content $C$ was used by us to fit the data

$$V = a - b \cdot \phi - d \cdot c, \tag{1}$$

where a, b, and d are constants. The best fit (at 40 MPa confining pressure) is

$$\begin{aligned} Vp\ (km\ /\ s) &= 5.59 - 6.93 \cdot \phi - 2.18 \cdot C \\ Vs\ (km\ /\ s) &= 3.52 - 4.91 \cdot \phi - 1.89 \cdot C\ . \end{aligned} \tag{2}$$

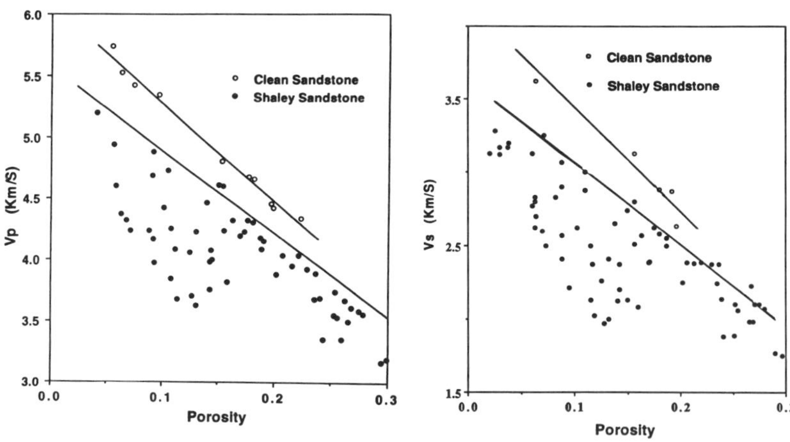


Figure 1. Measured (a) (b) shear velocities versus porosity for 75 sandstone samples at confining pressure Pc = 40 MPa and pore pressure Pp = 1.0 MPa.

The differences between the measured and predicted Vs values versus porosity are shown in fig. 2.

In the few clay free sandstones the measured velocities are higher than predicted by about 5% to 10%. Small amounts of clay in sandstones thus significantly soften grain contacts, most likely due to clay particles situated between grains. Because the clay particles' surface area is so large, even a small amount of clay can cover the entire pore surface area, including grain contacts.

## 3. Velocity-Porosity Relation in Uncemented Rocks

Velocities in uncemented sediments were measured in mixtures of sand and clay (Yin et. al, 1988) and analyzed using a simple geometrical/mechanical model (Marion, 1990). The samples were prepared by mixing clean, well-sorted Ottawa sand grains with pure room-dried kaolinite fine particles. A 0.5 mole solution of sodium-chloride was used to saturate the samples. Porosity and velocity values were obtained at pressure ranging from 10 to 50 MPa and at 1 MPa pre pressure. Porosity at room pressure was determined from samples' dry and saturated weights.

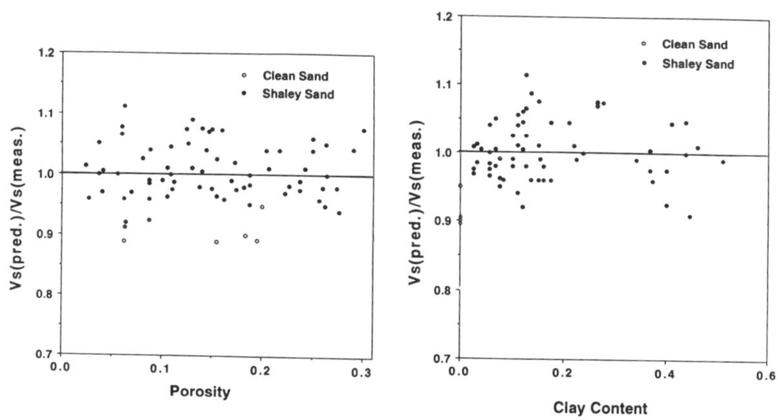


Figure 2. Linear fit between shear velocity $V_S$ versus porosity and clay content in sandstones (at $P_C$ = 40 MPa and $P_p$ = 1.0 MPa).

The results reveal systematic variations of porosity and velocity with clay content: the measured porosity shows a minimum at room pressure when clay weight fraction is 20%, this shifting to higher clay content with increasing confining pressure (fig. 3). In contrast, the compressional velocity (fig. 4) shows a maximum, at approximately 40% clay fraction. The shear velocity values also show variation with clay content, although they are more complex than either *Vp* or porosity. First there is sizable (up to 20%) decrease of velocity upon the addition of only a few percent clay, especially at elevated confining pressure. Second, a broad maximum in *Vs* is observed like *Vp* near 30%-40% clay (by weight).

POROSITY. The pore volume of our binary mixture can be analyzed in terms of the porosities of its two end members, ie., clean sand $\phi_s$ and pure clay $\phi_{sh}$ respectively (McGeary, 1961; Furnas, 1928; Westman and Hugill, 1930; Clarke, 1979).

Case 1: c < $\phi_s$. Here clay particles fit within the sand pore space, and the porosity of $\phi$ of the mixture decreases linearly with $C$

$$\phi = \phi_s - c\ (1 - \phi_{sh}) \quad for\ c < \phi_s. \tag{3}$$

Equation 3 holds until clay entirely fills the sand pore space when the sample porosity becomes

$$\phi = \phi_s \cdot \phi_{sh} \qquad for\ c = \phi_s.. \tag{4}$$

Case 2: c >$\phi_s$. The addition of clay can be accomplished here only by separating the sand grains, so that the porosity of the mixture is given by

$$\phi = c\,\phi_{sh} \qquad for\ c > \phi_s. \tag{5}$$

Case 1: $c < \phi_s$. When the clay particles are part of the pore-filling material and do not bear the load, velocity should increase with $C$ as the pore-filling moduli increase. Velocities are given by

$$V_p = \sqrt{\frac{K + 4G/3}{\rho}} \qquad V_s = \sqrt{\frac{G}{\rho}} \tag{6}$$

where K, G, and ρ — the bulk modulus, the shear modulus, and the density of the sand-clay mixture, respectively — are calculated using Gassmann's (1951) relations, and $\rho = (1-\phi_s)\rho_s + c(1-\phi_{sh})\rho_c + (\phi_s - c(1-\phi_{sh}))\rho_w$ where $\rho_s$, $\rho_c$ and $\rho_w$ are the densities of sand grains (quartz), clay (including bound water) and the saturating fluid respectively; $K_b$, $K_f$, and $K_s$ are the bulk moduli of the solid frame; the fluid filling the pore space, and the frame forming mineral respectively; $G_h$ is the shear modulus of the frame. We assume that the pore filling material has a vanishing shear modulus.

Case 2: $c > \phi_s$.. Sand grains are suspended in the shaley matrix, and the bulk and shear moduli obey the isostress model (Reuss 1929)

$$\frac{1}{K} = \frac{c}{K_{sh}} + \frac{1-c}{K_s} \qquad and \qquad \frac{1}{G} = \frac{1}{G_{sh}} + \frac{1-c}{G_s} \tag{7}$$

where $K$ and $G$ denote the bulk and shear moduli respectively and subscripts sh and s refer to saturated shale and quartz grains respectively. The density ρ is given by $\rho = (1 - c)\rho_s + c(1-\phi_{sh})\rho_c + c\phi_{sh}\rho_w$, Velocities from equation (6) were obtained using the measured values of $K_b$, $G_b$, $K_f$, $f_s$, and $f_{sh}$ and the bulk modulus, $K_f$, of the pore-filling material is given by

$$\frac{1}{K} = \frac{1 - \phi_{pf}}{K_1} + \frac{\phi_{pf}}{K_2} \tag{8}$$

where the bulk modulus of clay $K_1$ was determined experimentally. The porosity of the pore-filling material is

$$\phi_{pf} = \frac{\phi_s - c(1 - \phi_s)}{\phi_s} \tag{9}$$

The bulk and shear moduli of pure saturated clay, $K_{sh}$ and $G_{sh}$ were determined from P and S velocity measurements on saturated pure shale.

## 4. Critical Porosity and Elastic Bounds

Figure 5 summarizes the division, based on porosity and clay content, of the system sand-clay-brine into three domains — shaley sands, sandy shales, and suspensions.

The controlling parameters are simply the porosity of the clean sand $\phi_s$, and the porosity of the pure clay $\phi_{sh}$, and the critical porosity $\phi_c$, usually equal to $\phi_s$, at which clay particle exactly fill the sand pore space. Consequently, the shape and size of the domains is somewhat dependent on confining and pore pressures.

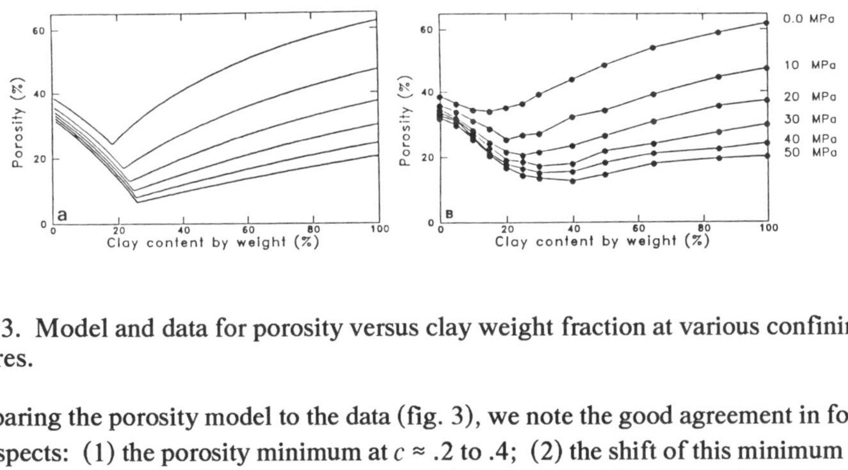


Figure 3. Model and data for porosity versus clay weight fraction at various confining pressures.

Comparing the porosity model to the data (fig. 3), we note the good agreement in four main aspects: (1) the porosity minimum at $c \approx .2$ to $.4$; (2) the shift of this minimum to higher clay content with increasing pressure; (3) the amount of porosity changes with pressure; and (4) the compaction behavior of sandy shale and shaley sand. This agreement suggests that the bimodal grain model is very adequate to describe porosity of sand-clay mixtures.

VELOCITY. Our calculation of compressional velocity versus clay content (fig. 4) is also based on the distinction between shaley sand ($c < \phi_s$) and sandy shale ($c > \phi_s$.).

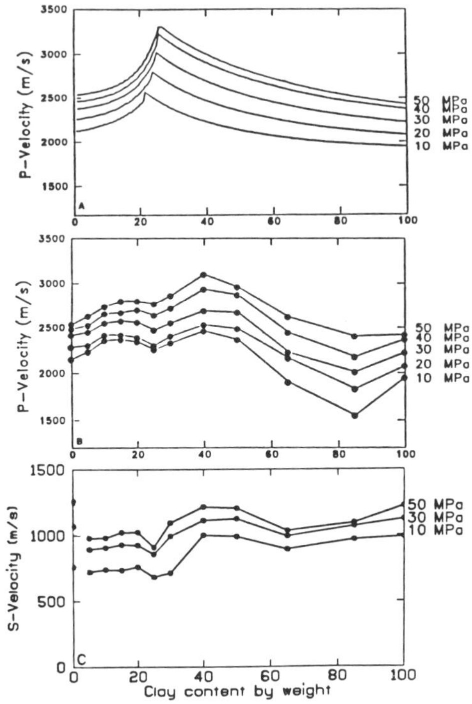


Figure 4. Velocities versus clay weight fraction at various confining pressures. Model results reproduce the peaks in velocities observed in unconsolidated sand-clay mixtures.

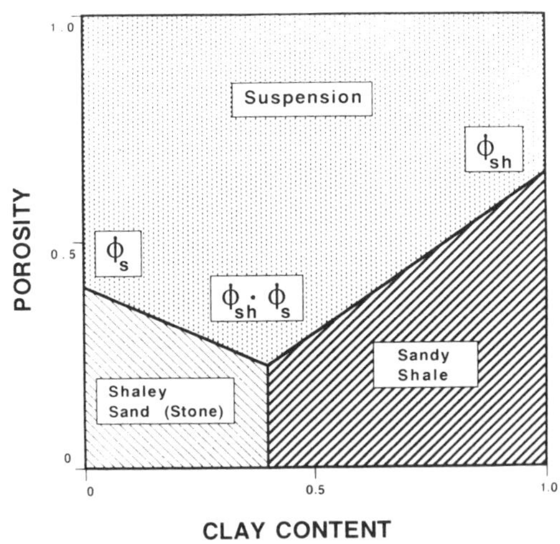


Figure 5. Approximate boundaries between sands, shales, and suspensions.

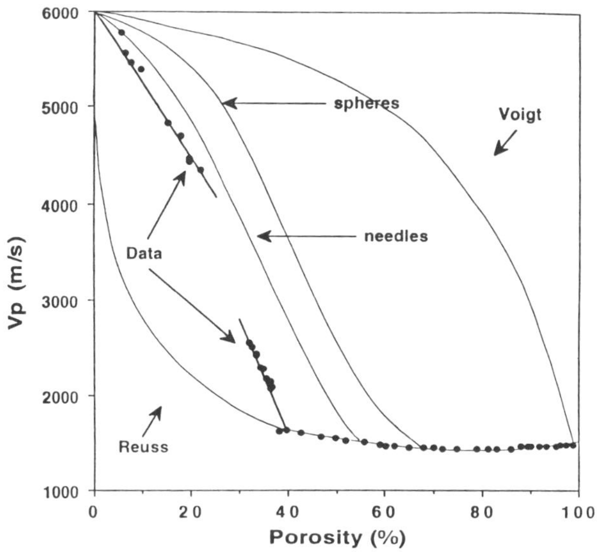


Figure 6. Velocity-porosity relation in clastic sediments without clay. Two transitions are important: (1) from suspension to load bearing grain matrix at ~ 39% porosity, and (2) from cemented to uncemented sand. Note that Voigt, Reuss, and the self-consistent spherical or needle-like pore models all fail to mimic the data.

Figure 6 combines all our data for compressional velocity versus porosity ranging from 2% to 85%, after removing the effects of clay. As shown, none of the existing models for the elasticity of this system are adequate predictors of velocities over the entire range of porosity. The Voigt (1928) and Reuss (1929) average are too far apart and too far from the data. Although the self-consistent approximations for spherical pores and especially needles come closer to the data, they still fail to match it, nor do they account for the critical porosity $\phi_C$,below which grains are load-bearing and above which grains are in suspension, at around 39%. Instead, these models yield a critical porosity around 55% or 60%. In clastic sediments, this porosity — both in the lab and in-situ — is most commonly around 36%-40%. In the absence of pore-filling clays, the compressional velocity versus porosity obeys almost a linear relation with porosity below $\phi_C$ — a relation which is surprisingly simpler than most theoretical models.

We propose here a modified Voigt average which specifically incorporates the critical porosity $\phi_C$. We assume that in a suspension, when $\phi > \phi_C$, the moduli closely obey the Reuss average, in which the stress in the solid and fluid is the same. In contrast, for load-bearing systems, in which $\phi < \phi_C$, we consider the moduli to follow a Voigt average of the two end components clean quarts at $\phi = 0$, and a critical suspension of grains in brine at $\phi = \phi_C$.

The standard use of Voigt and Reuss averages for a two component system is based on the relation between velocities and effective bulk and shear moduli $\overline{K}$ and $\overline{G}$ and system density $\overline{\rho}$ where

$$V = \sqrt{\frac{\overline{M}}{\overline{\rho}}} \tag{10}$$

with $\overline{M} = \overline{K} + \frac{4}{3}\overline{G}$ for $V_P$, and $\overline{M} = \overline{G}$ for $V_S$. The density is given by the usual average

$$\overline{\rho} = (1 - \phi)\rho_1 + \phi\rho_2 \tag{11}$$

where subscripts represent the properties of the pure end members quartz and clay. The Voigt and Reuss moduli are given respectively by

$$\overline{M}_v = (1 - \phi)M_1 + \phi M_2 \qquad \overline{M}_R = \left(\frac{1 - \phi}{M_1} + \frac{\phi}{M_2}\right) \tag{12}$$

The corresponding calculated compressional velocity, for a sand/water binary system is shown in fig. 7 together with our data. For $\phi<\phi_C$, the data is obviously unrelated to these averages, whereas for $\phi>\phi_C$, the data coincides with the Reuss average. At $\phi=\phi_C$, the effective moduli are $\overline{M}_c$.

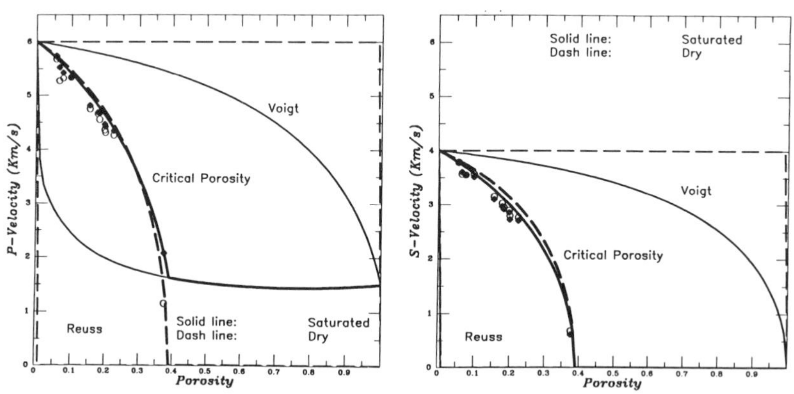


Figure 7. Comparison between computed modified Voigt and Reuss and measured (a) compressional; and (b) shear velocities in clean sandstones and sand.

We define next a modified porosity

$$\overline{\phi} = \phi / \phi_c \qquad 0 \leq \phi \leq \phi_c \tag{13}$$

when $\phi = 0$ then $\overline{\phi} = 0$, and when $\phi = \phi_c$, then $\overline{\phi} = 1$. Whereas the density still follows equation (11), the modified Voigt and Reuss, respectively

$$\overline{M}_v = (1 - \overline{\phi})M_1 + \overline{\phi}\,\overline{M}_c \qquad \overline{M}_R = \left[\frac{1 - \overline{\phi}}{M_1} + \frac{\overline{\phi}}{M_c}\right]^{-1} \tag{14}$$

where

$$\overline{M}_c = \left[\frac{1 - \overline{\phi}_c}{M_1} + \frac{\overline{\phi}_c}{M_2}\right] . \tag{15}$$

The agreement between the data and the computed velocities is amazingly good. Although a detailed analysis of this agreement is still needed, it implies that to a first approximation the packing of sand grains, cemented or uncemented is close to the stiffest one possible. it implies that whatever diagenetic processes transform sands into sandstones, involving porosity reduction and cementation, they do not greatly affect their stiffness relative to the modified Voigt bound.

## 5. Conclusion

The velocity data for cemented, uncemented, and suspended sands with clays and brines, spanning porosity, clay and brine from $\phi$=0 to 1.0, appear to follow surprisingly simple relations. For porosity below the critical porosity $\phi_C$, velocities in sandstones and sand are simply and to a first order linearly related to clay content. Furthermore, both Vp and Vs velocities follow the modified Voigt averages quite closely. In suspensions, for $\phi > \phi_C$, velocities follow very closely the Reuss average.

## 6. Acknowledgements

This work was supported by the Office of Naval Research under contract N00014-84-K-0560.

## 7. References

Birch, F. (1960) 'The velocity of compressional waves in rock to 10 kilobars, 1', J. Geophys. Res. 65, 1083-1102.

Clarke, R.H. (21979) 'Reservoir properties of conglomerates and conglomeratic sandstones', AAPG Bull. 63, 799-803.

Furnas, C.C. (1928) 'The relation between specific volume, voids, and size composition in systems of broken solids of mixed sizes', U.S. Bur. Mines, Rept. Investigations 2894, 1.10.

Gassman, F. (1951) 'Elastic waves through a packing of spheres', Geophysics 16, 673, 685.

Han, D., Nur, A., and Morgan, D. (1986) 'Effect of porosity and clay content on wave velocity in sandstones', Geophysics 51, 2093-2107.

Marion, D. (1990) 'Acoustical, Mechanical and Transport Properties of Sediments and Granular Materials', Ph.D. Thesis, Stanford Univ.

McGeary, R.K. (1961) 'Mechanical packing of spherical particles', J. Am. Ceram. Soc. 44, 513-522.

Reuss, A. (1929) 'Berechnung der Fliessgrenze von Mischkristallen aufgrund der Plastizitätsbedingung für Einkristalle', Zeitschrift fur Angewandte Mathematik and Mechanik 9, 49-58.

Voigt, W. (1928) 'Lehrbuch der Kristallphysik,' Teubner, Leipzig, Macmillan, New York.

Westman, A.E.R., and Hugill, H.R. (1930) 'The packing of particles', J. Am. Ceram. Soc. 13, 767-779.

Yin, H., Han, D.H., and Nur, A. (1988) 'Study of velocity and compaction on sand-clay mixtures', Stanford Rock and Borehole Project, Vol. 33.

# SHEAR WAVE ATTENUATION IN UNCONSOLIDATED LABORATORY SEDIMENTS

BURLIE A. BRUNSON
*Planning Systems Incorporated*
*7925 Westpark Drive*
*McLean, Virginia 22102*

ABSTRACT. Shear wave attenuation measurements were made using ceramic bimorph transducers to excite transverse vibrations in a cylindrical column of unconsolidated sediment. Three different water-saturated sediments were used in an attempt to determine the effects of grain shape and sorting on the frequency dependence of attenuation. The mean grain size of the sediments was held constant while the grain shape and size distributions were varied. The sediment assemblages used in the attenuation measurements included an unsorted-angular quartz sand, a well-sorted angular quartz sand, and well-sorted spherical glass beads. The unsorted sand showed the greatest attenuation over the measurement frequency range of 1 to 20 kHz. The well-sorted sand and the glass beads showed generally lower attenuation with the beads being the least lossy propagation medium. All three sediments showed evidence of viscous attenuation due to fluid-to-grain relative motion. This mechanism leads to a non-linear relationship between attenuation and frequency.

## 1. Introduction

In recent years, researchers have shown a great deal of interest in the propagation of transverse or shear waves through sedimentary materials. A subject of particular interest to acousticians and geophysicists is the frequency dependence of shear wave speed and attenuation in fluid saturated porous media.

This paper represents a brief summary of the results of some laboratory measurements of shear wave attenuation in water saturated medium grain sands and glass beads as a function of frequency, grain size sorting, and grain shape. It is intended to provide additional data to those researchers investigating the validity of the assumption that fluid-to-grain relative motion results in a non-linear relationship between attenuation and frequency. The work was sponsored by the United States Navy Office of Naval Technology and the Office of Naval Research.

## 2. Background

In a recent review of the subject of attenuation of sound in marine sediments, Kibblewhite [1] presented a comprehensive summary of the attenuation literature. In this review, the two opposing views of the frequency dependence of attenuation in marine sediments were presented. The first view, long held by Hamilton [2], is that attenuation can be considered to be a linear function of frequency for the entire range of frequencies generally considered in underwater acoustics (a few Hertz to a few megaHertz). The alternative view, based on the theoretical work of Biot [3-6] as applied to marine sediments by Stoll [7], holds that the interaction of the sediment frame and the pore fluid during passage of an acoustic wave leads to a nonlinear relationship between attenuation and frequency.

The linear versus nonlinear controversy was the motivation for the present research. This paper presents a short description of that research. A detailed description of the experimental and theoretical investigations have been compiled as a doctoral dissertation [8].



*J. M. Hovem et al. (eds.), Shear Waves in Marine Sediments*, 141–147.

## 3. Measurement Approach

To investigate the controversy over the frequency dependence of attenuation in marine sediments, an experiment was designed to measure shear wave attenuation. The approach was intended to insure that those physical properties of sediments thought to be important factors in shear wave attenuation would be controlled and measured during the conduct of the experiment. Factors considered included: frequency, grain size distribution, grain shape, porosity, permeability, and bulk density.

## 4. Propagation Media

In order to determine the effects of grain sorting and shape on the frequency dependence of shear wave attenuation, three assemblages were selected. The first assemblage was an unsorted angular quartz sand. The second assemblage was a well-sorted collection of the same angular quartz sand. The third assemblage, representing an "ideal" sediment, was a uniform collection of spherical quartz glass beads. All three assemblages had approximately the same mean grain size, 0.38 mm; thus, the differences were in the degree of sorting for the sands and the shape for the sand and beads. Those differences manifested themselves in both the physical properties and acoustical properties of the sediments. Table 1 summarizes the physical properties of the materials used as propagation media.

**TABLE 1. Sediment physical properties** ***(d is diameter, n is porosity, k is permeability, and ρ is dry density)***

| Sediment | Statistic | d (mm) | n (%) | k ($10^{-7}cm^2$) | ρ ($g/cm^3$) |
|---|---|---|---|---|---|
| Angular Sand (Unsorted) | Mean<br>SE | .37<br>.14 | 43.2<br>0.7 | 7.25<br>.68 | 1.51 |
| Angular Sand (Sorted) | Mean<br>SE | .38<br>.05 | 43.2<br>1.2 | 9.08<br>.74 | 1.51 |
| Spherical Beads (Sorted) | Mean<br>SE | .38<br>.05 | 35.5<br>0.4 | 11.43<br>1.51 | 1.56 |

### 4.1 MEASUREMENT PROCESS

Shear wave measurements were conducted using the system shown in Figure 1. Transverse vibrations (shear waves) were excited using a patented shear wave transducer consisting of a piezoceramic bimorph bender element mounted as a cantilever on a brass slug. The element is polarized such that it will vibrate in a plane normal to its axis when a differential voltage is applied across its electrode faces. The transducer element was mounted in an acrylic cylinder from which it was mechanically isolated to prevent excitation of compressional waves. The transverse motion of the bender was coupled to the sediment through a silicon rubber face. This flexible membrane serves to increase the active surface area while providing a good impedance match to the propagation medium. A similar transducer was mounted in the bottom of the acrylic tube containing the sediment. This transducer served as the shear wave receiver.

The measurement technique has been described in a previous paper [9], but will be summarized here for completeness. Once the measurement apparatus is set up, measured amounts of water and sediment are added to the acrylic tube. The mixture is vibrated to insure repeatable packing and excess water is removed through the drain tube. The shear wave projector is inserted and allowed to rest on the upper surface of the sediment column. An oscillator is used to impress a sinusoidal voltage across the faces of the bender element. The frequency of the oscillations is varied to produce measurements of shear wave amplitude at various frequencies under static physical conditions. The transverse vibrations are received, amplified, filtered, and displayed on an oscilloscope from which the amplitude of the received voltage may be determined. This amplitude, expressed in deciBels (relative to one volt), together with the sediment propagation path length is recorded for each frequency of interest. When all desired frequencies have been sampled, the projector is removed and the process is repeated for a new propagation path length. Multiple independent runs using different transducer pairs were performed to ensure the reproducibility of the attenuation estimates using the measurement technique described. After multiple runs, the received voltage versus path length data for a given frequency is fit using a least squares linear regression algorithm. The slope of the resultant linear fit yields the attenuation coefficient in deciBels per meter.

A basic assumption inherent in the measurements was that the measurement technique would result in plane wave propagation along the tube axis once away from the near field of the projector. In general, this was the case. Beyond a separation of approximately one acoustic wavelength, the amplitude of the transverse vibrations diminished at a rate directly proportional to the separation, that rate varying, of course, with frequency. The small amount of scatter displayed when the data was fit with a straight line attests to the validity of the plane wave assumption. [8,9]

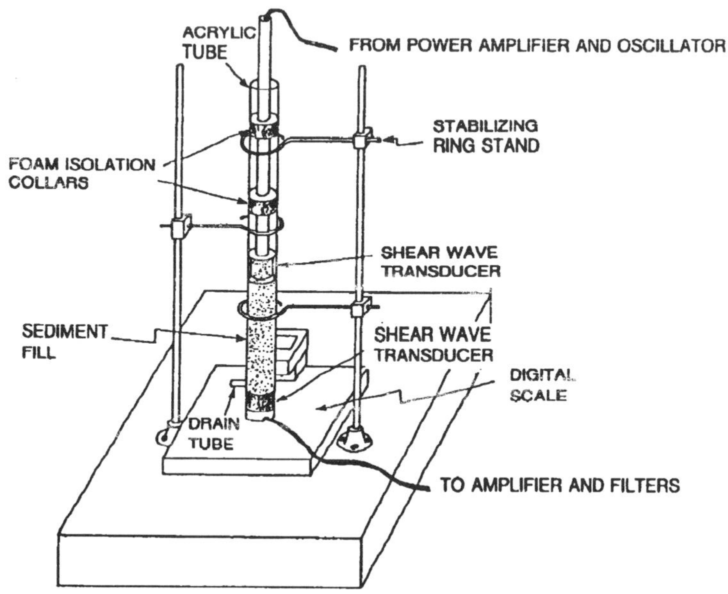


**Figure 1. Shear wave measurement system**

## 4.2 MEASUREMENT RESULTS

Table 2 is a compilation of the attenuation measurements performed using the measurement process described above. Attenuation in deciBels per meter are shown for each of the three propagation media for each of 17 discrete frequencies from 1 to 20 kiloHertz. In addition, a 90 percent confidence interval is provided to indicate the uncertainty in the measured values.

**TABLE 2. Shear wave attenuation $\alpha_S$ versus frequency in water saturated laboratory sediments**

| Frequency (kHz) | Sand (Unsorted) | | Sand (Sorted) | | Beads (Sorted) | |
|---|---|---|---|---|---|---|
| | $\alpha_S$ (dB/m) | 90% CI (dB/m) | $\alpha_S$ (dB/m) | 90% CI (dB/m) | $\alpha_S$ (dB/m) | 90% CI (dB/m) |
| 1.0 | 7.2 | 3.0 | 7.8 | 2.9 | 9.5 | 2.4 |
| 1.5 | 16.2 | 3.3 | 7.9 | 2.5 | 15.2 | 3.7 |
| 2.0 | 39.3 | 2.6 | 29.3 | 3.4 | 17.2 | 3.2 |
| 2.5 | 73.4 | 5.2 | 53.0 | 6.0 | 24.5 | 2.5 |
| 3.0 | 81.0 | 3.7 | 64.5 | 5.6 | 32.2 | 2.7 |
| 3.5 | 85.7 | 4.2 | 66.9 | 4.5 | 45.7 | 2.0 |
| 4.0 | 97.0 | 9.1 | 67.9 | 3.2 | 53.3 | 2.5 |
| 5.0 | 117.0 | 10.8 | 83.9 | 3.4 | 66.6 | 3.4 |
| 6.0 | 129.4 | 8.8 | 95.2 | 4.4 | 78.2 | 3.1 |
| 7.0 | 154.3 | 8.3 | 103.2 | 3.9 | 91.7 | 2.2 |
| 8.0 | 174.2 | 13.1 | 117.0 | 3.7 | 105.6 | 4.6 |
| 10.0 | 201.3 | 12.0 | 133.0 | 3.4 | 119.0 | 6.6 |
| 12.0 | 226.4 | 15.4 | 153.2 | 3.0 | 135.1 | 6.5 |
| 14.0 | 240.9 | 18.3 | 163.7 | 5.9 | 158.2 | 6.2 |
| 16.0 | 262.0 | 26.8 | 189.0 | 7.3 | 183.3 | 9.4 |
| 18.0 | 274.8 | 38.9 | 214.8 | 8.7 | 194.0 | 14.0 |
| 20.0 | 280.9 | 24.7 | 242.6 | 11.7 | 216.6 | 11.6 |

## 4.3 GRAIN SIZE SORTING EFFECTS

Figure 2 displays measured attenuation versus frequency for shear waves propagating in the two angular sand assemblages. Recall that the two sands have approximately the same mean grain size. The sorted sand is a subset of the unsorted sand derived by sieving. Since the measurements were performed on both sediments under similar conditions of packing, saturation, and stress, the differences in attenuation should be wholly attributable to grain size sorting. From the figure, we can see that both sands exhibit similar behavior below 2.5 kHz with the unsorted sand showing significantly higher attenuation. Above 2.5 kHz, both show a marked change in the frequency dependence. The sorted sand appears to show a rapid transition toward a 0.5 frequency exponent while the unsorted sand exhibits a frequency exponent of approximately 0.75 between 2.5 kHz and 8.0 kHz decreasing to 0.53 beyond. The sorted sand does not exhibit the same degree of slope change with increasing frequency. This may be due to a smaller, more consistent distribution of interstitial pore sizes resulting in a more consistent balance between the viscous and grain-to-grain friction loss mechanisms, an observation consistent with conclusions reached by other investigators [7, 10]. The greater loss exhibited by the unsorted sand may be attributable to the complicated interaction of grain-to-grain friction losses coupled with the presence of significant viscous losses due to grain-to-fluid relative motion at higher frequencies than in the well-sorted sediments. Overall, however, the two sand assemblages may be well represented by power curve fits to the data with both exhibiting a frequency to the 0.7 power dependency above 2.5 kHz.

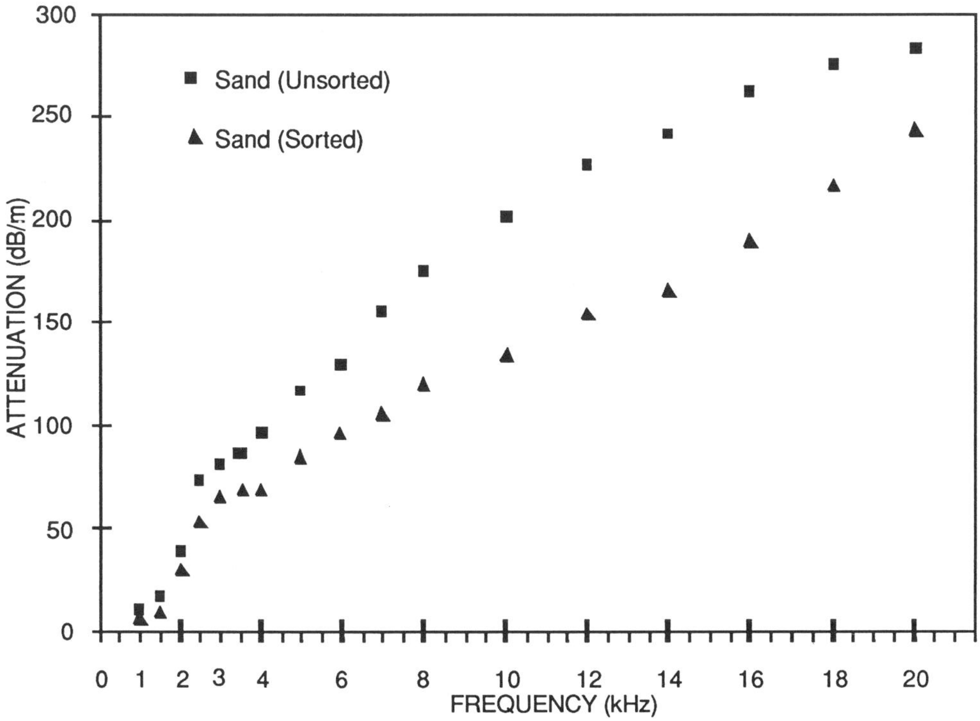


**Figure 2. Shear wave attenuation for unsorted and sorted angular sands**

## 4.4 GRAIN SHAPE EFFECTS

The effects of grain shape on the physical and acoustic properties of laboratory sediments can be observed by comparing the measurement results for the well sorted angular sand and the well sorted glass beads. The physical properties are shown in Table 1. While the two assemblages have nearly identical grain size distributions, the beads show much lower porosity and higher permeability. Table 2 and Figure 3 portray the attenuation data for each assemblage. At first glance, the two attenuation versus frequency plots appear quite similar. The major differences occur between 1.5 and 3.0 kHz. Here the angular sand exhibits a steep rise in the attenuation rate while the beads rise uniformly. This difference in behavior is consistent with the dominance of viscous losses attributable to grain-to-fluid motions in the sand while grain-to-grain losses dominate in the beads. Thus, the differences in the frequency dependence of attenuation observed in the two sediments at low frequencies may be due to the dominance of different loss mechanisms in each sediment. Once beyond 4.0 kHz, the two assemblages exhibit similar frequency dependences. Power curve fits show both to exhibit exponents of frequency between 1.0 and 0.5. Since the various loss mechanisms are additive, indications are that a combination of grain-to-grain friction, characterized by a linear frequency dependence, and grain-to-fluid losses, characterized by a 0.5 frequency exponent, is at work. The exponent for the beads is closer to 1.0 indicating a dominance of frame losses. The angular sand, on the other hand, shows an exponent closer to 0.5. Observations of higher attenuation in the dry beads indicate that grain-to-grain losses may be more significant than in the angular sand.

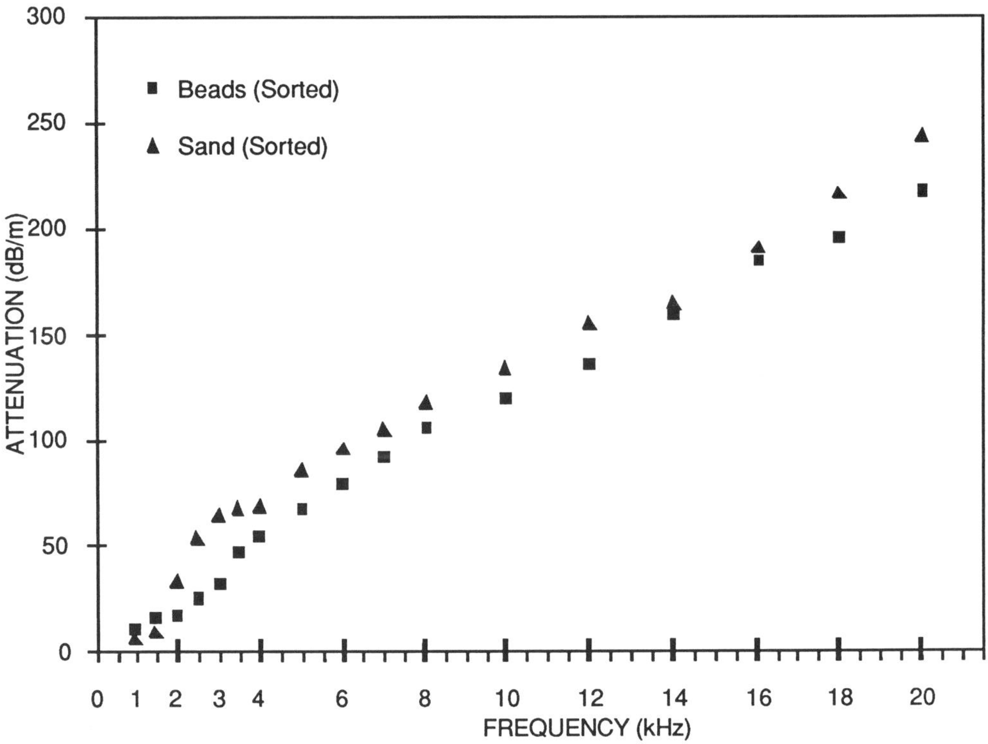


**Figure 3. Shear wave attenuation for sorted angular sands and sorted spherical beads**

## 5. Summary

Shear wave attenuations were measured over a significant frequency range (1-20 kHz) for three unconsolidated sediment assemblages chosen to indicate the effects of grain size sorting and grain shape on the frequency dependence of attenuation. Ancillary measurements under the same packing and stress conditions as the attenuation measurements have yielded physical properties appropriate to a test of the ability of the Biot-Stoll model to account for the attenuations observed.

The data led to the conclusion that the effect of poor grain size sorting (the unsorted sand case) is to increase the magnitude of losses consistent with grain-to-fluid relative motion. This may be primarily due to the increased specific surface area of the smaller grains. An additional effect is the apparent extension of the frequency range over which, according to Biot, the attenuation due to viscous interaction transitions from a frequency squared dependence to a dependence proportional to the square root of frequency. The actual frequency dependence is complicated by the presence of a significant degree of grain-to-grain friction at the lower and middle frequencies. The net result is a more gradual transition in the unsorted assemblage as more of the interstitial pores meet the conditions consistent with a particular viscous loss mechanism. As frequency increases, the effects of grain sorting become less important and the loss curves for the two angular sediments become parallel indicating a similar frequency dependence or attenuation mechanism at work in the saturated sediment.

The data from the sorted assemblages indicate that the primary effect of angularity on attenuation is to increase the magnitude of fluid-to-grain losses at lower frequencies. This effect may be related to the observed decrease in permeability probably caused by the presence of smaller more tortuous paths of flow and higher specific area of the angular assemblage. Once the frequency is high enough for both the sand and the beads to be in the higher frequency regime described by Biot theory, the attenuation differences diminish. Here the interaction of boundary layer viscous losses and grain-to-grain friction result in attenuation proportional to frequency raised to a power between 0.5 and 1.0. Higher dry frame losses observed in the beads cause the beads to exhibit a frequency exponent closer to 1.0 while the moderate frame losses of the angular sand lead to high frequency attenuation dominated by viscous boundary layer losses and a frequency exponent nearer to 0.5, an observation consistent with Biot theory.

None of the three sediments exhibited truly linear frequency dependence across the entire frequency range, although linear regression fits to the data produced reasonable results, particularly for the "ideal" spherical beads.

This work underscores the need to perform additional attenuation measurements under carefully controlled conditions with accompanying physical properties measurements to permit the sorting out of attenuation mechanisms at work in the specimens being examined.

## 6. References


1. Kibblewhite, Alick C., (1989) "Attenuation of Sound in Marine Sediments: A Review with Emphasis on New Low-Frequency Data", J. Acoust. Soc. Am. 86, 716-738.
2. Hamilton, E. L., (1987) "Acoustic Properties of Sediments", in *Acoustics and The Ocean Bottom*, edited by A. Lara-Saenz, C. Ranz Guerra, and C. Carbofile, Consejo Superior de Investigaciones Cientificas, Madrid
3. Biot, M. A., (1956) "Theory of Propagation of Elastic Waves in a Fluid-Saturated Porous Solid. I. Low Frequency Range", J. Acoust. Soc. Am. 28, 168-178.
4. Biot, M. A., (1956) "Theory of Propagation of Elastic Waves in a Fluid-Saturated Porous Solid. II. Higher Frequency Range", J. Acoust. Soc. Am. 28, 179-191.
5. Biot, M.A., (1962) "Mechanics of Deformation and Acoustic Propagation in Porous Media", J. Appl. Phys. 33, 1482-1498.
6. Biot, M. A., (1962) Generalized Theory of Acoustic Propagation in Porous Dissipative Media", J. Acoust. Soc. Am. 34, 1254-1264.
7. Stoll, Robert D., (1989) *Sediment Acoustics*, Lecture Notes in Earth Sciences:26, Springer-Verlag, Berlin.
8. Brunson, B. A., (1983) "Shear Wave Attenuation in Unconsolidated Laboratory Sediments", Thesis for Ph.D. degree, Oregon State University, School of Oceanography.
9. Brunson, B. A. and Johnson, R. K., (1980) "Laboratory Measurements of Shear Wave Attenuation in Saturated Sands", J. Acoust. Soc. Am. 68, 1371-1375.
10. McLeroy, E. G. and DeLoach, A., (1968) "Sound Speed and Attenuation from 15 to 15,000 kHz, Measured in Natural Sea Floor Sediments", J. Acoust. Soc. Am. 44, 1148-1150.

# SHEAR WAVE VELOCITIES OF GLACIO-MARINE SEDIMENTS: BARENTS SEA

S.K. BREEDING
*MAR, Inc.*
*101 N.W. Interchange*
*Bay St. Louis, MS 39520 USA*

D.A. DUNN
*University of Southern Mississippi*
*Department of Geology*
*Southern Station, Box 5044*
*Hattiesburg, MS 39406-5044 USA*

T.H. ORSI
*Texas A & M University*
*Department of Oceanography*
*College Station, TX 77843 USA*

ABSTRACT. Shear wave velocities were computed for surficial, glacio-marine sediments of the Barents Sea using three methods; Hamilton's equations, the Biot-Stoll model, and the Bryan-Stoll model. Comparisons are made between the model predictions and the *in situ* data. The predictions from the Biot-Stoll and the Bryan-Stoll models are in excellent agreement with the *in situ* data.

## 1. Introduction

Most, if not all, marine sediments possess sufficient rigidity to sustain shear wave propagation (Hamilton (1971)). Generation of shear waves by conversion of compressional waves at seismic discontinuities within the sea floor is an important energy loss mechanism. This loss is more critical in bottom-limited environments, such as shallow epicontinental seas where shallow-water paths interact numerous times with the ocean bottom (Akal (1980)). A thorough understanding of the characteristics and magnitude of shear waves and their application in geoacoustic models is therefore critical for accurate prediction of acoustic propagation in tactically important shallow water environments.

For various reasons, it is not always possible to obtain *in situ* measurements of shear wave velocity in marine sediments and, although laboratory instrumentation has improved (Brunson and Johnson (1980); Matthews (1982)), only a limited number of measurements have appeared in the



*J. M. Hovem et al. (eds.), Shear Waves in Marine Sediments*, 149–156.

literature (e.g., Richardson *et al.* (1991)). Other approaches for determining shear wave velocities include the Hookean model (Hamilton (1971)), the Biot-Stoll sediment model (Stoll and Kan (1981)), and the Bryan-Stoll model (Bryan and Stoll (1988)).

Most studies of the surficial Barents Sea have dealt with the spatial distribution of sediment types (Wright (1974); Bjorlykke *et al.* (1978); Elverhoi and Solheim (1983)), and little is known about their elastic characteristics. The purpose of this paper is to present and compare values of shear wave velocity for sediments from the Barents Sea calculated using Hamilton's empirical method, the Biot-Stoll sediment model, the Bryan-Stoll model, and *in situ* data.

## 2. Regional Setting

The Barents Sea is a shallow epicontinental sea north of North Cape (Norway) and the Kola Peninsula (Soviet Union), and between the islands of Novaya Zemlya and the Svalbard Archipelago (Figure 1). Marine geological and geophysical investigations indicate that the Barents Sea was extensively glaciated during the Pleistocene Epoch (Vorren *et al.* (1988)). Sediment

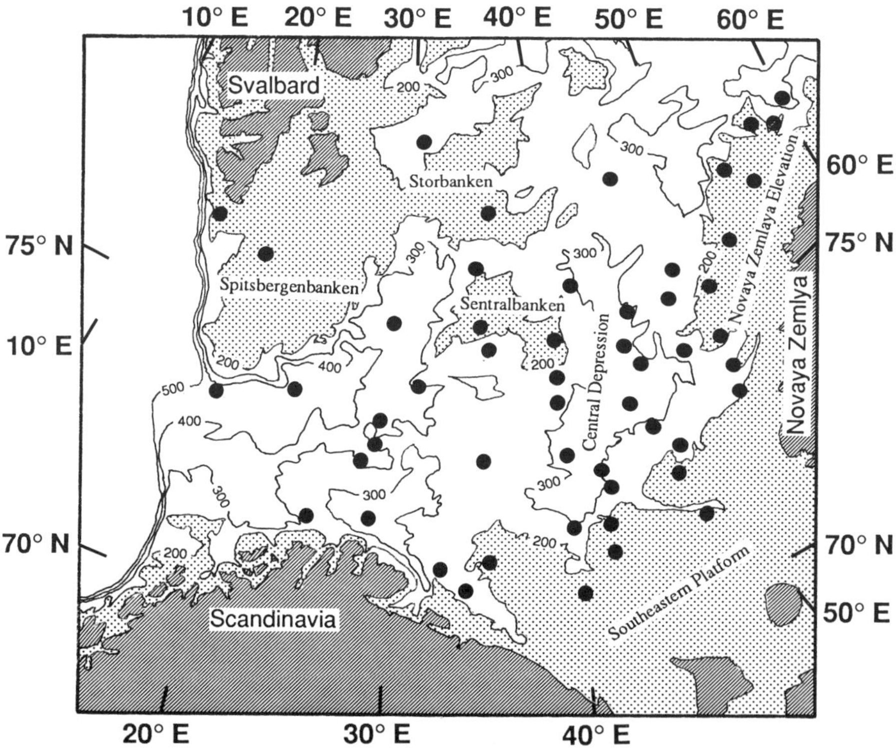


Figure 1. Physiographic setting of the Barents Sea. The locations of cores examined in this study are indicated by dots.

accumulation rates are quite low, because of the limited supply of clastic particles, erosive effects of sea level regressions, and bypassing of the shelf by sediments during extreme glaciations (Herman (1974)). Sand-silt-clays and silty clays predominate, and are generally depth-segregated, with coarser components found on topographic highs and the fine-fraction occupying lows.

## 3. Technical Approach

### 3.1. METHODS

Fifty-four gravity cores were collected from the Barents Sea in mid-1987, by the U.S. Naval Oceanographic Office (Figure 1). Once in the laboratory, the sediments were subjected to a series of geotechnical tests following the recommended procedures of American Society for Testing and Materials (1964).

Compressional wave velocities were determined parallel to bedding on the unsplit cores prior to geotechnical analyses using a pulse technique. The initial laboratory velocities were corrected to *in situ* conditions using tables of sound speed in sea water (U.S. Naval Oceanographic Office (1966)). Grain size distribution was determined for each sample and samples were classified using the nomenclature of Shepard (1954). Wet bulk density and porosity were calculated for each sample using measurements of water content, percent dry weight, and the specific gravity of the mineral solids. Wet bulk densities were corrected to *in situ* conditions (Hamilton (1971)) to account for the change in interstitial water density due to increased hydrostatic pressure. Porosities were corrected for salt residue following Hamilton (1974).

The laboratory data was used as input to the three models. The calculations discussed in the present study are restricted to the upper 10 cm of the sediment column.

### 3.2. ANALYTICAL METHODS USING HAMILTON'S EMPIRICAL APPROACH

Mathematical derivations of the elastic constants have been reviewed by Hamilton (1971; 1974). Of the three inputs are necessary to compute the elastic constants, sediment bulk density and compressional wave velocity measurements are commonly made. Shear wave velocity measurements are much less commonly made during traditional laboratory analyses. Hamilton (1971) suggested that bulk modulus, as defined by Gassman (1951), be estimated and used in the absence of measured shear wave velocity. The calculation requires an empirical estimate of the frame bulk modulus that is critical to the accuracy and validity of the results. To compute bulk moduli in the present study, we used inputs of aggregate, frame, and pore water bulk moduli. For the fine-grained fraction, an aggregate bulk modulus of 3.5 dynes/cm$^2$ x $10^{11}$ was used, and 4.0 dynes/cm$^2$ x $10^{11}$ for the coarser samples (Stoll and Kan (1981); Holland and Brunson (1988)). Frame bulk moduli were computed using regression equations for natural marine sands: $\log \kappa_f = 2.70932 - 4.25391(\eta)$; and silty clays: $\log \kappa_f = 3.73580 - 4.25075(\eta)$ (Hamilton (1971); Ogushwitz (1985)), where $\kappa_f$ is the frame bulk moduli and $\eta$ is the porosity corrected for salt residue. Pore water moduli ($\kappa_w$) were determined by the relationship, $\kappa_w = \rho V_p{}^2$, where $\rho$ is density and $V_p$ is compressional wave velocity. As noted by Hamilton (1971), all of the computed elastic properties are approximations, therefore, all input data have margins of error, some errors are known and some are not.

### 3.3. ANALYTICAL METHODS USING BIOT-STOLL'S SEDIMENT MODEL

The Biot-Stoll model requires 13 sediment, pore fluid, and frame properties; grain bulk modulus, grain density, porosity, pore fluid density, pore fluid bulk modulus, pore fluid viscosity, structure factor, permeability, pore size, the real and imaginary part of the shear modulus of the frame, and the real and imaginary part of the bulk modulus of the frame (Holland and Brunson (1988); Molinelli and Brunson (1985)). An estimate of 4.0 x $10^{11}$ dyne/cm$^2$ was used for the grain bulk modulus. Measured values of grain density and porosity were used as model inputs. *In situ* water depth, temperature, and salinity were used to calculate the the pore fluid density, bulk modulus, and viscosity. Estimates for the structure factor were used based on the grain size (1.25 for sands and 3.0 for clays). Permeability was determined from measured grain size distribution and porosity, using the Kozeny-Carman method. The pore size was determined using the hydraulic radius and mean specific surface which was calculated from the permeability. The shear modulus of the frame were determined using Stoll's stress procedure. The bulk shear modulus of the frame was calculated from the shear modulus and an estimate of poisson's ratio ranging from 0.2 for sands to 0.25 for clays.

### 3.4. ANALYTICAL METHODS USING BRYAN-STOLL'S MODEL

The Bryan-Stoll model is an empirical model which relates sediment dynamic modulus ($\mu$) to mean effective stress (p') and void ratio ($e$) with the following relationship: $\mu = \mu_o\,(p')^n \exp(\tau e)$, where $\mu_o$=2526 atm, n=0.5, and $\tau$=-1.5 (Bryan and Stoll (1988)). It was developed using many laboratory measurements of confining pressure, dynamic shear modulus, and published values of void ratio. The model predicts overburden pressure, shear modulus, shear wave velocity, and density, requiring only inputs of void ratio and incremental depth.

### 3.5. IN SITU DATA

The *in situ* shear wave velocity data was taken from eight sites located off the coast of LaSpezia, Italy, within diving depths. The measurements were taken using a pulse technique composed of one transmitter and two receivers placed to a depth of 30 cm below the sediment-water interface (Richardson *et al.* (1991)). The *in situ* shear wave data represent a variety of sediment types from a bottom-limited terrigeneous environment.

## 4. Results

Although the sediments studied range from gravel to clay, most classify as silty clay (42.6%) and sand-silt-clay (29.6%). Compressional wave velocity could not be measured for three samples because they contain quantities of gravel. Table 1 summarizes the average values of the salient parameters for the sediment samples.

The shear wave velocities obtained by the models described above are illustrated as a function of void ratio, porosity, and density in Figures 2-4, respectively. Figures 2-4 also include the *in situ* data from Richardson *et al.* (1991). In all three figures, the Biot-Stoll, the Bryan-Stoll, and the *in situ* data values for shear wave velocity are in good agreement with each other and are within the range of expected shear wave values. Bear in mind that the *in situ* data were measured at 30 cm, whereas the model predictions were for 10 cm. This would suggest that the model predictions would actually be slightly higher than the *in situ* data, had they been compared to data at 10 cm. Using Hamilton's early approach on these sediment samples results in a wider range of shear wave velocity values that are considerably higher than the other results.

The calculations for bulk modulus in Hamilton's method are critical to the accuracy and validity of the resulting shear wave velocities. In calculating the bulk modulus, we used an estimate for the aggregate bulk modulus. In addition, since there was not an empirical equation for the frame bulk modulus for sand-silt-clays, the equation for natural marine sands was used. This could account for the discrepancies in these results. It should be noted that this approach is from Hamilton's early work, over twenty years ago, and since then, the field of geoacoustic modeling has evolved.

Excluding the results from Hamilton's early method, Figure 2 illustrates the shear wave velocity monotonically decreasing as the void ratio increases. Similarly, the shear wave velocity decreases as the porosity increases (Figure 3). Conversely, an increase in density results in an increase in shear wave velocity (Figure 4).

TABLE 1. Sediment properties for measured parameters used as inputs for model predictions.

| Sediment Type | Number of Samples | Porosity (%) | Density (gm/cc) | Vp (m/sec) | Void Ratio (%) |
|---|---|---|---|---|---|
| Clay | 2 | 73.85 | 1.45 | 1418 | 2.77 |
| Silty Clay | 23 | 68.30 | 1.54 | 1434 | 2.25 |
| Sandy Clay | 1 | 66.20 | 1.57 | 1432 | 1.88 |
| Clayey Silt | 1 | 62.30 | 1.64 | 1482 | 1.60 |
| Sand-Silt-Clay | 16 | 55.70 | 1.75 | 1492 | 1.35 |
| Clayey Sand | 5 | 47.24 | 1.89 | 1535 | 0.89 |
| Silty Sand | 2 | 47.90 | 1.88 | 1550 | 0.90 |
| Sand | 1 | 39.80 | 2.01 | 1647 | 0.69 |

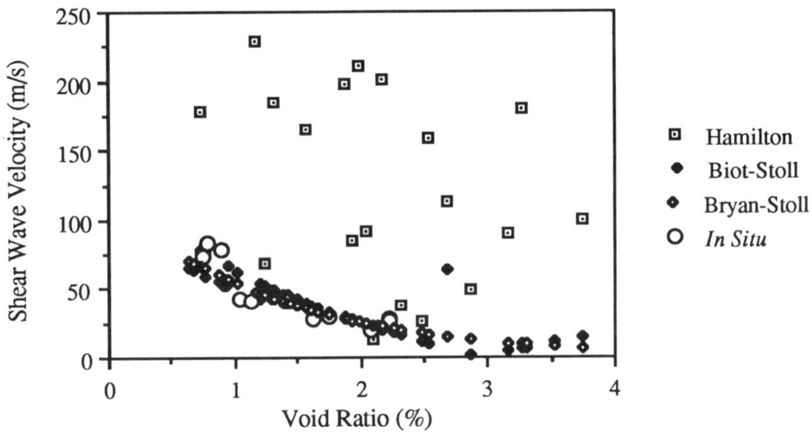


Figure 2. Shear wave velocity versus void ratio for model predictions and *in situ* data.

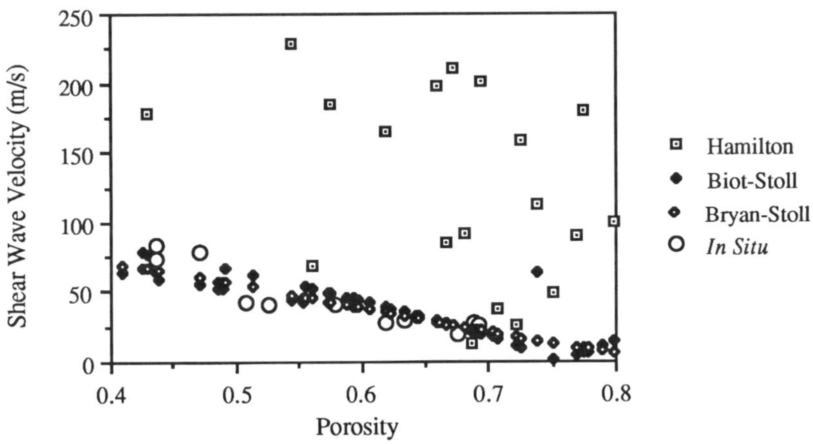


Figure 3. Shear wave velocity versus porosity for model predictions and *in situ* data.

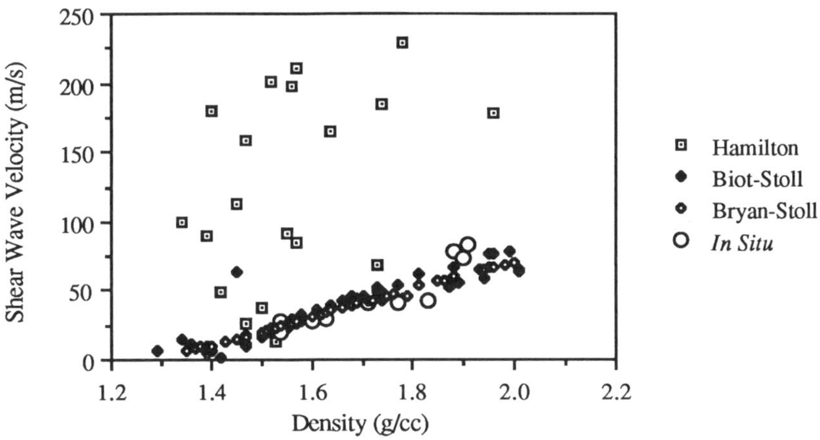


Figure 4. Shear wave velocity versus density for model predictions and *in situ* data.

## 5. Conclusions

The Biot-Stoll and the Bryan-Stoll models give realistic values for shear wave velocities in bottom-limiting environments such as the Barents Sea. Results from these models are slightly higher than but agree well with the *in situ* shear wave velocities. Definite trends in the results from these models exist as a function of void ratio, porosity, and density. Using Hamilton's earlier approach results in a wider range of values that are considerably higher than the results using the other methods. Therefore, it is suggested that a current model such as the Biot-Stoll or the Bryan-Stoll model be used whenever it is not possible to measure shear wave velocity *in situ* .

## Acknowledgements

We thank the U.S. Naval Oceanographic Office, and specifically the NAVOCEANO Marine Geological Laboratory, for providing the core data. C. Ingram is gratefully acknowledged. M. Richardson, J. Newcomb, and C. Holland are gratefully acknowledged for their comments and suggestions. D. McCaffrey is appreciated for providing graphical support.

## References

Akal, T., "Sea floor effects on shallow-water acoustic propagation," In: *Bottom-Interacting Ocean Acoustics*, W.A. Kuperman and F.B. Jensen, Plenum Press, New York, 557-575 (1980).

American Society for Testing and Materials, *Procedures for Testing Soils*: ASTM, Philadelphia, PA, Committee D-18, 4th Edition, 540 p. (1964).

D. Bjorlykke, B. Bue and A. Elverhoi, "Quaternary sediments in the northwestern part of the Barents Sea and their relationship to the underlying Mesozoic bedrock," *Sedimentology* **25**:227-246 (1978).

B.A. Brunson and R.K. Johnson, "Laboratory measurements of shear wave attenuation in saturated sands," *Journal Acoustical Society America* **68**:1371-1375 (1980).

G.M. Bryan and R.D. Stoll, "The dynamic shear modulus of marine sediments," *Journal Acoustical Society America* **83**:2159-2164 (1988).

A. Elverhoi and A. Solheim 1983, *Physical environment western Barents Sea, 1:1,500,000: Surface sediment distribution*: Oslo, Norway, Norsk Polarinstitutt Skrifter Nr. 179, 20 pp. (1983).

F. Gassman, "Über die elastizität poröser medien," *Vierteljahrsschr. Naturforsch. Ges. Zürich* **96**:1-23 (1951).

E.L. Hamilton, "Elastic properties of marine sediments," *Journal Geophysical Research* **76**:579-604 (1971).

E.L. Hamilton, "Prediction of deep-sea sediment properties: State-of-the-art," In: *Deep-Sea Sediments: Physical and Mechanical Properties*, A.L. Inderbitzen, ed., Plenum Press, New York, 1-43 (1974).

Y. Herman, "Topography of the Arctic Ocean," In: *Marine Geology and Oceanography of the Arctic Seas*, Herman, Y., ed., Springer-Verlag, New York, 73-81 (1974).

C.W. Holland and B.A. Brunson, " The Biot-Stoll sediment model: an environmental assessment," *Journal Acoustical Society America* **84**:1437-1443 (1988).

J.E. Matthews, "Shear wave velocity measurements in marine sediments," Geo-Mar. Lett. **2**:215-217 (1982).

E.J. Molinelli and B.A. Brunson, "PHYSED: Physical Sediment Model Software," Technical Manual, Planning System, Inc. (1985).

P.R. Ogushwitz, "Applicability of the Biot theory. I. Low-porosity materials," *Journal Acoustical Society America* **77**:429-440 (1985).

M.D. Richardson, E. Muzi, L. Troiano, and B. Miaschi, "Sediment shear waves: a comparison of *in situ* and laboratory measurements," In: *The Microstructure of Fine-Grained Sediments - From Mud to Shale*, R.H. Bennett, et al., eds., Springer-Verlag, New York, 403-415 (1991).

F.P. Shepard, "Nomenclature based on sand-silt-clay ratios," *Journal Sedimentary Petrology* **24**:151-158 (1954).

R.D. Stoll and T.K. Kan, "Reflection of acoustic waves at a water-sediment interface," *Journal Acoustical Society America* **70**:149-156 (1981).

U.S. Naval Oceanographic Office, *Handbook of Oceanographic Tables*: NAVOCEANO Special Publication No. SP-68, 302-318 (1966).

T.O. Vorren, M. Hald, and E. Lebesbye, "Late Cenozoic environments of the Barents Sea," *Paleoceanography* **3**:601-612 (1988).

P.L. Wright, "Recent sediments of the southwestern Barents Sea," *Marine Geology* **16**:51-81 (1974).

# ROCK ACOUSTICS: RELEVANCE OF THE POROUS VISCOELASTIC MODEL.

Patrick N.J. RASOLOFOSAON
*I.F.P. (B.P. 311)*
*92506 Rueil-Mamaison Cedex*
*France*

ABSTRACT. It is well established that **the acoustic behaviour of rock masses of infinite extension is well described by a linear viscoelastic theory.** However, in the presence of interfaces, such as a borehole wall in acoustic logging experiments, macroscopic fluid flow effects as described by Biot's poroelastic theory are superimposed on the viscoelastic phenomena. In other words, in these cases the poroelastic model and the viscoelastic model must be used jointly to correctly describe the experimental results. More precisely, **Biot's poroelastic effects are larger than (or the same order of magnitude as) classical viscoelastic effects when wavefronts encounter physical discontinuities (interfaces) between different propagation media** (as is the case for reflected-transmitted waves, surface or interface waves, guided waves...) **and when hydraulic exchange is possible through these interfaces. In these cases, it is necessary to describe the acoustic behaviour of rocks by means of a "poroviscoelastic" model which is a synthesis of Biot's poroelastic model and the classical viscoelastic model.** These theoretical predictions are confirmed by many recent experimental results corresponding to different propagation geometries.

## 1. Introduction

The purpose of this paper is to show the relevance of the "poroviscoelastic " model (PVM), a synthesis of the viscoelastic model (VM) and Biot's poroelastic model (BPM), for describing wave propagation in rocks, especially for propagation geometries involving discontinuities between different media. We first give a quick overview of the VM and BPM, delineating their main features and experimental corroborations. We then introduce the PVM which allows us to directly compare, from a theoretical point of view, the magnitude of viscoelastic effects and poroelastic effects in different propagation geometries. Next we show the good agreement between experimental results and the predictions of the PVM. Finally we discuss the practical consequences of these fundamental results.



*J. M. Hovem et al. (eds.), Shear Waves in Marine Sediments*, 157–166.

## 2. Overview of the viscoelastic model and Biot's poroelastic model

### 2.1. THE VISCOELASTIC MODEL

The literature on the VM is particularly plentiful. More specifically, in rock acoustics the reader is refered to Bourbié et al. (1987) for an extensive review both of theory and experiment. To summarize, one can say that from a rheological point of view viscoelastic media exhibit a delayed response to mechanical solicitations compared with elastic media which respond instantaneously. This implies that viscoelastic moduli, are no longer real and constant, as elastic moduli, but are complex value and frequency dependent. Consequently, in viscoelastic media, wave propagation is accompanied by attenuation and dispersion. To quantify the amount of attenuation/dispersion in the medium, we introduce a parameter Q, called the quality factor, which is inversely proportional to the energy dissipated per wavelength or during a cycle of vibration. For instance, highly attenuating media have low Q whereas non-dissipative or nearly elastic media have high Q values. Three classical viscoelastic models are commonly used in the geophysical literature. These are the standard linear solid (SLS) model, the nearly constant-Q (NCQ) model and the constant-Q (CQ) model. The attenuation, in terms of $Q^{-1}$, exhibits a relaxation peak in the SLS model, remains constant in a wide frequency band in the NCQ model, and is constant and frequency independent in the CQ model. Attenuation extrema are associated with a maximum of the velocity dispersion. The predictions of these models are in good agreement with many experimental results (see Bourbié et al., 1987 for further details).

### 2.2. BIOT'S POROELASTIC MODEL

The theory of wave propagation in porous media, often called poroelastic theory, was pioneered by Biot (1956a, 1956b). This theory describes the coupling between hydraulic and acoustic properties of a fluid-saturated porous medium. Its relevance and popularity rests on two main facts:

- First, one of its most fundamental predictions, namely the propagation of a second compressional P2-wave, which does not exist in elastic media, has been corroborated experimentally by Plona (1980).
- Secondly, the theory predicts a direct correlation between the attenuation/dispersion of the waves and the porosity/permeability of the propagation media. These latter parameters are fundamental for the exploitation of oil reservoirs.

Biot's pioneer work spawned a prolific number of theoretical works dealing with different propagation geometries (Geertsma and Smit 1961, Deresiewicz and Rice 1962, Deresiewicz 1962, Gardner 1962, Rosenbaum 1974, etc.). A unified physical description of poroelastic effects has been developed by Rasolofosaon (1987b). To summarize, one can say that poroelastic effects are important when wavefronts encounter physical macroscopic discontinuities, as is the case for transmitted/reflected waves, surface/interface waves or guided waves, and when the fluid transfer is free through these interfaces (open interface). Many authors corroborate the importance of hydraulic boundary conditions on the experimental results (Dunn 1987, Winkler 1987, Rasolofosaon 1987a, 1988a, 1988b).

## 3. Comparison between viscoelastic and poroelastic effects. The poroviscoelastic model.

### 3.1. COMPARISON BETWEEN VISCOELASTIC AND POROELASTIC EFFECTS

The main differences between the VM and BPM are illustrated by Figs. 1, 2 and 3. Figs. 1 and 2 show the influence of viscoelasticity and poroelasticity, respectively on normal incidence reflection and transmission at a liquid / solid interface. In Fig. 1, for the VM (SLS model), the deviations of the P-wave viscoelastic transmission and reflection coefficients from the respective elastic coefficients are plotted as functions of a normalized frequency. The normalization frequency Fref is the SLS relaxation frequency. In the case studied the fluid is perfect (no dissipation) and the solid is viscoelastic. In the VM, the solid velocity increases with frequency (especially around Fref for the SLS model) which results in an increase of the impedance contrast. Consequently the reflected energy increases and, due to energy conservation, the transmitted energy diminishes. As expected, the higher the attenuation (i.e. the lower Qp), the larger the effect. The results concerning BPM are different, see Fig. 2. We have plotted the normalized deviation from elasticity of the P-wave reflection coefficient (Fig. 2a), of the P1-wave transmission coefficient (Fig. 2b) and of the P2-wave transmission coefficients (Fig. 2c) as functions of a normalized frequency. The normalization frequency is equal to Biot's critical frequency which is inversely proportional to the permeability (Biot 1956a, 1956b). The elastic moduli of the reference elastic material are the zero frequency limit of the poroelastic moduli. Two hydraulic boundary conditions are analyzed:

- the sealed interface: no fluid flow between the porous medium and the neighbouring fluid.
- the open interface: free fluid transfer.

One of the main results is the following: when the fluid transfer is prevented (sealed interface), the P2-wave energy critically decreases and, due to energy balance, the energy of all the other ordinary waves (P1 and S) notably increases. The higher the porosity and/or the permeability, the larger the effects. These theoretical predictions have been experimentally corroborated by Rasolofosaon (1987a, 1988a, 1988b, 1989). A comparison between the influences of viscoelasticity and poroelasticity is sketched in Fig. 3.

It appears that the theoretical predictions of the VM and BPM are quite different. In many propagation geometries these two models must be used jointly to correctly describe the experimental results (Rasolofosaon 1987b). For this reason we introduce the PVM.

### 3.2. THE POROVISCOELASTIC MODEL

The idea of unification of the VM and BPM is not new since Biot himself suggested it (Biot 1956c, 1962). R.D. Stoll has developed and extensively used this model to study the acoustical properties of sediments (see Stoll 1989 for an extensive review). This model was adapted to rock acoustics (Rasolofosaon 1987b) and a complete energy formulation has been recently proposed (Rasolofosaon 1991). The latter analysis adapted to different propagation geometries is illustrated in Fig. 4. The viscoelastic part of the PVM is an SLS model and the poroelastic part of the model corresponds to a medium permeability clean sandstone (k =100 millidarcy, porosity= 20%). The upper part of Fig. 4 shows a typical attenuation curve with two relaxation peaks corresponding respectively to viscoelastic relaxation around 100Hz (chosen arbitrarily), and poroelastic relaxation around 100kHz. The relative heights of the two peaks allow a direct comparison between the magnitude of the viscoelastic effects and that of the poroelastic effects. The lower part of Fig. 4 displays a table of the viscoelastic quality factors (for P- and for S-waves) to be injected in the PVM to induce the same amount of poroelastic and viscoelastic attenuation/dispersion. Three cases are reported: the infinite medium (P1- and S-waves), the solid half-space (Rayleigh wave) and the liquid/solid interface (Stoneley wave). The main results are the following. First the poroelastic attenuation/dispersion is small in infinite media (Q>100). In the presence of macroscopic discontinuities (semi-infinite medium or liquid/solid interface) the poroelastic attenuation/dispersion increases drastically especially when fluid exchange occurs (Q<8!}. This infers that the viscoelastic part of the model takes into account most of the energy dissipation in the volume of the media of propagation whereas the poroelastic part of the model takes into account most of the energy conversion at the macroscopic boundaries, especially when the interface is open. Hence, most of the poroelastic attenuation/dispersion is due to P2-wave energy leaking from the interfaces. This wave is not observable in ordinary rocks because of its diffusive nature but the effect of its generation on the remaining waves can be drastic. These are the main reasons why the VM and BPM are complementary and must be used jointly (PVM) in certain geometrical configurations. Fig. 5 illustrates these results.

## 4. Relevance of the poroviscoelastic model (experiment versus theory)

This section reports experimental support for the PVM. The first experiment, illustrated in Fig. 6, consists in exciting Lamb's modes in rock plates immersed in water (Rasolofosaon 1987a) at different angles of incidence. Figs. 6a and 6b show respectively the experimental set-up and the experimental data vs. the PVM predictions. The two limit hydraulic boundary conditions (open interface and sealed interface) are analyzed. Three sandstones of comparable porosity (around 20%) and of various permeabilities (k=12; 230; 3800 millidarcy) are studied. This experiment shows firstly that the angular resonance peaks in the three rocks are very smooth which implies that the energy dissipation (by viscoelastic effects) in the plate is large. This is a classical result in all resonance scattering methods. Secondly, the experiment shows that the change of the hydraulic boundary condition induces a radical change in the resonance curves. More precisely the effect of preventing fluid transfer is to sharpen the resonance curves. The higher the permeability, the larger the effect. This is a typical poroelastic effect which cannot be described by classical VM only, whereas the experimental results are in good agreement with PVM predictions, see Fig. 6b.
The second experiment, illustrated by Fig. 7a, consists in generating surface waves at a liquid/rock interface (Rasolofosaon 1988a). We will focus our attention on the results concerning the Stoneley wave (Fig. 7b) which is the most affected by the poroelastic effects following Rosenbaum (1974). On Fig. 7b, the differential attenuation (i.e. the difference between the "open interface" attenuation and the "sealed interface" attenuation) of this wave is plotted versus the permeability.The main result is that, contrary to the case of (visco)elastic media, the differential attenuation is non-zero and permeability-dependent. It exhibits a maximum for intermediate permeability. These are typical poroelastic effects which are well discribed by BPM. The viscoelastic attenuation, which is evidently the same on a given rock in both interface hydraulic conditions, has been cancelled in the differential attenuation parameter. Thus another result which does not appear on Fig. 7b is the complementary viscoelastic attenuation actually observed on Stoneley waves when the interface is sealed. This effect cannot be accounted for with BPM only. Finally, this experiment again shows the necessity of using jointly the VM and BPM to correctly describe the experimental observations. Many other references report various experimental results leading to the same conclusions (Dunn 1987, Winkler et al. 1987, Rasolofosaon 1989)

## 5. Discussion and conclusions

It has been shown that for rock acoustics, the PVM must be used when dealing with macroscopic boundaries (as in the various problems of transmitted/reflected, surface/interface, guided waves) and when fluid transfer is possible through the interfaces. Otherwise a simple VM will suffice. In other words we have pointed out the critical influence of the hydraulic boundary condition. More precisely, poroelastic effects can be neglected as soon as hydraulic transfer through the interfaces is prevented.

In practical applications (borehole acoustic logging, sediment acoustics, etc.) the poroelastic part of the model carries much of the information about petrophysical parameters such as porosity/permeability. But part of this information is also contained in the viscoelastic part of the model. So one question naturally arises: Is it possible to extract these petrophysical informations from acoustic data? This question has been dealt with by Rasolofosaon (1988c) who concludes that in spite of the encouraging previous fundamental results, it seems that the inversion problem is plagued by major pitfalls of three main types:

- Many other extrinsic factors can strongly, and sometimes unpredictably affect the recorded acoustic signals (interface roughness, variation of the hydraulic and mechanical properties of the formations close to the interfaces, borehole condition in acoustic logging, etc.).
- Most previous analyses are based on studies of wave attenuation and/or dispersion. However the correct measurement of these parameters in the field, often under weakly controlled conditions, is quite tricky (see Bourbié et al. 1987 for a critical analysis).
- Viscoelastic and poroelastic attenuation/dispersion are difficult to separate in order to obtain an estimation of the fundamental petrophysical parameters.

In conclusion, despite of recent efforts in understanding the physics of wave propagation in rocks and in designing new tools of investigation (in acoustic logging or in underwater acoustics for instance), the inversion of acoustic data in terms of petrophysical parameters such as hydraulic parameters is not yet routine. As a consequence it is recommended that inversion problems should not be hastily resolved unless seriously considering these realities. In writing this, I do not intend to sow discouragement but to emphasize the challenge we are currently facing.

**Acknowledgements**

This research was done at the Rock Physics Laboratory of the Institut Français du Pétrole. The author gratefully acknowledges M. Masson for experimental help and Dr Jacquin, Dr Lucet, Dr Zinszner and M. Pouet for fruitful and enthusiastic discussions. Anonymous reviewers are also acknowledged for their helpful comments and suggestions.

## References


Biot, M.A., 1956a, Theory of propagation of elastic waves in fluid-saturated porous solid, I: Low-frequency range, J. Acous. Soc. Am., **28**, 168-178.

Biot, M.A., 1956b, Theory of propagation of elastic waves in fluid-saturated porous solid, II: Higher frequency range, J. Acous. Soc. Am., **28**, 179-191.

Biot, M.A., 1956c, Theory of deformation of porous viscoelastic anisotropic solid, J. Appl. Phys., **23**, 1482-1498.

Biot, M.A., 1962, Mechanics of deformation and acoustic propagation in porous media, J. Appl. Phys., **27**, 1482-1498.

Bourbié, T., Coussy, O., Zinszner, B.E., 1987, Acoustics of porous media,Gulf publishing Co, Houston, Texas.

Deresiewicz, H., 1962, The effect of boundaries on wave propagation in a liquid-filled porous solid, IV: Surface waves in a half space, Bull. Seismol. Soc. Am., **52**, 627-638.

Deresiewicz, H., Rice, J.F., 1962, The effect of boundaries on wave propagation in a liquid-filled porous solid, III: Reflection of plane waves at a free boundary, Bull. Seismol. Soc. Am., **52**, 595-625.

Dunn, K.J., 1987, Sample boudary effect in acoustic attenuation of fluid-saturated porous cylinders, J. Acous. Soc. Am., **81**, 1259-1266.

Gardner, G.H.F., 1962, Extensional waves in fluid-saturated porous cylinders, J. Acous. Soc. Am., **34**, 36-40.

Geertsma, J., Smit, D.C., 1961, Some aspects of elastic wave propagation in fluid-saturated porous solids, Geophysics, **26**, 169-181.

Plona, T.J., 1980, Observation of the second compressional wave in a porous medium at ultrasonic frequencies, Appl. Phys. Lett., **36**, 259-261.

Rasolofosaon, N.J.P., 1987a, New experimental of the influence of permeability on acoustic wave propagation, 57th SEG Annual International Meeting in New Orleans, paper S12.3

Rasolofosaon, N.J.P, 1987b, Acoustic wave propagation in porous media -Interface effects-(Experiments and theory), Doctorat-ès-Sciences thesis, University of Paris VII.

Rasolofosaon, N.J.P., 1988a, Influence of permeability on the attenuation of bulk and interface acoustic waves (Experiment vs. theory), 58th SEG Annual International Meeting in Anaheim, paper S10.3

Rasolofosaon, N.J.P., 1988b, Importance of interface hydraulic condition on the generation of second compressional wave in porous media, Appl. Phys. Lett. , **52**, 780-782.

Rasolofosaon, N.J.P., 1988c, Influence of permeability on the attenuation of acoustic waves (Experiment vs. theory) and critical analysis of the inversion of acoustic data in terms of permeability, SEG/EAEG Research Workshop in Reservoir

Geophysics, Dallas, Texas.

Rasolofosaon, N.J.P., 1989, Reflections on reflectors, 59th SEG Annual International Meeting in Dallas, paper RP1.1

Rasolofosaon, N.J.P., 1991, Plane acoustic waves in linear viscoelastic porous media - Energy, Particle displacement and physical interpretation, J. Acous. Soc. Am., **89**, 3.

Rosenbaum, J.H., 1974, Synthetic microseismograms -Logging in porous formation, Geophysics, **39**, 14-32.

Stoll, R.D., 1989, Sediment acoustics, Springer-Verlag, Berlin

Winkler, K.W., Plona, T.J., Froelich, B., Liu, H.L., 1987, Borehole Stoneley waves and permeability -Laboratory results, 57th SEG Annual International Meeting in New-Orleans, paper BHG 1.4.

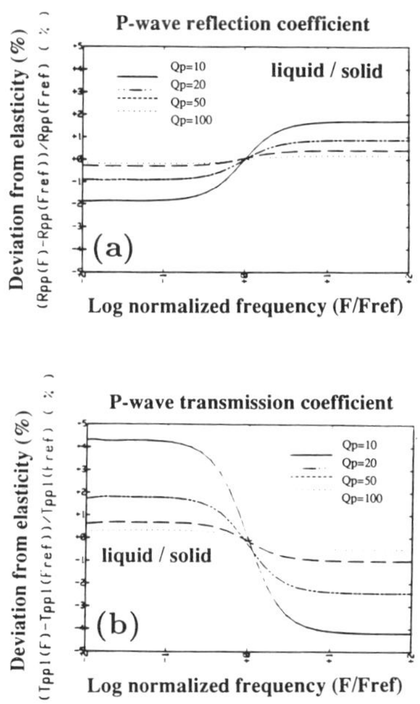


Fig. 1: Normalized deviations (from elasticity) of the P-wave (a) reflection and (b) transmission coefficients as functions of normalized frequency. Liquid / solid interface. Normal incidence. Parameter Qp. Viscoelastic model = SLS. Normalization frequency Fref is SLS relaxation frequency.

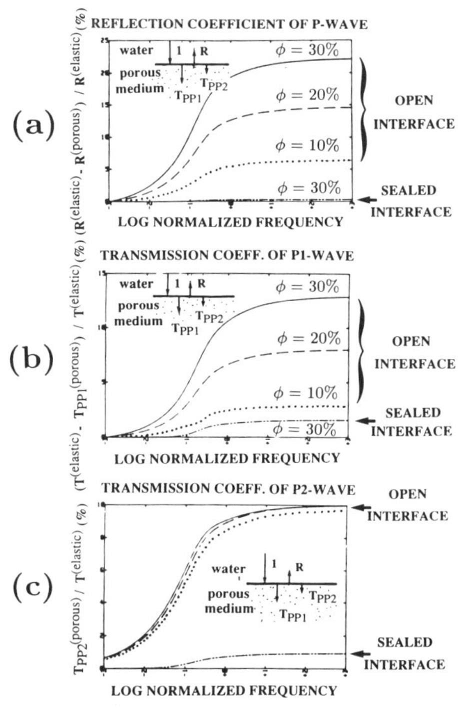


Fig. 2: Normalized deviations (from elasticity) of (a) the P-wave reflection coefficient, (b) the P1-wave transmission coefficient and (c) the P2-wave transmission coefficient as functions of normalized frequency. Normal incidence. Parameter porosity. Normalization frequency is Biot's critical frequency.

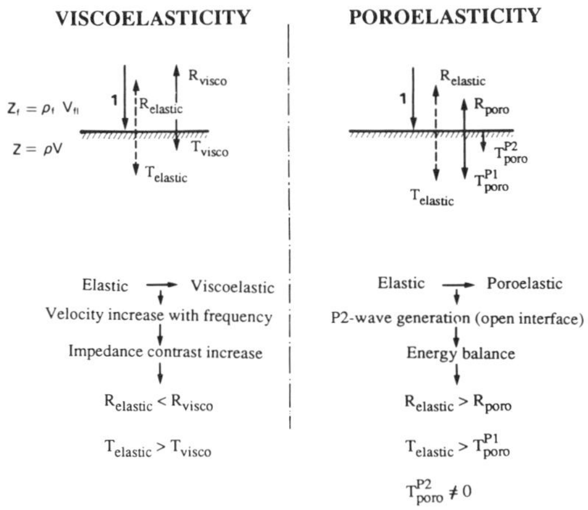


Fig. 3: Comparison between the influences of viscoelasticity and poroelasticity on transmission and reflection.

$Q_P^{visco}$ and $Q_S^{visco}$ inducing equal poroelastic attenuation and viscoelastic attenuation in the poroviscoelastic model:

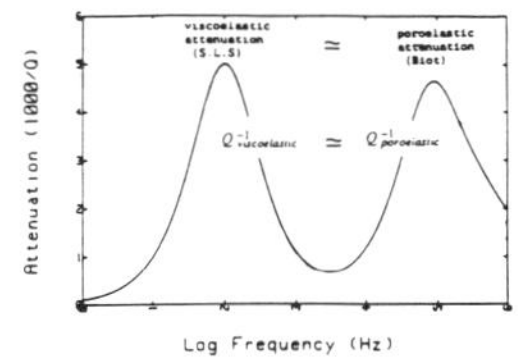


| | infinite porous medium (P1 and S waves) | semi-infinite porous medium (Rayleigh wave) | liquid/porous media - open interface - (Stoneley wave) |
|---|---|---|---|
| $Q_p^{visco}$ | 250 | 20 | << 8 |
| $Q_s^{visco}$ | 100 | 50 | - 8 |

Fig. 4: Viscoelastic attenuations to be injected in the poroviscoelastic model to induce equal viscoelastic attenuation and poroelastic attenuation. Cases of infinite porous medium, semi-infinite porous medium and liquid / porous medium interface.

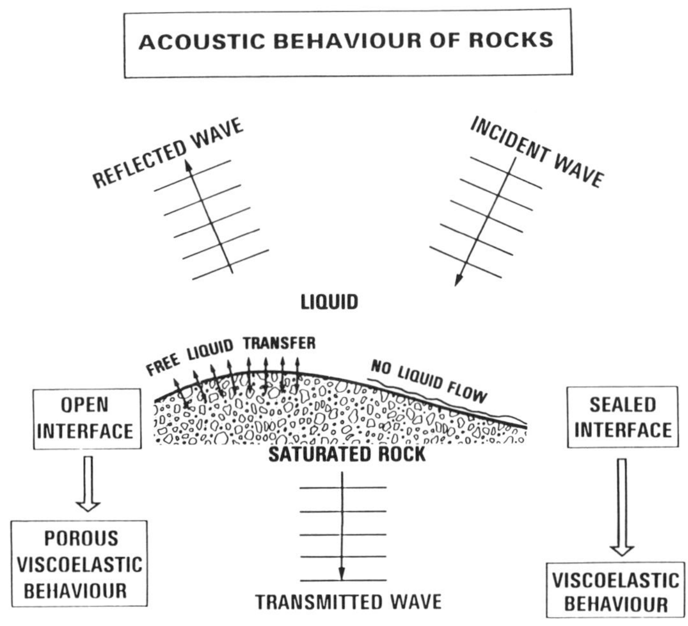


Fig. 5: Sketch summarizing the acoustic behaviour of rocks in presence of interfaces.

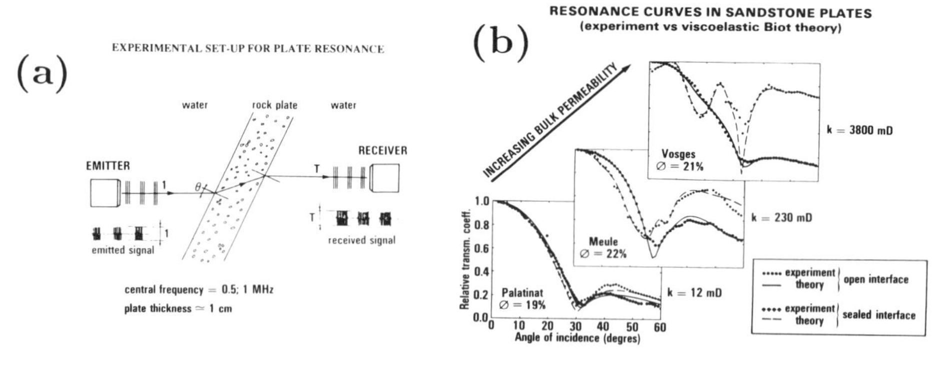


Fig. 6: (a) Experimental set-up and (b) experimental results vs. theory (poroviscoelastic model) on Lamb's modes resonances in rock plates.

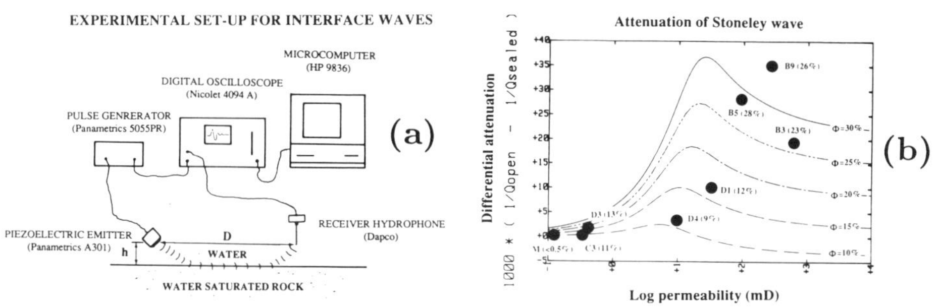


Fig. 7: (a) Experimental set-up and (b) experimental results vs. theory (poroviscoelastic model) on surface waves propagating along water / rock interfaces.

# INFLUENCE OF STRESS STATE AND STRESS HISTORY ON ACOUSTIC WAVE PROPAGATION IN SEDIMENTARY ROCKS

R.M. HOLT, E. FJAER, A.M. RAAEN AND C. RINGSTAD
*IKU (Continental Shelf and Petroleum Technology Research Institute)*
*N-7034 Trondheim*
*Norway*

ABSTRACT: Effects of currently and previously applied stresses on P- and S-wave velocities in sedimentary rocks (sandstones, chalks, shales) have been studied. Acoustic velocities have been measured on core samples in two orthogonal directions simultaneously under varying (isotropic, anisotropic) conditions of applied stress. Also, a series of model experiments has been performed, where synthetic sandstones were formed under stressed conditions, and where acoustic velocities were measured during and after unloading. The experiments show that the sensitivity to stress generally is largest at low stress levels, in particular below the maximum stress experienced previously. The symmetry of the stress field is to a large extent reflected in the wave propagation. Anisotropic unloading leads to an acoustic anisotropy which remains after the unloading is completed. The results underline that laboratory acoustic measurements should be performed under proper stress conditions. The unloading effect may help to quantify the *in situ* stress field.

## 1. Introduction

It is well established in the literature that sound velocities depend on the state of stress. For instance, Wyllie *et al.* (1958) and King (1966) found that sound velocities in sandstones and in unconsolidated sands increase with increasing external hydrostatic stress, and that the velocity increase becomes gradually smaller as the stress is increased. Nur and Simmons (1969) showed with granite specimens that an acoustic anisotropy developed as an anisotropic state of stress was applied. Similar observations have been made later also in sandstones (e.g. Holt and Fjaer, 1987; and Sayers *et al.*, 1990). Engelder and Plumb (1984), and later Teufel (1989), found a significant anisotropy, measured on unloaded cores at unconfined conditions. They related this to an anisotropic *in situ* stress state.

The motivation behind the present study was two-fold: i) possibly obtaining *in situ* stress information from acoustic data recorded in the field; and ii) obtaining guidelines for performing laboratory tests or correcting laboratory data, in order to compare with acoustic borehole log or seismic data recorded under different stress conditions. Since most of our experimental work actually has been related to problems associated with drilling for and production of petroleum, we have chosen to focus on relatively weak sedimentary rocks. Such materials have not been covered to any significant extent in the literature.

In this Paper, we shall show effects of the stress symmetry and of the stress level. We shall also see, through a simple model experiment, that not only the current state of stress but also the stress history has an influence on the resulting velocities and the velocity anisotropy. The results will reveal the importance of performing acoustic laboratory measurements under proper stress conditions.



*J. M. Hovem et al. (eds.), Shear Waves in Marine Sediments*, 167–174.

## 2. Experimental Details

The experiments were performed in a standard triaxial set-up. The test samples are cylinders with diameters ~ 38 mm and lengths ~ 75 mm. During each test, velocities of P- and S-waves propagating both axially and radially may be measured together with axial and radial stresses, pore pressure, and axial and radial strains. The acoustic waves are generated by Panametrics broadband ultrasonic transducers with a central frequency of 0.5 MHz. Transducers which measure axially propagating waves are mounted in the top and bottom pistons, and transducers which measure radially propagating waves are mounted near the central plane of the samples (see Fig. 1). Corrections to measured arrival times have been made both axially (due to transit times in pistons) and radially (due to pressure dependent transit times in the sleeve).

Samples from three different lithologies have been investigated:

Sandstone: Dry samples of a weak, 25% porosity, Triassic outcrop ("Red Wildmoor") sandstone.
Chalk: Water-saturated samples of a 43% porosity Upper Campanian outcrop chalk. The samples were saturated with tap water which had been stored with crushed chalk for about a week, to prevent chemically induced disintegration of the chalk samples.
Shales/clays: Weald Shale (16% porosity and 34% clay content) and Fullers Earth (57% porosity and 99% clay content). 3% brine was used for drilling and saturation before testing.

Sintered plates and filter drains were used for distributing pore fluid across the end faces and side walls when shale samples were tested. Pore pressures were measured at the top and bottom pistons.

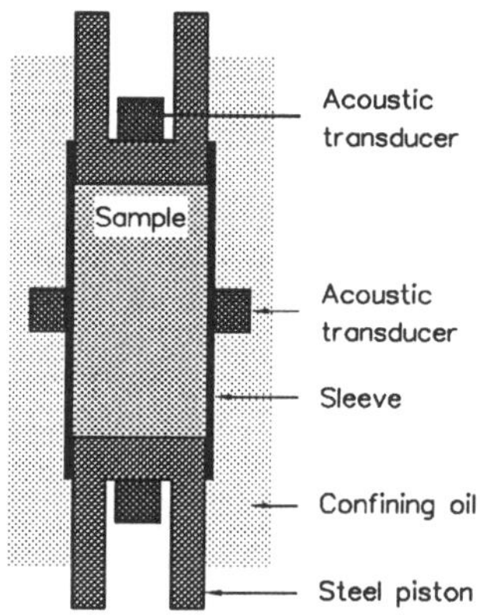


*Fig. 1: Sketch showing transducer positions in triaxial cell.*

## 3. Experimental results

In this Section, we shall present experimentally determined P- and S-wave velocities (with different directions of propagation and polarization) in weak sedimentary rocks, obtained during hydrostatic and triaxial loading, and unloading.

### 3.1 SANDSTONES

Fig. 2a shows measured acoustic velocities vs. pressure during hydrostatic loading. All velocities are seen to increase with increasing hydrostatic pressure and the increase is most pronounced at low pressures. The trends found here are typical of most sandstones investigated previously (e.g. King 1966).

Figs. 2b and 2c show acoustic velocities vs. strain in two triaxial tests. At low confining pressure (Fig. 2b), there is initially an increase in the velocity of the axially propagating P-wave (and the axially polarized S-wave), while the radially propagating P-wave (and the radially propagat-

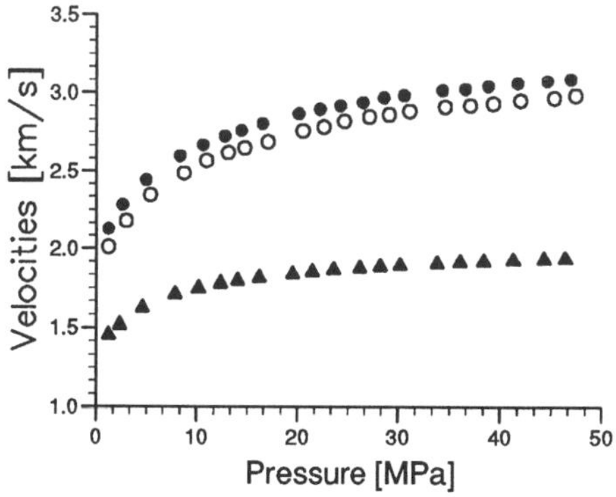


*Fig. 2a: Hydrostatic loading.*

*Figs. 2a-c:*
*Acoustic velocities in dry outcrop ("Red Wildmoor") sandstone (sample axis normal to bedding). The stress-strain behaviour is shown for the triaxial tests.*

- ● *Axial P-waves*
- o *Radial P-waves*
- ▲ *S-waves (radial propagation and axial polarization, or vice versa).*
- Δ *S-waves (radial propagation and tangential polarization)*

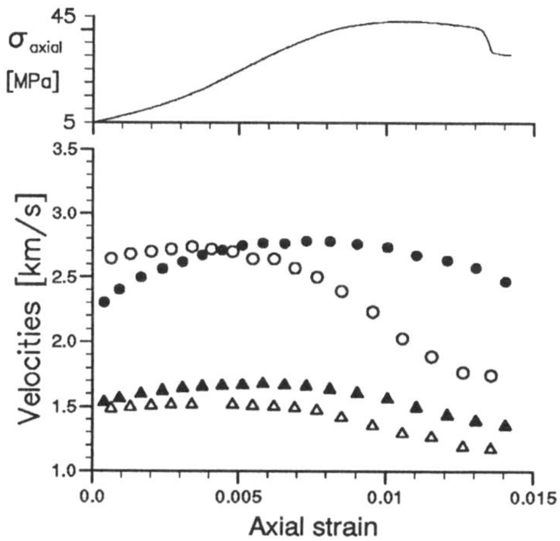


*Fig. 2b: Triaxial loading, $\sigma_{radial}$ = 5 MPa.*

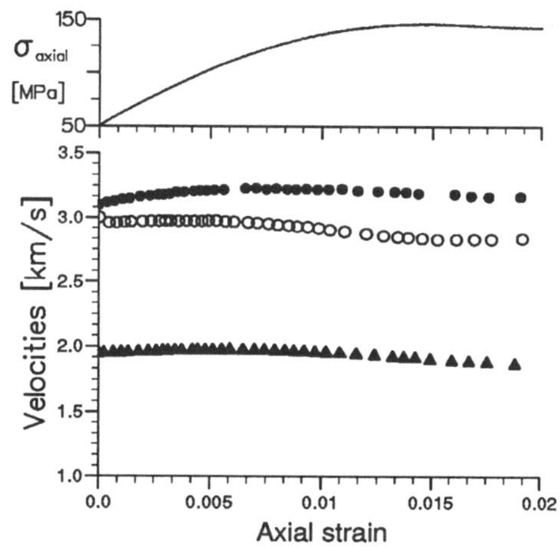


*Fig. 2c: Triaxial loading, $\sigma_{radial}$ = 50 MPa.*

ing and tangentially polarized S-wave) is rather insensitive to the stress change. Then, at a stress level corresponding to deviation from linearly elastic ("yield"), the wave velocities associated with radial propagation and/or polarization start to decline significantly. After failure, all velocities decrease slowly as a function of strain. This behaviour is much less prominent at high levels of confining pressure (Fig. 2c).

## 3.2 CHALK

Fig. 3a shows acoustic velocities as a function of hydrostatic pressure. All velocities are seen to be relatively stress insensitive during initial loading, but start to decrease when the hydrostatic stress level exceeds approximately 7.5 MPa. As loading continues, the velocities reach a minimum value and then start to increase. Observe that the initial anisotropy in P-wave velocities is preserved during the experiment.

The acoustic behaviour of chalk during triaxial loading is shown at low and high confining

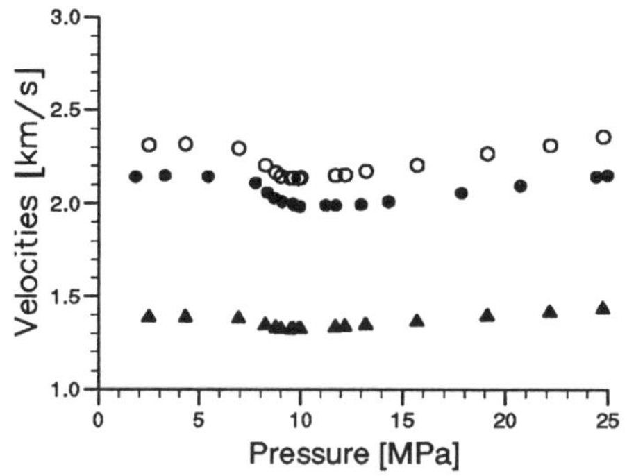


*Figs. 3a-c:*
*Acoustic velocities in brine saturated outcrop chalk (no visible bedding). The stress-strain behaviour is shown for the triaxial tests.*

- ● *Axial P-waves*
- o *Radial P-waves*
- ▲ *S-waves (radial propagation and axial polarization, or vice versa).*

*Fig. 3a: Hydrostatic loading.*

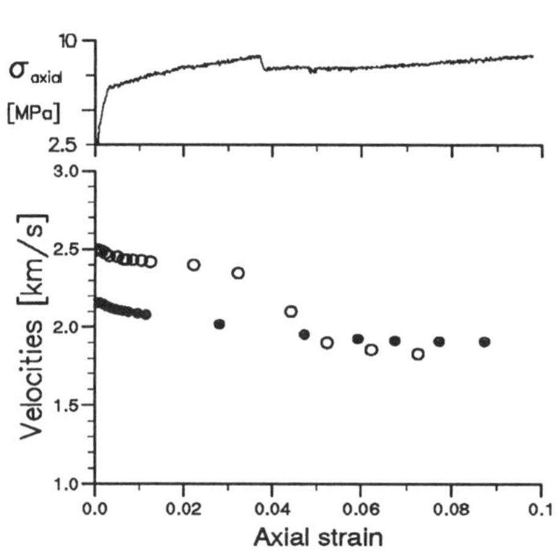


*Fig. 3b: Triaxial loading, $\sigma_{radial}$ = 2.5 MPa.*

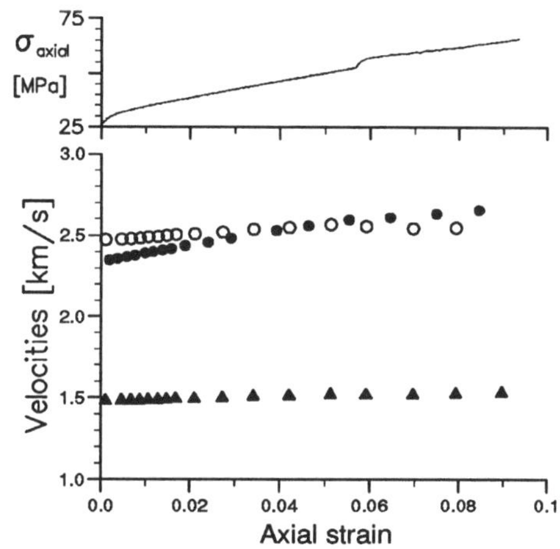


*Fig. 3c: Triaxial loading, $\sigma_{radial}$ = 25 MPa.*

pressures in Figs. 3b and 3c. In Fig. 3b, the failure of the specimen results in a strong decrease of the radial velocity, whereas the axial is less affected. Fig. 3c is the continuation of the experiment in Fig. 3a, and shows that all velocities are increasing during triaxial loading. The increase is, however, clearly anisotropic, such that the velocities of waves propagating (or being polarized) in the axial (maximum stress) direction increase faster than the others. Notice that the static stress strain curve in Fig. 3c shows no sign of failure; just a continuous trend of strain-hardening.

## 3.3 SHALES AND CLAYS

Sound velocities measured during drained hydrostatic loading for brine saturated Weald Shale and Fullers Earth are shown in Figs. 4a and 4b. In Weald Shale, the velocities are found to increase with increasing stress (Fig.4a), similar to what has been found in other low-porosity shales (Jones and Wang, 1981; Johnston, 1986). In Fullers Earth, the acoustic velocities are rela-

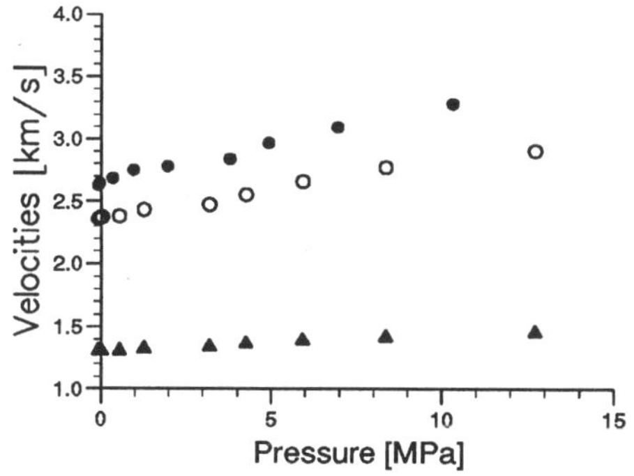


*Fig. 4a:*
*Drained hydrostatic loading of Weald Shale (sample axis parallel to bedding).*

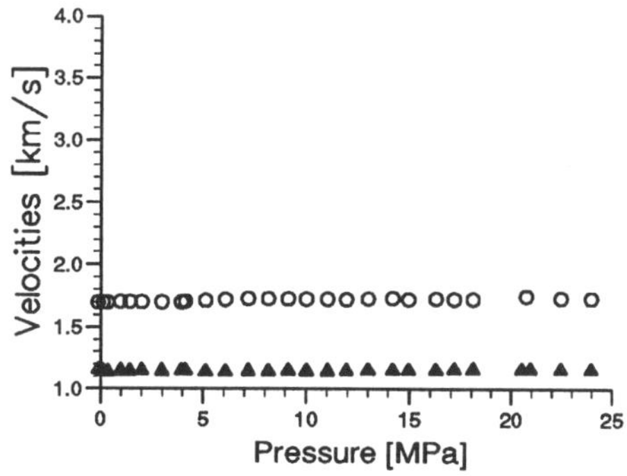


*Fig. 4b:*
*Drained hydrostatic loading of Fullers Earth (sample axis normal to bedding).*

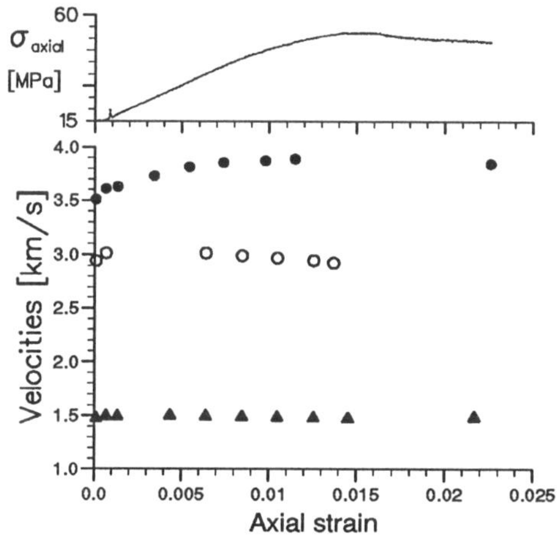


*Figs. 4a-c:*
*Acoustic velocities in brine-saturated outcrop shales. The stress-strain behaviour is shown for the triaxial test.*

- ● *Axial P-waves*
- o *Radial P-waves*
- ▲ *S-waves (radial propagation and axial polarization, or vice versa).*

*Fig. 4c:*
*Undrained triaxial loading of Weald Shale, (sample axis normal to bedding).*
*Initial pore pressure = 2 MPa;*
$\sigma_{radial} = 15$ *MPa.*

tively insensitive to stress changes (Fig.4b).

Following the drained hydrostatic loading, the specimens were kept in a state of constant external (isotropic) stress and constant pore pressure for a period of abut 20 hours. During this period, a significant volumetric deformation occurred due to consolidation. Table 1 shows the relative increase in P-wave velocity in Weald Shale during consolidation at 5 different stress levels. The results show that even for a constant effective stress, the velocity increases due to the compaction of the samples.

The acoustic behaviour is also monitored during undrained triaxial loading. Significant velocity changes are detected only in Weald Shale (Fig.4c). An acoustic anisotropy similar to that observed in sandstones (Figs. 2b and 2c) occurs.

*Table 1: P-wave velocity changes during consolidation at constant stress.*

| Velocity change $\Delta v_p / v_p$ [%] | Effective stress change $\Delta\sigma'$ [MPa] | Volumetric strain $\Delta\varepsilon_v$ [%] | Effective confining pressure $\sigma_{radial} - p_{pore}$ [MPa] |
|---|---|---|---|
| 13 | 0 | 2.3 | 3 |
| 16 | 0 | 1.6 | 8 |
| 17 | 0 | 1.6 | 10 |
| 4 | 0 | 0.3 | 13 |
| 31 | 0 | 3.5 | 23 |

### 3.4 SYNTHETIC SANDSTONES: EFFECTS OF UNLOADING

We shall now describe a simple model experiment which was performed to investigate the effect of a previously experienced stress state on acoustical parameters. In the experiment, which is described in more detail in Rathore *et al.* (1989), and further in Fjaer *et al.* (1990), synthetic sandstones are formed under stressed conditions. A sand and epoxy mixture (volume ratio 25:1), is subjected to axial stress under conditions of no lateral strain for a period of time sufficient for the cementing epoxy to cure. The resulting material becomes a porous structure of epoxy cemented sand. The state of stress is relatively isotropic during formation of the synthetic. When the axial stress is released, however, the radial stress will decrease at a lower rate than the axial stress. The result is the development of significant P- and S-wave anisotropies during unloading. These anisotropies remain after the specimens have been completely unloaded. The wave velocities are smaller for propagation (or polarization, in the case of S-waves) in the axial (previous maximum stress) direction. The velocity anisotropy is found to increase and the velocity magnitudes to decrease with increasing forming stress. The specimens are more compacted and hence have a lower porosity when formed at high stresses. Thus, the sound velocities tend to increase with increasing porosity, in contrast to the normal correlation.

## 4. Discussion

### 4.1 SYMMETRY OF THE STRESS STATE

Acoustic velocities in rocks depend not only on the magnitude of the stress the material is subjected to, but also on the symmetry of the stress field. This means that it is important to specify the direction of propagation and polarization relative to the principal axes of stress when discussing acoustic behaviour.

Consider first the case of isotropic loading, as shown in Figs. 2a, 3a and 4a for sandstone, chalk and shale, respectively. Acoustic velocities in orthogonal directions follow the same main pattern in a given specimen. In the sandstone and Weald shale samples, the velocities show a monotone increase, whereas in the chalk sample the behaviour is more complicated. In both cases there is a difference between the axial and radial velocities that remains throughout the loading. The behaviour can be explained by assuming a pressure dependent and a pressure independent contribution to the anisotropy. For the sandstone and shale, one may think of the pressure dependent part in terms of closing cracks. In chalk, the behaviour can be explained by a breaking of bonds (pore collapse) giving a velocity decrease, followed by a compaction phase which again stiffens the material. The pressure independent difference in the velocities may be related to inherent lithological anisotropy caused by oriented grains.

Proceeding next to anisotropic loading, the direction of propagation relative to the major principal stress is important. Considering the sandstone sample at a relatively low confining stress (Fig. 2b), one sees how the axial velocity increases prior to failure, whereas the radial velocity is less sensitive. After failure the radial velocity decreases strongly, with a smaller response seen in the axial velocity. The behaviour may again be explained in terms of microcracks in the material. In the pre-failure phase of the experiment, the stress applied to the sample results in a closing of microcracks normal to the sample axis, and has less effect on parallel cracks. This means that the path for the axial wave is improved, while that of the radial is less affected. As failure is approached, microcracks parallel to the sample axis start to open, and are reflected in a stronger decrease of the radial than the axial velocity. A similar behaviour is seen in the chalk sample at low confining pressure (Fig. 3b). Again, the radial behaviour is most sensitive to the failure process in the material.

## 4.2 STRESS REGIME

The response of the acoustic velocities to stress changes generally becomes smaller as the overall stress level is increased. This is seen in the isotropic loading of sandstone as a levelling out of the velocities as function of stress, Fig. 2a. Further, this is also seen by comparing the sandstone results in triaxial tests at low and high confining stresses, Figs. 2b and 2c. Although the behaviour in principle is the same for the high and low stresses, the response is smaller at high stresses. In chalk, similar effects are observed. Again, one may use the microcrack picture to qualitatively explain the results. For increasing isotropic stress, one expects the rate of crack closure to decrease as the number of open cracks decreases. This results in less sensitivity at high stresses. For triaxial loading, the number of microcracks normal to the sample axis, which have not closed in the hydrostatic phase of the experiment, is smaller at high confining stresses. This results in a smaller increase in the axial velocity in the initial triaxial phase.

## 4.3 STRAIN DEPENDENCE

Although the objective of this paper is to study the dependence of acoustic velocities on stress, we should not forget that the velocities probably respond to strain more than to stress. The significance of this distinction appears in situations where there is not a one-to-one relationship between stress and strain, for instance during consolidation, or in loading-unloading situations. The velocity increase during consolidation (at constant stress) of Weald Shale (Table 1) supports the hypothesis of strain being the prime variable.

## 4.4 STRESS HISTORY

The difference in behaviour between the high stress and the low stress regimes may be related to the previously experienced state of stress. This is reflected in the difference between the overconsolidated Weald Shale and the normally consolidated Fullers Earth claystone.

The simple model experiments described in Section 3.4 also confirm the influence of stress history on acoustic velocities: First, the velocities are functions of the stress during formation (which also was the peak stress experienced by the material). Second, the significant velocity anisotropy of the unloaded cores seems to reflect the anisotropic stress state during unloading, which illustrates a stress path effect on the velocities. The anisotropy and the velocity decrease can be explained by opening of oriented microcracks during unloading.

## 5. Conclusions

Acoustic wave velocities of sedimentary rocks are sensitive to changes in stress, in particular at stress levels below the maximum previously experienced stress. This implies indirectly that stress effects should be more pronounced in low porosity than high porosity geomaterials. The symmetry of the stress field (isotropic, anisotropic) is reflected in the wave propagation. These conclusions appear to be valid for sandstones, shales and chalk. Experiments with synthetics formed under stressed conditions have identified a memory of the previously applied maximum stress, and of the stress conditions during unloading.

The experiments underline the importance of reestablishing *in situ* or close to *in situ* stress conditions in the laboratory when acoustic measurements are performed for comparison with field data. The change in anisotropy as well as in velocity magnitudes when reloading a core sample can assist in finding *in situ* stress field orientation and stress magnitudes.

**Acknowledgements**
The authors would like to acknowledge the partial financial support for this work through the FORMEL consortium by Conoco Norway, Elf Aquitaine Norge, Fina Exploration Norway, Norsk Hydro and Statoil (for the work on sandstones), through Fina as responsible for the Mechanical Properties subproject in the North Sea Chalk Research Programme, through Statoil's British-Norwegian research cooperation (for the work on shales), and through NTNF (Royal Norwegian Council for Scientific and Industrial Research).

**References**

**Engelder, T. and Plumb, R.** (1984) "Changes in in situ ultrasonic properties of rock on strain relaxation", Int. J. Rock Mech. Min. Sci. & Geomech. Abstr. 21, 75-82.

**Fjaer, E., Holt, R.M., Rathore, J., and Åhlén, B.** (1990) "Acoustic anisotropy related to anisotropic stress conditions during formation and unloading", pres. at 4th Int. Workshop on Seismic Anisotropy, Edinburgh July 1990

**Holt, R.M. and Fjaer, E.** (1987) "Acoustic behaviour of sedimentary rocks during failure", in "North Sea Oil and Gas Reservoirs", ed. J.Kleppe et al., Graham & Trotman, pp.311-16.

**Jones, L.E. and Wang, H.F.** (1981) "Ultrasonic velocities in Cretaceous shales from the Williston basin", Geophysics 46, 288-97.

**Johnston, D.H.** (1986) "Shale properties at temperature and pressure", Oil & Gas J., Nov.3, pp 60-65.

**King, M.S.** (1966) "Wave velocities in rocks as a function of changes in overburden pressure and pore fluid saturants", Geophysics, Vol 31, pp 5073.

**Nur, A. and Simmons, G.** (1969) "Stress-induced velocity anisotropy in rock: an experimental study", J. Geophys. Res. 74, 6667-74.

**Rathore, J.S., Holt, R.M. and Fjaer, E.** (1989) "Effects of stress history on petrophysical properties of granular rocks", in Wahab Khair (ed.) Rock mechanics as a guide for efficient utilization of natural resources, 76572. AA Balkema, Rotterdam.

**Sayers, C.M., van Munster, J.G. and King, M.S.** (1990) "Stress-induced ultrasonic anisotropy in Berea sandstone", Int. J. Rock Mech. Min. Sci. & Geomech. Abstr. 27, 429-36.

**Teufel, L.W.** (1989) "Acoustic emissions during anelastic strain recovery of cores from deep boreholes", in Wahab Khair (ed.) Rock mechanics as a guide for efficient utilization of natural resources, 269-278. AA Balkema, Rotterdam.

**Wyllie, M.R.J., Gregory, A.R. and Gardner, G.H.F.** (1958) "An experimental investigation of factors affecting elastic wave velocities in porous media", Geophysics, 23, 459-93.

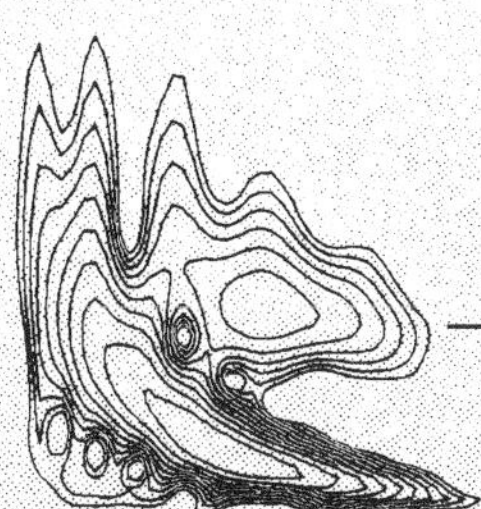

# Section 2

## *Field experiments and interpretation*

# A SUMMARY OF DREA OBSERVATIONS OF INTERFACE WAVES AT THE SEABED

DAVID M.F. CHAPMAN and PHILIP R. STAAL
*Defence Research Establishment Atlantic*
*Box 1012, Dartmouth, Nova Scotia*
*Canada, B2Y 3Z7*

ABSTRACT. During the past ten years or so, DREA scientists have observed interface waves in shallow water in the frequency range 2 Hz - 20 Hz over a variety of seabed types: sedimentary rock (chalk), unconsolidated sediments (sand and gravel), and crystalline bedrock (granite). In this paper we summarize our interface wave experiments and discuss how we used the observations to improve our understanding of low frequency acoustic propagation in shallow water. We have found that (a) the analysis of interface wave propagation is a good means of estimating seabed geo-acoustic parameters, including shear speed; and (b) over crystalline bedrock seabeds, interface waves in the frequency band 3 Hz - 6 Hz can be the dominant mode of propagation to ranges of 40 km or more.

## 1. Introduction

About 10 years ago, DREA scientists were involved in a research programme in shallow water acoustics that continues to this day, attempting to understand how the seabed influences acoustic propagation in the frequency range 2 Hz - 2000 Hz. Propagation models based on normal mode theory in stratified fluid media seemed to explain much of our transmission loss data [Ellis and Chapman (1980)]. This work was presented at a SACLANTCEN conference on bottom-interacting ocean acoustics [Kuperman and Jensen (1980)] at which we learned that shear waves in the seabed

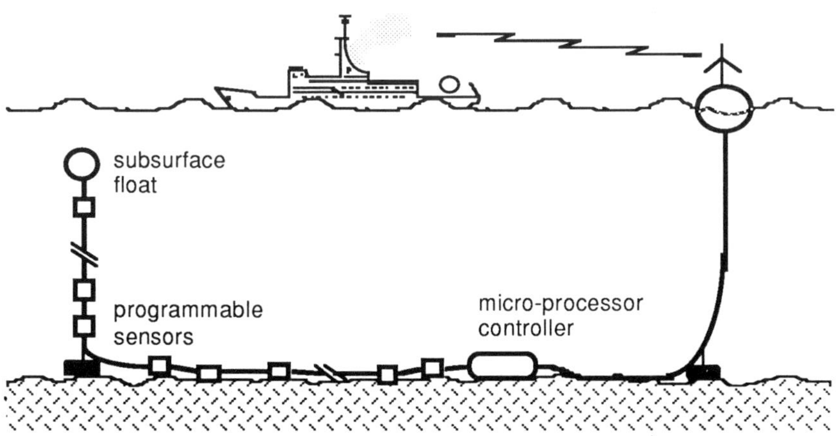


*Fig. 1. Typical deployment of the Hydra array on the seabed.*



*J. M. Hovem et al. (eds.), Shear Waves in Marine Sediments*, 177–184.

could be important in shallow water acoustics. We also learned that a special kind of seismo-acoustic wave could be observed at the water/sediment interface if the seabed supported shear waves: the *interface wave* or *Scholte wave* [Rauch (1980) and other papers]. Since that time, there have been several similar conferences at which interface wave research has been reported [Pace (1983), Akal and Berkson (1986), Lara-Sáenz *et al.* (1987), and Merklinger (1987)].

The very next year, in 1981, we performed a set of experiments over a chalk seabed [Staal (1983)] that showed high transmission loss below 300 Hz that could not be modelled using fluid media alone. This compelled us to consider shear waves as a loss mechanism. In the process, we not only verified that shear waves were indeed the loss mechanism [Chapman and Ellis (1983), Ellis and Chapman (1985)], we also discovered interface waves in the signals from hydrophones lying on the seabed.

Since then, we have paid special attention to the role of shear waves in shallow water propagation and we have observed interface waves in several of our experiments. This paper presents a summary of our observations of acoustic interface waves at the seabed. To keep the paper brief, we have omitted many details; we direct those readers interested in the details to the original sources listed at the end of the paper. Previously unpublished results are presented in Figs. 4, 5, and 7.

## 2. Interface Waves in Sedimentary Rock (Chalk)

Although geophones are commonly used in such experiments, we have successfully observed interface waves with our shallow-water hydrophone array *Hydra* [Staal (1987)]. A typical deployment of Hydra is illustrated in Fig. 1. The sensors on the seabed provide data with a good

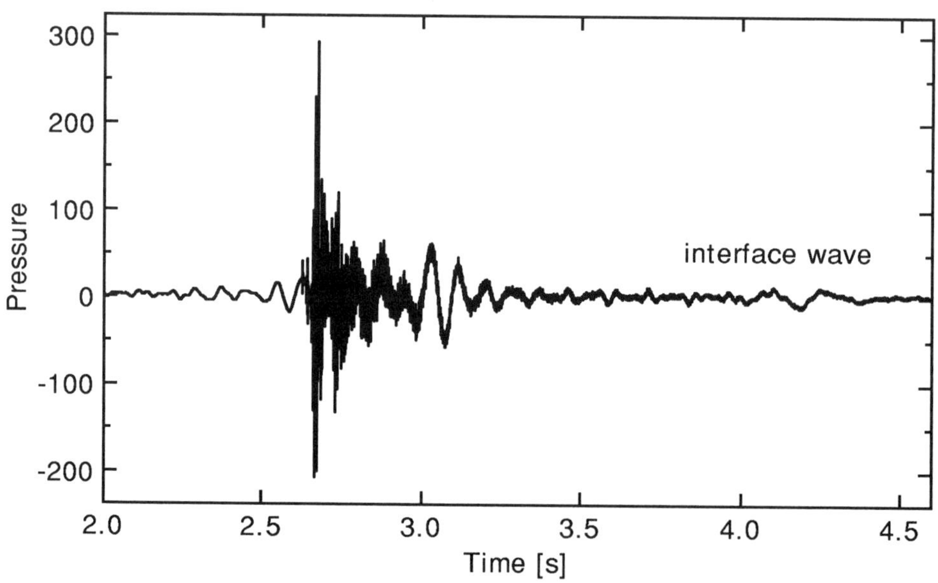


*Fig. 2. Signal received from an explosion at 5 km range over a chalk seabed, including an interface wave at 4 s.*

signal-to-noise ratio down to about 1 Hz and are capable of sensing the component of the interface wave that travels as a compressional wave in the water.

While analysing experimental data on transmission loss over a chalk seabed [Staal (1983)], we saw some weak late-arriving wavelets in the signals from the hydrophones on the seabed. Fig. 2 shows the signal received from the detonation of a 0.45 kg explosive at a range of 5 km; in order of arrival, we see the low-frequency compressional wave in the seabed, the high-frequency water wave, the dispersed arrivals of the various shallow water normal modes, and – at about 4 s – the interface wave. Several of these interface waves were observed in the data, but only at ranges less than 5 km.

From a time-of-flight analysis of these arrivals, we estimated the speed of the interface waves in chalk to be 1000 m/s ± 150 m/s. Subsequently, we used this estimate to specify a value for shear speed in chalk, required by acoustic propagation models that include compressional-to-shear wave conversion at the seabed as a propagation loss mechanism [Ellis and Chapman (1985)].

It was clear from the data that the interface wave in chalk was not a significant long-range propagation mechanism for underwater sound. However, the observation of interface waves at short ranges did provide valuable information regarding the geo-acoustic characteristics of the seabed that govern longer-range propagation at higher frequencies.

## 3. Interface Waves in Unconsolidated Sediments (Sand and Gravel)

Having discovered interface waves in our hydrophone data, we decided to try to observe them with geophones. We borrowed an ocean bottom seismometer (OBS) from the Bedford Institute of Oceanography and modified it to permit real-time ship-board recording of the signals from two geophones (vertical and horizontal orientation) and a hydrophone. The Terrain Geophysics Branch of the Geological Survey of Canada assisted by constructing and firing the source arrays, each consisting of 6 detonators spaced 25 m apart along a set of detonation wires.

Experiments involving the generation and detection of interface waves in seabeds of sand and sand and gravel were conducted in 1982 from the Canadian Forces ship HMCS CORMORANT [Dunphy (1982)]. The OBS and a source array were deployed over the side of the ship and the

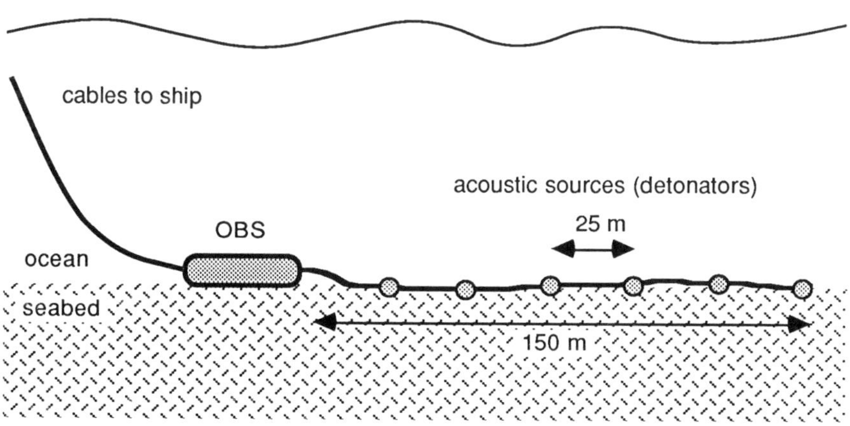


*Fig. 3. Deployment of an ocean bottom seismometer and a source array.*

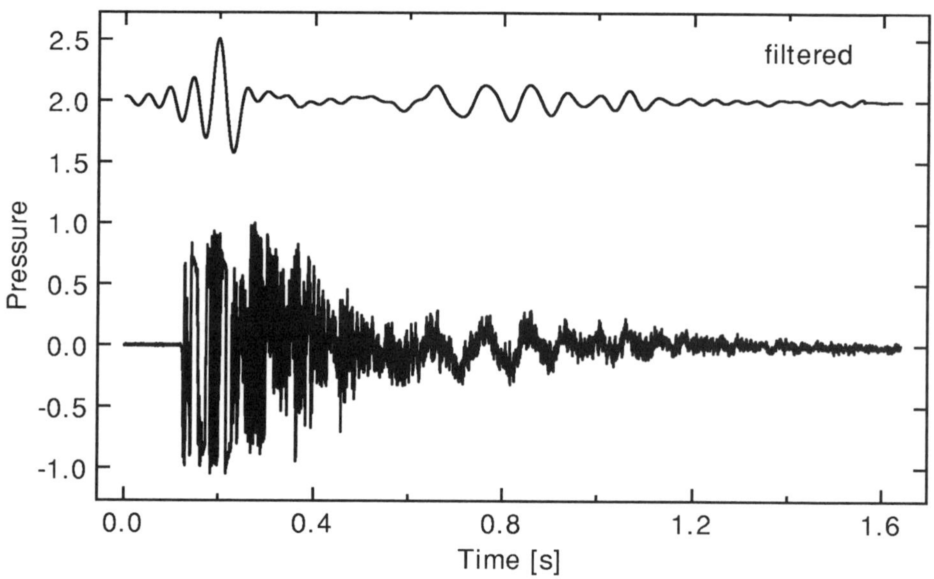


*Fig. 4.* *Signal received from an explosion at 100 m range over a sand seabed, including an interface wave at 0.8 s. The upper trace is a low-pass filtered version of the lower trace.*

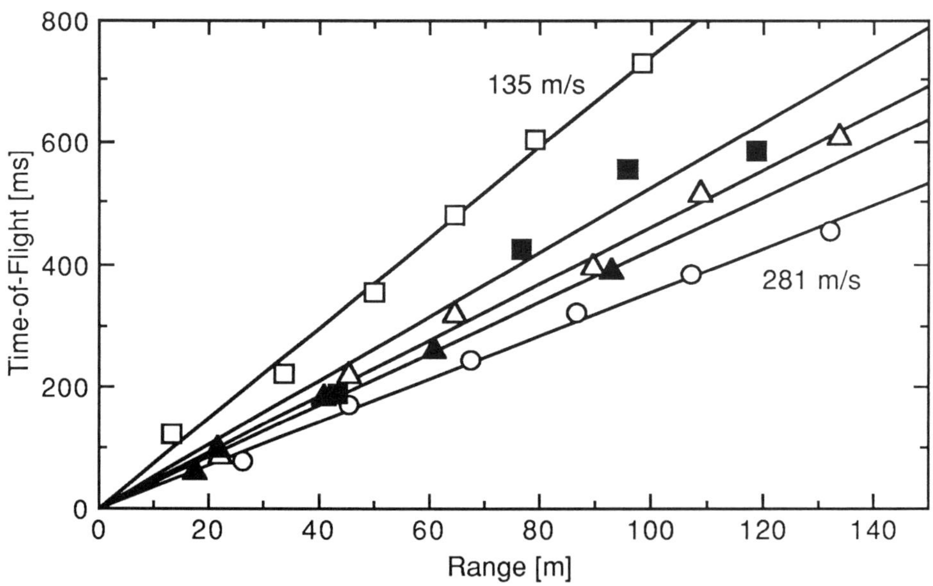


*Fig. 5.* *Time-of-flight data for interface waves over sand and sand and gravel seabeds. Each deployment is denoted by a distinct symbol type. Open symbols: sand seabeds; closed symbols: sand and gravel seabeds. Straight lines: least-square fits to the data.*

whole system was allowed to drag on the bottom, pulled by the drifting ship, until the array straightened out, as illustrated in Fig. 3. Then, each detonator was fired sequentially from the far end at 5-10 s intervals and the signals received at the OBS were recorded on the ship.

An example of the hydrophone signals is shown in Fig. 4, along with the same data low-pass filtered at 20 Hz. The high-speed compressional wave in the seabed arrives before the water wave, and a dispersed interface wave is evident at 0.8 s. Using the signals from the geophones, we plotted hodographs of vertical displacement vs. horizontal displacement as a function of time and were able to verify the elliptical motion of the seabed associated with the passage of the interface wave, as observed by Rauch (1980) and others.

The time-of-flight data for several different deployments is shown in Fig. 5. There is considerable spread in the data, partly due to uncertainties in the experimental method, but also due – we believe – to the intrinsic variability of the geo-acoustic properties of the seabed. The data collected indicate that the interface wave speed varied from 135 m/s to 281 m/s in sand and sand and gravel seabeds. In such materials, the shear speed is about 10% higher than the interface wave speed.

We did not continue interface wave experiments with OBS sensors, as we found their deployment cumbersome and the results only partially satisfactory. We have not seen interface waves using the Hydra array on a sand seabed, even when the source-receiver range is only a few kilometres.

## 4. Interface Waves in Crystalline Bedrock (Granite)

In contrast to our observations of interface waves in sand and chalk, we found that interface waves in granite were easy to generate and receive at ranges up to several tens of kilometres. In 1983 we conducted a propagation experiment with the Hydra array over a rough granite seabed with little or no sediment cover [Staal and Chapman (1986)]. Fig. 6 shows signals received from explosions at 11 ranges from 5 km to 38 km. We have aligned the water wave arrivals so that a vertical line through the figure represents the speed of the water wave; other lines at various inclinations represent slower wave speeds.

At all ranges, the transmission loss data associated with this experiment showed high loss in the 10 Hz - 100 Hz band. Moreover, the best propagation was observed in the 3 Hz - 6 Hz band associated with the interface wave.

As an aside, we have investigated several hypotheses regarding the 10 Hz - 100 Hz band of poor propagation, which is not predicted by a simple geo-acoustic model of water over smooth granite. At first we thought that the poor propagation was caused by scattering at the very rough granite seabed [Purcell (1986)]. A subsequent experiment in 1985 at the same location [Staal *et al.* (1987)] showed little correlation between the high transmission loss and the degree of bottom roughness. A more complex geo-acoustic model that includes shear waves in the thin sediment layer on top of the granite [Hughes *et al.* (1990)] produces satisfactory agreement between theory and experiment. Interestingly, the properties of the sediment layer are all-important for propagation in the 10 Hz - 100 Hz band, but have negligible influence on the interface wave, which seems to propagate as if the sediment were not present.

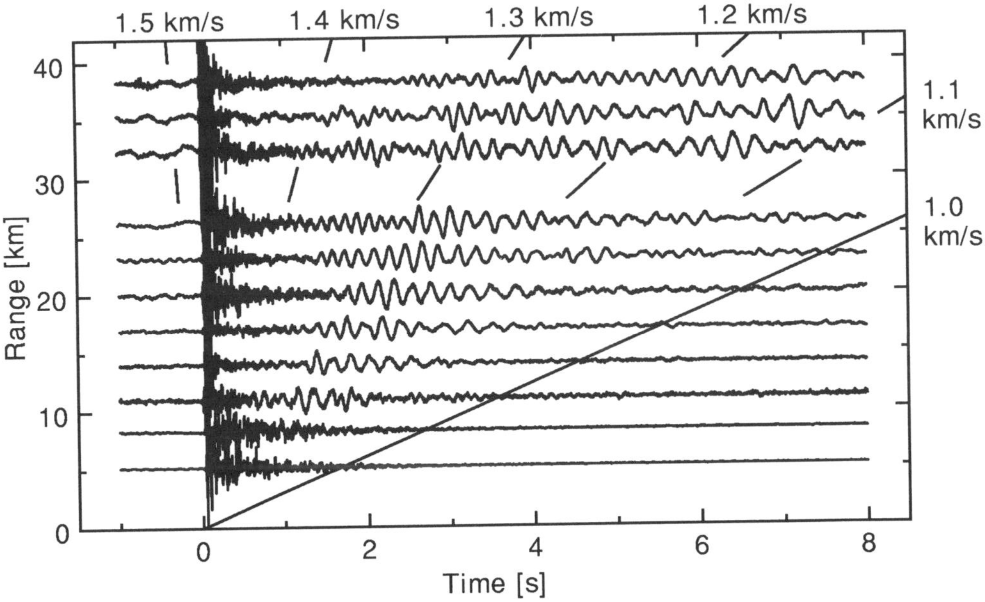


*Fig. 6.* *Signals, including interface waves, received from explosions at various ranges over a granite seabed.*

## 5. Dispersion of Interface Waves

The favourable propagation of interface waves in a granite seabed presents an opportunity to study the *dispersion* of the waves, that is, the dependence of the phase speed and the group speed upon frequency [Rauch (1980)]. In the time domain, dispersion appears as time-spreading of a waveform as it travels away from the source. Time-spreading of the interface wave in granite can be seen in Fig. 6: simply compare the duration of the waveforms at different ranges.

Dispersion in a waveguide is most noticeable when the acoustic wavelength in water is of the same order as the water depth, that is, when $Hf/c \approx 1$, in which H is the water depth, f is the wave frequency , and c is the wave speed. When $Hf/c \gg 1$, that is, at high frequencies or in deep water, we call the interface wave the *Scholte wave*. When $Hf/c \ll 1$, that is, at low frequencies or in shallow water, we call the interface wave the *Rayleigh wave*. (Some experts regard the low-frequency limit of the interface wave to be the *pseudo-Rayleigh wave*.) Between these frequency limits, the interface wave is dispersive, having a group speed between the Scholte speed and the Rayleigh speed, or less.

Dispersion in the granite seabed can also be illustrated by a grey-scale plot of intensity versus group speed and frequency as shown in Fig. 7. (These data come from the 1985 experiment.) The experimental data show arrivals from two distinct modes, and possibly a third. Superimposed on the experimental data are theoretical group speed curves for the first three normal modes of the three-layer water/sediment/granite geo-acoustic model used by Hughes *et al.* (1990) to model transmission loss in the same environment. The calculations were performed with the SAFARI

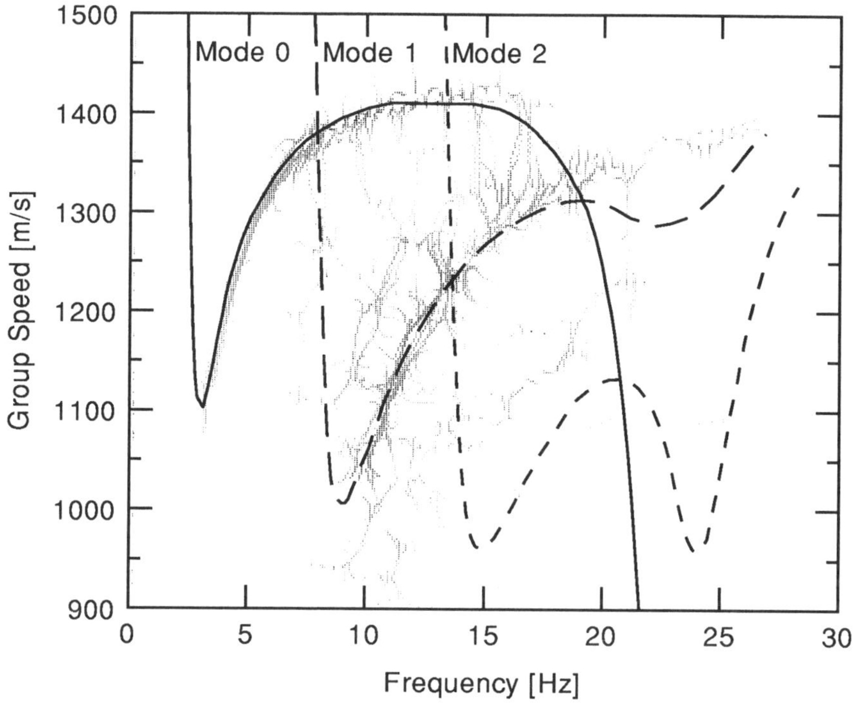


*Fig. 7.* *Spectrum level vs. group speed and frequency for the low-frequency portion of signals received from explosions over a granite seabed, with theoretical group speed curves superimposed for the interface wave (Mode 0) and the next two higher modes.*

code [Schmidt (1988)]. Previously, Staal *et al.* (1987) achieved reasonable agreement with a two-layer water/granite model, but the present model gives better agreement at the higher frequencies.

## 6. Conclusions

The study of interface waves at the seabed has been a small but important part of the shallow water acoustics research programme at DREA over the last decade. We have observed interface waves in the frequency range 2 Hz - 20 Hz over a variety of seabed types: sedimentary rock (chalk), unconsolidated sediments (sand and sand and gravel), and crystalline bedrock (granite). Although not the focus of our interest, the study of interface waves has improved our understanding of low frequency propagation. We have found that (a) the analysis of interface wave propagation is a good means of estimating seabed geo-acoustic parameters, including shear speed; and (b) over hard bedrock seabeds, interface waves in the frequency band 3 Hz - 6 Hz can be the dominant mode of propagation to ranges of 40 km or more.

## Acknowledgements

The authors thank Dave Heffler of the Bedford Institute of Oceanography (Dartmouth, N.S.) for the loan of the OBS and Jim Hunter and his team from the Terrain Geophysics Branch, Geological Survey of Canada (Ottawa, Ontario) for providing sources for the OBS experiments. We also thank Kevin Dunphy for his assistance on the OBS experiments during his term at DREA as a Co-op Student from Dalhousie University (Halifax, N.S.).

## References

Akal, Tuncay, and Jonathan M. Berkson (eds.) (1986) *Ocean Seismo-Acoustics*, Plenum Pub. Corp., New York, pp. 643-652.

Chapman, D.M.F., and Dale D. Ellis (1983) 'Geo-acoustic models for propagation modelling in shallow water', Canadian Acoustics **11**, 9-24.

Dunphy, Kevin (1982) 'Scholte Wave Velocity Observations', unpublished student work term report, DREA, December 1982.

Ellis, Dale D., and D.M.F. Chapman (1980) 'Propagation loss modelling on the Scotian Shelf: Comparison of model predictions with measurements', in [Kuperman and Jensen (1980)], pp. 541-555.

Ellis, Dale D., and D.M.F. Chapman (1985) 'A simple shallow water propagation model including shear wave effects', J. Acoust. Soc. Am. **78**, 2087-2095.

Hughes, Steven J., Dale D. Ellis, David M.F. Chapman, and Philip R. Staal (1990) 'Low-frequency acoustic propagation loss in shallow water over hard-rock seabeds covered by a thin layer of elastic-solid sediment', J. Acoust. Soc. Am. **88**, 283-297.

Kuperman, William A., and Finn B. Jensen (eds.) (1980) *Bottom-Interacting Ocean Acoustics*, Plenum Pub. Corp., New York.

Lara-Sáenz, A., C. Ranz-Guerra, and C. Carbó-Fité (eds.) (1987) *Acoustics and Ocean Bottom*, Consejo Superior de Investigaciones, Madrid.

Merklinger, H.M. (ed.) (1987) *Progress in Underwater Acoustics*, Plenum, New York.

Pace, N.G. (ed.) (1983) *Acoustics and the Sea-Bed*, Bath University Press, Bath, UK.

Purcell, Anthony J. (1986) 'Propagation of Scholte Waves and Normal Modes along a Rough Interface', DREA Contractor Report CR/86/408, April 1986.

Rauch, Dieter (1980) 'Experimental and theoretical studies of seismic interface waves in coastal waters', in [Kuperman and Jensen (1980)], pp. 307-327.

Schmidt, H. (1988) 'SAFARI: Seismo-Acoustic Fast field Algorithm for Range-Independent environments - User's Guide', SACLANTCEN REPORT SR-113.

Staal, P. (1983) 'Acoustic propagation measurements with a bottom-mounted array', in [Pace (1983)], pp. 289-296.

Staal, Philip R., and David M.F. Chapman (1986) 'Observations of interface waves and low-frequency shallow-water acoustic propagation over a rough granite seabed', in [Akal and Berkson (1986)], pp.643-652.

Staal, Philip R., David M.F. Chapman, and Pierre Zakarauskas (1987) 'The effect of variable roughness of a granite seabed on low-frequency shallow-water acoustic propagation', in [Merklinger (1987)], pp. 485-492.

Staal, P. R. (1987) 'Use and evolution of a modular digital hydrophone array', in *Proceedings of IEEE Oceans '87*, pp. 161-166.

# OBSERVATIONS OF THE RELATIVE CONTRIBUTIONS OF WATERBORNE AND SEDIMENT PATHS TO THE RECEIVED ACOUSTIC SIGNAL

HASSAN B. ALI and L. DALE BIBEE
*Naval Oceanographic and Atmospheric Research Laboratory*
*Stennis Space Center, MS. 39529-5004*
*USA*

ABSTRACT: Examples of the relative contributions of waterborne and sediment paths to acoustic propagation are presented using results obtained in a recent experiment conducted off the Oregon Coast by the Naval Oceanographic and Atmospheric Research Laboratory. The spectral energy contributions of the ground and waterborne paths to the received hydrophone signal are compared for several source-receiver separations, water depths, and frequency bands. The ground path is shown to increase in significance with decreasing frequency and, in some situations, to be as efficient a propagation path as the high-frequency waterborne path.

## Introduction

The propagation of sound in the sea has been of interest for at least five decades. The first significant underwater acoustic measurements to be documented were probably the shallow-water experiments conducted by Worzel and Ewing [1] in late 1943 and early 1944; their results were placed in a theoretical framework by Pekeris [2] in his pioneering work on acoustic waveguides. Contributions to the more general problem of the propagation of elastic waves in layered media have been made by several investigators. The classic works by Ewing, Jardetzky, and Press [3] and Brekhovskikh [4] are especially noteworthy. In recent years, a great deal of emphasis has been placed on numerical implementations of the classical theory of seismo-acoustic propagation. As a result, several useful numerical models are available to handle particular problems.

These early investigators were keenly aware of the partitioning of energy between the waterborne path and the seafloor ("ground") path, particularly at low frequencies. In spite of this, and the large number of experiments conducted since, both by acousticians and seismologists, partitioning remains poorly quantified. In part this resulted from limitations, particularly in dynamic range, of earlier recording instruments. Partitioning received little attention by either group because it belonged to that nebulous region lying between their interests. Seismologists concentrated mainly on deep-earth structures, for which waterborne signals are of relatively little value, while underwater acousticians have until recently emphasized frequencies for which bottom penetration is less significant, and narrow band data, from which it is difficult to isolate subseafloor paths. In recent years, the interest in low frequency and very low frequency (VLF, < 20 Hz) acoustic propagation has focused attention on energy partitioning between waterborne and subseafloor paths. At low frequencies (particularly VLF), the relatively small sediment attenuation allows



*J. M. Hovem et al. (eds.), Shear Waves in Marine Sediments*, 185–193.

significant penetration of acoustic energy into the seafloor, thereby necessitating inclusion of the effects of the seafloor on the propagation of the energy.

In this paper a first order empirical measure of spectral energy distribution is proposed which attempts to compare the relative contribution to the received signal in the water column of deeply penetrating energy with energy propagating through the water column and shallow sediments. Such a measure is useful, as it allows one to evaluate the importance of the deep-earth structure in predictions of VLF propagation. This parameter, which is described fully in the following section, is expected to vary with details of the environment and experimental geometry, but to be sufficiently robust to be characterized by general environmental parameters such as water depth, sediment thickness, and/or stratigraphic and tectonic setting. Based on this measure, the paper provides examples of energy partitioning observed in data obtained from recent propagation measurements conducted by the Naval Oceanographic and Atmospheric Research Laboratory (NOARL).

### Measure of the Relative Contributions of the Propagation Paths

To clearly establish the basis for the measure of partitioning used in this paper, a typical response to an underwater explosion ("shot") is discussed. Figure 1 presents the measured response of a water column hydrophone to a 10 km-distant shot. The length of the record shown here is approximately 6 seconds. Characteristic features of the response are labelled. The low-frequency arrival preceding the waterborne wave—i.e., the ground wave—represents energy travelling at a speed greater than that of the waterborne wave; hence, this is energy propagating in the bottom

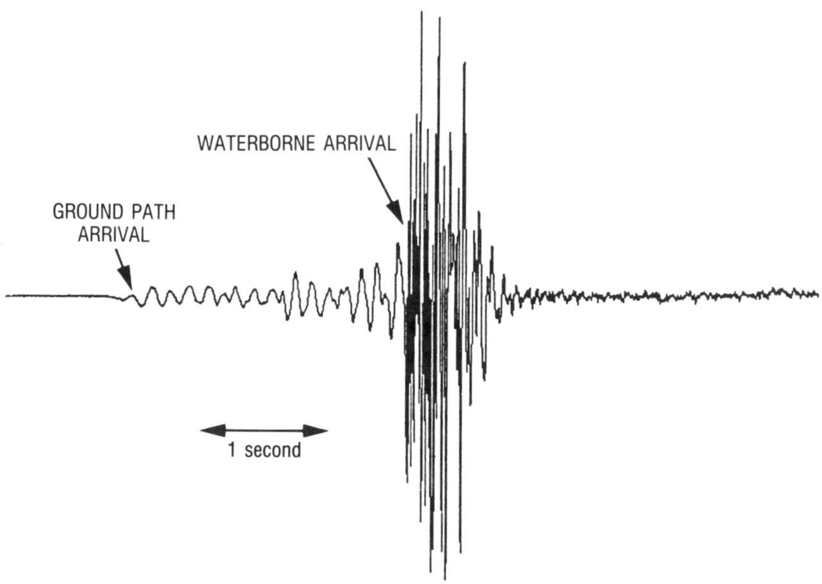


*Figure 1. Hydrophone response to a 10-km distant explosion (55-pound TNT), showing ground wave and water path arrivals.*

for a large part of its travel path. The frequencies contained in the ground wave are generally very low (here predominantly less than 10 Hz). The higher frequency components of the signal have been attenuated during deep penetration into the bottom layers. Generally, the high-speed ground waves will be dominated by compressional body waves and shear body waves converted to and from compressional body waves at the seafloor or subbottom layers.

Hence, analysis of the ground wave can provide information on the geoacoustic properties of the bottom and subbottom strata. In most of the shot records the ground wave appears to consist of several (two or three) packets of energy commonly referred to as distinct arrivals or seismic phases. This is often indicative of several bottom strata of different geoacoustic properties [5]. However, a simple correspondence of phases with subbottom layers is not necessarily correct since multipath propagation will typically generate more phases than layers. Based on vertical component geophone data obtained on a nearby ocean bottom seismometer (OBS), we believe the early part of the ground wave arrival in Figure 1 is energy returning from below acoustic basement. Acoustic basement in this area consists of a lithified sediment (clastic sedimentary rocks, consisting of sandstones, claystones, and siltstones) 1500 m below the seafloor. The later higher amplitude ground wave arrivals probably represent energy returning from depths shallower than this. By including all of these multiple phases under the category of ground path contribution, we claim to produce a parameterization which is more stable and less dependent on the details of the propagation path.

The arrival of the high-frequency waterborne acoustic wave is followed by a region in which waterborne and late-arriving ground waves interfere and eventually die away. These late-arriving ground waves can be attributed to multipath propagation in the slow-speed shallow sediments.

Based on the preceding, a reasonable first-order measure of deeply penetrating paths is provided by the total spectral energy in the time series comprising the high-speed, low-frequency ground waves preceding the waterborne arrivals. The spectral energy in the time series comprised of subsequent arrivals, beginning with the acoustic water wave, represents energy propagating predominantly in the water column and upper sediment layers of the sea bottom. Hence, the ratio of the two spectral energy terms is a useful parameter to assess the relative contributions of the two paths.

It is worthwhile pointing out that we are *not* considering the problem of partitioning of physical energy at the seafloor into the various modes of propagation. To solve that problem, an array of receivers that defines the full wavefield on both sides of the interface would be required. We are postulating a first-order measure of the relative contributions of deep-earth structures and waterborne/shallow-sediment paths to the received acoustic signal at a point in the field. The quantitative measure of the contributions is based on the *spectral energy* [the square of the modulus of the Fourier transform; units: $(Pa/Hz)^2$] of the corresponding time series.

## Experimental Arrangement/Data Acquisition

The results discussed below were obtained in a VLF experiment conducted in 1987 off the coast of Oregon. The test arrangement is shown schematically in Figure 2. A 16-element vertical hydrophone string was deployed in water depths of 183 m and 373 m. In addition, several OBS were deployed at various locations on the seafloor. Using explosives, air guns, and a cw sound source (fundamental frequency, 15 Hz), several upslope and cross-slope lines were completed. The 16 hydrophone outputs were multiplexed and telemetered via a UHF link to a nearby receiving ship. The received signal was then passed to an expanded video cassette recorder (EVCR)

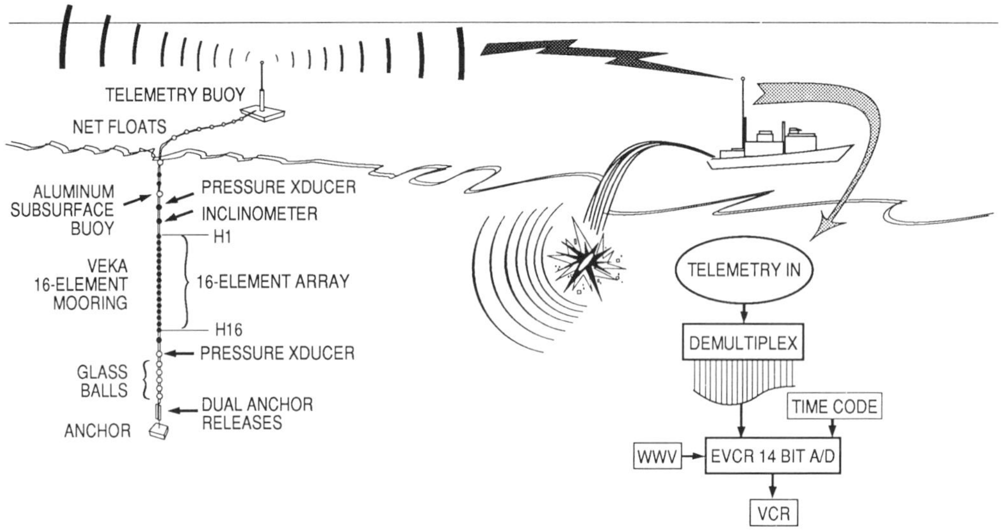


*Figure 2. Test arrangement and hydrophone data acquisition system of the VLF87 experiment.*

and converted to a VHS format video signal. The video signal was written on video tape with a standard VCR. Timing information in the form of IRIG-B time code was also recorded on the videotape. A common time base was provided by the WWV radio broadcast. The WWV signal was fed into the EVCR and digitized, multiplexed and recorded on video tape along with the acoustic data and IRIG-B signal. Subsequently (in the laboratory) the digitized data stored on the VCR tapes were played back through the EVCR to obtain the original 16 analog channels. Following low-pass filtering (100 Hz), the analog signals were digitized on a MASSCOMP computer at a sampling rate of 400 Hz. The digitized data were then written as multiplexed integers onto a 6250 bpi 9-track tape.

In this paper we present the response of the top hydrophone (H1), located 293 m below the water surface and 80 m above the bottom, to shots dropped along a 120-km long line beginning in shallow water (190 m), passing the hydrophone string, and terminating in deeper water (2740 m). Figure 3 shows the bathymetry along the shot-line and the locations of the hydrophones (marked by "x") and several of the shots (marked by circles). The shots consisted of 55-pound TNT charges and were detonated at depths ranging from approximately 60 to 90 m.

## Frequency Partitioning of the Propagation Paths

Considered first is the role of frequency selectivity in determining the relative importance of the ground and water paths. Figure 4 shows the response of hydrophone 1 to a 10 km-distant shot. The unfiltered data are shown along with the signal bandpass filtered into three different frequency bands. The sediment refraction (ground path) and water path are readily distinguishable on the basis of frequency content. In particular, the increase in relative importance of the ground wave with lower frequencies is quite evident. High-frequency propagation in the ground path is

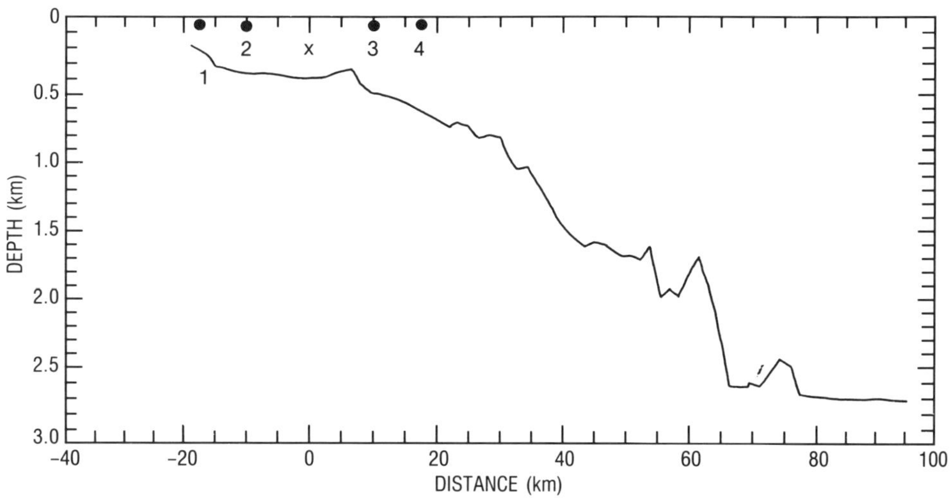


*Figure 3. Bathymetry along shot line 1.*

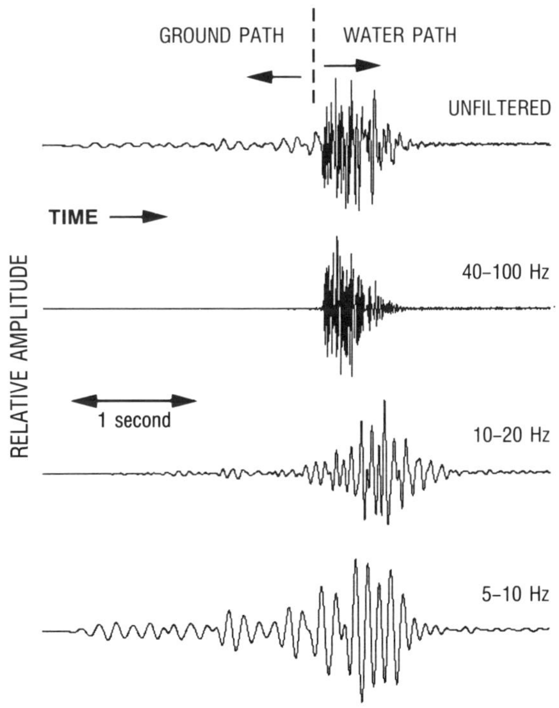


*Figure 4. Frequency selectivity of the propagation paths.*

significantly attenuated, but is clearly dominant in the water path. Above 40 Hz, for example, the ground wave is not even discernible in this particular signal.

### Range-Dependent Effects on the Relative Contributions of the Propagation Paths

Considered next is the spectral energy ratio as a function of source-receiver distance. It is expected, a priori, that the relative energy between the water and ground paths will be influenced by changing distance from the sound source in any realistic ocean environment. In addition to obvious geometrical spreading of the acoustic waves, changes in water depth result in modifications of the modal structure of the water column acoustic energy and therefore in the overall amplitude of water path phases. Varying sediment properties and bottom roughness influence the amplitude of the bottom interacting portion of the water path arrival. The amplitude of the ground path as a function of range is governed predominantly by the subbottom compressional (and to a lesser extent shear) velocity structure. The depths and magnitudes of major discontinuities between sedimentary layers and the velocity gradients within those layers exert strong controls on the focusing of energy emerging from the seafloor at different ranges. Scattering of energy by subbottom seismic velocity heterogeneities will result in redistribution of the emerging energy to different ranges and energy losses associated with scattering of energy out of the sagittal plane.

Examination of the range-dependence of the data reveals that the ground wave becomes undiscernible above background noise at ranges exceeding approximately 20 km on either side of the hydrophone—that is, independent of whether the source was in shallow water or deep water. The waterborne arrival, on the other hand, was clearly visible even at the maximum source distance (approximately 100 km). Figure 5 shows the ratio of spectral energy in the ground wave

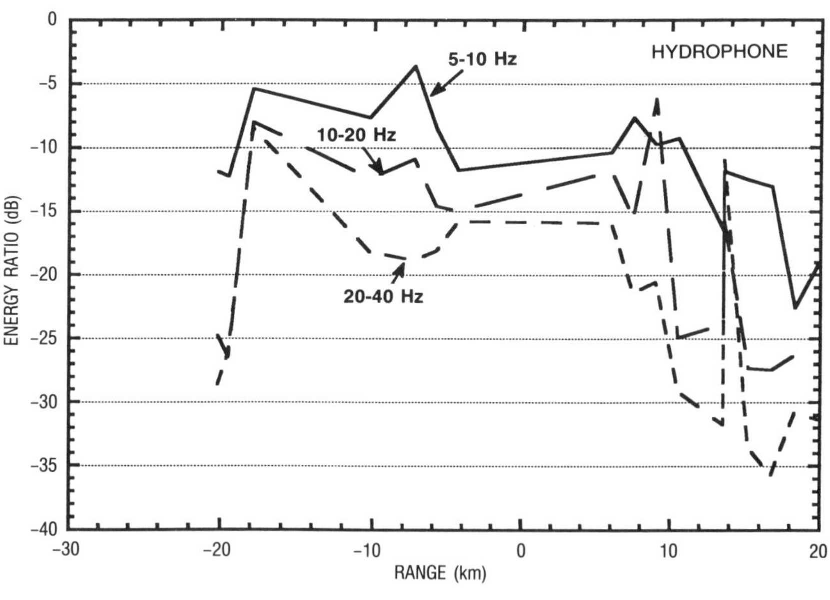


*Figure 5. Ratio of energy in the ground wave to energy in the waterborne path as a function of range and frequency.*

to energy in the waterborne path as a function of range, in several frequency bands. As expected, the lower frequency band contains more ground energy. Relative ground energy drops off both with increasing frequency and, in this case, with range. Referring to Figure 3, it is seen that the greater ranges correspond to deeper water and areas of increasing bottom slope. The peak occurring in shallow water, at approximately 7 km from the source (-7 km), may be due to energy returns from acoustic basement. This is strongly suggested by the record of seismograms at several ranges (record section) obtained from a nearby vertical geophone. The peak at +7 km may be associated with the obvious "abrupt" change in bathymetry at that location (Figure 3).

On a nearby vertical seismometer sensor, which is more sensitive to deeply penetrating compressional energy because of its vertical directivity, ground waves can be observed to at least 60 km. It is believed that scattering of ground wave energy from subbottom heterogeneities plays a major role in the ground wave energy losses in this environment. Based on sparker profiles performed during the experiment, and on published, interpreted seismic profiles [6, 7], it is clear that the Oregon continental margin is characterized by a complicated bottom and subbottom structure involving faulted and folded sedimentary strata. A similar experiment was conducted in 1985 in the more "benign" environment (passive margin) off Cape Fear, North Carolina. In that experiment, the ground wave arrivals from similar shots were very clear at distances of at least 60 km and were significantly larger than observed in this experiment [8]. The character of the ground waves was also different in the two areas: the signals were more impulsive in the East Coast data while they were more distributed in time with longer wave trains in the Oregon experiment. This observation is consistent with an increased scattering contribution in the Oregon data set that would tend to degrade long range propagation of the ground wave.

Figures 6, 7, and 8 show comparisons of the spectra of energy observed in ground wave and "waterborne" paths for several source locations. In particular, referring to Figure 3, the source was located at: point 4 (range, 18 km; water depth, 692 m) for Figure 6; point 2 (range, 10 km; water depth, 390 m) for Figure 7; and point 1 (range, 18 km; water depth, 250 m) for Figure 8. Therefore, by comparing the results in these three figures, one sees the effect of water depth (and range) on the relative contributions of waterborne and ground paths. With the source in deep water, one would expect relatively good waterborne propagation over a broad frequency range. With the source in shallow water, however, increased bottom interaction will remove energy from the water path. Figures 6-8 confirm this expectation. There is a significant decrease in the energy in the waterborne path between 692 m (Figure 6) and 250 m (Figure 8). At the lower frequencies, much of the waterborne loss in the shallow-water region may be due to conversion into ground wave energy. For example, at approximately 5 Hz the ground wave energy in shallow water (250 m, Figure 8) is almost 10 dB higher than the ground wave energy in deeper water (692 m, Figure 6), whereas the corresponding waterborne energy is approximately 10 dB lower in shallow water. However, without a detailed analysis of the path of the ground wave one cannot say, unequivocally, that the gain in ground path energy in this particular instance is due to the loss in the waterborne path. While the loss in waterborne energy at the higher frequencies is also likely to be caused by conversion into ground waves, scattering and anelastic attenuation within the seafloor prevent this energy from reemerging as an observed ground wave.

A general observation from all the figures, independent of range, is the relatively high levels of ground wave energy at the lower frequencies. The ground wave component at location 2 (Figure 7) is especially dominant. The results at location 3 (not shown here) show a similar high level for the ground wave. Considering only geometrical spreading, one *would* expect the levels of the ground wave at locations 2 and 3 (source-receiver ranges approximately 10 km) to be approximately 3 or 4 dB higher than those observed at locations 1 and 4 (source-receiver ranges

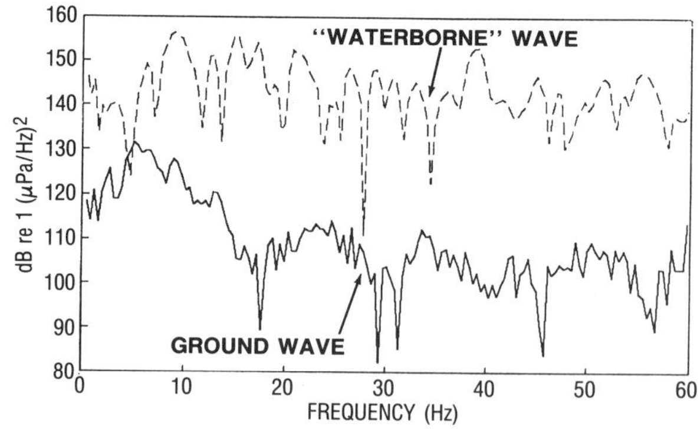


*Figure 6. Spectral energy partitioning: source-receiver separation, 18 km; water depth at source, 692 m.*

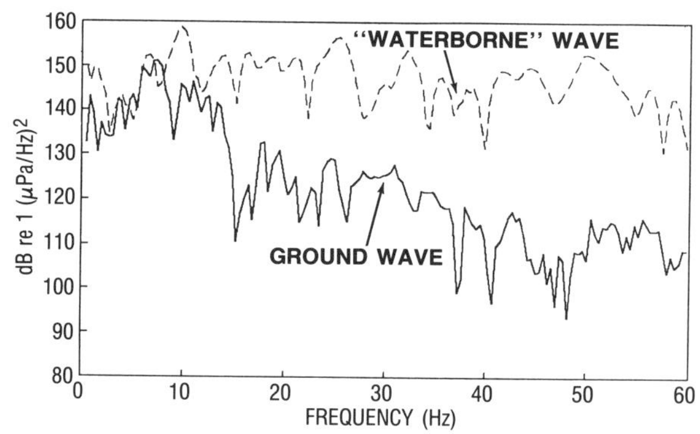


*Figure 7. Spectral energy partitioning: source-receiver separation, 10 km; water depth at source, 390 m.*

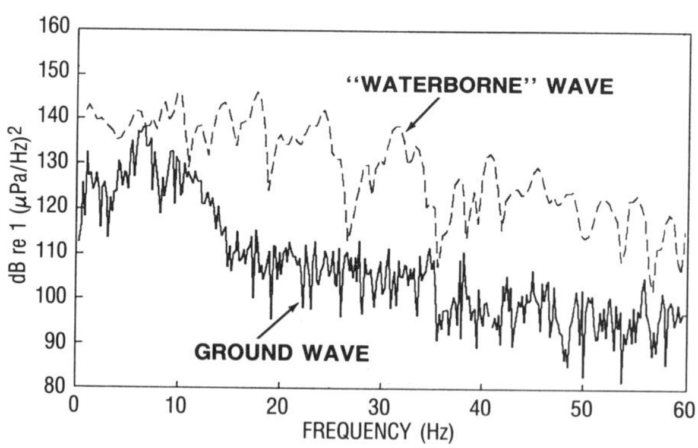


*Figure 8. Spectral energy partitioning: source-receiver separation, 18 km; water depth at source, 250 m.*

approximately 18 km). Clearly, however, the differences in ground wave responses at the two ranges are far too great to be attributable to spreading losses alone. It is noted that the ground wave response at –7 km (not shown here) is even more dominant than that at location 2. As pointed out in connection with Figure 5 , the increase in ground wave response at –7 km may be attributable to energy returns from acoustic basement.

## Conclusions

Based on a first order measure derived from the spectral energy content of a time series, examples of the relative contributions of sediment and waterborne paths have been presented using hydrophone data obtained in a recent NOARL experiment conducted off the Oregon Coast. Analysis of the results suggests the following conclusions for this data set:

1. On the hydrophone, the ground wave arrival was detectable at ranges up to 20 km from the source; the waterborne arrival was detectable to ranges in excess of 100 km.
2. The ground path is significant at VLF frequencies (< 20 Hz) and, in some instances, is as efficient as the high-frequency waterborne propagation.
3. The ground path contribution is enhanced if the sound source is located in shallow water and degraded for a deep-water source. The converse is true for the waterborne contribution.
4. The character of the ground wave arrivals suggests the presence of several subbottom layers of differing geoacoustic properties in the Oregon continental margin and a dominance of scattering losses from complicated subbottom structures. This is consistent with geologic surveys of the area.

## References


1. Worzel, J. L. and Ewing, M. (1948). 'Explosion sounds in shallow water', in Propagation of Sound in the Ocean, the Geological Society of America, Memoir 27, Boulder, Colorado.
2. Pekeris, C. L. (1948). 'Theory of propagation of explosive sound in shallow water', in Propagation of Sound in the Ocean, M. Ewing, J. L. Worzel, and C. L. Pekeris, the Geological Society of America, Memoir 27, Boulder, Colorado.
3. Ewing, W. M., Jardetzky W. S., and Press, F. (1957). Elastic Waves in Layered Media. McGraw-Hill, New York.
4. Brekhovskikh, L. M. (1960). Waves in Layered Media. Academic Press, New York.
5. Herring, C. 'Transmission of explosive sounds in the sea', in Physics of Sound in the Sea. Part 1: Transmission, Research Analysis Group, Committee on Undersea Warfare, National Research Council.
6. Snavely, P. D. Jr., Wagner, H. C., and Lander, D. L. (1980). 'Geologic cross section of the central Oregon continental margin', Geologic Society of America Map and Chart Series, MC-28J.
7. Kulm, L. D. and Fowler, G. A. (1974). 'Oregon continental margin structure and stratigraphy: A test of the imbricate thrust model', in The Geology of Continental Margins, C. A. Burke and C. L. Drake (eds.), Springer-Verlag, New York.
8. Ali, H. B., Bibee, L. D., Authement, M. J., Stephen, R. A., and Becklehimer, J. L. (1987). 'Low-frequency seismo-acoustic propagation in a sloping ocean environment: Measured results and numerical predictions', Naval Oceanographic and Atmospheric Research Laboratory, (formerly NORDA), Stennis Space Center, MS, NORDA Tech. Rep. 220.

# SHEAR-WAVE ANISOTROPY IN MARINE SEDIMENTS AROUND BRITAIN FROM SURFACE SOURCES.

COLIN MACBETH,
*Edinburgh Anisotropy Project,*
*British Geological Survey,*
*Edinburgh, UK. EH9 3LA.*

ABSTRACT. Land-based seismometer recordings of shallow underwater explosions in offshore areas exhibit, in some cases, distinctive long, slow (< 1.0 km/s), dispersed surface waves. These phases can be directly related to interfering modes of interface waves, with energy propagating within the poorly-consolidated, low velocity marine sediments. Rotated three-component seismogram data generated from underwater explosions in the Firth of Forth, the Clyde in Scotland, and the Strait of Dover, show significant energy on all three components of motion. The main features of this behaviour cannot be explained by anisotropic sediments exhibiting transverse isotropy with a vertical axis of symmetry. The recordings may be explained by an azimuthally anisotropic sediment layer, which results in a coupling of the Rayleigh and Love wave modes (except in a plane of symmetry). An alternative explanation may be lateral heterogeneities along the propagation path, such as a variation in the thickness of the sediment layer.

## 1. Introduction

This study is concerned with seismic data generated by underwater explosions around the British Isles, which display a remarkably well developed long (sometimes over 40 cycles), slow, surface wavetrain. These long coda are irregular in shape (which may be attributed to interfering modes), and appear to be strongly dispersed (group velocity dispersion of 1.8km/s/Hz). They have an apparent narrow bandwidth (1.4Hz), with a peak frequency of 1.2 Hz. The group velocity of these waves ranges typically from 0.4 to 1.5 km/s, which indicates that the wave energy must penetrate both the sea-bed sediments, and a portion of the underlying solid rockmass. These waves are similar to interface waves recorded on ocean bottom seismometers, which propagate along the sea floor from explosive charges (Rauch, 1986). The waves have similar character and duration at neighbouring seismometer locations, even when the seismometers are embedded in different geological environments. Consequently, it is unlikely that the observed wave character can be attributed to local, or site-dependent receiver effects. These long coda appear to require certain critical conditions for their generation. They are usually produced when underwater explosions are detonated



*J. M. Hovem et al. (eds.), Shear Waves in Marine Sediments,* 195–202.

within estuary regions, or in regions where there is sufficient thickness of unconsolidated sediment material to propagate the seismic energy at a particular frequency, and transfer the waves to land-based seismic stations. There is a critical distance from the explosion to the sea-land boundary, beyond which these waves will not propagate.

The dispersion of a selection of single-component recordings of these particular waves, assuming an isotropic earth structure, has been studied in an earlier paper by MacBeth and Burton (1988). It was concluded that the general character of the seismograms, although recorded on land, could be modelled by simple synthetic seismograms in which the dispersion of the waves was governed predominantly by the properties of a low wavespeed, isotropic sediment layer, and to a lesser extent by the surrounding water layer, and underlying solid rockmass. The waves thus appear to fit the accepted behaviour of interface waves commonly used to determine the shear properties of sediments (Jensen and Schmidt, 1986).

Previous studies have shown the value of using three-component recordings in the interpretation and examination of the sequence of modes constituting these interface waves (Snoek, Guidi, and Michelozzi, 1986; Schmalfeldt, 1986). However, these investigations concentrate mainly on the vertical and radial directions of motion. Here, the vertical, radial, and transverse directions of motion are interpreted to demonstrate the extra information which can be extracted about the nature of the propagation path by considering the polarization of the waves.

## 2. Character of observed three-component records

### 2.1 GENERAL FEATURES

Figures 1(a) to (d) show three-component seismograms generated by underwater explosions, illustrating the typical character of the records. The seismograms are taken from data collected on a UK seismic network, run by the British Geological Survey in Edinburgh (Browitt, Turbitt and Morgan, 1985). Table 1 gives details of the explosion origin times and locations, and the seismometer sites.

Figures 1(a) and (b) show two sets of three-component records from station EDI, in the LOWNET section of the UK array, situated in the Midland Valley region of Scotland. The recordings are of underwater explosions located in the Firth of Forth, 25km from this recording station. The coda waves have been detected on land-based seismometer stations up to 65km away. The explosions are detonated at the sea bottom, in 24m of water. They are a regular occurrence in this area, and provide many excellent examples (at least 50 since 1979) of the coda waves. It is interesting to note that although these events all appear to occur within a small epicentral area (2km$^2$), the wavetrains can differ considerably in appearance from shot to shot. This implies that a time variant process, such as tides or currents, is dominant in shaping the seismograms. The longest coda waves have been observed at station TFO, which records an underwater explosion from the Dover Straits (Figure 1(c)), at an epicentral distance of 24km. Much of the recorded waveform has group velocities of less than 1km/s. The propagation path in this case is almost entirely beneath the seafloor. Figure 1(d) shows another recording from the Midland Valley region, but this time the propagation path is on the west coast of Scotland. The underwater explosion is detonated in the Firth of Clyde, the waves propagating beneath the seafloor,

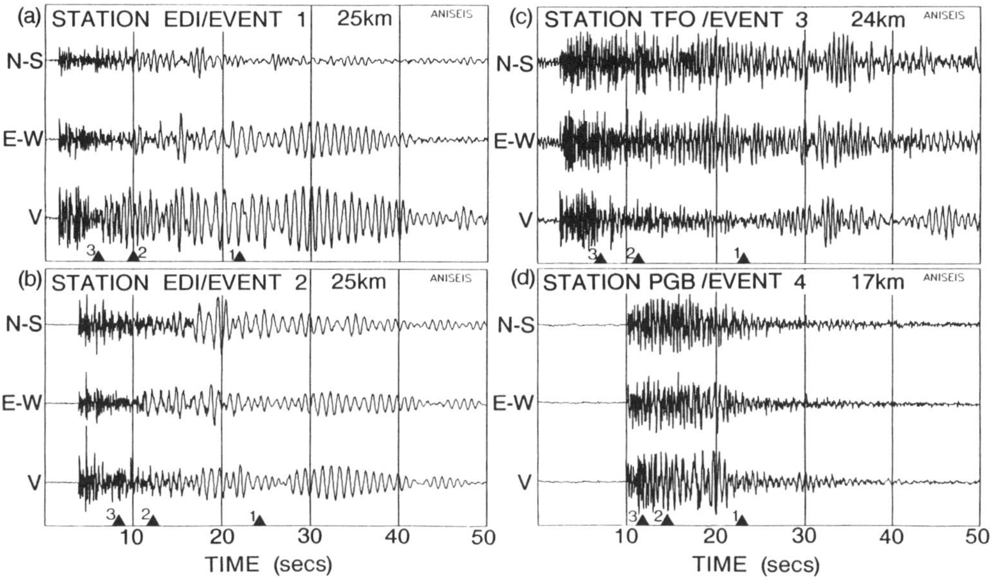


FIGURE 1. Selection of three-component recordings of underwater explosions around the UK. Triangles mark approximate wavespeeds of 1, 2 and 3km/s on the traces. The wavespeeds of the initial onset of the traces are: (a) 6.2 km/s; (b) 6.2km/s; (c) 6.2 km/s, and (d) 5.7km/s. Event numbers correspond to those in Table 1, from which relevant details of explosions can be found.

TABLE 1. Details of underwater explosions and seismic stations for which records are analysed.

| | location of explosion | lat. (°N) | long. (°W) | date | Time (GMT) | Station | Offset (km) |
|---|---|---|---|---|---|---|---|
| 1 | FIRTH OF FORTH | 56.1 | 3.0 | 08/2/79 | 12:10:18.4 | EDI | 25 |
| 2 | FIRTH OF FORTH | 56.1 | 3.0 | 16/7/90 | 12:59:57.6 | EDI | 25 |
| 3 | DOVER STRAITS | 51.0 | -1.4 | 04/5/90 | 10:15:19.4 | TFO | 24 |
| 4 | FIRTH OF CLYDE | 55.8 | 5.0 | 30/6/89 | 13:55:21.6 | PGB | 17 |

before making the transition into surface or interface waves on land. There is little coda wave energy below 1km/s. The number of available recordings of long-coda, low group velocity data around Britain are limited by the critical conditions for excitation and propagation, combined with the limited distribution of three-component seismometer stations, and their relatively recent deployment. Although other records of underwater explosions were found at other geographical locations, very few showed significant coda wave energy below a group velocity of 1km/s, and a character similar to those shown in Figures 1(a), (b) and (c).

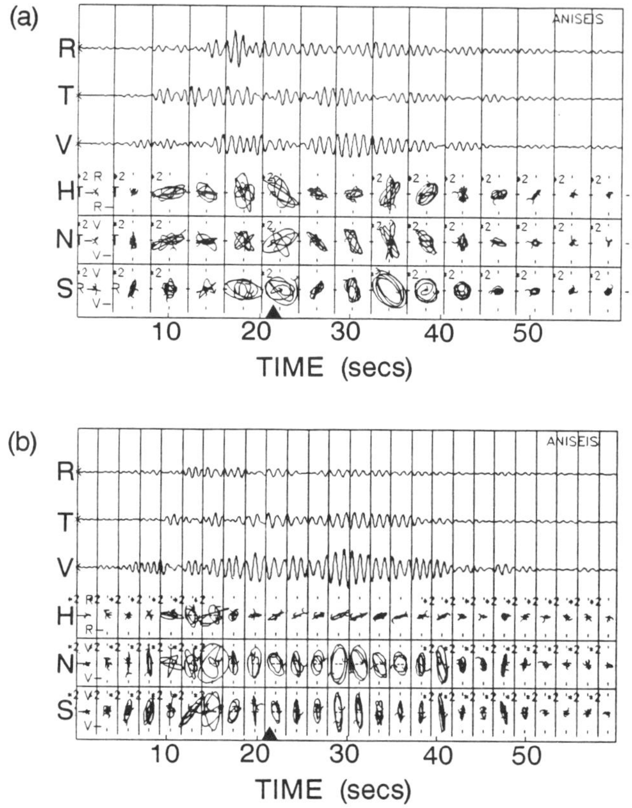


FIGURE 2. Observed particle motions projected into three orthogonal planes, for Events 1 and 2 (a and b) observed at station EDI. (Sagittal V-R, Normal V-T, Horizontal R-T). Observations are rotated and filtered about 1.2 Hz. The solid triangle marks 1km/s for reference.

## 2.2 OBSERVED POLARIZATION ANOMALIES

Figures 2(a) and (b) show three-component seismograms from events 1 and 2, rotated into radial and transverse directions, and band-pass filtered (0.8 - 1.6 Hz) with a Martin-Graham filter. The recordings show considerable energy in all three directions of the ground motion. The particle motions in the horizontal, sagittal, and normal planes display significant departures from pure Rayleigh or Love-wave behaviour (these departures are defined as polarization anomalies). The coda waves shown in Figure 2(a) cannot be attributed to a single mode of propagation, although it is possible to isolate a single mode over small 5s time-intervals. This is generally true of most of the recordings from other explosions detonated at this location, from 1978 to date. The recording in Figure 2(b) is unusual, as it consists of a single isolated mode with a distinctive particle motion.

## 3. Effects of anisotropy and heterogeneity on recorded waves

If the sediment and underlying layers are isotropic, one would expect coda waves generated from an explosive source to exhibit a motion which is purely Rayleigh-type, with no surface displacement in the transverse direction. One interpretation of the observed three-component displacement, is anisotropy in the near-surface sediments. However, marine sediments usually exhibit transverse isotropy with a vertical axis of symmetry (azimuthal isotropy) (Crampin, 1989). For this case the wave propagation would always be along a plane of mirror symmetry (Crampin, 1989), and Rayleigh and Love motions would be uncoupled. A possible departure from azimuthal isotropy, which would explain the observed effects, is the presence of aligned vertical parallel inclusions (Crampin and Atkinson, 1985) either in the sediments or just below. For waves propagating though material possessing this type of anisotropy, the radial, transverse, and vertical components of motion are coupled together. Instead of two families of distinct isotropic modes of propagation (Rayleigh and Love), these coalesce into one family of generalised normal modes. In general, the particle motion for these modes is no longer confined to the sagittal plane (as for Rayleigh waves) or the horizontal transverse direction (as for Love waves), but may take the form of an elliptical orbit, arbitrarily oriented in three-dimensions. Interface waves propagating along particular directions in an azimuthally anisotropic medium may have one (or more) of three distinctive particle motions, depending on the orientations of the plane or planes of anisotropic symmetry in the medium with respect to the direction of propagation (Crampin, 1975; Crampin, 1981). Figure 3 indicates the characteristic types. The particle motion may be either inclined (Figure 3(a)), tilted (Figure 3(b)), or sloping (Figure 3(c)), or a combination. Inclined motion refers to the case when the ellipse deviates from alignment with the great circle path. Tilting motion is defined by a rotation of the ellipse about an axis along the direction of propagation, and sloping motion describes a polarization ellipse rotated about the axis transverse to the direction of propagation (Crampin, 1975; Figure 3). The symmetry of the anisotropic system plays a crucial role in the nature of the recorded wave motion. The polarization patterns may provide information about the symmetry conditions, from which it is possible to estimate the orientation and type of anisotropic structure. For example, for the single mode in the coda waves of Figure 2(b), the radial and transverse directions of motion are in or out of phase with each other, and both have a definite phase shift (approximately 90°) with respect to the motion in the vertical direction. This particular type of polarization anomaly is inclined Rayleigh motion (as in Figure

3(a)), and hence indicates a horizontal plane of symmetry (Crampin, 1975; Kirkwood and Crampin, 1981). This suggests aligned vertical parallel inclusions, but does not exclude other more complicated anisotropic systems.

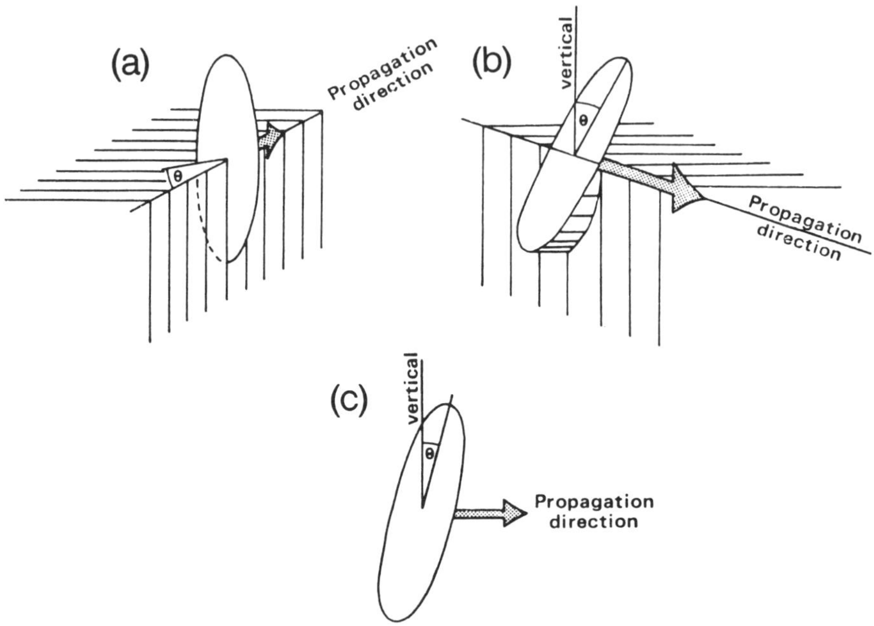


FIGURE 3. The three types of particle motions which can be exhibited by the generalized interface waves for anisotropic systems with different symmetry planes (after Crampin (1975)).

An alternative explanation to the behaviour of the coda waves may be scattering from lateral heterogeneities along the propagation path. Scattering causes a redistribution of energy between the individual modes representing the wavefield, the exact form of this interaction depending on the type, nature and scale of the heterogeneous region encountered by the waves. For transverse gradients in the seismic properties of the medium (such as a thinning of the sediment layer), there is a progressive net transfer of energy from the Rayleigh to the Love modes as the wavetrain propagates through the medium. The interaction may also occur at the source location, but would require a significant heterogeneity to generate the observed transverse displacement. As Rayleigh and Love waves arrive close together for structures dominated by a low velocity upper layer, then this effect may give the appearance of a single mode. A similar scattering effect is used to explain observations of Lg waves from underwater explosions by Kennett and Mykkletveit (1984), where there is a recorded displacement in the transverse direction as large as the vertical component, and an elliptical particle motion in the horizontal plane. Conversely, Maupin (1990) using similar data as these authors, has found that a small region of weak crustal anisotropy can also reproduce the observed effects. One indicator of the extent to which scattering plays a role is the limiting distance of propagation of the coda waves.

The resolution of the issue between anisotropy and heterogeneities requires the accurate observation of polarization patterns (particle motion in the three orthogonal planes of motion) over a range of azimuths. It may then be possible to detect the ordered change in the relative phase between the different directions of motion, which is diagnostic of some form of anisotropy, or some other variation related to a scattering process. Polarization anomalies interpreted in experiments like that of Snoek, Guidi, and Michelozzi (1986) would provide an indicator of the extent to which anisotropy contributes to the propagative character of these waves. For the present data it would be helpful to eliminate the effects of the sea-land boundary by using recordings from ocean bottom seismometers.

## 4. Conclusions

Land-based three-component recordings of slow, well-dispersed surface waves show significant energy on the transverse component of motion. The dispersion can be adequately explained by the properties of the low velocity sediment layer, the waves propagating with significant energy in the sediment layer. The three-component observations cannot be explained by isotropy or the usual azimuthal isotropy of marine sediments. One possible mechanism is azimuthal anisotropy caused by aligned vertical parallel inclusions in or below the sediment layer. Another explanation is scattering from lateral heterogeneities along the propagation path. More three-component observations at different azimuths are required to resolve the particular mechanism.

## Acknowledgements

I thank MACROC Ltd. and Applied Geophysical Software Inc. for approval to use the ANISEIS software. This work was supported by the Natural Environment Research Council, and is published with the approval of the director of the British Geological Survey (NERC). I would also like to thank David Greenwood of the British Geological Survey for his assistance in collecting, and preparing the data in this study.

## References

Browitt, C.W.A., Turbitt, T. and Morgan, S.N. (1985) 'Investigation of British earthquakes using the national monitoring network of the British

Geological Survey', in Earthquake Engineering in Britain, Thomas Telford, London, pp33-47.

Crampin, S. (1975) 'Distinctive particle motion of surface-waves as a diagnostic of anisotropic layering', Geophys.J.R.astr.Soc., 40, 177-186.

Crampin, S. (1981) 'A review of wave motion in anisotropic and cracked elastic media', Wave Motion 3, 343-391.

Crampin, S. and Atkinson, B.K. (1985) 'Microcracks in the Earth's crust', First Break 3, 3, 16-20.

Crampin, S. (1989) 'Suggestions for a consistent terminology for seismic anisotropy', Geophysical Prospecting 37, 753-770.

Jensen, F.B. and Schmidt, H. (1986) 'Shear properties of ocean sediments determined from numerical modelling of Scholte wave data, in T.Akal and J.M. Berkson (Eds), Ocean seismo-acoustics, NATO Conference Series IV:16, Plenum Press, New York and London, pp. 683-692.

Kennett, B.L.N. (1984) 'Guided wave propagation in laterally varying media - 1. Theoretical development', Geophys.J.R.astr.Soc., 79, 235-255.

Kennett, B.L.N., and Mykkeltveit (1984) 'Guided wave propagation in laterally varying media - II. Lg waves in north-western Europe', Geophys.J.R.astr.Soc., 79, 257-267.

Kirkwood, S.C., and Crampin, S. (1981) 'Surface-wave propagation in an ocean basin with an anisotropic upper mantle: numerical modelling', Geophys.J.R.astr.Soc., 64, 463-485.

MacBeth, C. and Burton, P.W. (1988) 'Surface-waves generated by underwater explosions offshore Scotland', Geophys.J.R.astr.Soc., 94, 285-294.

Maupin, V. (1990) 'Modelling of three-component Lg waves in anisotropic crustal models', Bull.Seism.Soc.Am., 80, 1311-1325.

Rauch, D. (1986) 'On the role of bottom interface waves in ocean seismo-acoustics:a review', in T.Akal and J.M. Berkson (Eds), Ocean seismo-acoustics, NATO Conference Series IV:16, Plenum Press, New York and London, pp. 623-641.

Schmalfeldt, B. (1986) 'A comparison of seismic and hydroacoustic measurements at very low frequencies in different shallow water areas, 653-662 in T.Akal and J.M. Berkson (Eds), Ocean seismo-acoustics, NATO Conference Series IV:16, Plenum Press, New York and London, pp. 653-662.

Snoek, M., Guidi, G., and Michelozzi E. (1986) 'Interface wave studies on the ligurian shelf using an OBS array:Experimental results and propagation models', in T.Akal and J.M. Berkson (Eds), Ocean seismo-acoustics, NATO Conference Series IV:16, Plenum Press, New York and London, pp. 663-672.

# REFRACTION MEASUREMENT OF SHEAR WAVE ANISOTROPY IN SHALLOW MARINE SEDIMENTS AND IMPLICATIONS FOR REFLECTION PROCESSING

PATRICIA A. BERGE, SUBHASHIS MALLICK, GERARD J. FRYER
*School of Ocean & Earth Science and Technology*
*University of Hawaii at Manoa*
*Honolulu, HI 96822, USA*
NOËL BARSTOW, JERRY A. CARTER, GEORGE H. SUTTON
*Rondout Associates, Incorporated*
*P. O. Box 224*
*Stone Ridge, NY 12484, USA*
JOHN I. EWING
*Woods Hole Oceanographic Institution*
*Woods Hole, MA 02543, USA*

ABSTRACT. In 1986, an on-bottom shear wave source and a multi-component on-bottom receiver with a hydrophone were used by Woods Hole Oceanographic Institution and Rondout Associates, Inc. to record a short ($< 200$ m) refraction profile near the United States Geological Survey's AMCOR drill hole 6011, off the coast of New Jersey. Synthetic seismograms match amplitudes and travel times for all components only if a transversely isotropic model is used. The data require a ~10-m-thick layer with ~16% shear wave anisotropy, overlying a 40-m-thick isotropic region. This model agrees with drill core data showing 10 m of silty clay overlying 40 m of sand. If a conventional reflection profile were recorded, the interval velocity would be overestimated by about 84% for $SV$-waves in the silty clay.

## 1. Introduction

Marine sediment cores routinely show anisotropy [1] resulting primarily from bedding, the interleaving of isotropic layers much thinner than a seismic wavelength [2]. A lamellate medium exhibits transverse isotropy with an axis of symmetry orthogonal to the layering [3, 4]. Describing the compressional and shear wave velocities in a transversely isotropic (TI) medium requires five independent elastic parameters [5], determined by recording compressional waves, and unconverted shear waves in two directions. Conventional marine seismic experiments may detect indications of anisotropy, but quantifying the anisotropy completely is impossible unless unconverted shear waves are recorded.

Recently, it has become technologically feasible for marine experiments to record unconverted shear waves by using multi-component ocean-bottom receivers and ocean-bottom sources [6, 7, 8, 9]. Berge et al. [10] have described the successful *in situ* measurement of transverse isotropy in shallow marine sediments, using an ocean-bottom shear source. In this paper, we briefly describe their key results and discuss implications for reflection seismology measurements recorded in anisotropic sediments.



*J. M. Hovem et al. (eds.), Shear Waves in Marine Sediments*, 203–212.

## 2. Background

In 1986 Rondout Associates, Inc. (RAI) and Woods Hole Oceanographic Institution (WHOI) used an ocean-bottom source and an ocean-bottom sensor to investigate the detailed seismic velocity and attenuation structure of shallow marine sediments off the coast of New Jersey [6, 7]. The shear wave source sled held two horizontally mounted 10-cubic-inch airguns that fired in opposing directions. The airgun blasts produced *P-SV* energy peaking at 10 Hz, while *SH* energy of about 20-30 Hz was transmitted into the sediments through the sled runners. (*P* designates elastic body waves with particle motion parallel to the direction of wave propagation in an isotropic medium. *SV* waves have particle motion in the sagittal plane and orthogonal to the direction of propagation. *SH* wave particle motion is orthogonal to the sagittal plane and to the direction of propagation). Three orthogonal geophones and a hydrophone were all mounted on a receiver toboggan. Record sections for radial horizontal (X), transverse horizontal (Y), and vertical (Z) motion were constructed by summing or differencing the airgun signals recorded on the geophones [6, 7].

A 200-m-long refraction profile and a shorter cross line were recorded in 21-m-deep water near the Atlantic Margin Coring Project (AMCOR) hole number 6011 (Fig. 1).

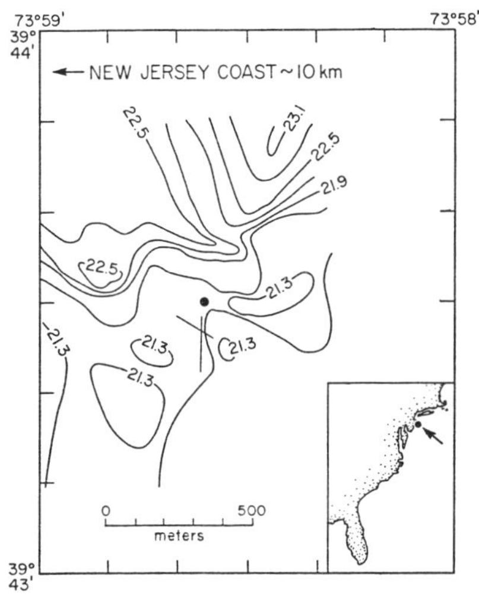


Figure 1. Location of the 1986 RAI/WHOI seismic experiment and nearby drill hole. Arrow and small dot in inset show map location. Large dot indicates AMCOR 6011 drill site [11]. Straight lines indicate seismic refraction profiles [6, 7, 10]. Only the north-south line is discussed in this paper. Curves show bathymetry in meters. Contour interval 0.305 m (1 foot).

Average receiver spacing was about 8 m. The seismic energy in the experiment sampled the sediments to a maximum depth of about 50 m. Drill cores collected at the 6011 site during the U. S. Geological Survey's AMCOR project [11] showed about 10 m of silty clay overlying about 40 m of sand. Geotechnical studies provided rough estimates of ~57% for the porosity and 2100 kg/m$^3$ for the density of the shallow sediments [12]. Silty clay is expected to be TI because of compositional layering at the grain scale [2].

Unfiltered multi-component record sections from the 200-m-long refraction profile recorded near the AMCOR 6011 site are shown in Fig. 2(a–d). Arrows indicate body wave arrivals, and interface waves in the pressure component data.

The pressure and vertical component data (Fig. 2c, d) show small amplitude *P* waves as first arrivals. On the pressure component, these arrivals are followed by Stoneley waves. The vertical component shows two sets of *SV* arrivals from fast deep and slower shallow

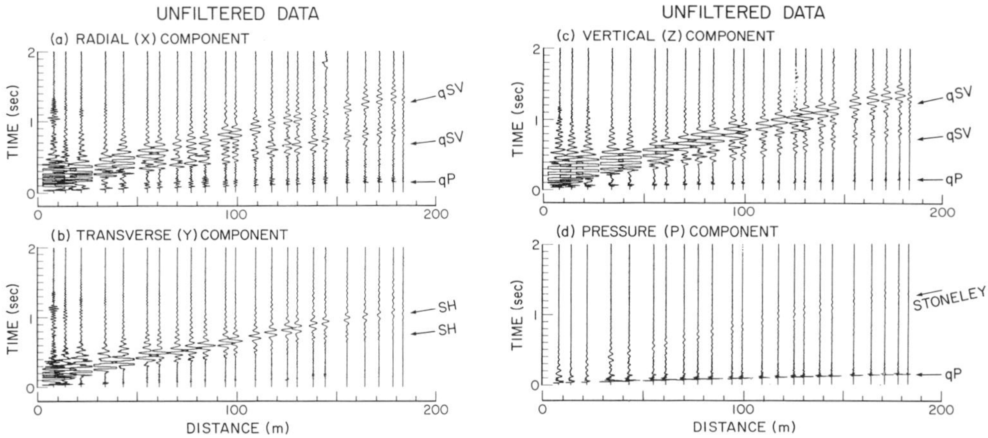


Figure 2. Unfiltered data record sections from seismic profile near AMCOR 6011 drill site. North on right; south on left. Times are unreduced; ranges in meters. (a) Radial horizontal (X) component. (b) Transverse horizontal (Y). (c) Vertical (Z). (d) Pressure (P).

layers, merging into Stoneley waves. The X component record section (Fig. 2a) is similar to the Z component record section. On the Y component record section (Fig. 2b), *SH* waves from deep and shallow layers merge into Love waves. A comparison of the X and Y component record sections shows that the second set of *SH* waves in the Y component data arrives earlier than the second set of *SV* waves in the X component data.

In attempts to model these data with isotropic models, Sutton et al. [6] and Carter et al. [7] could only match the data by invoking different structures for the X and Y components. Their modeling provided a starting point for anisotropic models [10].

## 3. Results and Discussion

### 3.1 ANISOTROPIC REFLECTIVITY MODELING

We expect silty clay to exhibit transverse isotropy with a vertical axis of symmetry. Only vertically travelling and horizontally travelling $P$ and $SV$ waves are pure-mode in such a medium. At other attitudes, particle motion is neither parallel to nor perpendicular to the direction of propagation. Hence, the waves are quasi-$P$ [13]. The $SH$ waves are pure-mode in all directions. In a TI medium, because of axial symmetry, the 21 independent elastic stiffnesses of general anisotropy reduce to five: $C_{11}, C_{13}, C_{33}, C_{44}$, and $C_{66}$ [5]. With this symmetry, the velocity of a horizontally travelling $SH$ wave is $\sqrt{C_{66}/\rho}$ where $\rho$ is density; the velocity of a horizontally travelling $qSV$ wave is $\sqrt{C_{44}/\rho}$ [14]. For transverse isotropy caused by thin layering, the stiffness $C_{66}$ must be larger than $C_{44}$ [4].

Our initial velocity model, based on the Sutton et al. [6] results, had a 10-m-thick TI layer overlying an isotropic half-space. A density of 2100 kg/m$^3$ was assumed. The code ANIVEC* [15] was run on a Cray supercomputer to produce synthetic seismograms to match all components of the data. Final synthetic seismograms were obtained after 153 iterative adjustments of the initial model [10].

---

*ANIVEC is a trademark of Geo-Pacific Corporation.

Table 1 presents the top 15 m of the final model Berge et al. [10] obtained for the 200-m-long profile recorded near the AMCOR 6011 drill site. Corresponding synthetics and filtered data record sections are displayed in Fig. 3(a–c). The modeling and synthetics were discussed in detail by Berge et al. [10]; we will summarize their main results below before discussing implications of shear wave anisotropy for seismic reflection data.

The X and Z component synthetic seismograms matched all significant phases in the data for traces further than one shear wavelength (~15–30 m) from the source (Fig. 3a, c). Closer to the source, the reflectivity program computed incorrect seismograms because of non-linear effects. A low value for $C_{44}$ in the top 10 m of the sediments in the model (Table 1), followed by a gradient and higher $C_{44}$ in the deeper part of the model, produced two sets of large amplitude $qSV$ arrivals merging into smaller amplitude Stoneley waves. Steep filter slopes produced the low frequency ringing artifact in the synthetics at short ranges. Because the profile was only about one $qP$ wavelength long, detailed modeling of the $qP$ arrivals was not attemped; however, the arrival times and amplitudes in the X and Z component synthetics matched the data fairly well for $qP$. A constant shear wave quality factor of $Q_s = 15$ was used in the top 10 m of the sediments. The deeper part of the model had $Q_s = 60$, and the compressional wave quality factor was $Q_p \approx \infty$. (The quality factor is proportional to the ratio of the peak strain energy stored to the energy lost in a cycle. A high quality factor denotes low attenuation, while a low quality factor implies high attenuation of seismic energy in the medium).

The synthetic seismograms for the Y component record section (Fig. 3b) showed small amplitude refracted $SH$ waves as the first arrivals. These were followed by larger amplitude $SH$ arrivals from the shallowest part of the model, merging into Love waves. The Y component data required a gradient in $C_{66}$ at the top of the model, and $C_{66} > C_{44}$ for the top 10 m of the sediments (Table 1).

Other non-unique models would also fit the data, but the data require that all such models must exhibit transverse isotropy, because the second set of $SH$ waves seen in the Y component data have shorter travel times than the corresponding $qSV$ waves in the X and Z component data [10]. No single isotropic model could explain the observed data for the vertical and both horizontal components.

For the top 10 m of sediments, the model (Table 1) has $C_{66} > C_{44}$, consistent with requirements for transverse isotropy caused by thin layering [4], as expected for silty clay [2]. The slowness surfaces and ray velocity surfaces (wavefront shapes) for the top 10 m are shown in Fig. 4(a, b). For the deeper, isotropic layers of the model, slowness and ray velocity surfaces are spherical.

**Table 1.** TI model used to generate reflectivity synthetic seismograms [10]

| $C_{11}$ | $C_{13}$ | $C_{33}$ (GPa) | $C_{44}$ | $C_{66}$ | depth† (m) |
|---|---|---|---|---|---|
| 2.25 | 2.25 | 2.25 | 0.0 | 0.0 | 0.0–21.0 |
| 5.544 | 5.376 | 5.544 | 0.0526 | 0.0669 | 21.0–23.5 |
| 5.544 | 5.376 | 5.544 | 0.0526 | 0.0683 | 23.5–26.0 |
| 5.544 | 5.376 | 5.544 | 0.0526 | 0.0700 | 26.0–28.5 |
| 5.544 | 5.376 | 5.544 | 0.0526 | 0.0714 | 28.5–31.0 |
| 5.586 | 5.376 | 5.586 | 0.0792 | 0.0864 | 31.0–33.5 |
| 5.586 | 5.366 | 5.586 | 0.110 | 0.110 | 33.5–36.0 |

† Source and receiver on bottom at 21 m depth. All depths are in meters below sea level.

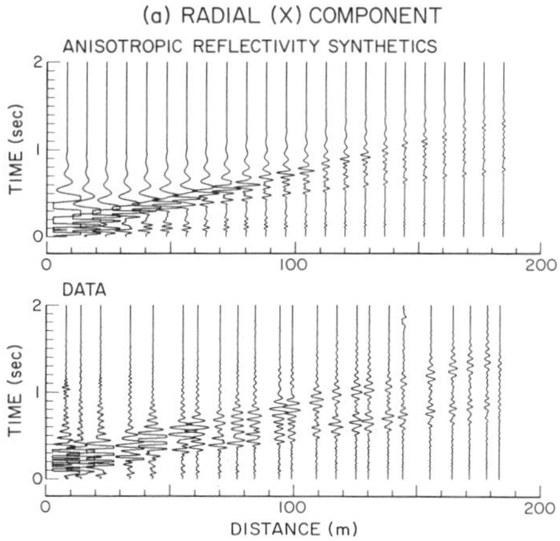


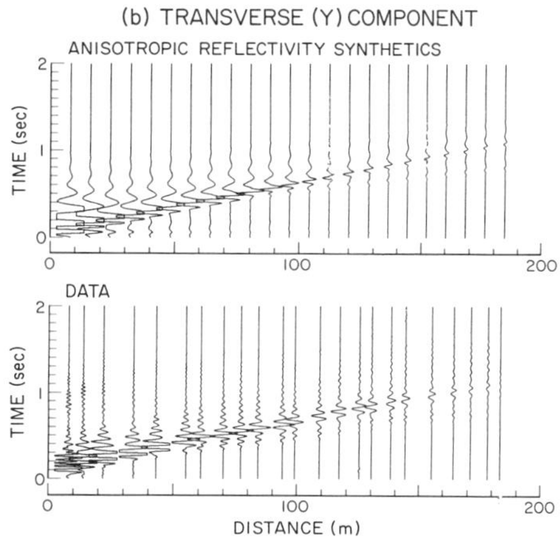


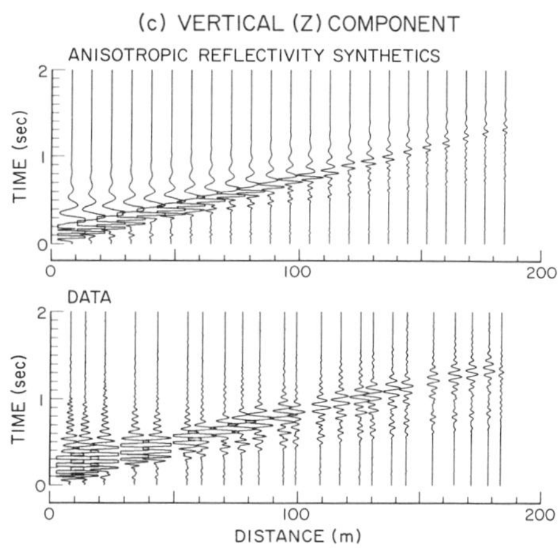


Figure 3. Anisotropic reflectivity synthetics compared to filtered data for the seismic profile near AMCOR 6011 drill site. Data and synthetics were filtered using a 4–25 Hz bandpass filter. Unreduced times in seconds, ranges in meters. (a) Synthetics (top) and data (bottom) for X component. (b) Y component. (c) Z component [10].

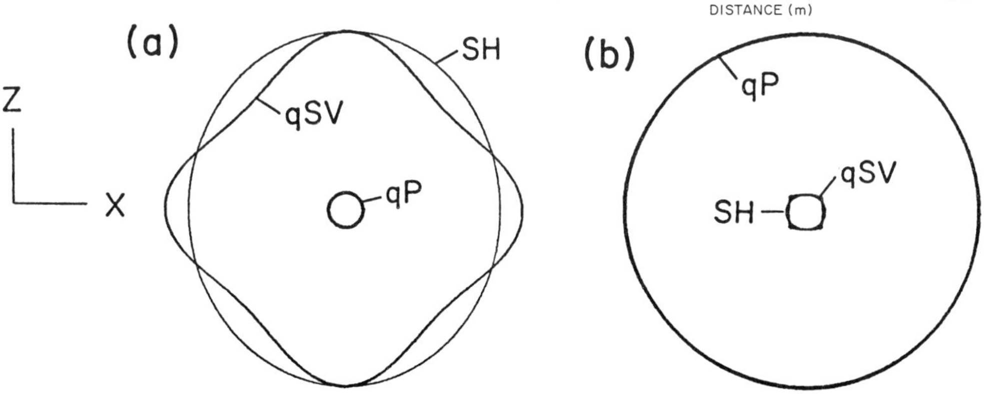


Figure 4. Cross-sections of phase slowness and ray velocity surfaces through $Y = 0$ plane. Since ray velocity is proportional to distance a wave travels in unit time, ray velocity surfaces show wavefront shapes. (a) Slowness surface, average for top 10 m of sediments for model in Table 1. (b) Ray velocity surface, average for top 10 m of sediments [10].

In general, for a single homogeneous layer, the travel time of a reflected ray is related to the layer thickness $h$, the source-receiver separation $x$, the ray velocity $V_{\mathrm{ray}}$, and the ray angle $\theta_{\mathrm{ray}}$ by

$$t^2(\theta_{\mathrm{ray}}) = \frac{4h^2}{V_{\mathrm{ray}}^2(\theta_{\mathrm{ray}})} + \frac{x^2}{V_{\mathrm{ray}}^2(\theta_{\mathrm{ray}})} \tag{1}$$

[16, 17, 18]. From equation (1), the slope of the $t^2$ versus $x^2$ curve is

$$\frac{d(t^2)}{d(x^2)} = \frac{1}{V_{\mathrm{ray}}^2} - \frac{t^2}{V_{\mathrm{ray}}^2}\frac{dV_{\mathrm{ray}}^2}{dx^2} = \frac{1}{V_{\mathrm{ray}}^2}\left[1 - \frac{2\cos^2\theta_{\mathrm{ray}}}{V_{\mathrm{ray}}}\frac{dV_{\mathrm{ray}}}{d\sin^2\theta_{\mathrm{ray}}}\right] \tag{2}$$

[18].

If the layer is isotropic, $V_{\mathrm{ray}}(\theta_{\mathrm{ray}})$ is constant and equation (1) is a straight line with slope $1/(V_{\mathrm{ray}}^2)$. In the isotropic case, for a $P$ or $S$ wave the normal moveout or stacking velocity $V_{\mathrm{stack}}$ equals the ray velocity $V_{\mathrm{ray}}$.

If the layer is TI with a vertical axis of symmetry, the *SH* wavefronts will have elliptical cross-sections in the sagittal plane, and equation (1) will be

$$t^2 = \frac{4h^2}{V_{SH_{\mathrm{vert}}}^2} + \frac{x^2}{V_{SH_{\mathrm{horiz}}}^2} \tag{3}$$

[16, 19, 20] where the constants

$$V_{SH_{\mathrm{vert}}} = \sqrt{C_{44}/\rho} \qquad \text{and} \qquad V_{SH_{\mathrm{horiz}}} = \sqrt{C_{66}/\rho} \tag{4}$$

[14] are the vertical and horizontal *SH* ray velocities.

Because $V_{SH_{\mathrm{vert}}}^2$ and $V_{SH_{\mathrm{horiz}}}^2$ are constants, equation (3) is a straight line with slope $1/\left(V_{SH_{\mathrm{horiz}}}^2\right)$. Thus the *SH* stacking velocity in a transversely isotropic layer is

$$V_{SH_{\mathrm{stack}}} = \sqrt{\frac{C_{66}}{\rho}} = V_{SH_{\mathrm{horiz}}} \tag{5}$$

for any ray angle $\theta_{\mathrm{ray}}$. Note that the *SH* stacking velocity is equal to the *SH horizontal* ray velocity, not the *SH vertical* ray velocity. Therefore the *SH* vertical ray velocity in the layer, sometimes called the "interval velocity," cannot be determined from the stacking velocity. If the *SH* vertical ray velocity were estimated by incorrectly assuming that it equaled the stacking velocity, an error of magnitude $\sqrt{(C_{66}/C_{44})}$ would result in the estimated value.

Although the *SH* waves in a transversely isotropic medium are elliptical, the $qP$ and $qSV$ waves are not. The $qSV$ waves in particular can have very complicated shapes that are not even approximately elliptical. The dependency of the ray velocities on the ray angles is not simple in the case of $qP$ or $qSV$ waves [13, 21, 22]. For either the $qP$ or the $qSV$ waves, equation (1) will be a curve with a different slope, given by equation (2), for every ray angle [19, 23]. Thus unlike the *SH* waves, the $qP$ and $qSV$ waves require different stacking velocities for every ray angle along the reflection spread. In practice, the curve given by equation (1) is fit with a straight line tangent at $\theta_{\mathrm{ray}} = 0°$, and the stacking velocity is determined from the slope of that line, i.e., from the slope of the curve near $\theta_{\mathrm{ray}} = 0°$, using the first term only for the right side of equation (2) [18]. Thus for short reflection spreads, the appropriate stacking velocities for ray angles near the vertical axis are

$$V_{P_{\rm stack}} = \sqrt{\frac{C_{44}}{\rho} + \frac{(C_{13}+C_{44})^2}{(C_{33}-C_{44})\rho}} = \sqrt{\frac{C_{33}}{\rho}}\,[1+2\delta]^{1/2}\,, \tag{6}$$

and

$$V_{SV_{\rm stack}} = \sqrt{\frac{C_{11}}{\rho} - \frac{(C_{13}+C_{44})^2}{(C_{33}-C_{44})\rho}} = \sqrt{\frac{C_{44}}{\rho}}\left[1+2\frac{C_{33}}{C_{44}}(\epsilon-\delta)\right]^{1/2}, \tag{7}$$

where $\epsilon$ and $\delta$ are functions of the elastic stiffnesses, as defined by Thomsen [18].

The $qSV$ stacking velocity in equation (7) differs from the $SH$ stacking velocity given in equation (5). Also, the $qP$ and $qSV$ stacking velocities in equations (6) and (7) are different from the respective vertical ray velocities

$$V_{P_{\rm vert}} = \sqrt{\frac{C_{33}}{\rho}} \qquad \text{and} \qquad V_{SV_{\rm vert}} = \sqrt{\frac{C_{44}}{\rho}} = V_{SH_{\rm vert}} \tag{8}$$

[14]. Incorrectly assuming that the vertical ray velocities equal the stacking velocities would lead to errors of magnitude $\sqrt{1+2\delta}$ and $\sqrt{1+2(C_{33}/C_{44})(\epsilon-\delta)}$ when estimating the $qP$ and $qSV$ ray velocities, respectively [24].

The $qP$ and $qSV$ stacking velocities in equations (6) and (7), determined at $\theta_{\rm ray} = 0°$, can only be used for short reflection spreads with small ray angles. Unless the $qP$ and $qSV$ wavefronts are nearly elliptical or spherical, different stacking velocities will be required at large ray angles. For some transversely isotropic media, equation (7) will not provide a useful stacking velocity even for small ray angles, because the $qSV$ wavefront shape and ray velocity vary too rapidly with $\theta_{\rm ray}$ at the vertical axis [25]. But studies of many known transversely isotropic media suggest that equations (6) and (7) can usually be applied for short reflection spreads recorded in sedimentary rocks for ray angles up to $\sim 40°$ [25, 26].

We can use equations (5), (6), and (7) to find the stacking velocities that would be used for a short reflection spread recorded in the silty clay layer at the AMCOR 6011 drill site. Comparing the stacking velocities to the ray velocities for the transversely isotropic silty clay provides an example of the magnitude of the differences between these velocities for common marine sediments. This illustrates the importance of distinguishing between stacking velocities and interval velocities in an anisotropic layer.

From Table 1, the elastic stiffnesses in the 10 m silty clay layer have values of $C_{11} = C_{33} = 5.544$ GPa, $C_{13} = 5.376$ GPa, $C_{44} = 0.0526$ GPa, and average $C_{66} \approx 0.0692$ GPa. Using $\rho = 2100\,\text{kg/m}^3$ [12] with these values for the elastic stiffnesses in equations (5), (6), and (7), we find that for the silty clay layer,

$$V_{P_{\rm stack}} = 1606 \text{ m/s}, \qquad V_{SV_{\rm stack}} = 291 \text{ m/s}, \qquad \text{and} \qquad V_{SH_{\rm stack}} = 181 \text{ m/s}. \tag{9}$$

From equation (8) we find that the vertical ray velocities are

$$V_{P_{\rm vert}} = 1625 \text{ m/s} \qquad \text{and} \qquad V_{SV_{\rm vert}} = 158 \text{ m/s} = V_{SH_{\rm vert}}. \tag{10}$$

In the silty clay layer, we have $\delta = -0.0113$ and $\epsilon = 0$, from Table 1 and Thomsen's [18] definitions of $\epsilon$ and $\delta$. It is not surprising to find that the stacking velocity for the $qP$ waves in the silty clay is almost the same as the vertical ray velocity, since the difference depends on $\sqrt{1+2\delta}$, and $\delta$ is very small for this material. But for the $qSV$ waves, the stacking velocity and the ray velocity differ by a factor of

$$\sqrt{1+2(C_{33}/C_{44})(\epsilon-\delta)} = 291/158 = 1.84. \tag{11}$$

Here the difference is large because $C_{33}/C_{44} \sim 100$. This ratio is large because the compressional wave velocity is nearly ten times the shear wave velocity for the shallow marine sediments, and the ratio of the elastic stiffnesses is proportional to the square of the ratio of the seismic velocities. The quantity calculated in equation (11) is not very sensitive to the particular velocity model we use, since any velocity model that fits the data must have compressional wave velocities about an order of magnitude greater than the shear wave velocities, for the shallowest sediments.

Equation (11) shows that the stacking velocity is 84% larger than the vertical ray velocity, for *qSV* waves in the silty clay. Clearly the stacking velocity for a seismic reflection profile is not a good approximation of the vertical ray velocity in this TI medium. Large errors result if the shear wave interval velocity is estimated using the *qSV* stacking velocity.

Fig. 5 shows cross-sections of the *qSV* and *SH* wavefronts for the silty clay layer from Table 1, along with the ellipse that would approximate the *qSV* wavefront if the stacking velocity given by equation (7) were used for seismic reflection data. The *qSV* stacking velocity is proportional to the horizontal axis of the ellipse [16, 19, 24]. Note that the *qSV* stacking velocity for the silty clay is larger than the horizontal *SH* ray velocity, which is, in turn, larger than the horizontal *qSV* ray velocity. If the *qSV stacking* velocity were confused with the horizontal *qSV ray* velocity, one could erroneously conclude that the horizontal *qSV* ray velocity was larger than the horizontal *SH* ray velocity, and that the anisotropy was not due to thin layering.

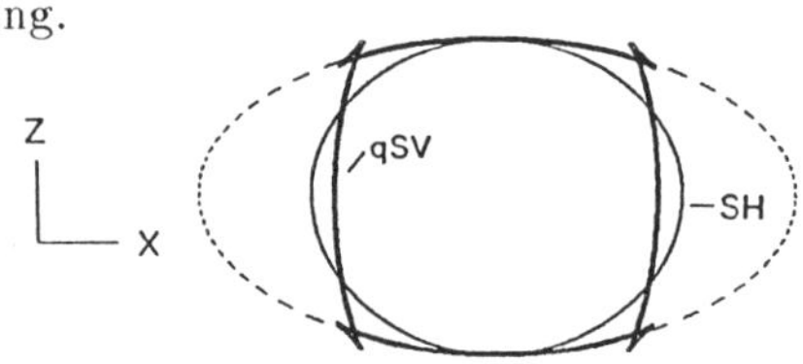


Figure 5. Cross-sections of shear wavefront shapes and stacking velocity ellipse approximation, through $Y = 0$ plane. Solid lines indicate *SH* and *qSV* ray velocity surfaces (wavefronts) for top 10 m of model in Table 1. Dashed line indicates ellipse approximation for *qSV* wavefront, found using *qSV* stacking velocity near vertical axis.

## 4. Conclusions

The simplest model that produces a good fit to the data from the refraction profile recorded near the AMCOR 6011 drill site requires an anisotropic layer for the top 10 m of the sediments [10]. This layer corresponds to the silty clay found in the drill hole [11]. Using

$$k = \frac{V_{\max} - V_{\min}}{V_{\text{mean}}} \tag{12}$$

[27, 28, 29], the shear wave anisotropy in the top 10 m of the sediments is 12 – 15% [10].

Ignoring anisotropy in marine data may lead to systematic errors in interpretation, including overestimates of depth and errors in Poisson's ratio [30, 31], incorrect estimates of interval velocities from stacking velocities, and disagreement between seismic reflection and seismic refraction models. The shear waves will show these errors more than the compressional waves. The anisotropy can be exploited, however, to yield information about the geologic structure, since the type of anisotropy often indicates its cause.

In a TI medium, an isotropic model can always be found to fit the travel times of the compressional waves [24, 32]. Additional information such as shear waves, polarities, amplitudes, or layer depths is required to detect the anisotropy [24]. Experiments using hydrophones alone cannot detect anisotropy in TI sediments. Future experiments to measure anisotropy in marine sediments and the oceanic crust must use on-bottom shear sources and multi-component on-bottom receivers, to record unconverted shear waves. Combined seismic reflection and refraction experiments would provide better control for estimating all five elastic stiffnesses for TI sediments.

## 5. Acknowledgments

This work was supported by the Office of Naval Research through contracts with the University of Hawaii, with Rondout Associates, Inc., and with Woods Hole Oceanographic Institution. The NATO SACLANTCEN provided a travel grant so that the paper could be presented at the Symposium on Shear Waves in Marine Sediments. School of Ocean and Earth Science and Technology contribution number 2354.

## 6. References

1 Bachman, R. T. (1983) 'Elastic anisotropy in marine sedimentary rocks,' J. Geophys. Res. 88, 539–545.

2 Carlson, R. L., Schaftenaar, C. H. and Moore, R. P. (1984) 'Causes of compressional-wave anisotropy in carbonate-bearing, deep-sea sediments', Geophysics 49, 525–532.

3 Postma, G. W. (1955) 'Wave propagation in a stratified medium', Geophysics 20, 780–806.

4 Backus, G. E. (1962) 'Long-wave elastic anisotropy produced by horizontal layering', J. Geophys. Res. 67, 4427–4440.

5 Love, A. E. H. (1920) A Treatise on the Mathematical Theory of Elasticity, 3rd ed., Cambridge University Press, Cambridge.

6 Sutton, G. H., Ewing, J. I., Dubois, D. L., Barstow, N. and Carter, J. A. (1987) 'Shear velocity/attenuation models for shallow marine sediments: offshore New Jersey', Eos, Trans. Am. Geophys. Union 68, 354.

7 Carter, J. A., Sutton, G. H., Barstow, N. and Ewing, J. I. (1987) 'Synthetic modeling of shear waves in shallow marine sediments', Journal of the Acoustical Society of America 82, suppl. 1, S112.

8 Akal, T., Schmidt, H. and Curzi, P. (1986) 'The use of Love waves to determine the geoacoustic properties of marine sediments', in T. Akal and J. M. Berkson (eds.), Ocean Seismo-Acoustics, Low-Frequency Underwater Acoustics, NATO Conference Series IV: Marine Sciences 16, Plenum Press, New York, pp. 841–852.

9 Gehrmann, T., Gimpel, P. and Theilen, F. (1984) 'Marine shear wave profiling', 54th Annual International Meeting, Society of Exploration Geophysicists, Abstracts and Biographies, 592–594.

10 Berge, P. A., Mallick, S., Fryer, G. J., Barstow, N., Carter, J. A., Sutton, G. H. and Ewing, J. I. (1991) '*In situ* measurement of transverse isotropy in shallow-water marine sediments', Geophys. J. Int. 104, 241–254.

11 Hathaway, J. C., Schlee, J. S., Poag, C. W., Valentine, P. C., Weed, E. G. A., Bothner, M. H., Kohout, F. A., Manheim, F. T., Schoen, R., Miller, R. E. and Schultz, D. M. (1976) 'Preliminary summary of the 1976 Atlantic Margin Coring Project of the US Geological Survey', U. S. Department of the Interior Geological Survey Open File Report No. 76–844.

12 Richards, A. F. (1978) 'Atlantic Margin Coring Project 1976. Preliminary report on shipboard and some laboratory geotechical data', U. S. Department of the Interior Geological Survey Open File Report No. 78–123.
13 Auld, B. A. (1973) Acoustic Fields and Waves in Solids, Vol. 1 and 2, John Wiley and Sons, New York.
14 Stoneley, R. (1949) 'The seismological implications of aelotropy in continental structure', Monthly Notices of the Royal Astronomical Society, Geophysical Supplement 5, 343–353.
15 Mallick, S. and Frazer, L. N. (1990) 'Computation of synthetic seismograms for stratified azimuthally anisotropic media', J. Geophys. Res. 95, 8513–8526.
16 Levin, F. K. (1978) 'The reflection, refraction, and diffraction of waves in media with an elliptical velocity dependence', Geophysics 43, 528–537.
17 Byun, B. S. (1984a) 'The characteristics of reflection traveltimes for some layer models', Geophysics 49, 119–123.
18 Thomsen, L. (1986) 'Weak elastic anisotropy', Geophysics 51, 1954–1966.
19 Levin, F. K. (1979) 'Seismic velocities in transversely isotropic media', Geophysics 44, 918–936.
20 Byun, B. S. (1982) 'Seismic parameters for media with elliptical velocity dependencies', Geophysics 47, 1621–1626.
21 Kraut, E. A. (1963) 'Advances in the theory of anisotropic elastic wave propagation', Rev. Geophys. 1, 401–448.
22 Musgrave, M. J. P. (1970) Crystal Acoustics, Introduction to the Study of Elastic Waves and Vibrations in Crystals, Holden-Day Inc., San Francisco.
23 Byun, B. S. (1984b) 'Seismic parameters for transversely isotropic media', Geophysics 49, 1908–1914.
24 Helbig, K. (1983) 'Elliptical anisotropy– Its significance and meaning', Geophysics 48, 825–832.
25 Berge, P. A. (1991) 'Short Note on estimating *SV*-wave stacking velocities for transversely isotropic solids', submitted to Geophysics.
26 Levin, F. K. (1989) 'Short Note on *SV*-velocities from *P-P* and *P-SV* data for transversely isotropic solids', Geophysics 54, 1336–1338.
27 Birch, F. (1960) 'The velocity of compressional waves in rocks to 10 kilobars: Part 1', J. Geophys. Res. 65, 1083–1102.
28 Birch, F. (1961) 'The velocity of compressional waves in rocks to 10 kilobars: Part 2', J. Geophys. Res. 66, 2199–2224.
29 Babuška, V. (1984) '*P*-wave velocity anisotropy in crystalline rocks', Geophys. J. R. Astr. Soc. 76, 113–119.
30 Fryer, G. J. and Miller, D. J. (1986) 'Effects and consequences of transverse isotropy in the seafloor', in T. Akal and J. M. Berkson (eds.), Ocean Seismo-Acoustics, Low-Frequency Underwater Acoustics, NATO Conference Series IV: Marine Sciences 16, Plenum Press, New York, pp. 589–597.
31 Fryer, G. J., Miller, D. J. and Berge, P. A. (1989) 'Seismic anisotropy and age-dependent structure of the upper oceanic crust', in J. M. Sinton (ed.), Evolution of Mid Ocean Ridges, Geophysical Monograph 57, IUGG 8, Am. Geophys. Union, Washington, D. C., pp. 1–8.
32 Miller, D. J. (1987) 'Transverse isotropy: some consequences for travel time inversion and models of the oceanic crust', MS thesis, University of Hawaii.

# SPATIAL VARIABILITY IN GROUND MOTION: EFFECTS OF MATERIAL HETEROGENEITY IN SEAFLOOR SEDIMENTS

MICHAEL SNOEK(*) and DIETER RAUCH(**)
* *M E E R , Kilmuckridge, Ireland*
** *U S E A S.p.A, La Spezia, Italy*

ABSTRACT. Seismo-acoustic experiments in the Ligurian Sea were used to test the potential of interface waves for probing the lateral fine structure of apparently regularly layered sediments. Two triaxial Ocean Bottom Seismometers (OBS) were deployed in the center of a radially shaped shot pattern. Charges were fired at up to 5 ranges for 16 azimuths off Tellaro and 2 ranges and 72 azimuths off Viareggio. Angular distribution of the peak-energy group velocity was correlated with polarization of radial particle motion. These results were then compared to profiles obtained from a high resolution seismic survey. We observe interface wave-propagation anomalies with deviation of up to 37% for the velocity and $23^0$ for the wave polarization for the Tellaro area. Deviation of up to 27% for the velocities were recorded for the Viareggio site. Comparison of the dispersion behaviour of observed and synthetic data established a shear-velocity/depth relationship in which $c_s$ increased from 100 m/s at the seafloor to 250 m/s at a depth of 30 m below the seafloor.

## 1. Introduction

Several recent shallow-water studies (e.g. [1]) have shown a surprising variation and azimuthal dependency of seismo-acoustic wavefield propagation. For a thorough understanding of the physical processes responsible for this variability, it is necessary to establish a quantitative relationship between the propagation medium, i.e. seafloor and subseafloor materials, and the observed ground motion characteristics.

Although anisotropy phenomena are commonly predicted and observed on land [2–4], direct seismic anisotropy data have become available for the marine environment only recently. Ocean Bottom Seismometers (OBS) have been deployed yielding data confirming anisotropic behaviour in the crystalline substructure of the upper oceanic crust. In a step further [5,6], 3-component seismometers were clamped to a borehole to investigate the spatial variability of propagation of the seismic wavefield. These experiments demonstrate shear wave splitting which was related to cracks in the upper crust basalts. Shear wave splitting was then modelled and the results used to estimate crack orientations in rocks [7]. Cracks in sediments do not exist; however, one finds instead, structured bedding as a function of changes in sedimentary processes, e.g. vertical grading of material, and preferred direction of sediment



*J. M. Hovem et al. (eds.), Shear Waves in Marine Sediments*, 213–219.

deposition due to currents, confirmed by grain orientation. Anisotropy measurements made on cores of consolidated sediments and sedimentary rocks show higher velocity parallel to bedding than perpendicular to bedding [8]. These studies indicate that anisotropy can considerably alter bottom loss estimates. It was also demonstrated [9] that anisotropic effects increase with depth, confirming previous empirical predictions [10].

Results from an experiment on the Ligurian Shelf NW of La Spezia indicated considerable azimuthal dependency of the seismic energy propagating in interface wave modes [11]. It appeared that the observed effects were caused by small scale lateral inhomogeneities. These results, as well as observations of previous interface wave propagation experiments, made us decide to conduct a series of experiments in a more controlled environment, i.e. at a site with *a priori* knowledge of the layering structure, sediment properties and interface wave propagation characteristics.

In this paper, we concentrate on the lateral variability in group velocity and horizontal hodograph polarization as a function of the sediment structure. In a second step we infer from this data the underlying shear wave velocity profile.

## 2. Physical Background

The realistic seafloor/subseafloor is a highly variable medium, formed by 'regular' as well as 'stochastic' sedimentation processes. Scholte-type interface waves as well as Love waves are valuable tools for investigations of these phenomena, since these waves are extremely sensitive to sediment lateral inhomogeneity and variability. Their dispersion behaviour further allows one to deduce the degree of stratification of the material. Interface waves are trapped in surficial bottom layers, and their elliptic particle motions are strongly polarized in the sagittal plane, i.e. the plane formed by the source-receiver direction and the normal to the interface. Velocity and polarization variations are caused by lateral changes of the geoacoustic parameters, i.e. material heterogeneity.

The term anisotropy has been used in an ambiguous way in the literature, leading to some confusion. In general the term 'anisotropy' characterizes the fact that values of certain physical parameters of a material depend on the direction of observation. Such anisotropic behaviour can be of very diverse origins. Anisotropy may be an intrinsic property of the entire propagation medium and thus a basic (or microscopic) quality characterizing each elementary cell of the material. On the other hand it may also be an 'extrinsic', 'effective' or 'apparent' quality caused by the action of superimposed external forces, the macroscopic fabric or texture of the solid and finally its inhomogeneous composition from bodies of different material and shape.

## 3. Experiment

We deployed two triaxial OBSs with horizontal spacing of about 30 m. The position, orientation, and the coupling of the sensor packs to the ground were controlled by divers. Each sensor pack contained three orthogonally mounted 1 Hz TELEDYN S-500 seismometers and one externally mounted hydrophone. Each OBS was connected to a surface radio buoy, which transmitted the signals to the ship. Data were recorded online using a MicroVAX and simultaneously onto an analogue tape as back-up. Explosive charges ranging between 180–1440 g were fired electrically on the seafloor. This procedure allowed us to determine exact position and timing of the explosion, while providing sufficient surface wave energy to probe this particular seafloor at ranges up to 4.2 km from the sensors. Seismic energy was generated on five circular profiles with radii of 550–1850 m and an angular spacing of $22.5^0$ at the Tellaro site (water depth at OBSs: $H = 17$ m). Seismic energy was geneated at the Viareggio site ($H = 21$ m) on two circular profiles of 800 and 1200 m radii with an angular spacing of $5^0$ for the 800 m and $10^0$ for the 1200 m range.

Seismic surveys were performed using an UNIBOOM in the frequency range 1–10 kHz. Under the given conditions (shallow water environment) the UNIBOOM system can normally only resolve the sediment structure for the first 12–15 m, since deeper reflectors are generally masked by multiples.

Sediment patches at the Tellaro site, show a predominant NW–SE orientation with a general slight increase in sediment thickness with water depth. The profiling plots suggest horizontally-layered, finely-stratified material with some indications of lateral variability. In the southern part of the Tellaro site, elongated sediment packages were oriented with the same general NW–SE trend. There are also extended areas, where no reflectors can be seen, probably due to gaseous sediments.

Sedimentary sequences in the Viareggio area are the result of two distinct sedimentation processes. From land towards the sea sediment wedge thickness increases. On this NE–SW trend is superimposed an NW–SE trend of increased sediment thickness which is the result of sediment discharge from the Magra river. Seismic profiles generally indicate a fairly homogeneous bottom material. Sediment homogeneity was indicated by previous interface wave experiments.

Sediment properties were determined *in situ* and from laboratory measurement on sediments collected fom the region. Sediments were collected using a 1.80 m long gravity corer (five positions) as well as shorter 30.0 cm (6.1 to 12.0 cm inside diameter) PVC hand-operated corer. Laboratory shear wave velocity measurements were made using bimorph ceramic bender transducer/receivers, and at six positions *in situ* shear and compressional wave measurements were made [12]. At the Tellaro site two propagation media with different S-wave velocities were evident. Shear-wave velocities range from 33–58 m/s for mud and 72–76 m/s for sand. A sand/silt layer of thickness roughly 0.4–0.6 m was randomly distributed, over roughly 6.0 m thick layer of mostly silty material. Thin layers sand were also found within the thick silty layer. Shear-wave velocities at Viareggio site were low, ranging from 25 m/s at the seafloor to 60 m/s at 2 m below the seafloor.

## 4. Data Analysis

In the present study we determine the azimuthal (or lateral) variation of interface wave group velocity and the horizontal deflection angle ($\beta$) of the recorded particle hodographs ($\beta > 0$ or $< 0 =$ deflection to the right or left as seen from top). To extract the velocity information from the recorded data we applied the multiple filter technique [13]. The results of this computation are displayed in a GABOR-Matrix, a normalized contoured plot of the interface wave energy as a function of group velocity and frequency.

The azimuthal distributions of interface wave group velocity were determined for different

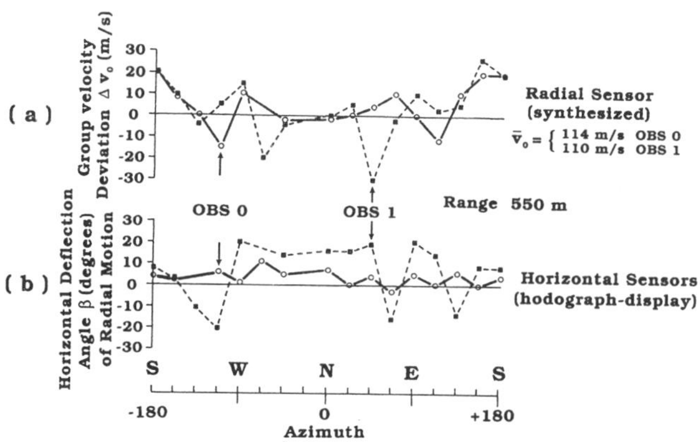


Figure 1. (a) Interface wave group velocity deviation, and (b) deflection angle for OBS 0 and OBS 1 as function of profile azimuth for the range 550 m at the Tellaro test site.

ranges for both OBS. We calculated the mean velocity for each range, from the values of the weighted center $v_0$ of the 98% energy contour. The deviations from these values for the 550 m range at the Tellaro site are plotted as a function of azimuth (Fig. 1a). We observe that the trends for both OBS are the same but the magnitude of the deviations differ. Maximum differences in group velocities between adjacent extrema for OBS 1 are 37%, whereas maximum deviations of only 30% were calculated for OBS 0.

The more densely sampled data collected off Viareggio show more moderate azimuthal velocity-variations, up to 27%, mainly due to the longer ranges. Interface wave velocity as a function of range and azimuth for different components as well as for different modes are presented in Fig. 2. The large scale trends observed basically reflect the sediment layer geometry. Radial and vertical small scale material variabilities are responsible for the fine structure. The trend of higher interface wave velocities in easternly direction may reflect the fact that the sediment wedge is thinning out towards the coast, allowing the wave to sense subbottom material of higher shear rigidity. Lower velocities were observed when the wave predominantely travels in lower density sediment wedges. These trends are consistent for both ranges (800 m and 1200 m) for both sensors at the Viareggio site. The fundamental mode ($M_0$) shows a smoother behaviour because over longer pathlengths the higher frequency content of the signal is damped out, leaving the lower frequencies which sense deeper into the ground and do not 'feel' the small scale material variations. This effect is also responsible for the slightly higher velocities observed at the 1200 m range.

Polarization analysis is another method to classify a wavetype and to infer wave propagation parameters. Figure 1b displays the horizontal deflection angle ($\beta$) of the radial particle motion as a function of profile azimuth for OBS 1 and OBS 0 at the 550 m range for the Tellaro site. To obtain a stable and representative bearing deviation angle ($\beta$), projected hodographs had to be averaged over a few periods of maximal signal amplitude because all three motion-components are affected by background noise and charges on the seafloor excite weak Love waves generating unwanted cross-deflections.

If the measured velocity distributions (Fig. 1a) vary smoothly with increasing azimuth

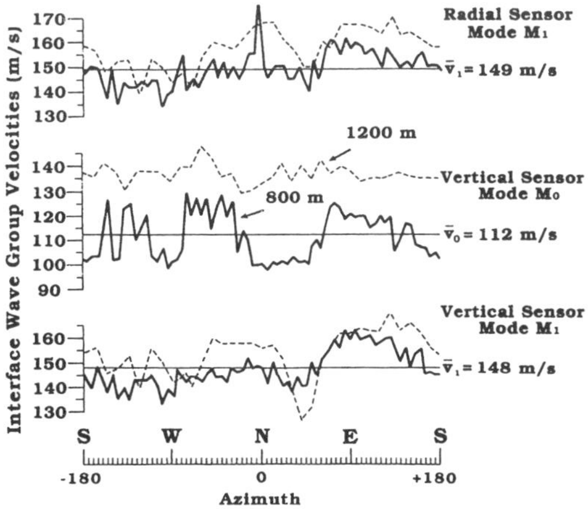


Figure 2. Interface wave velocities for the vertical and radial component (OBS 1) as function of profile azimuth and two ranges (800 m and 1200 m) at the Viareggio test site.

their rises or decays should correspond unequivocally to right (+) or left (−) deflections of the radial particle motion plotted below (Fig. 1b). Due to the above mentioned oscillatory fine-structure such a simple correlation can merely be attributed to a few intervals. Data from OBS 0 appear to be attenuated and biased. If sensor coupling problems are excluded, this difference may be explained if OBS 0 was located on a 'slower' patch of sediment which caused all crossing rays to bend slighty into the lens of 'softer' material. The prevalent right-deflection recorded by OBS 0 may be caused by the particular shape of this patch.

A clear relationship between Scholte wave data and layering structure could not be established suggesting that strong changes of shear-rigidity were usually not accompanied by similar changes in the compressional modulus.

## 5. Model Inversion

The multiple filter technique (MFT) analysis of interface waves provides us with group velocity versus frequency data and gives a qualitative idea of the dispersive character of the propagation medium, i.e. an indication of the layered structure of the sediments. SAFARI [14] was used as a modelling algorithm to parameterize the material for comparison of observed and synthetic data. Starting from an initial geoacoustic model based on all available environmental data, a fitting of the measured group-velocity curves was approached iteratively. We first calculated layer thicknesses from the vertical seismic profiles obtained on a previous cruise. We then assumed that the velocity profile remained the same over the observed range, which is not strictly true as our measurements have indicated. We compared the dispersion behaviour of the vertical component only for one azimuth at the Tellaro site, obtaining a 'representative' shear velocity profile, where the shear wave velocity increased from 100 m/s at the seafloor to 250 m/s at 30 m below the seafloor (Fig. 3).

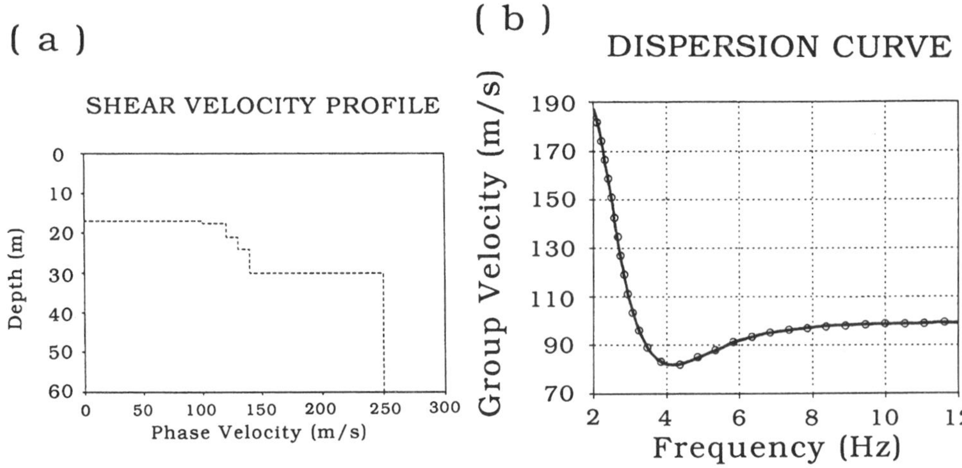


Figure 3. (a) Final input profile for the S-wave velocity (SAFARI model) and (b) resulting group velocity curve (solid line) fitting the observed data.

## 6. Conclusion

We have shown that in some shallow water areas spatial variability of seismic propagation in the upper seafloor is quite high. We are faced with velocity and directivity (polarization) anomalies, which, with our present knowledge, can only be specified between source and receiver. Since the magnitude of the polarization anomalies decreases with range, suggesting that the effects on the propagation of surface waves average out, we can assume that the actual 'events' responsible for the effects observed are of limited size. From the distance between OBSs ($\sim$ 30 m) and the data obtained from densely spaced core-samples collected at the Tellaro site, we conclude that the lower detection bound for sediment patches is $\sim$ 25–35 m. Such lateral extensions correspond roughly to half a wavelength at the peak-energy frequency ($\sim$ 3 Hz) of our 'probing' Scholte wave pulses.

It is evident that the propagation medium, which seems to be simply stratified in the seismic P-wave profiles, has a complicated shear velocity structure, as revealed by interface waves. We thus have a useful tool for probing the upper seafloor, and investigating anisotropy phenomena or material heterogeneity. Our findings demonstrate, that in real world environments, models based on a horizontally layered, laterally invariant propagation medium are not adequate.

**References**


[1] Macbeth, C. and Burton, P.W. Surface waves generated by underwater explosions offshore Scotland, *Geophysical Journal*, **94**, 1988: 285–294.

[2] Stoneley, R. The seismological implication of aeolotropy in continental structure, *Monthly Notices Roy. Astron. Soc. Geophys. Suppl*, **5**, 1949: 343–353.

[3] Crampin, S. Distinctive particle motion of surface waves as a diagnostic of anisotropic layering, *Geophys. J. R. astr. Soc.*, **40**, 1975: 177–186.

[4] Crampin, S. A review of the effects of anisotropic layering on the propagation of seismic waves, *Geophys. J. R. astr. Soc.*, **49**, 1977: 9–27.

[5] Stephen, R.A. Seismic anisotropy observed in upper oceanic crust, *Geophys. Res. Lett.*, **8**, 1981: 865–868.

[6] Stephen, R.A. Seismic Anisotropy in the Upper Crust at DSDP Site 504B, *In:* Akal, T. and Berkson, J., *ed.* Ocean Seismo–Acoustics, Plenum Press, 1986: 599–607.

[7] Crampin, S. Evaluation of anisotropy by shear-wave splitting, *Geophysics*, **50**, 1985: 142–152.

[8] Oakley, D.W. and Vidmar, P.J. Acoustic reflection from transversely isotropic consolidated sediments, *J. Acoust. Soc. Am.*, **73**, 1983: 513–519.

[9] Bachman, R.T. Acoustic anisotropy in marine sediments and sediment rocks, *J. Geophys. Res.*, **84**, 1979: 7661–7663.

[10] Hamilton, E.L. Sound velocity and related properties of marine sediments, North Pacific, *J. Geophys. Res.*, **75**, 1970: 4423–4446.

[11] Snoek, M., Guidi, G. and Michelozzi, E. Interface Wave Studies on the Ligurian Shelf Using an OBS Array: Experimental Results and Propagation Models, *In*: Akal, T. and Berkson, J., *ed.* Ocean Seismo–Acoustics, Plenum Press, 1986: 663–672.

[12] Richardson, M.D., Muzi, E., Troiano, L. and Miaschi, B. Sediment shear waves: A comparison of in situ and laboratory measurements, SACLANTCEN SM-210, SACLANT Undersea Research Centre, 1989.

[13] Dziewonski, A., Bloch, S. and Landisman, M. A technique for the analysis of transient seismic signals, *Bull. Seism. Soc. Am.*, **59**, 1969: 427–444.

[14] Schmidt, H. and Jensen, F.B. A full wave solution for propagation in multilayered viscoelastic media with application to Gaussian beam rflection at fluid-solid interfaces, *J. Acoust. Soc. Am.*, **77**, 1985: 813–825.

# IMPLICATIONS OF DEEP-WATER SEISMOMETER ARRAY MEASUREMENTS FOR SCHOLTE WAVE PROPAGATION

L. DALE BIBEE
*Naval Oceanographic and Atmospheric Research Laboratory*
*Code 362*
*Stennis Space Center, MS 39529*

LEROY M. DORMAN
*Geologic Research Division*
*Scripps Institution of Oceanography*
*La Jolla, CA 92093*

ABSTRACT: A field exercise was conducted in March 1990 to make measurements of Scholte wave propagation characteristics in a deep ocean environment. Signals from a series of bottom explosive shots were recorded on an array of ocean bottom seismometers. Clear Scholte phases were observed on the vertical seismometers to ranges of 1.25 km, but were attenuated to noise levels by 2 km range. Collocated hydrophones did not detect the Scholte wavetrain even at the closest ranges. The ratio of pressure (μPa) to vertical ground velocity (nm/s) was 68 dB in time windows dominated by body waves, but only 20 dB in windows dominated by the Scholte wave. Group velocities were low (30-100 m/s) and showed considerable variability despite the expected uniformity of the seafloor in this abyssal environment.

## Introduction

During the past twenty-five years numerous studies of seismic surface waves (Scholte waves) on the seafloor have been conducted. Ali and Schmalfeldt (1984), Jensen and Schmidt (1986), and Snoek (1990) presented summaries of previous field studies. Of these previous studies, the majority have been conducted in shallow-water environments. Snoek(1990) lists only three studies in water depths greater than 1000 m. Davies (1965) made measurements in several locations with a bottom hydrophone string and bottom charges. Whitmarsh and Lilwall (1982) report observations on ocean bottom seismometers (OBS) using a weight drop technique to generate Scholte waves, and Sauter et al. (1986) analyzed 3 seismograms from an OBS recording of seafloor shots. In part, the shallow-water interests stem from the importance of seismic phases to total acoustic propagation in these environments where the water layer is too thin to act as an effective waveguide. The proximity of sea-surface sources to the seafloor in shallow water should also provide enhanced excitation of seismic surface waves at the seafloor. The logistic difficulty of conducting deep-water, small-aperture experiments has also restricted the types of experiments that have been conducted in deep-water environments. In general however, the sedimentary processes and sediment types on continental shelves differ from those in the deep ocean and care must be taken in using the numerous shallow-water data sets to address deep-water issues.



*J. M. Hovem et al. (eds.), Shear Waves in Marine Sediments*, 221–230.

Given the exponential decay of the fundamental-mode wave functions away from the seafloor for surface waves, little direct excitation of these wave types would be expected from near-surface sources in deep water. However, a growing body of evidence suggests that a large component of the signals and noise recorded on seismometers on the seafloor is propagating in a Scholte wave mode. Tuthill et al. (1980) suggested that the spectra and particle motion of seafloor noise recordings indicate surface-wave contributions. More recently, Schreiner and Dorman (1990) reported evidence from seafloor array coherence measurements that Scholte propagation is dominant in deep-sea ambient, seismic noise. Dougherty and Stephen (1988) have shown evidence of Scholte wave particle motions in the coda of synthetic seismograms generated using numerical finite difference techniques in deep-water environments. Recordings of shallow explosions during recent VLF experiments at NOARL show characteristically longer coda on seismometer sensors as opposed to collocated hydrophones (Figure 1). The long time duration of these coda and the relatively large ratio of particle motions to pressure are believed to be indications of surface-wave contributions to the recorded signal.

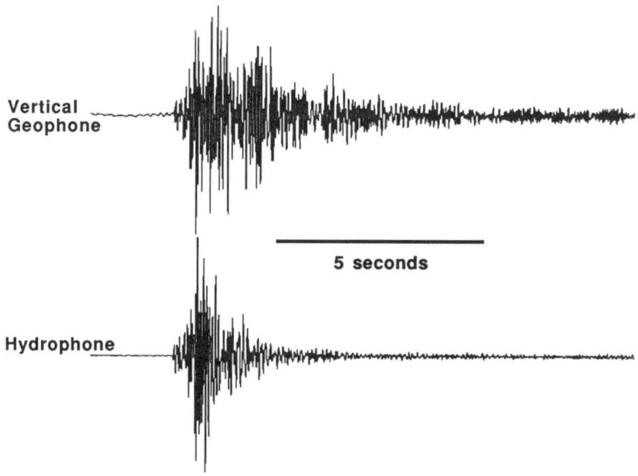


FIGURE 1. OBS recordings of near surface explosive shots typically show an extended coda for geophone records as opposed to recordings from a collocated hydrophone. One working hypothesis is that the energy in the coda is propagating as scattered Scholte waves. The shot was recorded during a study of the Oregon continental margin.

All of these reports of surface waves in deep water are thought to be the result of "indirect" excitation by scattering of body wave energy at or below the seafloor. Even relatively flat seafloor has been shown to produce significant backscatter from small scale roughness, and recent data have shown high variability in subbottom parameters below seemingly uniform seafloor (Rowe and Gettrust, 1990, Gettrust et al., 1988). These scatterers may act as secondary point sources near the interface which couple efficiently into surface-wave propagation modes.

At NOARL, an effort has been initiated this year to study the physics of generation and propagation of these deep-water seismic interface waves. Because the scattered field has a distributed source, traditional single station techniques for isolating surface-wave modes are not viable. Measurements of signals with substantial deep-water arrays are required. Two self-contained 12 node seismic arrays are currently under construction for NOARL field tests in 1991 and will be used in conjunction with "penetrator" sub-bottom seismometers and more traditional OBS. In preparation for that test, an experiment was conducted in March 1990 using OBS and on-bottom explosive shots to gather baseline data. Initial results of analysis of those data are reported here.

## Experiment

The experiment was conducted as an adjunct to a physical oceanographic experiment known as the Surface Waves Processes Program (SWAPP). The goals were not only to conduct a seismic surface-wave experiment, which is the focus of this paper, but also to measure ambient seismic and acoustic noise during a time when the sea surface was being extensively instrumented to measure the properties of the air-sea interface. The data of interest for this report were a series of explosive shots detonated at the seafloor and recorded on OBS/ hydrophones.

Figure 2 shows the location of the experiment, approximately 500 km west of Point Conception, California. The area is underlain by seafloor which is 37 million years old and exhibits abyssal-hill topography typical of much of the Pacific Ocean basin. The seafloor morphology consists of lineated troughs and hills, probably fault bounded, with a relief of about 100 m and a wavelength of a few kilometers across the strike of the lineation. Approximately 75 m of sediment overlie the basaltic basement in a fairly uniform thickness layer, but local thickness variations do occur (Figure 3). The majority of the experiment described in this report was conducted on a shelf over which the seafloor depth varied by only about 30 m from its average value of 4700 m.

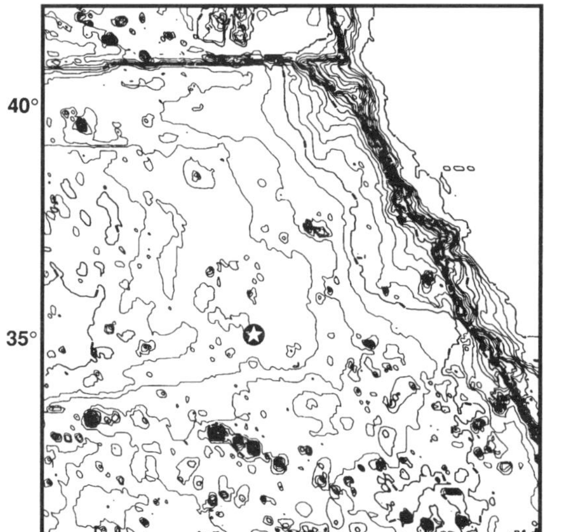


FIGURE 2. The experiment was located directly west from Pt. Conception, CA at the point indicated by the star. Note that the site is well away from the continental margin and not in the major fracture zone to the south.

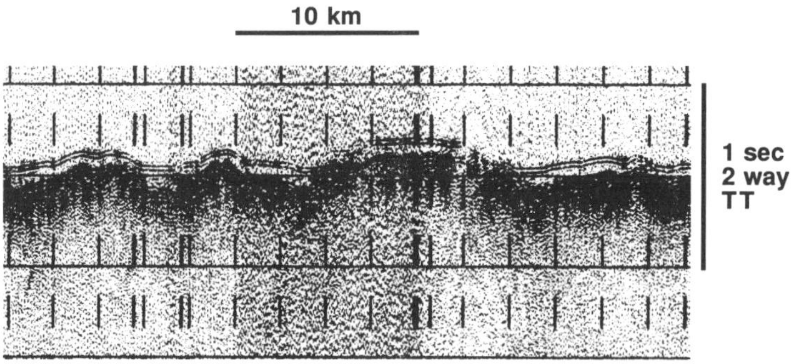


FIGURE 3. A sparker seismic reflection record over the area of the experiment shows 75 m of sediment overlying volcanic basement.

Each of the seven shots consisted of 23 kg of TNT detonated on the sea bottom at ambient pressure with an electric blasting cap and booster. A small expendable pressure case with a timer and firing circuit controlled the detonation. The rather large charge size was required to generate adequate low frequency energy since the high confining pressure forces the bubble pulse frequency of the explosion to more than 100 Hz.

Several receivers were used in the experiment, including a prototype tethered geophone array from Scripps Institution of Oceanography that was undergoing engineering tests and a NOARL vertical hydrophone array. The receivers used in the part of the study described here were NOARL OBS. Four instruments were deployed for the noise phase of the experiment; however, one expended its recording capacity prior to the active phase of the experiment. Three instruments (labeled 33, 34, and 36) recorded the seafloor shots. The OBS are upgraded versions of instruments original designed at Scripps Institution of Oceanography. While the mechanical design remains similar, the electronics have been completely redesigned at NOARL to incorporate a more modern micropower computer with a 13 bit A/D and gain-ranging analog electronics. The data are recorded on two optical WORM disk drives. Three-component, 1 Hz, L-4 seismometers are used as sensors along with a hydrophone and a two-axis acoustic current meter. A high precision, computer-compensated, quartz crystal is used as a time base.

## Data Reduction and Analysis

After transcription to a standard data format, the data were displayed as time series (Figure 4). The compressional wave traveling along a direct path from the source to the receiver dominates the early part of the seismograms. A series of later energy pulses at 6.2 second intervals represents energy traveling as a compressional body wave but traversing the water column multiple times after reflections off the sea surface and seafloor. On the vertical geophone a low-frequency, slow, dispersed wavetrain typical of Scholte wave propagation is observed. The hydrophone shows no evidence of the same wave train, although the body waves are clearly seen. While the hydrophone has somewhat lower response than the vertical geophone at low frequency, the reduction in size of the Scholte wave is much larger than this difference. A small precursor to the direct arrival is observed which represents energy refracted through the volcanic basement at velocities near 4.2 km/s.

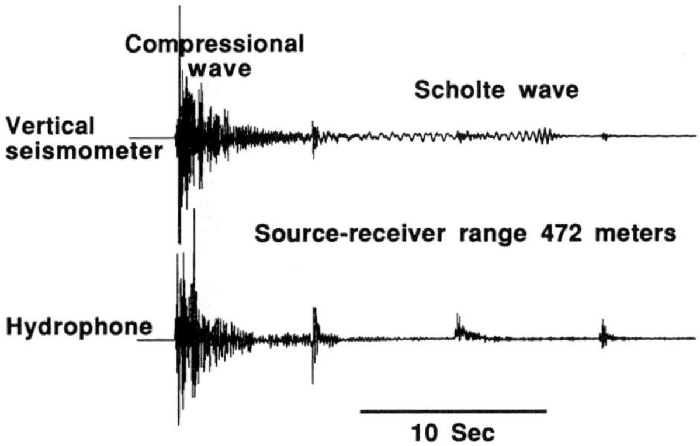


FIGURE 4. A time series plot shows the various phases recorded on the OBS. Note the lack of Scholte wave energy on the hydrophone sensor.

The particle motion of the dispersed wave in the saggital plane (vertical vs. radial) is elliptical as is typical of surface-wave propagation.

Since both instruments and shots were dropped freely from the surface, the position of each on the seafloor was not precisely known initially. The shot detonation time was also poorly known since the expendable timers were not of high precision. Fortunately, the geometry of the experiment, with a small bottom aperture and deep water, was ideal for determining the detonation time and source-receiver ranges from the relative arrival times of the direct wave and water-column multiples. The arrival time of the sea-surface reflection was only weakly dependent on the range but strongly dependent on the detonation time. An iterative procedure was used whereby the detonation time was calculated from the surface reflection and range was then calculated from the direct wave. These ranges were then fed back to generate more precise detonation times and the process was repeated iteratively until convergence was achieved. Finally, all of the ranges were combined in a least square procedure to calculate shot and receiver locations. The times are estimated to be in error by less than 20 ms, and the range errors are less than 30 m.

To isolate the dispersive characteristics of the surface wave, the multiple-window spectral techniques described by Landisman et al. (1969) were applied. Displays of spectral power as functions of discrete frequency and group velocity (referred to subsequently as dispersion matrices) were computed. Each window was four wavelengths long and was tapered with a Parzen window function to reject side-lobe contamination. The instrument response was removed from the computed spectra so that power level could be displayed in quantitative units of pressure (uPa) or particle velocity (nm/s). Figure 5 shows an example of one of these dispersion matrices plotted in three dimensions. The dispersed Scholte wave is responsible for the high power at group velocities between 30 and 100 m/s and is most prominent at frequencies between 1 and 4 Hz. In addition, high-power "lines" at particular group velocities are present. These non-dispersed signals originate from the late-arriving, body-wave, water-column multiples that appear in the windows for low group velocity. The apparent broadening of the "line" at low frequencies results from the longer window used at those frequencies and is a measure of the resolution in group velocity of the multiple-window technique.

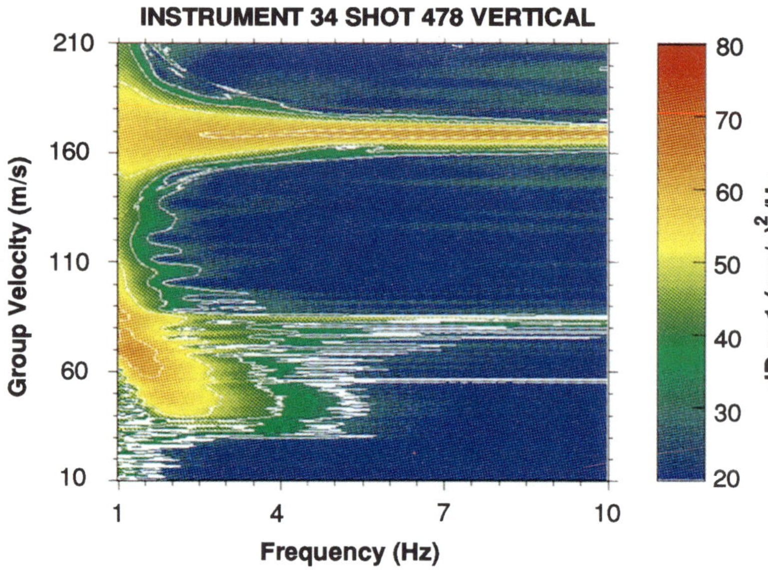


FIGURE 5. A display of a dispersion matrix for one of the shot-receiver pairs. Each element of the matrix is represented by a colored pixel which codes the amplitude of this value. The dispersed energy between 30 and 100 m/s is the Scholte wave. The line at 170 m/s is a body-wave, water-column multiple. The offset between source and receiver is 1075 m.

A dispersion curve can be fit to the high power elements of the dispersion matrix. The data forming this curve can then be inverted for shear-velocity structure in the seafloor; this is the subject of another paper in this volume (Schreiner et al., this volume). Our paper focuses on the quantification of properties of the seismic waves themselves, rather than the bottom properties as such. It is therefore of interest to compare the signal characteristics at different ranges and at different sensors. This is achieved by first extracting from the dispersion matrix the power levels along the dispersion curves.

## Discussion

Dispersion matrices were calculated for the four closest shots received by the three OBS, covering ranges between 345 m and 2100 m. Clear Scholte phases were seen on the vertical geophone for all shots at less than 1250 m range. At 2110 m only a trace of the Scholte phase was seen, and would not have been recognized if not keyed by the other receivers. For three of the four shots, the dispersion relations look similar but have minor variations (Figure 6). Only the fundamental mode could be resolved. Group velocities at frequencies above 3 Hz were approximately 35-40 m/s; below 3 Hz these increased to between 50 m/s and 90 m/s at 1 Hz. The variability between seismograms was most apparent at 1 Hz. At this frequency, the wavelength is comparable to the sediment thickness and basement begins to have an effect. The observed variability could be indicative of basement velocity variations or of minor variations in sediment thickness. Future work is planned to resolve these possibilities by correlating sediment reflection data with the variability in group velocity.

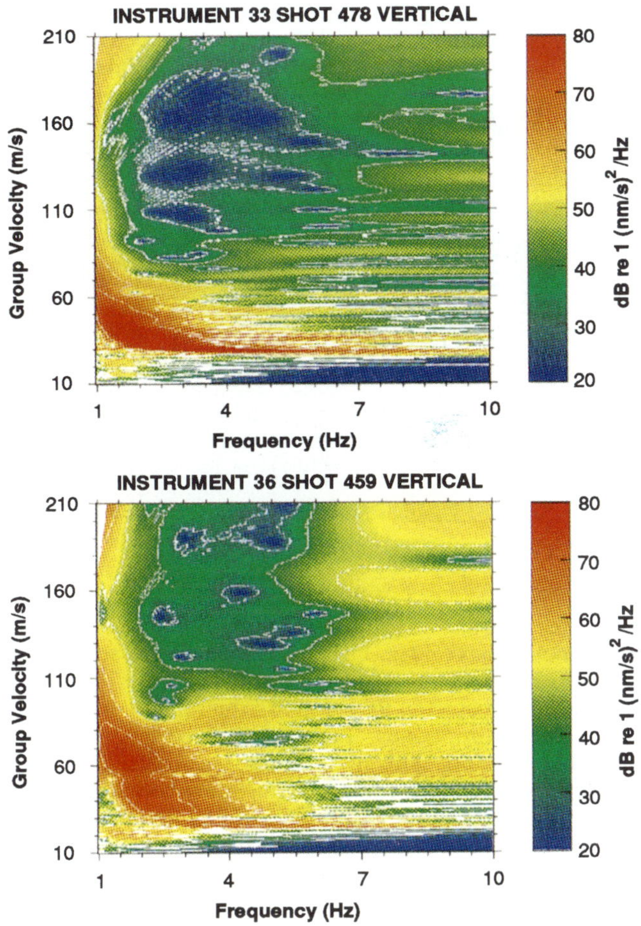


FIGURE 6. Two dispersion matrices at similar range are shown. The group velocities at 1 Hz are significantly different. The source-receiver offset was 386 m for the upper plot and 345 m for the lower plot.

The fourth shot (Figure 7) shows significantly higher group velocities (~70 m/s) at frequencies above 3 Hz for all receivers. This is evidence for large variability in the shallow-sediment, shear-wave velocity over a rather small area (less than 2 km square). Correlation of this variability with bathymetry and morphology is planned for the future.

By extracting the power along the dispersion curve, we can examine the range dependence of energy propagating to a set of receivers from a given shot. The uncertainties of source-signature variability from different shots is thus avoided. The Scholte phases are found to be highly attenuated with range. On shot 459 at 2.5 Hz, 14 dB decay between ranges of 345 m and 868 m is observed; only 4 dB would be predicted by cylindrical spreading. The remaining loss caused by scattering or anelastic attenuation is approximately 0.019 dB/m at 2.5 Hz, about 0.3 dB/wavelength. This requires a sediment S-wave Q of approximately 30. However, for shot 478 at 2.5 Hz, the additional loss exceeding geometric spreading is approximately 0.05 dB/m, requiring Q of approximately 12. Thus even among seismograms with similar dispersion curves the measured attenuation values are quite variable.

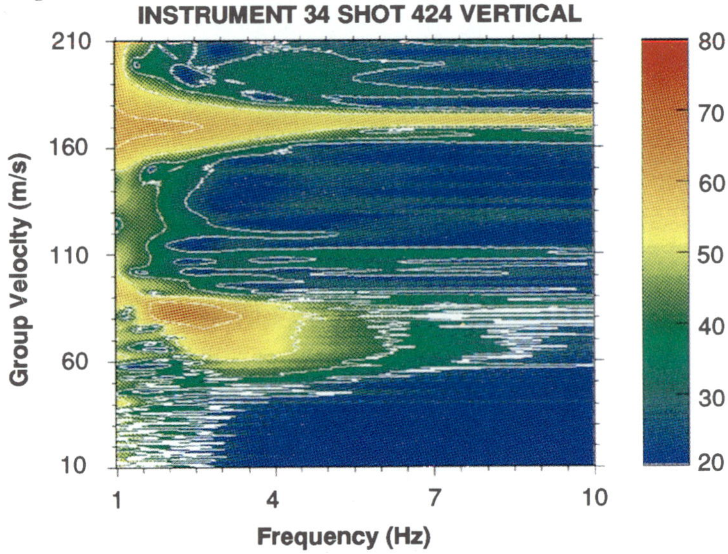


FIGURE 7. This shot shows group velocities that are twice those of the other shots at frequencies above 3 Hz. Offset between source and receiver is 1095 m.

A final question addressed in this study is the relative magnitude of Scholte wave effects on geophones and hydrophones. Figure 4 showed the lack of any obvious Scholte phases on the hydrophone record. This hydrophone was located less than 1 meter from the interface, well within a wavelength for all frequencies considered here. The computed dispersion matrix for the hydrophone also shows no evidence of the Scholte phase, even though it clearly shows the late arriving body wave multiples. Figure 8 shows the power extracted along the predicted, dispersion curve for both the hydrophone and geophone. The units for the two curves are different (one is pressure while the other is ground velocity).

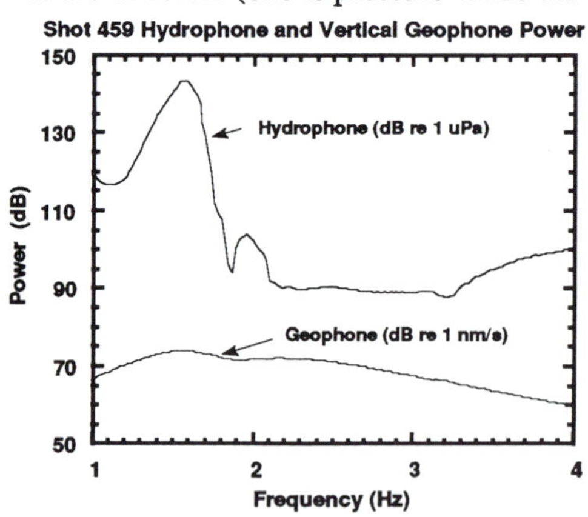


FIGURE 8. Ratios of hydrophone to geophone power are shown. The peak in hydrophone power at 1.5 Hz is due to a body-wave water-column multiple.

For a plane wave passing vertically through a homogeneous acoustic medium, the ratio between pressure and vertical particle velocity is simply the acoustic impedance of the medium (~64 dB for water in the units chosen). This simple relation does not hold for more complicated propagation scenarios. In Figure 8, a large peak in hydrophone power at approximately 1.5 Hz can be seen that does not occur on the geophone. At this frequency, the ratio is approximately 68 dB. The dispersion matrix plots indicate that at this frequency the dispersion curve crosses one of the "lines" generated by the water-column multiple. The high ratio of pressure to velocity (comparable to the expected body-wave value) is thus easily explained. Above this frequency, the pressure values are much lower (comparable to ambient noise), while the velocity values remain high (dominated by Scholte energy) resulting in a ratio of about 20 dB. Since the hydrophone value could be due to noise, this is a maximum value. A preliminary model-run using the SAFARI algorithm (Schmidt, 1988) indicates that the expected ratio for a similar Scholte wave is about 37 dB. While this calculated value is higher than that observed, it shows a large reduction from the body-wave value of 64 dB and provides a theoretical basis for the lack of observed Scholte waves on the hydrophone. Closer correspondence between the modeled and observed values may be achieved with a refined model of bottom parameters.

## Conclusions

Observations of Scholte waves on a deep-water seismometer array indicate considerable variability in both dispersion characteristics and in propagation losses over a 2 km area. This requires shear property variations that are surprisingly large given the abyssal sedimentary environment, an environment which would not be expected to be as dynamic as shallow environments. Higher density arrays of the type planned for NOARL's 1991 experiment are required to delineate these variations.

Ratios of pressure to vertical particle velocity for collocated hydrophone and geophone sensors are much smaller for Scholte waves than for body waves in this deep-water, soft-bottom environment. This implies that geophones should be the sensor of choice to measure Scholte waves for geoacoustic parameter inversion. Similarly, Scholte waves are much more likely to affect ambient noise and signal characteristics on geophones than on hydrophones, even if the two sensors are collocated.

## Acknowledgments

The authors thank the captain and crew of the USNS DESTEIGUER and the USS NAVAJO for providing competent and professional operational support at sea. William Everard, William Eslick, and Jeff Becklehimer provided valuable assistance in data collection at sea. Steve O'Hara provided preliminary model calculations. Hassan Ali suggested numerous improvements that were incorporated into the manuscript. This work was supported by NOARL Program Element 601153N, H. E. Morris, Program Manager. NOARL contribution PR91:004:362.

## References

Ali, Hassan B., and Schmalfeldt, Bernd (1984) 'Seismic sensing of low-frequency radiated ship noise', SACLANT ASW Research Centre Report SR-77.

Davies, D. (1965) 'Dispersed Stoneley waves on the ocean bottom', Bulletin of the Seismological Society of America, 55, 903-918.

Dougherty, Martin E., and Stephen Ralph A. (1988) 'Seismic energy partitioning and scattering in laterally heterogeneous ocean crust', PAGEOPH, 128,195-229.

Gettrust, J. F., Grimm, M., Madosik, S. and Rowe, M. (1988) 'Results of a deep-tow multichannel survey on the Bermuda Rise', Geophysical Research Letters, 15, 1413-1416.

Jensen, Finn B., and Schmidt, Henrik (1986) 'Shear properties of ocean sediments determined from numerical modelling of Scholte wave data', in Tuncay Akal and Jonathan M. Berkson (eds.), Ocean Seismo-Acoustics, Plenum Press, New York, pp. 683-692.

Landisman, M., Dziewonski, A., and Sato, Y. (1969) 'Recent improvements in the analysis of surface wave observations', Geophysical Journal of the Royal Astronomical Society, 17, 369-403.

Rowe, M. M. and Gettrust, J. F. (1989) 'DTAGS deep towed seismic survey of the Blake-Bahama Basin', NORDA Tecnical Note 449.

Sauter, A. W., Dorman, L. M., and Schreiner, A. E. (1986) 'A study of sea floor structure using ocean bottom shots and receivers', in Tuncay Akal and Jonathan M. Berkson (eds.), Ocean Seismo-Acoustics, Plenum Press, New York, pp. 673-681.

Schmidt, H. (1988) 'SAFARI, Seismo-acoustic fast field algorithm for range independent environments: Users guide, SACLANT ASW Research Centre Report SR-113.

Schreiner, Anthony E., and Dorman, L. M. (1990) 'Coherence lengths of seafloor noise: Effect of ocean bottom structure', Journal of the Acoustical Society of America, 88, 1503-1514.

Schreiner, Anthony E., Dorman, L. M., and Bibee, L. D. (this volume) 'Shear wave velocity structure at two deep water sites in the Pacific Ocean'.

Snoek, M. (1990) 'Interface-wave propagation studies: An example of seismo-acoustic propagation in non-homogeneous materials', SACLANT ASW Research Centre Report SM-229.

Tuthill, Jonathan D., Lewis, Brian T. R., and Garmany, Jan D. (1980) 'Stoneley waves, Lopez Island noise, and deep sea noise from 1 to 5 Hz', in G. H. Sutton, B. T. R. Lewis, J. Ewing, F. K. Duennebier, B. Iwatake, and J. D. Tuthill (eds.), Lopez Island Ocean Bottom Seismometer Intercomparison Experiment, Hawaii Institute of Geophysics Report No. HIG-80-4.

Whitmarsh, R. B. and Lilwal, R. C. (1982) 'A new method for determination of in-situ shear-wave velocity in deep-sea sediment', Geophysics, 47, 1745.

# SHEAR WAVE VELOCITY STRUCTURE FROM INTERFACE WAVES AT TWO DEEP WATER SITES IN THE PACIFIC OCEAN

A. E. SCHREINER, L. M. DORMAN
*Marine Physical Laboratory and Geological Research Division*
*Scripps Institution of Oceanography, 0215*
*University of California, San Diego*
*La Jolla, CA 92093-0215*
*U.S.A.*

L. D. BIBEE
*Naval Oceanographic and Atmospheric Research Laboratory*
*Code 362*
*Stennis Space Center, MS 39529-5004*
*U.S.A.*

ABSTRACT. The shear wave velocity structure of two deep ocean sites has been studied by detonating explosives on the ocean floor and examining the slowly moving Stoneley waves that were generated at the ocean floor interface. At the southern California site, the structure was found to be uniform insofar as the relatively sparse sampling can resolve. At the central California site, the velocities and velocity gradients were found to be variable over the space of 1 or 2 km. At both sites, the shear wave velocity at the seafloor is very low, between 27 and 32 m $s^{-1}$. The shear wave velocity gradient at the southern California site in the top 1 m of the sediments is 5.2 m $s^{-1}$ $m^{-1}$, and 3.1 m $s^{-1}$ $m^{-1}$ in the next 100 m. At the northern site, the velocity gradient for the top 1 m varies between 5.0 m $s^{-1}$ $m^{-1}$ and 8.0 m $s^{-1}$ $m^{-1}$, and for the next 20 m it varies between 4.5 m $s^{-1}$ $m^{-1}$ and 7.0 m $s^{-1}$ $m^{-1}$.

## 1. Introduction

The most effective way to make indirect measurements of the shear wave velocity on the inaccessible deep sea floor is to excite and record Stoneley/Scholte waves in the sedimentary waveguide. The dispersive wave trains are well separated in time from the other energy propagating out from the source, and the velocity dispersion is most strongly controlled by the shear wave velocity structure in the top 100 or so m of the sediments. Observations of Scholte wave dispersion have been used extensively to study the sea floor in shallow water (e.g. Rauch, 1986).

Because of the low propagation velocity of Stoneley waves and their evanescent nature, it is necessary to place the source energy and the recording sensors on the ocean floor. At oceanic depths this has proven to be a challenge. We have been making such measurements over the past 6 years. The first experiment followed the procedure of Davies (1965) and was done with explosives contained in pressure cases and detonated with a timer (Sauter et al., 1986). This was quite effective but limited the size of the explosives to 5 kg, larger amounts require prohibitively large pressure cases. With much experimentation, we found a suitable combination of detonators and



*J. M. Hovem et al. (eds.), Shear Waves in Marine Sediments*, 231–238.

boosters that could generate an explosion at ocean ambient pressure for the deep ocean. This method was used successfully with 23 and 45 kg charges for a tomographic survey of a seamount (Hildebrand et al., 1989). On that occasion, the shot location and timing was critical so the explosives were lowered on a wire, and detonated on command from the surface.

In March of 1990, as part of a larger ambient noise experiment known as SWAPP (Surface WAve Propagation Processes), ocean bottom shots were again set off, this time using timers. The SWAPP experiment was performed approximately 500 km from the central California coast, in water of 4700 m depth (Figure 1). Four ocean bottom seismometers (OBS) were deployed by the Naval Oceanographic and Atmospheric Research Laboratory (NOARL). Eight ocean floor shots were exploded to distances up to 3 km from the OBS's. A large variety of interface waves were recorded at different ranges and azimuths. Clear Stoneley waves were recorded up to a distance of 1.5 km. An 18 element telemetered seismometer array was also in operation during part of the shooting but was too distant to detect Stoneley waves from the one event it recorded.

Our aim in this paper is to compare the inferred shear wave velocity structure at the central California site with the structure at a southern California site (Sauter et al., 1986) (Figure 1 at the CIRCUS site), and discuss the variability observed at the central California site. One reason for studying the upper structure of the deep ocean floor is that velocity structure of the sediments has been found to be an important factor in the propagation characteristics of seafloor noise (Schreiner and Dorman, 1990; Dorman et al., 1991).

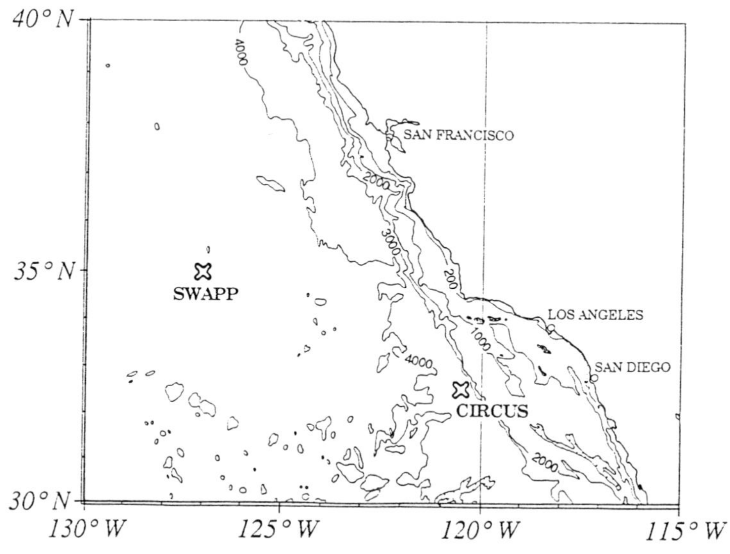


Figure 1. Coastline and bathymetry of the eastern Pacific ocean off California, showing the CIRCUS and SWAPP experiment sites.

## 2. Data Set

In the southern California experiment, 6 shots of 2 kg were fired on a north-south line at ranges of 0.41 to 3.5 km from the receivers (Figure 2a). Interface waves were visible to a distance of 1.35 km (Figure 3a). In the central California experiment, 7 shots of 23 kg were fired at ranges of 0.41 to 3.2 km and a variety of azimuths around the receivers (Figure 2b). Interface waves were visible to a distance of 1.6 km (Figure 3b). There are a total of 15 records showing interface waves. The dipersed waves were most clearly visible on the vertical seismometers; all the figures present the vertical component.

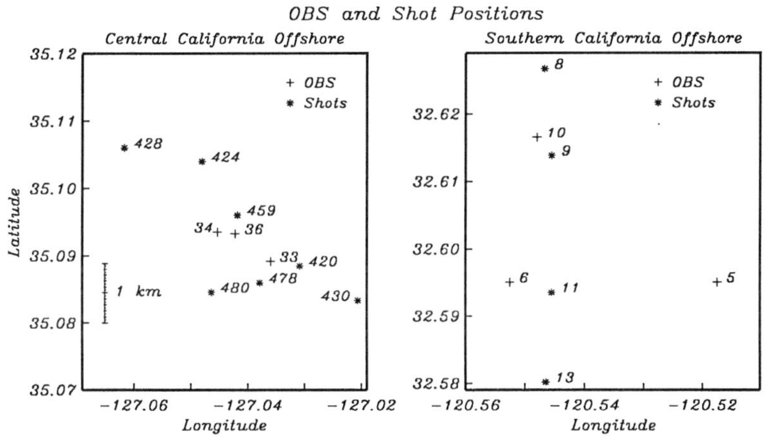


Figure 2. Relative locations of explosive sources and ocean bottom seismographs.
a) CIRCUS site, b) SWAPP site.

The three records from the CIRCUS site with visible interface waves all show the same dispersion (Figure 4a). At the SWAPP site, there was more variability; four representative time-frequency decompositions are shown in Figure 4b-e, and labeled I through IV. The differences in the dispersion curves are not obvious, with the exception of type IV (Figure 4e). However, closer examination reveals subtle differences in the velocities which, as will be described in the next section, can result in velocity models which are quite distinct.

Larger explosions were used in the SWAPP experiment in an effort to increase the visibility of the dispersed waves. However an ten-fold increase in shot size yielded only a 25% increase in distance over which the interface waves were detected. At ocean floor pressures, the bubble pulse frequency for a 23 kg shot is on the order of 90 Hz, and the source spectrum falls off sharply below the bubble frequency. The main interface wave energy is between 0.5 and 8 Hz, far below the source peak, so the explosive source is not being used efficiently. The ambient noise level at the SWAPP site was also substantially higher than the CIRCUS site due to the operations of three ships.

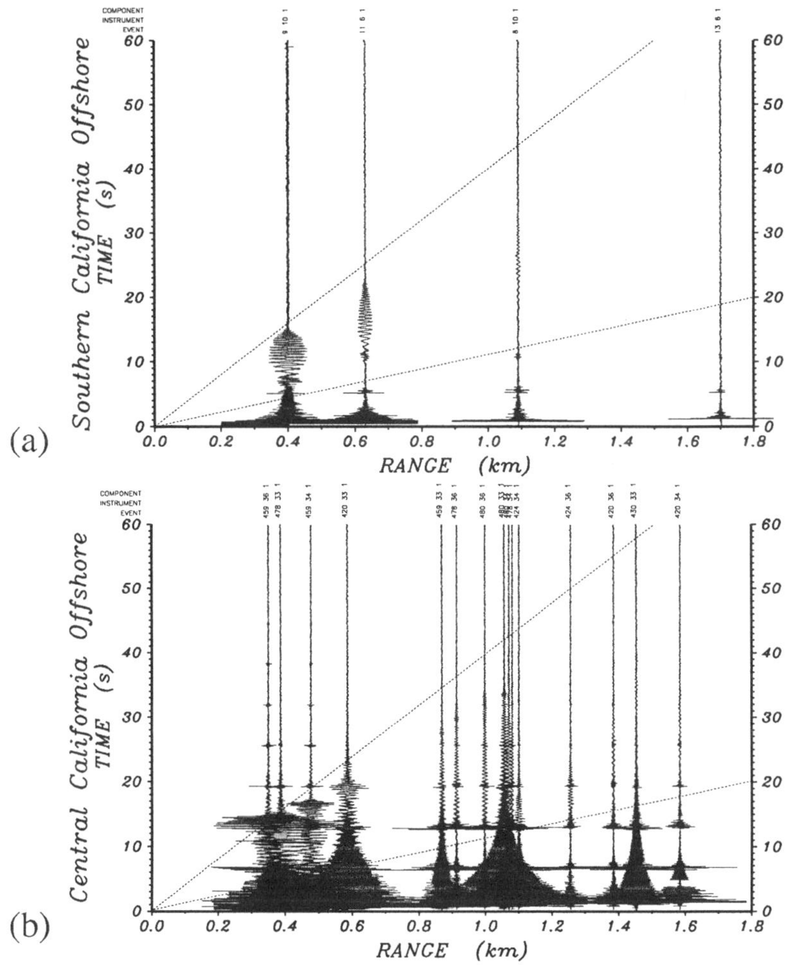


Figure 3. Seismograms for the vertical components of the OBS showing any combination of source and receiver within a separation of 1.8 km. The upper and lower dotted lines show group velocities of 25 m s$^{-1}$ and 90 m s–1 respectively.
a) CIRCUS experiment, b) SWAPP experiment.

## 3. Discussion

The interface wave energy was analyzed by applying a time-frequency decomposition of the dispersed wave trains via the multiple frequency window method (Dziewonski et al., 1969). Models of the upper sediments were constructed by the iterative comparison of computed dispersion curvest with the measured group velocity dispersion. The dispersion curves were calculated with a locked seismic mode method using propagator matrices (Gomberg and Masters, 1988).

The iterative modeling of the seismic dispersion required many runs of the synthetic seismogram program with incremental changes in the model parameters. The mode computation program accepts models with constant velocity layers. Many thin layers are required to represent smooth models, so to reduce the number of free parameters, the velocity model was specified in terms of a small number (5) of constant gradient layers and converted automatically to a large number (60) of constant velocity layers.

As a general rule, the maximum observed slowness depends on the minimum shear velocity in the model, at the ocean floor. The part of the dispersion curve between the highest observed frequency and the elbow in the curve is determined by the shear velocity gradient in the top 1 to 2 m of the sea floor. The slope of the lower part of the dispersion curve is mostly determined by the shear velocity gradient of the next 20 m of the model. The velocity gradient below 20 m has relatively little effect on the dispersion curve in the observed frequency range except for some narrow peaks in the higher order modes. Of course these features of the dispersion curve are not entirely independent and some trade-off may be required, and there are cases where this particular parametrization can not yield a model that fits the observed dispersion exactly. The dispersion curves are very insensitive to the compressional wave velocity structure and it was thus fixed at 1.51 km $s^{-1}$. The curves are somewhat more sensitive to the density but still to a relatively small degree and it was fixed at 1.4 Mg $m^{-3}$ for the sediments. For both sites, late arriving interface waves are visible out to about 1.5 km. However the high frequency (> 2 Hz) part of the wave train is attenuated within the first 800 m of travel. This part of the dispersion curve provides the strongest constraint on the shear velocity at the seafloor and the gradient in the upper few meters.

Some ambiguity in the models exists because the computed dispersion is unconstrained outside of the frequency range 0.5 to 6 Hz. Our data show a relatively small part of the dispersion curve.

The dispersion curves from the southern California data are all quite similar to each other and are shown by a representative example (Figure 4a). A model with a seafloor shear wave velocity of 30.5 m $s^{-1}$, a velocity gradient of 5.2 m $s^{-1}$ $m^{-1}$ in the upper 1 m, and a velocity gradient of 3.1 m $s^{-1}$ $m^{-1}$ in the next 100 m yielded at good fit to the observations. The second overtone appears to have been excited.

Four characteristic dispersion curves for mode 0 are visible in the central California offshore data, labeled I, II, III and IV (Figures 4b,c,d and e respectively). The majority (11 out of 15) of the interface waves generated at the central California site are fit reasonably well by a single velocity model (type I). The shear wave velocity at the seafloor is 29 m $s^{-1}$ with a gradient of 8.0 m $s^{-1}$ $m^{-1}$ for the upper 0.8 m. The gradient is reduced to 7.0 m $s^{-1}$ $m^{-1}$ for the next 20 m, and then becomes unresolved for the next 50 m. The paths that are fit by this model are in the west-central part of the site. The event shown in Figure 4b, although classified as type I does not match the computed curve exactly. It was not possible to find a model that would fit the high frequency and low frequency parts of the dispersion curve simultaneously. The reduction of the velocity model to a small number of constant gradient layers may be too much of a restriction.

The type II dispersion is displayed by three records with paths in the south-central part of the site. The model that fits these events has an even lower seafloor velocity of 27 m $s^{-1}$. The gradient is slightly higher at 8.2 m $s^{-1}$ $m^{-1}$ for the upper 1 m, then continues with a high gradient of 8.0 m $s^{-1}$ $m^{-1}$ for the next 20 m.

The type III dispersion is in evidence for two records, both from paths in the southeast part of the area. The best model for these events has a higher seafloor velocity of 32 m $s^{-1}$, and a lower gradient of 5.8 m $s^{-1}$ $m^{-1}$ in the upper 1.8 m, then a low gradient of 4.5 m $s^{-1}$ $m^{-1}$ for the next 20 m.

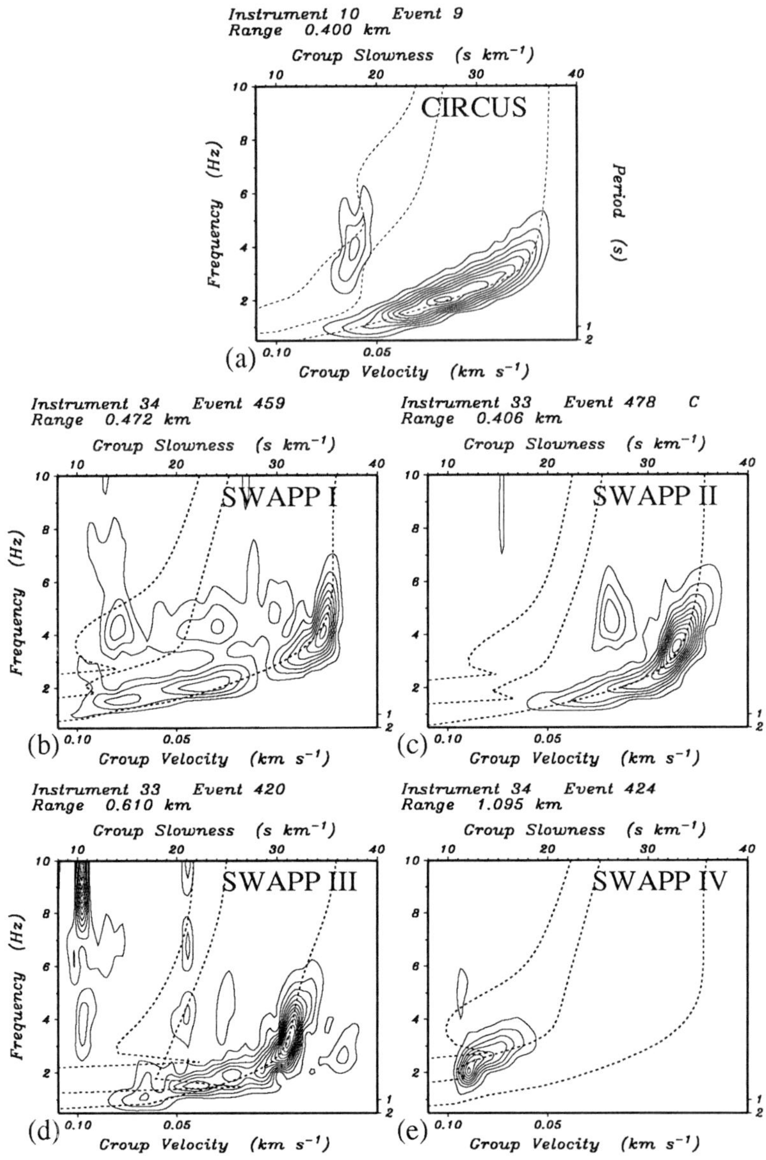


Figure 4. Group velocity dispersion analyses. The contours show amplitude of the time-frequency decomposition of the seismograms. The dotted lines are synthetic dispersion curves for modes 0 through 2 for the best fitting models.

a) CIRCUS site, b) SWAPP site type I, c) SWAPP site type II, d) SWAPP site type III, e) SWAPP site type IV.

The type IV dispersion curves (Figure 4e) of shot/receiver paths 424/34 and 424/36 could be the first overtone from the type I model. Either shot 424 did not excite the fundamental mode, or the shear wave velocity structure in the vicinity of shot 424 is very different from elsewhere in the area, or both may be true. At this point it is difficult to distinguish between the alternatives. OBS 33 is too far to record an interface wave from shot 424, otherwise it would have been productive to examine the difference between shot/receiver pair 424/33 and 459/33 because they are nearly aligned. It would also be useful to examine arrivals from shot 428 to any of the OBS because it is the only other shot to the northwest of the array; but again, the ranges are too great to observe an interface wave.

Higher order modes were apparently not excited at the central California site, with the possible exception of shot 424. Some of the records show energy which has the appearance of higher modes, but at this time, the models predict first and second overtones at very different velocities than are observed. It is difficult to see how a very different model could be constructed to fit the fundamental mode and the other energy as well.

By sorting the dispersion types by the location of their associated paths, a consistent picture of the variation of sediments properties at the site emerges (Figure 5). The dashed lines separate zones which correspond roughly to the different propagation characteristics of Figures 4b-e (except dispersion type IV is shown as zone I'). The shear wave velocity at the seafloor goes from 29.5 m $s^{-1}$ in the northwest and central area to 32 m $s^{-1}$ in the southwest part. The velocity gradient for the upper 20 m likewise changes from 7.0 m $s^{-1}$ $m^{-1}$ in the northwest, to 5.0 m $s^{-1}$ $m^{-1}$ and then 4.5 m $s^{-1}$ $m^{-1}$ in the southeast. The boundaries between these areas are not precisely constrained but appear to run in a north-south direction. Reflection records for a much wider area also suggest that the structural trend is north-south.

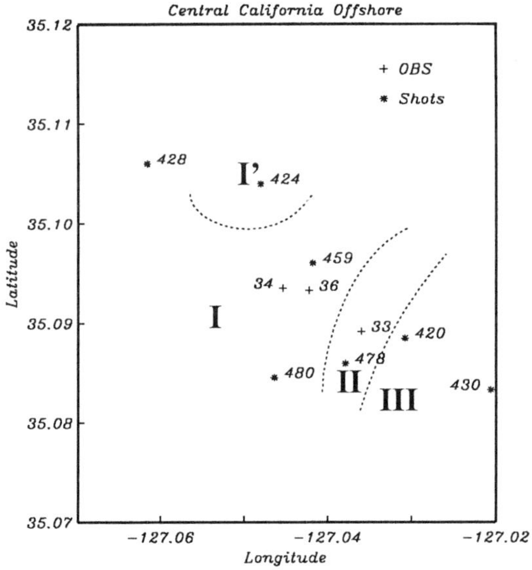


Figure 5. Map of the SWAPP site showing division of the seabed into different zones with different propagation characteristics.

## 4. Summary

The shear wave velocity structure of the upper sediments on the ocean floor is effectively determined from the dispersion analysis of interface waves. The generation and recording of such interface waves requires the placement of both source and receiver on the ocean floor because the energy is closely bound to the water-sediment interface. Despite the operational difficulty of placing explosives and receivers on the deep ocean floor, the results from the analysis of the interface waves are worth the effort. With enough paths a high resolution map of the upper sediments can be derived. Dispersion analysis of generated interface waves at two sites in the Pacific ocean have shown that substantial variability on the shear wave velocity structure may occur over short distances.

## References

Davies, D. (1965) 'Dispersed Stoneley waves on the ocean bottom', Bull. Seis. Soc. Am., 55, 903-918.

Dorman, L. D., Schreiner, A. E., and Bibee, L. D. (1991) 'The effects of shear velocity structure on seafloor noise statistics', this issue.

Dziewonski, A., Bloch, S., and Landisman, M. (1969) 'A technique for the analysis of transient seismic signals', Bull. Seis. Soc. Am., 59, 427-444.

Gomberg, J. S. and Masters, T. G. (1988) 'Waveform modeling using locked-mode synthetics and differential seismograms: applications to determination of the structure of Mexico', Geophys. J., 94, 193-218.

Hildebrand, J. A., Dorman, L. M., Hammer, P. T. C., Schreiner, A. E., and Cornuelle, B. D. (1989) 'Seismic tomography of Jasper Seamount' Geophys. Res. Lett., 16, 1355-1358.

Rauch, D. (1986) 'On the role of bottom interface waves in ocean seismo-acoustics: a review', in Akal, T. and Berkson, J. M. (eds.), Ocean Seismo-acoustics, Plenum Press, New York, pp. 623-641.

Sauter, A. W., Dorman, L. M., and Schreiner, A. E. (1986) 'A study of sea floor structure using ocean bottom shots and receivers', in Akal, T. and Berkson, J. M. (eds.), Ocean Seismo-acoustics, Plenum Press, New York, pp. 673-681.

Schreiner, A. E. and Dorman, L. M. (1990) 'Coherence lengths of seafloor noise: effect of ocean bottom structure', J. Acoust. Soc. Am., 88, 1503-1514.

# THE EFFECTS OF SHEAR VELOCITY STRUCTURE ON SEAFLOOR NOISE

LEROY M. DORMAN, ANTHONY E. SCHREINER
*Marine Physical Laboratory*
*Scripps Institution of Oceanography*
*University of California, San Diego*
*San Diego, CA 92093-0215*

L. DALE BIBEE
*Naval Oceanographic and Atmospheric Research Laboratory*
*Code 362*
*Stennis Space Center, MS 39529-5004*

ABSTRACT. The "microseism peak" and adjacent "noise notch" are prominent features of the earth's background noise spectrum between 0.08 Hz and about 5 Hz. They are caused by the combination of a peak in the source (ocean wave) spectrum and and by the presence of peaks in the spatial mode Q of the 4 gravest surface (and interface) wave modes. The peaks in the mode Q occur at a phase velocity of 1.5 km $s^{-1}$ and are due to the high Q of the water layer. The depressed mode Q at very low frequencies is caused by the low Q of the asthenosphere. The frequency of the low sides of the peaks in mode Q is strongly influenced by the ocean depth and the frequency of the high frequency side is controlled by the shear velocity structure of the shallow sediments.

## 1. Introduction

Seismic noise on the sea floor is very low at frequencies below 0.08 Hz, rising to a peak near 0.15 Hz and falling precipitously to very low values in the 5-10 Hz region (Webb and Cox, 1986; Sutton and Barstow, 1990; Schreiner and Dorman, 1990). This pattern is recognizable in recordings of pressure and in both vertical and horizontal motions of the sea floor. The amplitude ratios (or variance ratios) vary with frequency, horizontal components being distinctly noisier above about 0.4 Hz (Dorman and Schreiner, 1991).

By analyzing the dispersion of interface waves from explosions on the sea floor, we determined the shear velocity structure at DSDP site 469 (Sauter and others, 1986, Sauter, 1987) That site is in a deep-water sedimentary basin just off San Diego (see Figure 1). The surficial shear velocity is 30 m $s^{-1}$ and its gradient is $3.6 s^{-1}$. Later we reported measurements of the spatial and frequency dependence of the coherence of vertical seismic motion and showed that these data implied that the background noise field consists, to a significant degree, of interface waves. We showed that the excitation of the lower modes of the interface waves was consistent with scattering of acoustic energy from the nonlinear wave-wave interaction of ocean windwaves and swell into the seafloor waveguide (Schreiner, 1989, Schreiner and Dorman, 1990, Dorman and others, 1991).



*J. M. Hovem et al. (eds.), Shear Waves in Marine Sediments*, 239–245.

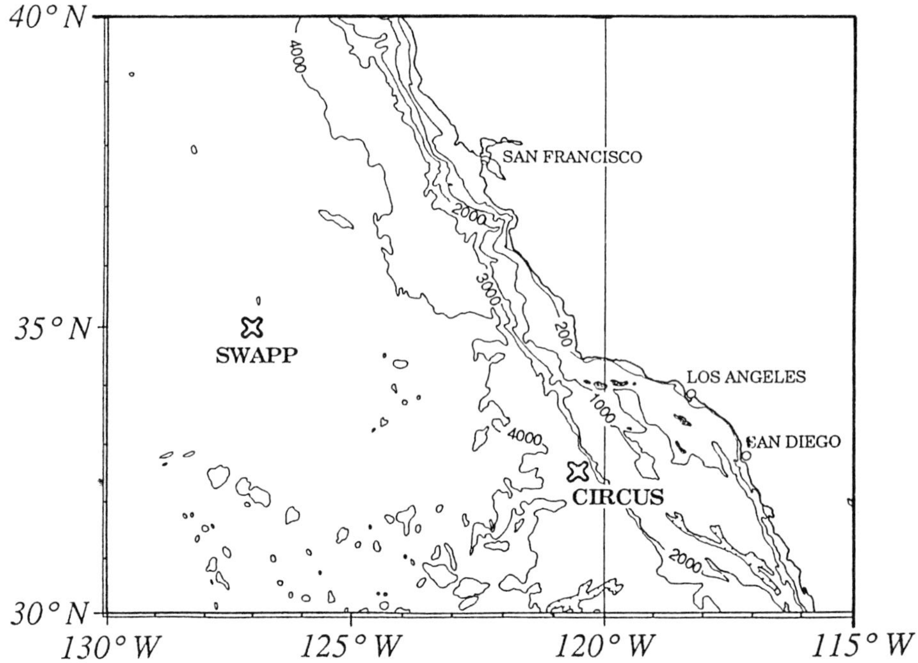


Figure 1. The experimental site is at the location of DSDP site 469, at 3800 meters water depth just past the Patton Escarpment which marks the edge of the California Borderlands. The sediment thickness at the drill site is 400 meters and thins toward the edges of the pond. The location marked SWAPP is the site of a different interface wave experiment.

The spatial and frequency dependence of the coherence shows that higher modes of interface waves dominate in the 0.8-4 Hz frequency range and that the propagation appears to be fundamental mode Rayleigh propagation at lower frequencies. We later used spectral ratios to refine the excitation pattern of the higher modes (Dorman and Schreiner, 1991). It thus appears that background noise can be well approximated by summing a small number of the gravest modes of surface waves (or interface waves).

## 2. Factors Affecting Noise Excitation

We can consider the physics of sea floor noise in three parts, generation or excitation, coupling, and propagation. Little is known about the coupling although scattering is certainly important at high (Stoneley wave) frequencies. We focus on the other two factors in this paper.

The most prominent source of sea floor noise at low frequencies is thought to be the nonlinear wave-wave mechanism first proposed by Miche and extended by Longuet-Higgins (1950) and by Hasselmann (1960). This theory shows that wavenumber spectral components of swell or wind waves traveling in opposing directions can generate acoustic waves with high phase velocity which can reach the floor of the deep ocean. The motions of the swell can only excite the sea floor at much lower frequencies than treated here.

Although the theory is widely accepted, it is difficult to exploit because the necessary frequency-wavenumber spectra of swell and wind waves are very difficult to measure. The amplitude of the acoustic wave generated by the nonlinear mechanism is proportional to the integral

$$I(\omega) = \int_{-\pi}^{\pi} G(\Theta,\omega) G(\Theta+\pi,\omega) d\Theta.$$

The directional factor $G(\Theta,\omega)$ at some frequency $\omega$ has only rarely been measured, (see for example Tyler and others, 1974) by radio wave scattering or by laboriously digitizing stereo photographs. In the absence of data, it is common to assume the $\cos^q$ dependence suggested by Longuet-Higgins and others (1963). Webb and Cox (1986) show that that this can, at times, produce a good fit over part of the microseism peak.

The remaining factor is that of propagation and we will argue next that that the propagation physics leaves an easily recognizable imprint on the observations. The size of the region of the ocean surface contributing to the noise observed at some location, assuming uniform excitation for simplicity, will be determined by the rate of attenuation of the propagation process. If attenuation is small, the region of excitation is large and we can expect large amplitudes. Conversely, if attenuation is high, the effective source region will be small and noise amplitudes will be low. The actual amplitudes observed will, of course, be the composite of the several transfer functions involved but each will leave its mark on the final product. By this argument, we see that the attenuation is a critical part of the propagation process.

Since it is the radius of the source region which is of greatest interest to us, we should examine the spatial Q of mode propagation in the ocean and the solid earth beneath. In dispersive media, where the phase and group velocity are different, temporal and spatial experiments will measure different values, whose ratios are the same as the the ratios of the phase and group velocity (see Aki and Richards, 1980, chapter 7). To make this calculation we must have a model and we use a laterally homogeneous flat earth.

### 2.1 THE VELOCITY STRUCTURE

The Stoneley wave analysis cited earlier provided the shear velocity of the upper few tens of meters. Seismic refraction experiments on oceanic crust of similar age (Spudich and Orcutt, 1980, Spudich and others, 1978) provide a model for the next 7 km. The rest of the lithosphere was taken from Orcutt and Dorman, 1977 and then a few hundred km of asthenosphere was taken from the PREM (Dziewonski and Anderson, 1981).

### 2.2 THE ATTENUATION STRUCTURE

Observation of the attenuation of the Stoneley waves constrained the shear Q in the upper 5 meters to the 22-45 range and the sediment shear Q below that was 100-110. Shear Q in the crust was taken to be 125 and upper mantle Q to be 250. Shear Q in the asthenosphere drops to 90-110. Figure 2 shows the model parameters.

## 3. Mode Calculations

Mode wavefunctions were calculated using the formalism of Woodhouse (1980) as implemented by Gomberg and Masters (1988). This calculation produces the stress-displacement vector in the solid as well as the phase and group velocities and attenuation as a function of frequency for the required modes.

The most interesting property for our purposes is the attenuation of the various modes. Figure

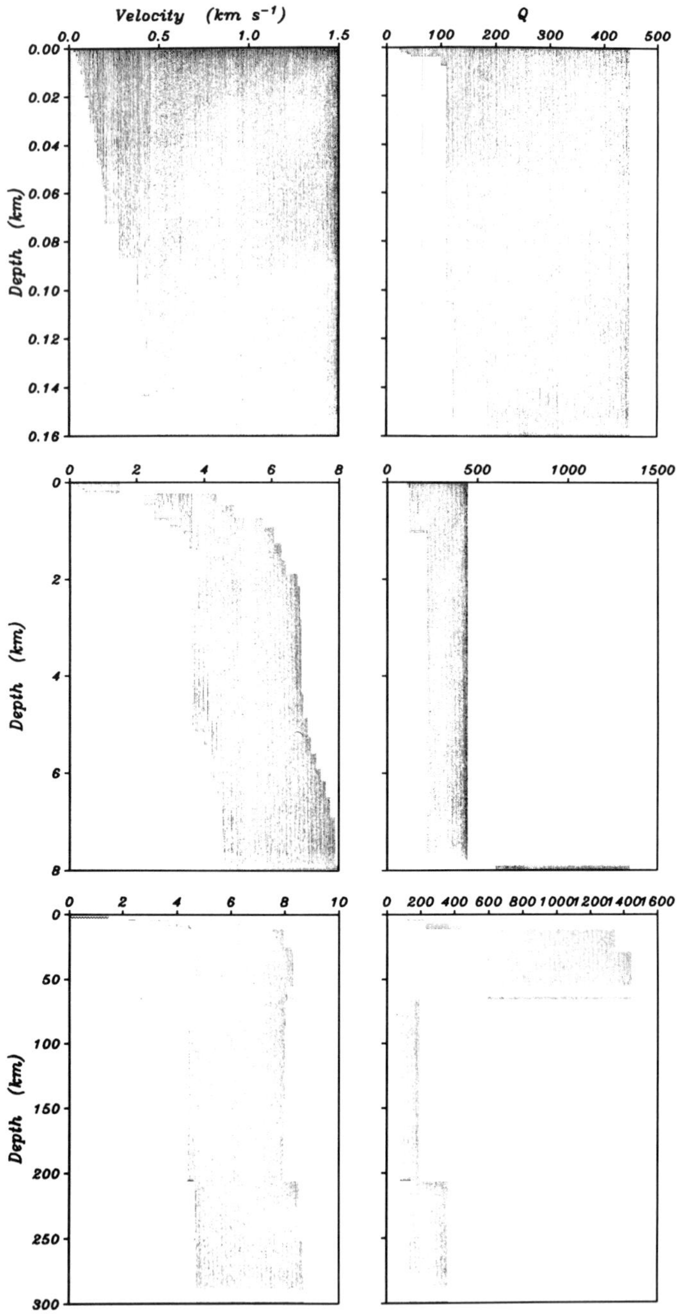


Figure 2. The oceanic structural model at three depth scales for velocity and Q; compressional wave values are dark gray, and shear wave values are light gray.

3a shows the spatial Q for the first 4 modes. The analysis of Dorman and Schreiner (1990) indicates that modes 0-3 are all that are important, at least at the DSDP 469 (CIRCUS) site. Examining each mode individually, we see that each mode has a prominent peak in Q, with the fundamental mode (0) having a peak at the lowest frequency and the higher modes having peaks at successively higher frequencies. In Figure 3b these same curves are replotted, but with phase velocity as the abscissa. Note that the peaks have been brought into alignment at a phase velocity of 1.5 km $s^{-1}$, so we can identify the high Q peak with the water layer of the model, which has high Q (effectively infinite compared to the solid earth layers).

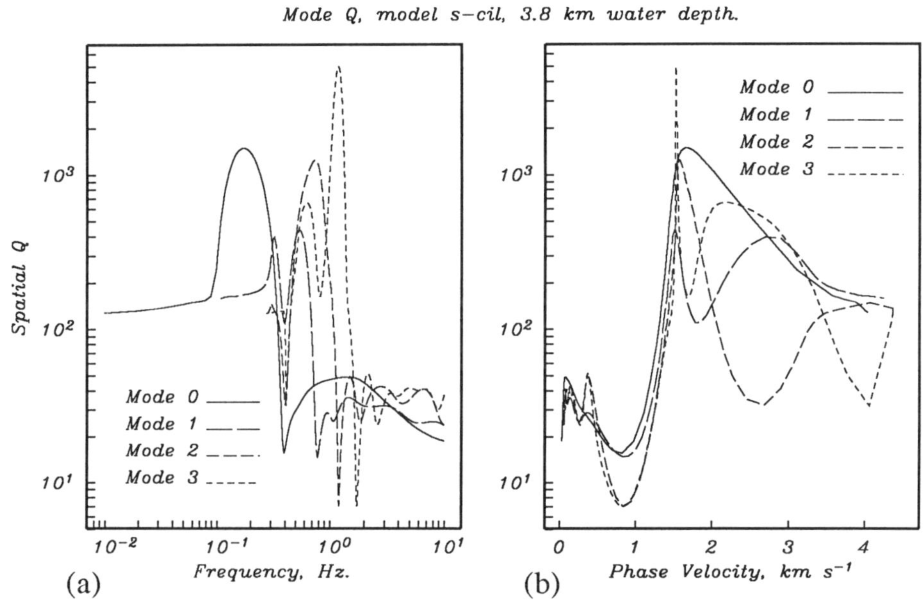


Figure 3. Spatial Q for the 4 gravest interface wave modes. (a) Q plotted as a function of frequency. (b) Q plotted as a function of phase velocity. Each mode has a peak at 1.5 km $s^{-1}$, the velocity of sound in water.

In Figure 4 we show spatial Q as in Figure for a model in which the thickness of the water layer is reduced by half. This reduces the peak Q reached by all the modes (because there is less high-Q material in the model) and raises the frequency of the low-frequency "edge" of the mode 0 peak. This happens because the low Q asthenosphere is, in terms of travel time, closer to the surface. The existence of the "noise notch" is thus, in part, a consequence of the presence of the asthenosphere. Other planets, such as Mars, which have cold interiors and do not permit plate tectonics, would not exhibit a "noise notch".

At the high frequency side of the graph, we see that the high side of the mode 3 peak corresponds to the steepest part of the microseism peak. The very low Q value to the right are a consequence of the low sediment Q and the very low group velocities of this part of the dispersion curve. Lower group velocity, of course, means that a energy takes a longer time to go a given distance.

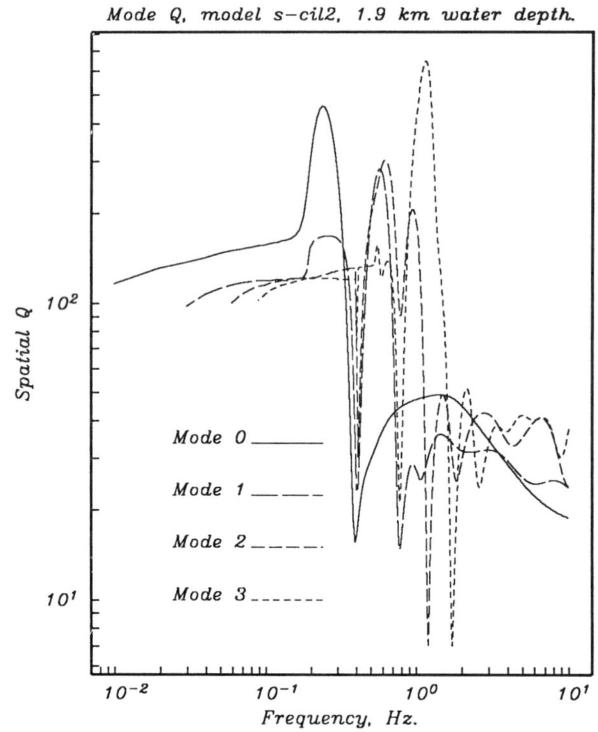


Figure 4. The same as figure 3a except that the thickness of the water layer is reduced from 3.8 km to 1.9 km.

The variance spectrum of wind waves typically falls off with frequency as $f^5$ or $f^6$ because of a phenomenon called saturation. Tall waves break and redistribute their variance to lower frequencies. Because of this, there is less excitation at the higher frequencies and the tall mode 3 peak in figure 3 is muted. In an ocean whose water depth is variable and whose seismic structure contains inhomogeneities and and thickness variations, the sharp peaks and troughs will be smoothed and the peaks will blend.

## 3. Conclusion

The "microseism peak" and "noise notch" exists because of the same physical properties in the earth which allow plate tectonics. The lower boundary of the peak is determined by the average depth of the oceans. The high-frequency side of the microseism peak is controlled by the properties of the upper few tens of meters of the sediments of the sea floor.

## References


Aki, K. and Richards, P. G., in *Quantitative Seismology: Theory and Methods,* W. H. Freeman and Co., San Francisco, Calif., 1980.

Dorman, L., Schreiner, A., Bibee, L., and Hildebrand, J., Deep-water sea-floor array observations of seismo-acoustic noise in the eastern Pacific and comparisons with wind and swell, in *Proceedings of conference "Natural Physical Sources of Underwater Sound"*, Kluwer, to appear, 1991.

Dorman, L. M. and Schreiner, A. E., On the Depth Dependence of Seismic Noise in the Sea Floor, *manuscript,* 1991.

Dziewonski, A. M. and Anderson, D. L., Preliminary Reference Earth Model, *Phys. Earth Planet. Inter., 25,* 297-356, 1981.

Gomberg, J. S. and Masters, T. G., Waveform modeling using locked-mode synthetic and differential seismograms: application to determination of the structure of Mexico, *Geophysical J., 94,* 193-218, 1988.

Hasselman, K. A., A statistical analysis of the generation of microseisms, *Rev. Geophys. Space Phys., 1,* 177-210, 1963.

Longuet-Higgins, M. S., A theory on the origin of microseisms, *Phil. Trans. Roy. Soc., Series A, 243,* 1-35, 1950.

Orcutt, J. A. and Dorman, L. M., An Oceanic Long Range Experiment, *J. Geophys., 43,* 257-263, 1977.

Sauter, A. W., Dorman, L. M., and Schreiner, A. E., A study of sea floor structure using ocean bottom shots, in *Ocean seismo-acoustics, low frequency underwater acoustics,* edited by T. Akal and J. M. Berkson, vol. 16, Plenum Press, New York, N.Y., 1986.

Sauter, A. W., Studies of the upper oceanic sea floor using ocean bottom seismometers, PhD. thesis, 100 pp., University of California, San Diego, La Jolla, Ca., 1987.

Schreiner, A. E., Observations of Seafloor Ambient Noise With an Ocean Bottom Seismometer Array, PhD. thesis, 111 pp., University of California, San Diego, La Jolla, Ca., 1989.

Schreiner, A. E. and Dorman, L. M., Coherence lengths of seafloor noise: effects of ocean bottom structure, *J. Acoust. Soc. Am.,* 1990.

Spudich, P. K. and Orcutt, J. A., Petrology and porosity of an oceanic crustal site: results from wave modeling of seismic refraction data, *J. Geophys. Res., 85,* 1407-1433, 1980.

Spudich, P. K. P., Salisbury, M. H., and Orcutt, J. A., Ophiolites found in oceanic crust?, *Geophys Res. Letts., 5,* 341-344, 1978.

Sutton, G. H. and Barstow, N., Ocean-bottom ULF Seismo-acoustic Ambient Noise .002 to .4 Hz, *J. Acoust. Soc. Am., 87,* 2005-2012, 1990.

Tyler, G. L., Teague, C. C., Stewart, R. H., Peterson, A. M., Munk, W. H., and Joy, J. W., *Deep Sea Res., 21,* 989-1026, 1974.

Webb, S. C. and Cox, C S., Observations and modeling of seafloor microseisms, *J. Geophys. Res., 91,* 7343-7358, 1986.

Woodhouse, J. H., Efficient and stable methods for performing seismic calculations in stratified media, in *Proceedings of the International School of Physics Enrico Fermi Course,* vol. 78, pp. 127-151, North Holland, Amsterdam, Netherlands, 1980.

# CONCURRENT OBSERVATIONS OF DIRECTIONAL SPECTRA OF OCEAN SURFACE WAVES AND MICROSEISMS FROM AN OCEAN SUBBOTTOM SEISMOMETER (OSS) ARRAY.

TOKUO YAMAMOTO and THOMAS NYE
University of Miami
Geo-Acoustics Laboratory (UM-GAL)
4600 Rickenbacker Causeway
Miami, Florida 33149
and
DEAN GOODMAN
UM/GAL Japan
Nakajima, Ishikawa, Japan

ABSTRACT. Analysis of spatially and temporally coherent seafloor motion is conducted using data from an OSS array deployed by the University of Miami's Geo-Acoustics Laboratories. Data from the Atlantic Generating Station 7-point array are used to measure the directional spectra of seafloor microseisms. The directional spectra of ocean surface waves are measured concurrently with a single OSS using the Maximum Entropy Method. Directional spectra phase velocities are found to be consistent with Scholte wave phase velocities calculated for the study area using the geological data measured by the Bottom Shear Modulus Profiler (BSMP) method (Trevorrow and Yamamoto, 1990). Scholte wave microseisms become strong during the period of two opposing seas and the propagation directions are in between the direction of propagation of the opposing ocean gravity wave spectra. During periods of non-opposing sea states, microseisms are weak and spatially incoherent. These observations are consistent with the theory for the origin of microseisms by Longuet-Higgins (1950) and Hasselmann (1963) - that they are a result of nonlinear interactions of opposing ocean gravity waves, and that the microseisms travel in a direction represented by the sum of the wave number vectors of the opposing ocean gravity waves.

## 1. Introduction

Documentation of the mechanisms of microseims evolution is the purpose of this paper. The energy transfer from the ocean waves to the seafloor creates a 'double frequency' seafloor motion, (Hasselman, 1963). This double frequency effect has been observed by University of Miami Geo-Acoustics Laboratory experiments in shallow



J. M. Hovem et al. (eds.), Shear Waves in Marine Sediments, 247–256.

water (Goodman et al., 1989) and relatively deeper water (Trevorrow et al., 1988). From measurements of the particle motion, the direction of wave propagation, and the phase velocity, Goodman et al. (1989) concluded that the dboule freqency microseims propagate as Scholte waves.

Figure 1 shows two averaged power spectra of ocean floor seismic fluctuations from OSS 306 in shallow water, 12.5 meters deep. (OSS units are buried 0.5 meters below the seafloor). The two spectra were averaged over a two hour time period on Aug. 20th (E1) and Aug. 21st (H1). The seismic channels show two peaks in the spectra. The large peak centered near 0.15 Hz is the response of the seabed to the passing ocean surface waves (Yamamoto et al., 1978). The smaller peak (centered at 0.5 Hz and present only during the E1 time period) is due to the 'double frequency' microseisms. The water depth acts as a physical low pass filter to the water wave induced (single frequency) seismic energy, while it has little or no effect on the double frequency seismic energy. Due to the width of the spectra, single frequency microseisms and double freuquency microseisms may coexist at the same frequency band (approximately 0.1–0.3 Hz). Since the energy level of the single frequency microseisms is much higher than that of the double frequency microseisms, it is difficult to identify the latter in the overlapping frequency band. Therefore, one must pay special attention to frequencies above the hydrodynamic cutoff in the study of the evolution of 'double frequency' microseisms. For this experiment the cutoff frequency is approximately 0.3 Hz.

The two seismic spectra indicate that the 'double frequency' microseism activity changes quickly with time. Contemporaneous data from an array of OSS's was analyzed to test whether the propagation of the microseisms was of a coherent or incoherent nature. Knowledge of the phase and level of spatially coherent microseisms were used to estimate the directional spectra of the disturbance based on the Maximum Likelihood Method described by Goddman et al. (1989). The aperture of the array was tuned to measure microseisms and therefore was not conducive to finding estimates of the directional spectra of the ocean surface gravity waves. An estimation of this spectra was found using measurements from a single OSS and the Maximum Entropy Method of directional resolution called the Buried Ocean Wave Directional Spectrometer (BOWDS) method (Nye et al., 1990a). The contemporaneous state of the ocean surface wave spectra measured by BOWDS was correlated with the microseisms directional spectra measured by the OSS array. This relationship can help to shed some light on the appropriate generating and propagating mechanisms for microseisms in shallow ocean waters.

## 2. Time History Observations of Winds, Ocean Surface Waves & Microseisms

A 7-point OSS array was deployed at the Atlantic Generating Station

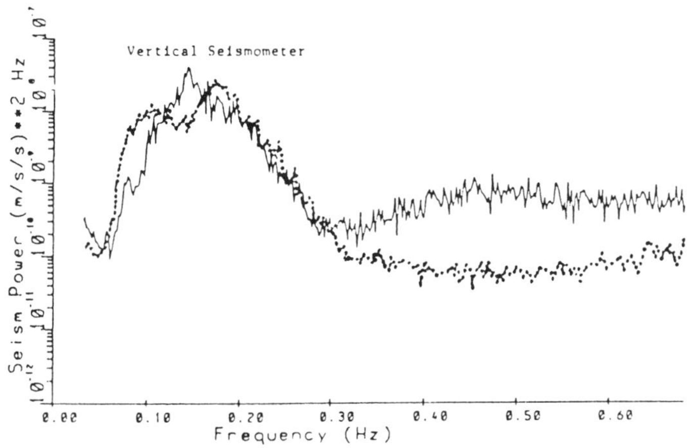


Figure 1. Comparison between vertical seismic power spectra for E1 (Aug 20, 00:20 hrs.) and H1 (Aug 21, 03:20 hrs). Spectra are averaged over two hours.

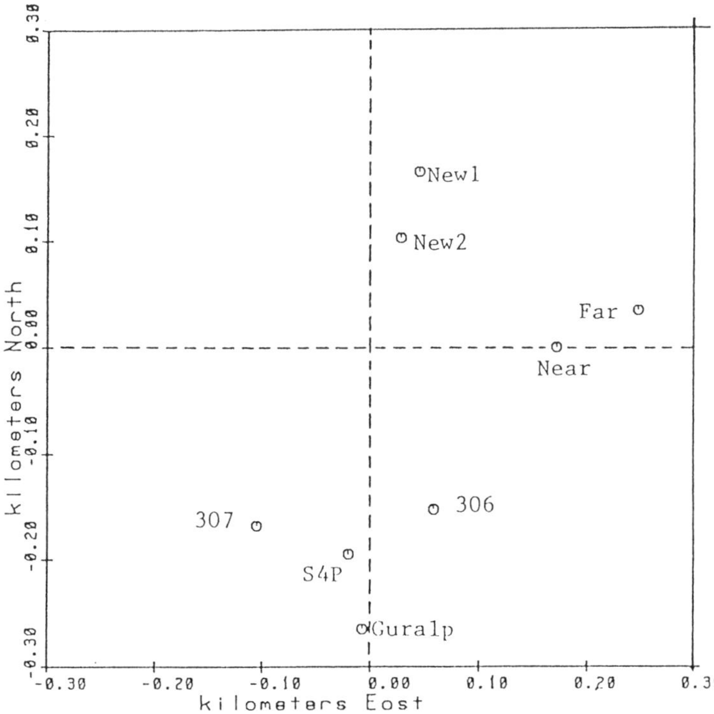


Figure 2. Location of BSMP/OBS experiments in August, 1989. Origin of East/North coordinates is N 39o 28.72', W 74o 14.62'. Water depth is 12 meters.

experimental site as shown in Figure 2. Element S4P had a pressure gauge and a current meter and therefore is not an OSS package. Time history observations of contemporaneous winds, ocean surface waves and microseisms were conducted using this OSS array. The OSS instrumentations, the method of deployment, the methods of data acquisition and analysis were the same as in Goodman et al. (1989) and Turgut et al. (1989).

### 2.1 WINDS

The time history observations of the wind speed and wind direction together with visual observations of sea state were made during the period 1989.08.17, 23:00 EDT to 08.22, 08:00 EDT (Nye et al., 1990b). It should be noted that after a 24 hour period of NE 25 knot winds, a wind shift to SW 20 knots took place on August 19. The SW wind prevailed thereafter. This timely wind shift provided an excellent opportunity to document the mechanisms of microseisms generation through contemporaneous observations of ocean gravity waves and microseisms directional spectra.

### 2.2 OCEAN SURFACE GRAVITY (WATER) WAVES

The directional spectra of ocean waves were measured by processing the data every two hours from a single OSS by the BOWDS method (Nye et al., 1990a). The directional spectra of ocean waves measured at four different instances during the observation period are shown in Figure 3. In the figures, the wave direction is the bearing in degrees of the direction which waves propagate *towards* as defined by the Meteorological Society. The significant wave height $H_s$ is also given in each figure. OSS "Far" was used to calculate the directional spectra during the time period A2 (Aug. 18. 17:45 to 19:45) because only this OSS was deployed at this time. Because the pressure signal of OSS "Far" was clipped, the wave height was estimated using the vertical seismic channel and a measured transfer function between the vertical seabed acceleration and the bottom pressure (Nye et al., 1990b). The spectra for the time periods E1 (Aug. 20, 00:20–02:20), F3 (Aug. 20, 14:20–16:20) and H1 (Aug. 21, 03:20–05:20) were measured by OSS "306". Pressure and all three orthogonal seismic channels of this OSS functioned.

The series of directional spectra A2, E1, F3, and H1 vividly show the changes of the sea wave directional spectrum during the observation periods and reflect the wind condition changes. Due to the NE 25 knot winds of several hours' duration prior to the observation, the ocean wave spectrum "Far A2" shows a strong single sea moving toward S–SW. As shown in the spectrum 306 E1, (Aug. 20, 00:20–02:20), the wind shift from NE 25 knots to SW 20 knots several hours prior to this time period created bi–directional sea conditions which were conductive to generating microseisms. Because the wind direction remained steady after the wind shift for at least 24 hours, the bi–directionality of

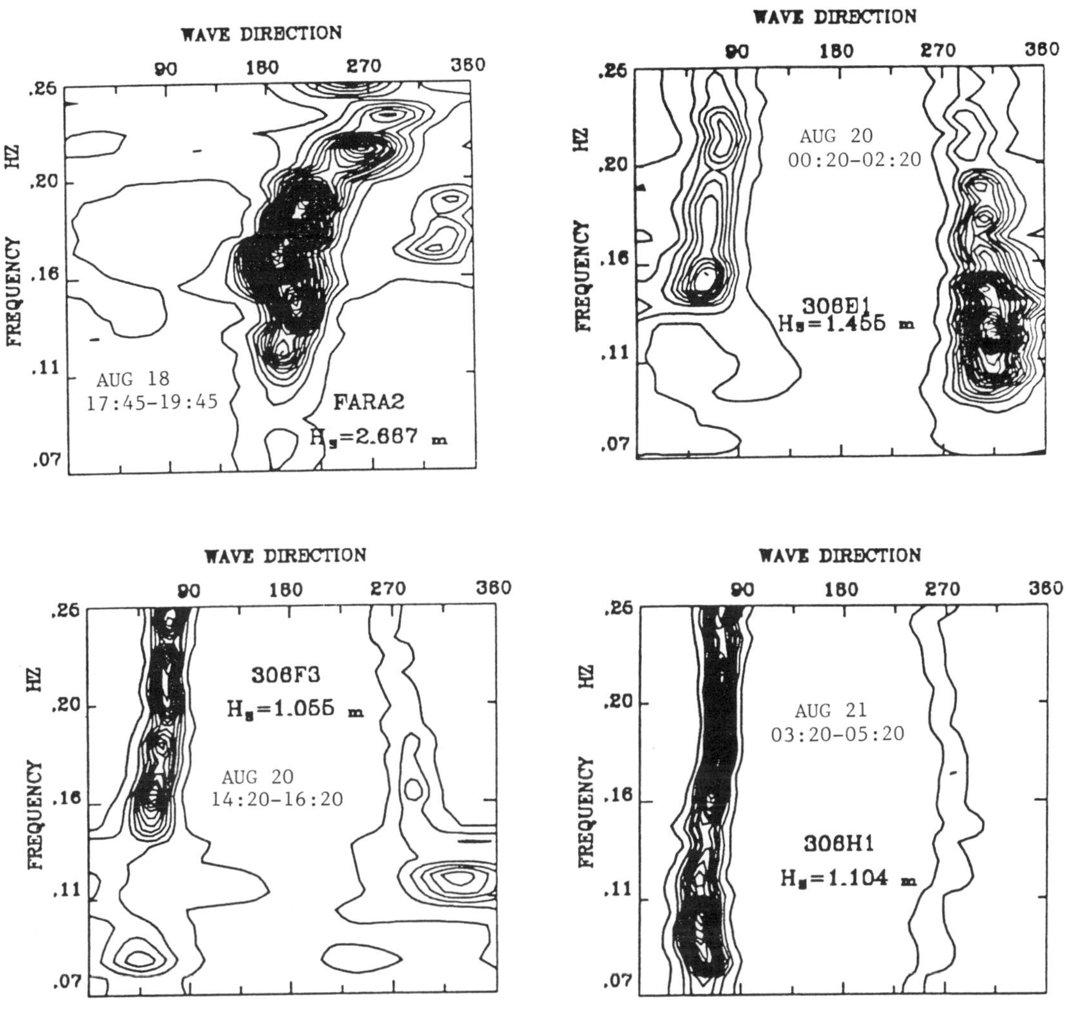


Figure 3. Directional spectra of ocean surface gravity waves, at four different time periods. Wave energy contour interval is 0.10 $m^2$/Hz/rad.

the sea diminished in time as depicted in spectra F3 and H1. The dramatic difference in the microseisms energy spectra for the time periods E1 and H1 is shown in Figure 1.

## 2.3 MICROSEISMS

Contemporaneous measurements of spatial coherence for the ocean wave states (E1, F3, H1) are shown in Figure 4(a–c). The spatial coherence between vertical seismometers was measured between OSS units 306 and New2 (range = 258 m) for the three data sets. This figure indicates that the highest coherence in the microseismic band (1–3 seconds) occured during time period E1 when the ocean wave directional spectra depicts strong opposing ocean wave components. Figure 4b indicates less coherence in the microseismic band as the level of opposing ocean wave state diminishes. For time period H1, figure 4c shows no measureable coherence in the microseismc band. Directional spectra of the time window August 20, 00:20–06:19 EDT, is shown in figure 5. Only four OSS elements; New2, Near, 306, and 307 were active during this time span and thus, used in the calculation of the directional spectra shown in figure 5. This contour plot represents the directional spectra at the discrete frequency of 2.26 seconds (which represents a 4 frequency bin average). The x–axis is phase velocity in m/s, while the y–axis represents azimuthal direction that the waves propagate *towards*. This data indicates that primary phase velocities between 200–400 m/s are the most likely velocities of seafloor microseisms in the study area. The direction that these microseisms propagate *towards* is approximately $20^{\circ}$, which is near the vector sum of the two arms of the bi–directional spectra of the ocean surface above. The small bimodal response toward $200^{\circ}$ is a result of the sidelobes of the four element array used in the calculation and is not real microseism energy. The directional spectra corresponding to the weak bi–directional sea state was computed by Goodman (1990). This indicated less directional resolution compared to the peak directions shown in Figure 5. This is an indication of the weakening state of the opposing sea waves, or conversely, the directional irresolve of the microseisms signify the end to the bi–directional sea state. The time window during the uni–directional sea state (H1) shows there to be no spatial coherence; thus, Maximum Likelihood Method directional spectra cannot be estimated.

A Haskell–Thompson matrix method was used to estimate the propagation of the Scholte wave mode through the seismic structure obtained from the BSMP measurements (Trevorrow and Yamamoto in this volume). A homogeneous elastic halfsace, which included a 400 m/s shear wave and a 2400 m/s compressional wave was assumed below the measured structure. The dispersion of the Scholte wave was found for the first mode only. The higher modes calculated were above the highest shear wave velocity in the profile and thus not a Scholte wave mode (Goodman 1990). The single Scholte wave mode discovered was only slightly dispersive below 2 seconds, and was nondispersive above 2 seconds.

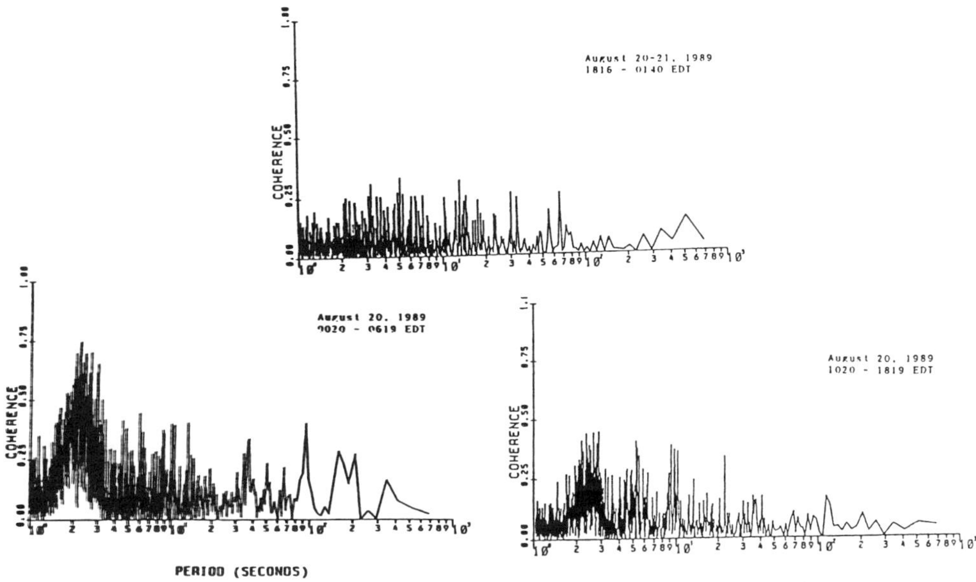


**Figure 4. Time history of spatial coherence measured between OBS 306 and NEW2 (range = 258 m) during the 4-point array study.**

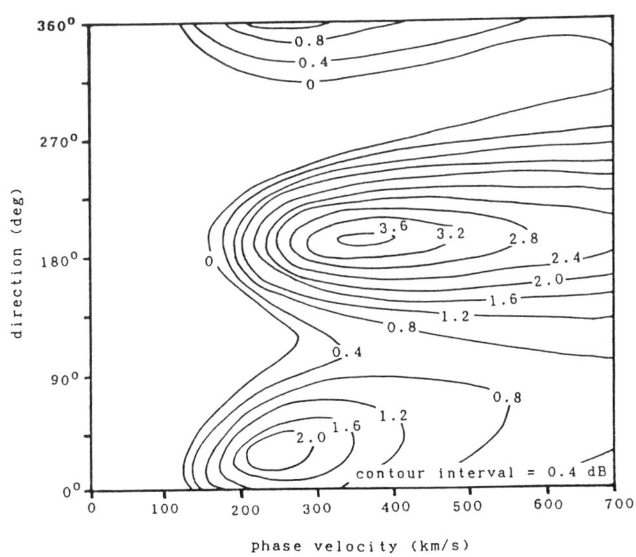


Figure 5. Directional spectra of microseisms (Aug 19–20, 1989, 2200–0700 EDT). Contours represent relative dB power predicted by the Maximum Likelihood Method. Scholte wave period = 2.26 seconds.

For the wave period of 2.26 seconds, the propagation velocity of the Scholte wave mode from the dispersion relation was 376 m/s. The measured propagation velocity of the microseism Scholte wave was found to be 365 m/s at its maximum energy density (see Figure 5). The measured velocities are reasonably close to modeled phase velocities of the Scholte wave mode.

A spectral decomposition technique was applied to analyze the 3–component seismic data collected from this study (Goodman, 1990). The purpose of this analysis was to determine the relative energy levels of Scholte type motion to that of Love wave type motion of the seabed. The total coherent energy in Scholte or vertically polarized (PSV) waves to that of Love or horizontally polarized (S + SH) waves was computed and is shown in Figure 6 for the microseisms, and gravity wave bands. The average value estimated for E PSV/S+SH across the microseims band is approximately 7.0 indicating that PSV–Scholte wave motions dominated SH–Love wave motions for this data set. The large ratios observed in the ocean wave band (Figure 6) confirm that the seabed motion was primarily vertically polarized as expected from the theory (Yamamoto et al. 1978).

The ellipticity of the spatially coherent microseisms is of interest. The measured ellipticity (dots) as compared with that found from applying the Haskell–Thompson method (line) is shown in Figure 7. The modeled and measured ellipticities are reasonably close.

## 4. Conclusions

The corroborative data for Longuet–Higgins' (1950) mechanism for microseism generation was obtained by a 7–point buried seismometer array. Ocean gravity (water) waves are shown to have significant opposing wave energies by the single OSS, MEM directional spectra estimation. The results show that the microseismic energy propagated as Scholte wave and occurred only when there was a bi–directional sea and that the resulting spectra propagated in a direction near the vector sum of the two arms of the opposing seas. These results are consistent with Longuet–Higgins' theory. The measured phase velocities and particle motions of microseisms were in reasonably close agreements with model predictions.

## Acknowledgements

This work was supported under Office of Naval Research contract numbers N00014–89–J–1146 and N00014–90–J–1019. Special thanks are due to Mark Trevorrow for kindly assisting us in data processing and spectral computations. We would also like to thank Mark Trevorrow, Altan Turgut, Humberto Guarin, Andrew Rogers and Koichi Ando for their assistance during the sea cruises using the R/V Cape Henlopen.

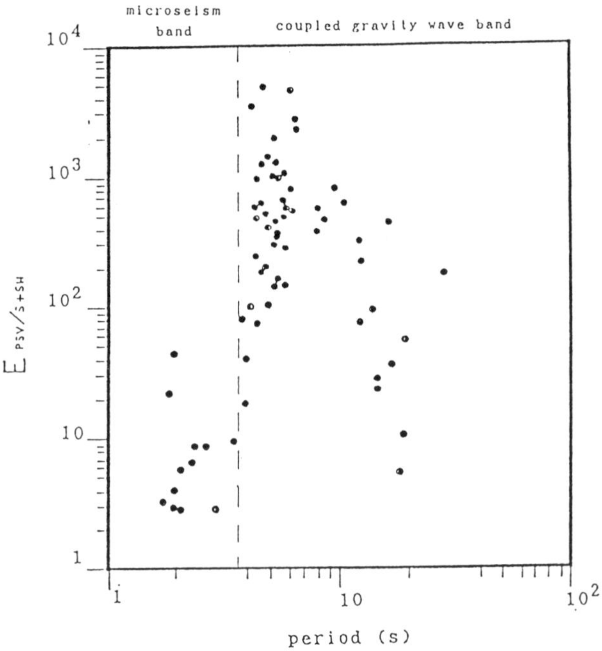


Figure 6. The ratio of coherent vertically polarized waves to coherent horizontally polarized waves.

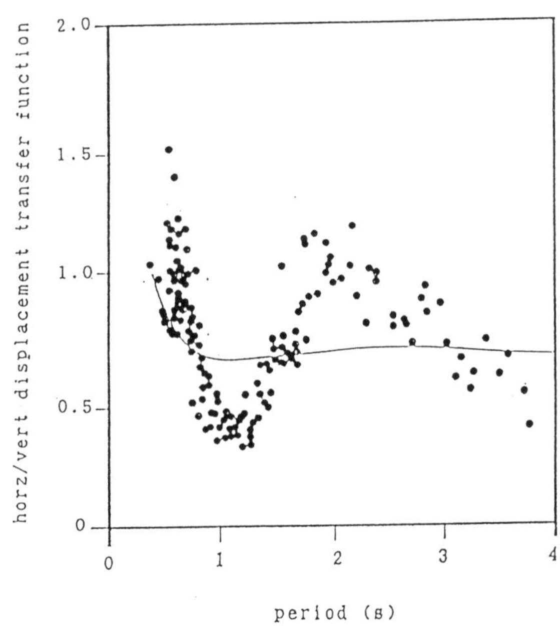


Figure 7. Comparison of measured ellipticities and those values determined from applying the Haskell-Thompson matrix method to the known geophysical profiles at the AGS study area.

## 4. References


4.1 Goodman, D., (1990) 'Analysis and observations of spatially coherent seafloor microseisms', Ph.D. Dissertation, University of Miami, RSMAS, Div. of Applied Marine Physics, Aug. 1990.

4.2 Goodman, D., T. Yamamoto, M. Trevorrow, C. Abbott, A. Turgut, M. Badiey, and K. Ando, (1989) 'Directional spectra observations of seafloor microseisms from an ocean-bottom seismometer array', J. Acoust. Soc. Am., 86(6), 2309–2317.

4.3 Hasselmann, K.A., (1963) 'A statistical analysis of the generation of microseisms' Rev. Geophys. 1, 177–210.

4.4 Longuet-Higgins, M.S. (1950) "A theory of the origin of microseisms', Philos. Trans. R. Soc. London, Ser., A243, p1–35.

4.5 Nye, T., Yamamoto, T., and Trevorrow, M., 1990. 'Measurements of the directional spectra of shallow water waves using the maximum entropy principle and a single ocean bottom seismometer', J. Atmos. Ocean Tech. V.7, No. 5, p 781–791.

4.6 Nye, T., Yamamoto, T., Trevorrow, M., Turgut, A., Rogers, A., and Goodman, D. (1990) 'Summary of 1989 single OBS directional spectra of ocean surface gravity wave measurements from New Jersey shelf and Miami sites', RSMAS Tech. Report 90–002, GAL Report 1010, University of Miami.

4.7 Trevorrow, M., Yamamoto, T., Turgut, A., Abbott, C., Badiey, M., Goodman, D. and Ando, K. (1988) 'High resolution bottom shear modulus profiler experiments on the New Jersey Shelf, summer 1987', RSMAS tech. rep. 88–002, University of Miami.

4.8 Trevorrow, M. and Yamamoto, T. (in press 1990) 'Sedimentary shear modulus and shear speed profiles from a gravity wave inversion', Proc. NATO conference on SHEAR WAVES .

4.9 Turgut, A., Yamamoto, T., Trevorrow, M. and Goodman, D., (1989), 'An OBS array for observations of directional spectra' Current practices and new technology in Ocean Engineering, Vol. 13, 1988.

4.10 Yamamoto, T., Koning, H.L., Sellmeijer, H., and Van Hijum, E. (1978) 'On the response of a poro-elastic bed to water waves', J. Fluid Mech. 87, 193–206.

# WAVE PROPAGATION IN A BOREHOLE

M.N. TOKSÖZ and C.H. CHENG
*Earth Resources Laboratory*
*Department of Earth, Atmospheric, and Planetary Sciences*
*Massachusetts Institute of Technology*
*Cambridge, MA 02139, U.S.A.*

ABSTRACT. In full waveform acoustic logging, a pressure source generates various types of elastic waves propagating down the borehole. These waves, because their wavelengths are of the order of the borehole radius, are guided waves. Depending on the velocity of propagation, these guided waves are either normal modes, like the Stoneley wave, where all the energy is contained around the borehole, or leaky modes, like the P leaky mode, where some of the energy is converted into body waves and radiates away from the borehole into the formation. Both the normal and leaky modes are significantly affected by the P and especially the S wave velocity of the formation. Other physical parameters affecting their propagation are attenuation, permeability and anisotropy. With proper processing techniques in combination with inversion algorithms, we can obtain estimates of the S wave velocity even in soft marine sediments where no refracted S wave arrival exists on the full waveform logs.

## 1. Introduction

Since its introduction in the late 1970's, full waveform acoustic logging has held the promise of providing shear wave velocity measurements in a borehole. The first generation monopole tool, where the source radiation is axisymmetric with respect to the borehole axis, can indeed provide a direct shear wave velocity measurement using the refracted shear head wave and/or psuedo-Rayleigh wave arrival time. However, this is only possible in "hard" or "fast" formations where the shear wave velocity is higher than the acoustic velocity of the borehole fluid. This is clearly not the case for marine sediments. In this paper we outline the approaches to determine the in situ shear wave velocity of soft marine sediments through borehole acoustic logging.

### 1.1 WAVE PROPAGATION IN A BOREHOLE

A lot of work has been done studying the problem of elastic wave propagation in a fluid-filled borehole, starting with Biot (1952). A comprehensive description is given in Toksöz et al. (1984). There are compressional and shear head waves which are refracted along the borehole boundary; the latter exist only in formations where the shear wave velocity is higher than the acoustic velocity of the borehole fluid, and thus will not exist in soft marine sediments. Associated with these head waves are resonances, the P leaky mode and the psuedo-Rayleigh wave. These exist because the wavelengths are comparable to the size of the borehole. Thus they have cutoff frequencies below which they cannot exist. The P leaky mode is so called because of its continuous conversion into shear waves radiating away from the borehole boundary. This is necessary to satisfy the boundary



*J. M. Hovem et al. (eds.), Shear Waves in Marine Sediments*, 257–265.

conditions. In contrast, the pseudo-Rayleigh wave is a properly guided normal mode, with no leakage of energy. Phase velocity is bounded by the shear wave velocity of the formation; its amplitude decays away from the boundary. However, since it is a result of the resonances of the shear waves and the borehole, just like the refracted shear waves, the pseudo-Rayleigh wave does not exist in formations made up of soft sediments.

A final wave type is the Stoneley wave. It exists because of the interface between the solid formation and the liquid borehole. Its amplitude decays away from the borehole boundary in both the solid and the liquid. It does not have a low frequency cutoff. At high frequencies, the curvature of the borehole surface becomes insignificant and the Stoneley wave velocity approaches that of the Scholte wave velocity of a liquid half-space over a solid half-space. In soft sediments, the phase velocity of the Stoneley wave depends significantly on the shear wave velocity of the formation (Cheng et al., 1982; Cheng and Toksöz, 1983).

## 2. Shear Wave Velocity Determination in Soft Sediments

Because of the lack of a refracted shear head wave and the associated pseudo-Rayleigh wave, direct detection of shear wave velocity from full waveform acoustic logs in soft marine sediments is impossible. We have to resort to indirect methods and to rely on the amplitude of the P leaky mode and the phase velocity of the Stoneley wave.

### 2.1 SHEAR WAVE VELOCITY FROM STONELEY WAVE VELOCITY

As mentioned above, the Stoneley wave is an interface wave propagating along the borehole fluid-solid interface. Its velocity approaches that of the Scholte wave velocity at high frequencies and thus is controlled by the shear wave velocity of the formation. Figure 1 shows the partition coefficient of the Stoneley wave in soft sediments (after Cheng et al., 1982). The partition coefficient is the normalized partial derivative of the phase velocity of the Stoneley wave with respect to the body wave velocities. From Figure 1 we can see that a change in the shear wave velocity of the formation will cause a change in the Stoneley wave velocity of nearly the same magnitude. By determining the phase velocity of the Stoneley wave, and knowing other relevant parameters such as the formation P wave velocity and density, the borehole fluid density, the borehole radius, and the frequency of the Stoneley wave, we can invert for the shear wave velocity of the formation. Examples are given in Zemanek et al. (1984) and Stevens and Day (1986).

More recently, however, it has been demonstrated that the phase velocity of the Stoneley wave is significantly affected by the permeability of the formation, especially in high permeability zones (Williams et al., 1984; Cheng et al., 1987,among others). Thus it is not possible to estimate shear wave velocity from Stoneley wave velocity in high permeability formations without taking the permeability into account. In addition, in soft sediments the excitation of the Stoneley wave is restricted to low frequencies (less than 5 kHz). Thus in a lot of the earlier generation full waveform tools, the Stoneley wave was not excited because these tools have center frequencies between 10 to 15 kHz. Under such conditions, no Stoneley wave was identifiable in the full waveform log. An alternative way of estimating shear wave velocity is to use the amplitude information contained in the P head wave and P leaky mode.

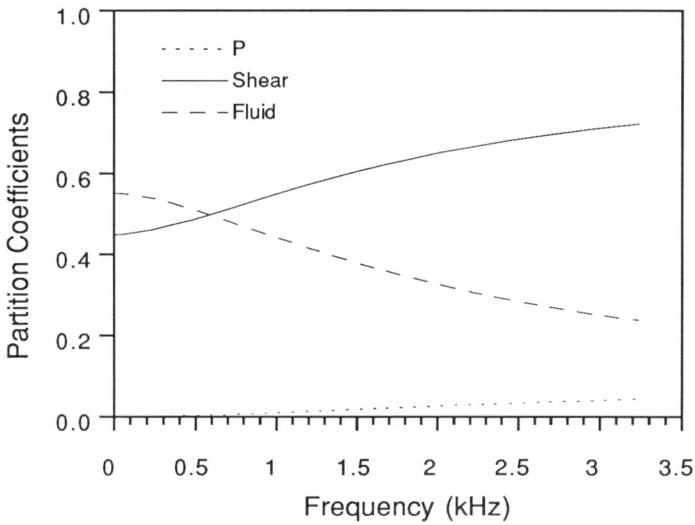


Figure 1. Partition coefficients (normalized partial derivatives) versus frequency for the Stoneley wave in soft sediments (after Cheng et al., 1982).

## 2.2 SHEAR WAVE VELOCITY FROM P WAVE AMPLITUDE

There are two major factors affecting the amplitude of the P head wave and P leaky mode propagating in a borehole. The first and most easy to understand is the formation P wave attenuation ($1/Q_p$). The second is the constant conversion and subsequent leakage to shear waves at the borehole fluid-solid boundary (Figure 2). This conversion depends on the Poisson's ratio (equivalently, the P- to S-wave velocity ratio) of the formation. Examples of the variation in the P wave amplitude as a function of formation Poisson's ratio can be found in Toksöz et al. (1985).

Cheng (1989) presented a method of inverting the P wave amplitude for formation shear wave velocity and P wave attenuation. The algorithm consists of windowing the P wave and P leaky mode from the full waveform logs recorded at a pair of receivers, taking the Fourier spectral amplitude ratio of the two waveforms, and inverting the spectral ratio for the shear wave velocity and P wave attenuation using a standard non-linear damped least-squares inversion algorithm. The forward model used for the inversion is the P wave and P leaky mode generated by a branch cut integration algorithm (Tsang and Rader, 1979). This forward method is quite fast when compared to a discrete wavenumber summation type algorithm. The parameters needed for input are the P wave velocity and density of the formation and the borehole fluid, the borehole radius, and the attenuation of the borehole fluid. The algorithm was tested on both synthetic and real data. Please refer to Cheng (1989) for a detailed description of the algorithm.

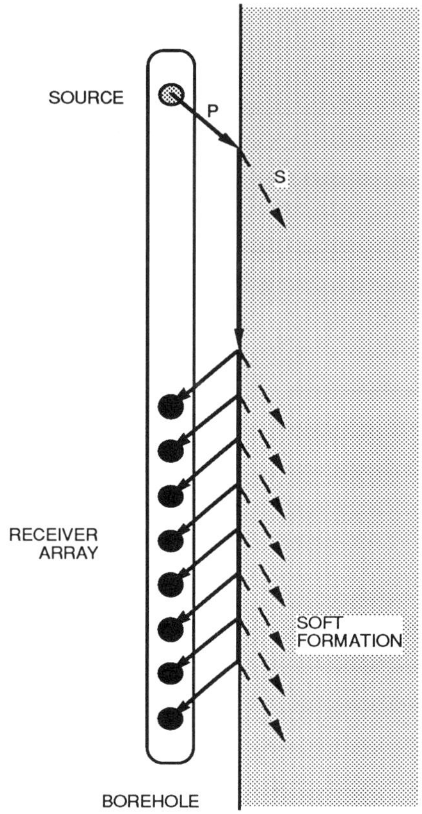


Figure 2: Schematic figure of conversion of P head wave along the borehole boundary and re-radiation into the formation as shear wave.

## 2.3 EXAMPLE FROM DSDP SITE 613

We have applied the method of Cheng (1989) to full waveform acoustic logging data collected on DSDP Leg 95 at site 613. The site is located near the Baltimore Canyon Trough on the eastern North American continental slope. Water depth at the site is 2300 m. Three major lithological units were logged. Lithologic Unit 1, the uppermost layer, recorded clay rich terrigenous deposition. Below a major unconformity, Lithologic Units 2 and 3 were comprised of siliceous carbonates with low clay content (Wilkens et al., 1986). The major difference between Lithologic Unit 2 and Lithologic Unit 3 is a phase change in the silica component from radiolarian and diatom tests composed of opal-A (Unit 2) to porcellanite cement composed of opal-CT (Unit 3). The phase change of silica resulted in a 15-20% decrease in porosity and an increase in bulk density of sediments across the Unit 2 – Unit 3 boundary. The boundary between Units 2 and 3 is the widely recognized reflector Horizon $A^C$ seen in seismic data from much of the western Atlantic basin (Tucholke, 1981). Logging began at 120 mbsf (meters below sea-floor) in Unit 1, proceeded into Unit 2 from 279 mbsf to 442 mbsf, and finished at a bottom hole depth of 570 meters in Unit 3. Data recording problems spoiled much of the data from Unit 1, so the analysis will be restricted to the two carbonate units.

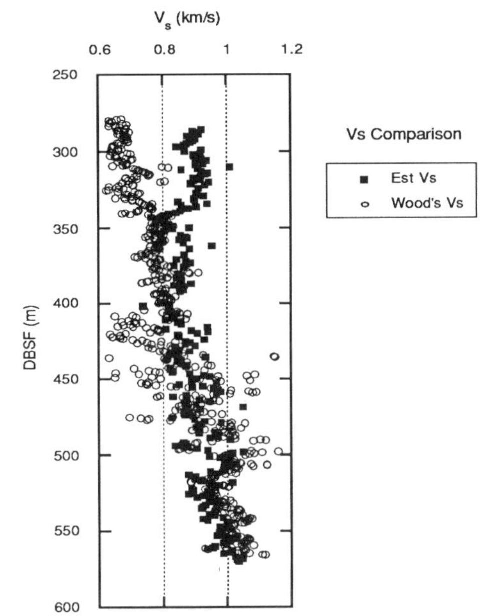


Figure 3: Shear wave velocities from inversion (filled squares, Cheng, 1989) of full waveform acoustic log data and those estimated by taking the difference between log $V_p$ and calculated $V_p$ from Wood's (1930) equation (open circles) for DSDP site 613.

Figure 3 shows the results of the inversion. Also shown is the shear wave velocity estimated by taking the difference between the observed P wave velocity and the P wave velocity calculated using Wood's (1930) equation given the constituents and the porosity. Wood's equation assumes that there is no rigidity to the system, and thus this method gives the theoretical upper bound of the shear wave velocity in semi-consolidated sediments. The inverted shear wave velocity is less than or equal to this estimate from about 340 mbsf and below. In general, the shear wave velocity falls in the range of 0.8 to 1.0 km/sec, consistent with recent laboratory measurements in similar marine sediments (Manghnani, personal communication).

There are several possible reasons why the inverted shear wave velocities are higher than the Wood's shear wave velocities at shallower depths (above 340 mbsf): the first reason is that as the velocities get slower, the travel times get longer, and since the tool is recording only a fixed time window, not all of the waveforms at the far receiver were being recorded at the shallow depths. Another possible reason is as the shear wave velocity gets lower at these shallower depths, less and less energy is being converted into shear waves radiating away in the formation. In the limit of zero shear wave velocity, the borehole becomes a liquid-liquid waveguide and no shear wave conversion will take place. Thus the sensitivity of the inversion to the low shear wave velocities is small, and thus the inversion results above 340 mbsf are not trustworthy. They are included here for completeness.

Another area where the inverted shear wave velocities are higher than the Wood's velocities is around the Unit 2 – 3 transition at 442 mbsf. There are fine scale heterogeneities around the transition zone, and both porosity and P wave velocity logs show rapid variations. Some of these variations can be attributed to cycle skipping. Thus the Wood's velocities are less trustworthy in this zone, and the inverted velocities are probably more reliable.

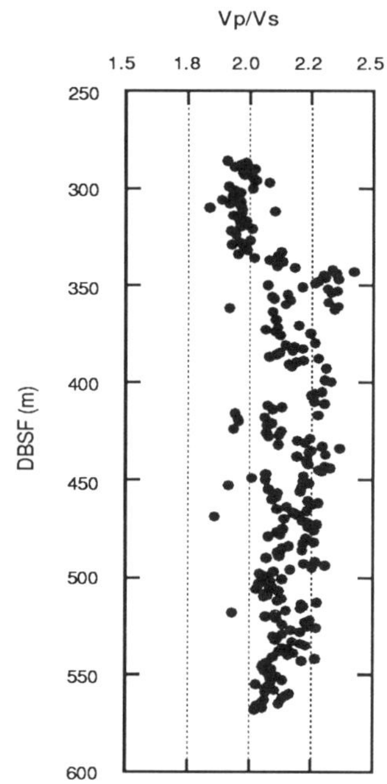


Figure 4: The Vp/Vs ratio for DSDP site 613. The Vp is from the conventional sonic log, the Vs from inversion of full waveform acoustic log data.

Figure 4 shows the $V_p/V_s$ ratio for Units 2 and 3. The ratio starts from around 2.4 at 340 mbsf to about 2.1 at 560 mbsf. This is consistent with the consolidation of the sediments with depth. As discussed in the previous paragraphs, the results above 340 mbsf are probably in error.

Interestingly, the results through the Unit 2 – 3 transition region did not show large variations as in the Wood's shear wave velocities. This implies the latter variations are probably due more to variations in the porosity log rather than the P-wave velocity log.

Another result from the inversion algorithm is the P wave quality factor $Q_p$, which is proportional to the inverse of the attenuation. Figure 5 shows the $Q_p$ derived from the inversion of the full waveform data the same time as the estimation of $V_s$. Also shown are the independent measurements of Goldberg et al. (1985) from the first cycle of the waveform only. The two agree very well over the range of 340 to 560 mbsf. More significantly, both show a decrease in $Q_p$ or an increase in attenuation as a function of depth. It is believed that this is due to the fact that as consolidation begins, more and more calcite grains are coming into contact, thus increasing the energy lost due to grains sliding or rubbing against each other. In contrast, in relatively unconsoliated sediments, the chief mechanism of attenuation is the grains moving or rotating in the liquid matrix, thus creating viscous attenuation effects. Thus in these sediments the grain boundary effects are seen to contribute more and more to P wave attenuation as consolidation occurs.

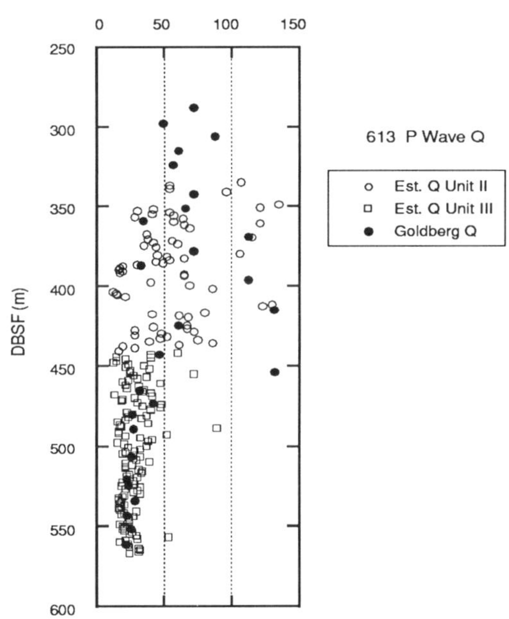


Figure 5: $Q_p$ obtained from the inversion of full waveform acoustic log data and those by Goldberg et al. (1985). Note the decrease in $Q_p$ with depth.

## 3. Conclusions

In this paper we reviewed the acoustic waves generated in a borehole by a full waveform acoustic logging tool. In soft sediments, there is no refracted shear head wave and the associated shear wave normal mode (pseudo-Rayleigh wave), thus a direct determination of formation shear wave velocity under such conditions is impossible. We presented two alternative ways of estimating shear wave velocity from the full waveform logging data: an inversion based on the Stoneley wave velocity and one based on the P wave and P leaky mode amplitude. Both are shown to be applicable to real data. However, any indirect method suffers from the need to be able to know and model the field condition adequately. The new generation of direct shear wave logging tools being developed should provide accurate direct measurements of shear wave velocity in marine sediments.

## 4. Acknowledgements

The authors appreciate discussions with Roy Wilkens and Murli Manghnani of the Hawaii Institute of Geophysics. The work was supported by the Full Waveform Acoustic Logging Consortium at M.I.T. and by National Science Foundation Grant No. OCE87-10226 (CHC). Parts of the computing were performed at the Pittsburgh Supercomputing Center.

## 5. References

Biot, M.A. (1952) 'Propagation of elastic waves in a cylindrical bore containing a fluid', J. Appl. Physics, 23, 997–1005.

Cheng, C.H. (1989) 'Full waveform inversion of P waves for Vs and Qp', J. Geophys. Res., 94, 19619–19625.

Cheng, C.H. and Toksöz, M.N. (1983) 'Determination of shear wave velocities in "slow" formations', Trans. SPWLA 24th Ann. Logging Symp., Paper V.

Cheng, C.H., Toksöz, M.N. and Willis, M.E. (1982) 'Determination of in situ attenuation from full waveform acoustic logs', J. Geophys. Res., 87, 5477–5484.

Cheng, C.H., Zhang, J. and Burns, D.R. (1987) 'Effects of in-situ permeability on the propagation of Stoneley (tube) waves in a borehole', Geophysics, 52, 1279–1289.

Goldberg, D., Moos, D. and Anderson, R.N. (1985) 'Attenuation changes due to diagenesis in marine sediments', Trans. SPWLA 26th Ann. Logging Symp., Paper KK.

Stevens, J.L. and Day, S.M. (1986) 'Shear velocity logging in slow formations using the Stoneley wave', Geophysics, 51, 137–147.

Toksöz, M.N., Cheng, C.H. and Willis, M.E. (1984) 'Seismic waves in a borehole — a review', in M.N. Toksöz and R.R. Stewart (eds.), Vertical Seismic Profiling: Advanced Concepts, Geophysical Press, pp. 256–275.

Toksöz, M.N., Wilkens, R.H. and Cheng, C.H. (1985) 'Shear wave velocity and attenuation in ocean bottom sediments from acoustic log waveforms', Geophys. Res. Lett., 12, 37–40.

Tsang, L., and Rader, D. (1979) 'Numerical evaluation of the transient acoustic waveform due to a point source in a fluid-filled borehole', Geophysics, 44, 1706–1720.

Tuchulke, B.E., (1981) 'Geologic significance of seismic reflectors in the deep western North Atlantic basin', SEPM Spec. Publ., 32, 23–37.

Wilkens, R.H., Schreiber, B.C., Caruso, L. and Simmons, G. (1986) 'The effects of diagenesis on the microstructure of Eocene sediments bordering the Baltimore Canyon Trough', Init. Repts. DSDP, 95, Washington (U.S. Govt. Printing Office), 527–548.

Williams, D.M., Zemanek, J., Angona, F.A., Dennis, C.L. and Caldwell, R.L. (1984) 'The long space acoustic logging tool', Trans. SPWLA 25th Ann. Logging Symp., Paper T.

Wood, A.B., (1930) A Textbook of Sound, G. Bell and Sons, Ltd., London.

Zemanek, J., Angona, F.A., Williams, D.M. and Caldwell, R.L. (1984) 'Continuous acoustic shear wave logging', Trans. SPWLA 25th Ann. Logging Symp., Paper U.

# APPLICATION OF SHEAR WAVE MEASUREMENTS IN BOREHOLES

R. KANESTRØM
*Norsk Hydro a.s*
*P.O.Box 4313*
*N-5028 Bergen*
*Norway*

ABSTRACT. Measurements of shear waves in boreholes have a number of applications in seismic prospecting. Shear waves combined with compressional waves permit theoretical rock models to be defined for different lithologies. The $Vp/Vs$ ratio is demonstrated to be a good indicator for sand and shale, and the dynamic elastic moduli can be computed for different types of rock. The shear wave velocity is essential for a successful use of amplitude versus offset data and for an optimal use of amplitudes during the stacking process. The different aspects are demonstrated by real data examples in the paper. Systematic measurements and use of shear waves in boreholes will contribute to a significant advance in seismic prospecting.

## 1. Introduction

Combined use of P- and S-waves will contribute to an improvement in petroleum exploration. The P- and S-velocities can be determined from seismic surface measurements, but at the present time the accuracy and vertical resolution is too poor in most cases for an effective use in lithological prediction. S-wave exploration seismology has been less successful offshore than onshore because the source and receivers are located in a water layer. Consequently, the generation and recording of the S-waves are dependent on the P to S conversion and on the S to P conversion. However, the S-wave velocity is a very important parameter even if we are dealing only with the P-wave response. The angle-dependent reflectivity of P-waves is a function of contrasts in P-wave velocity, density and S-wave velocity between two media.

The continuously-moving S-wave velocity well logging sonde will advance the techniques of lithological prediction and seismic inversion considerably. In this paper I will demonstrate some applications of S-wave measurements in boreholes, and also its importance in offshore exploration.

## 2. Method

### 2.1 THE THEORETICAL ROCK MODEL

Numerous theoretical models developed during the past 30 years may be used to compute the effective elastic moduli of a two phase medium, given the elastic moduli of the individual components. We have chosen to use the Kuster-Toksöz model (Kuster and Toksöz (1974a,b), Toksöz et al. (1976), Toksöz and Cheng (1978)) since



*J. M. Hovem et al. (eds.), Shear Waves in Marine Sediments*, 267–274.

it allows ellipsoidal inclusions rather than spherical ones and since, in addition, the results of Kuster-Toksöz models seem to agree reasonably well with those of laboratory experiments and sonic logging. Consideration of the Kuster-Toksöz model has shown a distinction in behaviour for rocks characterized by equant pores, i.e. pores of high aspect ratio from those dominated by cracks.

To obtain the theoretical rock model I have used the full-waveform sonic log data, the density log, results from the core analysis (mainly porosity and grain density) and information about the pore fluid, formation pressure and lithology. Figure 1 shows the log-derived $Vp$ and $Vs$ data as a function of porosity. The lower part of Figure 1 gives the parameters for the theoretical rock model which fits the observed P- and S-velocities. The P- and S-velocities calculated from the rock parameters are represented by curves in the upper part of Figure 1.

The advantage of establishing a theoretical rock model for a given condition is that the velocities can be predicted for other conditions.

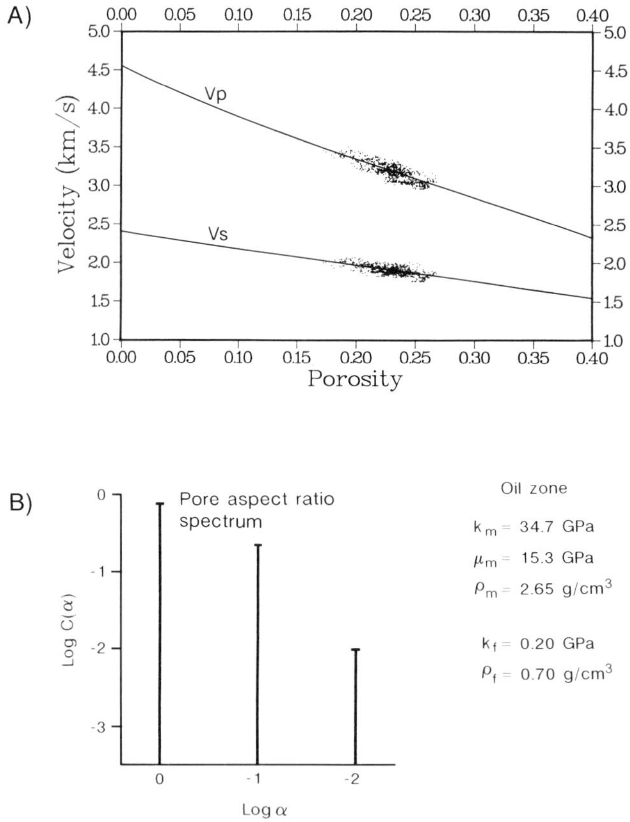


Figure 1. A) Observed P- and S-velocities (dots) in an oil well. Curves represent the calculated velocities based on the parameters given in B). B) Reservoir rock parameters.

## 3. Results

### 3.1 SAND-SHALE

Figure 2 shows an example of $Vp$ and $Vs$ measurements in a borehole. As seen from the figure, the lithology alternates between sand and shale with small calcite cemented zones and a few coal layers. It is not possible to predict the lithology from the values of the P- and S-velocities. However, the $Vp/Vs$-ratio is an effective indicator of sand and shale as shown in Figure 2.

The calcite cemented zones are not revealed by the $Vp/Vs$-ratio, but are easily identified by the high P- and S-velocities. The coal layers can be identified by the low P- and S-velocities in these zones.

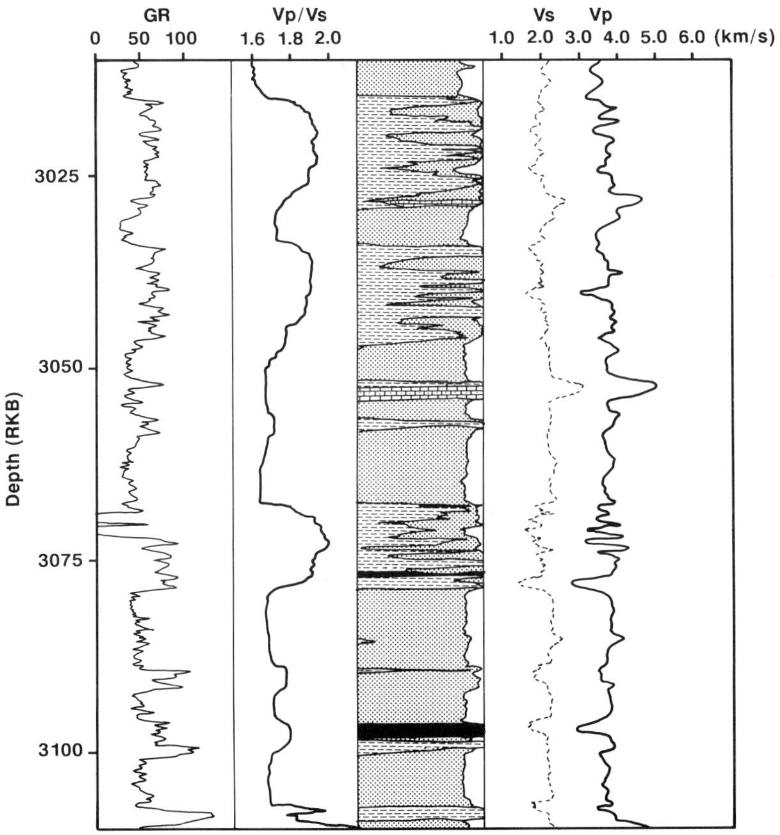


Figure 2. Observed P- and S- velocities. Note the change in $Vp/Vs$ with lithology. GR represents the gamma ray log.

3.2 THE DYNAMIC ELASTIC CONSTANTS

When both P- and S-velocities are recorded, the Poisson ratio can be computed. Incorporating the density log permits the shear modules and the incompressibility to be calculated. Figure 3 shows an example where we are going from the gas zone into the oil zone. The figure includes the Poisson ratio ($\sigma$), the shear modulus ($\mu$), the incompressibility ($k$) and the ratio $k/\mu$. As seen from the figure there is a marked change in $\sigma$ and $k$ over the gas-oil contact (GOC). Figure 3 shows that P- and S-waves can be combined to give a very precise definition of the GOC.

There is a gradual decrease in the shear modulus in the gas zone as we approach the GOC. The decrease in $\mu$ starts about 8 m above the GOC and becomes approximately constant in the oil zone. From the GOC the oil saturation decreases upwards in the gas zone. We interpret the decrease in $\mu$ as a "softening" effect caused by the oil in the pores.

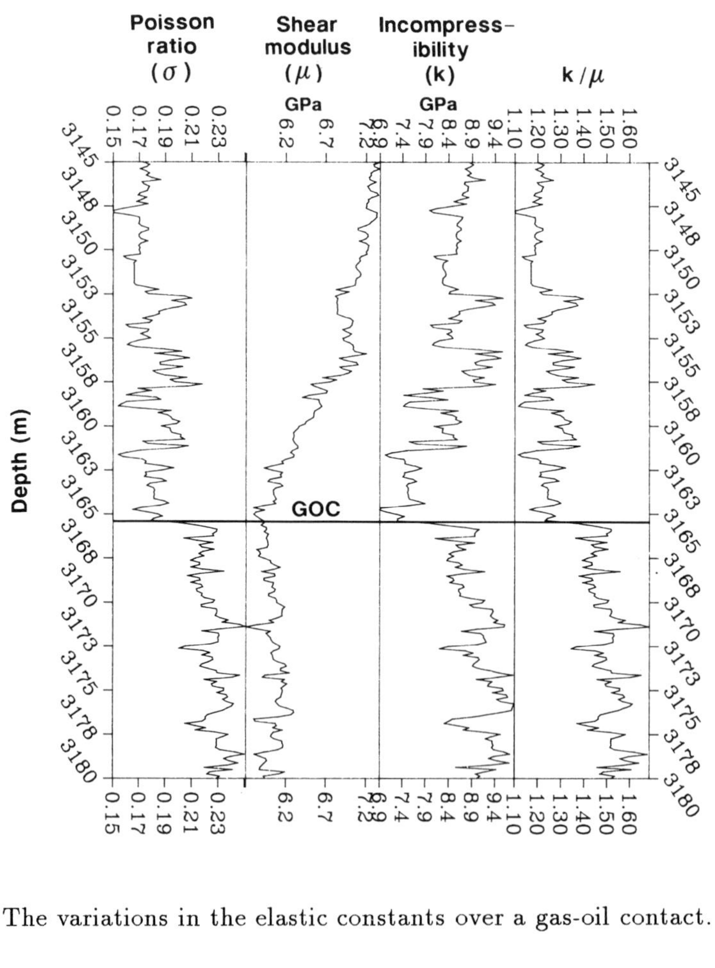


Figure 3. The variations in the elastic constants over a gas-oil contact.

### 3.3 VELOCITIES IN GAS AND OIL ZONES

It is usually believed that the velocity in the gas zone is lower than in the oil zone. This is not always true. Figure 4 shows an example where the P and S-velocities are higher in the gas zone than in the oil zone. We have made similar observations in other wells. The observed change in velocity across the GOC is not a result of the different pore fluids, but is essentially a result of the difference in lithology as shown in Figure 4. The difference in the elastic moduli over the GOC is believed to be a long term effect caused by the difference in pore fill.

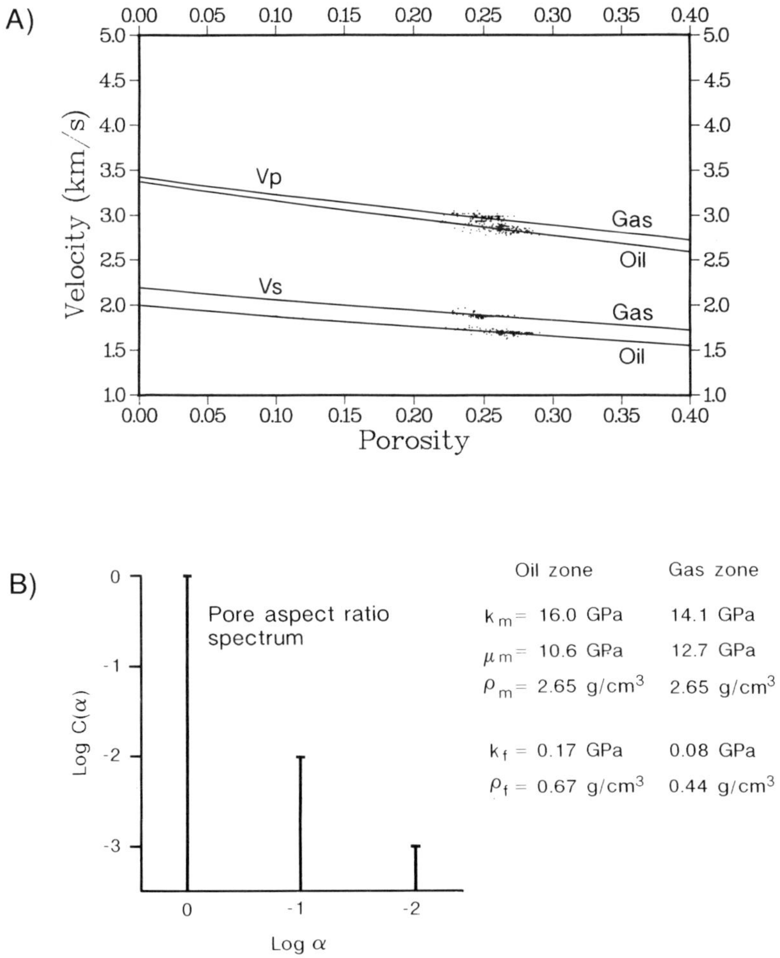


Figure 4. A) Observed P- and S-velocities (dots) in the gas and oil zone. Note that the velocities are higher in the gas zone than in the oil zone. B) Reservoir rock parameters.

3.4 SEISMIC RESPONSE FROM A GOC

As shown in Figure 4, the P- and S-velocities across the GOC may behave differently from what is usually believed. I have computed the seismic response from the GOC based on the P- and S-velocities and the density measured in two wells in the same field. The results which are shown in Figure 5 demonstrate that the seismic response from the GOC may change significantly from one well to another in the same field. It is very important to be aware of this problem when mapping the GOC from seismic data.

The results in Figure 5 demonstrate the need for S-wave measurements in boreholes. The computed amplitude versus offset (AVO) will assist in an optimal use of the seismic data during processing. Measurements of S-waves are an absolute requirement for successful seismic monitoring of the GOC during production.

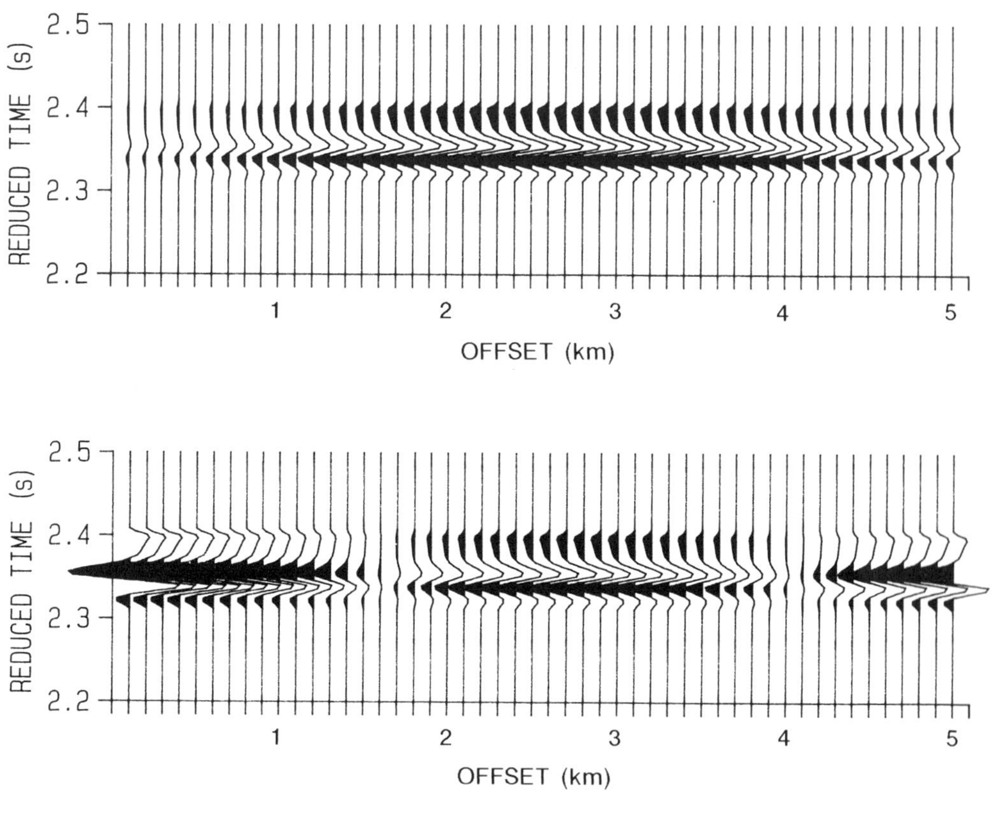


Figure 5. Calculated seismic response from the GOC based on borehole data at two different wells in the same field.

3.5 IMPROVEMENT OF SEISMIC SECTIONS

The results in Figure 5 show that conventional stacking need not always be a successful process. If we knew the AVO at the levels of interest, the stacking process could be performed in a more optimal fashion.

I have tested this on a seismic line running through a well. The top of the reservoir could hardly be interpreted from the conventionally processed data. The borehole data were analysed and the AVO computed from the top of the reservoir showed that offset values greater than about 1400 m should not be used in the stacking process. Figure 6 shows the original seismic section and the reprocessed data based on a priori information about the AVO from the top of the reservoir.

The data in Figure 6 are marine seismic data, however, the S-velocity is an important factor contributing to the AVO for P-waves.

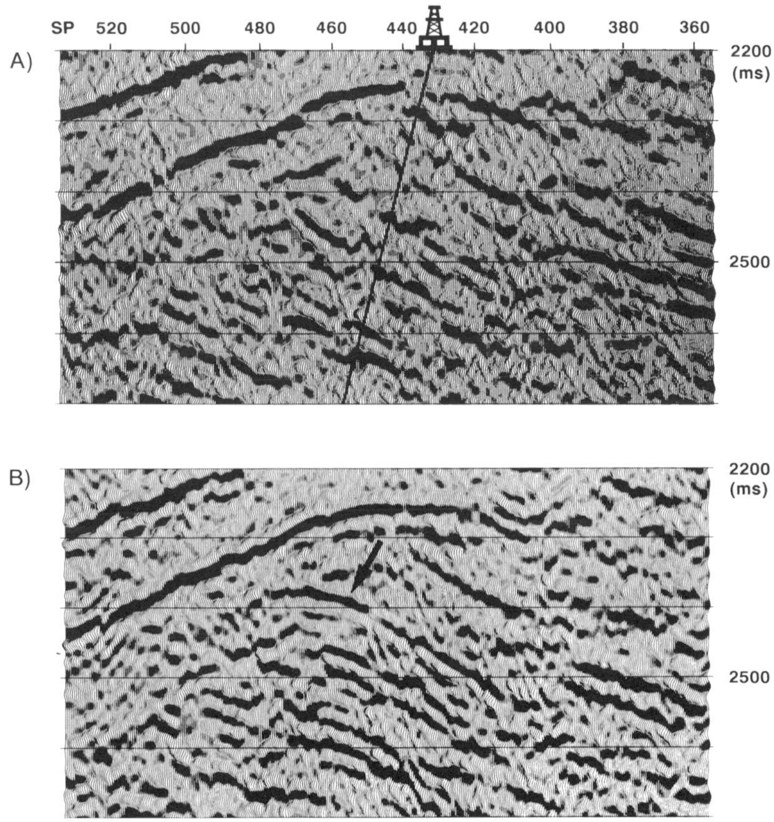


Figure 6. A) Seismic section from conventional seismic processing. B) Seismic section based on a priori information about the seismic response from the top of the reservoir.

## 4. Conclusion

The theoretical rock model is an essential tool for the application of AVO data and the interpretation of acoustic impedances derived from seismic data. To establish the theoretical rock model, shear wave velocities are required.

Shear wave measurements in boreholes have a wide range of application in seismic prospecting ranging from the calculation of the elastic constants of the formation through lithological prediction (including porosity and pore fill) to practical application in seismic procesing to improve the quality of P-wave sections. Systematic measurements and use of shear waves in boreholes will contribute to a significant advance in seismic prospecting.

## 5. References

Kuster, G.T. and Toksöz, M.N. (1974a) 'Velocity and attenuation of seismic waves in two-phase media: Part I theoretical formulations', Geophysics 39, 587-606.

Kuster, G.T. and Toksöz, M.N. (1974b) 'Velocity and attenuation of seismic waves in two-phase media: Part II experimental results', Geophysics 39, 607-618.

Toksöz, M.N. and Cheng, C.H. (1978) 'Modeling of seismic velocities in porous rocks and its application to seismic exploration', The Arabian Journal for Science and Engineering, Special Issue, 109-131.

Toksöz, M.N., Cheng, C.H. and Timur, A. (1976) 'Velocities of seismic waves in porous rocks', Geophysics 41, 621-645.

# EXPERIENCE WITH THE SEISMIC CONE PENETROMETER IN OFFSHORE SITE INVESTIGATIONS

G.DE LANGE
*Fugro-McClelland Engineers B.V.*
*P.O. Box 250*
*2260 AG Leidschendam*
*The Netherlands*

ABSTRACT. The seismic cone penetrometer has been used in offshore site investigations by Fugro-McClelland since 1987. The system consists of a hydraulically driven spring hammer on the seabed and a triaxial set of geophones incorporated in a conventional in-situ piezocone penetrometer. It is currently used in the Vertical Seismic Profiling mode. The seismic cone is pushed into the soil from the seabed or from the bottom of an advancing borehole. Shear waves were recorded in (glacio)marine sediments in the Norwegian Trench. It was observed that the amplitudes and velocities increased, and material damping decreased during standstill of the cone, which is attributed to pore pressure dissipation after cone penetration. Contrary to the shear wave energy, the compression wave energy may be transmitted through the tools. Experiments have shown that the decoupling of the seismic cone from the rods enables a full dynamic characterization of the soil in terms of moduli of rigidity and elasticity.

## 1. Introduction

Solving the problems of dynamic loading in offshore geotechnics has long relied on estimated or indirectly determined values of shear modulus. Experience on land had already indicated that the measurement of the shear wave velocity in-situ could serve as ground truth for laboratory tests on samples (Richart, 1975). It was therefore desirable to develop an economical method to measure the shear wave velocities in-situ in offshore site investigations. The incorporation of geophones in a cone penetrometer offered the possibility to combine standard geotechnical testing with shear wave testing in the same operation. The viability of the test was shown by Robertson et al. (1986), illustrated by comparisons with the well established crosshole test (ASTM, 1985).

The seismic cone penetrometer has been used in offshore site investigations by Fugro-McClelland since 1987 and a considerable amount of experience has been accumulated on the response characteristics of the system. This paper briefly describes the system. Comparisons with laboratory tests are shown and operational constraints and further developments are discussed.

## 2. Description of the seismic cone system

Standard cone penetration testing (CPT) employs a cone-tipped probe which measures



*J. M. Hovem et al. (eds.), Shear Waves in Marine Sediments*, 275–282.

resistance on the tip, friction on the side-wall and often pore pressure while the probe is pushed into the soil at a constant rate. Offshore the cone may be pushed into the soil with a steel rod from the seabed using the WHEELDRIVE SEACALF reaction frame. The test may also be performed at the bottom of a borehole using the WISON wireline penetrometer tool (Fig. 1). The WHEELDRIVE SEACALF employs four wheels which grip the test rods and when rotated, move them up or down. The amount of penetration depends on the soil conditions, but typically ranges from about 20m in dense sands and hard clays to between 30m and 60m in softer normally consolidated clays. Deviations from the vertical are monitored with an inclinometer built into the cone. The WISON system is normally used in conjunction with straight flush rotary drilling. At the required depth the WISON tool is lowered on its umbilical cable to the open bit where it is latched in the drill collar. The drillstring is set on the bottom and the WISON tool is activated to push the penetrometer into the soil, the reaction being provided by the SEACLAM reaction frame. In this way much greater depths can be achieved.

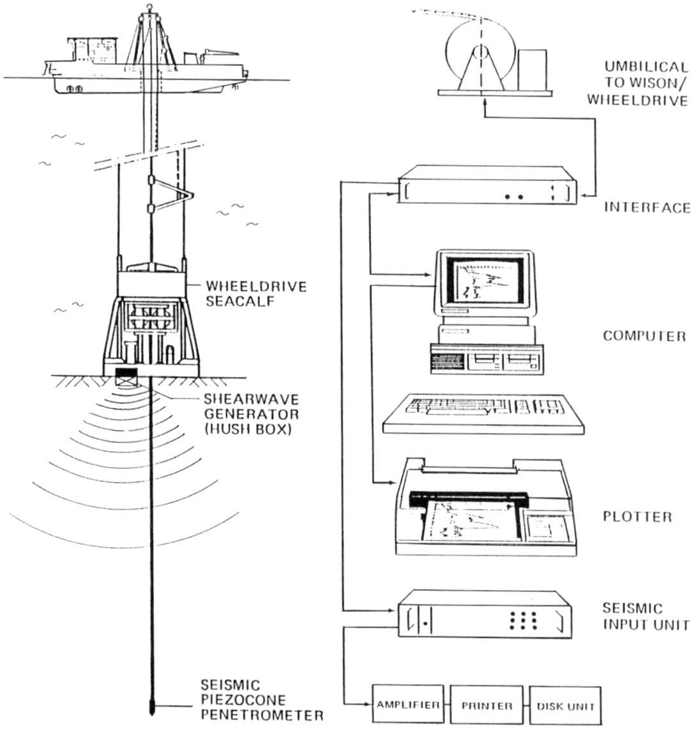


*Figure 1. Seismic cone in WHEELDRIVE SEACALF mode*

In the seismic CPT the principle of downhole vertical seismic profiling (VSP) is used, i.e. the shear wave source is located at the seabed and the shear wave is received at the seismic cone penetrometer. The shear wave source consists of a horizontally oriented, hydraulically activated, spring hammer mounted in a steel box (HUSHBOX). Vertical vanes protruding from the HUSHBOX provide lateral soil reaction for the hammer. The HUSHBOX is mounted in rubber sockets on the WHEELDRIVE SEACALF or SEACLAM frame. The hydraulic system allows the generation of an unlimited amount of blows enabling signal stacking.

The seismic cone penetrometer developed by Fugro-McClelland incorporates a triaxial set of geophones in a piezocone penetrometer. The triaxial receiver unit ensures that the seismic cone is independent of its orientation with respect to the direction of particle motion, which is perpendicular to the vertical plane through the cone and the centre of the HUSHBOX (true SH-wave).

When performing a seismic CPT, the seismic cone is pushed into the soil to the required depth and the HUSHBOX is activated, triggering the recording of the signal. This sequence is repeated at regular intervals, and thus a plot of shear wave travel time versus depth is obtained (Fig. 2). The resulting shear wave velocities can be correlated with the corresponding cone resistance profile. The soil density profile required for the calculation of the shear modulus may be interpreted from the CPT profile, or measured on samples, or measured directly in-situ with a nuclear density probe (Tjelta et al., 1985).

An important advantage of using the seismic cone with a dedicated SH-wave generator is the excellent coupling of the probe and the sediment and the lateral compliance of the cone-rod system ensuring an optimum sensitivity for SH-waves. This allows a small offset of the HUSHBOX (0.7m). Consequently, refraction of the shear wave can be ignored in the data reduction.

## 3. Response characteristics

An extensive site investigation, involving seismic cone tests, was performed in the Norwegian Trench to obtain soil design parameters for offshore production platforms (De Lange et al., 1990). The soil profile consists of normally consolidated very soft to firm marine clay overlying a sequence of overconsolidated sandy glaciomarine clays. The WHEELDRIVE SEACALF mode could be utilized down to about 40m below seabed. The WISON borehole mode was used down to 83m below seabed. Thanks to the high signal-to-noise ratio one blow was sufficient down to 40m depth. During the penetration with the WHEELDRIVE SEACALF rotation of the cone was observed, amounting to 45 degrees for every 15m of penetration. To analyse the strain amplitudes the signals from the orthogonal horizontal geophones were added vectorially, knowing from observation that no component was present in the sagittal plane. The strain amplitude $\gamma$ is obtained by deviding the particle velocity by the shear wave velocity (White, 1965), assuming that, thanks to its lateral compliance, the geophone response is equal to the soil particle velocity (Auld, 1980). The plot of $\gamma$ versus depth shows that below about 10m depth the strain amplitude produced by the HUSHBOX is smaller than $10^{-3}$ % (Fig. 3), the generally accepted upper limit below which the shear modulus has its maximum value $G_{max}$ (Richart, 1975). Consequently, the shear moduli calculated for the upper 10m are just under $G_{max}$.

A number of pore pressure dissipation tests were performed at different depths in a WHEELDRIVE SEACALF test enabling the analysis of changes in the response with time. When a cone penetrometer is pushed in a soil of low permeability (i.e. clay), excess pore pressures developed in a zone around the cone. During standstill of the cone the dissipation of this excess pore pressure is monitored in order to arrive at a measure of permeability of the soil. Shear waves were generated before and after dissipation which typically lasted for several hours. Shear wave travel times were generally one to two milliseconds less after dissipation. As the shear wave velocities were in the range of 250 m/s to 300 m/s, this indicates that the zone of excess pore pressure extends over a relatively short distance. Pre- and post-dissipation shear wave velocities were in the same range above 35m but below this depth increased with time by 12%, suggesting different dissipation characteristics.

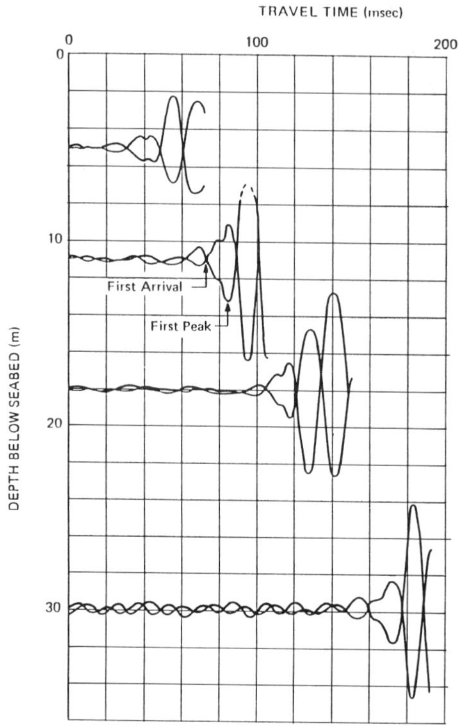


*Figure 2. Shear wave travel time versus depth below seabed. Note that shear wave amplitudes are normalized.*

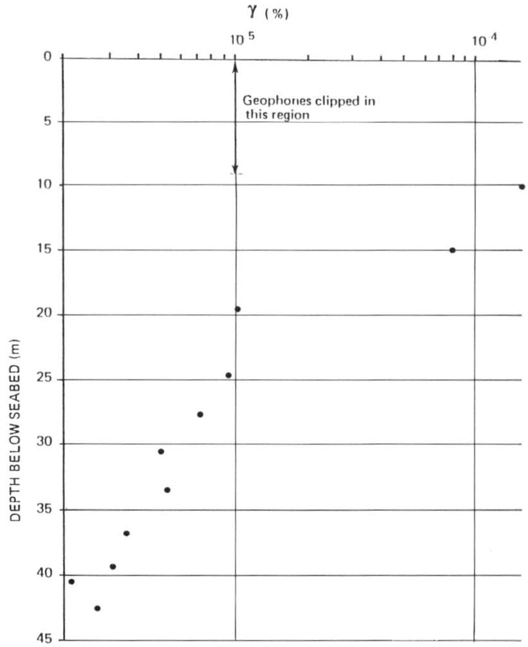


*Figure 3. Shear strain amplitude (γ) versus depth below seabed for WHEELDRIVE seismic cone test in the Norwegian Trench.*

## 4. Material damping measurement

An attempt was made to establish the material damping coefficients with the data obtained at this site. The attenuation coefficient (α) and its relation to the specific energy dissipation function (1/Q) and damping ratio (D) are given by the following equations (Richart et al., 1970):

$$\alpha = \frac{1}{R_2 - R_1} * \ln \frac{A_1 * R_1}{A_2 * R_2} \qquad (m^{-1})$$

$$\frac{1}{Q} = \lambda * \frac{\alpha}{\pi}$$

$$D = \frac{1}{\sqrt{4Q^2 + 1}} * 100 \qquad (\%)$$

in which $A_1$ and $A_2$ are the geophone amplitudes at depths below the source of $R_1$ and $R_2$ respectively, and $\lambda$ is the wavelength of the shear wave. Figure 4 presents a logarithmic plot of average normalized amplitude versus depth. It can be seen that after pore pressure dissipation a considerable increase in amplitude (10%-25%) is recorded with a corresponding lower material damping. Below 30.6m no post-dissipation damping could be calculated ($\alpha < 0$), indicating that differences in the rate or timespan of dissipation are major factors.

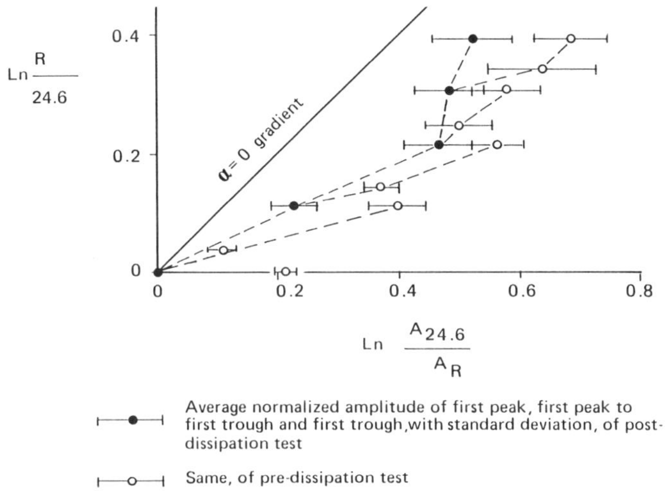


*Figure 4. Average normalized amplitude versus depth. The $\alpha$ = 0 gradient indicates geometrical damping. Smaller gradients indicate a material damping component ($\alpha > 0$).*

A summary of the material damping coefficients derived from the data presented in Figure 4, including the damping ratio obtained in a resonant column test in the laboratory is given in Table 1:

*TABLE 1. A comparison of material damping between 24.6 m and 30.6 m measured with the seismic cone and in the laboratory*

| | Before Dissipation | After Dissipation | Resonant Column at 26 m |
|---|---|---|---|
| $\alpha$ ($m^{-1}$) | 0.0955 | 0.0408 | |
| Q | 4.38 | 10.27 | |
| D (%) | 11.33 | 4.87 | 5.03 |

The comparison with the laboratory test suggests that the post-dissipation value is closer to the undisturbed in-situ value. Many of the problems associated with this pseudo-interval measurement would be solved by mounting a fixed spacing double set of geophones. With such a true interval measurement the influence of pore pressure dissipation on material damping and shear modulus can be more carefully monitored.

## 5. Comparison with laboratory measurements

A comparison with laboratory measurements of $G_{max}$ was performed by de Lange et al. (1990) (see Fig. 5). The results of resonant column tests and piezoceramic bender element tests performed on samples from the same sites at in-situ pressures indicate that, as the degree of consolidation increases with depth, the resonant column results are lower relative to the in-situ values of $G_{max}$. The bender element tests that were performed on-board ship immediately after sampling agreed very well with the seismic cone values. This indicates that the samples in the resonant column tests may have suffered some degree of disturbance (Athanasopoulos and Richart, 1983). The three methods combined provide a comprehensive testing package covering the entire range of load and strain levels of interest in dynamic geotechnical problems: the seismic cone provides $G_{max}$ at in-situ loading conditions, serving as ground truth for the laboratory tests, the bender element test provides the variation of $G_{max}$ with changing loading conditions, while in the resonant column test the variation of G with strain rate can be monitored.

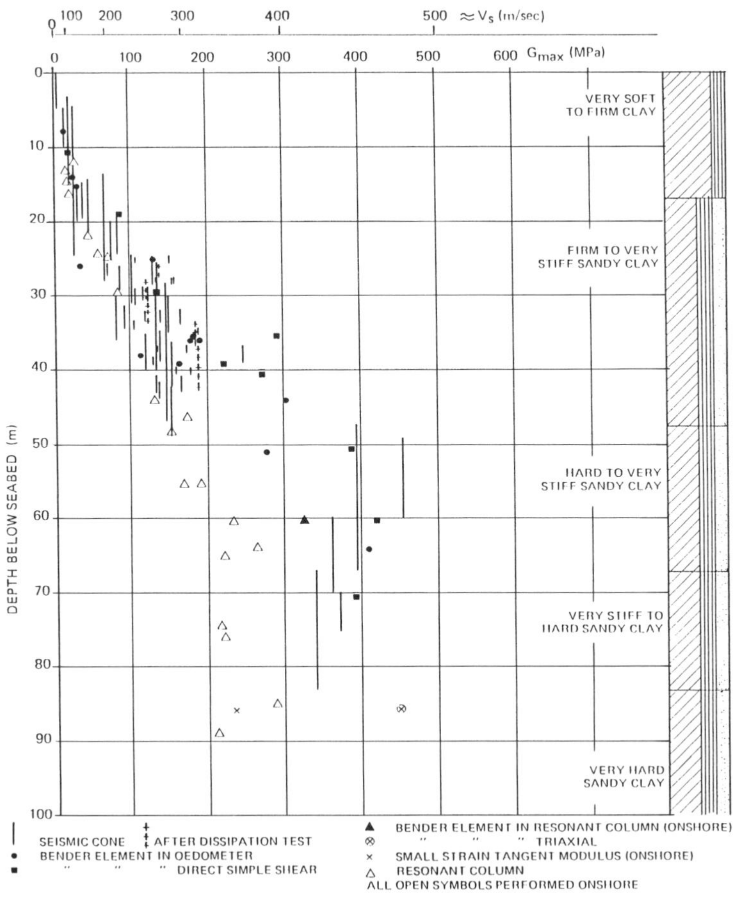


*Figure 5. Initial shear modulus versus depth below seabed*

## 6. Measurement of compression waves

Seismic CPT tests onshore and offshore have shown that compression (P) wave energy is easily transmitted through the push rods to the geophones, obliterating the "soil" arrivals. An example of this effect is shown in Figure 6. In this onshore CPT the P wave, which is generated by striking the ground vertically, is transmitted to the push rod by a layer of peat exhibiting a very high adhesion to the rod. The energy ratio of the waves travelling through the soil and through the rod is further deteriorated by the high damping characteristics of the peat. The problem was solved by inserting a small sliding section in the push rod just above the seismic cone, allowing the cone to move independently from the rod, thus eliminating the disturbing P wave. A reliable modulus of elasticity and Poisson's ratio could now be determined.

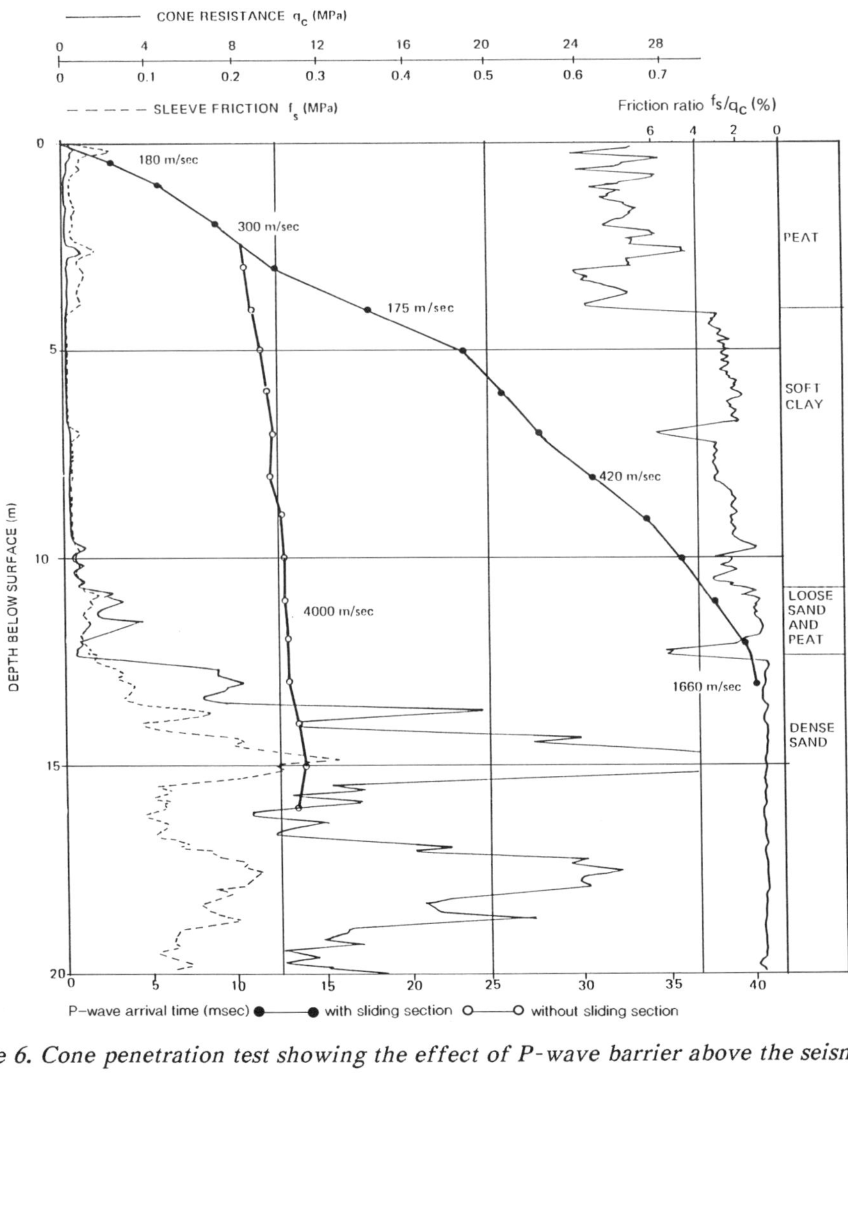


*Figure 6. Cone penetration test showing the effect of P-wave barrier above the seismic cone.*

## 7. Conclusions

The seismic CPT is a reliable in-situ instrument providing ground truth for laboratory dynamic soil tests. Experience in the field has shown that while the cone remains stationary the shear wave travel time and amplitude decrease, during excess pore pressure dissipation. The effect on the material damping coefficient is more significant and should be studied further with a fixed-spacing double geophone set. The seismic cone can easily be adapted to be sensitive to P waves.

## 8. Acknowledgements

I would like to thank Norsk Hydro, Statoil and NAM for their permission to use data from the site investigations sponsored by them. The work done by the Norwegian GeotechnicalInstitute and Nick Willet from Fugro-McClelland Ltd. on the previous study is greatfully acknowledged. Thanks are also due to Ronald van Steveninck for providing the amplitude data.

## 9. References

ASTM (1985) 'Standard test method for crosshole seismic testing', in *Annual Book of ASTM Standards*, Section 4, Vol. 04.08, Designation D4428, ASTM, Philadelphia, pp. 885-898.

Athanasopoulos, G.A.and Richart, F.E., Jr. (1983) 'Effect of stress release on shear modulus of clays', *Jl. of GeotechnicalEngineering*, ASCE, Vol. 109, No. 10, 1233-1245.

Auld, B. (1980) 'Cross-hole and down-hole $V_s$ by mechanical impulse', *Jl. of the GeotechnicalEngineering Division*, ASCE, Vol. 106, No. GT9, 1061-1062.

Lange, G. de, Rawlings, C.G. and Willet, N. (1990) 'Comparison of shear moduli from offshore seismic cone tests and resonant column and piezoceramic bender element laboratory tests', *Underwater Technology*, Vol. 16, No. 3, 13-20.

Richart, F.E., Jr. (1975) 'Some effect of dynamic soil properties on soil-structure interaction', *Jl. of the GeotechnicalEngineering Division*, ASCE, Vol. 101, No. GT12, 1197-1240.

Richart, F.E., Jr., Hall, J.R., Jr. and Woods, R.D. (1970) *Vibrations of Soils and Foundations*, Prentice Hall, Englewood Cliffs.

Robertson, P.K., Campanella, R.G., Gillespie, D. and Rice, A. (1986) 'Seismic CPT to measure in-situ shear wave velocity', *Jl. of GeotechnicalEngineering*, ASCE, Vol. 112, No. 8, 791-803.

Tjelta, T.I., Tieges, A.W.W., Smits, F.P., Geise,J.M. and Lunne, T.A. (1985) 'In-situ density measurements by nuclear backscatter for an offshore soil investigation', in *Proc. 17th Annual Offshore Technology Conference*, Paper No. OTC 4917, Vol. 1, pp. 201-206.

White, J.E. (1965) *Seismic Waves*, McGraw-Hill, New York.

# COMPARISON OF TECHNIQUES FOR SHEAR WAVE VELOCITY AND ATTENUATION MEASUREMENTS

T. G. MUIR*, T. AKAL, M. D. RICHARDSON**, R. D. STOLL***,
A. CAITI, and J. M. HOVEM
*SACLANT Undersea Research Centre*
*19026 Viale San Bartolomeo 400*
*La Spezia, Italy*

**Applied Research Laboratories, The University of Texas at Austin*
***Naval Oceanographic and Atmospheric Laboratory*
****Lamont-Doherty Geological Observatory, Columbia University*

ABSTRACT. In order to support ocean acoustic modelling and sonar performance prediction, it is necessary to acquire geoacoustic data on sediment structure. The direct measurement of shear wave velocity and attenuation within the sedimentary volume is a difficult and time consuming process. More rapid, remote measurements involving inversion techniques are evolving, but their evaluation requires "ground truth" data obtained by direct measurements. A comparison of various "in situ" methods used at SACLANTCEN is presented to illustrate the development of testbed sites where ground truth data is obtained. These methods include 1) near surface measurements with bimorph transducers, 2) downhole measurements with impact sources and geophone receivers, 3) explosive shot measurements with bottomed and buried seismographs, and 4) measurements with explosive sources and line arrays of geophones. The role of these methods in the evolution of rapid remote sensing techniques is discussed.

## 1. Introduction

Over the past 20 years, ocean acoustics research, as well as operational experience with sonar systems, has shown that the sea floor is the dominant feature for sound propagation loss in shallow water. Early SACLANTCEN experiments and analytical modelling efforts [1] identified shear wave conversion at the bottom as a dominant loss mechanism. The importance of bottom shear effects on propagation were found to be enhanced by summer thermal conditions (negative sound velocity gradients), increasing shallowness of the water column, and the nearness of the bottom shear velocity to the compressional velocity in the water [2]. Shear wave velocity and its dependence on depth within sediments appears to be the most important parameter in this process, while shear wave attenuation and compressional wave velocity and attenuation and sediment density are of lesser importance [3].

Ideally, ocean surveys done in support of acoustic modelling and sonar performance prediction, should utilize survey tools capable of sensing these parameters underway, in a rapid area search. Remote sensing techniques involving inversion methods are nonetheless being evolved with potential for rapid area survey of the surficial sediments. Evaluation and development of these



*J. M. Hovem et al. (eds.), Shear Waves in Marine Sediments*, 283–294.

techniques requires utilization of testbed sites where ground truth data on sediment geoacoustic parameters is known. Several types of instrument systems have been developed at SACLANTCEN to delineate bottom geoacoustic parameters at testbed sites in the Ligurian Sea, the Adriatic Sea, and the Norwegian Sea. The design and utilization of shear wave components of these systems is illustrated in this paper.

## 2. Direct, Near Surface Methods

A special system was developed to measure sediment geoacoustic parameters, including compressional and shear velocity and attenuation, a few tens of centimeters below the sediment-water interface. Since this system is treateded in a paper by Barbagelata, et.al., elsewhere in this proceedings, only a brief summary is given here.

### 2.1 PROBE HARDWARE

The main geoacoustic features of the system are shown in Figure (1). Shear and compressional wave probes are attached to a triangular frame that uses weights to force the probes into the sediment. In very shallow water, divers can be used to carefully guide the probes into the sediment, while in deeper water, typically a few hundred meters, a bottom-mounted sleeve system (not shown) allows the triangular frame to be guided into the bottom at normal incidence.

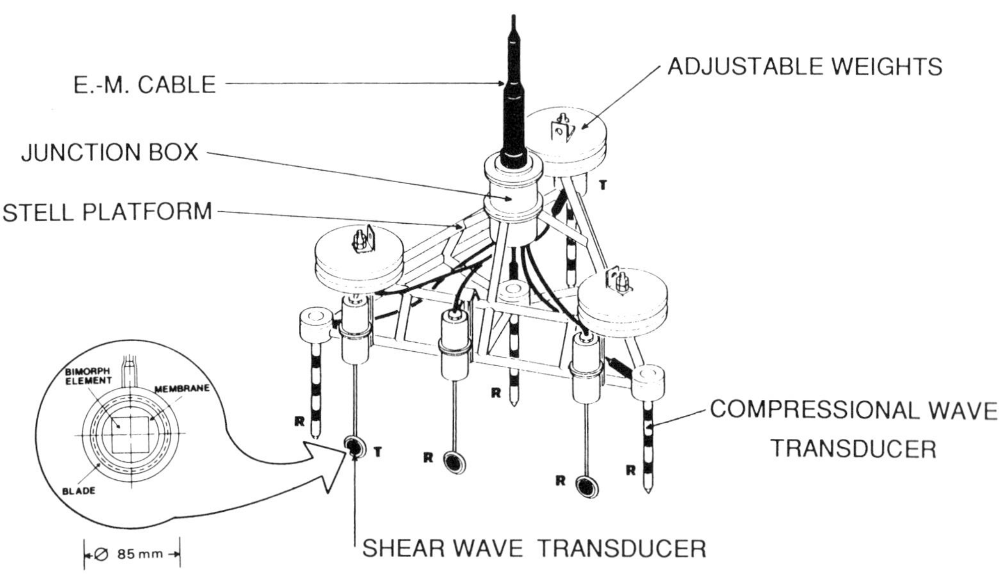


Figure 1. System for near surface geoacoustic measurements.

2.1.1 *Shear Wave Probes.* Transmit and receive elements consist of square bimorph bender elements mounted in a flexible membrane within a protective, blade-like, stainless steel ring, also shown in Figure (1). The element oscillation and orientation is such that horizontally polarized shear waves travel between the transmitter and two receivers at a mean sediment depth of 30 cm.

2.1.2 *Electronics.* Standard cw pulse techniques are used to transmit and receive shear wave signals over the indicated propagation paths. The signals are sine wave tones of up to 10 cycles per burst. The operating band of the shear wave system is 100 to 2,000 Hz. The shear wave transducers appear to be sensitive to impedance coupling to the sediment, which changes their resonant frequency from sediment to sediment.

2.2 MEASUREMENTS

2.2.1 *Procedures.* Shear wave measurements are made directly, from the ratio of measured transducer separation and pulse arrival time, which yields the horizontal shear velocity.

2.2.2 *Results.* Some sample shear wave pulse data are shown in Figure (2), for measurement of shear speeds, which are quite low near the surface. Also shown in Figure (2) is a horizontal beam pattern, taken by successive placement of a receiver at different azimuths, demonstrating the dipole-like radiation pattern of the bimorph projector radiating shear waves.

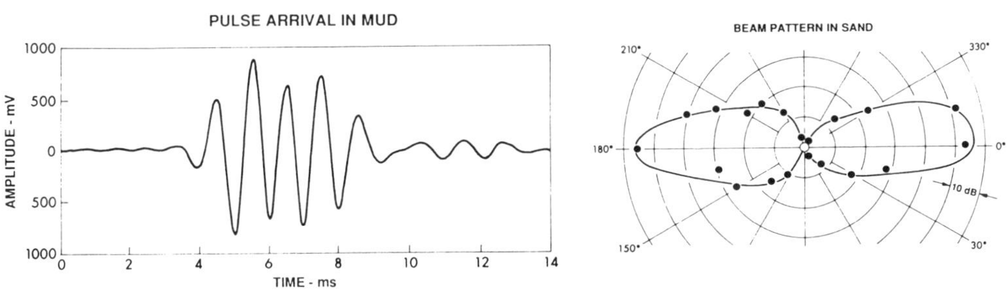


Figure 2. Sample shear wave data for sediments in the Ligurian Sea. The pulse arrivals in a sand bottom yielded 88.2 m/sec at a depth of 33 cm. The beam pattern was acquired at 525 Hz, in a sand sediment at 50 cm depth.

2.2.3 *Comments.* The near surface system described here is useful in support of high frequency applications requiring only near surface data as well as for low frequency, seismo-acoustic applications where the process of initializing the geoacoustic model of the sediment must begin. Ground truth data on the sediment structures can be quickly acquired near the surface and then followed up with more extensive measurements within the sub-bottom sediment structures.

## 3. Direct Downhole Methods

Ground truth measurements as a function of depth in the sediments can be made with boreholes, using either direct, single-hole, or hole-to-hole techniques, in support of inversion techniques involving refraction or reflection seismometry.

## 3.1 MEASUREMENT TOOLS

3.1.1 *Hydro-pneumatic Drilling Rig.* The creation of useful boreholes in unconsolidated sediments requires special procedures. A SACLANTCEN tool for making boreholes in shallow water is illustrated in Figure (3). It consists of a bottom-mounted tripod with a circular foot and a casing tube that guides a rather heavy, free-falling drill collar, some 6 m in length. Erosive cutting forces are provided by a combination of compressed air and pressurized water that act from nozzles within the drill collar, so as to minimize the disturbance of the natural sediments exterior to the tube. The hydrojet provides cutting power and the air jet provides lift for the sediment slurry. Once the hole is drilled, it can be plugged with a thin-walled plastic tube for downhole measurements. The tube is weighted and capped at the bottom, filled with seawater, and has a removeable cap at its top end enabling subsequent access for seismo-acoustic logging measurements. A drilled hole can also be used to implant seismometers and other instruments, intended for permanent deployment, as shown.

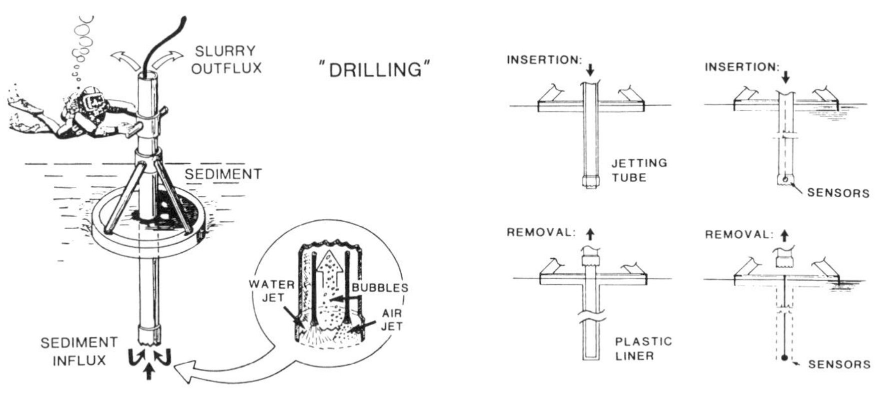


Figure 3. Hydro-pneumatic drilling rig.

3.1.2 *Downhole Shear Wave Measurement Tools.* The shear wave tools are illustrated in Figure (4), and consist of: a) a transmitter in the form of an electromagnetic mallet, suspended from a hydraulic clamp that attaches the device securely to the interior of the tube wall, and b) receivers in the form of moving coil geophones that are securely mounted within a hydraulic clamping device in both vertical and horizontal orientation. The mallet supplies a point force on the elastic tube wall, which results in a multi-polarized transient signal to the sediment, dependent on the boundary conditions. Vector components of the propagating shear wave are received and analyzed for velocity and attenuation parameters.

## 3.2 RESULTS

3.2.1 *Raw Data.* A time series and frequency spectrum for a shear wave signal received by a geophone with a natural frequency of 4.5 Hz are shown in Figure (5). Here, the source and receiver were both oriented in the radial direction at a depth of 4.5 m in the sediment, with a separation range of 2.9 m. The reaction of the dynamic system of source, tube, sediment, receiver, etc., is a pulsed, sinusoidal signal containing a few wavelengths, with a spectral peak around 100 Hz, and a minor lobe around 25 Hz.

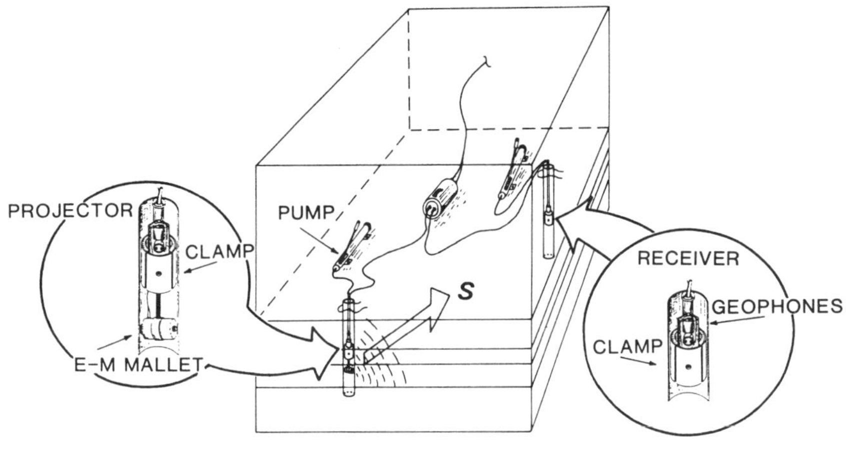


Figure 4. Downhole shear wave measurement system.

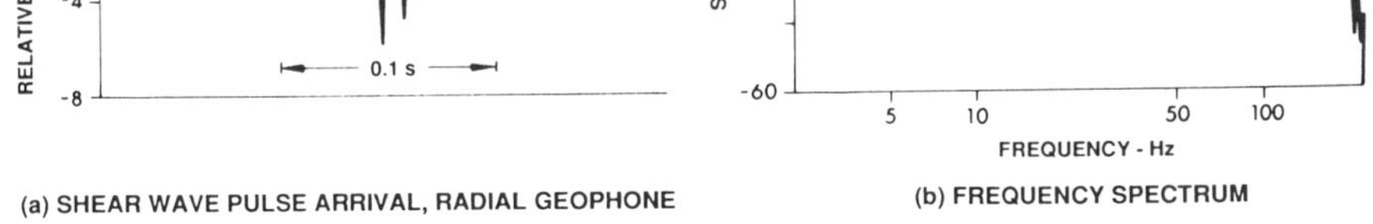


Figure 5. Radial, cross-hole shear wave signal, for a mud bottom in the Ligurian Sea.

3.2.2 *Shear Wave Velocity.* Figure (6) presents geoacoustic parameters as a function of depth. The cross-hole velocity data was acquired using four horizontally displaced boreholes. Here, the orientation of the source and receivers was transverse, the depth increments were 0.5 m, and the separation ranges were 1.7, 2.8, and 11.2 m from the source hole. It was possible to reduce the velocity data at the 100 Hz maximum in the spectrum and the resulting measurements (direct and differential) were made between various receiver holes. The scatter in the data (5 to 15 %) is indicated by the individual data points. It can be seen that the shear velocities are quite low, and increase with depth. Profiling data, also shown in Figure (6), show the sediments in this area to be layered, and in fact gas bearing.

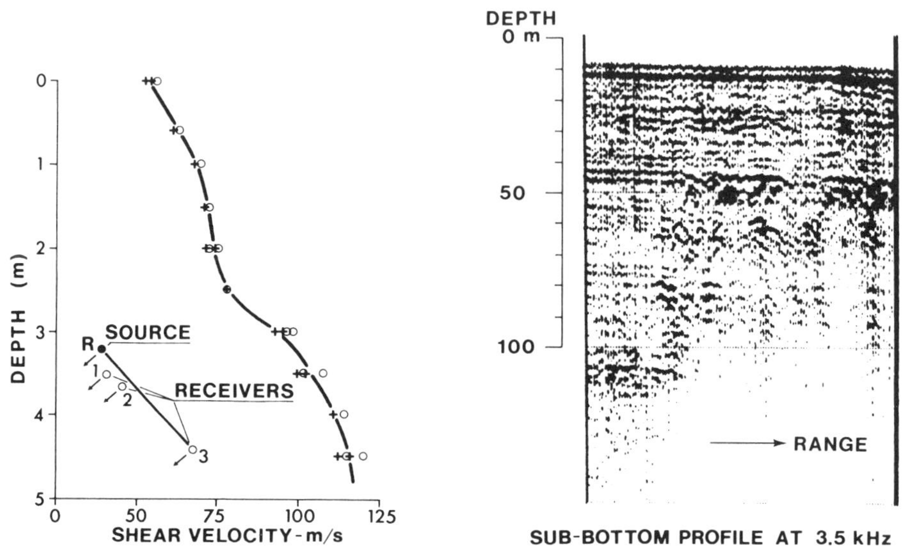


Figure 6. Shear wave velocity and sub-bottom profile in a mud bottom of the Ligurian Sea.

3.2.3 *Attenuation.* This downhole data ensemble was differentially analyzed to extract the attenuation vs depth vs frequency and the results are plotted in a quasi three-dimensional format in Figure (7). It can be seen that the attenuation decreases with both depth and frequency. For most of the depths, the frequency response is adequately modeled by a dependence on frequency to the power of 1.5, as is demonstrated by the data at a depth of 2.0 m, also presented in Figure (7). These are the first in situ measurements to show a nonlinear dependence of shear wave attenuation on frequency in unconsolidated sediments.

3.2.4 *Comments.* The direct, downhole system described here is useful in obtaining precise ground truth data vs depth at testbed sites. As a research tool, this method has provided new insight into shear wave processes. However, it is a time-consuming and expensive technique. The approaches discussed in the following sections involve remote sensing and inversion to obtain data on the geoacoustic parameters, covering greater areas in less time.

## 4. Point Source and Single Seismograph Methods

Approximately ten years ago, SACLANTCEN began a variety of seafloor experiments on seismo-acoustics, using a new digital, three-axis seismograph [4]. These included ambient noise and propagation loss measurements, as well as measurements on basic geophysics issues. A summary of recent work follows.

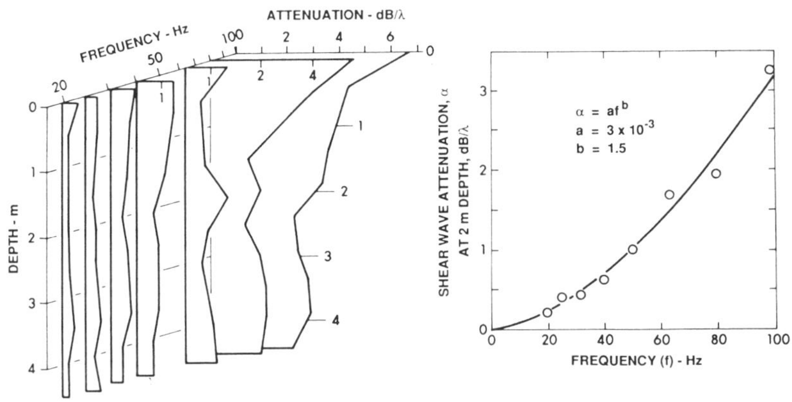


Figure 7. Shear wave attenuation for a mud bottom of the Ligurian Sea.

## 4.1 MEASUREMENT TOOLS

4.1.1 *Deployment Scenario.* A typical SACLANTCEN experiment in shallow water is sketched in Figure (8), showing an ocean bottom seismometer (OBS), an ocean sub-bottom seismometer (OSS), and a mechanically decoupled radio buoy for data telemetry to a receiving ship, such as R/V ALLIANCE, which is quiet, and is furthermore located out of the local noise field. Bottom and sub-bottom hydrophones and a vertical array of four hydrophone elements have also been used to simultaneously measure sound propagation loss in the water column, while seismic data is being acquired. A smaller vessel, R/V MANNING is used to detonate the explosive shot sources.

4.1.2 *Seismometers.* The OBS and OSS instruments are illustrated in Figure (9). The OBSs are disc shape and are of two three-axis designs; one employs 4.5 Hz moving coil geophones while the other uses 1 Hz geophones. Both use digital processing for amplification, gain ranging (to prevent saturation), and multiplexing. The OSS is in the shape of a small vertical cylinder, made by casting 4.5 Hz geophones along three axes in a polyurethane potting compound. One such unit has been permanently implanted at a depth of 6 m in a mud sediment in the Ligurian Sea and connected to a removeable electronics canister that rests on the seafloor.

4.1.3 *Shot Source.* Typically, TNT charges weighing 180 g are lowered to the bottom and electrically detonated with a blasting cap. A calculated, free field signature [5] and the low frequency component of the spectrum are shown in Figure (10), for a water depth of 16 m. The rise and decay of the initial blast is a small fraction of a millisecond, which produces kilohertz frequencies, well above the geophone and seismo-acoustic band. However, the first bubble pulse has a period of 78 msec, giving a spectral peak near 13 Hz, which likely provides the dominant source of seismo-acoustic signals.

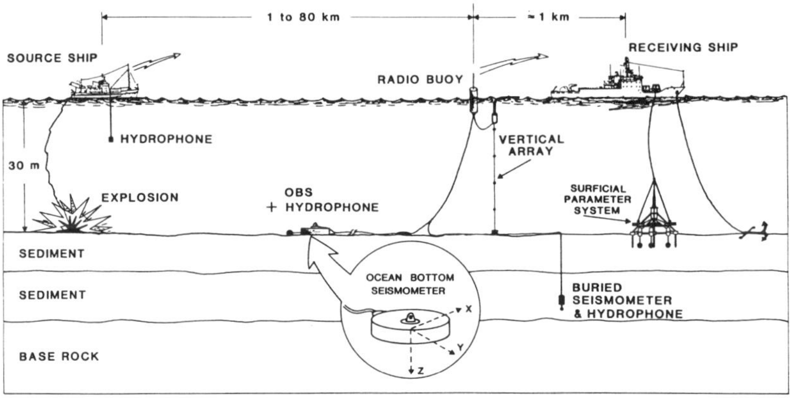


Figure 8. Seismo-acoustic measurement scenario.

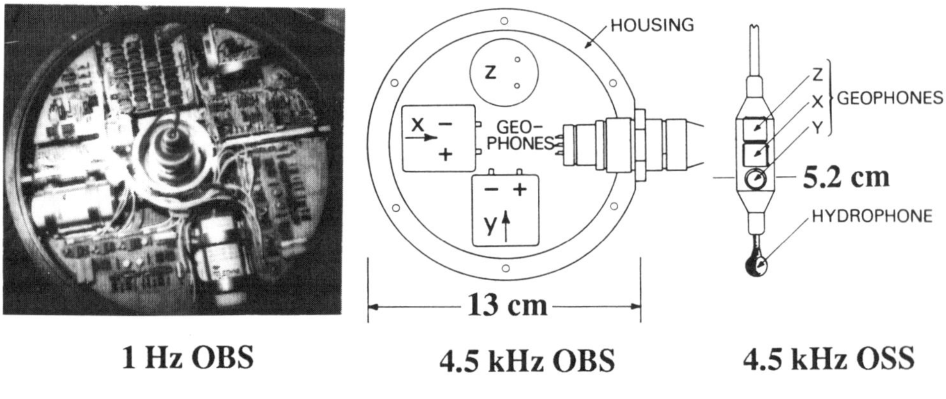


Figure 9. Seismometer systems (not to scale).

## 4.2 RESULTS

4.2.1 *Raw Data.* The results of a shot fired on the bottom in 16 m of water at a range of 350 m are shown in Figure (11). Here we compare vertical particle velocity from 4.5 Hz seismometers, one in an OBS on the bottom, the other in an OSS buried to a depth of 6 m in the aforementioned mud sediment testbed in the Ligurian Sea. Both the time series and the spectra have been normalized for display purposes, but the bottomed OBS geophone consequently required 3 dB more gain than its buried counterpart in the OSS. As a result, the buried vertical geophone is more sensitive to seismic signals of the interface type shown here. It is also subject to less ambient noise, typically 10 dB or so. Despite the maximum in the source signal at 13 Hz, the received spectra both peak at 3 - 4 Hz, obviously driven by the frequency selectivity of the interface wave phenomena and influenced by the low frequency roll-off of the 4.5 Hz geophones.

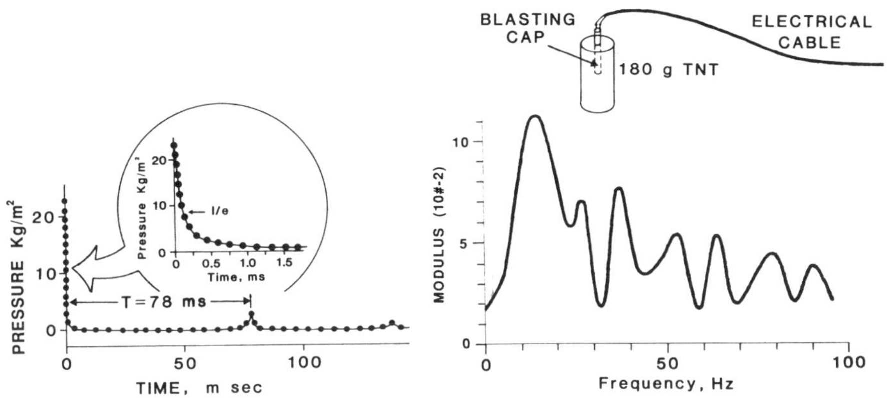


Figure 10. Free field shot signature and spectrum, calculated at a depth of 16 m.

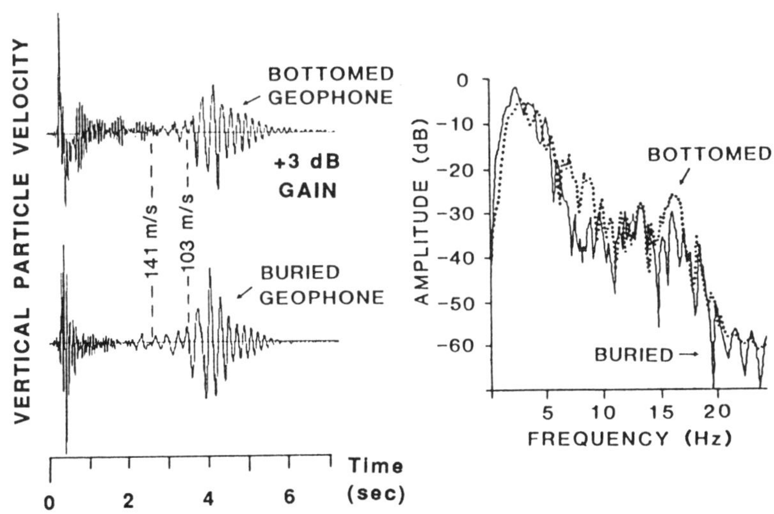


Figure 11. Comparison of buried and bottomed seismic signals from a bottom shot source.

4.2.2 *Inverted Data.* Seismo-acoustic measurements made as a function of range by the techniques described above were used to estimate sediment shear parameters through an iterated inversion process. The procedure involves two parts, described in detail in the paper by Caiti, et.al., in this proceedings. First, the shear velocity profile is estimated through a least-square fitting of the group velocity of the fundamental propagating mode, using a simplified Thomson-Haskell integration scheme as a forward model. Second, the SAFARI model [6] is used interactively to tune the shear wave velocity and attenuation profile to match computed and measured dispersion data, which are contour plotted in a Gabor matrix format. Here, we used the measured attenuation, dependent on the frequency to the power of 1.5, which is a deviation from standard practice with SAFARI calculations. The results are shown in Figure (12), which supports the existence of dispersive, bimodal interface waves at this site. The resulting range stack of synthetic seismograms is in reasonable agreement with those measured.

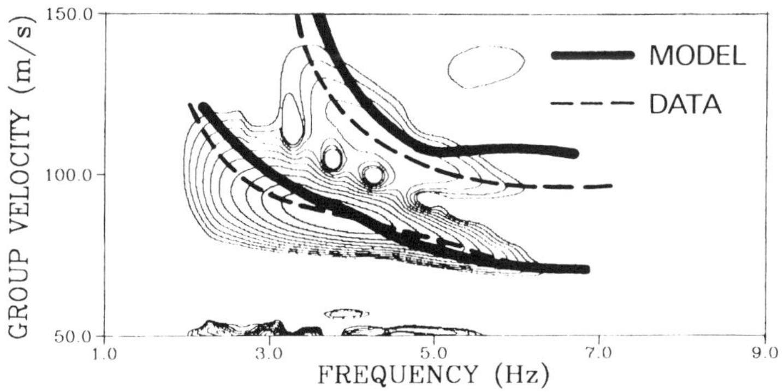


Figure 12. Comparison of dispersion data for the mud bottom at a Ligurian Sea testbed.

The resulting geoacoustic model for shear wave parameters is shown in Figure (13), along with the directly measured parameters. It can be seen that the direct and inverted data agree reasonably well, which is not surprising since they were forced to agree through fine tuning. The lowest mode appears to be a Sholte wave travelling in the near surface portion of the sediment, while the next (and weaker) mode is likely associated with a Stonely wave at an interface some 25 m deep.

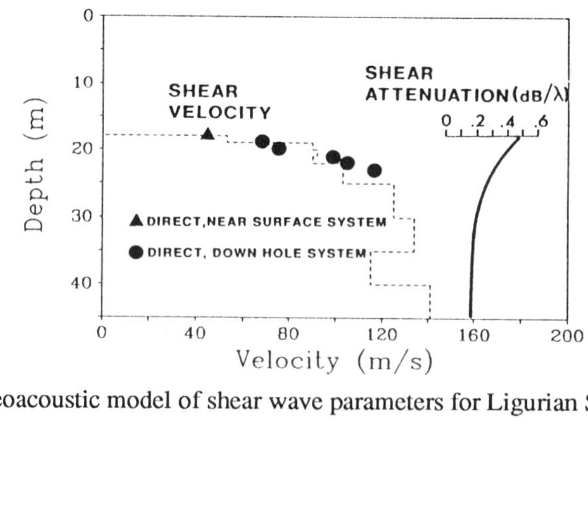


Figure 13. Geoacoustic model of shear wave parameters for Ligurian Sea testbed.

4.2.3 *Comments.* Although the work with the various methods discussed above was intended to help evaluate the inversion technique, it should be obvious from considering the extent of data coverage by the methods compared in Figure (13), that the inversion method has the potential to provide far more data. It is also non-invasive and quasi-remote, which for ocean surveys is important in that remote sensing techniques are comparatively less labor intensive and more efficient. Further, three axis seismograph receivers have the added advantage of adressing the vector field of seismic waves, enabling studies of horizontally polarized waves as well as hodograph time histories of particle orbits.

## 6. Future Systems

The aforementioned systems are useful for acquiring ground truth data, especially at testbed sites where only a small area needs to be delineated. If it is desired to survey larger areas of the ocean, even the small percent that constitute the shallow seas, it will be necessary to do so with mobile systems having higher search rate capability. A suite of survey instruments is invisioned for this purpose, and a hypothetical family of instruments is sketched in Figure (14). It consists of a bathymetric scanner that would delineate the bottom topography, a logging device that would provide direct seismo - acoustic ground truth at selected stationary intervals, a sub - bottom profiler to provide data on sediment stratification and layer impedance contrast, a wide angle reflectometer to provide near - surface sediment parameters from inversion of high frequency compressional wave data, and some type of sub - bottom device, tentatively labeled a "Scholte wave scanner", that would provide a geoacoustic model as a function of depth in the sediment from inversion of converted shear wave data. The last instrument represents a formidable challenge because the excitation function for seismic interface waves decays exponentially with distance from the bottom [4], and the noise introduced by hydrodynamic forces will likely be in the preferred seismo - acoustic band, at very low frequencies. The bottom towed sled instruments, discussed elsewhere in this proceedings by Davis, et.al., are a step in the right direction for increased survey rate.

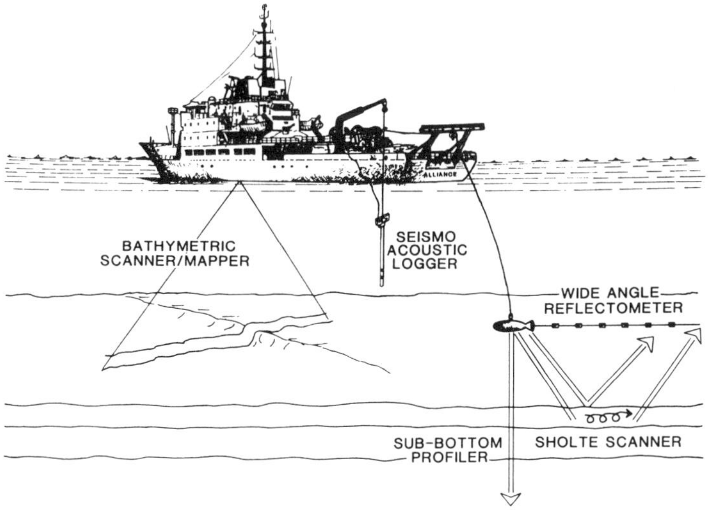


Figure 14. Hypothetical suite of bottom survey instruments for higher search rate applications.

**References**

[1] Akal, T., and Jensen, F. (1983), "Effects of the sea-bed on acoustic propagation," in N. G. Pace (ed.) *Acoustics and the Sea-Bed*, Bath University Press, Bath, U.K., pp. 225-232.

[2] Akal, T. (1980), "Sea floor effects on shallow-water acoustic propagation," in W. A. Kuperman and F. B. Jensen (eds.), *Bottom Interacting Ocean Acoustics*, Plenum Press, New York, pp. 557-576.

[3] Jensen, F. B. (1989), "Ocean Seismo-acoustic modeling: numerical methods," in Y. Desaubies, A. Tarantola, and J. Zinn-Justin (eds.) *Oceanographic and Geophysical Tomography*, Elsevier Scientific Publishers, New York, pp. 1-31.

[4] Rauch, D. (1980), "Experimental and theoretical studies of seismic interface waves in coastal waters," in W. A. Kuperman and F. B. Jensen (eds.), *Bottom Interacting Ocean Acoustics*, Plenum Press, New York, pp. 307-327.

[5] Personal communication, Jean Goertner, Naval Surface Weapons Center, September 1987.

[6] Schmidt, H. (1989),"SAFARI Users Guide," SACLANT Undersea Research Centre, La Spezia, Italy.

# SHEAR WAVE VELOCITY GRADIENTS IN NEAR-SURFACE MARINE SEDIMENT

MICHAEL D. RICHARDSON*, ENRICO MUZI, BRUNO MIASCHI, and FERDA TURGUTCAN
SACLANT Underwater Research Centre
Viale S. Bartolomeo 400
19026 La Spezia, Italy

*Seafloor Geosciences Division
Naval Oceanographic and Atmospheric Research Laboratory
Stennis Space Center, MS 39529-5004

ABSTRACT. Positive gradients in values of sediment shear wave velocity were measured from several shallow-water sites near La Spezia, Italy. Shear waves were generated and received using identical bimorph ceramic bender probes inserted into the upper 2 m of sediment. Shear wave velocity was measured directly using a pulse technique. Velocities ranged from 17 m/s near the sediment-water interface in high-porosity silty clays to over 100 m/s in lower porosity sands and silty sands. Shear wave gradients of 4 to 17 $s^{-1}$ were measured in the upper 2 m of sediments at the muddy sites, and a higher gradient of 85 $s^{-1}$ was measured at the hard-packed sandy site. Velocities at the muddy sites were generally less than predicted, but measurements and predictions were in agreement for sandy sites. Comparisons suggest that sediment shear modulus is controlled by more than the process of consolidation in near-surface, fine-grained sediments but is adequately modelled in sands.

## 1. Introduction

Measuring and interpreting values of near-surface shear wave velocity are complicated by two factors: the poorly defined relationships between shear wave velocity and sediment type, and the strong vertical gradients of sediment shear modulus that result from overburden pressure [1,2]. Laboratory consolidation tests, downhole-crosshole direct measurements, and inversion of time arrivals of seismic and surface waves have provided much-needed information about vertical gradients of shear wave velocity. Using data collected by these measurement techniques, both Hamilton [2] and Bryan and Stoll [3] have developed empirical relationships to define vertical gradients of shear wave velocity (or shear modulus) in several types of sediment. Empirical relationships between shear wave velocity and physical properties of sediments have also been established for fixed depths in near-surface marine sediments [4,5,6]. These measured values of shear wave velocity near the surface are often lower than those predicted by the empirical relationships of Hamilton and Bryan and Stoll.

The upper few meters of sediments include a zone of rapid changes in the physical, geotechnical, and chemical properties of sediments. These changes result from intense biological activity and from active physical (storms, waves, currents, erosional and depositional events) and chemical processes [7]. One might expect that shear wave velocity (i.e., shear modulus or



*J. M. Hovem et al. (eds.), Shear Waves in Marine Sediments, 295–304.*

rigidity) values and gradients would differ from idealized physical models, laboratory measurements made on disturbed or artificial sediments, or low-frequency measurements that integrate physical properties over long pathlengths. We present the vertical gradients of values of shear wave velocity, measured in the upper 2 m, for a variety of sediment types near La Spezia, Italy. These gradients are compared to predictions of empirical models to learn more about the physical mechanisms that control shear wave propagation in near-surface marine sediments.

## 2. Methods

Sediment shear wave velocity was measured using a pulse technique [4,8]. Transmit and receive probes were constructed from identical 31.75 $mm^2 \times 0.48$ mm thick bimorph ceramic benders. The ceramic bender elements were potted in a stainless steel ring with soft silicone rubber (hardness = 35 Shore A) to allow relatively unrestricted bender movement. A thin covering of much harder polyurethane resin (hardness = 80 Shore A) held the ceramics in place and provided a tough coating to protect the ceramics during insertion into the sediment. Shear wave probes were attached to the ends of 2.4-m hollow steel pipes. During deployment, probe orientation and distance below the sediment-water interface were controlled by scuba divers using a PVC frame (Fig. 1). Shear wave velocity was measured at 10-cm depth intervals and over 30-cm and 70-cm pathlengths parallel to the sediment-water interface. Shear waves were generated as a 2- to 6-cycle sine wave pulsed every 0.5 s. Driving frequency (200–2000 Hz) and driving voltage (100–230 V p-p) depended on the varying mechanical load of sediments on the compliant bender ceramic face, the sediment shear wave velocity and attenuation, and the pathlength between the receive and transmit probes. Received signals were amplified with 40-dB preamplifiers mounted in receiver probe heads. The signals were band-pass filtered and recorded with a digital waveform recording oscilloscope. Shear wave velocity was calculated from the time delay and receiver-transmitter separation.

Gradients in values of shear wave velocity were measured at seven sites (Fig. 2). At least two deployments of the measurement system were made at each site except at Venere Azzura, where hard-packed sandy sediments resisted probe penetration. Duplicate 12-cm (inside diameter) gravity core samples were collected at each site for sediment physical property determination. Dry-sediment density was determined with a helium pycnometer. Sediment porosity, void ratio, and wet density were calculated from weight loss of the sediment after it was dried in an oven at 105°C for 48 hours and from the measured dry density [9]. Sediment grain size distribution was determined by dry sieving for the gravel- and sand-sized particles and by using a Micromertics Sedigraph for the silt- and clay-sized particles [10].

## 3. Results

Sediments at Diga, Portovenere, and Viareggio consisted of primarily silt- and clay-sized particles (Table 1). Values of sediment porosity decreased only slightly with depth in the sediment, primarily in response to increased overburden pressure and not from changes in sediment grain size. Values of in situ shear wave velocity increased from a range of 20 to 25 m/s near the sediment-water interface (10–30 cm) to a range of 40 to 50 m/s at 2-m depth

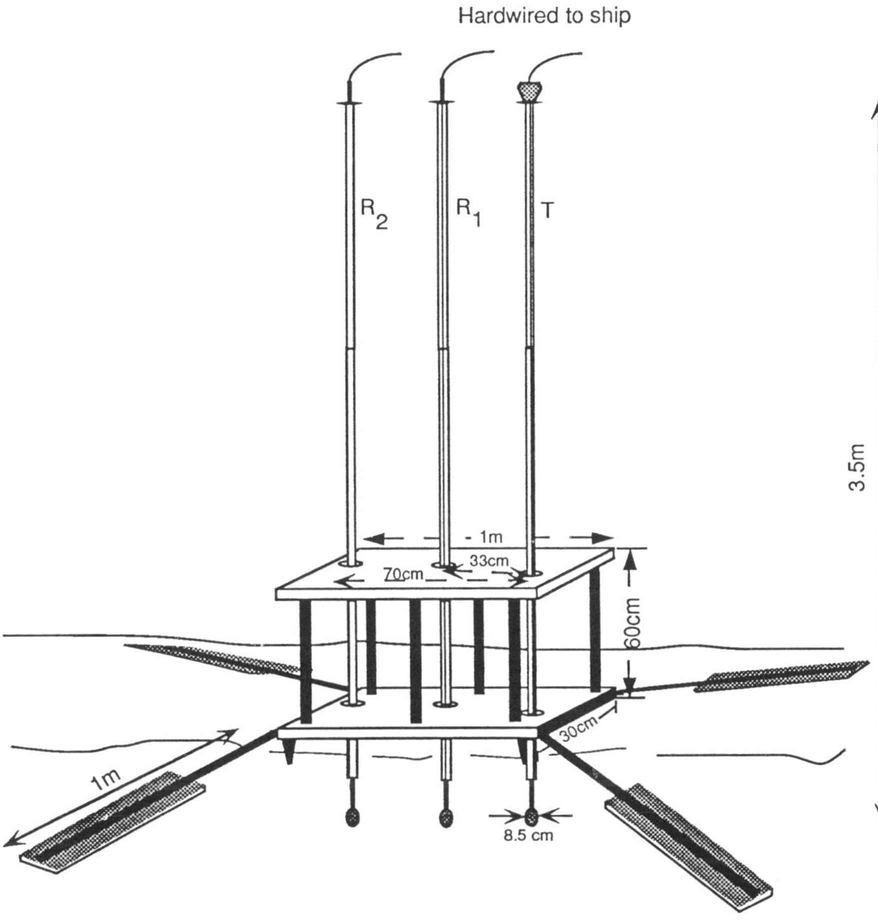


Figure 1. PVC measurement frame.

in the sediment (Fig. 3). The sediments at Boa Dragaggio and Turf consisted of interlayered fine sands and silty clays. The fine sand layers were particularly evident at 120 to 180 cm depths at the Turf site and had lower porosity and higher void ratio values than the silty clay layers. Shear wave velocity in the finer grained silty clay sediment layers increased from a range of 30 to 40 m/s in near-surface sediment to nearly 60 m/s at 2 m, but sandy sediments had higher values of shear wave velocity that ranged from 70 to 113 m/s. The sediments at Venere Azzura were well-sorted, fine, hard-packed, low-porosity sands. Near-surface values of shear wave velocity (56–100 m/s) and the gradient (85 $s^{-1}$) were much higher than at the muddy sites. Values of shear wave velocity at Santa Teresa increased from about 25 m/s near the surface to a range of 40 to 50 m/s at 1.6 m. The upper 70 cm of sediment at Santa Teresa consisted of high-porosity clays (range of mean grain size 8.5–9.8$\phi$). From 80 to 160 cm, the sediments contained large amounts of sand- and gravel-sized particles (up to 60–85% by weight

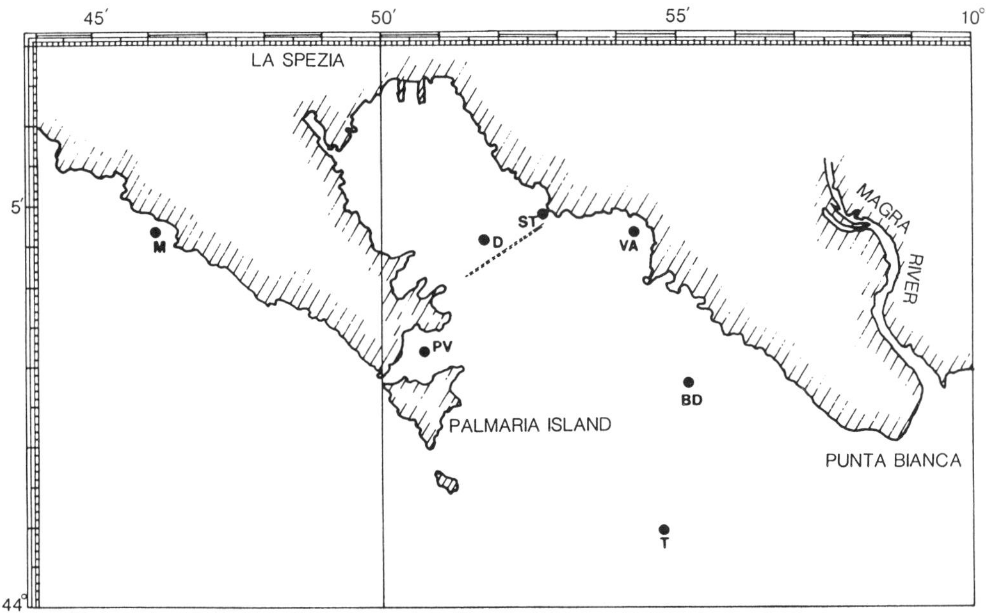


Figure 2. Location of sampling sites: Diga (D), Santa Teresa (ST), Portovenere (PV), Boa Dragaggio (BD), Turf (T), Venere Azzura (VA). Viareggio site (43°48.62'N,10°07.16'E) was 33 km southeast of Palamaria Island.

Table 1. Summary of environmental conditions (mean and range) for the upper 2 m of sediments at seven sampling sites in the vicinity of La Spezia, Italy

| Site | Depth (m) | Date | Porosity (%) | Mean Grain Size (ϕ) |
|---|---|---|---|---|
| Diga | 7 | 20 April 89 | 63 (61-65) | 10.0 (9.9-10.1) |
| Santa Teresa | 10 | 21 April 89 | 54 (36-67) | 5.3 (0.7-10.0) |
| Portovenere | 12 | 30 April 89 | 63 (62-67) | 9.9 (9.8-10.1) |
| Boa Dragaggio | 14 | 4 July 89 | 57 (53-64) | 8.2 (5.3-9.7) |
| Turf | 18 | 5 July 89 | 52 (44-59) | 7.4 (4.0-10.0) |
| Venere Azzura | 7 | 7 July 89 | 45 (42-49) | 3.4 (2.6-4.2) |
| Viareggio | 22 | 10-11 July 89 | 58 (45-66) | 9.1 (8.1-9.6) |

of whole and broken molluscan shells and rock fragments varying in size and shape). Mean grain size (0.7–3.6ϕ) and porosity (mean 48%, range 36–61%) were significantly different from the upper 70 cm.

In summary, two trends in values of shear wave velocity were evident: a positive gradient with depth in the sediment, and higher values of shear wave velocity associated with coarser

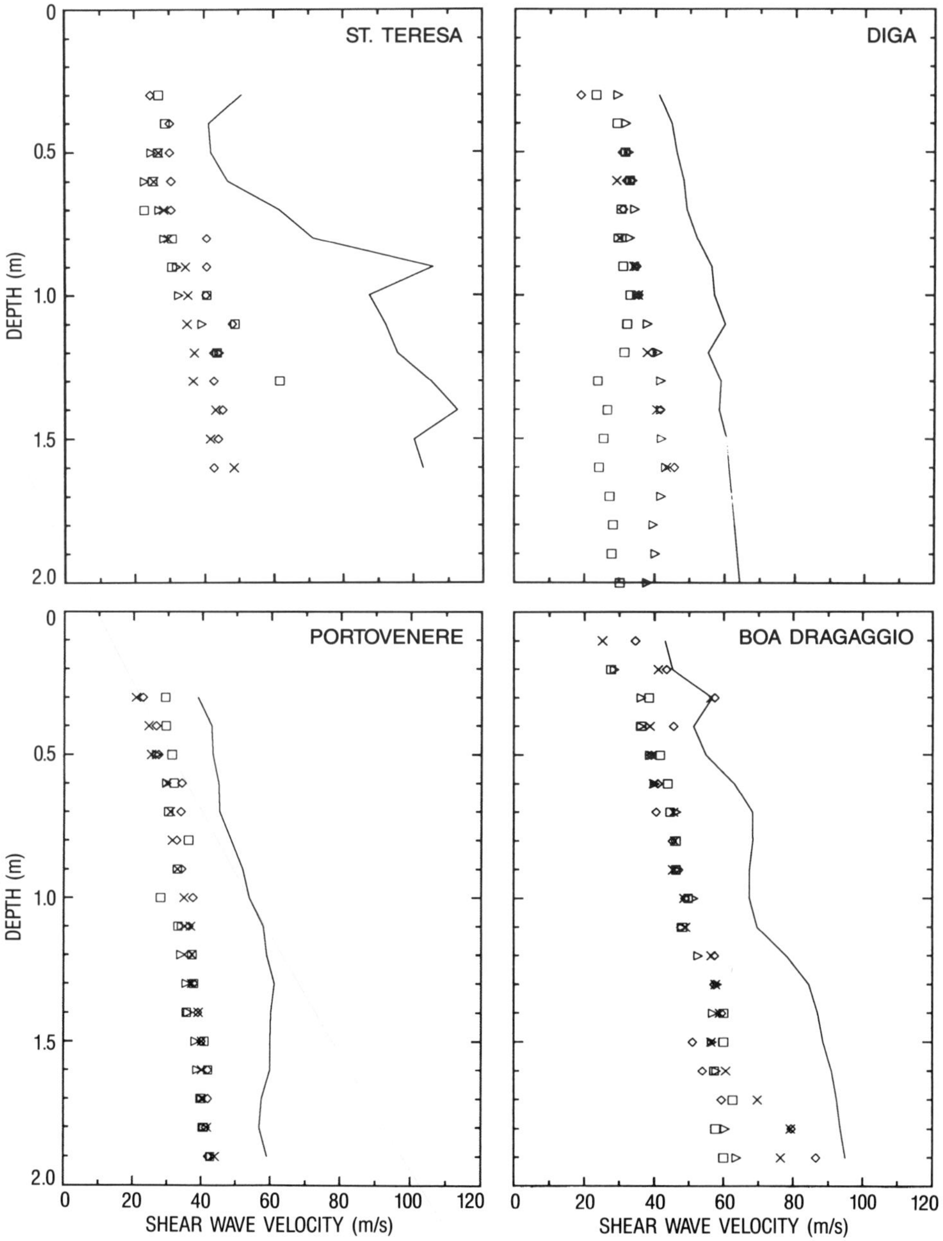


Figure 3. Depth distribution of values of shear wave velocity for seven locations in the vicinity of La Spezia, Italy. Measured velocities (symbols) are compared to shear wave velocities predicted using the Bryan and Stoll model (lines).

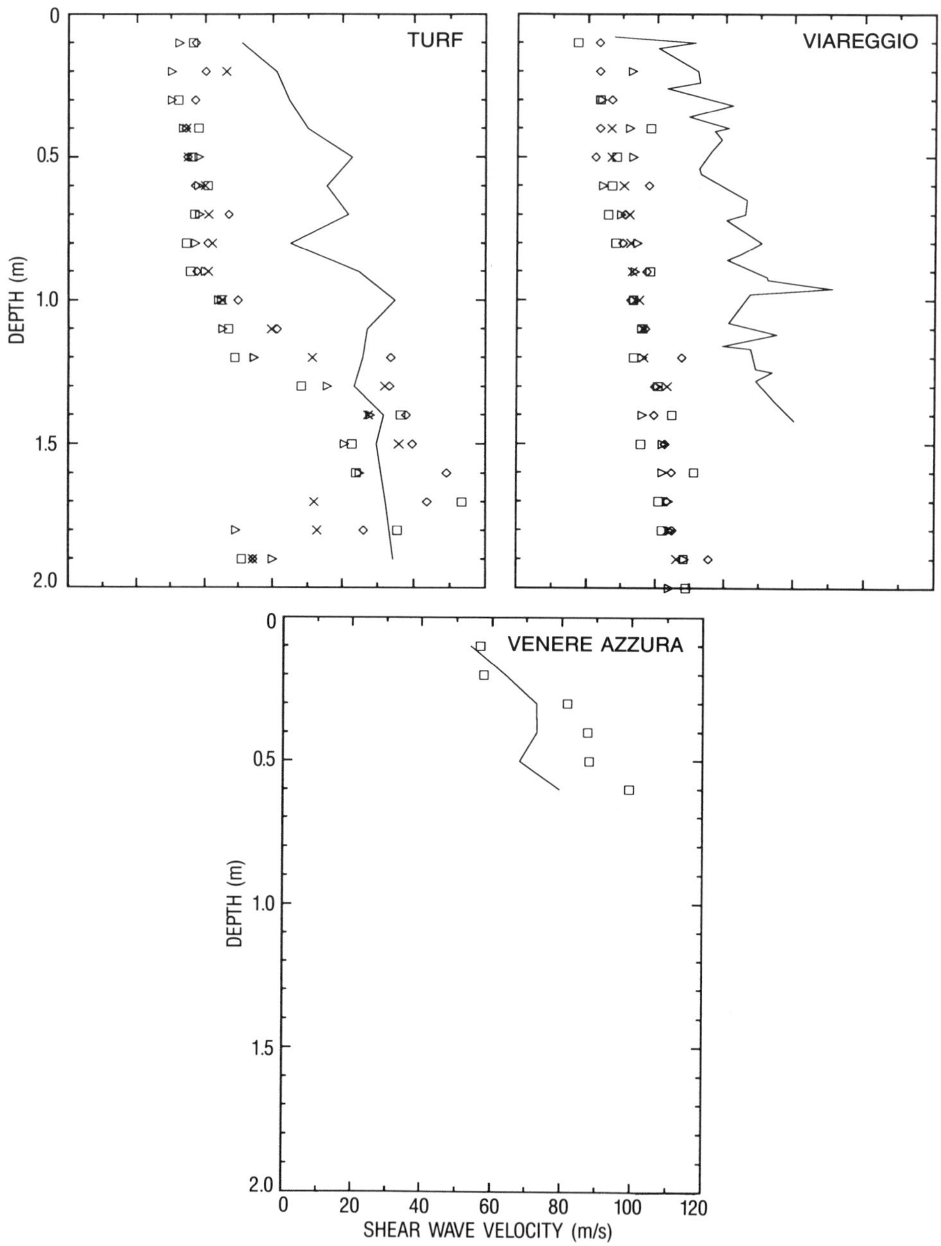


Figure 3. (continued).

grained, lower porosity sediments. Both trends are qualitatively in agreement with published measurements and models.

## 4. Discussion

Over the past 12 years, SACLANTCEN investigators have measured near-surface shear and interface wave velocities at the same La Spezia locations sampled for this study [11-15]. These studies confirm the existence of low shear wave velocities near the surface that are similar to those we measured. Near-surface vertical gradients of shear wave velocity were determined by crosshole measurement techniques and by inversion of interface wave dispersion curves collected using either ocean-bottom seismometers or line arrays of bottomed geophones. These gradients were also similar to the data we have presented.

Shear wave velocity gradients can be predicted for the upper few hundred meters of sediment using “poro-viscoelastic” models [1] or empirical relationships [2,3]. We did not measure enough sediment characteristics to predict gradients of shear wave velocity from the full application of poro-viscoelastic models; therefore, we will restrict our data comparisons to predictions from empirical relationships. Hamilton [3], in a compilation of values of shear wave velocity determined by mostly in situ techniques, suggests the following regressions for calculating shear wave velocity ($V_s$, m/s) versus depth (D, m):

$$V_s = 128D^{0.28} \text{ for sands} \tag{1}$$

$$V_s = 116 + 4.65D \text{ for silty-clays} \tag{2}$$

Bryan and Stoll [3] summarized the effects of mean effective stress (p') and void ratio ($e$) on sediment dynamic modulus ($\mu$) with the following expression:

$$\mu = a\,(p')^n \exp(b\,e)\,, \tag{3}$$

where $a = 2526$ atm, $n = 0.45$, and $b = -1.5$. The formula was based on 494 concurrent low-strain laboratory measurements of dynamic shear modulus, confining pressure (range 14–700 kPa), and void ratio (range 0.35–2.5) compiled from the literature. Overburden pressure, calculated from sediment density and void ratio, is assumed to be 0.67 times the mean effective stress [16]. Shear wave velocity can then be calculated from sediment dynamic modulus and sediment wet density ($\rho$, kg/m$^3$):

$$V_s = (\mu/\rho)^{0.5}\,. \tag{4}$$

Both empirical relationships predict higher near-surface shear wave velocity values than were measured in muddy sediments (Diga, Portovenere, and Viareggio). Hamilton's values of 116 to 125 m/s for silty-clay sediments overestimate measured shear wave velocities by 100 to 500% and predictions from the Bryan and Stoll model are roughly 50% higher than the measured values (Fig. 3). In contrast, both Hamilton's and Bryan and Stoll's shear wave velocity estimates and our measurements for sand (Venere Azzura) generally agreed. At the Venere Azzura site, predictions from the Bryan and Stoll model averaged 13% lower than measured shear wave velocities, and Hamilton's predictions were 19% higher than measured shear wave velocities. At the Turf and Boa Dragaggio sites, shear wave velocities estimated by the Bryan and Stoll model were higher than measured values in muddy sediment layers and nearly the same in sand layers. Hamilton's predictions may not be appropriate for these layered sediments. At the Santa Teresa site, predictions from the Bryan and Stoll model were higher than measured shear wave velocities, especially from 80 to 160 cm where sediments contained large amounts of sand- and gravel-sized particles. The mixture of silty clay with sand- and gravel-sized shells and gravel did not have the same shear modulus as would a homogeneous sediment with similar grain size and void ratio.

The near-surface shear wave velocities in soft sediments were lower than predicted. This decrease may result from any one or a combination of the following factors: (1) low in situ values of shear modulus that may result from near-surface physical, chemical, or biological processes; (2) differences in laboratory, in situ probe, and seismic inversion measurement techniques, including frequency, strain amplitude, strain rate and time, stress history, and effective stress conditions; (3) shear wave velocity anisotropy as a result of differential vertical and horizontal effective stress; (4) heterogeneous distribution of sediment properties such as at Santa Teresa; and (5) disturbance of sediments during insertion of the probes into sediments or during collection of sediment samples. The differences in measured and predicted values of shear wave velocity should be investigated in combined laboratory and field experiments where effective stress is measured and/or controlled.

Stoll [17] suggests increasing the parameter $a$ by 30 to 150% in order for formula 3, that was derived from laboratory measurements, to be valid for in situ field conditions. This correction is equivalent to 14 to 58% increase in shear wave velocity. As can be seen in Figure 3, most in situ values of shear wave velocity were actually lower than predicted by formula 3.

We therefore recommend that near-surface shear wave velocity be estimated using a predictive relationship that combines the results of this study and those of Richardson et al. [4]. In situ shear wave velocity, at 30 cm depth in the sediment, can be estimated using the following empirical relationship [4]:

$$V_s = 57.6/e \,, \tag{5}$$

where $e$ is the void ratio. If we assume that shear wave velocity increases to the power of 0.3 with depth (a good approximation of our measured data), then shear wave velocity gradients in the upper 2 m of sediment can be predicted by the following:

$$V_s = (85/e)D^{0.3} \tag{6}$$

where D is depth in meters below the sediment-water interface.

This predictive relationship applies only to sites that consist of a single sediment type and should not be applied to layered or mixed sediments, such as were found at Turf, Boa Dragaggio, and Santa Teresa. Continued investigation of near-surface shear wave velocities may extend the applicability of this relationship to other sediment types and improve the predictive accuracy of formula 6. Physical and geoacoustic properties were not measured on the same sediment sample; therefore, we were restricted to describing general trends between sediment physical and geoacoustic properties and could not calculate empirical relationships such as those of Bryan and Stoll [3].

The predictive relationship developed by Bryan and Stoll [3] should be used to estimate shear wave velocity of sediments deeper than the first few meters. This relationship includes provisions to account for layering by including the effects of changing void ratio, as well as increases in effective stress. Continued analysis of field and theoretical data should improve the accuracy of this empirical relationship [16]. Hamilton's predictive relationships do not account for vertical changes in sediment type, which reflect changing sedimentary processes, and should be used only in the absence of knowledge of the subbottom distribution of sediment properties.

## 5. References


[1] Stoll, R. D. (1989) Sediment Acoustics, Lecture Notes in Earth Sciences, Springer-Verlag, New York, 153 p.

[2] Hamilton, E. L. (1976) 'Shear wave velocity verses depth in marine sediments: a review', Geophysics 68, 985-996.

[3] Bryan, G. M. and Stoll, R. D. (1988) 'The dynamic shear modulus of marine sediments', J. Acoust. Soc. Am. 83, 2159-2164.

[4] Richardson, M. D., Muzi, E., Troiano, L., and Miaschi, B. (1990), 'Sediment shear waves: a comparison of in situ and laboratory measurements', in R. H. Bennett, W. R. Bryant and M. H. Hulbert (eds.) Microstructure of Fine-Grained Sediments, Springer-Verlag, New York, pp. 403-415.

[5] Huws, D. G., Davis, A. M. and Bennell, J. D. (1991) 'Mapping of the sea bed via in situ shear wave (SH) velocities,' in J. M. Hovem, M. D. Richardson, and R. D. Stoll (eds.), Shear Waves in Marine Sediments, Kluwer Academic Publishers, Dordrecht, The Netherlands.

[6] Jones, S. A. and Jago, C. F. (1991) 'Small-scale in situ measurements of S-H velocity in surficial sedimentary deposits: localised textural and biological controls', in J. M. Hovem, M. D. Richardson, and R. D. Stoll (eds.), Shear Waves in Marine Sediments, Kluwer Academic Publishers, Dordrecht, The Netherlands.

[7] Richardson, M. D. and Young, D. K. (1980) 'Geoacoustic models and bioturbation', Mar. Geol. 52, 205-218.

[8] Barbagelata, A. Richardson, M., Miaschi, B, Muzi, E., Guerrini, P., Troiano, L. and Akal, T. (1991) 'ISSAMS: an in situ sediment acoustic measurement system', in J. M. Hovem, M. D. Richardson, and R. D. Stoll (eds.) Shear Waves in Marine Sediments, Kluwer Academic Publishers, Dordrecht, The Netherlands.

[9] Kermbon, A., Gehin, C., Blavier, P., and Tonarelli, B. (1969) 'Acoustic and other physical properties of deep-sea sediments in the Tyrrhenian Abysall Plain', Mar. Geol. 7, 129-145.

[10] Briggs, K. B. and Richardson, M. D. (1984) 'Physical and acoustic properties of surface sediments from the Venezuela Basin: a data report', Naval Ocean Research and Development Activity, Stennis Space Center, MS, NORDA Technical Note 238, 282 p.

[11] Rauch, D. (1980) 'Experiments and theoretical studies of seismic interface waves in coastal waters', in W. A. Kuperman and F. B. Jensen (eds.) Bottom-Interacting Acoustics, Plenum Press, New York, pp. 307-327.

[12] Akal T., Curzi, P. V., and Michelozzi, E. (1984) 'Geoacoustic and geotechnical properties of sediments: a pilot study', Mem. Soc. Geol. Italy 27, 411-421.

[13] Snoek, M. and Rauch, D. (1991) 'Spatial variability in ground motion: effects of material pheterogeneity in sea-floor sediment', in J. M. Hovem, M. D. Richardson,

and R. D. Stoll (eds.), Shear Waves in Marine Sediments, Kluwer Academic Publishers, Dordrecht, The Netherlands.

[14] Caiti, A., Akal, T., and Stoll, R. D. (1991) 'Determination of shear wave velocity profiles by inversion of interface wave data', in J. M. Hovem, M. D. Richardson, and R. D. Stoll (eds.), Shear Waves in Marine Sediments, Kluwer Academic Publishers, Dordrecht, The Netherlands.

[15] Muir, T. G., Akal, T., Richardson, M. D., Stoll, R. D., Caiti, A., and Hovem, J. (1991) 'Comparison of techniques for shear wave velocity and attenuation measurements', in J. M. Hovem, M. D. Richardson, and R. D. Stoll (eds.), Shear Waves in Marine Sediments, Kluwer Academic Publishers, Dordrecht, The Netherlands.

[16] Stoll, R. D., Bryan, G. M., Flood, R., Chayes, D., and Manley, P. (1988) 'Shallow seismic experiments using shear waves', J. Acoust. Soc. Am. 83, 93-102.

[17] Stoll, R. D. (1991) 'Shear waves in marine sediments-bridging the gap from theory to field applications', in J. M. Hovem, M. D. Richardson, and R. D. Stoll (eds.), Shear Waves in Marine Sediments, Kluwer Academic Publishers, Dordrecht, The Netherlands.

# ISSAMS: AN IN SITU SEDIMENT ACOUSTIC MEASUREMENT SYSTEM

A. BARBAGELATA, M. RICHARDSON*, B. MIASCHI,
E. MUZI, P. GUERRINI, L. TROIANO and T. AKAL

*SACLANT Undersea Research Centre*
*Viale San Bartolomeo 400*
*19026 La Spezia, Italy*

**Seafloor Geosciences Division*
*Naval Oceanographic and Atmospheric Research Laboratory*
*Stennis Space Center, Mississippi 39529-5004*

ABSTRACT. An *In Situ* Sediment Acoustic Measurement System (ISSAMS) has been designed and tested for measurement of near surface sediment geoacoustic properties in water depths up to 200 m. Shear wave velocity is measured over pathlengths of 33 and 71 cm at a sediment depth of 30 cm. Compressional wave velocity and attenuation is measured over 58 and 100 cm pathlengths at sediment depths of 10, 20 and 30 cm. Preliminary tests that have lead to the current design are described to justify the current measurement configuration. System characteristics are presented and a strategy for further development and improvement of the ISSAMS system is outlined. Representative data collected from the Adriatic Sea are presented and compared to laboratory and seismic measurements. These results demonstrate the accuracy and usefulness of ISSAMS.

## 1. Introduction

Knowledge of sediment geoacoustic properties is of fundamental importance for prediction of stability of marine slopes, consolidation behaviour of sediments, strength of marine foundations, scattering and propagation of acoustic energy from and within sediments.

Near surface sediment geoacoustic properties have been measured *in situ* using probes deployed by scuba divers and research submersibles and remotely from surface ships; between and down boreholes using explosive and various vibratory techniques; with various seismic reflection and refraction techniques; from direct path and inversion measurement of surface waves; and in the laboratory using geotechnical techniques related to the stress/strain behaviour of sediments. Application of these techniques, especially relating to shear wave velocity and attenuation measurements, is covered by various authors in this volume [1].

It was our intent to develop a remote method to measure shear and compressional wave velocities and attenuations in surficial marine sediments. It is within this surficial zone (upper 50cm of sediment) that the most active and rapid diagenetic changes in sediment properties occur [2]. The system presented is capable of making rapid, direct *in situ* measurements of sediment geoacoustic properties in the upper 30cm of shallow-water marine sediments. We have included information on early development stages to justify the current measurement configuration and present concepts for further development.



*J. M. Hovem et al. (eds.), Shear Waves in Marine Sediments*, 305–312.

## 2. ISSAMS Characteristics and Development

### 2.1 GENERAL

ISSAMS has passed through four of the five planned developmental stages. During the first stage, individual probes were tested to determine beam pattern, frequency response, signal attenuation, and coupling characteristics with the sediment. During stage 2, we tested for potential acoustic coupling between probes and the deployment frame, and determined penetration resistance of the probes into the sediment. The scuba diver deployed third stage was used to test the measurement system in a range of sediments commonly found on the continental shelf, including soft muds, hard packed sands, and loose gravels [3]. The current fourth stage is a diver-independent shallow water geoacoustic measurement system which is hardwired to surface electronics via a 300m electro-mechanical cable. We also present requirements for a depth-independent fifth stage which will be connected to surface electronics via a single coaxial cable.

### 2.2 COMPRESSIONAL WAVE PROBES

Transmit and receive probes consist of three identical radial-poled ceramic cylinders mounted on a hollow steel pipe (Fig.1). The ceramic cylinders are 25mm outside diameter x 25mm long and have 3.2 mm thick walls. Ceramic elements are attached to the stainless steel pipe

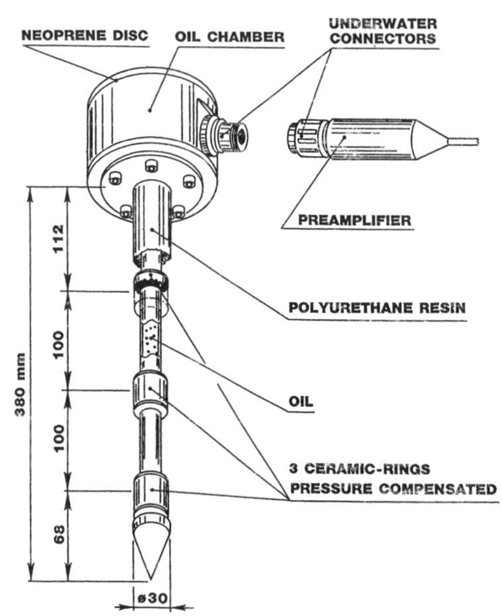


Figure 1. Compressional wave receiver probe.

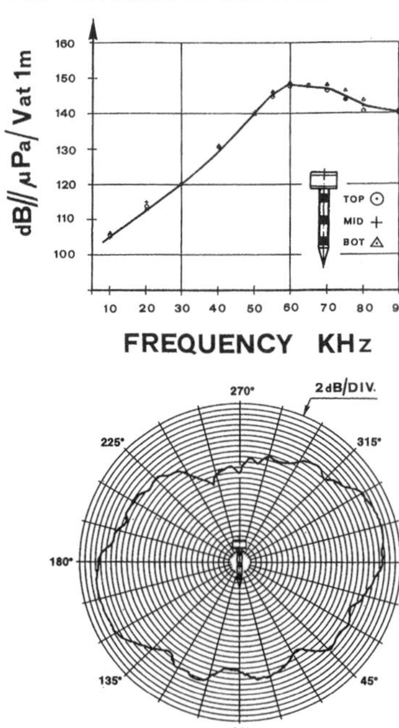


Figure 2. Vertical directionality and receiving sensitivity, for compressional wave probes in fresh water.

by epoxy resins and are pressure compensated through connections to an oil filled reservoir [4]. The probes are potted in hard polyurethane resin for protection. Receive probes include a 40dB gain pre-amplifier (Fig.1). The probes have a maximum sensitivity at 60 kHz, with omnidirectional characteristics in the horizontal plane and a $40^0$ (3dB) beamwidth in the vertical plane at 60 kHz (Fig.2).

Coupling between probes and sediments during diver deployment was good with strong signals transmitted through muddy and sandy sediments over path lengths up to 2 meters. Coupling between probes and frame was negligible therefore no additional decoupling material was used for the next configuration of ISSAMS.

During deployment of ISSAMS, individual ceramics are located 10, 20, and 30cm below the sediment-water interface and measurements are made over 9 separate paths with 58 or 100 cm probe separations (Fig.3).

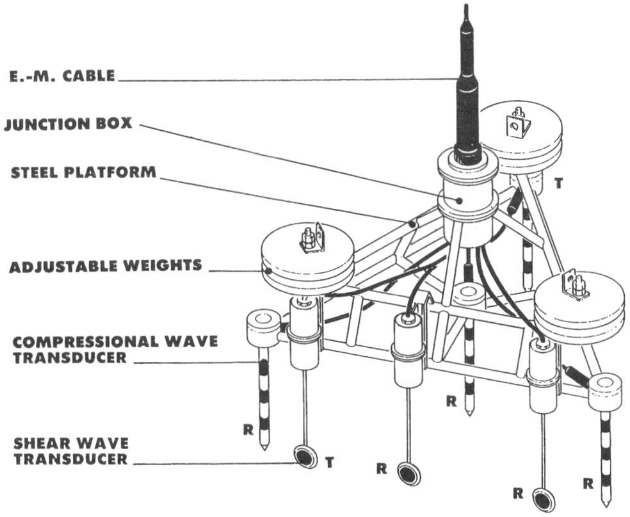


Figure 3. ISSAMS Stage 3 measurement system.

## 2.3 SHEAR WAVE PROBES

Transmitters and receivers are identical 31.75 mm square x 0.6 mm thick 'Bimorph' ceramic benders (Fig.4). The ceramics are potted in a stainless steel ring with soft silicone rubber to allow relatively unrestricted bender movement. A thin covering of a harder polyurethane resin holds the ceramics in place and provides a tough coating to protect the ceramics during insertion into the sediment (penetration depth 30cm). The receive probes include a 40dB gain pre-amplifier (Fig.3). The horizontal beam pattern of the probes illustrates the property of maximum sensitivity perpendicular to the direction of particle motion (Fig.5).

Coupling between probes and sediment, during hand-deployment by scuba divers, was good. Shear waves propagated over distances up to 2 meters in both sands and muds (Fig.6). However, some energy was also coupled into the test-frame, complicating time-delay measurements. Hence probes in the third stage system were decoupled from the frame by potting in a cylinder of silicone rubber (Fig.3).

Shear wave measurements are made over 33cm and 71cm pathlengths approximately 30cm below the sediment-water interface. Measurement frequency varies with sediment type due to the varying mechanical load imposed on the compliant bender ceramic by different sediment types, which results in a shift in resonance frequency. Shear waves propagating through sediments at 20 to 100m/s and 250 to 1500 Hz have wavelengths between 5 and 10cm.

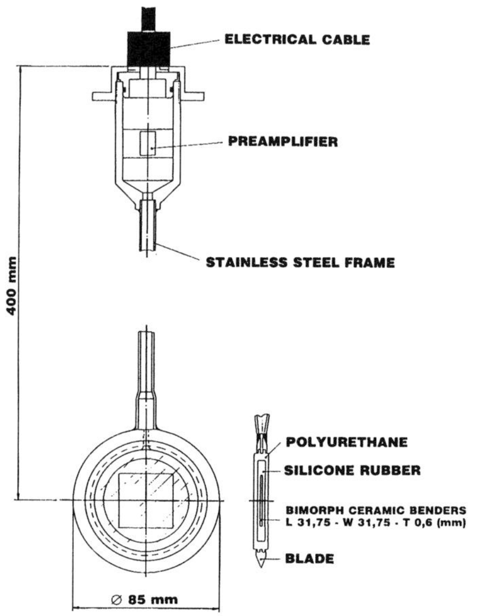


Figure 4. Shear wave receiver probe.

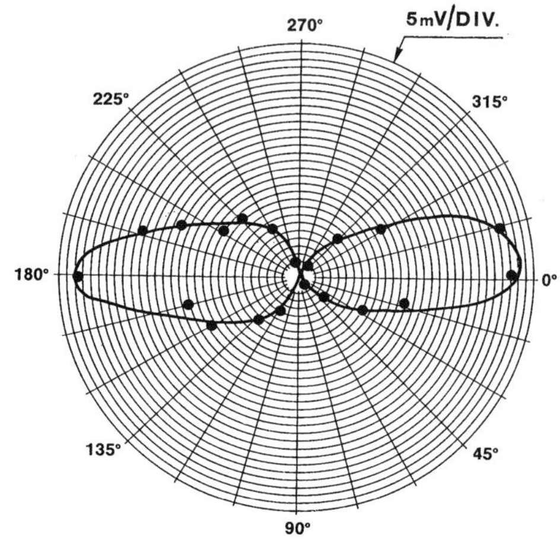


Figure 5. Beam pattern of shear wave probe.

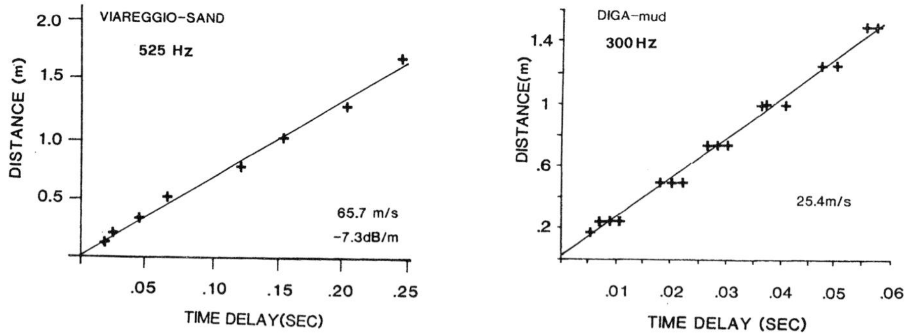


Figure 6. Shear wave velocity from distance-time delay measurements.

## 2.4 ELECTRONICS

Electronic configuration is the same for compressional and shear wave measurements (Fig.7). Both shear and compressional wave ceramic elements are driven with a pulsed sine wave. Shear waves are generated as a 6-cycle sine wave with 500ms repitition rate. Driving frequency (100-2000 Hz) and driving voltage (100-230Vp-p) vary depending on coupling, sediment shear wave velocity and attenuation, and path length between receiver and transmitter. The resonant frequency in muddy sediment is lower (200-500 Hz) than that of sandy sediment (500-1500 Hz). Compressional waves are generated as a 10-cycle sine wave with 100 ms repitition rate. Driving frequency is a constant 60 KHz and the driving voltage (30-100Vp-p) varies depending on path length, attenuation and sediment type. Pre-amplified and filtered receiver signals are monitored and recorded with a digitizing oscilloscope.

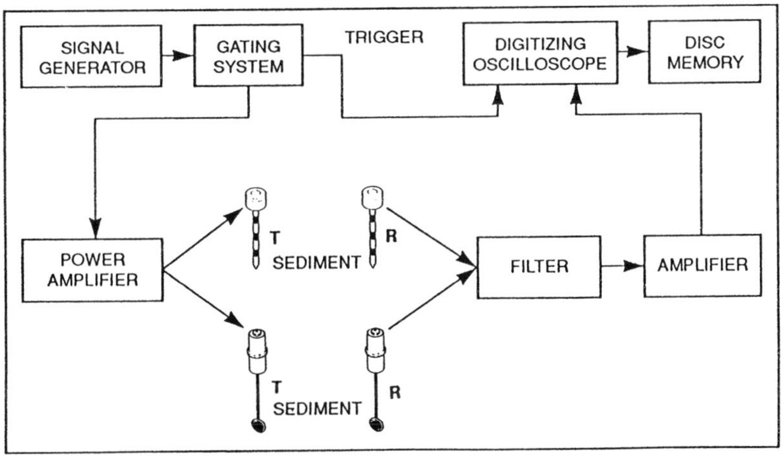


Figure 7. Block diagram of compressional and shear wave measurement system electronics. Preamplifiers (40dB gain) are not shown.

## 2.5 DATA CALCULATIONS

Compressional wave probes are calibrated in the water column, a few meters above the bottom. Examples of received signals propagated through sediment are presented (Fig.8). Sediment compressional wave velocity ($V_p$, m/s) is calculated from the travel time differences between water and sediment ($\Delta t$ , sec.):

$$V_p = \frac{C_w}{1 - (\Delta t C_w / d)}$$

where $C_w$ is the measured or calculated acoustic water velocity (m/s) and $d$ is the distance between probes. $C_w$ is calculated from temperature, salinity, and depth while $d$ is determined from $C_w$ and the travel time in water. Compressional wave attenuation is calculated as $20 \log_{10}$ of the ratio of received voltage between probes in seawater and sediment. Attenuation is then extrapolated to a l-m pathlength and expressed as dB/m.

Shear wave velocity is calculated from the time delay and measured distances between transmit and receive probes. Examples of received signals are presented (Fig.9).

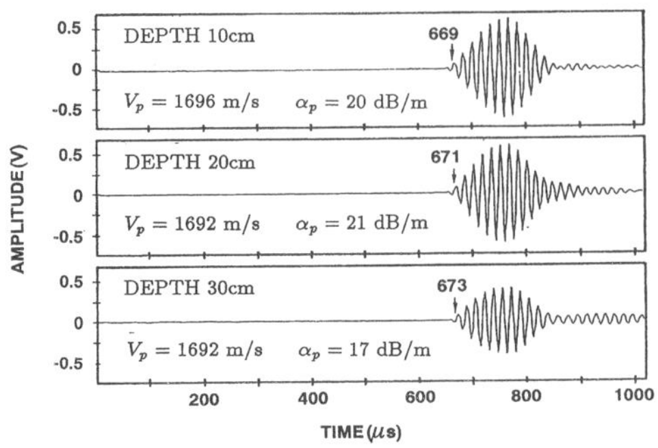


Figure 8. Examples of received compressional wave signals from a subtidal fine sand.

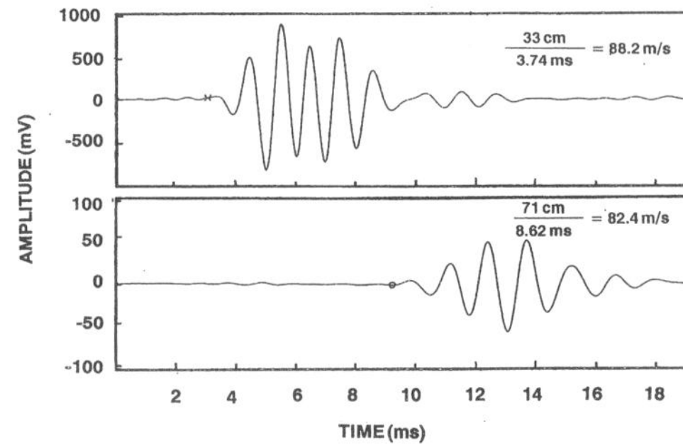


Figure 9. Examples of signals recorded from two shear wave receivers in hard packed fine sand.

## 3. ISSAMS

### 3.1 CURRENT CONFIGURATION

The current version of ISSAMS (stage 4) can be deployed without the aid of divers in water depths up to 200m (Fig.10). An external frame guides the probes into the sediment with minimal disturbance. The system also allows retraction of the probes prior to retrieval of the platform, thus providing protection for the delicate probes.

ISSAMS is positioned just above the sediment-water interface for calibration of compressional wave probes and then lowered to the sediment surface. The external frame is released from the rest of ISSAMS and the weight of the triangular platform together with the internal shaft drives compressional and shear wave probes into the bottom. After measurements are completed, the triangular frame and probes are released from the weights and shaft and a spring pulls the probes out of the sediment. Retrieval of ISSAMS compresses the spring and cocks release mechanisms and ISSAMS is ready for additional deployment without return to the surface. The current version of ISSAMS is hardwired to the surface via an electro-mechanical cable containing 16 twisted conductor pairs.

### 3.2 ADRIATIC DATA

ISSAMS was recently tested as part of a seismo-acoustic experiment in the Central Adriatic Sea (Table 1). In situ shear wave velocities were higher than laboratory measurements made on cores collected from the same locations [5]. Preliminary estimates of interface wave velocities from travel time data, collected at the same locations were nearly the same as shear wave velocities (R. D. Stoll, personal communication). Values of *in situ* shear wave velocity and porosity were in general agreement with the empirical predictive relationships developed by Richardson et. al., [3] during testing of the early developmental stages of ISSAMS.

*In situ* and laboratory values of compressional wave velocity were not significantly different at the muddy site but at the sandy site laboratory measured values of compressional wave velocity were 20-30 m/s higher than *in situ* measured compressional wave velocities. The dispersive behaviour of compressional waves in sediments could account for these differences [6], as could a reduction of sediment water content during core collection. When attenuation values are corrected for frequency, laboratory measured values are higher than in situ values (probably the result of disturbance of sediments during coring).

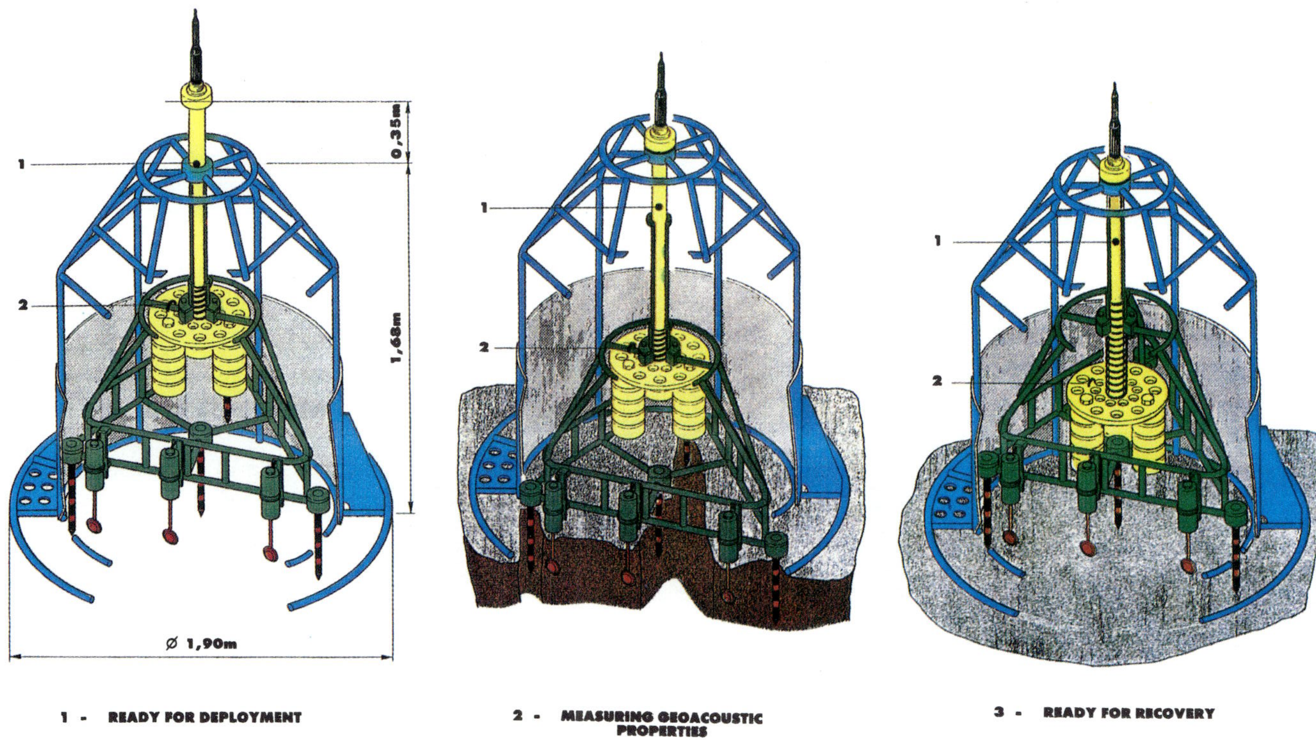


Figure 10. Deployment sequence for ISSAMS Stage 4 system. Release mechanism (1) allows separation of the external frame from probes, shaft and weights. Release mechanism (2) allows separation of the probes from the shaft and weights.

**Table 1.** A comparison of laboratory and *in situ* values of compressional wave velocity ($V_p$) and attenuation ($\alpha_p$) and shear wave velocity ($V_s$) from two Adriatic Sea sediments. Interface wave velocities were determined by R.D.Stoll from a preliminary analysis of stacked travel times from bottom geophone array.

| Sediment type | Sand | Mud |
|---|---|---|
| Porosity (%) | 37 | 68 |
| Mean grain size ($\phi$) | 3.5 | 8.6 |
| Depth (m) | 59 | 62 |
| **In Situ Measurements** | | |
| $V_p$ (m/s) | 1557-1568 | 1467-1486 |
| $\alpha_p$ (dB/m @ 60 KHz) | 21-33 | 0-5 |
| $V_s$ (m/s) | 78-82 | 27-31 |
| **Laboratory Measurements** | | |
| $V_p$ (m/s) | 1580-1604 | 1468-1487 |
| $\alpha_p$ (dB/m @ 200 KHz) | 105-146 | - |
| $\alpha_p$ (dB/m @ 400 KHz) | - | 11-94 |
| $V_s$ (m/s) | 50 | 15 |
| **Seismic Measurements** | | |
| $V_s$ (m/s) | 115 | 30 |

## 3.3 FUTURE DEVELOPMENT OF THE ISSAMS SYSTEM

To allow deeper deployment it is necessary to modify ISSAMS to operate from a lighter twin-conductor cable. The following additional sensors should also be included: a CTD package for calibration of the compressional wave transducers; sediment temperature probe; sediment resistivity probe to estimate the porosity; pore pressure sensor to estimate sediment disturbance.

Finally, inclusion of a second shear wave transmitting probe would increase the number of shear wave velocity measurements per deployment, and may yield values of shear wave attenuation without the need for consistent probe-sediment coupling.

## 4.0 References

[1] Hovem, J.M., Richardson, M.D., and Stoll,R.D. (1991) "Shear Waves in Marine Sediments" Kluwer Academic Publishers, Dordrecht.

[2] Richardson, M.D. (1986) "Spatial Variability of Superficial Shallow-water Sediments Geoacoustic Properties" Akal, T. and Berkson, J.M. (eds.) Ocean Seismio- Acoustics, Plenum Press, London, 527-536.

[3] Richardson, M.D., Miaschi, B., Muzi, E. and Troiano, L., 1990 "Sediment Shear Waves: A comparison of in situ and laboratory measurements" SM-210 SACLANTCEN Report.

[4] Allman, R.L. (1974) "Equipment for In Situ Measurements of Sound Velocity and Attenuation." NUC TP 434, Naval Undersea Center, San Diego CA. 74 pps.

[5] Richardson, M.D., Curzi, P.V., Muzi, E., Turgutcan, F. and Akal, T. (1990) "A Generic Geoacoustic Model of the Adriatic Sea" SACLANTCEN Report.

[6] Stoll, (R.D. 1990) "Sediment Acoustics" Lecture Notes in Earth Sciences, Springer Verlag, New York 153 pps.

# SMALL-SCALE IN SITU MEASUREMENTS OF S-H VELOCITY IN SURFICIAL SEDIMENTARY DEPOSITS: LOCALISED TEXTURAL AND BIOLOGICAL CONTROLS.

S.E. JONES AND C.F. JAGO
*School of Ocean Sciences*
*University College of North Wales*
*Menai Bridge*
*United Kingdom*

ABSTRACT. Using piezoelectric ceramic bender transducers, SH velocity ($V_s$) has been determined *in situ* in the upper 80mm of saturated intertidal sediments. Variability was measured on a range of spatial scales and over a seasonal cycle. The full data set revealed primary variability controlled by bulk textural composition, the depositional environment and the benthic infaunal community. $V_s$ decreased with fining mean grain size, which was associated with increasing porosity and fines content, but a significant degree of unexplained second order variability remained. This was investigated in sands with predominantly uniform bulk textural properties, where $V_s$ was related to variation in packing configuration caused by local and seasonal perturbation of a uniform sediment framework, in response to hydrodynamic and biological processes. There was inverse dependence on porosity and fines content and direct dependence on coarse shell fractions. Furthermore, burrowing organisms (*Arenicola* and *Corophium*) reduced $V_s$, while tube-building organisms (*Lanice*) increased it.

## 1. Introduction.

The acoustic S-wave velocity ($V_s$) of an unconsolidated sedimentary deposit is controlled primarily by the structural properties of the sediment framework: in particular by the number and rigidity of the intergranular contacts, across which the S-wave disturbance must pass. It is therefore related to porosity, effective stress, grain mineralogy and shape, the degree of interlocking of the grains and the nature of the intergranular forces: frictional for sands, cohesive for clays, with adhesive biological secretions affecting both.

Surficial sediments are defined here as comprising the upper 10-20cm of the sea bed. They are subject to active, spatially and temporally variable, hydrodynamic, biological and geochemical processes, resulting in complex, heterogeneous structures which must be assessed *in situ*. Measurement of surficial $V_s$ is of obvious importance in sediment geoacoustic application [1,2]. In addition, there is great potential for its application as a tool for indirect *in situ* assessment of structural properties of surficial sediments, since it is controlled by factors which in turn affect erodibility, permeability, and consolidation [3]. These are of fundamental importance in the study of biogeochemical cycling within the marine environment.

Until recently, there have been very few direct *in situ* measurements of $V_s$ in surficial marine sediments [4]. However, building on recent developments in



*J. M. Hovem et al. (eds.), Shear Waves in Marine Sediments*, 313–320.

piezoelectric transducer technology for application in laboratory test cells [5], a technique for small scale *in situ* $V_s$ measurement has been devised and successfully applied in a range of saturated intertidal sediments [6]. The results from this set of measurements form the subject of this paper.

## 2. Materials and Methods.

All sampling was performed in intertidal sediments, on their exposure by the ebbing tide, in order to maintain conditions of full saturation. This was essential, as $V_s$ is sharply increased on drainage of the sediment, due to increased effective stress at the intergranular contacts [7]. A series of contrasting locations was selected, within which localised variability was assessed. Sites were marked out (typically $1m^2$) and sampled, either at equal intervals along a predetermined line or guided by observed variation in macrofaunal density. The maximum extent of any location was 200 x 50m. In addition, a set of five sites within one location was monitored at monthly intervals over a period of one year to assess seasonal variability.

$V_s$ was measured using piezoelectric bender elements as SH transducers, which were inserted vertically into the sediment to a depth of 40mm (Figure 1). Between two and six replicates were obtained per site. The transducers were self-supporting, to avoid interference by acoustic transmission through a rigid housing. A battery-powered Oyo Sonic Viewer Model 5217A generated repeated delta-function ('spike') excitations for the transmitter, the received signal being displayed and stacked on its integral digital oscilloscope. This produced high quality signals at the resonant frequency of the transducer-sediment system, which varied between 300 and 1500Hz

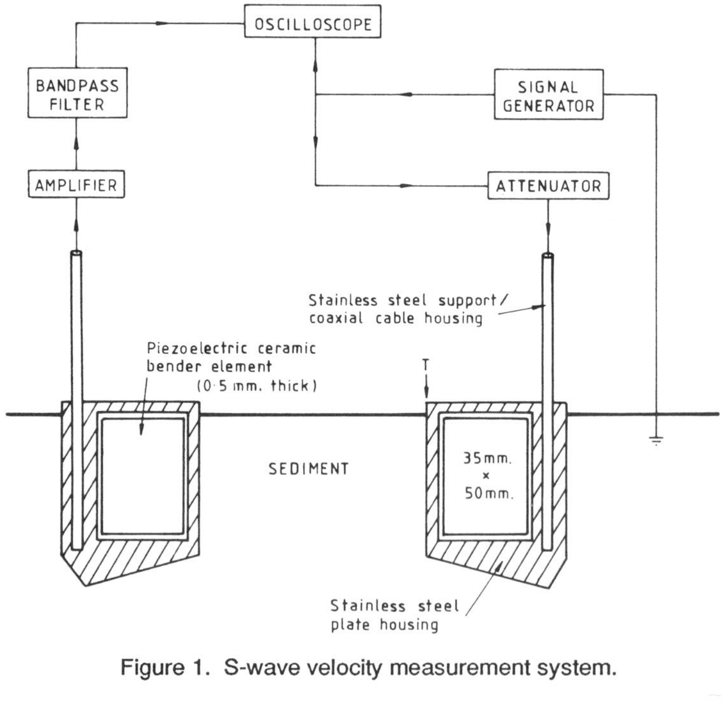


Figure 1. S-wave velocity measurement system.

in the saturated surficial deposits studied. Two transmitter-receiver separations were employed, fixed at 50mm and 125mm (+/- 2mm). Signal arrival-time could be influenced by sediment properties down to a maximum of 80mm from the surface. In vertically homogeneous sediments, $V_s$ was obtained from the difference in the two arrival times. In some deposits, however, marked vertical heterogeneity was observed, resulting in significant curvature of the arrival-time/probe-separation relationship. In order to characterise the surface layer, $V_s$ was calculated in these cases from the signal onset time and the source-receiver separation. Strongest heterogeneity was observed in fine-grained, cohesive sediments and in deposits with texturally contrasting sublayers.

Two or three replicate 5cm ($100cm^3$) sediment cores were collected at each site for textural analysis and determination of porosity. Sediment was washed free of salt, dried for 24 hours at at 110°C and weighed. Porosity was then calculated from the dry sediment weight and the sediment grain density (obtained from tabulated values for quartz, carbonate and clay minerals)

Standard procedures for textural analysis were adopted [8]. Silts and clays (fines) were separated out by wet-sieving: sands were dry-sieved; fines were disaggregated and subjected to pipette analysis. The overall grain size distribution was then computed and grain size statistics obtained [9]. Carbonate content was determined from fractional weight loss after acidification by excess 10% HCl.

Several locations were selected because they exhibited a range in number per unit area of a single benthic macrofaunal species [10]. *Arenicola marina*, a sedentary deposit feeding polychaete, was assessed by counting its distinctive coiled faecal cast. *Lanice conchilega*, a tube-building suspension feeding polychaete, was assessed by counting its characteristic protruding, shell-coated, fringed tube. *Corophium*, a burrowing suspension feeding amphipod, was assessed by disturbance of the deposit and counting exposed burrows, or by sieving $300cm^3$ of sediment through a 710mm screen and counting entrapped organisms.

## 3. Results and Discussion.

### 3.1. VARIABILITY IN TEXTURAL AND STRUCTURAL CHARACTERISTICS.

A variety of scales of variability has been assessed using this data set. The highest variability in all parameters was encountered over large spatial scales, encompassing primary or first order variation in textural characteristics and packing configuration controlled by the nature of the sediment source, the depositional environment, and the benthic infaunal community. Figure 2 illustrates the broad trend downwards in $V_s$ with fining mean grain diameter, which is associated with increasing porosity and fines content, in agreement with other work [1,3]. A considerable degree of scatter was observed, however, indicating that for a given mean grain size there is a range of packing configurations possible, controlled by composition of the grain size distribution, depositional structures and biological activity. This is illustrated by the data from a laboratory study of natural sands [3] included in the figure. The bars correspond to the range in $V_s$ obtained during compaction of a given sample between high and low porosity limits.

The nature of this observed scatter has been investigated for a subset of the data including all sands (containing up to 5% fines), but not including the sharply contrasting mudflats. Further measurements are required in silts and muds before a full combined analysis is possible.

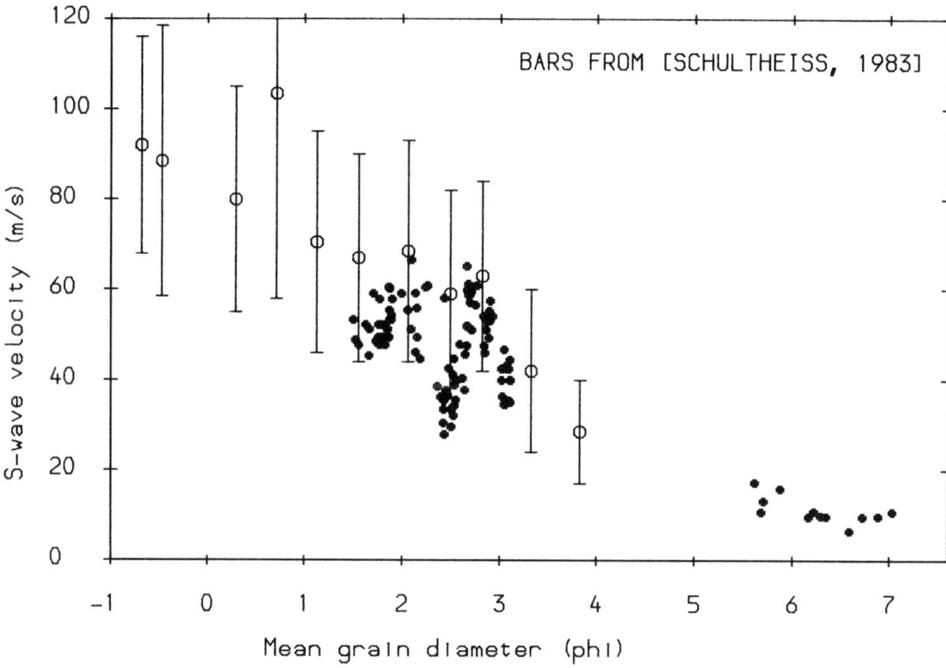


Figure 2. Observed variation in $V_s$ with mean grain diameter. Comparison between *in situ* and laboratory data.

For each parameter, within-location coefficients of variation were averaged to yield an estimate of medium-scale spatial variability (10-200m). The overall coefficient of variation encompasses larger spatial scales (>500m). Within-site replication enabled an estimate of average small-scale spatial variability (0.1-1m) for porosity and $V_s$, although this includes both natural spatial heterogeneity and random sampling error. Seasonal variability was obtained from the coefficients of temporal variation measured during the seasonal study. Figure 3 illustrates these different scales of variability for selected parameters. Macrofaunal counts have been omitted because their distributions resulted in coefficients of variation of greater than 100%: they showed the highest variability of all parameters over localised, seasonal and large scales.

Large scale spatial variability was associated with a wide range in bulk textural composition, including variation in mean grain diameter from 1.25-3.25 phi and in carbonate content from 5-30%. Over localised or medium spatial scales, bulk textural variability was drastically reduced, with a high proportion of the grain size distribution, including the mode grain size, remaining uniform. Localised second order textural variability was introduced via the admixture of relatively small (and highly variable) proportions of a coarse subpopulation. This coarse population generally contained an enhanced proportion of shell fragments, causing higher localised variability of carbonate content. Fines content also formed a highly variable subpopulation at some locations. With the exception of the macrofaunal counts, these second order textural parameters were the most variable over all spatial and temporal scales.

Porosity and $V_s$ also exhibited much lower degrees of variability over medium spatial scales than overall. It is therefore possible to interpret localised variability in the packing configuration of sands in terms of minor perturbation of a texturally and structurally uniform sedimentary framework. This perturbation may be due to the admixture of small proportions of coarse or fine particles, to depositional structures or to macrofaunal activity, and leads to second order variability of porosity and $V_s$.

Seasonal textural variability was even lower than localised variability, indicating that the textural signature of a deposit reflects a highly localised, time-averaged response to the hydrodynamic environment and benthic infaunal community. In contrast, porosity and $V_s$ exhibited comparable localised and seasonal variabilities. This suggests that packing configuration is more sensitive than textural composition to seasonal variation in hydrodynamic and biological processes.

The relative variability of different parameters is also of interest. Over localised scales, mode grain size was the least variable, followed by porosity and mean grain size. $V_s$ followed, then the second order textural parameters and finally the macrofaunal counts. Over large scales, porosity was the least variable, the remaining hierarchy remaining unchanged. However, over a seasonal cycle, and at individual locations with very low textural variability, porosity was more variable than the bulk textural parameters. This demonstrates that porosity is relatively insensitive to variation in absolute particle size, being more influenced by grain shape and size distribution, and also by seasonal and localised hydrodynamic and biological processes. $V_s$ was always more variable than both porosity and mean grain size, indicating sensitivity to both packing configuration and absolute size. This was

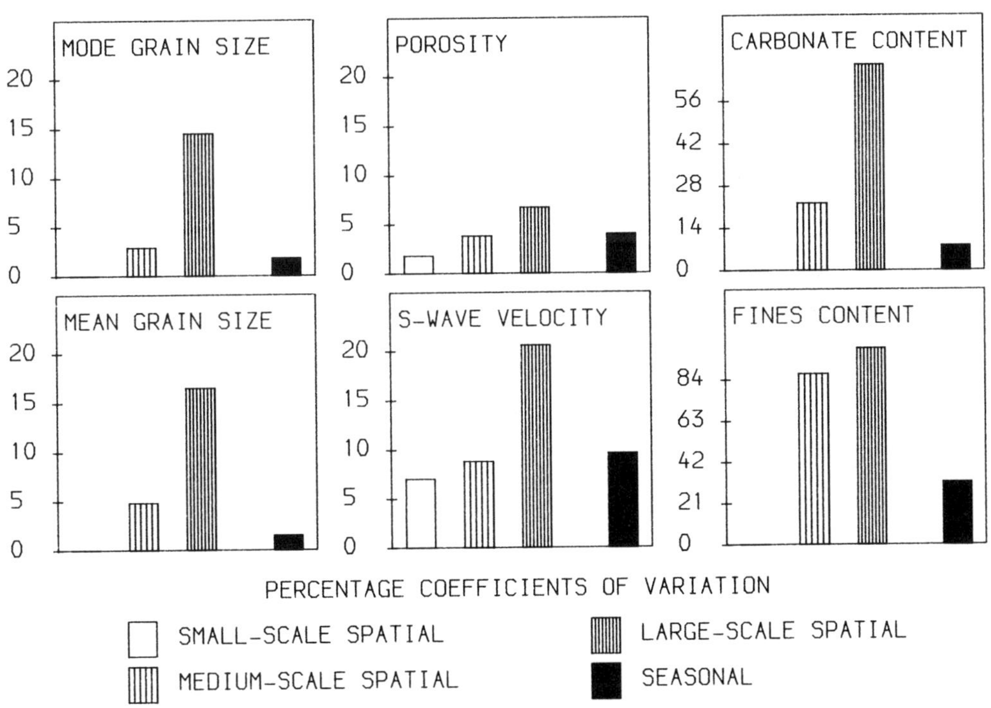


Figure 3. Relative scales of variability. Comparison of observed percentage coefficients of variation over a range of spatial and temporal scales.

expected because $V_s$ should be influenced by factors affecting the number and nature of intergranular contacts, which include size, shape, and degree of interlocking of the sediment grains in addition to porosity.

### 3.2. TEXTURAL AND BIOLOGICAL CONTROLS ON SHEAR WAVE VELOCITY.

At individual locations, statistical analysis of measured parameters revealed a number of controls on $V_s$ which can be interpreted in terms of their second order effect on properties of the bulk sedimentary framework. Simple direct relationships were not universally identified, due to sampling error, interaction between textural and biological variables and the influence of factors which could not be quantified (such as hydrodynamically derived structures and microbiological effects). However, some location-specific correlations (to 95% significance level) were obtained, from which examples have been presented in Figure 4. The error bars correspond to mean within-site standard deviations and therefore incorporate both small-scale spatial heterogeneity and random sampling error.

For a given sediment texture, $V_s$ was inversely related to porosity [3]. Thus in the Taf Estuary, where 90% of the sampled grain size distribution was uniform, a strong negative correlation was obtained (Figure 4(a)). For a given porosity, $V_s$ is affected by second order textural variability which influences intergranular contacts. Thus in the Cefni Estuary (Figure 4(b)), a small increase in fines content significantly reduced $V_s$, without affecting porosity. This is interpreted as due to reduction of friction at the intergranular contacts by fine particles within the sedimentary framework, or by organic matter associated with the fines. In contrast, variation in proportion of a coarse, shelly subpopulation at Lligwy Beach (Figure 4(c)) increased $V_s$, probably due to enhanced intergranular friction associated with angular shell fragments.

Biological activity was also found to significantly affect $V_s$. Burrowing and feeding activity by *Arenicola* causes extensive liquefaction and mixing of surficial sediments, and coating of particles by mucopolysaccharide secretions. Generally *Arenicola* was found to reduce $V_s$. Figure 4(d) illustrates the effect of seasonal variation in *Arenicola* activity at one site in the Cefni Estuary. *Corophium* maintains small U-shaped fluid-filled burrows which can be present in very high densities: these also reduced $V_s$, as shown in Figure 4(f). *Lanice*, however, constructs rigid, shell-lined vertical tubes: these acted to increase $V_s$ (Figure 4(e)).

Many of these relationships, although significant, indicate a high degree of scatter. This can be partially explained by sampling error, but is also influenced by variation of additional textural and biological controls. Multiple regression analysis [11] has been applied to identify linear combinations of partial controls (to 95% significance), resulting in improved degrees of explained variability ($R^2$). For example, at Llanddwyn Beach, $V_s$ was increased independently by both *Lanice* and the coarse subpopulation ($R^2$ = 56%). In the Taf Estuary, $V_s$ was reduced independently by both *Corophium* and porosity ($R^2$ = 79%), indicating that the effect of *Corophium* is not simply one of increasing porosity. At Lligwy Beach, $V_s$ was increased by the coarse subpopulation and reduced by *Arenicola* activity ($R^2$ = 65%).

Multiple regression has also been used to investigate larger scale controls, encompassing wider ranges in textural parameters and porosity. Over the 150 sandy intertidal sites monitored, $V_s$ was independently related to carbonate content, fines content, porosity and *Arenicola* activity ($R^2$ = 70%). For a given porosity, an increased proportion of shell fragments increased $V_s$, admixture of fines reduced $V_s$, and bioturbation by *Arenicola* also reduced $V_s$. These are in good agreement with controls identified at individual locations.

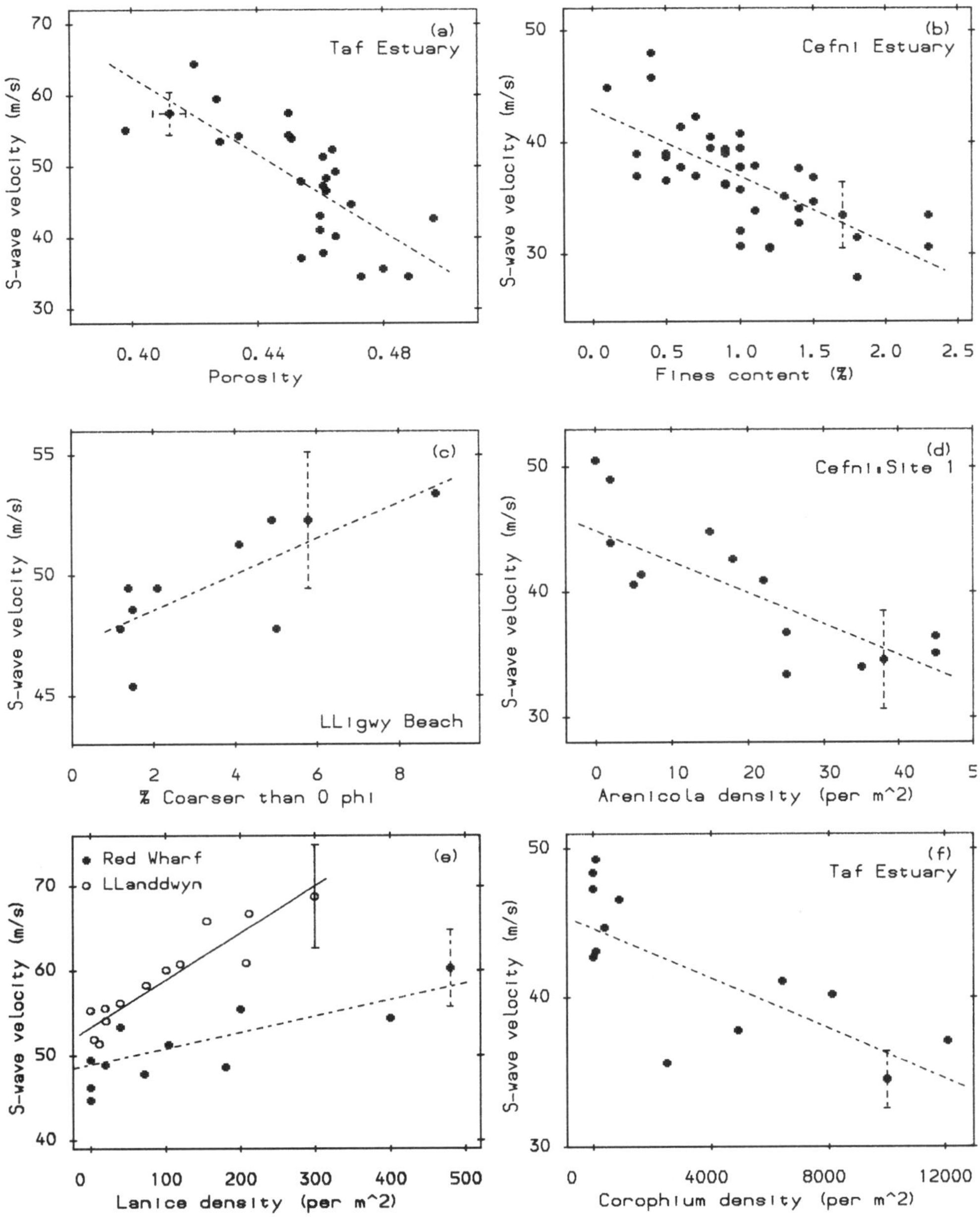


Figure 4. Controls on V*s*.
Straight lines represent least squares regression:
bars represent average within-site coefficient of variation.

## 4. Conclusions.

Surficial $V_s$ is both spatially and temporally variable, in response to a complex combination of textural, hydrodynamic and biological controls. Even where bulk textural variability is very low, $V_s$ is significantly affected by localised and seasonal variability of textural composition, and of biological and hydrodynamic processes. This indicates sensitivity to packing configuration which confirms its potential as a technique for assessment of sediment structural properties. It also suggests that simple predictive equations based on bulk textural parameters provide only a first order approximation to $V_s$, and that direct *in situ* assessment remains the most desirable means of determination.

## 5. Acknowledgements.

This work was supported by a NERC research contract using facilities provided by the School of Ocean Sciences, University College of North Wales. We gratefully acknowledge the assistance of Dr J.D. Bennell (UCNW), who designed and constructed the *in situ* SH transducers, and Dr. M.D. Richardson (NOARL), who contributed critical advice during the early stages of the project.

## 6. References.

[1] Hamilton, E.L., 1980. Geoacoustic modelling of the sea floor. *J. Acoust. Soc. Am.* 68:1313-1340.

[2] Richardson, M.D., 1985. Spatial variability of surficial shallow water sediment acoustic properties, in: *Ocean Seismo-acoustics* (T. Akal and J. Berkson, eds.) Plenum, New York.

[3] Richardson, M.D. and Young, D.K., 1981. Geoacoustic models and bioturbation. *Mar. Geol.* 38:205-218.

[4] Schultheiss, P.J., 1983. *The influence of packing structure on seismic wave velocities in sediments.* UCNW Marine Geology Report No. 83/1.

[5] Bennell, J.D. and Taylor Smith, D., 1991. A review of laboratory shear wave techniques and attenuation measurements with particular reference to the resonant column, in: Hovem, J.M., Richardson, M.D. and Stoll, R.D., (eds.), *Shear Waves in Marine Sediments.* Kluwer Academic Publishers, Dordrecht (this volume).

[6] Jones, S.E., 1990. *Geophysical properties of marine sediments: textural and biological controls.* Thesis submitted for PhD, University College of North Wales, U.K.

[7] Wu, S., Gray, D.H. and Richart, F.E., 1984. Capillary effects on dynamic modulus of sands and silts. *J. Geotech. Eng.* ASCE, 110(9):1188-1203.

[8] Carver, R.E., 1971. *Procedures in Sedimentary Petrology.* Wiley Interscience, New York.

[9] Folk, R.L. and Ward, W.C., 1957. Brazos river Bar, a study in the significance of grain size parameters. *J. Sedim. Petrol.* 27:3-26.

[10] Eltringham, S.K., 1971. *Life in Mud and Sand.* The English Universities Press.

[11] Sokal, R.R. and Rohlf, F.J., 1981. *Biometry.* 2nd Edition, W H Freeman & Co., San Francisco.

# IN SITU MEASUREMENTS OF SHEAR-WAVE VELOCITY IN OCEAN SEDIMENTS

**R.B. Whitmarsh, P.R. Miles**
**Institute of Oceanographic Sciences Deacon Laboratory, Wormley, Godalming, Surrey, GU8 5UB, U.K.**

ABSTRACT This paper presents shear-wave results from the Northeast Atlantic and from the Norwegian Sea. In the first case a link was established between shear-wave velocity and sediment lithification from constraints imposed as part of the modelling of compressional-wave synthetic seismograms and in the second case bottom shots and OBS's were used to generate and detect shear-waves directly. These observations add to the meagre number of shear-wave velocity measurements made in oceanic depths and in particular indicate the part played by shear-waves in causing energy to be reflected within the sediment column.

## Introduction

Knowledge of the shear-wave velocity structure in ocean sediments (i.e. in water depths greater than a few hundred metres) is important for the calculation of low-frequency long range sound propagation. In particular this is true because of the potential coversion of compressional to shear-wave energy and the subsequent strong attenuation of this energy (Beebe & Holland, 1986). A second order effect is the large transverse isotropy (shear-wave anisotropy of upto 26 %) which occurs in strongly layered sediments (Fryer & Miller, 1986; Whitmarsh & Lilwall, 1982). In addition, because of the link between dynamic and static shear moduli there is interest in using shear-wave velocities to estimate shear strength for a variety of engineering purposes (Abbis, 1981). Lastly shear-wave velocity is a sensitive indicator of sediment rigidity, and therefore of induration and lithification (Whitmarsh et al., 1990), and may provide geological information without drilling.

Shear-wave velocity in an ocean sediment depends critically on the inherent rigidity formed by the physical interlocking and chemical bonding of mineral grains. In deep sea sediments, particularly those close to the sea bed, the rigidity is low (as indicated by shear-wave velocities of a few tens of metres per second at the sea bed; Davies, 1965; Whitmarsh & Lilwall, 1982) and these grain-to-grain connections are probably fragile and are unlikely to withstand the physical changes involved in coring or drilling and in recovery to sea surface temperatures and pressures. Thus there is a strong case for making shear-wave measurements in situ.

Earthquakes are strong sources of shear-wave energy but by their nature are unpredictable in time and limited in spatial distribution. Most shear-wave measurements in ocean depths have been made by controlled-source techniques whereby shear-waves and/or Scholte waves are generated directly by sources on the sea bed (e.g. Davies, 1965; Whitmarsh & Lilwall, 1982). Sometimes the conversion of compressional wave energy, from sources in the water, to shear-waves at the sediment-basement interface (White & Stephen,1980) has been used to compute the average shear-wave velocity within the sediment layer from the so-called S-P time measured by an OBS (ocean bottom seismograph; e.g. Lewis & McClain, 1977). In rare cases direct measurements have been possible with probes from a submersible (Hamilton et al., 1970). In general it is apparently not possible to use downhole logging tools, for example in Ocean Drilling Program boreholes (Anon.,1986), unless the shear-wave velocity of the borehole surroundings exceeds the compressional-wave velocity of the borehole fluid.

This paper presents shear-wave results from the Northeast Atlantic and from the Norwegian Sea. In the first case a link was established between shear-wave velocity and sediment lithification from constraints imposed as part of the modelling of compressional-wave synthetic seismograms. In the second case bottom shots and OBS's were used to generate and detect shear-waves directly. These observations add to the meagre number of shear-wave velocity measurements made in oceanic depths (Davies, 1965; Hamilton et al., 1970; Lewis & McClain, 1977; Whitmarsh & Lilwall, 1982; Au & Clowes, 1984; Toksoz et al., 1985; Sauter et al.,1986; Duennebier et al.,1987; Nagumo et al.,1980a, b; Sereno T.,1990; Shearer & Orcutt, 1986; Whitmarsh et al.,1990; McClain & Lewis, 1980).



*J. M. Hovem et al. (eds.), Shear Waves in Marine Sediments*, 321–328.

## Observations in the Northeast Atlantic

The data were collected during Discovery Cruise 161 in August-September 1986 (Whitmarsh, 1986). Digital ocean-bottom seismographs (DOBS; Kirk et al.,1982; Peal & Kirk, 1983) were used to record signals from an array of four 16-litre airguns fired every 2 minutes at a speed of 4 to 5 knots. Shot-to-DOBS ranges were calculated from water-wave travel times.

In the first area a series of three seismic refraction profiles was shot along a NE-SW line located about 200 km ESE of Madeira in a water depth of about 4350 metres (Fig.1; Whitmarsh et al., 1988, 1990). Here just one of the three seismic refraction profiles is described; the other two profiles show essentially the same features. The record section recorded by one of the DOBS at Point MM (Fig. 2) shows, besides the direct water-wave, basement arrivals with an apparent velocity of about 5 km/sec and clear sediment arrivals with an apparent velocity of about 2.5 km/sec. Close inspection reveals that the sediment arrivals actually fall into three groups (8.5-10.3, 10.5-13 and beyond 13 km). Each group is delayed in time by roughly 0.2 sec with respect to the previous group so that the

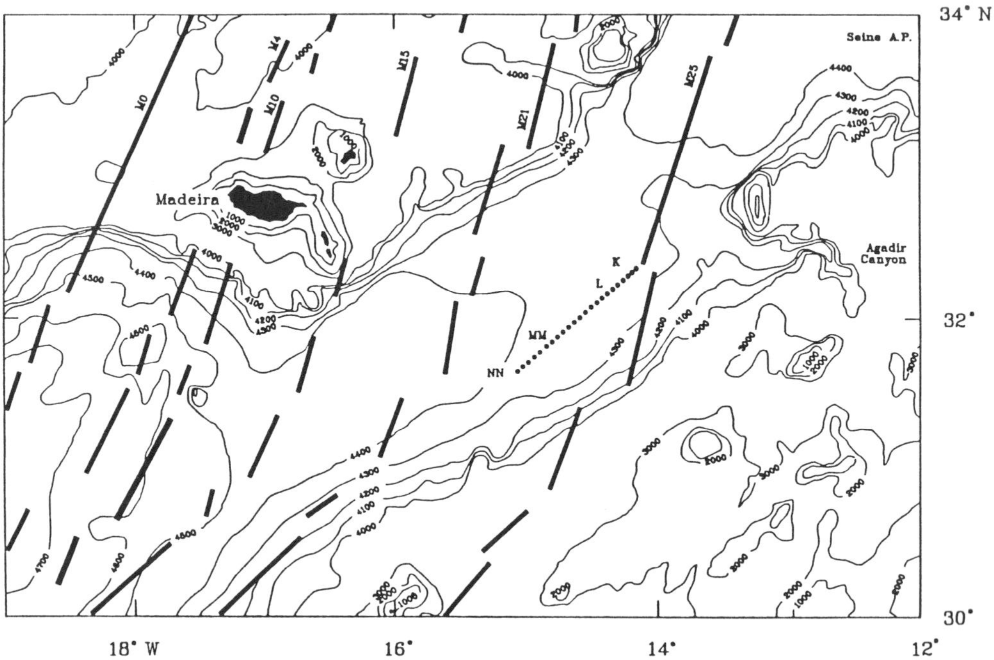


Figure 1. Bathymetric chart of the work area off Madeira (contours in metres). Dotted line denotes OBS profiles with measurements at points L, MM and N; thick lines are magnetic isochrons (Klitgord & Schouten, 1986).

groups appear to be en echelon on the record section. The same delays are seen on the record section of the adjacent DOBS and, within a few hundred metres, at the same range so that it may be concluded that they are functions of the velocity/depth structure and are not the result of lateral inhomogeneities. In fact the amplitude and shape of these arrivals, observed at about the same range, are closely similar. Similarly the travel times of corresponding peaks and troughs at the same range vary by up to 0.02 sec between the two DOBS indicating a lack of significant heterogeneity within the sediments.

One possible explanation for the en echelon sediment arrivals is that they are multiply-reflected diving rays. In this case the apparent velocity should decrease, as range increases, on passing from one set of arrivals to the next. This hypothesis was tested objectively, using the pair of DOBS as an array to estimate the apparent velocity of arrivals from each shot, by cross-correlating between traces; this process showed that indeed the apparent velocity behaved as predicted (Whitmarsh et al., 1990). Travel time modelling was constrained by the sediment travel times discussed above as well as by independent estimates, from near vertical incidence reflections, of 0.45 and 0.95 km thick layers with interval velocities of 1.9 and 2.4 km/sec respectively within the sediments. The thickness of the upper layer precluded the possibility of the multiples being reflected at the sea bed. Thus reflection within the sediments was suspected and modelling indicated that this had occurred at the interface between the upper and lower sediments. The final model was checked using synthetic seismograms (Fuchs & Mueller, 1971; Fig. 2).

In order to obtain sufficiently strong multiple phases a substantial contrast in shear-wave velocity is required across the intrasediment reflecting horizon. This is illustrated in Fig.3 where a Poisson's ratio of 0.25 was assumed for the lower sediments whereas Vs in the upper sediments must be about 0.25 km/sec and certainly less than 0.5 km/sec. No realistic changes in density can produce the required effect instead. Although, following Helmberger et al.,1979 rather high Q factors (Qa=500, Qb=250) were assumed when calculating the synthetic seismograms no significant amplitude change was detected if Q was reduced by a factor of five (Whitmarsh et al., 1988).

Also in 1986 we shot a number of lines with airguns and DOBS over the continental margin west of Iberia. Here four out of seven record sections showed the same sort of en echelon sediment arrivals as were seen near Madeira and these DOBS form a group over the Iberia Abyssal Plain

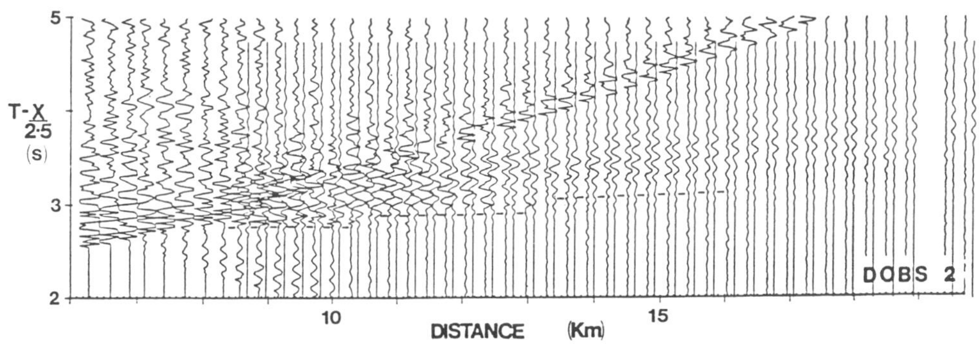


Figure 2. Unfiltered vertical geophone record section recorded at Point MM. Reduction velocity 2.5 km/sec. Synthetic seismograms, drawn as shorter lines, were calculated for a shear-wave velocity in the upper sediment layer of 250 m/sec (see text). The heavy dashed lines indicate the sediment arrivals.

(Fig.4). It is not possible with our data to demonstrate that the en echelon behaviour is not due to lateral inhomogeneities because the DOBS were deployed singly. However lateral homogeneity is a reasonable assumption in view of the character of the seismic reflection profiles associated with the lines. Travel time modelling required the same sort of intrasediment reflector as was required near Madeira. The computed depths of this reflector lie very close to an angular unconformity visible in coincident reflection profiles (Fig.5). Synthetic seismogram modelling also confirmed the need for an intrasediment reflector with a relatively strong shear-wave velocity contrast. Here the upper layer has a mean velocity closer to 100 m/sec and certainly less than 250 m/sec.

The intrasediment reflector can be traced laterally some 220 km northeast to DSDP Site 398 where it was cored at a depth of 380 m (Sibuet, Ryan et al.,1979). The reflector has been linked to a number of physical changes in the cores, principally a transition from ooze to chalk (Whitmarsh et al

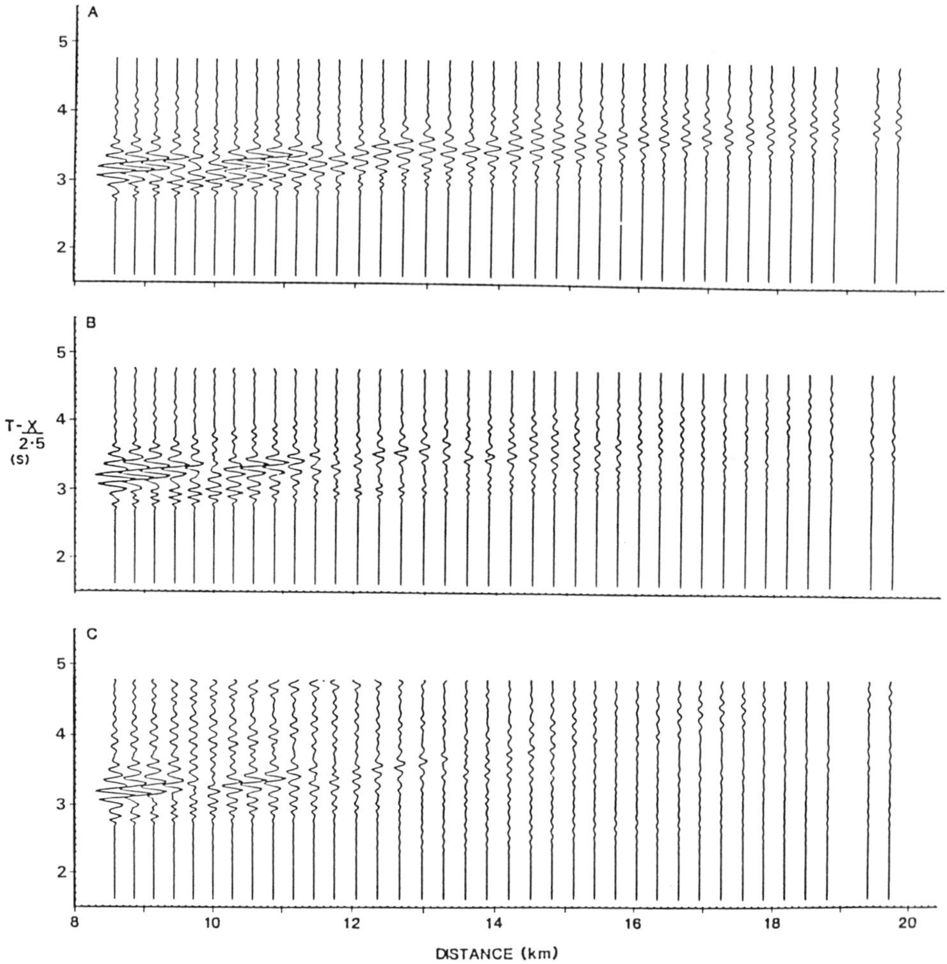


Figure 3. Three sets of synthetic seismograms based on the best compressional wave velocity/depth model for Point MM. The synthetics differ in the shear-wave velocity assumed for the sediments above the intrasediment reflector as follows; (a) 100 m/sec (b) 250 m/sec (c) 500 m/sec. The figure shows that a shear-wave velocity of about 250 m/sec is required to fit the observed record section in Figure 2.

1990; Sibuet, Ryan et al., 1979). Such an increase in induration is expected to lead to a strong increase in shear-wave velocity as we have inferred. Off Madeira only a rather tenuous link can be made to a DSDP borehole (a sea bed sill intervenes) but even here there is a transition from clay to claystone at a depth which is within the range of depths of the intrasediment reflector there. Thus there is some evidence to suggest that the increase in lithification which occurs at a depth of a few hundred metres in ocean sediments is associated with an important increase in shear-wave velocity. At the same time this horizon is an important reflector of compressional-waves at wide-angles.

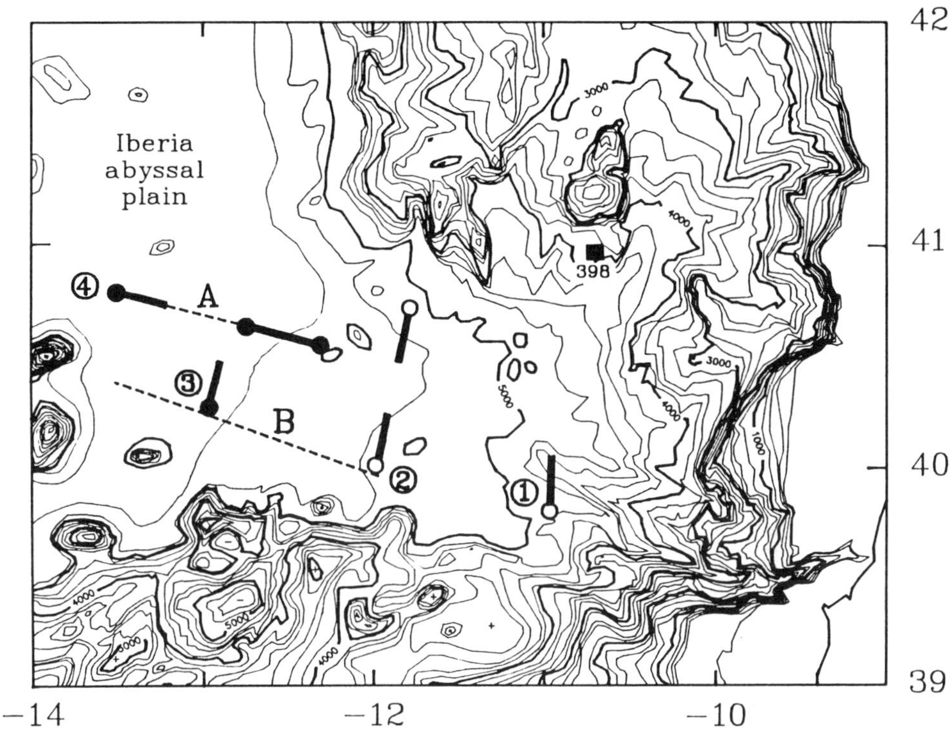


Figure 4. Location and numbering of airgun refraction profiles in the eastern Iberia Abyssal Plain which recorded en echelon sediment arrivals (dots) and those which did not (open circles). Square denotes DSDP Site 398.

## Observations in the Norwegian Sea

In the summer of 1989 we conducted a series of experiments with bottom shots and DOBS along a NW-SE line across the Norwegian Sea with the objective of determining the velocity structure of the sediments in a variety of geological environments. Bottom shots were chosen as sources because they enable sediment arrivals to be observed as first arrivals by an OBS to greater ranges than is possible with surface sources alone.

The DOBS and 10 kg shots were deployed using an acoustic navigation system with sea bed beacons, which gave the relative position of the ship to an accuracy of about 3 m, along lines upto 4000 m long. The relative positions of shots and DOBS on the sea bed were determined to an accuracy of better than 22 m from the arrival time difference of the direct and surface-reflected water waves. The technique has been described in detail by Kirk et al, in press.

Figure 6 shows a record section, reduced to 500 m/sec, for one of the lines at 64°15'N,2°17'W. Between the high amplitude Scholte waves and the strong high frequency water-borne energy a series of low frequency wavelets can be discerned. These appear to be correlatable from trace to trace. Travel time modelling strongly suggests that they are shear-waves which have undergone multiple reflection at the sea bed. This interpretation is supported by the calculated hodochrons also shown on the record-section from which multiples upto fifth-order can be picked out. Multiples of a given order die out with increasing range in part because the reflection coefficient at the sea bed steadily decreases as the angle of incidence increases. Synthetic seismograms have yet to be calculated to check this interpretation. The corresponding compressional-wave travel times have also been modelled and the two velocity profiles are shown in Fig.7. Although the increase in

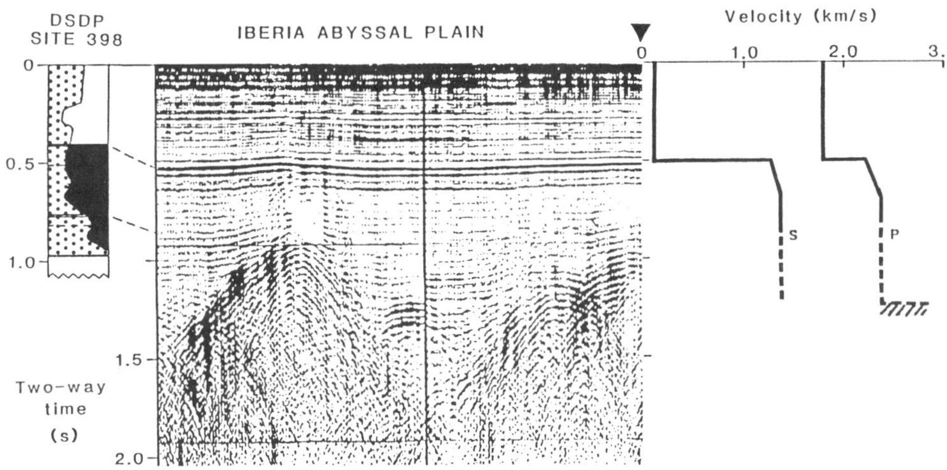


Figure 5. Seismic reflection profile along the refraction profile recorded by DOBS 1 at the west end of Line 4. The velocity structure was derived from synthetic seismogram modelling. Note the near coincidence of the abrupt velocity increase in the model at 0.5 sec twt with a reflector seen on the reflection profile. A summary lithological column from DSDP Site 398 is shown for comparison. White = ooze; black = chalk; stipple = non-carbonate fraction.

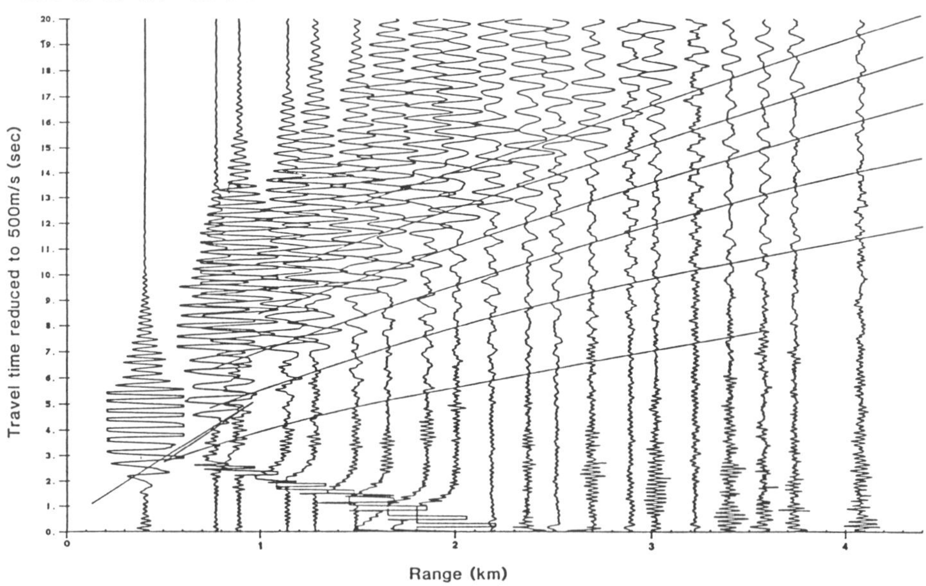


Figure 6. Vertical geophone record section from Site 1 filtered low-pass 7 Hz and reduced to a velocity of 500 m/sec. Calculated hodochrons based on the model in Figure 7 are shown for ray paths which have been multiply reflected at the sea bed (see text).

compressional-wave velocity at 240 m depth can be closely correlated with a reflector on a coincident reflection profile the same is not true of the smaller increase in shear-wave velocity at 125 m depth.

## Conclusions

The above results from recent work by the Institute of Oceanographic Sciences Deacon Laboratory shed light on two different aspects of the influence of shear-waves on seismic propagation within ocean sediments. In both cases however there is strong evidence to suggest that, for different reasons, the shear-wave velocity structure has an important effect on the trapping of both compressional and shear-wave energy within the upper part of the sedimentary column. In the first

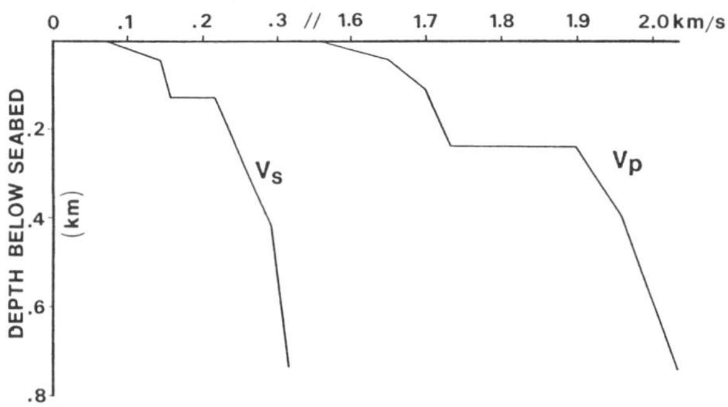


Figure 7. Computed velocity/depth profiles for compressional and shear-waves at Site 1 based on travel time modelling.

case compressional-wave energy is multiply reflected from an intrasediment reflector caused by a sharp increase in shear-wave velocity associated with an increase in sediment lithification and in the second case the sea bed itself has been shown to be an important reflector of shear-waves in the presence of a strong shear-wave velocity gradient.

## Acknowledgements

We thank the Admiralty Research Establishment Portland for support in carrying out this work at sea and ashore.

## References

Abbis C.P. (1981) 'Shear wave measurements of the elasticity of the ground', Geotechnique, 31, 91-104.

Anonymous (1986) 'Wireline Logging Manual', Ocean Drilling Program, Lamont Doherty Borehole Research Group, Palisades, New York.

Au, D. and Clowes, R.M. (1984) 'Shear-wave velocity structure of the oceanic lithosphere from ocean bottom seismometer studies', Geophys. J.R. astro. Soc., 77, 105 - 123.

Beebe, J.H. and Holland, C.W. (1986) 'The effect of inconsolidated sediment rigidity on low frequency acoustic propagation', in T. Akal, J.M. Berkson (eds.), Ocean Seismo-acoustics. NATO Conf. Series, Series IV Marine Sciences 16, 207 - 215.

Davies, D. (1965) Dispersed Stonely Waves on the Ocean Bottom Bull. Seismol. Soc. Amer. 55, 903.

Duennebier, F.K., Leinert, B., Cessaro, R., Anderson, P. and Mallick, S. (1987) 'Controlled-source seismic experiment at Hole 581C'. Init. Rept. DSDP, 88, 105 - 125.

Fryer, G.J. and Miller, D.J. (1986) 'Effects and consequences of transverse isotropy in the seafloor' in, T. Akal and J.M. Berkson (eds.) Ocean Seismo-acoustics. Plenum Press, New York., 589 - 597.

Fuchs, K. and Muller, G. (1971) 'Computeration of synthetic seismograms with the reflectivity method and comparison with observations'. Geophys. Journal Astro. Soc., 23, 417 - 433.

Hamilton, E.L., Bucker, H.P. Keir, D.L. and Whitney, J.A. (1970) 'Velocities of compressional and shear waves in marine sediments determined in situ from a research submersible'. J. Geophys. Res. 75, 4039 - 4049.

Helmberger, D.V., Engen, G. and Scott, P. (1979) 'A note on velocity, density and attenuation models for marine sediments determined from multibounce phases'. J. Geophys. Res., 84, 667 - 671.

Kirk, R.E., Langford, J.J. and Whitmarsh, R.B. (1982) 'A three-component ocean bottom seismograph for controlled source and earthquake seismology'. Mar. Geophys. Res. 5, 327 - 341.

Kirk, R.E., Robertson, K., Whitmarsh, R.B. and Miles, P.R. (in press) 'A technique for conducting seismic refraction experiments on the ocean bed using bottom shots. Mar. Geophys. Res.

Klitgord, K.D. and Schouten, H. (1986) Chapter 22. Plate kinematics of the central Atlantic in, P.R. Vogt, B.E. Tucholke (eds.) The western North Atlantic region. Geology of N. Amer. Vol. M Geol. Soc. Am., 351 - 378.

Lewis, B.T.R. and McClain, J. (1977) 'Converted shear waves as seen by ocean bottom seismometers and surface buoys. Bull. Seismol. Soc. Amer. 67, 1291 - 1302.

McClain, J.S. and Lewis, B.T.R. (1980) 'A seismic experiment at the axis of the East Pacific Rise', Mar. Geol., 35, 147 - 169.

Nagumo, S., Ouchi, T. and Koresawa, A. (1980) 'Seismic velocity structure near the extinct spreading centre in the Shikoku Basin, N. Philippine Sea'. Marine Geology 35, 135 - 146.

Nagumo, S. Kasahara, J. and Koresawa, A. (1980) 'Large Poisson's ratio and low S-wave velocity within the Japan trench inner wall toe'. Marine Geology 35, 129 - 133.

Peal, K.R. and Kirk, R.E. (1983) 'An event recording ocean bottom seismograph'. IEEE proceedings of 3rd working symposium on oceanographic data systems, IEE Computer Society Reprint, pp 5.

Sauter, A.W., Dorman, L.M. and Schreiner, A.E. (1986) 'A study of sea-floor structure using sea floor shots and receivers, in Ocean seismo-acoustics symposium', June 10 - 14 SACLANT ASW Research Centre, NATO (eds.) T. Akal and J.M. Berkson. Conf. Ser. Ser. IV Mar. Sci. 16, 673 - 681.

Sereno, T.J. (1990) 'Numerical modelling of water-waves recorded by a sub-bottom seismometer in the Southwest Pacific', J. Geophys. Res. 95, 2575 - 2591.

Shearer, P.M. and Orcutt, J.A. (1986) 'Compressional and shear wave anisotropy in the oceanic lithosphere - the Ngendei seismic refraction experiment', Geophys. J. R. astr. Soc. 87, 967 - 1003.

Sibuet, J.-C., Ryan, W.B.F. and Shipboard Scientific Party. (1979) Initial Reports of the Deep Sea Drilling Project, Vol. 47, U.S. Govt. Printing Office, Washington D.C.

Toksoz, M.N., Wilkens, R.H. and Cheng, C.H. (1985) 'Shear wave velocity and attenuation in ocean bottom sediments from acoustic log waveforms'. Geophys. Res. Letts., 12, 37 - 40.

White, R.S. and Stephen, R.A. (1980). 'Compressional to shear wave conversion in oceanic crust. Geophys. J. Roy. astro. Soc. 63, 547 - 566.

Whitmarsh, R.B. (1986) R.R.S. Discovery Cruise 161. 'Portuguese ocean-continent boundary, plate boundaries west of Iberia and seismic structure of sediments off Madeira' I.O.S. Cruise Report No. 188, 35 pp.

Whitmarsh, R.B. and Lilwall, R.C. (1982) 'A new method for the determination of in-situ shear-wave velocity in deep-sea sediments', presented at Oceanology International '82, 2nd - 5th March 1982. Brighton, England. Paper No. 4.2, 21 pp. Spearhead Publications: Kingston-upon-Thames.

Whitmarsh, R.B., Miles, P.R. and Pinheiro, L.M. (1990) 'The seismic velocity structure of some NE Atlantic continental rise sediments: or lithification index? Geophys. J. Int., 101, 367 - 378.

# SEAFLOOR SHEAR WAVE VELOCITY DATA ACQUISITION: PROCEDURES AND PITFALLS.

A.M.Davis, D.G.Huws and J.D.Bennell
School of Ocean Sciences
University College of North Wales
Menai Bridge
Wales, U.K.

**Abstract**
The technological problems associated with making *in situ* measurements of seafloor sediment shear wave velocities have to date precluded their general use in offshore reconnaissance surveying or in site specific investigations. A number of researchers have reported experiments in which they have used a variety of signal generators, detectors and processing techniques to measure shear wave propagation phenomena. These seem to have met with varying degrees of success and may not have been taken-up by industrial users for a variety of reasons. Similar experiments have been carried out at the University College of North Wales (UCNW), with the ultimate aim being to produce an integrated geophysical system capable of evaluating seabed sediment velocities over a range of depths extending from the immediate surface layer up to 100 metres below the seabed surface. These experiments have shown that an ideal source is one which is coupled to the sediment surface and preferentially generates horizontally polarised shear waves. A matched detector is thus one which is horizontally responsive and oriented with its axis parallel to the source motion. While the prevailing noise conditions require that both the source and receivers are stationary, it is possible to collect data in a 'pseudo-underway' fashion whilst steaming slowly along a pre-determined survey track. In this way information on seabed properties can be obtained rapidly and in a cost effective manner. Dependent on the scale of measurement, this can be indicative of subtle changes in the surface sediment layer or can be used to determine the depth variation in physical properties.

## 1. Introduction

The use of shear wave velocity data in land-based investigations has increased several-fold in recent years. This can largely be attributed to the greater awareness of engineers to the value of the velocity as a basic input parameter in foundation design. While the obvious use is in dynamic design e.g. for vibrational loading studies, earthquake risk analysis etc., the recognition of the fundamental importance of low



*J. M. Hovem et al. (eds.), Shear Waves in Marine Sediments*, 329–336.

strain data in engineering evaluation is further extending the application of the seismically-derived elastic parameters.

Similar and additional uses will be found in the offshore environment. For example the water column itself imposes dynamic stresses on the seafloor which can ultimately cause liquefaction of the sediment body. This can occur either through water waves acting directly on the sediment or via stresses translated to the seabed through interaction with any structure founded on it. Clearly offshore engineering studies will benefit from the introduction of any new geophysical methods designed to obtain *in situ* physical property data.

The aim of this paper is to introduce measurement techniques developed by the authors at the University College of North Wales and designed to provide information relating to the seabed surface layer and to depths of a few tens of metres below the water-sediment interface. At the same time supplementary details on operational procedures/difficulties will be given as an aid to other potential investigators.

## 2. Background

Techniques for land-based shear wave data acquisition are well established. The seismic refraction method remains the favoured technique for an initial appraisal of any site for engineering purposes, as it allows data to be collected inexpensively and without any disturbance to the site. It can give indications of *in situ* velocity structure/gradients from which the depth variation of the physical properties can be inferred. Once the main site investigation is underway a more accurate evaluation of the subsurface material properties can be made using any available boreholes e.g. through crosshole shooting, seismic tomography etc.

In the marine environment, the presence of the water column obviously imposes constraints on the use of any of the standard land-based methods of investigation: instrumentation must be specially designed for operation on the seabed, and deployment methodology given due consideration. A number of authors have concluded that shear wave refraction shooting offers the best approach to obtaining *in situ* shear wave velocity data for seafloor sediments. In 1974 Schwarz and Conwell reported a seismic refraction experiment using an S wave source coupled to the seabed and claimed a maximum penetration of 100 metres below the mudline. More recently Stumpel et. al. (1984) presented results from measurements in the Baltic Sea where a 'sheargun', essentially an airgun modified to produce a horizontal blow, was used to obtain information in gas-bearing sediments where compressional wave surveys had failed. Both groups of authors expressed their preference for SH (horizontally-polarised) wave generation because of the tendency for the vertically-polarised shear wave (SV) to convert to a compressional wave at a boundary. Stoll et. al. (1988) also used airguns deployed in an unconventional manner to obtain shear wave velocity data as a function of depth in seabed sediments. The authors report good agreement between *in situ* velocities and those predicted by geoacoustic models.

Essentially the seismic refraction method relies on the measurement and analysis of waves critically refracted from boundaries across which there is a velocity increase. For seafloor sediments, which are often characterised by a velocity gradient (Taylor Smith, 1986), the depth variation in physical properties can be inferred from analysis of travel time data obtained for a range of source receiver offsets. For engineering purposes, where the zone of interest is relatively shallow, large energy sources are not required; this is particularly true if a repeatable source is used in conjunction with signal stacking.

## 3. The UCNW Seabed Refraction Systems

### 3.1. Basic Design Criteria

The shear wave signal generator was designed to simulate the action of the 'horizontal traction source' commonly used for land-based work. With this, an impact made by a 10kg. hammer on the end of a weighted wooden plank can be detected over surface distances of a few hundred metres (Davis and Schultheiss, 1980). For the marine system, the two controllable factors influencing the horizontal range and hence penetration depth of the propagating wave, were identified as the energy imparted by the horizontal blow and the coupling of the source with the seabed surface. While in the first instance the new instrumentation was being developed to obtain information relating to the shallow subsurface layers (therefore requiring short surface shot to detector spreads), future requirements for a greater depth sensing capability were also borne in mind when designing the seismic sources.

In order to achieve a variable energy output the SH source design was based on the electro-magnetic principle, as used by Schwarz and Conwell (1974). An electro-magnetic hammer device (seismic slug) was constructed using a moving mass and coil system, and the energy varied by altering the number of turns on the coil and/or the amount of power fed to the coil. These seismic sources are impulsive devices which can be made to preferentially generate either shear or compressional waves dependent on their mode of operation. A horizontally-oriented slug produces horizontal shearing motion which is translated into the seabed if suitably coupled to the sediment surface. To achieve this coupling and facilitate towing along the seabed the slug is sledge mounted. To activate the source a high voltage is discharged instantaneously through the coil with the horizontal shearing motion resulting from the action of the moving bar hitting the end stops.

For SH measurements orientation of detectors relative to source motion is critical if signals are to be easily identifiable. For optimum S-wave detection a horizontally-mounted geophone with its axis at right angles to the profile line (i.e. axis parallel to impact direction) is needed. For marine observations the deployment procedure should aim to achieve this alignment, although the particle motion can be resolved from a 3-component record. Gimballed geophones are used to ensure that the transverse component is maintained in the horizontal plane. (Gimballed geophones, with a suitable frequency response range, are commercially available and can easily be adapted for marine refraction work).

For successful shear wave work, direct coupling of both the source and receivers to the seabed should be maintained during each test. This is most easily achieved via self-weight although coupling of receivers can be assisted in other ways (see 3.2).

### 3.2. Pseudo-underway System

For investigations of near surface layers a pseudo-underway system has been designed, constructed and is now fully operational. The seabed hardware consists of seismic sources, a focused electrode pad for measurement of the apparent formation factor **AFF**( Jackson 1975), and an integral towed array of gimballed geophones. The ship-board instrumentation includes electronics to drive the seismic sources, a multi-channel digital seismograph to record and display seismic signals, a digital oscilloscope for measuring voltage variations (for electrical resistivity measurements), and a system control box. A multi-core cable is used to feed electrical energy to the seabed instruments and transmit received signals back to the on-board recording instruments. While the sledge, and hence sources, are coupled to the seabed under self-weight, the coupling of the geophones is partly maintained by attachment to a continuous section of rubber matting (see fig.1). This also maintains the alignment of the geophones and ensures a constant spacing between source and receivers. A full description of the system can be found in Davis et. al. (1989).

'Profiling' is achieved by towing the system slowly along the seabed surface with point measurements made by paying out slack on the towing cable, thus rendering the seabed instruments stationary for a short period. A typical shear wave record is given in fig.2, with signals from a single shot recorded at 5 geophone stations. A detailed analysis of data for each shot point can be carried out to provide information on any initial velocity gradient or an average value of the surface layer down to the maximum penetration depth of the system (as defined by the spread length). For each measurement point, travel time plots constructed using geophone interval times, appear as straight lines if no gradient exists (as the receiver interval is a constant), or may be curved in the presence of a gradient. Subtle variations in the physical property distribution within the survey area can be highlighted by plotting the point measurements as traverses (e.g. fig. 3) or contouring if sufficient data are available (see Huws et.al. 1991). The velocity data shown in the figure relate to the upper 1-2 metres of seabed material at each measurement station, values being obtained from the initial slope of each time-distance plot.

### 3.3. Large-scale Refraction System

The large-scale refraction system, developed for measurement of physical properties down to a few tens of metres, uses a more powerful version of the basic electro-magnetic hammer device, with the sledge-mounted source deployed from an anchored ship and geophones deployed separately remote from the sledge. Experiments have been carried out using two different modes of received signal transmission. One utilises a direct cable link

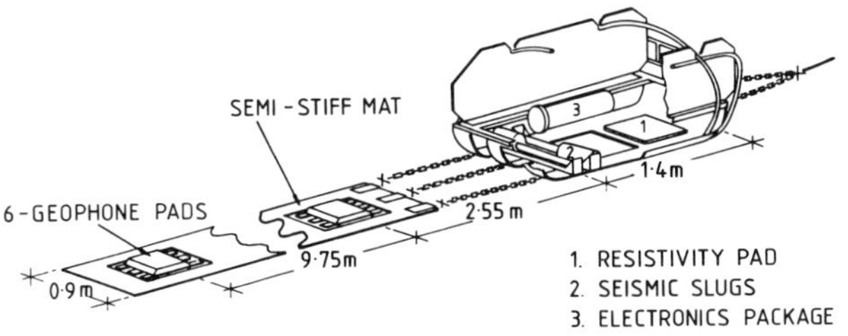


Fig. 1. The pseudo-underway seafloor geophysical system.

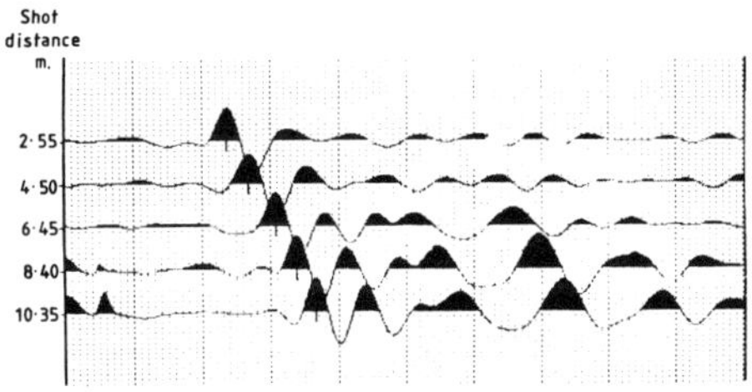


Fig. 2. Example shear wave record collected with the pseudo-underway system.

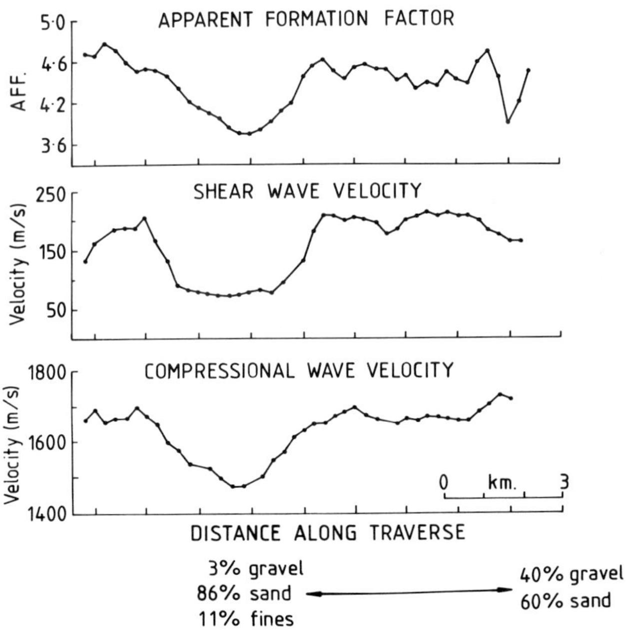


Fig. 3. Example data set for a traverse completed with the pseudo-underway system.

back to the ship from two 3-component geophones deployed on the seabed (fig. 4), and the other uses a single seabed geophone (mounted onto a free-fall penetrometer device) coupled to a sonobuoy, with data transmitted telemetrically to the ship-board instruments. Of the two, the direct cable link system proved more successful, the major problem with the sonobuoy system being obtaining data at close ranges (i.e. <50 metres offset) while guaranteeing that seabed instrumentation does not become entangled. Furthermore, with the sonobuoy system, signal recognition is complicated by the fact that there is no control on orientation of the horizontal sensors relative to the source motion. In contrast, provided the long offset cable is deployed correctly in relation to the prevailing tidal flow (i.e. to maintain a straight line between source and receivers) the geophone response can be optimised.

Using the configuration displayed in fig.4, the two geophones are deployed at a 'realistic' maximum offset, shown to be between 200 and 300 metres with this system. Tests are then carried out with signals detected by the two geophones and by an additional reference geophone mounted on the sledge adjacent to the source (to provide the shot instant and act as a noise monitor). On completion of an individual test the cable is hauled in a short distance and the test repeated. In this way a multi-channel record is built up and the refraction spread completed (fig.5). Apparent surface velocities and any velocity variation with depth are obtained by analysing the time-distance plots for first arrivals (using the signal propagating directly through the water to obtain the receiver offset).

While this procedure has significantly extended the depth capability of the original small-scale system described in 3.2, a full reverse-shot spread would need to be completed before a more rigorous interpretation could be attempted.

## 4. Discussion and Conclusions

In this paper the authors have outlined two methods for *in situ* shear wave velocity data acquisition and made additional recommendations relating to practical procedures. While both systems are relatively straightforward to operate, the success of the survey is largely dependent on deployment procedures, with particular attention needed to be given to prevailing oceanographic conditions. The data can then be used to map out variations in the physical properties of the seabed materials both laterally and with depth.

Results to date collected with the small-scale system have shown that the shear wave velocity is sensitive to subtle variations in seabed properties. As a pseudo-underway survey tool it offers a novel approach to investigating seabed properties and, used alongside the larger-scale refraction system, clearly holds much potential for use in site appraisal for engineering and other purposes.

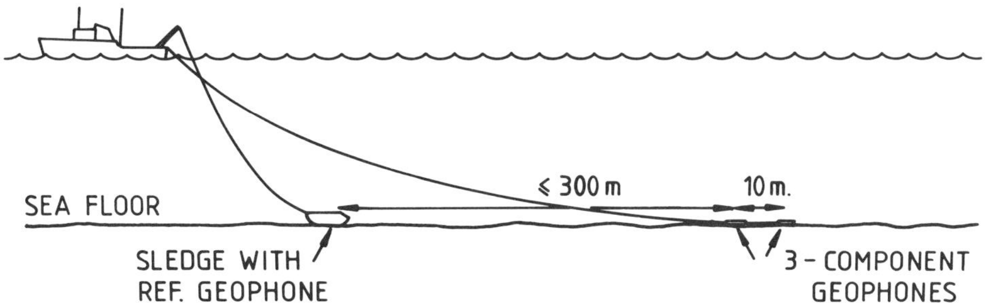


Fig. 4. Shear wave refraction configuration using a long offset geophone cable.

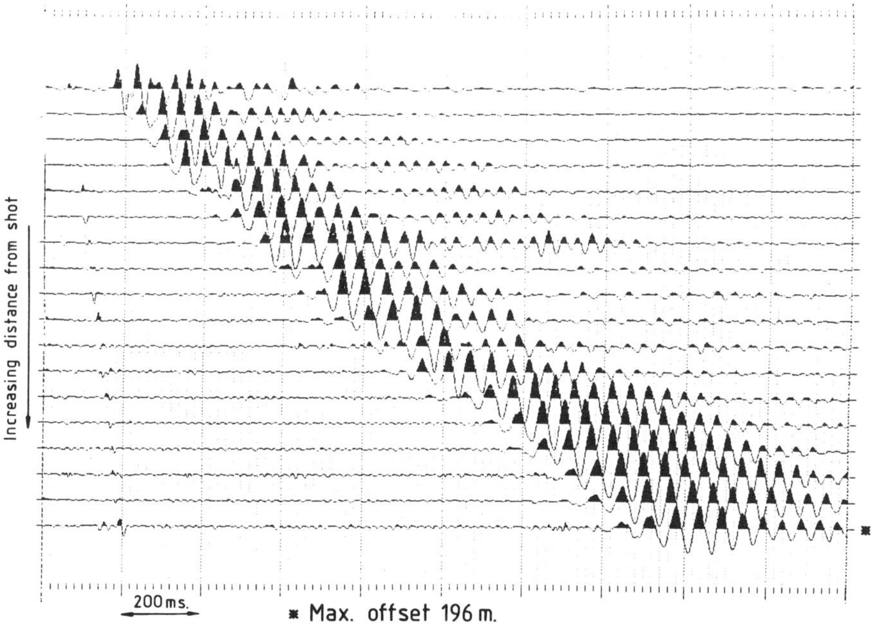


Fig. 5. Example multi-channel seismic record collected with SH wave source and long offset cable.

**Acknowledgments**

The shear wave systems were developed as part of an EC contract awarded by the Directorate-General for Energy. Inclusion of the resistivity pad onto the first system was by kind permission of the British Geological Survey and their help throughout the project is gratefully acknowledged.

**REFERENCES**

Davis, A.M. and Bennell, J.D. (1986) 'Dynamic properties of marine sediments', in Akal and Berkson (eds.) Ocean Seismo-acoustics, Plenum Press, New York,, pp.501-510.

Davis, A.M., Bennell, J.D., Huws, D.G. and Thomas, D.(1989) 'Development of a seafloor geophysical sledge', Marine Geotechnology 8, 99-109.

Davis, A.M. and Schultheiss, P.J. (1980) 'Seismic signal processing in engineering site investigation. A case history', Ground Engineering 13, 44-48.

Huws, D.G., Davis, A.M. and Bennell, J.D. (1990) 'Mapping of the seabed via (SH) velocity measurements', in Hovem, Richardson and Stoll (eds.) Shear Waves in Marine Sediments, Kluwer Academic Publishers.

Jackson, P.D. (1975) 'An electrical resistivity method for evaluating the *in situ* porosity of clean marine sands', Marine Geotechnology 1(2), 91-115.

Schwarz, S.D., and Conwell, F.R. (1974) 'A technique for the *in situ* measurement of shear wave velocities (Vs) for deep marine foundations', Offshore Technology Conference, OTC 2014, 755-762.

Stoll, R.D., Bryan, G.M., Flood, R., Chayes, D. and Manley, P. (1988) 'Shallow seismic experiments using shear waves', J.Acoust.Soc.Am. 83(1),93-102.

Stumpel, H., Kahler, S., Meissner, R. and Milkereit, B. (1984) 'The use of seismic shear waves and compressional waves for lithological problems of shallow sediments', Geophysical Prospecting 32, 662-675.

Taylor Smith, D. (1986) 'Geotechnical characteristics of the seabed related to seismo-acoustics', in Akal and Berkson (eds.) Seismo-Acoustics, Plenum Press, New York, 483-500.

**MAPPING OF THE SEA BED VIA**
***IN SITU* SHEAR WAVE (SH) VELOCITIES**

D. G. HUWS, A.M. DAVIS, J. D. BENNELL
*School of Ocean Sciences*
*Menai Bridge*
*LL57 5EY*
*United Kingdom.*

ABSTRACT. There is a distinct need both in industry and academia for a reconnaissance tool capable of accurately and reliably mapping the *in situ* properties of seabed sediments. A pseudo-underway seafloor sledge has been developed which measures *in situ* values of horizontally polarised shear wave velocities ($V_s$) and electrical resistivity (in the form of apparent formation factor - AFF). This system was used during an integrated geophysical survey in an area within the Firth of Clyde ,U.K. The results reveal variations in $V_s$ across the area which are mirrored in bathymetry, AFF data and grab sample information. In general, deeper water is characterised by the presence of muds of low $V_s$ (c.30 m/s) and low AFF, whilst shallower water sediments are predominantly sands or muddy sands of greater $V_s$ (60-90 m/s) and high AFF, suggesting a bathymetric control on sediment properties. Shear wave velocity is also found to vary with sediment depth. Both linear and discrete layer velocity-depth models were fitted to the travel time data. The linear velocity model best suited data from the mud-dominated zone, suggesting a gradual change in velocity related properties with depth. Sub-bottom profiler data give support to this since the zone in question is characterised by an apparent lack of acoustic reflectors. Velocity gradients are in agreement with previously published literature for similar sediments and depth ranges, being typically c. 17 /s.

## 1. Introduction

There is a distinct need, both in industry and academia, for a reconnaissance tool capable of accurately and reliably mapping *in situ* properties of the sea floor. Conventional site surveys usually involve measuring sediment properties from field collected cores. However, disturbance during collection, transport or storage of such samples (in the form of drainage or internal biogenic gas production), and during the measurement process itself, can lead to substantial errors in any subsequent laboratory reading (ARE Report, 1989).



*J. M. Hovem et al. (eds.), Shear Waves in Marine Sediments*, 337–343.

A pseudo-underway seabed towed sledge system has been developed at the University College of North Wales (UCNW) which can measure the shear (SH) wave velocity of the sea bed to a depth of 4-5 m. on a routine basis (Davis et al., 1989). Essentially, it consists of a shear wave source, seismic receivers, and a pad capable of measuring the electrical resistivity of the seafloor sediment to a depth of c.0.3 m. (Jackson, 1975).

This paper briefly explains how the measured data relate to the physical properties of marine sediments and also discusses the implications of results collected with the seafloor sledge system during a recent survey.

Simplistically, the shear wave velocity of a saturated marine sediment is a function of the sediment bulk density and dynamic shear modulus. The latter, in turn, is related to a host of sediment properties of which the most significant are the effective confining stress and the voids ratio (Hardin and Richart, 1963). Variations in shear wave velocity in the upper few metres of sediment are therefore quite significant since there usually occur marked increases in effective confining stress with overburden thickness at these depths.

From electrical resistivity readings, the apparent formation factor (AFF) of the substrate may be ascertained (Davis et al., 1989). This is empirically related to the inverse of sediment porosity by Archie's law via an exponent termed the cementation factor.

Mapping the spatial variability of shear wave velocity and AFF values will thus indirectly reveal trends in the sediment dynamic shear modulus and porosity distribution within a given area.

## 2. Sledge Survey Case History

### 2.1 METHODS

The survey area is situated in Irvine Bay, off the south west coast of Scotland (figure 1).

The underlying aim of the investigation was to map the sea bed sediment properties of the area via shear wave velocity and electrical resistivity measurements. Prior to the sledge investigation, the area was surveyed in detail using standard geophysical techniques (sub-bottom profiling, echo sounding, side scan sonar) and grab sampling. Background data from the area included one density log and four lithological core logs (ARE, Portland Report, 1989).

Details of sledge survey techniques are given by Davis et al. (this volume).

Shear wave signal quality was generally good throughout the survey. Given that over 300 sledge readings were taken within a relatively small area (12 sq. km.), the data are best represented in the form of contour plots.

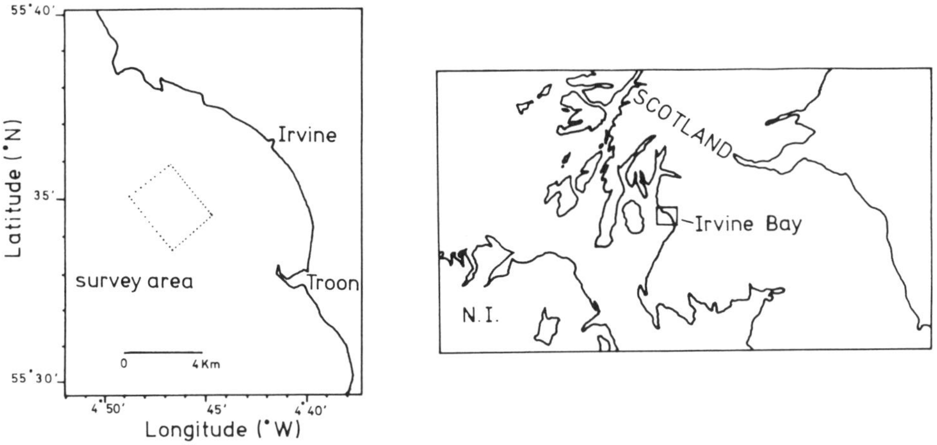


Figure 1. Position of Irvine Bay and the survey area.

## 2.2. RESULTS AND DISCUSSION

Initially a corrected bathymetric chart for the area was produced (figure 2a). The main feature to note is the general increase in water depth moving offshore (due SW) from <38 m. to >50 m. Other features include a "shallow" at the western tip of the area and a stretch of deep water orientated SW - NE in the central portion of the area.

Figure (2b) shows the AFF value distribution over the same area. Note that since, for a given cementation factor, AFF is inversely related to sediment porosity, higher AFF values imply lower porosity values and vice-versa. Conversion of the data to porosity was not considered appropriate since the variation in cementation factor within the area was not known. Laboratory techniques have been developed at UCNW to determine this factor for sand samples (Jackson, 1975).

The results of the particle size analyses performed on the grab samples are shown in figure (2c). Only fifteen samples were collected, therefore the resulting map should only be interpreted in a general sense. The data reveal that the area traverses the flank of a littoral platform consisting mainly of sand, through a transitional section of muddy sand and sandy mud, before finally entering a muddy zone in deeper water. This bathymetric control on sediment type is typical of the whole of the Firth of Clyde (British Geological Survey, 1985) and results from the increasing tidal energy regimes in the coast-proximal shallow water areas.

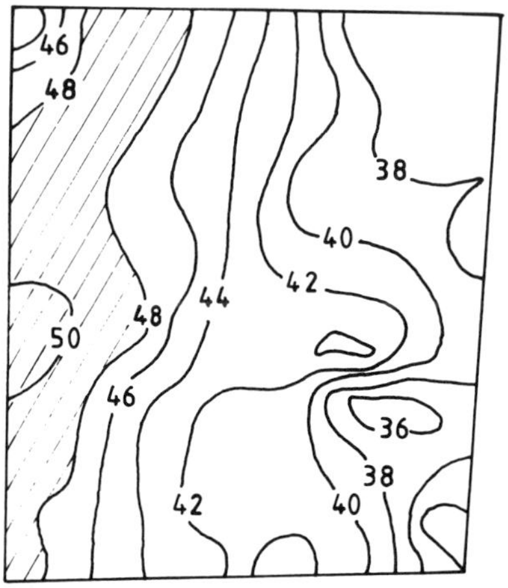


(a) Bathymetry of the survey area
(shading for >48 m. depth)

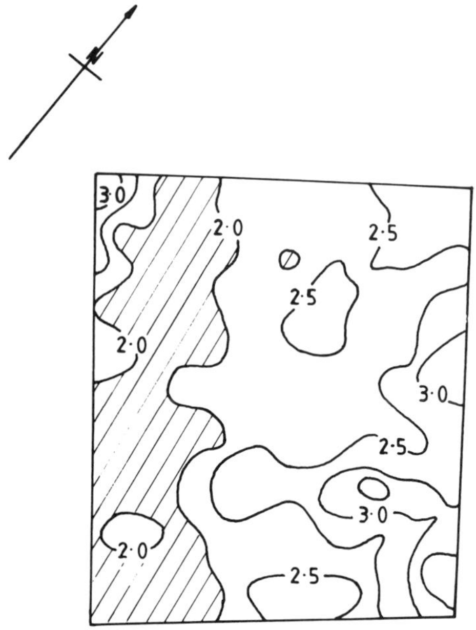


(b) Apparent formation factor
(shading for AFF < 2.0)

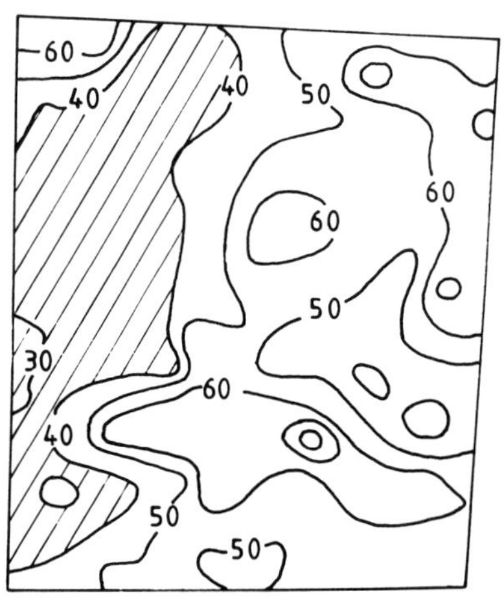


(c) Sea bed sediment distribution
(M-mud; sM-sandy mud; mS-muddy sand;
S-sand (British Geological Survey,
1985).

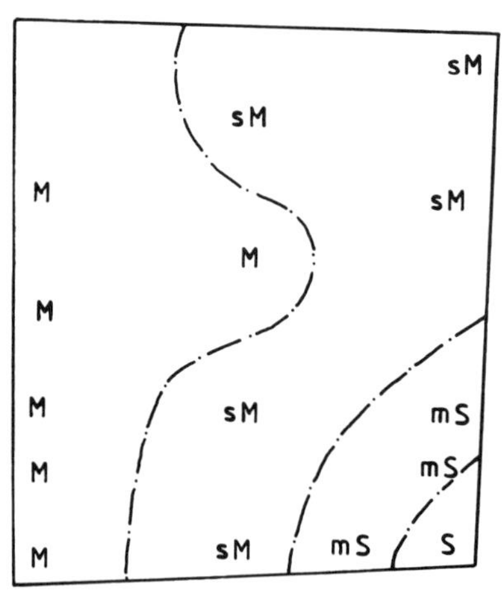


(d) Shear wave velocity in m/s
(shading for $V_s$ < 40 m/s)

Figure 2: Physical Property Distribution map for the Irvine Bay survey area

The shear wave time-distance plots reveal three distinct apparent velocity-receiver offset characteristics. Distinct changes in velocity at certain offsets is a typical feature of many plots (figure 3a), implying the presence of one or more sub-bottom refractors. There are also data in which a gradual increase in apparent velocity with offset occurs (figure 3b). Both of these plot-types result from some form of velocity-depth ($V_z$) structure. A minority of readings showed no such variation.

Two ($V_z$) models were applied to the data. The model used by Morris and Abbis (1979) divides the sub-bottom into a series of discrete horizontal layers of varying thicknesses and increasing velocity. In its simplest form, it is equatable to a two layer case and as such the model could be fitted to most of the recorded data.

The second model assumes a linear increase of velocity with depth (Abbiss, 1981) such that

$$(V_z) = V_o + kz$$

where $V_o$ = the extrapolated velocity at the surface (m/s)
$k$ = the velocity gradient ( /s)
$z$ = sediment depth (m)

This ($V_z$) structure leads to a time-distance curve given by

$$T = \frac{2}{k} \text{ arc sinh } \frac{kx}{2V_o}$$

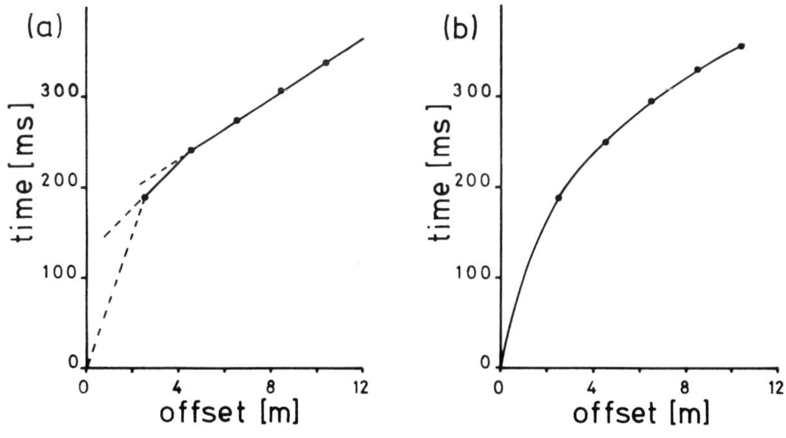


FIGURE 3: Typical time-distance plots from the Irvine Bay Survey Area, showing (a) distinct changes; and (b) a gradual increase in apparent velocity with offset.

The linear model best fit the time-distance data collected from the area of muddy sediment. This suggests that the (Vz) relationship is governed mainly by smoothly varying confining pressures and voids ratios, and not by possible changes in sediment material, as implied by the discrete layer model. This postulation is supported by the sub-bottom profiling data which is characterised by an apparent lack of distinct acoustic impedance contrasts (reflectors) in the mud-dominated areas.

For the muddy area, a typical linear velocity-depth function would have a theoretical extrapolated seabed surface velocity of 10 m/s and a velocity gradient of 17 /s for the depth range of 1 to 5 m. below the sea floor. This compares well with velocity gradients quoted previously in the literature for clays and silts (Hamdi, 1981, Richardson et al., this volume).

Given this depth variation in shear wave velocity, a standard interpretation procedure was adopted to allow all velocities to be represented on one map. Since, for the purposes of this survey, the zone of greatest interest was the sea bed surface, the velocity between the two geophones nearest the source was chosen as this standard. This represents an average velocity value for a sediment depth of c. 1.1 m. The contour plot generated with these values is shown in figure (2d).

Values of Vs range from less than 30 m/s to greater than 90 m/s. However, clear correlations are evident among figures 2a-d. High shear wave velocities are generally associated with shallow water, lower porosity sands whilst lower shear wave velocities typify deeper water, higher porosity muds. In order to demonstrate this, the zones in which shear wave velocities are generally less than 40 m/s, water depths greater than 48 m. and apparent formation factors less than 2.0 have been shaded (fig.2). On closer inspection, the interelationships between water depth, mapped seabed sediment types, apparent formation factor and shear wave velocities are quite marked. Minor bathymetric features also reveal themselves as small anomalies in shear wave velocity and AFF, demonstrating the rigid bathymetric control over sediment properties within the area. Correlation between mapped sediment type and the other parameters are not exact due to the paucity in sampling data.

Using density log data from collected from the area, it would become possible to extend the interpretation procedure so as to calculate values for small strain shear moduli using one of the standard solutions to the general wave equation. Given more data for the variation of density with depth, it would ultimately be possible to produce contour maps of predicted small strain shear moduli for any depth up to 5 m. below seabed.

## 3. Implications and Conclusions

Engineering estimates and geoacoustic models (e.g. the Biot-Stoll model (Stoll, 1986)) rely upon *in situ* parameter estimates to provide a representative description of a sediment body. Shear wave velocity (and its variation with depth) is one such parameter and it is a required input for the majority of geoacoustic models (Hamilton, 1980).

Velocities are often ascertained indirectly via interface wave modelling or they are predicted from other parameters using empirical relationships. However, it is infinitely preferable to use actual values collected *in situ* and the pseudo-underway seafloor sledge system has proved itself to be capable of mapping shear wave velocity variations of the sea bed sediments down to depths of 4 to 5 m. on a routine basis. As such, it "plugs a hole" in the field of directly measureable sea bed acoustic properties.

## 4. Acknowledgements

The sledge system was originally developed as part of an E.C. contract awarded by the Energy Directorate. The Admiralty Research Establishment (Portland) contributed to the funding of the cited marine survey at Irvine Bay and also provided background survey literature. The electrical resistivity pad design and electronic circuitry was developed and produced by Dr. Peter Jackson (British Geological Survey, Keyworth). The authors also wish to acknowledge the substantial support given by the technical department at Menai Bridge.

## 5. References

Admiralty Research Establishment Report (1989) 'Analysis of cores from ARE (UTH2) sea bed survey', ARE Portland, UTH Tech. Memo, 23/89(C).

Abbiss, C.P. (1981) 'Shear wave measurements of the elasticity of the ground', Geotechnique, 31(1), 91-104.

British Geological Survey (1985) 'Seabed sediments and Quaternary geology', Clyde Sheet 55°N - 06°W, 1:250 000 Series.

Davis, A.M., Bennell, J.D., Huws, D.G., Thomas, D. (1989) 'Development of a seafloor geophysical sledge', Marine Geotechnology, 1(2), 91-115.

Davis, A.M., Huws, D.G., Bennell, J.D. (1991) 'Seafloor shear wave velocity data acquisition : procedures and pitfalls', in Shear Waves in Marine Sediments, J.M. Hovem, M.D. Richardson and R.D. Stoll (eds.), Kluwer Academic Publishers, Dordrecht, (this volume)

Hardin, B.O. and Richart, R.E. (1963) 'Elastic wave velocities in granular soils', Proc. Am. Soc. Civil Eng., 89, 33-65.

Hamdi, F.A.I. (1981) 'Variations of seismic velocity with depth in near surface marine sediments and their engineering significance', PhD. Thesis, Univ. of Wales (U.K.).

Hamilton E.L. (1980) 'Geoacoustic modeling of the sea floor', J. Acoust. Soc. Am., 68(5), 1313-1340.

Jackson, P.D. (1975) 'An electrical resistivity method for evaluating the in situ porosity of clean marine sands', Marine Geotechnology, 1(2), 91-115.

Morris, D.V. and Abbiss, C.P. (1979) 'Static modulus of Gault clay predicted from seismic tests', Ground Engineering, 12(8), 44-50.

Stoll R.D. (1986) 'Acoustic waves in marine sediments', in T. Akal and J.M. Berkson (eds.), Ocean Seismo-acoustics, Plenum Press, N.Y. and London, 417-434.

# OBSERVATIONS OF SPLIT SHEAR-WAVES FROM YOUNG OCEAN CRUST

N. JONES, M.C. SINHA, R.B. WHITMARSH[1]
*Department of Earth Sciences*
*University of Cambridge*
*Cambridge CB3 0EZ, U.K.*

ABSTRACT. Pressure waves generated by an airgun source give rise to converted shear-waves at the sea-floor interface. Such shear-waves, turning through young oceanic crust, have been recorded by an array of digital OBS's. Particle motion analysis shows that shear-wave splitting has occurred during their passage through the in-situ rockmass; the polarization of the *faster* component generally lies in a N-S direction, parallel to the ridge-axis.

## 1. Introduction

In April 1986, a detailed seismic refraction survey was carried out jointly by Cambridge University and the Institute of Oceanographic Sciences to investigate anisotropy in young ocean-crust (1-3Ma). The chosen area lay on the eastern flank of the Mid-Atlantic Ridge (MAR) at 40° 15′ N, 29° 20′ W, about 30km east of the median valley. Gloria sonographs over this area reveal highly elongate, parallel targets striking 010° N; these have been interpreted as inward facing faults and fissures (Searle and Laughton, 1977). In conjunction with the OBS work, a Deep Tow seismic reflection profile was shot normal to ridge axis across the survey site and on as far as the eastern summit of the median valley. This revealed a series of rotated fault blocks, of average width 2km (Joppen, 1989), bounded by the normal faults seen by Gloria.

Neither the bathymetry (Fig. 2), the magnetics, nor Gloria (I.O.S. long-range sidescan sonar system) showed any signs of fracture-zone related discontinuities; the nearest fracture-zone was the Kurchatov some 40km north of the survey site. This was an important factor as fracture-zones usually represent anomalous crustal features (White et al., 1984).

Bearing in mind the uniformity of the observed tectonic fabric and the closeness of the median valley, it was thought that a fairly uniform stress regime might exist and, hence, the possibility of investigating seismic anisotropy was favourable (White and Whitmarsh, 1984; Stephen, 1981,1985).

[1]I.O.S. Deacon Laboratory, Brook Road, Wormley, Surrey GU8 5UB, UK



*J. M. Hovem et al. (eds.), Shear Waves in Marine Sediments*, 345–352.

## 2. Experiment Geometry

The experiment consisted of 629 shots from an airgun array into a number of digital OBS's positioned in such a way as to ensure a broad coverage in azimuth and range for shot-receiver pairs. The survey geometry is illustrated in Fig. 1. The shot and receiver positions were obtained from the I.O.S. 5 KHz acoustic navigation system, together with satellite fixes and the ship's DR system. Final shot and receiver positions were obtained to within a few tens of metres accuracy (~ 100m in the worst cases) by using an optimisation technique based on the best satellite fixes and ranges calculated from water-wave arrival times. Figure 2 is a plot of the bathymetry over the site compiled from standard precision-echosounder data and Gloria sonograms.

The airgun array consisted of five guns (160, 300, 466, 700 and 1000 cu.in. chambers) fired every two minutes and towed at a speed of 4.5 knots, giving an average shot spacing of 270m. Two types of digital OBS were employed, namely the Cambridge DOBS (Duschenes et al., 1985) and the I.O.S. DOBS (Kirk et al., 1982). Basically these were 4-component instruments recording the three channels of a geophone package and a separate hydrophone signal for a dynamic range of over 110dB and a bandwidth of 0.1Hz to 64Hz. The instruments were pre-programmed to record over 15 second time-windows at sampling rates of 256Hz for the Cambridge DOBS and 128Hz for the I.O.S. DOBS.

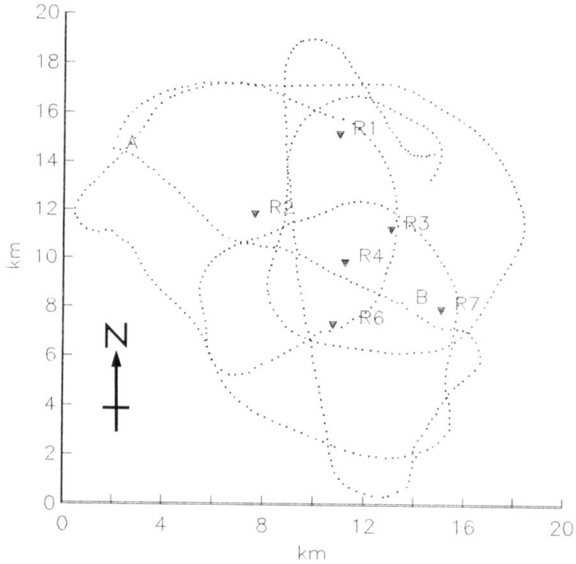


Figure 1: Shooting-track. Triangles represent receiver locations, dots represent shot-points. Origin of rectangular grid:- 40° 10′ N, 29° 17′ W.

## 3. Data

A sample section of range-sorted data for the three geophone channels is shown in Fig.3, representing a portion of the shooting-track having fairly uniform azimuth with respect to receiver R7. To aid interpretation of this section, a mute has been applied at the water-wave onset leaving a clear picture of the two preceding phases, namely the P- and S-waves. A reduction velocity of 3.2kms$^{-1}$ emphasizes the divergent nature of the secondary phase with respect to the P-wave onset. Basic comparisons of arrival times for the two phases along each trace yields a Poisson relation where $(Vp/Vs)^2 \simeq 2.2$-$2.5$. Fourier analysis reveals a dominant 7Hz characteristic frequency for the S-wave arrivals and so, with a suitable low-pass filter, much of the pre-shear-wave noise can be removed. However, delineating S-wave arrivals accurately from such sections as illustrated in Fig. 3 can be hindered by the remnant effect of the P-wave arrival. To this end, the use of polarization diagrams (PD's) is required.

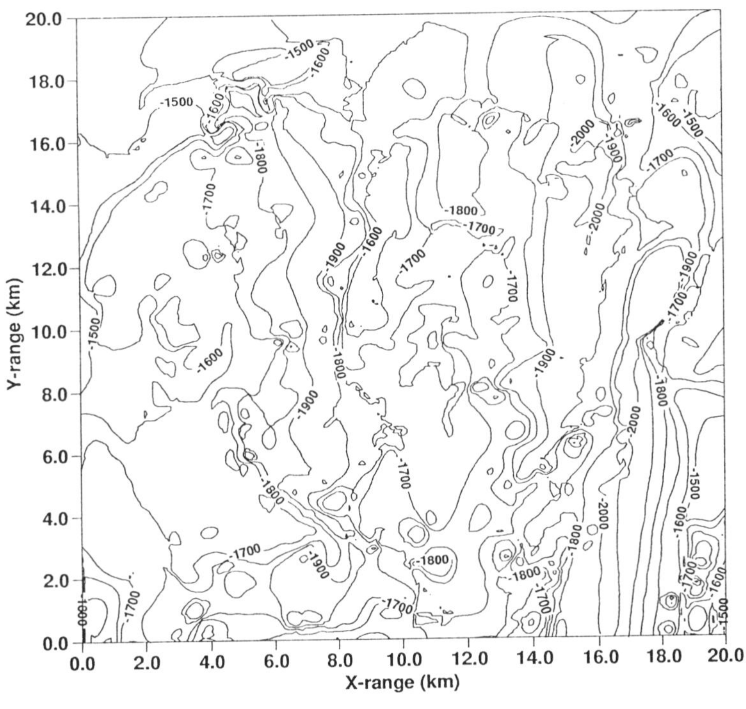


Figure 2: Bathymetry at survey-site. Origin at 40° 10′ N, 29° 17′ W; contours in metres.

A crude 1-D velocity-depth model, based on the assumption that the rock-mass is homogeneous and seismically isotropic, was calculated from P-wave travel-times picked on R7, for shot-points along section A-B (see Fig.1). The times were forward modelled using Maslov Asymptotic Ray Theory (Chapman and Drummond, 1982). A feature of the model is the non-existence of Layer 1 sediments, a fact born out by the Gloria sonograms. An initial velocity of 3.3 kms$^{-1}$ at the top of Layer 2 and a relatively high gradient of 1.3 s$^{-1}$ is estimated. Though a very simple model, the synthetic and observed arrival times showed good agreement. Further, low initial velocities and high gradients for young oceanic crust have been seen elsewhere on the MAR, eg. White and Purdy (1983).

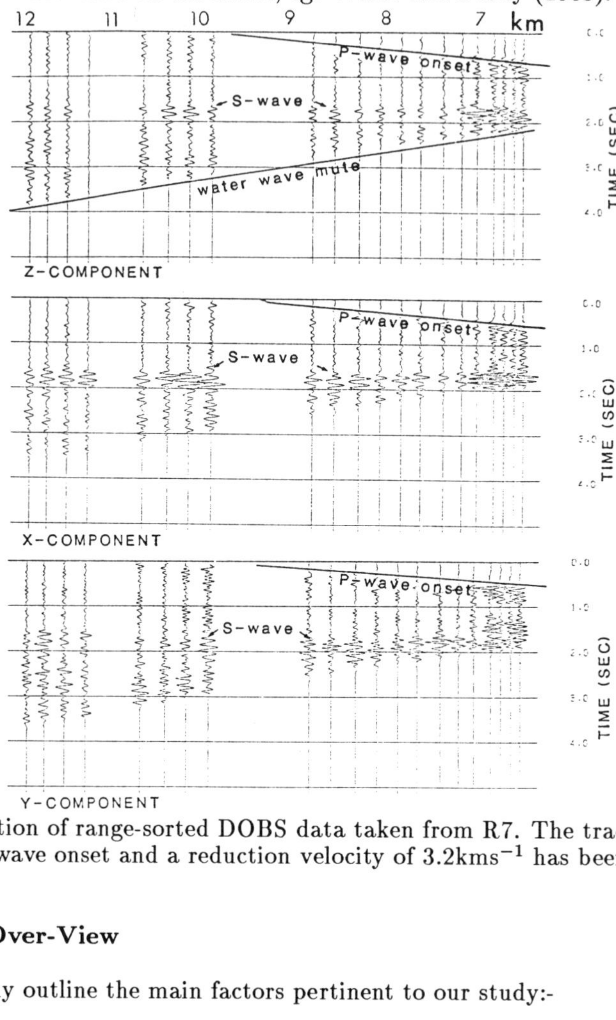


Figure 3: Section of range-sorted DOBS data taken from R7. The traces have been muted at the water-wave onset and a reduction velocity of 3.2kms$^{-1}$ has been applied.

## 4. Theory Over-View

Here we briefly outline the main factors pertinent to our study:-

- Following Crampin (1981,1984), we know that in an anisotropic medium, solution of the wave equation indicates that there are three body-waves for every direction of phase propagation. These waves have orthogonal polarisations and, generally, different velocities. They are termed the quasi-P, quasi-S1 and quasi-S2 waves. As the two qS-waves propagate through an anisotropic medium, they become separated in time. This effect constitutes the phenomenon of shear-wave splitting.

- Most of the observations of shear-wave splitting to date have been from VSP's and micro-earthquake studies where the ray-paths were relatively straight. For this experiment however, the ray-paths are highly curved and a steep velocity gradient is present. It is therefore possible that the polarization vectors of the qS phases rotate about the direction of propagation, see Shearer and Chapman(1988;1989a,b). This is known in turn to cause coupling between the two qS-pulses.

- Heterogeneity within the medium traversed by the shear-waves could account for splitting. Radial shear-wave energy can easily be converted to transverse motion by scattering effects; however, it is thought that this conversion would generally arrive later than the unscattered radial. Hence observations of initial transverse motion at the S-wave onset might be a good case for anisotropy.

- There are inherent problems of recording shear-waves at the free-surface (water/sea-floor interface). Booth et al. (1985) stress the importance of only considering shear-wave arrivals incident within the *shear-wave window*, defined as the solid angle bounded by the angle, $i_c = \sin^{-1}(Vs/Vp)$.

## 5. Methods

As mentioned earlier, as much accuracy as possible was needed to delineate the shear-wave onsets. To this end, we employed the traditional analytical process of examining particle motion diagrams, produced by rotated horizontal seismograms, for ray-paths of near-vertical incidence. Further, to aid visual identification of the initial S-wave motion, both the rotated seismograms and the particle-motion diagrams were colour-coded with respect to time.

We have a redundancy of data over various azimuths and ranges, however only a limited number of shot-receiver pairs were selected for particle-motion analysis. The selection factors depended on:

- Signal-to-noise ratio; obviously, due to the nature of the shooting track, many of the shear-wave onsets are obscured by the water-wave at close range or by P-wave noise.

- Whether the incidence angle was likely to be within the shear-wave window; cut-off range (8.0km) was calculated on the basis of our velocity depth model and our estimate of the shear-wave window, below which initial shear-wave motions were discounted. The angle of the onset in the vertical plane was also observed in the PD's.

- How well the geophone azimuth had been constrained. In order to achieve this, only shots with good signal-to-noise ratio for the water-wave onset were used. PD's of the first 5-10 samples of each water-wave arrival, from the horizontal geophones, were plotted and a best-fit straight line determined, giving the apparent shot azimuth. Having observed numerous apparent shot azimuths and knowing their true shot-azimuths to within 1°, we were able to estimate the orientations of the geophone azimuths.

It is important to see how consistent these estimates are over a range of shot-azimuths by plotting one against the other, since there are a number of factors which might cause scatter in the water-wave polarisation data eg. mismatched or tilted horizontal geophones, inaccurate source or receiver positions, etc., (Duennebier et al., 1987).

We were unable to obtain a consistent result for R10. Howerver we were able to deduce the geophone azimuth to within $\pm$ 6 ° for the other Cambridge instruments (R7,R11).

Knowing the geophone azimuth, we rotated the horizontals through a given angle for each shot under consideration so that the axes lie along the radial and transverse directions. Figure 4 is a typical example of this process.

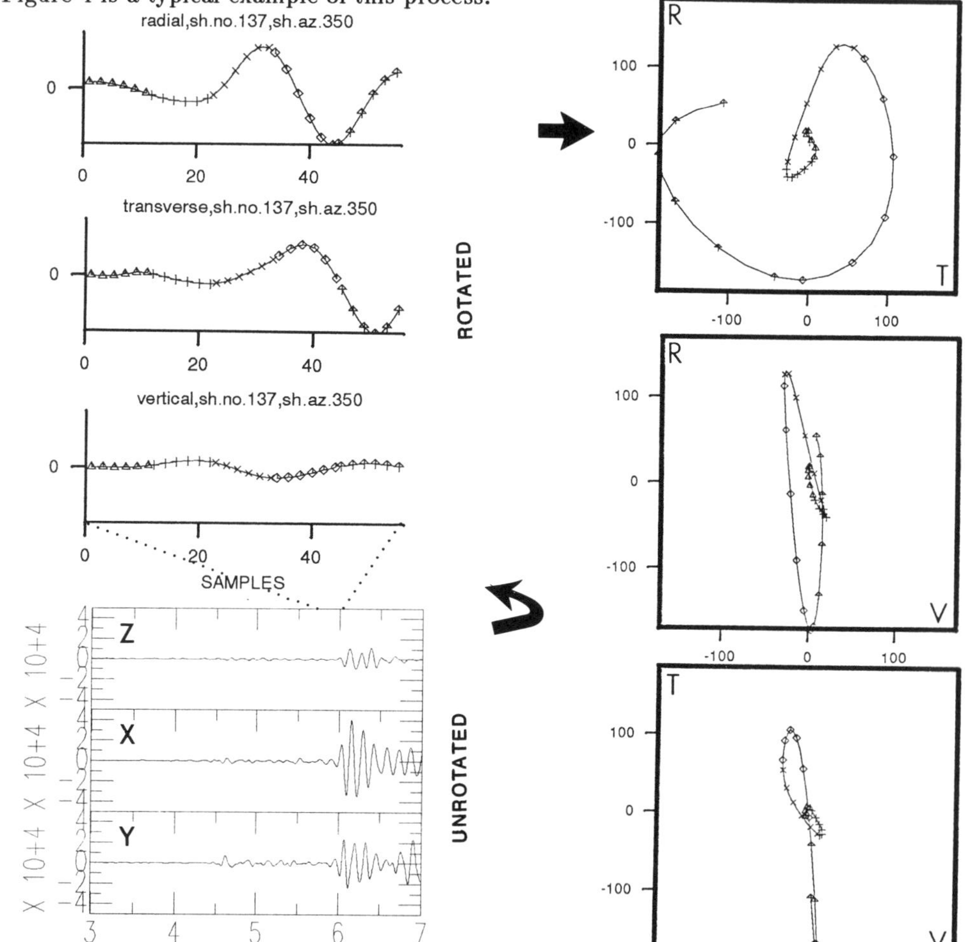


Figure 4: Example of shear-wave analysis

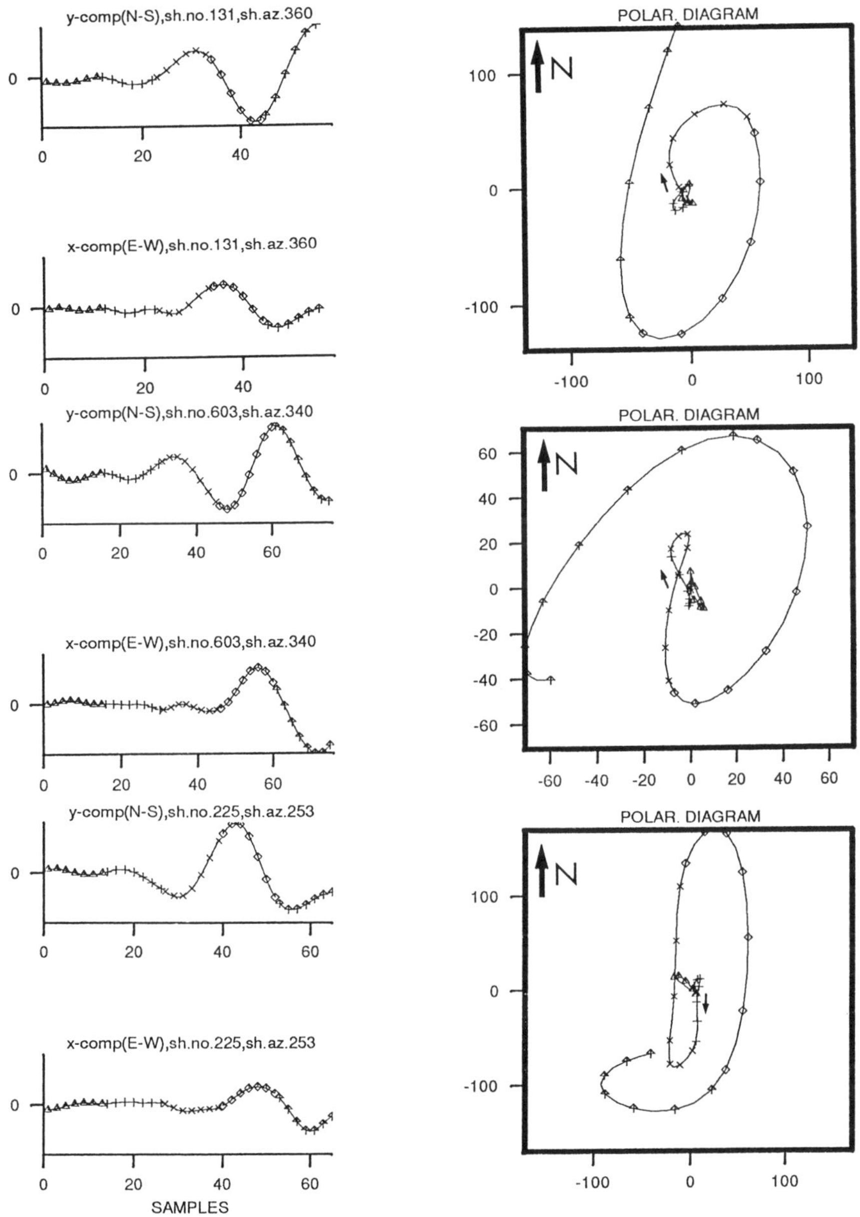


Figure 5: Examples of S-wave onsets from rotated seismograms.

## 6. Results and Conclusions

A compilation of PD's for shots at ranges greater than 8km is shown in Fig. 5. The diagrams illustrate portions of seismograms both pre- and post-S-wave onset. By iteratively glancing between the PD's and the seismograms, it becomes apparent that the initial onset is generally in the N-S direction. In fact this was the case for all the shear-wave onsets that we analysed. The fact that initial onsets occur both in radial and transverse directions over a range of azimuths, together with fairly uniform N-S orientations for all rotated seismograms suggests that we might have a case for anisotropy related shear-wave splitting.

## 7. References

Booth,D.C., Crampin,S., Evans,R. and Roberts, G., 1985, Shear-wave polarisations near the North Anatolian Fault-3. Geophys.J.R.astr.Soc., 91, 287-311.

Chapman,C.H. and Drummond,R., 1982, Body-wave seismograms in inhomogeneous media using Maslov asymptotic theory, Bull.Seis.Soc.Amer., 72, S277-S317.

Crampin,S., 1981, A review of wave motion in anisotropic and cracked elastic-media, Wave Motion, 3, 343-391.

Crampin,S., 1984, An introduction to wave propagation in anisotropic media, Geophys.J.R. astr.Soc., 76, 165-172.

Duennebier,F.K.,Anderson,P.N. and Fryer, G.J., 1987, Azimuth determination of and from horizontal ocean bottom seismic sensors, J.Geophys.Res.,92,3567-3572

Duschenes,J.D., Potts,C.G. and Rayner,M., 1985, Cambridge deep ocean geophone, Mar. Geophys.Res., 4, 437-450.

Joppen,M.,1989, Seismic Studies of Small Ocean Basins And Zones Of Crustal Accretion, Ph.D. Thesis, Univ. of Cambridge.

Kirk,R.E., Whitmarsh,R.B., and Langford,J.J., 1982, A three-component ocean bottom seismometer for controlled source and earthquake seismology, Mar.Geophys.Res., 5, 327-341.

Searle,R.C. and Laughton,A.S., 1977, Sonar studies of the Mid-Atlantic Ridge and Kurchatov Fracture Zone, J.Geophys.Res.,82,5313-5328.

Shearer,P.M. and Chapman,C.H., 1988, Ray tracing in anisotropic media with a linear velocity gradient, Geophys.J., 94, 575-580.

Shearer,P.M. and Chapman,C.H., 1989, Ray tracing in azimuthally anisotropic media- 1.Results for models of aligned cracks in the upper crust, Geophys.J., 96, 51-64.

Shearer,P.M. and Chapman,C.H., 1989, Ray tracing in azimuthally anisotropic media- 2.Quasi-shear wave coupling Geophys. J., 96, 51-64.

Stephen,R.A., 1981, Seismic anisotropy observed in the upper oceanic crust, J.Geophys.Res. Lett.,8,865-868.

Stephen,R.A., 1985, Seismic anisotropy in the oceanic upper crust, J.Geophys.Res., 90, 11383-11396.

White,R.S., Detrick,R.S., Sinha, M.C. and Cormier, M.H., 1984, Anomalous seismic crustal structure of oceanic fracture zones, Geophys.J.R.astr.Soc., 79,779-798.

White,R.S. and Purdy,G.M., 1983, Crustal velocity structure on the flanks of the Mid-Atlantic Ridge at 24° N, Geophys. J.R.astr.Soc, 75, 361-385.

White,R.S. and Whitmarsh,R.B., 1984, An investigation of seismic anisotropy due to cracks in the upper oceanic crust at 45° N, Mid-Atlantic Ridge, Geophys.J.R.astron.Soc.,79,439-467.

# Estimation of Shear-Wave Speed in Ocean-Bottom Sediments by Comparing Oblique-angle Reflectivity Measurements with Normal Incidence Data

T. M. McGee, Department of Exploration Geophysics, University of Utrecht
P.O.Box 80.021, 3508TA Utrecht, The Netherlands

## Abstract

Information concerning the speed of shear waves in sea-floor sediments is encoded in the way that the amplitude of water-bottom reflections varies with angle of incidence. The simplest theoretical model of this variation is described by the mathematical relationships for plane-wave reflection from a liquid/solid plane interface. These relationships are stated in terms of ratios of elastic parameters across the interface. Published measurements provide a basis for tabulating a set of values for these ratios and investigating the theoretical variation of amplitudes reflected from more-or-less realistic water/soil interfaces. This investigation suggests a theoretical nomogram which is useful for estimating shear speed if Rayleigh reflection coefficient and density contrast are known. The results of a field experiment are plotted on such a nomogram and found to be consistent with the type of soil known to be present at the reflection point. It is concluded that this type of analysis can be done in near-to-real time from a surface vessel operating in the conventional "common-reflection-point" profiling mode if the hydrophone array is approximately twice the water depth, or more, in length.

## Introduction

When acoustic waves propagating in water encounter the water bottom they are partially reflected and partially refracted. The reflected energy is purely compressional at all angles of incidence because the water does not support shear waves. The refracted energy is purely compressional at normal incidence but, if the bottom sediments exhibit rigidity, it is partly converted to shear at oblique angles of incidence. This conversion reduces the amplitude of the reflected wave. The amount of reduction can provide an estimate of the speed of shear waves (s-waves) in the water-bottom sediments if the angle of incidence, the geometry of the wavefront and sea-floor contrasts in acoustic impedance and the speed of compressional waves (p-waves) are known.

The angle of incidence may be determined from the recording geometry if the speed of sound in the water column is known as a function of depth. The wavefront geometry depends on the type of source and receiver employed but, if both are small compared to wavelengths of interest, it can be assumed to be spherical. If the sea floor is sufficiently flat that geometrical focusing and scattering are negligible, the impedance contrast may be determined from the rate of decay of water-layer multiples recorded on a receiver located near the source, a so-called zero-offset trace. The p-wave speed in most sea-floor soils is not very different from that in water. Synthetic models indicate that, for angles of incidence less than about 60°, the error incurred by assuming them to be equal is not large.

A nomographic scheme for estimating s-wave speed in sea-floor soils is described and the results of a field experiment show it to be feasible. Although the method relies on a number of assumptions, it has the advantage of being amenable to near-to-real time processing on a surface vessel operating in the conventional "common-reflection-point" profiling mode. It is therefore considered useful for general reconnaissance or for choosing sites at which to carry out detailed studies.



*J. M. Hovem et al. (eds.), Shear Waves in Marine Sediments*, 353–360.

**Basic Mathematical Relationships**

The reflection coefficient at a liquid/solid interface for incidence in the liquid is given by Ewing, Jardetsky and Press (1957). It is a function of the angle of incidence and can be written in the form

$$R(i) = \frac{adF\cos i - \sqrt{E}}{adF\cos i + \sqrt{E}} \tag{1}$$

for

$$F = (1-2G)^2 + 4(b/a)(1-G)\sqrt{EG} \quad , \quad E = 1 - a^2\sin^2 i \quad , \quad G = 1 - b^2\sin^2 i$$

where $i$ = angle of incidence,
$a$ = ratio of the p-wave speed in the solid to that in the liquid,
$b$ = ratio of the s-wave speed in the solid to the p-wave speed in the liquid
and $d$ = the ratio of the density of the solid to that of the liquid.

(It should be noted that the quantity $b/a$ is the ratio of s-wave to p-wave speed in the sediment.)

The angle of incidence varies from $i = 0°$ (normal incidence) to $i = 90°$ (grazing incidence). The reflection coefficient is real-valued for angles at which both $E$ and $G$ are non-negative. Since $a > b$, it is real for all $i$ if $a \le 1$. If $a > 1$, the coefficient is complex for angles greater than $i = \arcsin(1/a)$, the so-called *critical angle*. If $b > 1$, total reflection occurs for angles greater than $i = \arcsin(1/b)$. Total reflection means that the reflection coefficient is of unit magnitude and, if it is complex, only its phase varies. Varying phase results in the shape of the reflected pulse being different from that of the incident pulse.

Two special situations are of particular interest here. The first is normal incidence where eq.(1) simplifies to

$$R(0) = \frac{ad - 1}{ad + 1} \tag{2}$$

which is called the *Rayleigh reflection coefficient*. It is not affected by the shear properties of the sediment, being dependent only on the product $\{ad\}$ which is called the *acoustic impedance contrast* across the sea floor. The second is when there is no change in p-wave speed across the sea floor, i.e. $a = 1$. Then eq.(1) can be written

$$R(i) = \frac{dF - 1}{dF + 1} \quad \text{where} \quad F = (1-2G)^2 + 4b(1-G)\sqrt{G}\cos i \ . \tag{3}$$

It is probably rather common that this is at least approximately valid. Measurements showing $a \approx 1$ have been reported by Hamilton (1972) and Chapman et al (1983).

The ratios $a$, $b$ and $d$ are related by the well-known result of elastic wave theory that the speeds of p-waves and s-waves are functions incompressibility, rigidity and density. Since a liquid exhibits no rigidity, it is possible to write

$$b = .5\sqrt{3(a^2 - k/d)} \quad \text{or, equivalently,} \quad k = d\,(a^2 - \frac{4}{3}b^2) \tag{4}$$

where $k$ is the ratio of the incompressibility of the solid to that of the liquid. Since $k$ is positive, the second relationship of eq.4 shows that $b/a < .866$, approximately.

**Theoretical Investigations**

In order to establish a more-or-less realistic basis for theoretical calculations, parameter ratios for various categories of "synthetic" sea floors are compiled as Table 1. All ratios assume the liquid to have a p-wave speed of 1500 m/s and unit density. Values of $a$ and $d$ for the generic soil categories of sand, dense clay, silt and light clay are based on p-wave speed and density measurements reported by Hamilton (1972). These are used in eq.2 to calculate Rayleigh reflection coefficients. The values tabulated in the two other categories are based not as much on measurement as on an intention to "bracket" the values in the generic soil categories. The category "friable" sandstone is intended to represent sea floors that are more competent than the generic soils but not as competent as very hard bedrock. The category "gassy" mud is intended to represent very soft sea floors. The $b$ -values shown are calculated from eq.4 by assuming the indicated $k$ values. The tabulated values of $b$ are required for the theoretical investigations, but are irrelevent to the experimental results.

The amplitude (signed magnitude) and phase of the reflection coefficient for each category in Table 1 is plotted versus angle of incidence in fig.1. The phase curves for dense clay, silt, light clay and "gassy" mud are identically zero since their coefficients are real for all angles of incidence ( $a \leq 1$ ). Categories with $a > 1$ have a critical angle marked by a narrow peak. That for sand is $i \approx 60^\circ$ and that for "friable" sandstone is $i = 30^\circ$. (Such narrow peaks are not actually observed in the field because real sea floors do not behave as mathematical interfaces.) Above the critical angles, coefficients are complex and the phase curves are not zero. At angles greater than $i \approx 42^\circ$ the "friable" sandstone is totally reflecting ( $b > 1$ ) and its phase curve increases across a range of about $540^\circ$.

If a reflection coefficient is divided by its corresponding Rayleigh reflection coefficient it is said to be *normalized*. Normalized reflection amplitude curves for each category of Table 1 are shown in fig.2.

If values of $a$ and $R(0)$ are known, eq.2 may be solved for the value of $d$ and then eq.3 used to plot a family of curves whose members correspond to different values of $b/a$. If such a family is normalized, it comprises the *shear–speed nomogram* for the particular $a$-value and Rayleigh reflection coefficient. An example of such a nomogram drawn for $a = 1$ and $R(0) = +.2$ is shown in fig.3. It consists of curves for values of $b/a$ ranging from 0.0 (the minimum possible value) to 0.866 (the maximum possible value). The shape of the curves follows an orderly progression toward smaller normalized amplitudes as $b/a$ increases from .00 to .66 (dashed curves), but the direction of progression then reverses (solid curves). Thus the nomogram would yield a unique estimate only for soils such that $b/a < .665$. In Table 1 this value lies between dense clay and sand, implying that the nomogram does not discriminate between them.

The similarity between the dashed curves in fig.3 and the normalized amplitude curves for generic soil categories in fig.2 suggests that shear-speed nomograms may be useful for estimating the value of $b/a$. Of course, some error would be involved in cases of soils for which $a \neq 1$. The magnitudes of such errors for soils similar to those in Table 1 are illustrated in fig.4 ( $a > 1$ ) and fig.5 ( $a < 1$ ).

Fig.4 is a shear-speed nomogram drawn for $a = 1$ and $R(0) = +.39$. Plotted on it are normalized amplitude curves for "sand-clay" and "clay-sand", two "soils" contrived such that they exhibit $R(0)$ equal to the average of that for sand and dense clay in Table 1. The "clay-sand" has $b/a = .65$ and its curve corresponds to that in the nomogram. The "sand-clay" ( $b/a = .68$ ) curve does not correspond to the nomogram because it has $a > 1$. The difference is not pronounced for angles of incidence less than about 57°, however. In that region the "sand-clay" curve appears to closely follow the nomogram curve for $b/a = .66$. Thus an estimate based on the nomogram is about 3% too low.

Similarly, fig.5 is a nomogram for $a = 1$ and $R(0) = +.19$ (the average of silt and light clay in Table 1). Normalized amplitude curves for "silt-clay" ( $b/a = .49$ ) and "clay-silt" ( $b/a = .47$ ) are plotted on it. The "silt-clay" curve follows the nomogram correctly and the "clay-silt" curve plots very close to it for angles of incidence less than about $58^\circ$. In this case the nomogram estimate is about 4% too high.

These tests indicate that, if the Rayleigh reflection coefficient is known, a shear-speed nomogram drawn for it and $a = 1$ may be used to estimate values of $b/a$ to within 5% by determining normalized reflected amplitudes for angles of incidence up to about 60°.

| sediment type | a | d | R(0) | k | b | b/a |
|---|---|---|---|---|---|---|
| "friable" sandstone | 2.00 | 2.20 | +.63 | 2.20 | 1.50 | 0.75 |
| sand | 1.15 | 2.04 | +.40 | 1.00 | 0.79 | 0.69 |
| dense clay | 1.00 | 2.20 | +.38 | 1.00 | 0.64 | 0.64 |
| silt | 1.00 | 1.72 | +.26 | 1.00 | 0.56 | 0.56 |
| light clay | 0.97 | 1.30 | +.12 | 1.00 | 0.36 | 0.37 |
| "gassy" mud | 0.33 | 1.01 | -.50 | 0.11 | 0.03 | 0.09 |

Table 1 - Parameter ratios used in theoretical investigations

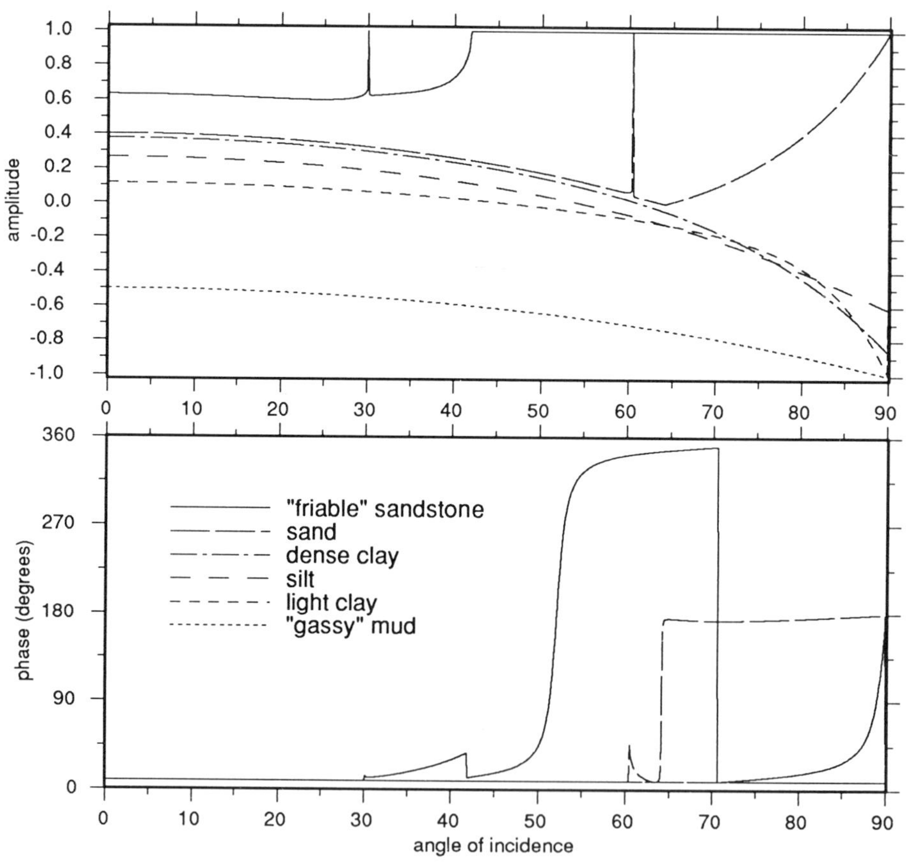


Fig.1 - Reflection coefficients for catagories in Table 1

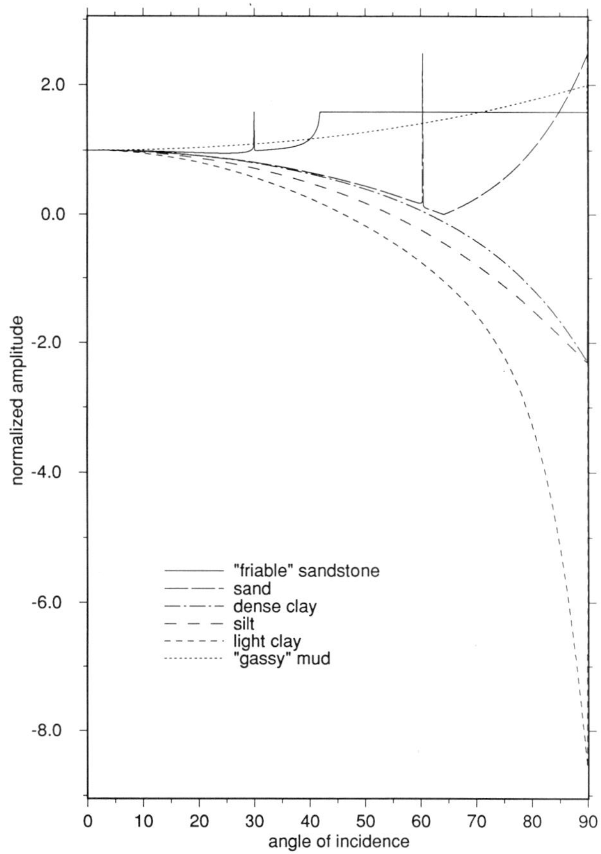


Fig.2 - "Typical" normalized sea-floor reflection coefficients

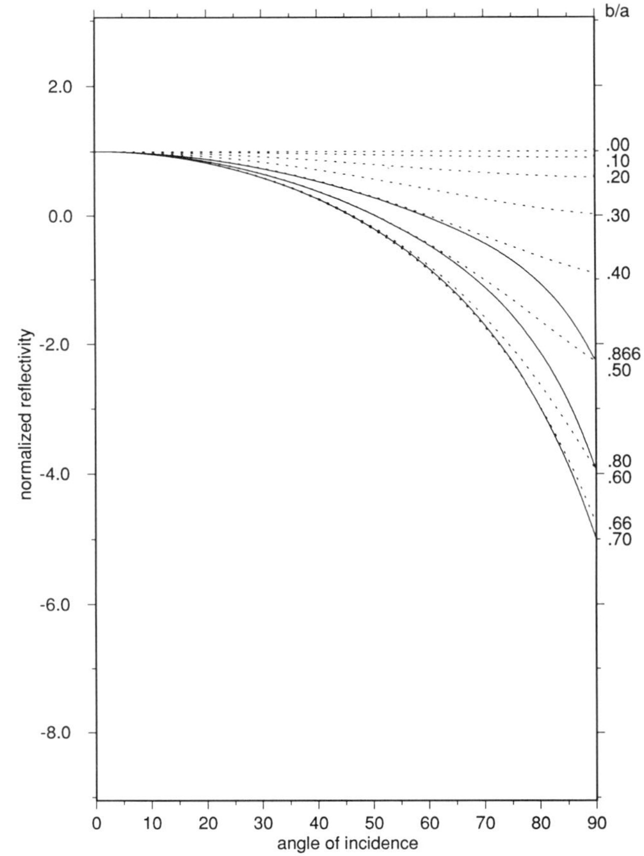


Fig.3 - Shear-speed nomogram drawn for a = 1.0 and R(0) = +.20

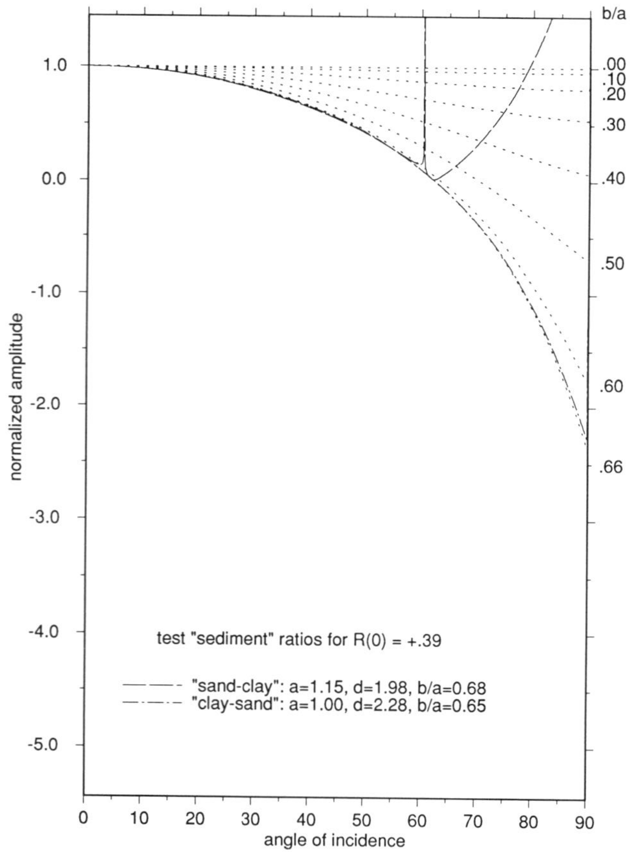


Fig.4 - Test of shear-speed nomogram for a>1

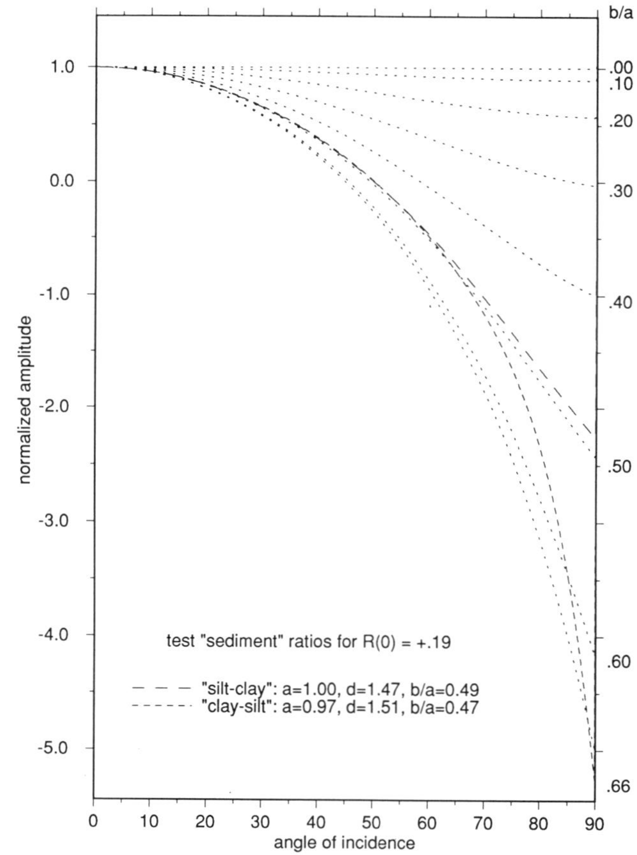


Fig.5 - Test of shear-speed nomogram for a<1

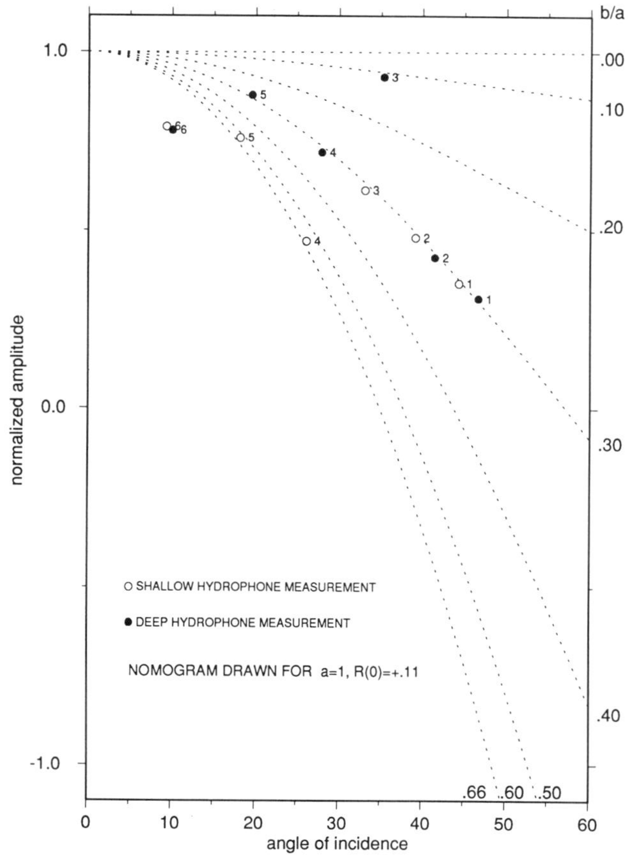


Fig.6 - Shear-speed nomogram showing results of field experiment

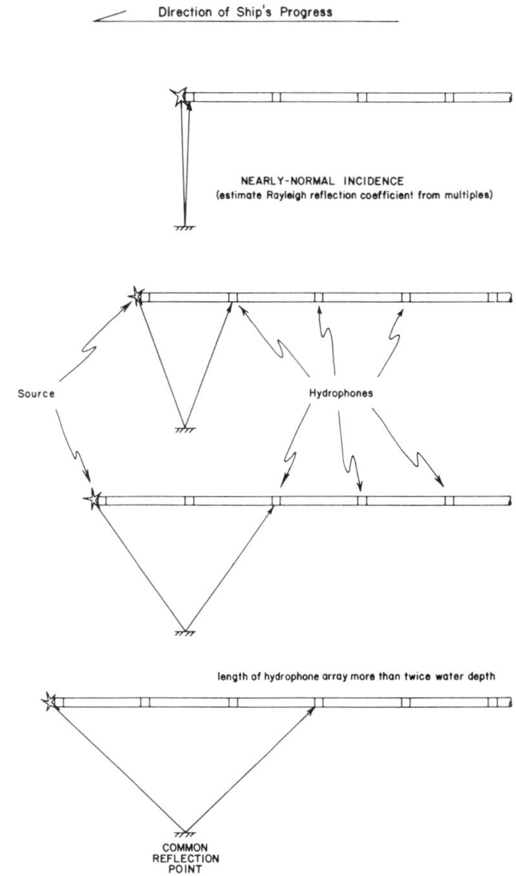


Fig.7 - "Common-reflection-point" profiling

## Experimental Results

The use of shear-speed nomograms to estimate values of $b/a$ from field data was tested in an experiment carried out in Sooke Basin on the southwest coast of Vancouver Island in British Columbia, Canada. Details of the experiment may be found in Thalassic Data Limited (1985). The acoustic source was a small airgun suspended immediately beneath the surface so that it would not create a bubble. Sea-floor reflections were recorded for angles of incidence from 0° to about 47° at two different hydrophone depths in water 22m deep. Amplitude ratios were measured by crosscorrelation. The correlation functions were nearly symmetrical, thus indicating minimal scattering loss. STD profiles verified that the speed of sound in the water column did not vary by more than 0.5% , so it was not considered necessary to correct the geometry for raypath bending in the water. Corrections were made for source strength variation and spherical wavefront divergence, however. Subbottom (3.5 kHz) profiles indicate the sea-floor sediment to be at least 3m thick and possibly containing gas. The upper half metre of sediment was sampled by divers and found to be a cohesive clayey silt such that $d = 1.25$ . Under the assumption $a = 1$ , this indicates $R(0) = .11$ . Fig.6 is a shear-speed nomogram drawn for $a = 1$ and $R(0) = .11$ and the corrected results are plotted on it. The majority of data points fall near a nomogram curve for about $b/a = .31$ which agrees well with the light clay category of Table 1.

## Conclusions

The theoretical tests indicate that shear-speed nomograms drawn for $a = 1$ and an appropriate value of $R(0)$ may be used to estimate the shear-to-compressional speed ratio in sea-floor sediments for which $b/a < .66$ . When drawing the nomogram, it is first necessary to estimate the Rayleigh reflection coefficient. This may be done by comparing the amplitudes of successive water-layer multiples at normal (or nearly-normal) incidence. The calculations are not formidable and could be done in near-to-real time.

When compared to values of $b/a$ in Table 1, the estimate demonstrated in fig.6 corresponds to the light clay category. This agrees with the soil found within the uppermost half metre at the reflection point. Thus it is concluded that the technique provides accurate estimates in at least some cases.

The estimation technique is especially well-suited to a surface vessel doing multichannel profiling in the "common-reflection-point" mode (illustrated in fig.7). One channel should be located near enough to the source that it can be considered to record normal-incidence data. The hydrophone array should be long enough that the channel furthest from the source records data at an angle incidence greater than 45° but less than 60° .

## Acknowledgements

Grateful acknowledgement is made to the Department of National Defence (DREP) of Canada for permission to use the reflection data set, to R. A. Spence Engineering Ltd. of Vancouver for analyzing the soil at the test site and to the Ministry of Defence (DGM/DWOO) of the Netherlands for financial support to present this paper.

## References

Chapman, N.R., Zelt, C.A., and Busch, A.E. (1983) Geoacoustic modelling of deep ocean abyssal plains. Acoustics and the Seabed, ed. N. G. Pace, University of Bath Press.

Ewing, W. M., Jardetsky, W. S., and Press, F. (1957) Elastic waves in layered media. McGraw-Hill.

Hamilton, E. L. (1972) Compressional-wave attenuation in marine sediments. Geophysics **37**, 620-646.

Thalassic Data Limited. (1985) Estimation of shear-wave speed in ocean-bottom sediments by comparing oblique-angle reflectivity measurements with normal incidence data. Canadian Department of National Defense No. 01SB.97708-5-5950.

# SEDIMENT $Q_\beta$ FROM SPECTRAL RATIOS OF CONVERTED SHEAR REFLECTIONS

Peter D. Bromirski, L. Neil Frazer, and Fred K. Duennebier
GG/SOEST/University of Hawaii at Manoa
2525 Correa Road
Honolulu, HI 96822

ABSTRACT. Spectral ratios of the basement-converted shear reflections PS and PSSS on refraction seismograms are used to estimate the sediment shear quality factor $Q_\beta$. The spectral ratio method is tested on a variety of synthetic seismograms computed using the constant-$Q$ dispersion relation of Kjartansson (1979). $Q_\beta$ estimates from synthetic seismograms were within 15% of the true $Q_\beta$ for all sediment models tested, both in the presence of added noise, and with the use of a complicated source function. Tests on synthetic seismograms computed from realistic sediment models, composed of alternating layers with high impedance contrast between them, indicate that apparent attenuation due to intrabed multiples is not a significant loss mechanism between 5 and 15 Hz. The method is applied to 9 Hz horizontal component airgun-OBS refraction data collected over 356 m of primarily high-porosity biosiliceous clay in 5467 m of water near 44°N, 160°E. Effective $Q_\beta$ for the sediment column was found to be 107 $\pm$ 15 ($\alpha = 0.255 \pm 0.03$ dB/$\lambda$) at 9 Hz.

## 1. Introduction

The importance of sediment shear parameters on acoustic propagation loss in the ocean has been identified by several studies (e.g. Vidmar, 1980; Hughes et al., 1990). Direct measurement of sediment shear attenuation parameters has been limited by the lack of satisfactory ocean bottom shear sources and by problems in ocean bottom seismometer (OBS) design (Sutton and Duennebier, 1987). Here the shear source problem is circumvented by analysing converted shear reflections from the top of the crust (basement) observed on OBS horizontal component refraction seismograms.

The OBS data used in this study were obtained in conjunction with the OSS IV Experiment on DSDP Leg 88 near Hole 581C. The data were recorded by an Hawaii Institute of Geophysics isolated sensor ocean bottom seismometer, OBS Y-220, from a large (30-liter) airgun towed by Soviet research vessel Dimitri Mendeleev at a depth of 20 m. Single-channel reflection data show pelagic sediments smoothly draped over ocean crust with a 2-way travel time of about 0.35 s. Holes drilled at Site 581 encountered roughly 350 m thick continuous pelagic siliceous clays with occasional chert bands becoming numerous near basement (Duennebier et al., 1987).

The $P + S$ travel time through the sediments is about 2.00 $\pm$ 0.02 s obtained from the



*J. M. Hovem et al. (eds.), Shear Waves in Marine Sediments*, 361–368.

$PS$ reflection on the horizontal data (Figure 1a). The time and distance constraints give an average sediment compressional velocity $\bar{c}_\alpha$ of about 1.7 km/s and an average shear velocity $\bar{c}_\beta$ of 0.20 $\pm$ 0.01 km/s. The identification of the $PSSS$ multiple shear reflection (Figure 1a) was the motivation for this study. The spectral ratio method was used since the $PS$ and $PSSS$ reflections are observed for the same source and receiver, and the shear paths are virtually the same. When the ratio of the amplitude spectra of these phases are taken, source and receiver effects cancel.

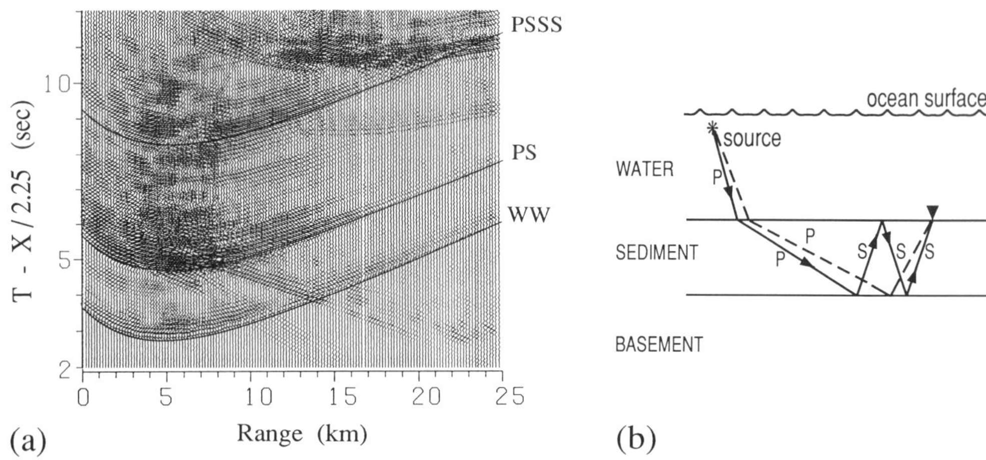


Figure 1. (a) Horizontal component OBS refraction data reduced to 2.25 km/s with travel-time curves for the water wave $WW$ and the $PS$ and $PSSS$ reflections for a 1-layer average sediment model. (b) Raypath diagram for the $PS$ (dashed) and $PSSS$ (solid) reflections.

## 2. Effective Attenuation

Attenuation measured from field data is termed effective attenuation ($\alpha_E$) as it includes contributions from both intrinsic ($\alpha_I$) and apparent ($\alpha_A$) attenuation. This can be expressed in terms of $\alpha$ or $Q$ as

$$\alpha_E = \alpha_I + \alpha_A \qquad \text{or} \qquad Q_E^{-1} = Q_I^{-1} + Q_A^{-1} \tag{1}$$

Either $Q_I$ or $Q_A$ may be frequency-dependent. In-situ measurements indicate that $Q_E$ is frequency-independent in the seismic band (Kanamori and Anderson, 1977).

Some attenuation mechanisms to account for intrinsic losses were presented in Toksöz and Johnston (1981). Intrinsic $\alpha_I$ is attributed to friction mechanisms and frame anelasticity while apparent $\alpha_A$ is the amplitude decay of a seismic pulse resulting from the scattering of energy by heterogeneities such as microbeds (O'Doherty and Anstey, 1971; Schoenberger and Levin, 1978; Menke, 1983). In apparent attenuation, energy is redistributed to other parts of the coda unlike $\alpha_I$ where energy is removed from the coda. However, its affect on spectral content is difficult to distinguish from intrinsic losses.

## 3. Spectral Ratios

The spectral ratio (SR) method uses the ratio of the amplitude spectra of different arrivals to estimate $Q$. Variations of the spectral ratio method have been used by Jacobson et al. (1981) and Jannsen et al. (1985) to determine compressional $Q_\alpha$ in marine sediments. The amplitude spectrum of a vertically traveling signal multiply reflected between $z = z_o$ and at $z = z_1$ (Figure 1b) is given by

$$|A_1(\omega)| = |A_o(\omega)G(z)R| \ \exp[-2\alpha(z_1 - z_o)]$$

where $G(z)$ is the geometrical spreading and $R$ contains reflection and transmission coefficients, assumed independent of frequency. In this case $z_o$ is the depth at the water/sediment interface and $z_1$ is the depth at the sediment/basement interface. We use a vertically incident converted shear phase at $z_o$ as our reference amplitude $A_o(\omega)$ corresponding to the $PS$ reflection (the dashed raypath in Figure 1b) while $A_1(\omega)$ corresponds to the $PSSS$ reflection. Thus, $A_o(\omega)$ includes source, receiver, and noise contributions which we assume are constant for each trace. We take the natural log of the ratio of the two spectra, and letting $2(z_1 - z_o) = T\bar{c}$ where $T$ is the travel time difference between successive multiples, $\bar{c}$ is the average shear wave velocity in the sediments, and $\alpha = \omega/2\bar{c}Q$, we have

$$\ln\left[\frac{A_1(\omega)}{A_o(\omega)}\right] = const - (\frac{T}{2}Q^{-1})\omega \tag{2}$$

in which *const* is a term independent of $\omega$ that includes contributions from spreading and and from reflection and transmission coefficients. Note that if $Q$ is independent of $\omega$, then equation (2) describes a straight line with slope $m = -\frac{T}{2}Q^{-1}$. The value of $m$ can be determined by a least squares fit to the spectral ratios. From this slope one obtains an effective $Q$ for any successive pair of reflected shear multiples.

Using Snell's Law for $c_\beta = 0.2$ km/s and $c_\alpha = 1.5$ km/s, the shear-wave incident angle at the water/sediment interface is less than 5°, resulting in a less than 1% increase in path length from vertical and negligible S to P conversion. Note that S to P conversion affects only the *const* term in (2). Also, our tests with synthetic data computed by an exact method (Mallick and Frazer, 1987) recover $Q_I$ exactly.

## 4. Numerical Tests on Synthetic Data

Since the data we are examining have a narrow bandwidth and since it has been widely observed that $Q$ depends very little upon frequency within the range of frequencies observed in seismic investigations (e.g., Strick, 1967; Kanamori and Anderson, 1977), we assume $Q$ to be independent of frequency and use the constant $Q$ dispersion relation of Kjartansson (1979) in the computation of reflectivity synthetic seismograms.

A variety of multi-layer sediment models were used to test the accuracy of the spectral ratio method. Intrinsic $Q_I$ for an $n$-layer constant-$Q$ sediment model is given by

$$Q_I = \frac{Z_T}{\bar{c}}\left[\sum_{j=1}^{n}\frac{z_j}{c_jQ_j}\right]^{-1} \qquad j = 1, n \tag{3}$$

where $Z_T$ is the total thickness, $\bar{c}$ is the average velocity for the stack of layers, and $z_j$, $c_j$, and $Q_j$ are the thickness, phase velocity, and $Q$, respectively, of the $j^{th}$ layer.

*Test parameters.* The synthetic traces were computed with a Nyquist frequency of 40 Hz matching the OBS data sampling interval of 0.0125 s. We selected 55-point wavelets to match the length of the *PS* phase coda in the OBS data. A 10% Hanning taper was applied prior to transforming to the frequency domain. Tests showed that a 256 point window padded with zeros gave the best results. In all spectra shown, zero dB is relative to 1 (digital unit)$^2$/Hz.

We smoothed the amplitude spectra before taking the spectral ratios using a $\frac{1}{4}, \frac{1}{2}, \frac{1}{4}$ smoothing function keeping the endpoints fixed. The airgun source was modeled using a modified form of the explosive source function presented in Spudich and Orcutt (1980). All tests were performed on synthetic traces at range 0.41 km to compare with data trace 3. Sediment thickness was 356 m.

TABLE 1. Model parameters used in computing the synthetic traces in Figures 2a and 2b.

| | | $c_\alpha$ | $Q_\alpha$ | $c_\beta$ | $Q_\beta$ (2a) | $Q_\beta$(2b) | $\rho$ | h(km) |
|---|---|---|---|---|---|---|---|---|
| water | | 1.5 | 10000 | 0.0 | 0 | 0 | 1.0 | 5.467 |
| sediments | L1a | 1.55 | 30 | 0.160 | 190 | 45 | 1.35 | 0.200 |
| | L1b | 1.75 | 30 | 0.295 | 225 | 70 | 1.75 | 0.156 |
| basement | | 4.25 | 300 | 2.25 | 250 | 250 | 2.65 | ___ |

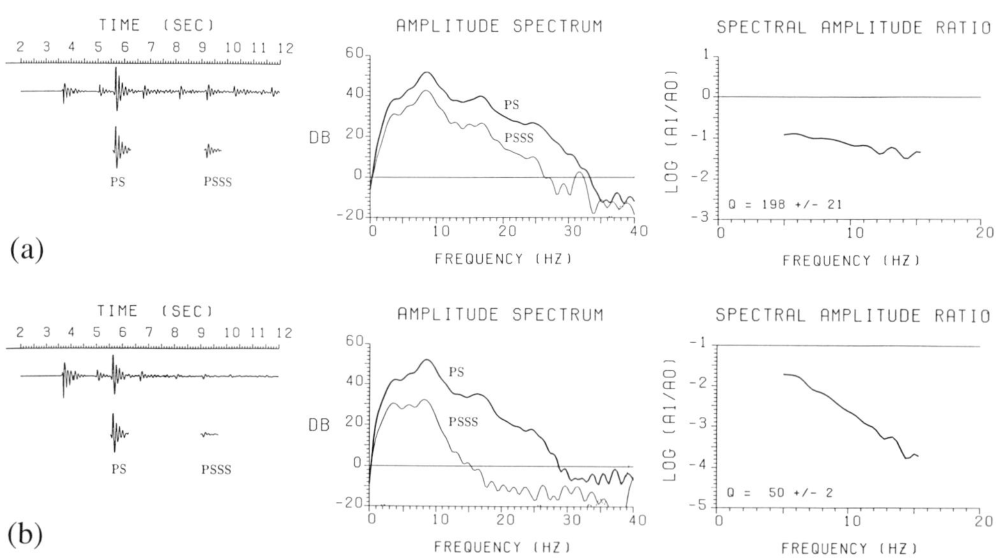


Figure 2. Synthetic traces with selected wavelets, their associated amplitude spectra and spectral ratio curves for model $Q_I$ of (a) 200 and (b) 50.

*2-layer sediment model.* Horizontal component synthetic seismograms (Figure 2a,b) were computed to test the accuracy of the method for high and low $Q_\beta$ for two 2-layer models (Table 1) that differ only in $Q_\beta$. These models give $Q_\beta \approx 200$ and $Q_\beta \approx 50$ using (3). The smoothed amplitude spectra of the wavelets selected for the SR tests along with their spectral ratios are shown in Figure 2. The $\pm$ refers to 95% confidence limits for the least squares fits to the spectral ratios, which give $Q_\beta$ estimates within 5% of the model values. The differences between the two models can readily be seen in the relative amplitudes of the *PS* and *PSSS* phases, their amplitude spectra, and the slope of the spectral ratio curves. The accuracy of these SR $Q_\beta$ estimates suggests that the implicit assumption of vertical raypaths is acceptable. These results also imply that layering and S to P conversion do not pose problems for this method.

*Apparent attenuation.* The importance of intrabed multiples in the apparent $Q_A^{-1}$ term in (1) can be estimated by making $Q_I$ large, which gives $Q_E^{-1} = Q_A^{-1}$. If apparent attenuation is small, then $Q_A$ is large and (1) reduces to $Q_E = Q_I$. Synthetic seismograms were computed for realistic models containing alternating layers of high impedance contrast with layer thicknesses of $\frac{1}{2}\ \lambda_\beta$, for an oscillating sediment velocity structure of $c_\beta = 150$ m/s and 250 m/s with a constant shear $Q_I = 150$. This model was designed to maximize the frequency of impedance contrasts while maintaining the travel-time and path-length constraints and the average $c_\beta = 200$ m/s. If apparent attenuation is significant, the measured $Q_E$ will be less than the model $Q_I$. The synthetic trace for this model is shown in Figure 3. Although the 95% confidence limits are larger than in the previous tests, the least squares fit to the spectral ratios gives a $Q_\beta$ estimate within 12% of the model value. This suggests that apparent attenuation from intrabed multiples is not significant at 9 Hz and that the $Q$ we are measuring is $Q_I$.

In addition to estimating the importance of $Q_A$, the long codas for this model also test the sensitivity of the method to interference problems due to source function length. This may explain why the 95% confidence limits are significantly larger than for the previous 2-layer sediment model tests.

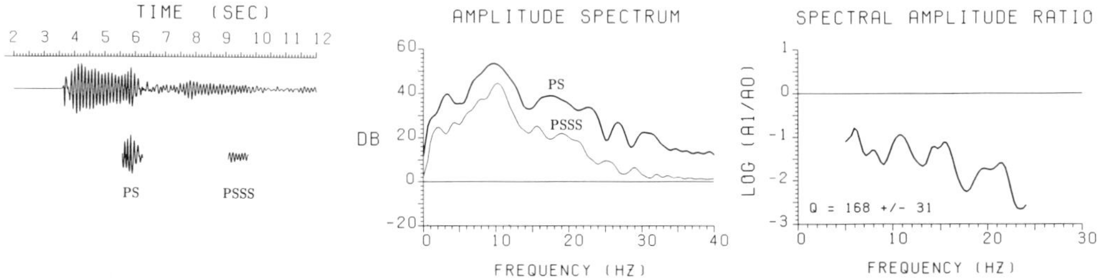


Figure 3. Synthetic trace with selected wavelets, their associated amplitude spectra and spectral ratio curve for a model with layer thickness of 1/2 a shear wavelength.

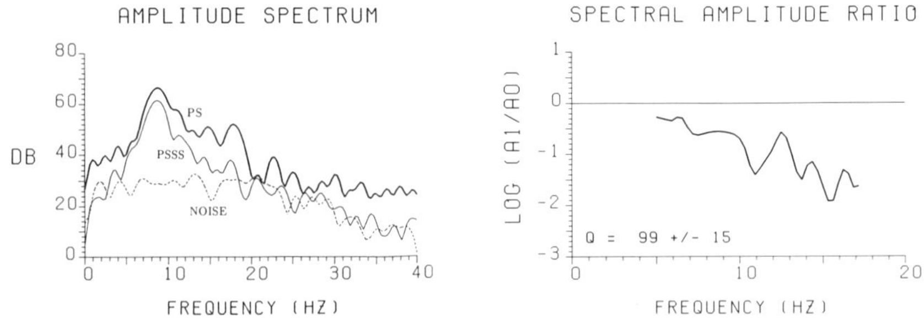


Figure 4. Amplitude spectra and spectral ratio curve for the wavelets selected in Figure 5 for data trace 3 at 0.41 km.

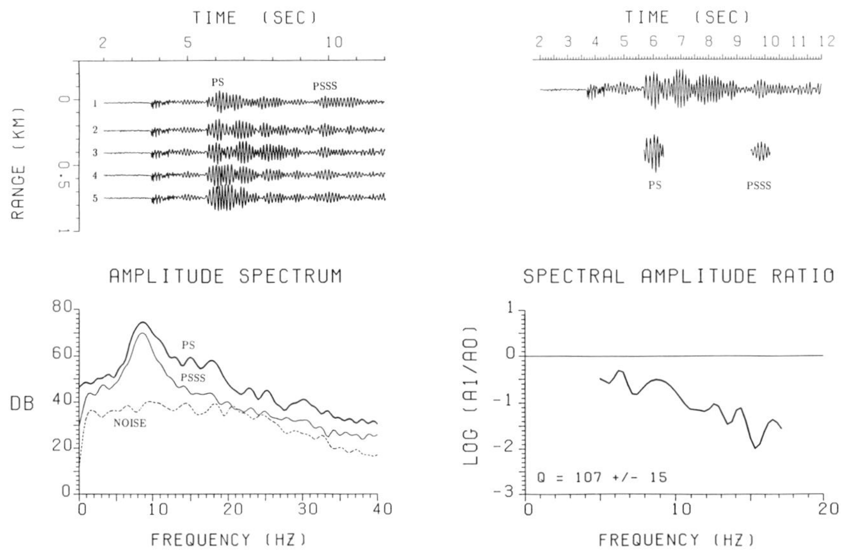


Figure 5. Lower figures are the average spectra and spectral ratio curve for the data traces 1-5 (upper left). Upper right is data trace 3 at 0.41 km with selected wavelets.

TABLE 2. Results of spectral ratio analysis of data traces 1-5.

| Range (km) | Travel Time (s) | Bandwidth (Hz) | $Q_\beta$ |
|---|---|---|---|
| 0.027 | 3.84 | [2,17] | $155 \pm 45$ |
| 0.238 | 3.92 | [3.5,17] | $134 \pm 35$ |
| 0.405 | 3.82 | [5,17] | $99 \pm 15$ |
| 0.576 | 3.82 | [2,13] | $101 \pm 32$ |
| 0.747 | 3.83 | [3,13] | $63 \pm 14$ |

## 5. Application to OBS Refraction Data

Next our method was applied to the OBS refraction data. Each wavelet window, position and length, was adjusted in order to obtain as smooth an amplitude spectrum as possible. After the mean was removed from the wavelets, the time-domain Hanning taper and smoothing functions were applied in the same way as for the synthetics. We selected the portion of the data prior to the first arrival starting at 2.0 s, with the same length as the wavelets used for the spectral ratios, as the noise estimate. We subtracted the noise power from the spectral power estimates after smoothing. If noise is included in the spectral ratios, then the $Q_\beta$ estimate will be higher than the true value.

We restricted our analysis to offsets less than 1.0 km. Since the S-rays have a near vertical path, the near traces are least contaminated by scattered energy and interference effects. The amplitude spectra of the wavelets selected for trace 3 at 0.41 km (Figure 4) show that the data have a dominant frequency of about 9 Hz and that noise levels restrict the useable bandwidth to about 5-15 Hz. A linear least squares fit to the spectral ratios was made for frequencies at which the spectral estimates for the $PS$ and $PSSS$ wavelets were at least 3 dB above the noise level. We obtained an effective $Q_\beta = 99 \pm 15$ for this trace.

Table 2 shows that, although there is considerable variation from trace to trace in the results for $Q_\beta$, the 95% confidence intervals of the first four traces overlap. To average path irregularities and possible interference effects, we took an arithmetic average of the spectral estimates for the 5 traces in Table 2 prior to obtaining the spectra and ratios seen in Figure 5. These gave $Q_\beta = 107 \pm 15$. Although it is not evident for the traces analysed, interference is probably present in these data to some degree and we suspect that interference of the crustal $S$ refraction with the $PSSS$ coda for the trace at 0.747 km is the reason this $Q_\beta$ estimate is significantly lower than the other traces. Synthetic seismograms computed for a single sediment layer using $Q_\beta = 100$ match the amplitude decay of the $PS$ and two successive shear multiples observed in the data.

## 6. Conclusions

Our spectral average $Q_\beta = 107 \pm 15$, corresponding to an attenuation of about 0.255 $\pm$ 0.03 dB/$\lambda$, is consistent with results from Scholte-wave studies (Jensen and Schmidt, 1986). Since we are assuming a minimum sediment travel path, i.e. perfectly vertical shear raypaths, and the actual path is somewhat longer, and since the reflection profiling indicates subhorizontal layering, and both interference and noise tend to steepen the slope of the spectral ratio curves, the spectral ratio method gives a lower bound for $Q_\beta$. We conclude that sediment $Q_\beta$ can be obtained to a precision of $\pm$ 15% or better in areas where there is significant P to S conversion at the sediment/basement interface and sediment thickness allows separation of shear-wave arrivals.

## 7. References

Duennebier, F.K., Lienert, B., Cessaro, R., Anderson, P., and Mallick, S. (1987) 'Controlled-source seismic experiment at Hole 581-C', *In* Duennebier, F.K., Stephen, R., Gettrust, J.F., et al., eds., *Init. Repts. DSDP, 88*, Washington (U.S. Govt. Printing Office), 105-125.

Hamilton, E.L. (1980) 'Geoacoustic modeling of the sea floor', *J. Acoust. Soc. Am., 77*, 1789-1799.

Hughes, S.J., Ellis, D.D., Chapman, D.M.F., and Staal, P.R. (1990) 'Low frequency acoustic propagation loss in shallow water over hard-rock seabeds covered by a thin layer of elastic-solid sediment', *J. Acoust. Soc. Am., 88*, 283-297.

Jannsen, D., Voss, J. and Theilen, F. (1985) 'Comparison of methods to determine Q in shallow marine sediments from vertical reflection seismograms', *Geophys. Prosp., 33*, 479-497.

Jacobson, R.S., Shor, G.G., Jr., and Dorman, L.M. (1981) 'Linear inversion of body-wave data - Part II: Attenuation versus depth using spectral ratios', *Geophysics, 46*, 152-162.

Jensen, F.B. and Schmidt, H. (1986) 'Shear properties of ocean sediments determined from numerical modeling of Scholte-wave data', *In* Akal, T. and Berkson, J.M., eds., *Ocean Seismo-Acoustics: Low Frequency Underwater Acoustics*, Plenum Press, New York, 683-692.

Kanamori, H. and Anderson, D.L. (1977) 'Importance of physical dispersion in surface wave and free oscillation problems: Review', *Rev. of Geophys. and Space Phys., 15*, 105-112.

Kjartansson, E. (1979) 'Constant Q-wave propagation and attenuation', *J. Geophys. Res., 84*, 4737-4748.

Mallick, S. and Frazer, L. N. (1987) 'Practical aspects of reflectivity modeling', *Geophysics, 52*, 1355-1364.

Menke, W. (1983) 'A formula for the apparent attenuation of acoustic waves in randomly layered media', *Geophys. J. Roy. Astr. Soc., 75*, 541-554.

O'Doherty, R.F. and Anstey, N.A. (1971) 'Reflections on amplitudes', *Geophys. Prosp., 19*, 430-458.

Schoenberger, M. and Levin, F.K. (1978) 'Apparent attenuation due to intrabed multiples, II', *Geophysics, 43*, 730-737.

Spudich, P.K.P. and Orcutt, J.A. (1980) 'Petrology and porosity of an oceanic crustal site: Results from wave form modeling of seismic refraction data', *J. Geophys. Res., 85*, 1409-1433.

Strick, E. (1967) 'The determination of Q, dynamic viscosity and transient creep curves from wave propagation measurements', *Geophys. J. Roy. Astr. Soc., 13*, 197-218.

Sutton, G.H. and Duennebier, F.K. (1987) 'Optimum design of ocean bottom seismometers', *Mar. Geophys. Res., 9*, 47-65.

Toksöz, M.N. and Johnston, D.H., eds. (1981) *Seismic Wave Attenuation*, SEG Geophysics reprint series No. 2, Tulsa.

Vidmar, P.J. (1980), 'Ray path analysis of sediment shear wave effects on bottom reflection loss', *J. Acoust, Soc. Am., 68*, 639-648.

# CONSTRAINTS ON SHEAR VELOCITIES IN DEEP-OCEAN SEDIMENTS AS DETERMINED FROM DEEP-TOW MULTICHANNEL SEISMIC DATA

JOSEPH F. GETTRUST and MARY M. ROWE
*Naval Oceanographic and Atmospheric Research Laboratory*
*Seafloor Geosciences Division, Geology/Geophysics Branch*
*Stennis Space Center, MS 39529-5004*
*USA*

ABSTRACT. A method is presented for constraining shear velocity estimates that is based on a frequency domain versus offset (or, equivalently, grazing angle). This technique emphasizes signal variability as a function of wavenumber and provides a means for estimating shear velocities when no shear phases or converted shear phases are observed. A model study demonstrates the technique, which is then applied to deep-tow multichannel data from the Bermuda Rise. These data support Poisson's Ratios of >0.48 within the upper 50 m of sediments on the rise; however, the estimated Poisson's Ratio for a turbidite stringer at ~20-m depth is ~0.46. Because the technique is compatible with marine multichannel seismic data, it provides a way to estimate the spatial variability of shear velocities at scales of ~20 m.

## Introduction

The Naval Oceanographic and Atmospheric Research Laboratory (NOARL) developed the Deep Towed Acoustic Geophysics System (DTAGS), an instrument that places both the acoustic source and the multichannel hydrophone array within several hundred meters of the seafloor in water depths up to 6 km. This geometry makes it possible to obtain detailed estimates of the structure and compressional velocities ($V_p$) within the upper 500 m of deep-ocean sediments (Gettrust et al., 1988; Bowles et al., 1990), as well as estimates of acoustic bottom loss as a function of grazing angle and frequency (Rowe and Gettrust, 1989).

With DTAGS data we can directly measure shear velocities ($V_s$), but only if significant $V_p$ to $V_s$ converted (PSP) phases occur within the sediments. With the data taken to date, PSP phases have not been observed. However, we will show that it is possible to place bounds on shear velocities within the upper sediments by examining reflectivity as a function of frequency and offset. This technique can be thought of as a frequency domain representation similar to Amplitude Versus Offset (AVO) studies now commonly used in seismic exploration. To emphasize the connection to AVO studies, we will refer to the reflectivity technique as Spectrum Versus Offset (SVO).

## Background

The SVO technique is an outgrowth of our efforts to constrain bottom loss estimates as a function of grazing angle. Often, these estimates have been made at discrete frequencies and produce attenuation curves that are impossible to relate to the environment. With broadband signals, the



*J. M. Hovem et al. (eds.), Shear Waves in Marine Sediments*, 369–378.

modulus of the frequency spectrum of the earth's response function versus grazing angle can be plotted, and that surface related to the sediment structure being sampled.

Assuming we have a known source signal, receiver transfer functions, and source-receiver geometries, we can estimate the transfer function of the subbottom as a function of frequency and grazing angle through the following:

$$\mathbf{B}(\omega,\ \phi) = \frac{\mathbf{S}(\omega)\,\mathbf{E}(\omega,\ \phi)\,\mathbf{G}(\omega,\ \phi)}{\mathbf{S}(\omega)\,\mathbf{G}(\omega,\ \phi)} \tag{1}$$

$$\mathrm{BA}(\omega,\ \phi) = 20.0\ \log_{10}(\ |\mathbf{B}(\omega,\ \phi|)\ , \tag{2}$$

where $\omega$ is frequency, $\phi$ is grazing angle, $S(\omega)$ is the frequency domain representation of the source signal, and $E(\omega, \phi)$ and $G(\omega, \phi)$ are the frequency and grazing angle representations of the earth and hydrophone groups, respectively. The earth response $BA(\omega, \phi)$ is the modulus of the band-limited transfer function of the sediments in decibels.

This approach has been shown by others (Dicus, 1976; Chapman, 1983; and Hastrup, 1986) to provide robust estimates of bottom attenuation. More important, Equation (2) allows one to relate the geoacoustic response to the earth's structure. This suggests that we may be able to use this representation to constrain estimates of $V_s$, even though we have not directly observed shear phases.

## Preliminary Model Tests

Range-independent forward modelling of seismic data requires estimates of $V_p$, $V_s$, density ($\rho$), and anelastic attenuation ($\gamma_p$ and $\gamma_s$), as a function of depth (z). As our data directly constrain only $V_p(z)$, we must estimate $V_s(z)$, $\rho(z)$, and both $\gamma_p(z)$ and $\gamma_s(z)$. Because travel paths of multichannel seismic reflection data that sample the uppermost sediments are relatively short, our model studies show that $\gamma_p(z)$ and $\gamma_s(z)$ are not dominant factors, and we use fixed estimates of those values appropriate for the geologic material sampled. Under these conditions, $\rho(z)$ and $V_s(z)$ remain as the free parameters to be fit.

If the scattering matrix is examined for seismic waves at an interface, then the transmission and reflection coefficients are functions of changes in $V_p$, $V_s$, and $\rho$. In the case of relatively small changes between two media, variation in $V_s$ usually is more important than variation in $\rho$ (Aki and Richards, 1980). As the SVO is composed of the scattering components of the sediment column sampled, we can expect that it will be more sensitive to changes in $V_s(z)$ than to changes in $\rho(z)$.

Because Poisson's Ratio ($\sigma$) often is used to relate $V_p$ to $V_s$ ratios and is independent of $\rho$, it provides a convenient basis for our numerical modelling. All computations were done with SAFARI (Schmidt, 1984). All reflectivity model runs were made using 181 frequencies and 97 grazing angles (covering the range 275 to 525 Hz and 45° to 90° grazing angle). Pulse code runs were made using 1024 wavenumbers and 4096 frequencies (250 to 650 Hz).

During this study, we computed SVOs for a simple model (Figure 1, Table 1) where we varied either $\sigma$ or $\rho$ within layer 3 (Table 2). We varied $\sigma$ within that layer (Table 2) over a wide range (0.25 to 0.5); for the purposes of this paper, we present results for cases where $\sigma$ within that layer varies from 0.49 to 0.40 (Figures 2 and 3). All SVO surfaces are dominated by interference

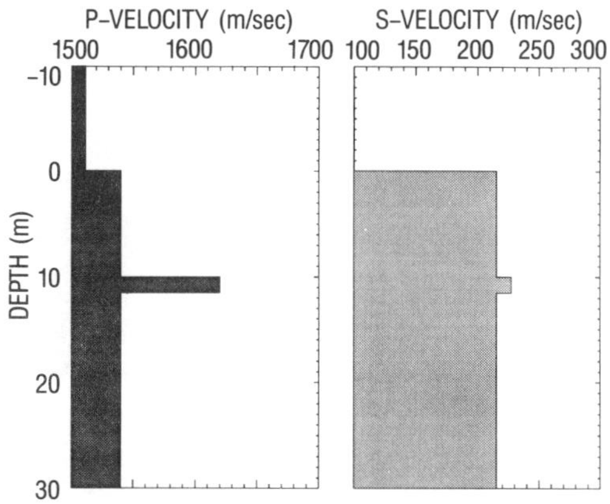


*Figure 1. The velocity-depth functions used to compute (SVO) functions presented in Figures 2, 3, and 4.*

TABLE 1. Parameters for simple test model.

| Depth (m) | $V_p$ (m/s) | $V_s$ (m/s) | $\gamma_p$ (dB/λ) | $\gamma_s$ (dB/λ) | Density (g/cm³) |
|---|---|---|---|---|---|
| 0 | 1511 | 0 | 0 | 0 | 1.00 |
| 0 | 1540 | 215 | 0.2 | 0.5 | 1.65 |
| 10 | 1620 | 227 | 0.2 | 0.5 | 2.00 |
| 11.5 | 1540 | 215 | 0.2 | 0.5 | 1.65 |

TABLE 2. Variation in $V_s$ and density of layer 3, Figure 1.

| Model Number | $V_p$ (m/sec) | $V_s$ (m/sec) | Poisson's Ratio | Density (g/cm³) |
|---|---|---|---|---|
| 1 | 1620 | 227 | 0.49 | 2.0 |
| 2 | " | 385 | 0.47 | " |
| 3 | " | 488 | 0.45 | " |
| 4 | " | 661 | 0.40 | " |
| 5 | " | 227 | 0.49 | 2.1 |
| 6 | " | " | " | 2.3 |

bands, where the frequency difference (~80Hz) between primary interference bands is related to travel times within the first elastic layer by $\frac{1}{\Delta f} = T$, where T = the two-way travel time within that layer.

In the noise-free seismic traces generated with these models (Figures 2 and 3), some differences in the amplitude of reflections from layer 3 are evident; however, we suggest that it would be much more difficult to interpret these data in conventional seismic sections than in their companion SVOs.

With $V_p$ fixed, examination of the SVO representation for $\sigma = 0.40$ (Figure 3) provides the quickest insight into an important source of the variability observed in $(\omega, \phi)$ space. Clearly,

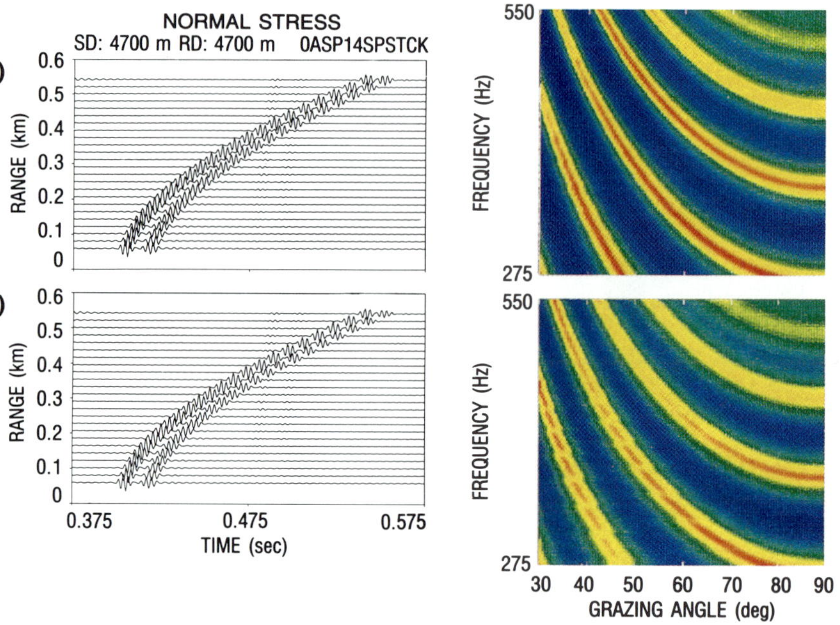


*Figure 2. (a) SVOs and seismograms used as a base model ($\sigma = 0.49$ for layer 3). (b) Results for the same base model but with $\sigma = 0.47$ in layer 3.*

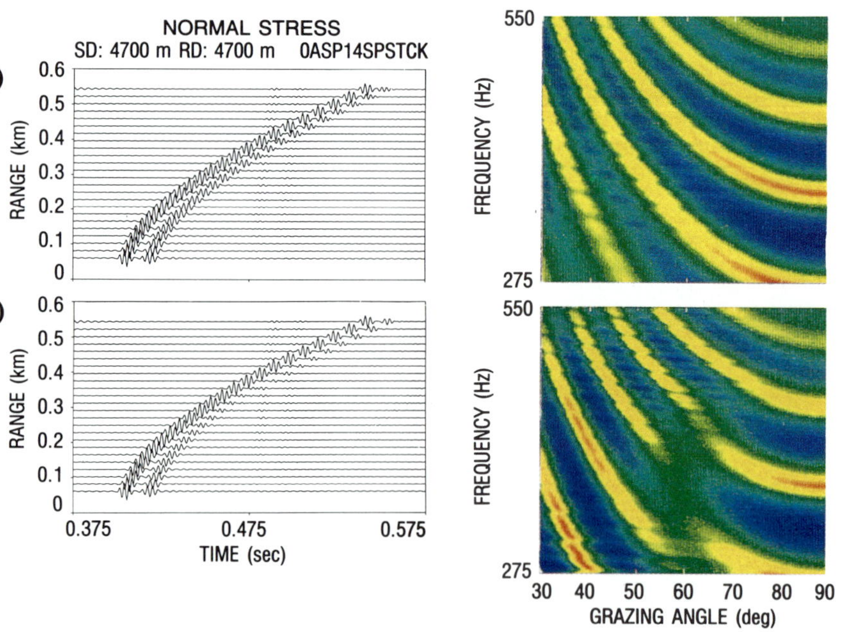


*Figure 3. Same as Figures 2(a) and 2(b), except the values of $\sigma = 0.45$ (a) and 0.40 (b).*

perturbation of interference bands due to increased $V_s$ in layer 3 (Table 2) is quite dramatic for $\sigma = 0.40$. The companion seismograms also show differences from those presented in Figures 2(a) and 2(b); however, the SVO representations provide significant additional information. Note the discontinuity in bands at ~70° grazing angle. This discontinuity occurs at the critical grazing angle (71°) for shear waves at the layer 2–layer 3 interface (see Table 2). This discontinuity is not as evident for larger values of $\sigma$ because smaller changes in $V_s$ across that interface move the critical angle to smaller grazing angles, while the decreased acoustic impedance mismatch at that interface reduces the amplitude of reflected shear phases.

If we fix $\sigma$ and vary only $\rho$ (Figure 4), band cutoffs of the type discussed above are not generated in the grazing angle range sampled (the critical grazing angle for $\sigma = 0.49$ is ~13°). Note that increasing the density by these factors does not produce the same perturbations in interference bands in the SVO surfaces. SVO plots for cases with decreasing density (not shown) produce similar results, although the mean value of the surface can change significantly.

We tested both increases and decreases in $\rho$ of up to 20%, and observed no band cutoff such as that noted above. Moreover, the change in the overall amplitude levels in $(\omega, \phi)$ space for large changes in $\rho$ provide further constraints on interpretation. This suggests that with the SVO representation, we have a means to discriminate (to some extent) between changes in $V_s$ and changes in $\rho$.

## Estimates Using DTAGS Observations

DTAGS data presented here were obtained in 1984 from the Bermuda Rise region where the system was towed at 500 to 800 m above the seafloor; the nominal water depth in the area is ~5 km. The source bandwidth is 250 to 650 Hz. The hydrophone array consists of 24 independent groups with a nominal group separation of 42 m. The maximum source–receiver offset is 1011.0 m (see Gettrust et al., 1988, for further details).

With our deep-tow configuration, the direct water-wave arrivals are not contaminated with surface or seafloor-reflected signals, and attenuation through the water is <0.01 dB at 525 Hz. Because the source is isotropic (±0.5 dB) the bottom interacting signal is the convolution of the source signal (estimated from the direct water-wave) with the sediment response and the (known) hydrophone group response. Therefore, after correcting for spherical spreading, we can estimate the bottom response function using Equations (1) and (2).

The geometry of the array and its elevation above the bottom constrained grazing angles to between 50° and 80°. Critical grazing angles for the sediment sound speeds observed in the data (Gettrust et al., 1988) are less than 40°. Grazing angles were determined from the (fixed) source–receiver offset and the elevation of the array above the bottom. Grazing angles for data presented here range from 50° to 86°. The maximum tilt of the array during the experiment was <11° from horizontal. Effects of the array tilt on the calculated attenuation values were determined to be on the order of $10^{-3}$ dB and, therefore, could be neglected. Computation of the hydrophone group response shows a smooth decrease in response of less than 0.2 dB over the frequency band and grazing angles sampled.

Here, we present a sample of these data that is based on five adjacent shots (each separated by ~15 m) that are averaged to obtain a smoothed SVO estimate:

$$\overline{\mathrm{BA}}(\omega, \phi) = \frac{1}{5}\sum_{i=1}^{5} \mathrm{BA}_i(\omega, \phi) \tag{3}$$

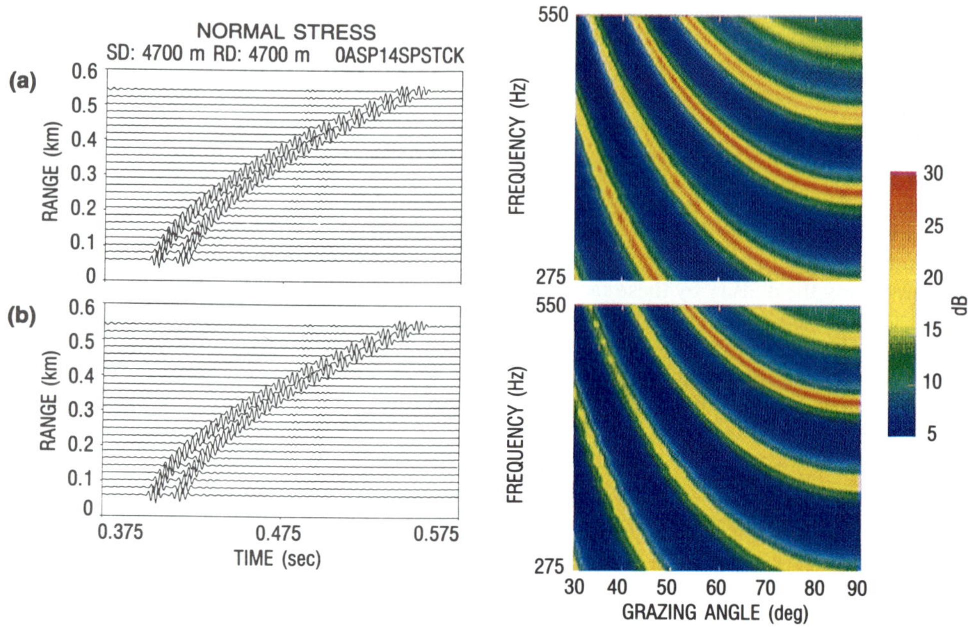


*Figure 4. Seismograms and SVOs using the same velocity-depth functions as the Figure 2(a) data, except the density in layer 3 has been increased by 5% (a) and 15% (b).*

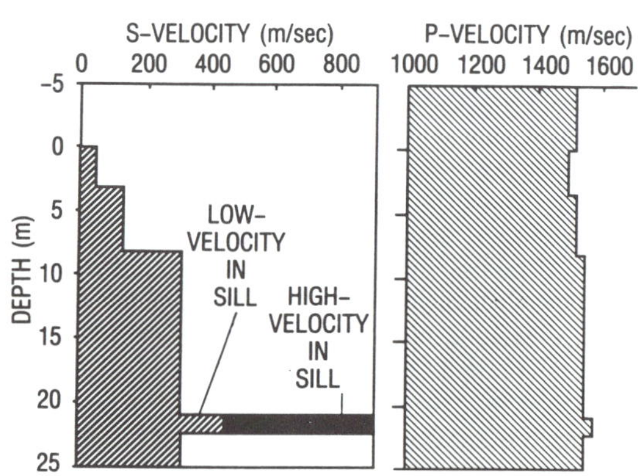


*Figure 5. The velocity-depth function for $V_p$ determined using conventional multichannel seismic techniques.*

Even though the data presented are from a "simple" area along the Bermuda Rise profiles, we know that lateral variability over the region sampled is significant at the source frequencies used (Gettrust et al., 1988). This problem is exacerbated by the fact that our navigation limits us to common shot point data; this limitation increases the variability within our sample, but does not preclude us from gaining significant insight into $V_s(z)$ in the uppermost sediments.

Figure 5 and Table 3 present the $V_p(z)$ function derived from the DTAGS multichannel data in the area where data used to generate the SVO presented in Figure 6 were taken. The velocity-depth function for $V_s$ was estimated by fitting numerical SVO modelling results (Figures 7 through 9) to an SVO plot for five contiguous shots (Figure 6) from a profile on the Bermuda Rise.

The $V_p(z)$ data are used to constrain our interpretation. The model results (see Table 3) that fit these averaged data (Equation 3) reasonably well are presented in Figure 7. The value of $\sigma$ used for the computation of the SVO presented in Figure 7 is ~0.48 for material above and below the thin sill at a ~21-m depth ($\sigma = 0.46$ within the sill). While lateral variability of elastic parameters within the sediments limit the accuracy of our fit, increasing our shear velocity estimates degrades the fit to actual data.

To demonstrate that $\gamma_p(z)$ and $\gamma_s(z)$ are not significant issues in this case, Figure 8 presents modelling results for the same velocity-depth function used to generate the SVO shown in Figure 7; however, in this case both $\gamma_p$ and $\gamma_s$ are identically zero. The only difference between the results presented here and those presented in Figure 7 is the smoothing afforded by finite

TABLE 3. Velocity-depth model for Bermuda Rise data.

| Depth (m) | $V_p$ (m/s) | $V_s$ (m/s) | $\gamma_p$ (dB/λ) | $\gamma_s$ (dB/λ) | Density (g/cm$^3$) |
|---|---|---|---|---|---|
| 0 | 1535 | 0 | 0 | 0 | 1.0 |
| 0 | 1510 | 50 | 0.2 | 0.8 | 1.1 |
| 3.1 | 1540 | 100 | 0.2 | 0.6 | 1.55 |
| 8.14 | 1560 | 300 | 0.2 | 0.6 | 1.65 |
| 21.3 | 1580 | 450 | 0.2 | 0.5 | 1.75 |
| 22.3 | 1560 | 300 | 0.2 | 0.5 | 1.65 |

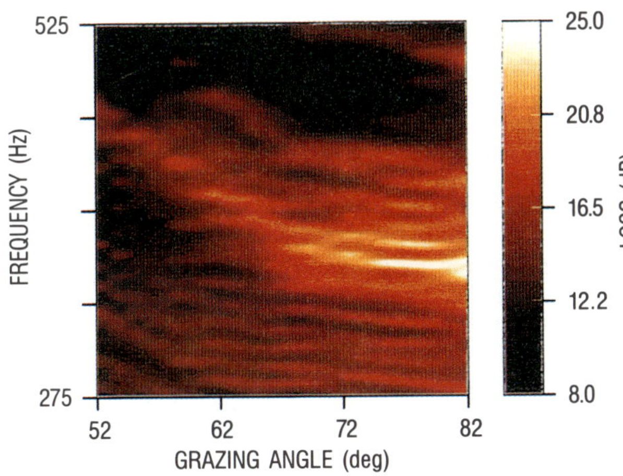


*Figure 6. SVO for five contiguous shots from a DTAGS profile on the Bermuda Rise. Note termination of interference band at 57° (480 Hz) which we relate to $V_s$ in the thin sill at 22 m depth (Figure 5).*

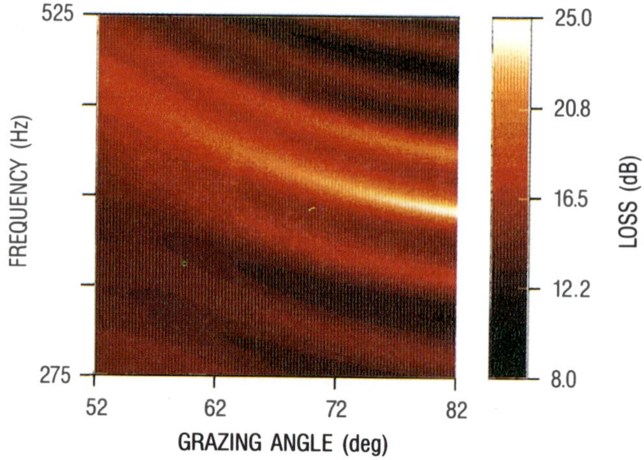


*Figure 7. "Best fit" SVO for these data (see Figure 6). The depth to reflection horizons and estimates of $V_p$ are fixed by the observations.*

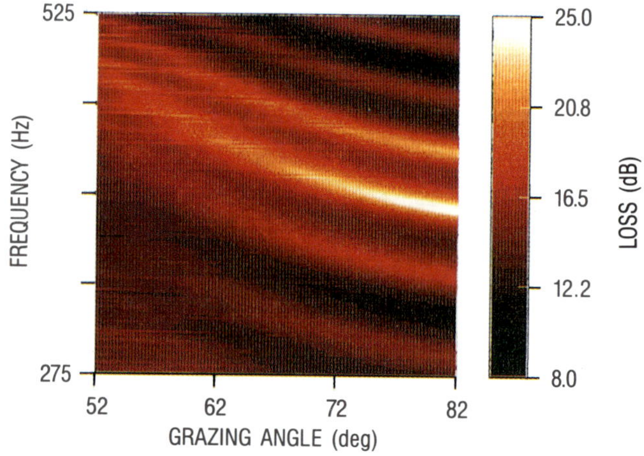


*Figure 8. SVO used to generate Figure 7 except $\gamma_p$ and $\gamma_s$ are set to zero.*

anelastic attenuation. Comparison of the data presented in Figures 7 and 8 clearly shows that the effect of finite attenuation is simply to smooth the $(\omega, \phi)$ surface.

The effect of increasing our estimate of $V_s$ in the thin sill (Figure 5) is shown in Figures 9a and 9b. In these cases, the value of $\sigma$ within the sill has been decreased to 0.40 (Figure 9b) and 0.25 (Figure 9b). Neither SVO presented in this figure fit the observations as well as that presented in Figure 7. A comparison of the SVOs for these "hard" sill cases with those for the target data (Figure 6) and "best fit" model (Figure 7) show that it is possible to constrain estimates of $V_s$ even within thin layers.

With $\sigma$ close to its upper limit of 0.5 throughout most of our "best fit" model, it also is interesting to determine how well an acoustic model fits these observations. The results for an acoustic model ($V_s$ identically zero) are shown in Figure 10. The data shown here demonstrate that $V_s$ must be considered for geoacoustic predictions even in the frequency range 275 Hz to 525 Hz. A comparison of results for the acoustic model with both the target data and models with finite $V_s$ values (Figures 6 and 7) shows that the results for the acoustic model differ dramatically from reality; clearly, acoustic models of the subbottom are not appropriate for estimating bottom loss in the low-frequency regime.

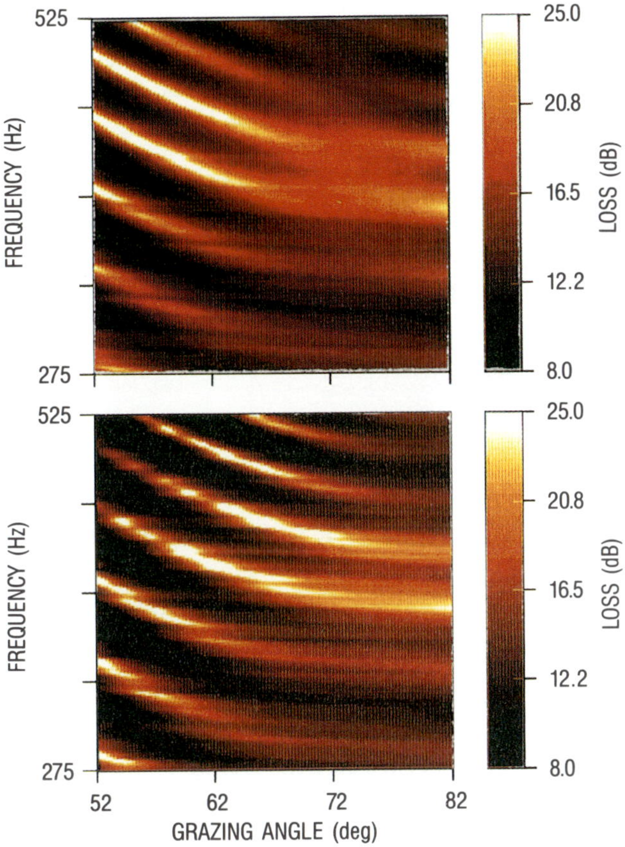


*Figure 9. SVO used to generate Figure 7, except that shear velocity in the thin sill was increased such that σ for that layer was decreased to 0.40 (a) and 0.25 (b).*

**Summary**

Interference bands can dominate scattering functions, even when scattering is from (relatively) thinly layered sills with thicknesses less than a wavelength. This banding was predicted by Spencer et al. (1977) for sedimentary sections that contained layers too thin to be resolved by the seismic section. Hastrup (1986) noted that where thin layers have a half-wavelength thickness the seismic waves are reflected incoherently, giving rise to a band of high attenuation across all grazing angles for the appropriate band of frequencies. Layers that have a quarter-wavelength thickness can give rise to anomalously low attenuation (Gilbert, 1980; Tyce et al., 1980; Hastrup, 1986).

In the work reported here, we examined the sensitivity of these interference bands to changes in $V_s$ and showed that it is possible to use this (SVO) representation to constrain estimates of $V_s$, even in relatively thin layers. Although this technique does not provide the resolution of, say, Scholte wave studies, it does allow us to constrain estimates of $V_s$ at any point along a profile. Given the lateral variability of $V_p$ that is found in deep-ocean sediments (Gettrust et al., 1988; Rowe and Gettrust, 1990) such techniques as SVO, with which the spatial variability of $V_s$ can be estimated, will significantly enhance our understanding of geologic processes (and resultant geoacoustic properties) in deep-ocean regions.

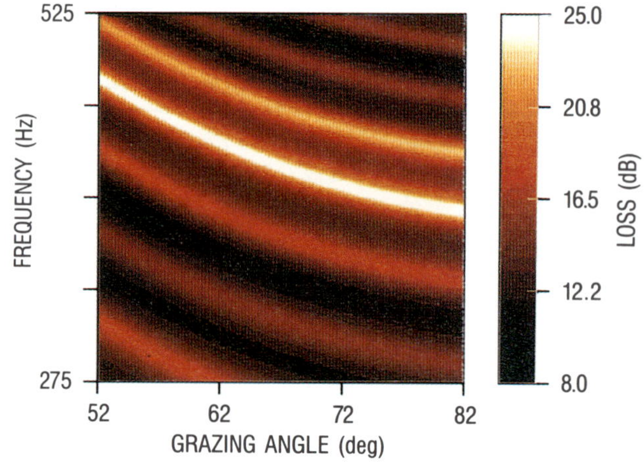


*Figure 10. SVO for an acoustic model equivalent to that used to generate our "best fit" (Figure 7) except $V_s$ is identically zero for all layers.*

**Acknowledgments**

This work was funded by NOARL Program Element 62435N, Dr. E. R. Franchi, Program Manager. NOARL contribution PR90:005:362.

**References**

Aki, K. and P. G. Richards (1980) 'Quantitative Seismology: Theory and Methods', v. 1, W. H. Freeman and Company, San Francisco, p. 154.

Bowles, F. A., J. F. Gettrust, and M. Rowe (1991) 'Geologic interpretations based on deep-tow single channel and multichannel seismic data from the Bermuda Rise', Marine Geology 96, 279-293.

Chapman, N. R. (1983) 'Modeling ocean-bottom reflection loss measurements with the plane-wave reflection coefficient', Journal of the Acoustical Society of America 73, 1601-1607.

Dicus, R. L. (1976) 'Preliminary investigations of the ocean bottom impulse response at low frequencies', US Naval Oceanographic Office, Stennis Space Center, MS, Technical Note 6130-4-76.

Gettrust, J. F., Grimm, M., Madosik, S., and Rowe, M. (1988) 'Results of a deep-two multichannel survey on the Bermuda Rise', Geophysical Research Letters 15, 1413-1416.

Gilbert, K. E. (1980) 'Reflection of sound from a randomly layered ocean bottom', Journal of the Acoustical Society of America 68, 1454-1458.

Hastrup, O. F. (1986) 'Acoustic bottom reflectivity', Naval Underwater Systems Center, New London, CT, NUSC Technical Document 8129.

Rowe, M. M. and Gettrust, J. F. (1990) 'DTAGS deep towed seismic survey of the Blake-Bahama Basin', Naval Ocean Research and Development Activity, Stennis Space Center, MS, NORDA Technical Note 449.

Schmidt, H. (1984) 'Modelling of pulse propagation in layered media using a new fast field program', in L. B. Felsen (eds.), Hybrid Formulation of Wave Propagation and Scattering, Martinus Nijhoff Publishers, Dordrect, The Netherlands.

Spencer, T. W., Edwards, C. M., and Sonnad, J. R. (1977) 'Seismic wave attenuation in nonresolvable cyclic stratification', Geophysics 42, 939-949.

Tyce, R. C., Mayer, L. A., and Spiess, F. N. (1980) 'Near-bottom seismic profiling: high lateral variability, anomalous amplitudes, and estimates of attenuation', Journal of the Acoustical Society of America 68, 1391-1402.

# CHANGES IN P AND S VELOCITIES CAUSED BY SUBDUCTION RELATED SEDIMENT ACCRETION OFF WASHINGTON/OREGON

B.T.R. LEWIS
University of Washington, WB-10
Seattle, WA, 98195, USA

ABSTRACT. Based on the results of a seismic refraction experiment at the deformation front of the Cascadia subduction zone it appears that the accreted uplifted sediments in the deformation front (with the exception of the top few tens of meters) may well have a lower velocity (higher porosity) than the original undeformed Cascadia Basin sediments. This could be the result of tectonically induced folding causing a disruption (microfracturing ?) of the sediments which in turn leads to a reduction of rigidity, an increase in permeability and porosity, and a decrease in seismic velocity.

## 1. Introduction and Objectives

The initial motivation for this work was derived from a study reported by Carson (1977). In this study Carson used porosities from cores (up to several meters long) and dredged rocks from the Washington margin to suggest that there was an apparent decrease in porosity of up to 50 % as one proceeded from Cascadia Basin across the accretionary prism. Subsequent work, based largely on ALVIN dives and reported by Kulm et al (1986), Moore et al (1990), Carson et al (1990), Lewis and Cochrane(1990), has supported the concept that dewatering accompanies subduction related accretion by identifying seafloor seeps of fluids from the sub-seafloor. Because seismic velocity is sensitive to porosity (Hamilton, 1978) it is reasonable to test this hypothesis by using standard marine seismic methods to measure insitu velocities. Bray and Karig (1985), Davis et al (1990), and Bangs et al (1990) have used velocities from Multichannel Seismic data (MCS) to infer porosity reduction across the Nankai, Cascadia and Barbados subduction zones. An early application of seismic methods using Ocean Bottom Seismometers (OBS) to the Cascadia subduction zone (Cochrane et al (1989)) was performed without the benefit of accurate multi-beam bathymetry, resulting in data that could not be related to specific geologic features and lacked the detail needed for definitive conclusions about dewatering. In 1987, with the benefit of accurate multi-beam bathymetric maps, we undertook another OBS seismic refraction experiment on the geologic features which had been studied in detail with ALVIN. The results of this experiment,together with data from a shear wave experiment, are reported here.

## 2. Experiment Location and Methods

Two experiments are described. In the first OBS's were deployed in the zone of initial deformation off Oregon and P wave velocities were measured and analyzed for indications of dewatering of accreted sediments. In the second we tested a specially designed acoustically telemetering OBS that was used with ALVIN to measure Stonely waves, and hence to infer the shear speed of the upper tens of meters of sediment.



*J. M. Hovem et al. (eds.), Shear Waves in Marine Sediments*, 379–386.

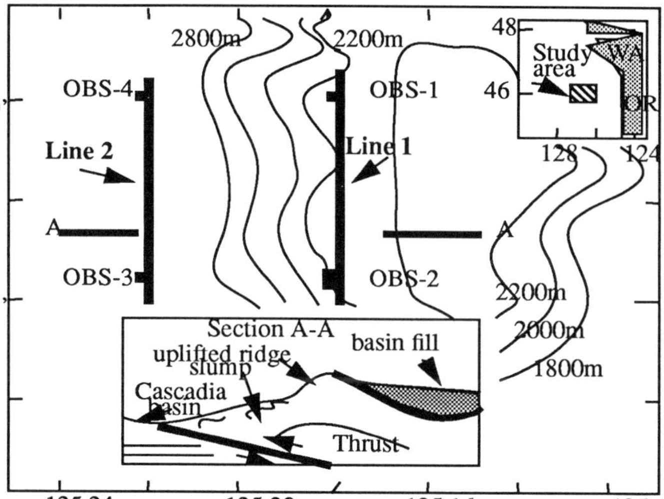


Figure 1. Location of the study area and refraction lines on the uplifted ridge (line 1) and on Cascadia Basin (line 2).

## 2.1. P WAVES

The objective of this experiment was to measure and compare P wave velocities on the undeformed Cascadia basin and on the first accreted ridge. A site off Oregon was selected on the basis of the considerable detailed data available from ALVIN dives and the logistic support available from Atlantis ll between ALVIN dives.

The location of the site and a geologic cross section are shown in figure 1. Uplift of at least 900 meters has occurred along a landward dipping thrust fault and it is notable that the primary evidence for dewatering is found at the top of the uplifted ridge, (Lewis and Cochrane, 1990, Moore et al, 1990). In this experiment we placed two University of Washington OBS's (Lister and Lewis, 1977 ) at the top of the ridge and two on Cascadia Basin, forming a rectangle, (see figure 1). The OBS locations were carefully selected so that the strike lines between OBS's would be at a constant depth, thereby removing effects of changing water depth on travel times. The source for seismic waves was a 300 cu in air gun fired at 10 sec intervals with the ship steaming at about 5 knots along the lines connecting the OBS's. During the air gun firing the SEA-BEAM system was running and this allowed a verification of the shot positions and depth variations along the shot lines. All the OBS's functioned properly. After the data were digitized it was found that the OBS's on each of the strike lines (lines 1 and 2 in figure 1) had essentially the same travel times and I then concentrated on analyzing the best data from each line, which was from OBS's 1 and 4.

## 2.2 S WAVES

Shear wave velocities reported here come from a test of a system which was designed to measure shear velocity in the upper few tens of meters of sediment using Stonely wave propagation and to be deployable from ALVIN. The system consisted of an OBS which was deployed by ALVIN and acoustically telemetered data from a vertical seismometer to ALVIN, together with an ALVIN deployable weight-drop source and a hydrophone to detect the sound of the weight drop. The operational concept was to deploy the OBS and then activate the source at several distances up to a about 100 meters from the OBS, with data being recorded in ALVIN on a tape recorder. The principal difficulty with realizing this concept was with navigation. The sonar on ALVIN was not usually able to locate the OBS and as a result we spent most of the time trying to find the OBS after it had been deployed. A successful test of the system was made on the Washington Margin in a geologically similar location to the uplifted ridge in figure 1. However we did not have SEA-BEAM coverage at this site and it is likely that the test was conducted in a slope- basin sediment pond and not on older uplifted sediments.

## 3. Results

### 3.1. P WAVES

Record sections of the vertical data from OBS-1 (uplifted ridge) and OBS-4 (Cascadia Basin) are shown in figure 2. The distance scale was determined from the water wave travel times and the

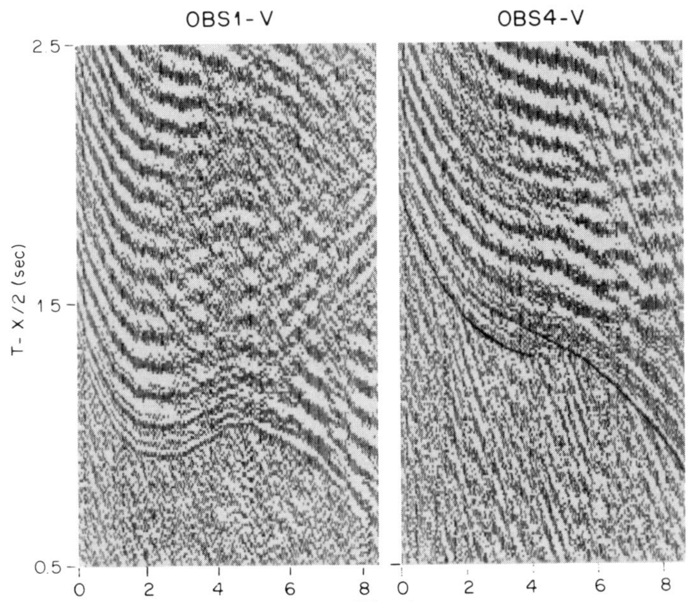


**Figure 2. Vertical component data from OBS-1 (uplifted ridge) and OBS-2 (Cascadia Basin)**

depth of the OBS. We derive from the data estimates of the velocity-depth function below lines 1 and 2. The data clearly show the cross- over from the water-wave to the refracted wave at a distance of about 4 km as well as in P-S conversions on the horizontal data (which are not shown here). To obtain a velocity-depth estimate a two-step process was used. In the first step I used a trial and error method in which the travel times of the data were matched to P-wave travel times (including triplications) computed for a velocity-depth model with a ray-tracing algorithm. In the second step tau-p values for the model were compared to tau-p bounds computed from the vertical component data by slant stacking (see Kappus et al, 1990 and Dorman 1979). After verifying that the model tau-p values lay within the data tau-p bounds, the tau-p bounds were extrapolated to the sea-floor using the model as the center of the bounds. Extrapolation was required because the data tau-p values are contaminated by the larger amplitude water-waves when the refracted and water-wave travel times are close together. Equipped with complete estimates of the tau-p bounds I then transformed the upper and lower tau-p bounds into bounds on the velocity-depth function with an inversion algorithm. The results of this two step process are shown in figures 3 and 4.

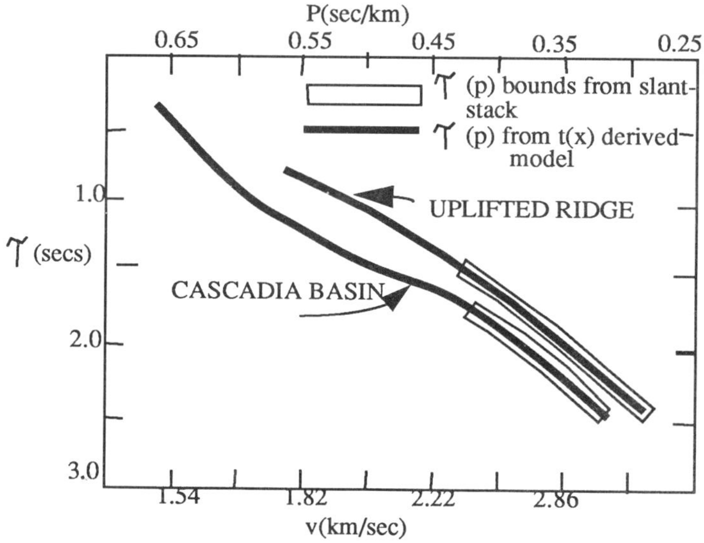


Figure 3. $\tau$ (p) bounds as measured from slant-stacked t(x) data, together with $\tau$(p) for a velocity model determined from the t(x) data.

There are two principal results from the velocity-depth solutions: a. The velocity of the upper 100 meters of Cascadia Basin is lower than that of the uplifted ridge. b. Below about 100 meters the uplifted ridge has a uniformly lower velocity than Cascadia basin, which is contrary to our starting hypothesis. That the uppermost sediments of Cascadia Basin have a velocity slightly lower than sea-water, and a high gradient below this, is evident from the data in figure 2. The possible causes of these results is taken up in the discussion section.

### 3.2. S WAVES

Stonely waves were recorded on the uplifted ridge on the Washington margin with the telemetering OBS and a weight-drop source. Two parameters in these data are poorly constrained.

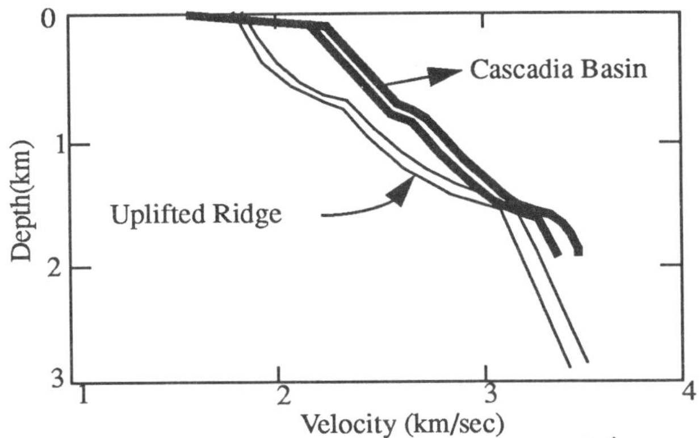


Figure 4. Velocity-depth bounds obtained from $\tau$ (p) bounds.

One is location of the test with respect to structure; there was no detailed bathymetry and the site could well have been over a slope -basin sediment pond. The other is the source to receiver distance. Distance was estimated from sonar targets but they may not be reliable. Consequently the absolute shear velocities are approximate and the geologic implications are not robust. In spite of these limitations a dispersed wave train from a weight-drop at an estimated distance of 20 meters was observed and group velocities measured from these data are shown in figure 5, together with a theoretical curve for the model shown in this figure. The shear velocities are very low at the sediment-water interface (50 m/s) and increase rapidly with depth, consistent with other sea-floor measurements ( Davies 1965, Tuthill et al 1981, Schreiner and Dorman, 1990).

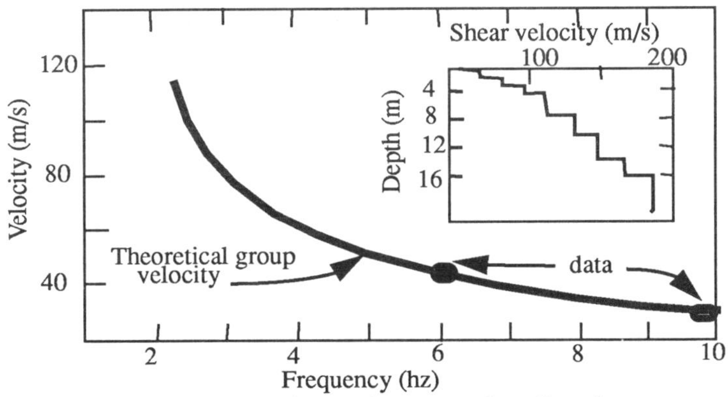


Figure 5. Shear velocity estimates from a sea-floor Stonely wave experiment.

## 4. Discussion

As stated in the introduction an original motivation for this work was the observation that samples of Cascadia Basin sea-floor had a much larger porosity than samples of the accreted and uplifted sediments This observation is clearly supported by the measured seismic velocities of the uppermost tens of meters, with the Cascadia values of 1.5 km/s being significantly lower than the ridge value of 1.8 km/s. This result is not that surprising because we are comparing very recently deposited turbidites (which have not yet reached an equilibrium compaction state) with much older sediments (<300 kyr, Kulm and Fowler 1974) What is surprising is that at depths greater than about 100 meters the Cascadia sediments have a higher velocity than the uplifted ridge sediments. If one interprets this result only in terms of porosity changes it implies that the older uplifted sediments have a higher porosity than those on Cascadia Basin, which is a possible, but counter- intuitive, conclusion.

In discussing alternative mechanisms that could cause these changes in velocity it is useful to recall that $Vp = ((k+4\mu/3)/\rho)1/2$, and $Vs = (\mu/\rho)1/2$, where k=bulk modulus, $\mu$=rigidity modulus, $\rho$=density, Vp=compressional velocity, Vs=shear velocity, and Vp, Vs are functions of pressure and temperature. It is noted that the very large gradient in shear velocity at the sea- floor must be due to the acquisition of rigidity as the sediments become consolidated and cemented and that of all the physical parameters $\mu$ will have the largest change with depth. Porosity affects the density $\rho$, but we note that the effect on velocity must be mostly through the elastic moduli, since increasing porosity causes decreasing density and this would cause an increase in velocity if k and $\mu$ remained constant. Empirical data (Hamilton 1978) shows that increasing porosity in fact causes a decrease in velocity, implying that the elastic moduli are more sensitive to porosity changes than density. k is mostly a function of the components of the sediment, and not how they are connected, whereas $\mu$ is a sensitive function of how components are connected. This line of reasoning can be used to generate a hypothetical shear velocity distribution as follows. If we assume that the density and bulk modulus vary a relatively small amount with depth (see ODP leg 110, 1988 for justifiction) and we use Vp and the density $\rho$ of the of the unconsolidated sediments at the sea-floor (where Vs ~ zero) as a measure of k, then we can calculate the rigidity modulus $\mu$ (and hence Vs) as a function of depth from Vp. This is a rather crude estimate of Vs but it serves as a reasonable starting point for comparison with observations of Vs and for a discussion of how tectonics may affect velocity. Values of Vs computed from this assumption for Cascadia Basin and the uplifted ridge are shown in figure 6 and we see, of course, that the curves follow the Vp patterns closely. In addition the Vs values at the sea-floor match closely the shear velocities obtained from Stonely wave experiments.

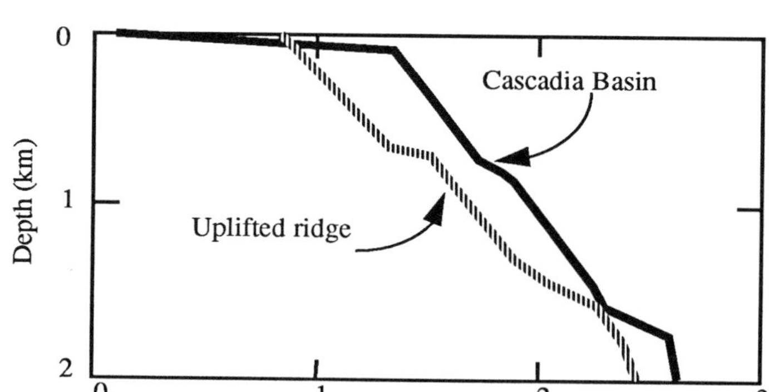


Figure 6. Shear velocity estimates from the P wave model

Returning now to a discussion of why the uplifted ridge sediments have lower velocities than Cascadia Basin, it seems there are initially two possibilities. First, the uplift of the sediments by almost 1 km will result in a decompression that reduces the velocity. The problem is by how much. Data on the effect of pressure on unconsolidated sediments are experimentally difficult to obtain and sparse. Discussions with workers in this field suggest that this effect is less than the observed change in velocity. Second is the possibility that tectonic uplift and the accompanying anticlinal folding has resulted in a disruption of the sediments, possibly by generation of microfractures and an increase in porosity, which causes a decrease in the rigidity modulus. This option is preferred at present because it would increase the bulk permeability of the sediments and it would favour increased water flow from the top of the anticline, a feature which is observed at this site. As the rocks age and dewatering and cementation proceed the velocity should finally increase to values above Cascadia Basin values. Although the seismic velocities observed at this site show that the accretionary process and associated dewatering is more complicated than at first assumed, they are providing constraints and insights on the process. Evidence for a similar decrease in velocity has been reported for the accretionary front of the Barbados Trench by Bangs et al 1990, who also allow that a lowered rigidity may be the cause.

## 5. Conclusions

Based on the results of this experiment it now seems clear that the inference of large amounts of dewatering in the accretionary front of the Cascadia subduction zone is the result of comparing samples of recent turbidites, that have not yet reached an equilibrium compaction state as a result of burial, with samples of much older sediments that have undergone compaction by burial and also undergone tectonic deformation. Therefore this inference may have little to do with tectonically induced dewatering. In fact I arrive at the opposite conclusion; that the most recently accreted and uplifted sediments may well have a higher porosity than the original Cascadia Basin sediments. This could be the result of the tectonically induced folding causing a disruption (microfracturing ?) of the sediments which in turn leads to a reduction of rigidity, an increase in permeability and porosity, and a decrease in seismic velocity.

## 6. Acknowledgements.

This work was funded by the National Science Foundation under grant OCE-8614474 and I wish to thank V. Kulm and B.Carson for their scientific and logistic support.

## 7. References

Bangs, N.L.B., G.K. Westbrook, J.W.Ladd and P. Buhl (1990) "Seismic velocities from the Barbados Ridge complex: Indicators of high pore fluid pressures in an accretionary complex", J. Geophys. Res., 95, 8767-8782.

Bray, C. J. and Karig, D. E. (1985) "Porosity of sediments in accretionary prisms and some implications for dewatering processes", J. Geophys. Res. 90, 768-778.

Cochrane, G. R., B. T. R. Lewis and K. J. McClain (1989) Structure and subduction processes along the Oregon-Washington margin, Pure and Applied Geophysics, 128, 767-800.

Carson, B. (1977) "Tectonically-induced deformation of deep- sea sediments off Washington and Oregon: mechanical consolidation", Marine Geology, 24, 289- 307.

Carson, B., E. Suess, and J. C. Strasser (1990) "Fluid flow and mass flux determinations at vent sites on the Cascadia margin accretionary prism". J. Geophys. Res., 95, 8891-8898.

Davies, D. (1965) "Dispersed Stonely waves on the ocean bottom", Bull. Seis. Soc. Am., 55, 903.

Davis E. E., R. D. Hyndman and H. Villinger (1990) "Rates of fluid expulsion across the northern Cascadia accretionary prism: constraints from new heat flow and multichannel seismic reflection data". J. Geophys. Res. 95, 8869-8891.

Dorman, L. M. (1979) "A linear relationship between earth models and seismic body wave data", Geophys. Res. Lett. 6, 132- 134.

Hamilton, E. L. (1978), "Sound velocity-density relations in sea-floor sediments and rocks", Jour. Acoust. Soc. Amer. 63, 366-377.

Kappus M. E., A. J. Harding and J. A. Orcutt (1990) "A comparison of tau-p transform methods", Geophysics, 55, 1202-1215.

Kulm, L. D. and Fowler, G. A. (1974) "Oregon continental margin structure and stratigraphy: a test of the imbricate thrust model", In: The Geology of Continental Margins (ed. Burk, C. A., Drake, C. L.) (Springer-Verlag, NY) pp. 261-283.

Kulm, L. D., Suess, E., Moore, J. C., Carson, B., Lewis, B. T., Ritger, S. D., Kadko, D. C., Thornburg, T. M., Embly, R. W., Rugh, W. D., Massoth, G. J., Langseth, M. G., Cochrane, G. R., and Scamman, R. L. (1986) "Oregon subduction zone: venting, fauna, and carbonates", Science 231, 561-566.

Lister C. R. B and B. T. R. Lewis (1977) "A direct recording ocean bottom seismometer", Mar. Geophys. Res., 3, 65-85.

Lewis, B. T. R. and G. C. Cochrane, "Relationship between the location of chemosynthetic benthic communities and geologic structure on the Cascadia subduction zone", J. Geophys. Res., 95, 8783- 8794.

Moore, J. C., D. Orange and V. Kulm, (1990) "Interrelationship of fluid venting and structural evolution: Alvin observations from the frontal accretionary prism, Oregon, J. Geophys. Res. 95, 8795-8809.

Schreiner A. E. and L. M. Dorman (1990) "A study of sea-floor noise using an array of seismomters", submitted to J. Acoust. Soc. Am.

Shipboard Scientific Party (1988) Proc. ODP, Initial Reports (Pl -A) 110: College Station, Tx (Ocean Drilling Program).

Tuthill J. D., B. T. R. Lewis and J. D. Garmany (1981) "Stonely waves, Lopez Island noise, and deep sea noise from 1 to 5 hz.", Mar. Geophys. Res., 95-108.

# COMPRESSIONAL AND SHEAR WAVE VELOCITIES IN THE UPPER CRUST

Orest Diachok, Stephen Wales, Ronald Dicus,
Fred Feirtag, David Shirley and John Siegel
*Naval Research Laboratory, Code 5120*
*Washington, D.C. 20375-5000 USA*

ABSTRACT. New reflectivity vs. grazing angle data at a number of thinly sedimented sites in the Pacific Ocean spanning 0.5 to 64 m.y. reveal that compressional and shear wave velocities, $V_p$ and $V_S$, of the uppermost crust increase with geologic age. Measurements between 40 and 64 m.y. reveal compressional and shear critical angles of 60 ± 3° and 9° ± 3° respectively. The corresponding compressional and shear speeds are 3090 ± 300 m/s and 1564 ± 15 m/s respectively. The relatively high estimated shear speeds are in sharp contrast to inferred shear speeds in 0.5 m.y. old crust, viz., 800 m/s, whereas the estimated compressional speeds in the older crust only slightly exceed $V_p$ in young crust, which is about 2400 m/s.

## 1. Introduction

Bottom interacting acoustic fields at small grazing angles in the deep ocean are primarily dominated by sediment-refracted energy when the sediment thickness is greater than about 100 meters, [1-4] and to a large extent by the upper crust when the sediment thickness is below about 100 meters. Consequently, detailed knowledge of upper crustal geoacoustic properties is essential for acoustic field prediction in thinly sedimented regimes. These include seamounts, substantial areas in the vicinity of mid-ocean ridges, and most significantly the majority of the Pacific Ocean.

Unconsolidated sediments on the ocean floor originate primarily from land (terrigenous), volcanoes and deceased oceanic plants and animals (pelagic). Deep trenches surrounding the Pacific Ocean limit terrigenous sediments from flowing into this ocean. As a result the mean sediment thickness in the central North Pacific is low, generally below 100 meters (based on Lamont-Doherty Geophysical Observatory data contoured by B. Hersey)[5]. Exceptions occur in areas in close proximity to land and volcanoes, and in the high biological productivity zone near the equator. As seismic-acoustic properties of the upper crust are age-dependent, it is noteworthy that the sediment thickness overlying crust as old as 150 m.y. in non-equatorial regions of the western Pacific[6] is generally less than 100 meters.

The most significant acoustic parameter affecting the reflection process at the upper crustal boundary at small grazing angles in the absence of a sediment layer is $V_S$. If $V_S$ is higher than the speed of sound in water, $V_W$, then total reflectivity occurs at grazing angles less than the critical angle. This type of behavior is illustrated in Figure 1, where $\theta_S = 9°$. If $V_S < V_W$, then there is no shear critical angle, and nearly total mode conversion to shear waves occurs accompanied by large interfacial reflection losses. In a previous paper[7] the results of a low frequency reflectivity vs. grazing angle experiment conducted at a sediment-free site in the vicinity of the East Pacific Rise (Project ROSE) were documented; estimated $V_p$ was shown to be



*J. M. Hovem et al. (eds.), Shear Waves in Marine Sediments*, 387–394.

consistent with a limited set of well-sampled seismic refraction measurements at a near-by site.[8] The following conclusion was drawn about the $V_s$:

> The interfacial $V_s$ in the upper crust at a geologically young, sediment-free Pacific site, was estimated to be 800 m/s, substantially lower than $V_w$.

This estimate, which is based on single hydrophone data, should be considered a lower bound. Array measurements of the coherent component of the reflection coefficient would probably have resulted in lower reflectivity at small grazing angles and higher inferred $V_s$.

All of the data analyzed in this paper are limited to the Pacific, where rms roughness is relatively small, to minimize effects of roughness on the reflection coefficient. The conclusions, however are expected to be applicable to other oceans.

The objectives of this paper are to 1) describe new measurements of the coherent component of the reflection coefficient vs. grazing angle and inferred $V_p$ and $V_s$ in older crust (40 - 64 m.y.) for comparison with previous measurements in very young crust; and 2) discuss the implication of these results on acoustic field prediction in thinly sedimented regions of the world's oceans.

In Section 2 of this paper the age dependence of interfacial $V_p$ and $V_s$ in thinly sedimented regions of the Pacific will be summarized. In Section 3 results of the reflectivity experiments will be described. Effects of boundary roughness and a thin sediment layer will be considered in Sections 4 and 5, respectively. Reflectivity data will be presented and discussed in terms of critical angles in Section 6. Wave field prediction will be considered in Section 7. Conclusions and recommendations for future research are outlined in Section 8.

## 2. The Nature of the Upper Crust

The upper crust is formed at mid-ocean ridges and is transported over periods of millions of years to great ranges. Over this period the density and velocities of the solid basalt component of this medium, measured on dredged samples from sediment-free sites at high frequencies and high pressures[9], decrease as a result of chemical alteration. Aging makes this material progressively "softer". Recently formed crust is cracked and porous. As it ages the cracks and pores are thought to be filled primarily by calcite and sediments from above and to a limited extent by minerals brought from below by circulating water.

As a result of these effects, the effective low frequency seismic properties of this composite are age-dependent. Estimates of the effective interfacial $V_p$ in thinly sedimented regions of the Pacific Ocean, extrapolated from well sampled seismic refraction data[10-13] and inferences from far field reflectivity vs. angle measurements (to be subsequently discussed in this paper), are shown as a function of geologic age in Figure 2. Since $V_p$ and $V_s$ may be anisotropic, data shown in this figure are restricted to measurements made parallel to the ridge, hypothetically the high speed direction. Excluded from this compilation are measurements made at ridge crests, fracture zones, thickly sedimented sites, near seamounts and those characterized by a low sampling rate. The data shown here suggests that $V_p$ and $V_s$ increase with crustal age, hypothetically due, at least in part, to the filling of cracks and pores.

The variance $\Delta V_p$, of young crust has been estimated by Toomey et. al[10] to be less than or approximately equal to ± 10%; the corresponding correlation lengths normal and parallel to the axis were estimated to be approximately 2 and 6 km respectively. These lengths are comparable to bathymetric correlation lengths of Pacific abyssal horsts (hills).

## 3. Reflectivity Measurements

New reflectivity vs. grazing angle measurements were made at 12 sites over the age range 40 to 64 m.y. The sediment thickness, measured with a water gun, was estimated to be less than 40 meters at all sites. To minimize scattering due to roughness all measurements (reported here) were made parallel to the dominant horst-grabben structure. Sea-beam maps were recorded to reveal seamount locations. The rms roughness, determined from the Sea-beam profiler, will be discussed in Section 4. The acoustic signals were detected on a 1100 m long array of hydrophones towed broadside to the propagation path at a depth of 300 m from the Canadian research ship, CFAV Endeavour. Standard Mk 61 2 lb underwater explosives were dropped from the USNS Desteiger every two degrees (grazing angles) and detonated at a depth of 240 m.

The data was tape recorded over a 5-300 Hz bandwidth and subsequently processed on a single hydrophone and coherently added (beamformed) to provide estimates of the total energy and the coherent component of the reflection coefficient vs. grazing angle. The data were filtered in third octave bands. Only the data centered around 8 Hz will be presented in this paper. After beamforming, reflection coefficients were computed essentially as described in reference 7. Details of the processing are described elsewhere.[14]

Prior to presentation and interpretation of the data, two effects which may confound the identification of critical angles will be discussed, viz. bottom roughness and the thin sediment layer.

## 4. Effects of Boundary Roughness

Scattering loss from a Gaussian randomly rough surface, a reasonable approximation for the basalt boundary[7], may be simply described and readily implemented with a theory developed by Eckart.[21] According to this theory, scattering loss obeys the relationship:

$$R = e^{-g^2}$$

where R = coherent energy reflection coefficient, g = 2kr sin θ, k = wave number, r = rms roughness, and θ = grazing angle. This model is considered applicable provided $r < \lambda << L$, where λ = wavelength, r = rms roughness and L = correlation length. For the experimental sites reported here, these parameters were derived from the Sea Beam vertical profiling beam. The data was spatially averaged over about 50km, thereby encompassing many abyssal hills. The resultant $r \approx 50$ m and $L \approx 2000$ m. At 8 Hz, λ = 191 m. Consequently, the Eckart equation is considered reasonable for this application.

According to this model the scattering loss in the vicinity of the compressional and shear critical angles is about 5 and 1 dB respectively. Consequently scattering effects may be expected to be significant for estimation of $\theta_p$ and negligible for estimation of $\theta_s$.

## 5. Effects of a Thin Elastic Sediment Layer

A thin sediment layer, generally less than 100 m, overlies the upper crustal boundary in the Northeast Pacific. Such sediment layers are known to be spatially variable in thickness (nominally ± 50%) on a scale of about 2 km normal and 6 km parallel to the dominant horst-grabben structure. The approximate $V_p$ and $V_s$ values of this medium are 1550 and 150 m/s respectively.

Reflectivity from the flat boundary of a fast ($V_s$ = 2500 m/s) elastic solid covered by a thin horizontally stratified elastic sediment layer has been theoretically investigated by Hawker[18-19] and Vidmar[20-22]. Vidmar demonstrated that a thin elastic layer can diminish the reflection coefficient by a few dB in the angular range $\theta_s < \theta < \theta_p$, and by tens of dB when $\theta < \theta_s$ at frequencies as low as 10 Hz. In contrast, a thin fluid sediment layer produced essentially no effect.

Hawker demonstrated that at sufficiently high frequencies, when $\lambda << H$ where H is the sediment thickness, the angle of the reflection minimum corresponds to mode conversion (according to Snell's law) of compressional to Stoneley waves at the sediment-rock interface. At lower frequencies (where $\lambda$ is comparable to H) the reflection minimum persists, but shifts to smaller grazing angles than predicted by the Stoneley wave phase speeds.

When $V_s$ is only slightly greater than $V_w$, e.g. when $V_w < V_s < 1600$ m/s, the interference minimum disappears, and the reflection process becomes transparent to the sediment layer. In older crust[13], where $V_s$ may be as high as 2400 m/s, interference nulls due to the thin sediment layer will undoubtedly affect the reflectivity, and thereby complicate estimation of the shear speed.

## 6. Estimation and Interpretation of $V_p$ and $V_s$

At the compressional critical angle $\theta_p$ the theoretical plane wave specular reflection coefficient equals unity in the absence of a thin sediment layer or rough boundary scattering. Rough boundary scattering and spherical wave effects diminish and angularly broaden the compressional critical angle peak.

Calculations that incorporate these effects agreed reasonably well with previously reported measurements on a single hydrophone at a 0.5 m.y. old sediment-free site. A comparably comprehensive data-theory comparison of the new measurements does not appear to be necessary for identification of critical angles. Measurements made at the 58 m.y. site, shown in Figure 1, clearly reveal compressional and shear critical angles at 59° and 9° respectively.

Inspection of critical angle measurements made at other sites in the 40-64 m.y. age range yields compressional and shear speeds graphically displayed in Figure 2. The relatively large scatter in $V_p$ may be attributed to 1) real spatial variability (Toomey et.al. estimated that $\Delta V_p < 10\%$ on a scale of 2 and 6 km along and across strike respectively in the vicinity of the East Pacific Rise), and 2) the mean slope of reflecting surfaces, averaged over a Fresnel Zone. A 1° departure from the horizontal translates into a 2° error in the critical angle measurement. The radial and transverse Fresnel zone dimensions in the vicinity of $\theta_p$ are nominally 1 km x 1 km, which are small compared to the correlation lengths of the horst and grabben geometry (6 and 2 km in the radial and transverse directions). Consequently significant contributions to $\Delta V_p$ may be expected from both effects. The radial and transverse Fresnel zone dimensions in the vicinity of

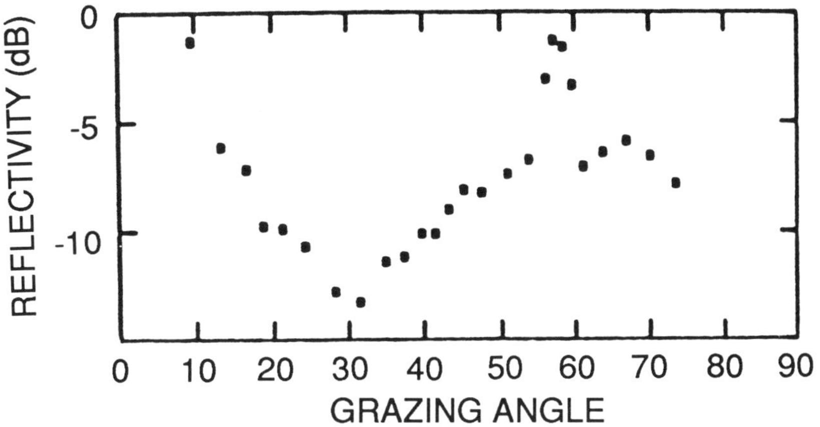


Figure 1. Reflectivity vs grazing angle at 8Hz at 59 m.y. old site

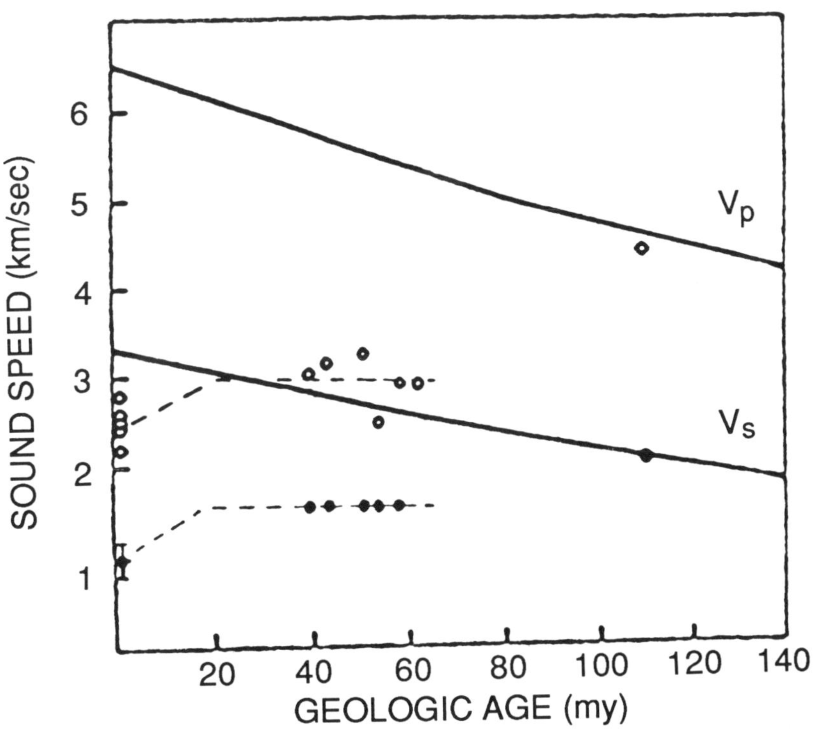


Figure 2. Seismo-acoustic (lower curves and data) and ultrasonic (——) measurements of compressional and shear speeds in sediment-free/thinly sedimented volcanic rock (basalt) vs. geologic age.

$\theta_s$ are nominally about 3 and 18 km respectively, which are relatively large compared to the correlation lengths of $V_s$ and the geometrical structure resulting in spatially averaged (over about 6 "grains") estimates of $V_s$. Consequently, the contributions of the two effects to the variance in the shear speed are relatively small. The difference in the measured variances in $V_p$ and $V_s$ is also due in part to the low sensitivity of $V_s$ to $\theta_s$ at small angles, and the high sensitivity of $V_p$ to $\theta_p$ at large angles. Although it is not clear from Figure 2, $V_s$ increases slightly, viz. from 1560 to 1580 m/s over the age range 40 - 64 m.y. In view of the variance in $V_s$ and the small number of samples, this trend may not be significant.

## 7. Acoustic Field Prediction

The angular region over which bottom interacting rays (modes) are required for acoustic field prediction is dependent on the amplitude, phase and coherence of reflected signals. Conditions under which the reflectivity and spatial coherence are high, and the phase may be expected to be predictable, occur at angles less than the shear critical angle in the 40 - 64 m.y. age range. At sufficiently low frequencies the reflectivity, coherence, and phase predictability is hypothetically high due to diminished rough boundary scattering losses and thin layer effects. In young crust ($\lesssim$ 20 m.y.) the lack of a shear critical angle diminishes the need for incorporating bottom interacting modes in acoustic field prediction. In much older crust interference effects due to the thin sediment layer will probably unpredictably scramble the phases of modes that interact with the crust at small grazing angles.

## 8. Conclusions

Reflectivity vs grazing angle data at a number of thinly sedimented sites spanning 0.5 to 64 m.y. reveal that compressional and shear speeds of the upper-most crust increase with geologic age, hypothetically a result of crack filling. Measurements between 40 and 64 m.y. reveal compressional and shear critical angles of 60° ± 3° and 9° ± 3° respectively. The corresponding compressional and shear speeds are 3090 ± 300 and 1564 ± 15 m/s respectively. The relatively high shear speeds are in sharp contrast to inferred shear speeds in 0.5 m.y. old crust, viz 800 m/s. The existence of shear critical angles in the older crust implies that prediction of acoustic fields in the ocean at very low frequencies in this age range requires incorporation of bottom interacting modes.

## 9. Recommendations for Future Research

a) Develop methods for probing the physical properties of both the basalt and the thin sediment layer at small grazing angles where multipath arrivals are not separable using modal decomposition and/or matched field processing. Develop analytical search algorithms for rapidly estimating geo-acoustic parameters.

b) Conduct acoustic sensing measurements of the upper crust with horizontal and/or vertical arrays particularly in ocean crust older than 65 m.y. and younger than 20 m.y. Test theoretical predictions that reflectivity from sufficiently old crust characterized by very high shear speeds ($\sim$ 2500 m/s) is affected by sediment-related interference effects. Refine estimates of the shear speed in young crust with towed array measurements.

c) Experimentally determine the magnitude of sound speed anisotropy, and adapt existing, and if deemed appropriate, develop new theories of reflectivity from anisotropic solids.

d) Incorporate effects of sub-bottom refracted modes, including boundary scattering upon re-entry into water.

## ACKNOWLEDGEMENTS

This work was supported by the Office of the Chief of Naval Research Geology & Geophysics Program, Code 1125GG, and the AEAS Program, Code 124A. The authors appreciate the encouragement from numerous OCNR program managers, particularly Dr. Randy Jacobsen, Mr. Ed Estalot and Dr. Leonard Johnson. The measurements reported here would not have been possible without the strong commitment of the management, and the high level of at sea expertise of the technical staff of the Defense Research Establishment Pacific (DREP). The authors are particularly grateful to Dr. Ross Chapman of DREP for his interest, enthusiasm, technical contributions and commitment to this project.

## REFERENCES

1. Hill, M. N. (1952) "Seismic Refraction Shooting in an Area of the Eastern Atlantic", Phil Trans. R. Soc. London A244:561.

2. Morris, H. E. (1970) "Bottom Reflection Loss Model with a Velocity Gradient", J. Acoust. Soc. Am. 48:1198.

3. Dicus, R. L. (1976) "Preliminary Investigations of the Ocean Bottom Impulse Response at Low Frequencies", Naval Oceanographic Office, NSTL Station, MS.

4. Hamilton, E. L. (1980) "Geoacoustic Modeling of the Ocean Floor", J. Acoust. Soc. Am. 68:1313.

5. Spofford, C. (1983) "Blug Overview", Science Applications Inc., McLean, VA.

6. Larson, R. L. and Pitman, W. C. (1985) The Bedrock Geology of the World, W. H. Freeman and Co., New York, N.Y.

7. Diachok, O., Dicus, R. and Wales, S. (1986) "Effects of Upper Crustal Geoacoustic Parameters on Low Frequency Sound", in Seismo-Acoustics, T. Akal and J. Berkson, Plenum.

8. Purdy, G. M. (1982) "The Variability in Seismic Structure of Layer 2 Near the East Pacific Rise at 12°N.", J. Geophys Res., 87, 8403.

9. Christensen, N. I. and Salisbury, M. H. (1973) "Velocities, Elastic Moduli and Weathering - Age Relations For Pacific Layer Basalts", Earth and Plan. Science Let. 19:461.

10. Toomey, D. R. et. al (in press) "The Three Dimensional Seismic Velocity Structure of the East Pacific Rise Near Latitude 9° 30".

11. Harding, A. J. et. al (1989) "Structure of Young Oceanic Crust at 13°N on the East Pacific Rise From Expanding Spread Profiles", J. Geophys Res., 94, 163.

12. Vera, E. E. and Mutter, J. C. (1988) "Crustal Velocities in the Rose Area of the East Pacific Rise: One Dimensional Travel Time and Expanded spread Profiles", J. Geophys. Res., 93, 6635.

13. Shearer, P. and Orcutt, J. (1985) "Anisotropy in the Oceanic Lithosphere-Theory and Observations from the Ngendei Seismic Refraction Experiment in the Southwest Pacific", Geophys. Jr. Astr. Soc. 80:493.

14. Dicus, R. L. et. al, in preparation.

15. Eckart, C. (1953) "The Scattering of Sound from the the Sea Surface", J. Acoust. Soc. Am. 25:566.

16. Lonsdale, P. unpublished results.

17. Lonsdale, P. (1977) "Regional Shape and Tectonics of the Equitorical East Pacific Rise", Mar Geophys Res., 3, 295.

18. Hawker, K. (1978) "Influence of Stoneley Waves on Plan-Wave Reflection Coefficients: Characteristics of Bottom Reflection Loss", J. Acoust. Soc. Am. 64:548.

19. Hawker, K. (1979) "The Existence of Stoneley Waves as a Loss Mechanism in Plane Wave Reflection Problems", J. Acoust. Soc. Am. 65:682.

20. Vidmar, P. (1980) "The Dependence of Bottom Reflection Loss on the Geoacoustic Parameters of Deep Sea (Solid) Sediments)", J. Acoust. Soc. Am. 68:1442.

21. Vidmar, P. (1980) "Ray Path Analysis of Sediment Shear Wave Effects on Bottom Reflection Loss", J. Acoust. Soc. Am. 68:639.

22. Vidmar, P. (1980) "The Effect of Sediment Rigidity on Bottom Reflection Loss in a Typical Deep Sea Sediment", J. Acoust. Soc. Am. 68:634.

# Sedimentary Shear Modulus and Shear Speed Profiles from a Gravity Wave Inversion

Mark V. Trevorrow and Tokuo Yamamoto
Univ. of Miami, RSMAS
4600 Rickenbacker Cswy
Miami, Florida, U.S.A. 33149

**Abstract:** Experimental techniques and inverse methods have been established to extract sediment shear modulus vs depth profiles from measurements of wave-induced seabed motion. Seabed sediments are modelled as a quasi-static, incompressible, layered elastic medium. An iterative eigenvalue expansion technique is used to extract the inverse. Using semi-empirical sediment models, it is possible to deduce sediment porosity and bulk density, and thus shear wave speed, from the shear modulus profile. The final results are supported by comparisons with borehole stratigraphy logs and shear wave refraction results at a site on the New Jersey Shelf.

# 1 Seabed Sediment Models:

As surface gravity water waves propagate through continental shelf waters (roughly <100m deep), they induce small motions of the seabed sediments through variations in hydrodynamic pressure on the seafloor. The layered seabed sediments can be seen to behave in a massless, incompressible, elastic manner in response to water wave-induced pressures (see Refs. 2-6 for a complete explanation of these modelling assumptions). Under these circumstances it is possible to predict realistically the seabed response to passing water waves if the sediment *shear modulus* at every depth is known. Conversely, through the use of geophysical inverse methods, it is possible to extract the sediment shear modulus with depth profile from measurements of seabed motion and wave-induced pressure. This new passive sediment remote-sensing method and instrumentation system has been called the Bottom Shear Modulus Profiler (henceforth BSMP).

Surface gravity waves begin to cause measurable seabed displacements when they reach water depths roughly equal to their wavelength. On the shallow continental shelf, this surface gravity wave-induced seabed motion is predominant in the frequency range 40 - 330 mHz, depending on the water depth. From experiments conducted over 5 years, we have found the typical seabed surface displacement to have a peak amplitude of the order of 0.1 mm at a frequency of 0.1 Hz (see Trevorrow *et al.*[4]).

The water waves are modelled using the familiar linear dispersion relation

$$\omega^2 = g\, k\, \tanh(\, k\, d\, ) \tag{1}$$



*J. M. Hovem et al. (eds.), Shear Waves in Marine Sediments*, 395–402.

In this case the wavenumber, $k$, is a unique function of angular frequency, $\omega$, gravity, $g$, and water depth, $d$. The sediment motion is coupled directly to the water wave and the has the same wavelength, with the seabed displacement and water surface displacement 180° out of phase. The sediment response exhibits the same prograde particle motion as the water waves. The driving force for the seabed displacement is the wave-induced bottom pressure, given by

$$P = \frac{a\, \rho_w\, g}{\cosh(\, k\, d\,)} \tag{2}$$

where $a$ is the water wave amplitude at the water surface and $\rho_w$ is seawater density.

As a first approximation, we shall assume the seabed sediments to be isotropic, lossless, and horizontally stratified. Modelling the seabed response comes through the integration of a *propagator matrix*, which has the form

$$\frac{d\underline{r}}{dz} = \begin{bmatrix} 0 & k & 1/G(z) & 0 \\ -k & 0 & 0 & 0 \\ 4G(z)\,k^2 & 0 & 0 & k \\ 0 & 0 & -k & 0 \end{bmatrix} \underline{r} \tag{3}$$

Where $G(z)$ is the sediment shear modulus profile, and the vector $\underline{r}$ is the displacement/stress vector whose components are: $r_1$ = u = horizontal seabed displacement, $r_2$ = w = vertical seabed displacement, $r_3 = \tau_{xz}$ = shear stress in the x-z plane, and $r_4 = \tau_{zz}$ = vertical normal stress.

The differential equations (3) can be integrated numerically given any shear modulus with depth profile, a wavenumber, and appropriate boundary conditions (see Yamamoto and Torii[5] and Trevorrow *et al.*[3,4] for details on solving this system). In addition, the authors have developed a semi-empirical sediment model (outlined in Yamamoto *et al.*[6]) which allow calculation of porosity, $\beta(z)$, from the shear modulus profile and vice versa.

Under usual circumstances, the densities of the pore fluid and the individual grains in a marine sediment are approximately known. The density of seawater varies only slightly about a value of 1025 $kg/m^3$, while the sediment grains typically found on the shallow continental margins have a density of approximately 2650 $kg/m^3$. Thus, we can define the bulk density by

$$\rho = \beta\rho_w + (1-\beta)\rho_s \tag{4}$$

where the subscripts $(w, s)$ refer to water and solid respectively.

From knowledge of both the shear modulus and sediment porosity, the shear wave speed is straightforwardly given by

$$c_s = \sqrt{\frac{G}{\rho}} \tag{5}$$

## 2 Inverse Methods

Knowing the shear properties of the sediments at every depth, it is possible to predict the seabed response to pressure forcing induced by water waves. It is also possible to calculate in the *reverse* direction; to extract the shear modulus profile of the sediments from measurements of the wave-induced pressure and the seabed response. The success of the forward theory is limited only by the accuracy of the physical model and the assumptions made in deriving it. This inverse calculation, by contrast, is mathematically non-exact due to the inevitable noise in the admittance data and the fact that the water waves have only a finite frequency bandwidth. This non-exactness manifests itself as an inherent uncertainty and resolution limit in the final result.

As is typical in geophysical problems, the relation between the shear modulus profile (model) and seabed admittance (data) is complicated and non-linear. The general, discrete, non-linear relation between a vector of data parameters, $\underline{d}$, and the vector of model parameters, $\underline{m}$, can be written conceptually as

$$\underline{d} = f(\underline{m}) \tag{6}$$

where $f()$ is some non-linear functional mapping model parameters into data. We then expand Eq. (6) in a Taylor series about some initial value $\underline{m}_0$

$$\underline{d} = f(\underline{m}) \approx f(\underline{m}_0) + \left[\frac{\partial f(\underline{m})}{\partial \underline{m}}\right] (\hat{m} - \underline{m}_0) + \ldots \tag{7}$$

We can replace $f(\underline{m}_0)$ with $\underline{d}_0$, the calculated admittance data from the initial shear modulus profile $\underline{m}_0$. Rearranging yields

$$\hat{m} = \underline{m}_0 + \left[\frac{\partial f(\underline{m})}{\partial \underline{m}}\right]^{-g} (\underline{d} - \underline{d}_0) \tag{8}$$

where the power '$-g$' denotes that we are specifying a *generalized* inverse of the derivative matrix, rather than a formal mathematical inverse. This formula is then used in an iterative fashion to cause $\underline{m}_0$ to creep towards $\hat{m}$. The derivative matrix is calculated approximately using a central difference scheme.

The generalized inverse of the derivative matrix $[\partial f(\underline{m})/\partial \underline{m}]^{-g}$ is at the heart of the inversion algorithm. Because of the presence of noise in the data and the simultaneous occurence of both *over-* and *under-determined* conditions in the model, the derivative matrix is fairly ill-conditioned. To handle the generalized inverse of such a matrix we use a Singular Value Decomposition (SVD). The SVD performs a spectral decomposition of the derivative matrix into a set of coupled eigenvalue-eigenvector matrices. The generalized inverse is readily calculated from the SVD matrices, and Eq.(8) becomes an eigenvalue expansion. Some noisy, higher-order solutions are discarded for stability. For full details on the solution of this inverse problem, and a discussion of resolution and uncertainty limits, see Trevorrow *et al.*[2].

## 3 Instrumentation:

The BSMP instrument packages, in both analog and digital radio-telemetry formats, are the products of five years of experience and improvements. The measurement system consists of two separate housings connected by a 2 m long umbilical cable, with connection to a surface buoy or ship for real-time recording or telemetry. The seismometer housing is a small Aluminum alloy cylinder (dia. = 30.5 cm, height = 25.5 cm) designed to contain three orthogonally mounted seismometers (either Teledyne-Geotech model S-510 or S-750, or Guralp CMG-3) and two pendulum tiltmeters. The other housing is a similar Aluminum alloy cylinder which contains electronics and serves as a support for a differential pressure sensor (either InterOcean WS200 or PME-109). The BSMP could be deployed in either of two configurations: plate mounted or buried. The buried configuration provided superior instrument/sediment coupling (see Trevorrow *et al.*[3]). Shallow burial of the seismometer housing is achieved by use of a hydraulic jet burial system.

In the analog configuration, the 3-component seismic and hydrodynamic pressure signals are relayed directly to the attending ship, where amplification, filtering and recording are done. The digital radio-telemetry BSMP configuration has the amplifiers, filters, digitizers, and FM Telemetry systems located in the electronics housing. The digitizing rate for these units is set at 4 Hz. Each BSMP channel has its own amplifier and bandpass filter (3.86-500 mHz pass) circuitry. Previous seismometer designs were limited to frequencies above 30 mHz due to rising internal noise levels at low frequencies. However, the Guralp instruments have better low-frequency performance due to capacitive plate design and feedback noise suppression circuitry.

## 4 Spectral Results:

The analysis of BSMP data follows standard spectral and transfer function estimation techniques. The desired frequency band for this analysis is in the range 0.002 to 1.0 Hz, which is best suited by digitizing the raw voltage time-series at 4 Hz and using a base FFT size of 4096 points (17 minutes). Then, we can define vertical *admittance*, $\eta(f)$, as

$$\eta(f) = \frac{\mathrm{w}}{a} = \frac{-\rho_w\, g}{4\pi^2\, f^2\, \cosh(kd)}\, \frac{S_{vp}(f)}{S_p(f)} \tag{9}$$

where $S_{vp}$ is the time-averaged cross-spectrum between vertical acceleration and pressure and $S_p$ is the time-averaged pressure power spectrum. Also, it has been found convenient to use vertical seabed *impedance*, $I(f)$ (Pa/m), which is the ratio of sea-bottom pressure to seabed displacement. Using our previous definition of admittance this becomes

$$I_v = -4\pi^2 f^2\, \frac{S_p}{S_{vp}} = \frac{\rho_w g}{\cosh(kd)\; \eta(f)} \tag{10}$$

Figure 1 shows the time-averaged vertical acceleration and pressure power spectra from experiments at the Atlantic Generating Station (AGS, $N39°\,28'$, $W74°\,16'$, Aug. 29, 1987). Approximately 6.5 hours of data went into the calculation of these spectra. Notice the large, broad-band energy peak in the frequency range 50 to 275 mHz that is present in both spectra. This is due to water wave energy, mostly low-frequency *swell.* The rapid cutoff of wave energy at 170 - 350 mHz is due to the fact that water wave induced pressures are only significant within one wavelength of the surface. In the vertical seismic spectrum, the increase in spectral levels at frequencies below 40 mHz is due to internal instrumental noise being boosted by the frequency response corrections. A more detailed discussion and summary of shallow water seabed seismicity is given in Trevorrow *et al.*[4]

Figure 2 shows the vertical seabed admittance magnitude and coherence spectra calculated from the data shown in the previous figure. Notice the range of very high coherence

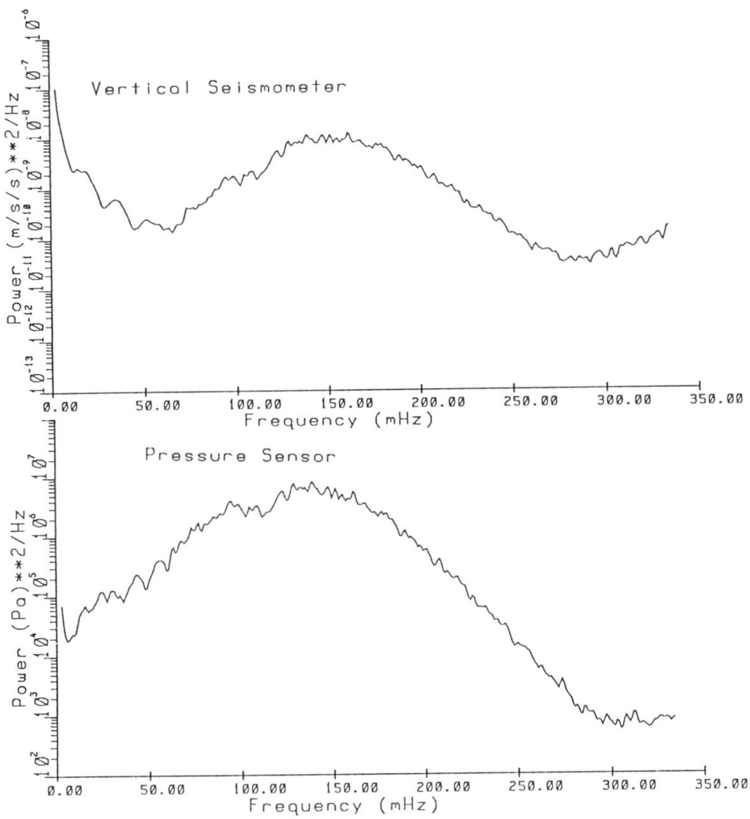


Figure 1. Vertical acceleration and pressure power spectra from AGS87-302. Water depth = 12 m.

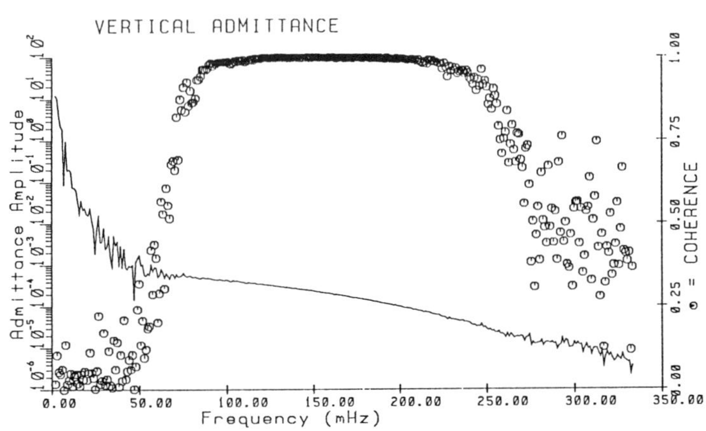


Figure 2. Vertical admittance and coherence spectra from AGS87-302 (data shown in Fig. 1).

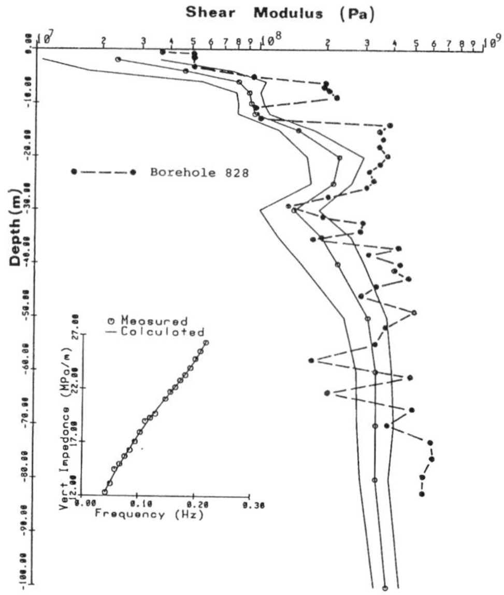


Figure 3. Comparison of inverted shear modulus profile from AGS87-302 with reference borehole 828. Inset shows experimental vs model seabed impedance match.

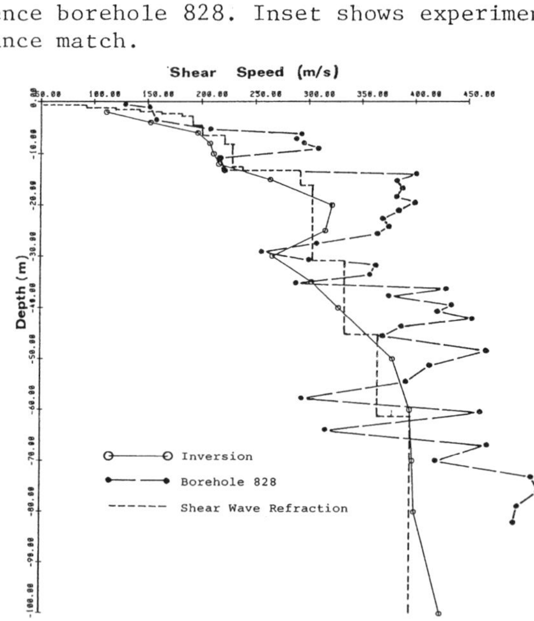


Figure 4. Comparison of invered shear wave speed profile from AGS87-302 (converted from shear modulus) with shear wave refraction result (from J. Ewing), and borehole 828 (converted from blow count).

($\approx$0.99) in the frequency range 60 to 270 mHz. This very high coherence implies that there is negligible noise contamination from ambient seismicity (microseisms) in this frequency band. This is the band of good admittance data useful in the inversion process.

## 5 Shear Modulus Profile Results:

Continuing with the example data set presented in Figures 1 and 2, the inversion results from AGS-87 302 is shown in Figure 3. A comparison profile is calculated from a Standard Penetration Test (SPT) borehole from the exact same location (SPT borehole data are contained in Miller and Dill[1]). The error bound on the inversion is $\pm 2$ standard deviations. Both the inversion and reference profiles clearly show the distinct hard layer beginning at 12-15 m and a softer silty-clay layer at 30 m. Notice that there is reasonable agreement between the inversion and the reference profile at all depths, although most of the variations in shear modulus shown in the reference occur on too fine a vertical scale to be resolved by the inversion.

An inset to Figure 3 shows the 20 admittance data points, converted to vertical Impedance, selected for use in the inversion. The calculated impedance curve fits smoothly through the data points, with a rms relative mismatch error of only 1.01 %. The longest wavelength available in this data set is 254 m (at 43 mHz), so a maximum profile depth of 100 m ($\approx \frac{1}{2}\lambda_{max}$) is appropriate. The shortest wavelength is 19.1 m (at 286 mHz), which restricts the minimum resolution depth to about 3 m.

Figure 4 shows a comparison of shear wave speed profiles for this AGS site under consideration. The inversion result and SPT borehole profiles are converted from shear modulus using the methods outlined in Yamamoto *et al.*[6]. Also shown is an independent shear wave speed profile from active refraction experiments conducted at the same site by J. Ewing and G. Purdy of WHOI and G. Sutton and J. Carter of Roundout Assoc. (J. Ewing personal communication). Again, there is good agreement in a depth-averaged sense between all three profiles, especially in resolving the hard layer at 12 m.

## 6 Discussion and Conclusions:

- The method of seabed shear modulus inversion using surface gravity wave-induced seabed motion can accurately estimate the depth-averaged magnitude and structure of sediment shear modulus. Further, the shear modulus profile can be converted to realistic shear wave speed profiles. This follows from direct comparisons of the BSMP results with sedimentary reference data and shear wave refraction experiments.

- In deeper waters, the higher frequency waves necessary to resolve fine details in the shear modulus structure near the seabed surface were absent from the coherent admittance data sets. This cutoff occurs when the water wavelength is approximately 1.3 times the water depth. The low frequency limit to coherent admittance data seems to be set by a decrease in surface gravity wave energy below approximately 50 mHz.

## Acknowledgements

This work was supported under several O.N.R. contracts, 1985-90. The authors are grateful for this support and the continued encouragements of Drs. Raymond Fitzgerald and Marshall Orr (Code 1125OA).

## 7 References:

1. Miller, H. J., and Dill, C., *Final Report, Geophysical Investigation of Atlantic Generating Station Site and Offshore Region*, for the Public Service Electric and Gas Company, Newark, New Jersey (1974).

2. Trevorrow, M., Yamamoto, T., Badiey, M., Turgut, A., and Conner, C., 'Experimental Verification of Seabed Shear Modulus Profile Inversions using Surface Gravity (water) Wave-induced Seabed Motion', Geophysical Journal, 93 (3), 419-436 (1988).

3. Trevorrow, M., Yamamoto, T., Turgut, A., Goodman, D., and Badiey, M., 'Very Low Frequency Ocean Bottom Seismic Noise and Coupling on the Shallow Continental Shelf', Mar. Geophys. Res., 11 (2), 129-152 (1989).

4. Trevorrow, M., Yamamoto, T., Turgut, A., and Goodman, D., 'Measurements of Ambient Seabed Seismic Noise below 1.0 Hz on the Shallow U. S. Continental Shelf', J. Acoust. Soc. Am., 86 (6), 2318-2327 (1989).

5. Yamamoto, T., and Torii, T., 'Seabed Shear Modulus Profile Inversions using Surface Gravity (water) Wave-induced Bottom Motion', Geophys. J. Roy. Astr. Soc., 85, 413-431 (1986).

6. Yamamoto, T., Trevorrow, M., Badiey, M., and Turgut, A., 'Seabed Porosity and Shear Modulus Inversion using Surface Gravity (water) Wave-induced seabed motion', Geophys. J. Int., 98 (1), 173-182 (1989).

# MEASUREMENTS OF COMPRESSIONAL WAVE AND SHEAR WAVE SPEEDS, ATTENUATION, PERMEABILITY, AND POROSITY IN MARINE SEDIMENTS

A. TURGUT and T. YAMAMOTO
*University of Miami, RSMAS*
*4600 Rickenbacker Cswy.*
*Miami, Florida 33149, U.S.A.*

ABSTRACT. Measurements of compressional wave and shear wave speeds and attenuation are performed in sandy marine sediments. Simple relations are derived to estimate porosity and permeability of marine sediments from the measured compressional and shear wave speeds and attenuation. The estimated porosity and permeability values are compared with direct core measurements. Agreement between estimated and directly measured values indicates that acoustic measurement of porosity and permeability is feasible.

## 1. Introduction

For the last three decades, theoretical studies and experiments in marine sediments have produced an expansion of knowledge about the sea floor and its effects on acoustic wave propagation in the ocean. To characterize geoacoustic properties of the seabed, systematic measurements of acoustic wave propagation in marine sediments are still needed, especially at low frequency covering rather broad bands. Establishing relations between acoustic wave propagation (dispersion, attenuation, wave speed) and the physical properties of the seabed (porosity, permeability, density, elastic moduli, and saturation level) will help to better understand the physical processes involved in the acoustic wave propagation in marine sediments.

In early studies, Stoll and Bryan [1] showed that acoustic wave propagation in marine sediments can be formulated by Biot's [2] theory for the saturated porous media. Toksöz *et al.* [3] presented systematic laboratory measurements of compressional and shear wave attenuation and velocity in dry and brine-saturated rocks. Tittman *et al.* [4] observed velocity dispersion correlated with intrinsic attenuation ($Q^{-1}$) for several porous rocks. Numerous examples of attenuation and wave speed measurements at discrete frequencies were reported by Hamilton [5] covering the frequency range 5 Hz to 1 MHz. A recent literature review of the attenuation measurements both in sediments and in porous rocks is given by Kibblewhite [6]. In this study, *in-situ* measurements of compressional wave speed and attenuation are presented within the frequency range of 1-30 kHz. Measurements of shear wave speed at low frequencies (<1 kHz) are also presented. Finally, porosity and permeability of marine sediments are estimated from acoustic measurements.



*J. M. Hovem et al. (eds.), Shear Waves in Marine Sediments*, 403–410.

## 2. Theory

In this section, frequency equations for compressional and shear waves are given following the Biot theory [2]. Approximate solutions for compressional wave velocity and attenuation are also given in order to predict porosity and permeability of marine sediments. For simple harmonic motion of compressional waves, Biot's equations of motion can be expressed as [7]

$$Hl^2\phi_1 + Cl^2\phi_2 = \omega^2(\rho\phi_1 + \rho_f\phi_2), \tag{1}$$

$$Cl^2\phi_1 + Ml^2\phi_2 = \omega^2(\rho_f\phi_1 + m'\phi_2), \tag{2}$$

where $H, C, M$ are Biot's elastic moduli, $l$ is wavenumber, $\phi_1$ and $\phi_2$ are scalar displacement potentials, $\omega$ is the angular frequency, $\rho$ is the bulk density, and $\rho_f$ is the density of pore fluid. The virtual mass density $m'$ is given by $m' = m - i\eta F(\kappa)/k_s\omega$, where $\eta$ is the dynamic viscosity, $k_s$ is the permeability, and $\kappa$ is the ratio of the pore size $a$ to the oscillatory boundary layer thickness, $\kappa = a(\rho_f\omega/\eta)^{1/2}$. Virtual mass $m$ is expressed as $m = (1+\alpha)\rho_f/\beta$, where $\alpha$ is the added mass coefficient and $\beta$ is porosity. $F(\kappa)$ is the viscous correction factor introduced by Biot [2] to treat the high frequency regime in which Poiseuille flow breaks down.

The frequency equation can be derived from Eqs.1 and 2 as

$$\begin{vmatrix} Hl^2 - \rho\omega^2 & \rho_f\omega^2 - Cl^2 \\ Cl^2 - \rho_f\omega^2 & m'\omega^2 - Ml^2 \end{vmatrix} = 0. \tag{3}$$

Similarly, a frequency equation is given for the shear wave as [2]

$$\begin{vmatrix} \mu l^2 - \rho\omega^2 & \rho_f\omega^2 \\ -\rho_f\omega^2 & m'\omega^2 \end{vmatrix} = 0. \tag{4}$$

The solution of the frequency equation (Eq.3) for the fast compressional wave can be approximated as [8]

$$l^2 \approx \frac{\omega^2(\rho m' - \rho_f^2)}{Hm' + M\rho - 2C\rho_f}. \tag{5}$$

For high Q values, the intrinsic attenuation $Q^{-1}$ and the propagation velocity $V$ can be written as

$$Q^{-1} = -\mathrm{Im}(l^2)/\mathrm{Re}(l^2), \qquad V^2 = \omega^2/\mathrm{Re}(l^2). \tag{6}$$

Using Eq.5, we obtain

$$V^2 = \mathrm{Re}\left(\frac{Hm' + M\rho - 2C\rho_f}{\rho m' - \rho_f^2}\right), \tag{7}$$

with the limiting cases

$$V_0^2 = H/\rho, \quad \text{and} \quad V_\infty^2 = (Hm + M\rho - 2C\rho_f)/(\rho m - \rho_f^2). \tag{8}$$

Also, using Eq.5, we can easily show that intrinsic attenuation can be defined as

$$Q^{-1}(\omega) \approx \left(\frac{V_\infty^2}{V_0^2} - 1\right) \Big/ \left(\frac{m_i}{A} + \frac{A}{m_i}\frac{V_\infty^2}{V_0^2}\right), \tag{9}$$

where $m_i$ is the imaginary part of $m'$ and $A = (\rho m - \rho_f^2)/\rho$.

The approximate solutions and the exact solutions of Biot's model for different types of marine sediments were compared and good agreement was found, especially for the sediments with less rigidity [7].

The compressional wave velocity is more sensitive to porosity than the grain bulk modulus $K_r$ and bulk modulus of the skeletal frame $K_b$, which are the only parameters with some uncertainties. The porosity can be calculated by using low frequency compressional and shear wave velocities and certain values of the grain bulk modulus and bulk modulus of the frame. Considering $K_r \gg K_b$ for most marine sediments, the low frequency limit in Eq.8 can be further simplified such that porosity can be expressed as

$$\beta = \frac{K_f(K_r - K)}{(K_r - K_f)(K - K_b)}, \tag{10}$$

where $K = H - \frac{4}{3}\mu = \rho(V_p^2 - \frac{4}{3}V_s^2)$ is the system bulk modulus and $K_f$ is the bulk modulus of the fluid. Using the measured velocity value at the high frequency limit and Biot's coefficients $C$ and $M$, one can also estimate the virtual mass coefficient $m$ in Eq.8. Finally, Eq.9 can be used for the permeability estimation. According to Biot's theory, location of the viscous attenuation peak is mostly determined by the permeability of the porous medium. Therefore, the permeability can be estimated if the frequency of the viscous attenuation peak is measured. Ignoring the effects of viscosity correction factor $F(\kappa)$, the location of the relaxation frequency (viscous attenuation peak) can be found as

$$f_r = \frac{\rho\eta}{2\pi(\rho m - \rho_f^2)k_s}\frac{V_0}{V_\infty}. \tag{11}$$

## 3. Experiment

Several tests were performed to collect the most reliable data throughout the experiment. Cross hole transmission tests were conducted to determine the compressional wave velocity and intrinsic attenuation. The source and receivers were buried 5 m deep and 1-3 m apart by a newly developed hydraulic jet burial system [9]. Deep burial of source and receiver is needed to avoid surface reflection interference with direct arrivals and to detect mostly body waves rather than surface waves. Typical source and receiver separations were 0.5 to 2 m for an acoustic transducer source and 2 to 5 m for a sparker source. The configuration of the cross hole transmission experiment is shown in Fig.1.

### 3.1. INSTRUMENTATION

*3.1.1. Source.* The most important characteristics of a source for acoustic wave generation are a) source level, b) repeatability, c) directionality, and d) frequency response. Two different types of sources, namely a single tip sparker and an omnidirectional acoustic transducer, were used in the experiments. The energy level of a single tip sparker varies from 300 to 600 Joules depending on the firing rate. The acoustic source level for the single tip sparker is about 190-210 dB re 1 $\mu$Pa @ 1 m. Excellent repeatability was observed from a single tip sparker for the same physical environments. Directionality of a single tip sparker showed the characteristics of a dipole source because of the asymmetric position of the electrode tips. The frequency response of a sparker could be varied from 200 to 5,000 Hz, depending on the energy level. In this particular experiment, signals were observed within the frequency range of 200 to 1500 Hz. Although these characteristics of a single tip sparker satisfied all the requirements for our experiment, the capability of changing pulse shape was limited. An ITC-1001 acoustic transducer was used to cover the higher frequencies (up to 30 kHz), and to control the pulse shape generation. Since the time signatures of each receiver are compared for the velocity and attenuation measurements, the frequency response of the source did not have to be known precisely. However, it is desirable to have a smoothly varying response in the frequency band of interest. The repeatability of the transducer was also checked in the saturated sediment, and perfect repeatability was observed.

*3.1.2. Receiver.* Benthos AQ-17 hydrophones were used to detect acoustic signals in the frequency range of 300 Hz to 30 kHz. In a fluid or fluid-like material (fluid-saturated sediment in our case), the most common acoustic receiver is a hydrophone. Hydrophones are usually made of a piezoelectric material which generates an electrical charge directly proportional to the excess applied pressure. Another property is that hydrophones are not required to be placed in a specific orientation with respect to the source (omnidirectional hydrophones). Special attention has to be paid in coupling the receivers to the ground to obtain accurate attenuation measurements.

*3.1.3. Signal Generation.* A Teledyne MINI SPARKER-253 power supply was used to generate a spark by applying a 7,000 Volt charge between the electrodes of a single tip sparker. A Textronix AFG-5001 arbitrary function generator and an ELGAR A501 power amplifier were used to drive the source transducer. The arbitrary function generator is state-of-the-art equipment which can be used to generate any arbitrary wave form by using mathematical functions or any discrete values. During the experiments, a two-cycle burst signal with the desired frequency and a one-cycle shifted sine wave, $[1 + \sin(x - \pi/2)]$, were generated. Output voltage ($\pm$ 10V) from the function generator was magnified by 20 dB using the A501 power amplifier.

*3.1.4. Signal Acquisition.* Two ONO-SOKKI CF-910 spectrum analyzers were used for real-time signal acquisition in both time and frequency domains. In addition to real time analysis, the collected data sets were recorded on magnetic FM tapes by using a TEAC SR-51 analog FM recorder. Then, selected data sets were digitized with a 12 bit A/D converter. Digitization rate was up to 64,000 samples per second, depending on the frequency range of interest. A block diagram of signal generation and data acquisition systems is given in Fig.2.

### 3.2. RESULTS FROM CROSS HOLE TRANSMISSION EXPERIMENTS

Time signatures from each hydrophone are compared for the travel time and attenuation measurements of compressional and shear waves. Typical 64-stack time signature of a Ricker wavelet ($f_0 \approx 1$ kHz) recorded from two hydrophones and their power spectra, impulse response function, phase, coherence, and transfer function are shown in Fig.3.

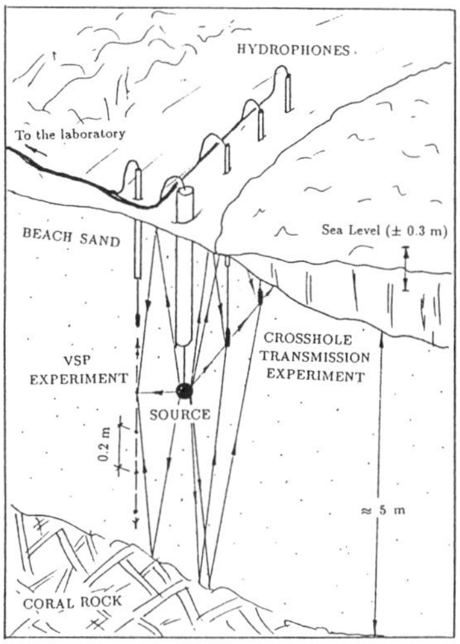


Figure 1: Configuration of the cross hole transmission experiment.

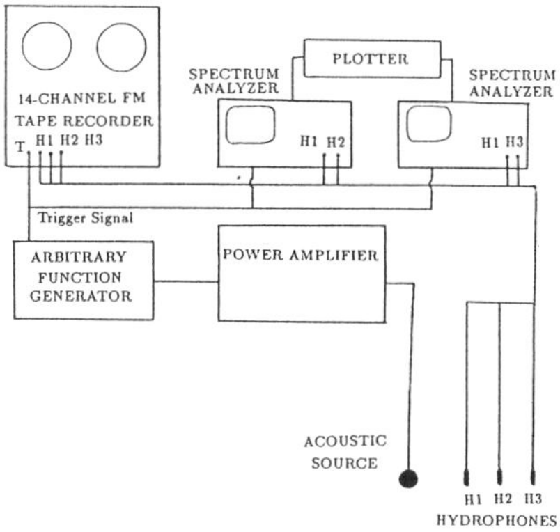


Figure 2: Block diagram of the signal generation and data acquisition systems.

The transfer function of two time signatures, $g_1(t)$ and $g_2(t)$ is given in frequency domain as

$$H(\omega) = G_2(\omega)/G_1(\omega), \tag{12}$$

where $G_1(\omega)$ and $G_2(\omega)$ are Fourier transforms of $g_1(t)$ and $g_2(t)$, respectively. The impulse response function $h(t)$ is the inverse Fourier transform of the transfer function $H(\omega)$. Any characteristic peak in the impulse response function represents a characteristic time delay between the two time signatures.

A specially designed single tip sparker source was used for frequencies lower than 1 kHz. Shear waves were also generated during the low frequency sparker experiments due to the asymmetric position of the electrode tips. Fig.4. shows the typical time signatures of the sparker signal and impulse response from the two hydrophones. Shear wave velocities measured from the impulse response function were approximately 110 m/s. The spectral ratio method is used to determine the frequency dependent compressional wave velocity and intrinsic attenuation. Compressional wave velocity and intrinsic attenuation can be calculated from the phase spectrum and transfer function using the following equations

$$V(f) = V_0 + \frac{(R_2 - R_1)f \text{ x } 360^0}{\theta}, \text{ and } Q^{-1}(f) = \frac{c(f)}{\pi f(R_2 - R_1)}\left(\ln\frac{R_1}{R_2} + \ln\frac{A_1(f)}{A_2(f)}\right), \tag{13}$$

where $R_1$ and $R_2$ are receiver distances from the source, $f$ is the frequency, $\theta$ is the phase in degrees, and $A_1(f)/A_2(f)$ is the amplitude ratio of the two signals. Fig.5 shows the comparison of measured compressional wave velocity and intrinsic attenuation with the theoretical predictions.

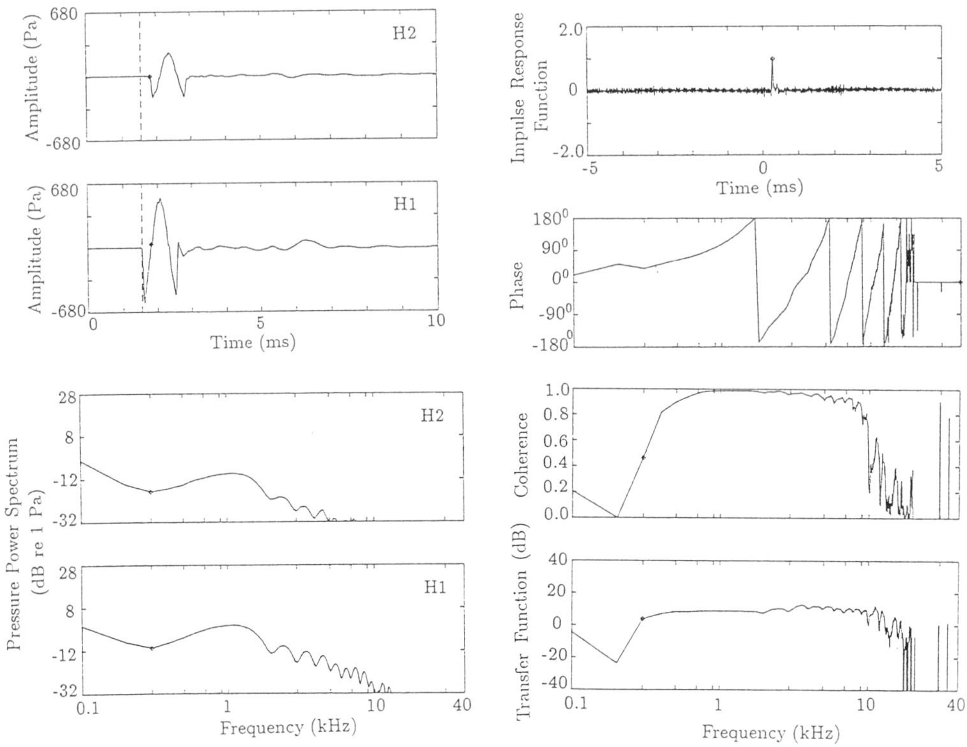


Figure 3: Acoustic signals recorded from two hydrophones and their spectra, impulse response function, phase, coherence, and transfer function.

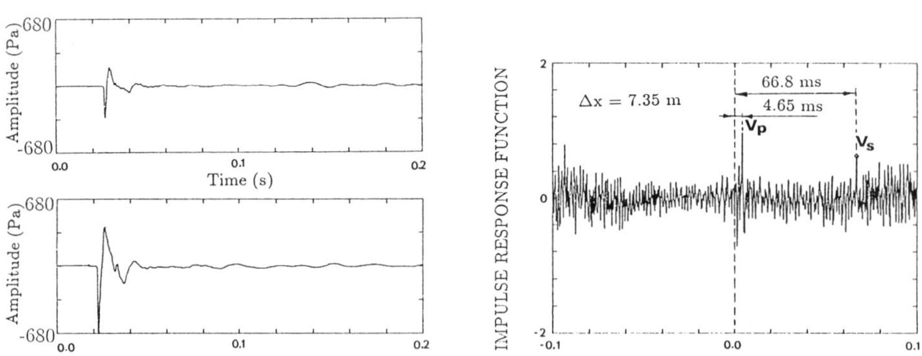


Figure 4: Low frequency sparker signals recorded from two hydrophones and the impulse response function.

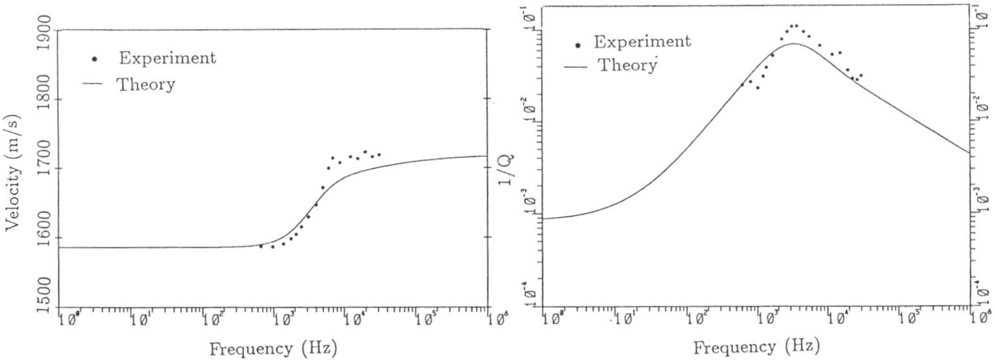


Figure 5: Comparison of measured compressional wave velocity and intrinsic attenuation with theory.

## 4. Experimental Results and Discussion

The frequency dependent intrinsic attenuation and corresponding dispersion in the compressional wave velocity were detected in sandy marine sediments. This was achieved by conducting the experiments in a rather homogeneous medium so that attenuation due to scattering is minimum. The porosity and permeability were calculated for the medium using Eqs. 10 and 11. A comparison of porosity and permeability calculated from acoustic measurements and direct core sample measurements is given in Table 1. In the direct core sample measurements, porosity was obtained from the measurements of grain density and bulk density of 100% saturated sediment, and the permeability was measured using a falling head permeameter. In Table 1, mean values and 95% confident intervals are given for five different core measurements. The following values were used in the acoustic prediction of porosity and permeability: low frequency compressional wave velocity $V_0$ =1585 m/s (wave speed measured at 1 kHz), high frequency compressional wave velocity $V_\infty$ =1725 m/s (wave speed measured at 30 kHz), shear wave velocity $V_s$ =110 m/s (obtained from sparker source data), relaxation frequency $f_r$ = 3900±200 Hz, the grain bulk modulus $K_r$ =3.0x$10^{10}$-5.2x$10^{10}$ N/m$^2$, bulk modulus of the skeletal frame $K_b = (1.0 - 4.0)\text{x}\rho V_s^2$, fluid bulk modulus $K_f$ =2.3x$10^9$ N/m$^2$. Porosity and permeability variations in the acoustic estimations are due to the uncertainties in the bulk modulus of the skeletal frame, the grain bulk modulus, and the viscous attenuation peak. The added mass coefficient is also estimated as $\alpha = 0.24$ from the high frequency compressional wave velocity in Eq. 8.

In Table 1, the consistency of estimated and directly measured porosity and permeability indicates that Biot's prediction of viscous losses for certain values of porosity and permeability is the main loss mechanism for sandy sediments within the frequency range of 1-30 kHz.

Table 1. Comparison of porosity and permeability values obtained from core measurements and acoustic measurements.

| | Core measurements | Acoustic measurements |
|---|---|---|
| Porosity | 0.44±.018 | 0.435-0.455 |
| Permeability ($m^2$) | $(1.75\pm0.21)\text{x}10^{-11}$ | $(1.63\pm0.08)\text{x}10^{-11}$ |

**Acknowledgements:** This work was supported under O.N.R. contract N0014-87-J-1116. The authors are grateful for this support and the continued encouragement of Dr. Marshall Orr (Code 1125OA).

**References:**

1. Stoll, R. D., and Bryan, G. M. (1970) 'Wave attenuation in saturated sediments', J. Acoust. Soc. Am. 47, 1440-1447.
2. Biot, M. A. (1956) 'Theory of propagation of elastic waves in a fluid saturated porous solid. II. Higher frequency range', J. Acoust. Soc. Am., 28, 179-191.
3. Toksöz, M. N., Johnston, D. H., and Timur, A. (1979) 'Attenuation of seismic waves in dry and saturated rocks, I. Laboratory measurements', Geophysics, 44 , 681-690.
4. Tittman, B. R., Nadler, H., Clark, V. A., and Ahlberg, L. A. (1981) 'Frequency dependence of seismic dissipation in saturated rocks', Geophysics, Res. Lett.', 8, 36-38.
5. Hamilton, E. L., (1980) 'Geoacoustic modelling of the seafloor', J. Acoust. Soc. Am., 68, 1313-1340.
6. Kibblewhite, A. C. (1989) 'Attenuation of sound in marine sediments: A review with emphasis on new low frequency data', J. Acoust. Soc. Am., 86, 716-738.
7. Turgut, A. and Yamamoto, T. (1990) 'Measurements of acoustic wave velocities and attenuation in marine sediments', J. Acoust. Soc. Am.,87, 2376-2383.
8. Geertsma, T., and Smit, D. C. (1961) 'Some aspects of elastic waves propagation in fluid saturated porous solids', Geophysics, 26, 169-181.
9. Turgut, A., Yamamoto, T., Abbott, C., Badiey, M., Trevorrow, M., and Goodman, D. (1987) 'High Resolution BSMP: A shallow water real-time OBS array', Technical Report No. 87-007, RSMAS, Univ. of Miami, FL.

# SEA-BED SHEAR MODULI FROM MEASUREMENTS OF TIDALLY INDUCED PORE PRESSURES

M. T. HURLEY
*School of Ocean Sciences*
*University College North Wales*
*Bangor, Gwynedd*
*United Kingdom*

P. J. SCHULTHEISS
*GEOTEK*
*Marley Lane*
*Haselmere, Surrey*
*United Kingdom*

ABSTRACT. A method for determining an average sea-bed shear modulus (G) of the top few metres of the sea-bed has been developed and tested. The method employs a simple mathematical model and in-situ measurements of the variation of tidally induced pore-pressures with depth. The model is based on Biot's theory, assuming a homogeneous sea bed with constant physical properties. We tested the model using data collected at Great Meteor East (GME), which is located on the Madeira Abyssal plain of the N Atlantic Ocean. We used in-situ measurements of tidally induced pore-pressures - collected using a pop-up pore-pressure instrument (PUPPI) - along with physical properties data from laboratory experiments on core-samples. We input these measurements into the model and obtained an average value of G = 20 MPa for the first 4 metres of the sea-bed. This value of G is greater than measurements obtained from alternative techniques used at GME. The reasons for these differences are: simplicity of the model; limitations of our in-situ pore-pressure measurements, and the uncertainties present in the alternative techniques. Therefore, we have demonstrated the feasibility of this method of determining G from tidally induced pore-pressure measurements within defined limits. We recommend improvements to the model and the pore-pressure measuring system PUPPI, which will increase the accuracy of shear moduli determined from this method.

## 1. Introduction

The way in which the sea-bed is affected by the passage of waves along the surface of the sea has been studied by a number of authors (e.g. Yamamoto et al., 1978, Mei and Foda, 1981, Okusa, 1985). Further, techniques have been developed to determine sea-bed properties through use of 2-D hydrodynamic models (Yamamoto and Torii, 1986). We examine a simple 1-D case where variations of pore-pressures in the sea-bed are caused by the effects of tides. We assess the feasibility of determining an average value of the shear modulus, over the upper few metres of the sea-bed, using in-situ measurements of tidally induced pore-pressures.



*J. M. Hovem et al. (eds.), Shear Waves in Marine Sediments*, 411–418.

## 2. A model for tidally induced pore-pressures

We use a simple model of the sea-bed to obtain an equation which describes the variation of tidally induced pore-pressures (p) with depth (z). The model is shown in Figure 1. The tide applies a cyclic 1-D hydrodynamic stress on the sea-bed of amplitude q and frequency $\omega$. The boundary condition at the top of the sea-bed is $p(z=0) = -q$. The sea-bed is assumed to have constant physical properties, which are: the porosity ($\phi$); the fluid bulk modulus ($K_f$); the fluid viscosity ($\eta$); the permeability (k); the shear modulus (G); the frame bulk modulus ($K_b$) and the constrained frame compressibility ($a_b$). The sea-bed is assumed to be thicker than the diffusivity wavelength $\lambda_\kappa$ and rests on an impermeable rigid basement.

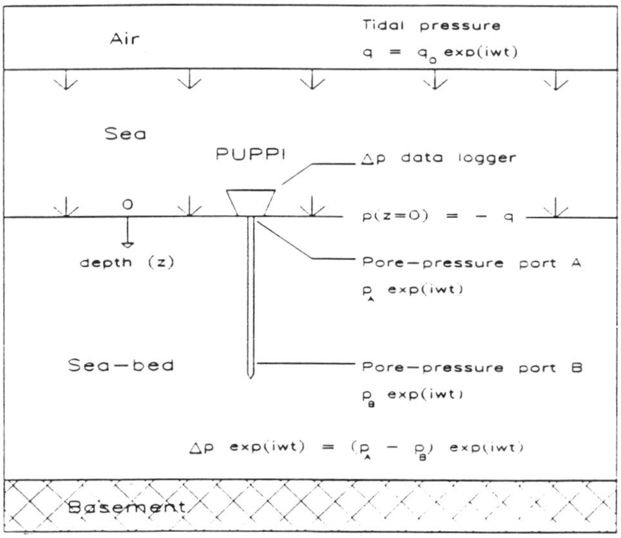


Figure 1. A schematic model showing the tidal pressures on and in the sea-bed and the differential pore-pressure instrument PUPPI

Using the model in Figure 1, the following equations for the variation of pore-pressure (p) with depth (z) in the sea-bed can be obtained with Biot's (1941, 1956) theory:

$$p\, e^{i\omega t} = P_1\, e^{i\omega t} + P_2\, e^{-cz}\, e^{i\omega t} \tag{1}$$

$$P_1 = -q\,\{K_f / \{\phi a_b\} + 1\}^{-1} \tag{2}$$

$$P_2 = -q\,\{1 - \{K_f / \{\phi a_b\} + 1\}^{-1}\} \tag{3}$$

$$c = \{\omega / \{2\kappa\}\}^{1/2} \tag{4}$$

$$\kappa = k / \eta\, \{\phi / K_f + a_b\}^{-1} = \omega \lambda_\kappa^{\,2} / \{8\pi^2\} \tag{5}$$

$$a_b = K_b + 4G / 3 \tag{6}$$

### 3. Modelling tidally induced pore-pressures

The variation of the tidally induced pore-pressure with depth can be calculated using Equations 1 to 6, providing that the parameters k, $K_f$, $\phi$, etc are known, and the simple model adequately describes the data. We will study how the average permeability and shear modulus of the sea-bed affect the variation of tidally induced pore-pressure with depth. We will consider the specific example of the sea-bed at a test site named Great Meteor East (GME), located on the Madeira Abyssal Plain in the North East Atlantic Ocean. The sea-bed at GME comprises over 200m thick unconsolidated turbidite and pelagic sediments, which overlay a basalt basement. The sea is over 5000m deep at GME, so there will be no free gas in the sea-bed and we will not have to consider partial saturation effects (Okusa, 1985).

Schultheiss and Gunn (1985) and Hurley (1989) have carried out a series of laboratory experiments on core samples of the sediments of GME. Average values of the physical properties for the sediments from the upper-few metres of the sea-bed at GME, from these studies, are shown in Table 1. Some of these properties, for example, $\phi$, $\eta$, $K_f$ and $\omega$, are known with reasonable accuracy (< 5%). The same cannot be said for k, $K_b$ and G. The permeability ranges over two orders of magnitude - due to physical and scaling differences introduced by the various laboratory measuring techniques. The shear modulus ranges over two orders of magnidute - due to laboratory technique dependent uncertainties and strong in-situ shear modulus gradients. Unlike k and G, we know of no direct measurements of $K_b$ (which is often called the "dry" bulk modulus) on very high porosity ($\phi \simeq 80\%$) unconsolidated sediments. We do have some $K_b$ and G measurements for high porosity ($\phi \simeq 40\%$) unconsolidated sands, which yield the relationship $K_b = 2.17G$. We assume that this relationship holds for the unconsolidated turbidite sediments from GME. Unfortunately, we cannot quantify the uncertainty associated with this assumption.

TABLE 1. Average physical properties of sediments from the first few metres of the sea-bed at GME

| | |
|---|---|
| Porosity ($\phi$) | 0.8 |
| Fluid viscosity ($\eta$) | $10^{-3}$ Pa.s |
| Fluid modulus ($K_f$) | $(2.4).10^{9}$ Pa |
| Tidal frequency ($\omega$) | $(1.454).10^{-4}$ Hz |
| Permeability (k) | $(5).10^{-15}$ $m^2$ |
| Frame bulk modulus ($K_b$) | 2.17G Pa |
| Shear modulus (G) | $(5).10^{6}$ Pa |

The parameters given in Table 1 can be used in Equations 1 to 6 to give predictions of the variation of the magnitude of the tidally induced pore-pressure (p) with depth (z) at GME. Using these data, we illustrate how variations in G and k affect these predictions in Figures 2 and 3. Figure 2 shows how p varies with z for different shear moduli for a constant permeability. Each curve tends to a different constant value at depth, depending on the shear modulus of the sea-bed. The greater the shear modulus is, the smaller the induced pore-pressure becomes. Figure 3 shows how p varies with z for different permeabilities for a constant shear modulus. All curves tend to the same asymptote, but at a different rate. The greater permeability is, the longer the decay takes. Further discussions on the character of wave induced pore-pressure versus depth curves are given by Yamamoto et al., 1978, Mei and Foda, 1981, Okusa 1985 and Hurley, 1989.

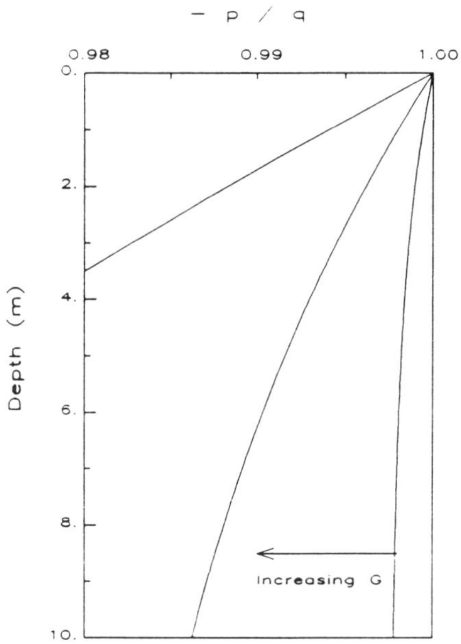


Figure 2. Predicted variation of tidally induced pore-pressures with depth at GME for a constant permeability ($k = (1.26).10^{-13}$ $m^2$) and for a range of shear moduli (G = 2, 20 and 200 MPa).

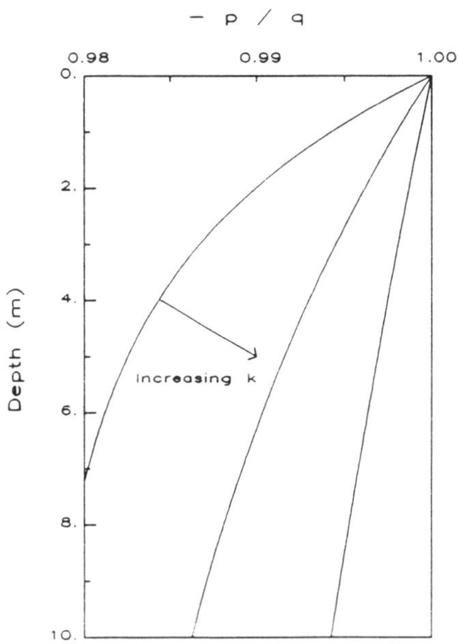


Figure 3. Predicted variation of tidally induced pore-pressures with depth at GME for a constant shear modulus (G = 20 MPa) and a range of permeabilities ($k = (1.26).10^{-14}$, $(1.26).10^{-13}$ and $(1.26).10^{-12}$ $m^2$).

## 4. Measurements of tidally induced pore-pressures

Schultheiss and McPhail (1985) have developed a pop-up pore pressure instrument (PUPPI). It is a simple device which measures the differential pore-pressure between a port ($p_A$) at the surface of the sea-bed, and a port ($p_B$) up to 6 metres below the surface (see Figure 1). PUPPI has been deployed over 20 times at GME. A typical record of the differential pore pressure ($\Delta p$) data collected from one deployment at GME is shown in Figure 4. Also shown in Figure 4 is a record of the variation in tidally-induced hydrodynamic stress (q) obtained from tide-gauge measurements taken at the same location. Note that the $\Delta p$ and q records have two frequency components: the diurnal tidal frequency and the modulating lunar spring-neap tidal frequency. Furthermore, the variations of $\Delta p$ and q with time are in-phase, but the amplitude of the $\Delta p$ record is two orders of magnitude smaller than that of the q record.

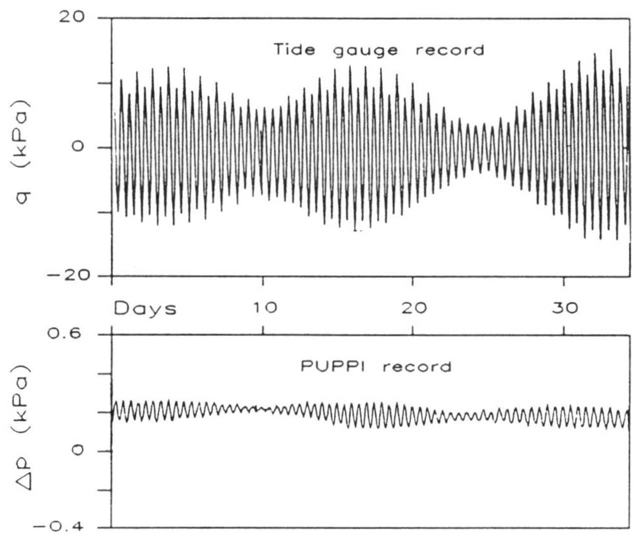


Figure 4. Measurements of tidal pressures on the sea-bed (q) and in the sea-bed (q - $\Delta p$) the sea bed at GME

## 5. Shear modulus and permeability from tidally induced pore-pressures

It is possible to use $\Delta p$ and q measurements, such as those shown in Figure 4, along with the data in Table 1, to determine the shear modulus and permeability of the sea-bed at GME. We have measurements of the tidally induced pore pressure ($p = (q - \Delta p) / q$) at depths of z = 2 m and z = 4 from 20 PUPPI deployments at GME. The average values (along with the spread) of these measurements are shown in Figure 5 as the dots. Also shown in Figure 5 is the best-fit tidally induced pore-pressure curve, where the permeability and shear modulus have been adjusted to give best agreement with pore-pressure measurements. The best-fit line is not exact, but does lie within the spread of the measured tidally induced pore-pressure data.

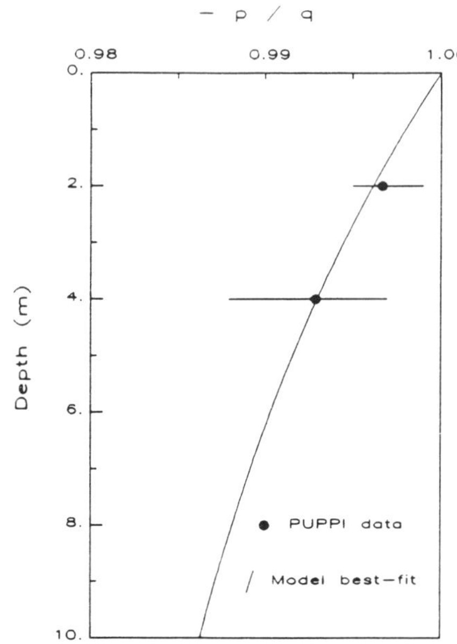


**Figure 5. Best-fit predicted and measured induced pore-pressures with depth at GME**

The best-fit shear modulus is G = 20 MPa and the best-fit permeability is $(1.26).10^{-13}$ $m^2$. Using the parameters shown in Table 1 along with these values for G and k in Equation 5 gives $\lambda_\kappa \simeq 70$m. Therefore, the assumption that the sea-bed is thicker than the diffusivity wavelength $\lambda_\kappa$ is valid. However, the best-fit values for G and k are significantly greater than the average values from laboratory measurements shown in Table 1. The reasons for these discrepancies require further explanation

The best-fit average sea-bed shear modulus is about an order of magnitude greater than laboratory measurements taken on samples from the upper 5 metres of the sea-bed at GME (Hurley, 1989). This discrepancy could be due to a number of factors, the most likely of which are: coring disturbance reducing G in the laboratory samples; inaccuracy of the assumed relationship between $K_b$ and G, and limitations of the model and the $\Delta p$ data. The reduction of G through coring disturbance is supported by the observation that G measurements on core samples are a factor of 4 smaller than G inverted from Stoneley wave velocities at GME (Whitmarsh and Lilwall, 1982). The problem of characterising the relationship between $K_b$ and G for very high porosity unconsolidated sediments could be addressed by laboratory resonant column experiments. However, reducing the limitations of the model and the $\Delta p$ data are obvious areas for improvement.

Using the simple model, we have two equations (from the two $\Delta p$ data values from two depths) and two unknowns k and G (with $K_b$ assumed). This model could be improved by letting the physical properties vary in a non-linear fashion with depth - as they certainly do in real life. However, increasing the complexity of the model is of no value unless we increase the number of $\Delta p$ data points. A simple remedy is to add further pressure ports to PUPPI. This would allow the variation of $\Delta p$ with depth to be better characterised. Note that the data-logger has insufficient temporal resolution to capture any phase-lag information.

The best-fit permeability is $(1.26).10^{-13}$ $m^2$, which is one to two orders of magnitude greater than core measurements on turbidite samples from GME. This discrepancy could be accounted for by the limitations of the model and the $\Delta p$ data, but it is interesting to note that scale differences between the laboratory and in-situ measurements may also contribute. Weaver and Schultheiss (1983) have discovered that open vertical worm-burrows up to two metres in length exist in the turbidite sediments from GME. On a length scale of metres, these burrows can increase the permeability of the the average sea-bed permeability by orders of magnitude. However, the burrows are not preserved in the laboratory samples which are of a length-scale of centimetres. Therefore, the PUPPI pore-pore pressure measurements - hence the best-fit permeability - will be affected by the worm burrows, whilst the laboratory permeability measurements will not. This point illustrates the importance of length-scale when evaluating the permeability of sediments, as is well appreciated for rocks.

## 6 Conclusions

We have developed a new method to determine the average value of the shear modulus (G) of the upper few metres of the sea-bed. The method uses a simple mathematical model and differential pore-pressure measurements obtained from PUPPI. Using this method we obtained average values of G = 20 MPa for the sea-bed at GME, which is slightly larger than G values obtained form alternative techniques. The method is limited by the simplicity of the model, the lack of a relationship between $K_b$ and G for very high porosity sediments and, most importantly, the number of pressure ports on PUPPI. Allowing for these limitations, we have demonstrated the feasibility of our method of determining G from tidally induced pore-pressure measurements. Improving the capability of the differential pressure measurements - by using more closely spaced sensors with greater pressure resolution - will increase the certainty of G and k values obtained from this method. Further improvements will be given by developing the model to account for the variations of G and k with depth.

## 7 References

Biot, M.A. (1941) 'General theory of three dimensional consolidation', J. Appl. Phys., 12, 155-164.

Biot, M.A. (1956) 'Theory of propagation of elastic waves in a fluid saturated porous solid, I. Low frequency range'. J. Acoust. Soc. Am., 28, 168-178.

Hurley, M.T. (1989) 'Application of Biot's theory to sea-bed sediments', PhD Thesis, University of Wales.

Mei, C.C. and Foda, M.A. (1981) 'Wave-induced responses in a fluid-filled poroelastic solid with a free interface - a boundary layer theory', Geophys. J. R. Astr. Soc., 66, 579-631.

Okusa, S., (1985) 'Wave-induced stresses in unsaturated submarine sediments', Geotechnique, 35, 517-532.

Schultheiss, P.J. and Gunn, D.E. (1985) 'The permeability and consolidation of deep-sea sediments', Institute of Oceanographic Sciences Report, 201.

Schultheiss, P.J. and McPhail, S.D. (1985) 'Direct indication of pore-water advection from pore-pressure measurements in Madeira abyssal plain sediments', Nature, 320, 384-350.

Weaver, P.P.E. and Schultheiss, P.J. (1983) 'Vertical open burrows in deep-sea sediments 2-m in length', Nature, 301, 329-331.

Whitmarsh, R.B. and Lilwall, R.C. (1982) 'A new method for the determination of in-situ shear wave velocity in deep-sea sediments', Oceanology Int., 4.2.

Yamamoto, T., Koning, H.L., Sellmeiher H., and Hijum, E.V. (1978) 'On the response of a poro-elastic sea-bed to water waves', J. Fluid Mech., 87, 193-206.

Yamamoto, T. and Torrii, T. (1986) 'Seabed shear modulus profile inversion using surface gravity (water) wave-induced bottom motion', Geophys. J. R. Astron. Soc., 85, 413-431.

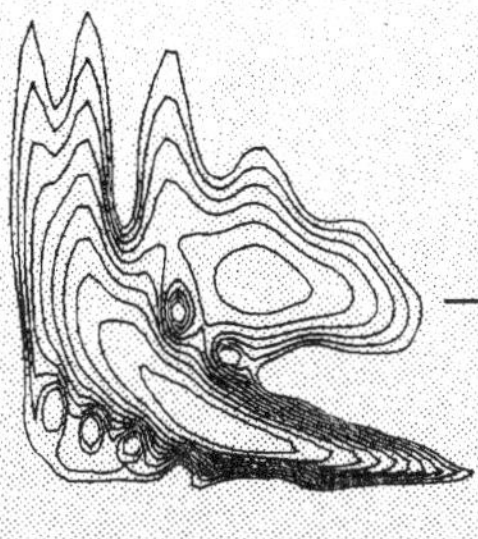

# Section 3

## *Modelling and inversion techniques*

# EXCESS ATTENUATION IN LOW-FREQUENCY SHALLOW-WATER ACOUSTICS: A SHEAR WAVE EFFECT?

FINN B. JENSEN
*SACLANT Undersea Research Centre*
*19026 La Spezia*
*Italy*

ABSTRACT. Broadband propagation data collected in a shallow-water area of the Barents Sea exhibits unusually high losses for frequencies below 200 Hz. Several geoacoustic models are tested with the SAFARI code in an attempt to construct a plausible bottom model (homogeneous or layered, fluid or elastic), which may account for the observed excess attenuation at low frequencies. It is shown that only an elastic bottom with a high shear speed near the water/bottom interface allows for a match between modeled and measured propagation loss data.

## 1. Introduction

The seabed is known to be the controlling factor in low-frequency shallow-water acoustics. A lossy seabed causes attenuation of waterborne sound due both to a critical-angle effect and to the attenuation of compressional waves in the seabed material. Early experimental results, however, indicated that low-frequency losses observed in some shallow-water areas could not be explained within the framework of a "fluid" sediment model. A natural extension was to include elasticity of sediment materials and thus allow for the coupling of acoustic energy into shear waves in the seabed. This additional degree of freedom in the physical model turned out to be able to account for the measured excess losses in some shallow-water areas, and with approximately the correct frequency dependence. However, the required shear speeds were found to be as high as 500–700 m/s in materials judged to be unconsolidated sands and gravel. Considering that *in situ* measurements in sandy materials indicate much lower shear speeds (100–300 m/s) within the upper few meters of the sediment layer, the above modeling results based on an acoustic inversion clearly have become controversial.

This paper reviews the modeling procedure which has led to the controversial shear speed results. The approach is the standard one of iterative forward modeling, where geoacoustic parameters are changed in a systematic way until an acceptable agreement is obtained between theory and experiment. Of course, we impose constraints on parameter variations to be consistent with either actual *in situ* measurements of selected parameters or with reported values in the literature. In any



*J. M. Hovem et al. (eds.), Shear Waves in Marine Sediments*, 421–430.

event, the scope of the acoustic inversion is to determine a *plausible* geoacoustic model for the area of interest.

## 2. The Acoustic Data

The variability of acoustic propagation in shallow water is illustrated in Fig. 1, where data sets from two different geographical areas are compared. The contoured quantity is transmission loss in decibels ($TL = -10\log[E_{1/3}(r,z)]$), where $E_{1/3}$ is the normalized spectral energy averaged over a 1/3-octave band. Both data sets cover a frequency range of 25–3200 Hz and propagation ranges up to 45 km.

The upper data set (Fig. 1(a)) was collected in the Barents Sea in just 60 m water depth. Note the high transmission losses recorded below 200 Hz, where energy levels fall off rapidly indicating that most of the acoustic energy emitted by the source is lost to the seabed. In contrast to this high-loss environment, we show in Fig. 1(b) a data set from the UK shelf area in 90 m water depth. Here propagation conditions are excellent over the entire frequency band. This second data set, which represents "normal" propagation conditions for thick sandy sediments, can be modeled without invoking high shear speeds in the seabed.

We shall here concentrate on the data set from the Barents Sea (Fig. 1(a)), which exhibits high losses for frequencies below 200 Hz. This low-frequency excess attenuation of waterborne sound must be attributed to seafloor effects, and we shall attempt to construct a plausible geoacoustic model (fluid or elastic) which accounts for the observed low-frequency losses.

## 3. Modeling Results

The forward modeling was performed with the SAFARI code [1], an accepted standard for solving wave-propagation problems in horizontally stratified viscoelastic media. The input parameters are wave speeds ($c_p, c_s$), wave attenuations ($\beta_p, \beta_s$) and density ($\rho$) in each layer, the layer thickness, source/receiver positions, and frequency. The code provides a full-spectrum coherent field calculation at the specified source frequency. In order to remove multipath interference effects from the narrowband modeling result, we subsequently apply a 1-km range smoothing window before comparison with the experimental data averaged over a 1/3-octave band.

The environmental information collected in the Barents Sea shows that the water depth is around 60 m along the entire track, and that the homogeneous water column has a *p*-wave speed of 1450 m/s. The bottom information is sparse, consisting of only two cores each with a penetration depth of just 60 cm. The core analysis shows that the upper part of the seabed consists of gravel intermixed with sand. The porosity is as low as 20–30 %, and the density is around 2.2 g/cm$^3$. With this information on hand we can estimate the *p*-wave speed from empirical information published by Hamilton [2] on velocity-porosity relationships for continental-shelf sediments. We find $c_p$=1800–2000 m/s, which agrees well with published data for sand/gravel materials. In the modeling we pick the lower value of 1800 m/s, which together with a density of 2.0 g/cm$^3$, are considered fixed values for the near-surface materials in our various geoacoustic models.

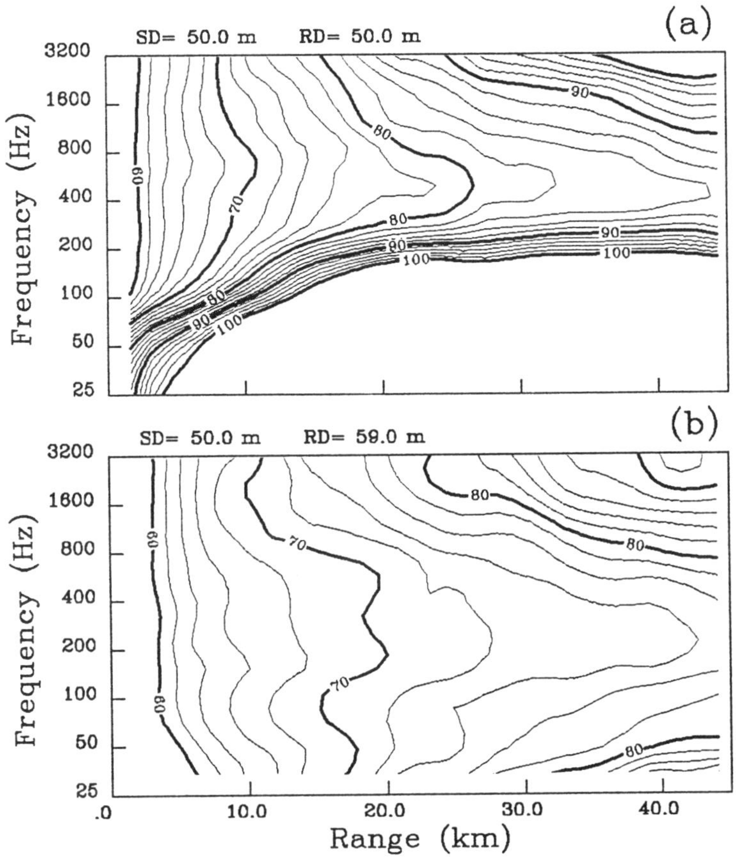


Figure 1. Measured 1/3-octave band propagation losses in two shallow-water areas. (a) Data from the Barents Sea in 60 m water depth exhibiting extreme losses at frequencies below 200 Hz (excess attenuation); (b) Data from the UK shelf area in 90 m water depth.

Table 1: Homogeneous fluid bottom

| Layer | Thickness (m) | Sound speed (m/s) | Attenuation (dB/$\lambda$) | Density (g/cm$^3$) |
|---|---|---|---|---|
| Water | 60 | 1450 | 0.0 | 1.0 |
| Substrate | $\infty$ | 1800 | 0.5 | 2.0 |

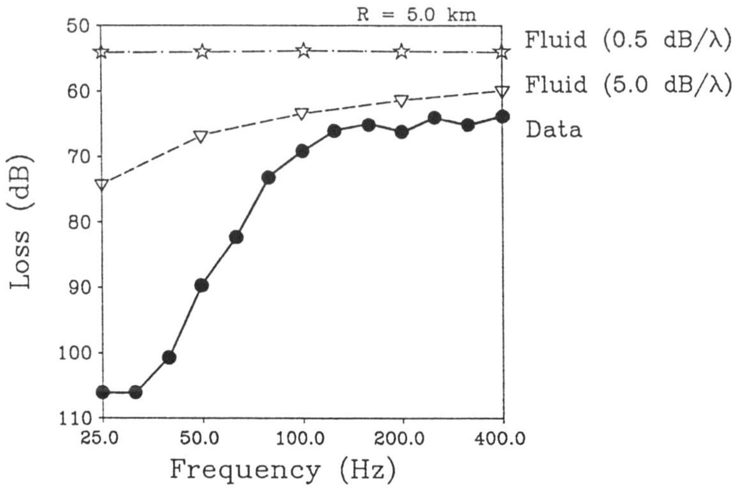


Figure 2. Comparison of measured propagation losses in the Barents Sea at a range of 5 km with model predictions for a homogeneous fluid bottom (Table 1). The two model results correspond to different $p$-wave attenuations.

### 3.1. HOMOGENEOUS FLUID BOTTOM

We first attempt to model the Barents Sea data with a homogeneous fluid bottom (Table 1). The only unknown parameter is the $p$-wave attenuation, which has been chosen to be 0.5 dB/$\lambda$. The literature [2,3] show almost an order of magnitude variation in attenuation values for different sediment types. Recent measurements by Stoll [3] indicate quite low $p$-wave attenuations in sandy sediments in the frequency range 25–400 Hz. Thus a value of 0.1 dB/$\lambda$ might be more realistic than the high-loss value chosen by us. On the other hand, we need considerable bottom loss to explain the experimental data, and $\beta_p$=0.5 dB/$\lambda$ is not an entirely unrealistic value.

The model result for a homogeneous fluid bottom is shown in Fig. 2 together with measured propagation losses at a range of 5 km. Clearly the fluid bottom does not provide sufficient loss at low frequencies, where the data are 50 dB below the model prediction. Not even by introducing the entirely unrealistic $p$-wave attenuation of 5.0 dB/$\lambda$ do we have sufficient bottom loss at low frequencies. Hence we may conclude that it is not possible to explain the observed losses in the Barents Sea within the framework of a (homogeneous) fluid bottom.

### 3.2. SEDIMENTARY BOTTOM WITH A SHEAR GRADIENT

We next consider the same fluid bottom described in Table 1, but now with the added complexity of shear in the sediment. While density and $p$-wave properties are known to have small depth gradients in unconsolidated shelf sediments, the shear speed increases rapidly with depth near the water/bottom interface. Representative

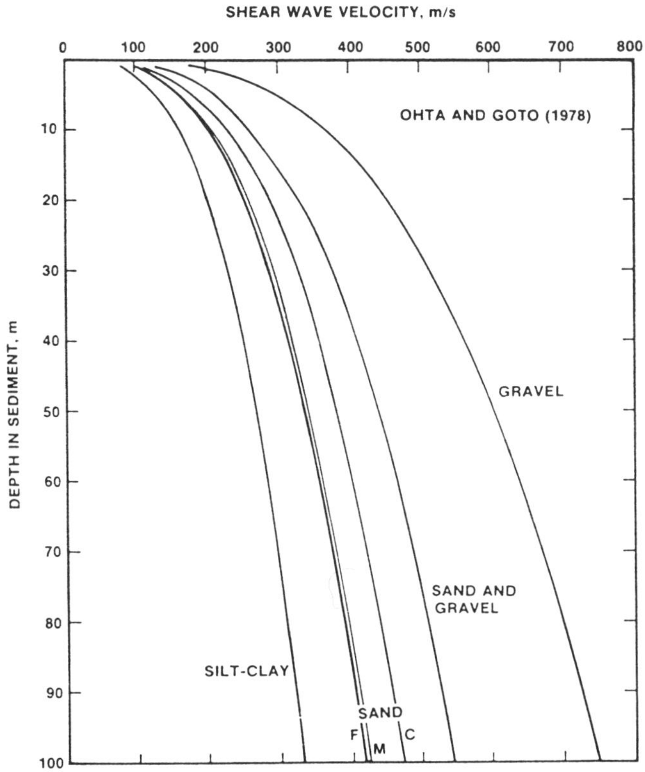


Figure 3. Shear-speed profiles in unconsolidated sediments (from Hamilton [2]).

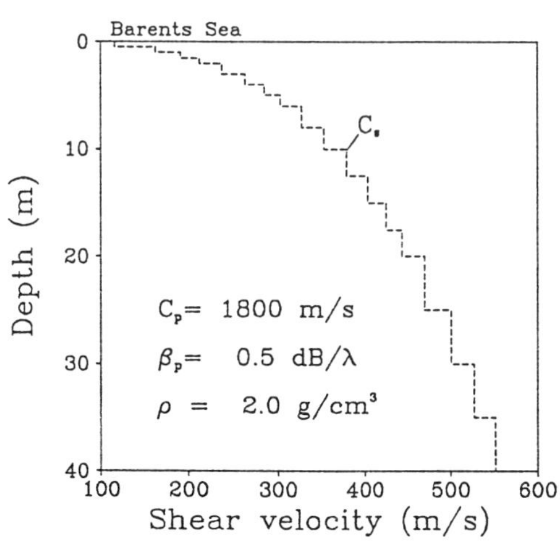


Figure 4. Geoacoustic model for sedimentary bottom (gravel) with a shear-speed gradient.

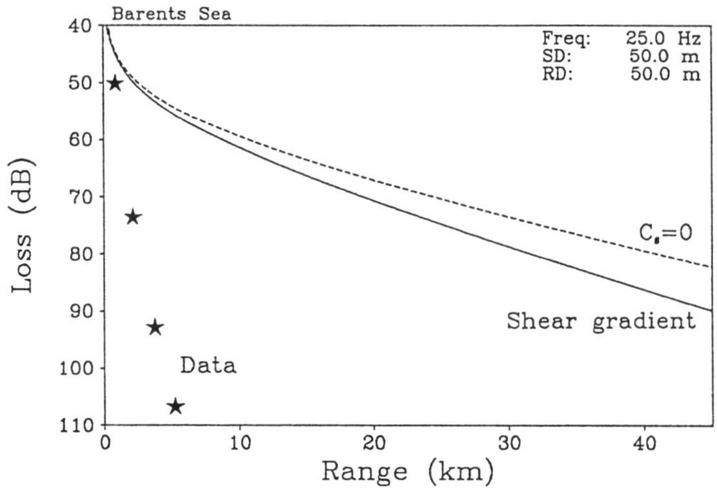


Figure 5. Comparison of measured propagation losses in the Barents Sea at 25 Hz with model predictions for a sedimentary bottom with a shear-speed gradient (Fig. 4).

shear gradients for different sediment types are given in Fig. 3 (after Hamilton [2]). Note that the shear speed is around 100 m/s near the seafloor.

In order to maximize shear losses in our modeling of the Barents Sea data, we select the curve for gravel in Fig. 3 to represent our shear speed variation with depth ($c_s = 178D^{0.312}$). Hence our geoacoustic model becomes as shown in Fig. 4, where the shear speed is seen to take a value of 120 m/s near the seafloor, increasing with depth to 550 m/s. The shear-wave attenuation is kept constant over depth at a value of 1.0 dB/$\lambda$.

The model result is shown in Fig. 5 (solid line) for a frequency of 25 Hz. For comparison we also show the predicted loss when shear is neglected alltogether (dashed line). We notice that even though shear does cause an additional loss of approximately 8 dB at 45 km, the effect is much too small to explain the experimentally observed losses. In fact, the additional loss caused by the shear gradient could have been obtained also by increasing the $p$-wave attenuation from 0.5 to 0.7 dB/$\lambda$. Hence, the effect of the shear gradient is within the uncertainty associated with the $p$-wave attenuation values, and consequently shear effects can be ignored in unconsolidated sediments. Of course, this statement is valid only to the extent that sediment shear speeds can be represented by the curves given in Fig. 3.

The reason for the small effect of shear in unconsolidated sediments is the low shear-speeds encountered just below the seafloor ($\sim$ 100 m/s). In fact, even at 25 Hz the important part of the shear profile is within the upper 5 m of the sediment where the mean shear speed is around 200 m/s. In the next section we show that much higher shear speeds are required near the water/bottom interface to account for the observed low-frequency losses in the Barents Sea.

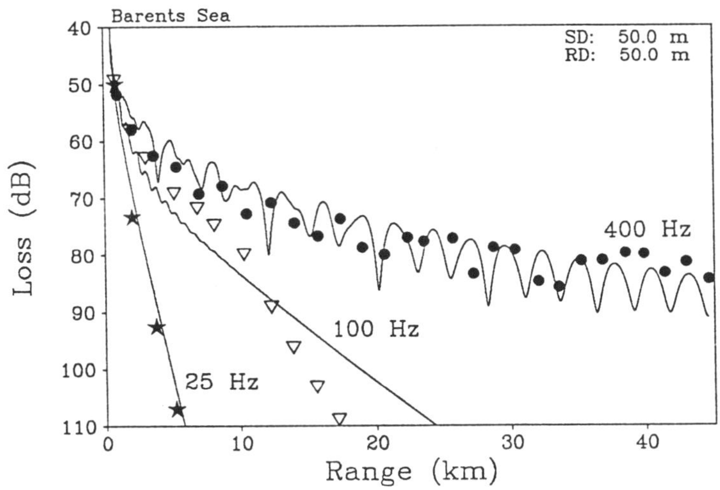


Figure 6. Comparison of measured propagation losses in the Barents Sea at 25, 100 and 400 Hz with model predictions for a homogeneous elastic bottom (Table 2).

## 3.3. HOMOGENEOUS ELASTIC BOTTOM

We now consider the bottom to be a semi-consolidated material with a high shear speed, i.e. we let $c_s$ be a free parameter to be determined so as to obtain agreement between modeled and measured low-frequency losses. An acceptable agreement was achieved with the parameters given in Table 2. Note that the required shear speed is 700 m/s, which leads to the model/data agreement shown in Fig. 6. We now have a strong coupling of acoustic energy into shear waves in the bottom resulting in very high propagation losses at 25 Hz. The 100 Hz data show an anomalous decay with range, which is not evident in the model result. This behavior is most likely due to lateral variations in the environmental properties not accounted for in the acoustic model. At 400 Hz the model/data agreement is again excellent.

Now, is a homogeneous elastic bottom with a shear speed of 700 m/s a plausible geoacoustic model for this area? Some support for this type of bottom model may be found in a recent study by Hughes *et al.* [4] of low-frequency propagation loss in various shallow-water areas at high latitudes. They report wave speeds in glacial till to be $c_p$=2000 m/s and $c_s$=800 m/s, which are values very similar to the ones used by us (Table 2). Hence, a plausible bottom model for the Barents Sea area is that we have a thick layer of glacial till (or similar material) with a shear speed of around 700 m/s.

## 3.4. LAYERED ELASTIC BOTTOM

The final geoacoustic model considered is an unconsolidated thin sediment layer on top of a solid substrate. Such a model was also considered by Hughes *et al.* [4] for

Table 2: Homogeneous elastic bottom

| Layer | Thickness (m) | Sound speed (m/s) | Attenuation (dB/$\lambda$) | Density (g/cm$^3$) |
|---|---|---|---|---|
| Water | 60 | 1450 | 0.0 | 1.0 |
| Substrate | $\infty$ | $p$ 1800 | 0.5 | 2.0 |
| | | $s$ 700 | 1.0 | |

Table 3: Layered elastic bottom

| Layer | Thickness (m) | Sound speed (m/s) | Attenuation (dB/$\lambda$) | Density (g/cm$^3$) |
|---|---|---|---|---|
| Water | 60 | 1450 | 0.0 | 1.0 |
| Sediment | 1.5 | $p$ 1800 | 0.5 | 2.0 |
| | | $s$ 140 | 1.0 | |
| Substrate | $\infty$ | $p$ 3700 | 0.5 | 2.5 |
| | | $s$ 1850 | 1.0 | |

several data sets. One indication that this model might be appropriate also for the Barents Sea track stems from independent geophysical data published by Faleide [5]. He reports that the bottom consists of sandstone (shale) covered by a thin layer (0.5–3 m) of glaciomarine sediments. Using directly the $p$- and $s$-wave properties given by Faleide [5] for sandstone, we settled on the layered elastic bottom model given in Table 3. Hence, we have a 1.5 m thick sediment layer with a low shear speed of 140 m/s overlying a fast substrate. This geoacoustic model is consistent both with our coring result (0.6 m of sand and gravel) and with the bottom model given in Ref. [5].

It is well-known that a layered bottom gives rise to reflection-loss resonances, which cause high propagation losses at particular frequencies. This phenomenon was treated in considerable detail by Hughes *et al.* [4], who showed that these resonances occur at frequencies where the layer thickness equals an odd multiple of $\lambda_s/4$ in the top layer. In our case we get high losses at $F$=23.3, 70, 117 Hz, etc., which is evidenced by the computed plane-wave reflection loss at the seafloor shown in Fig. 7. However, we do not expect to see sharp resonances in the experimental data, due both to the 1/3-octave band averaging and to lateral variations in layer thickness and material properties.

The model results for both the homogeneous and the layered elastic bottom are shown in Fig. 8. Note that both results compare well with the experimental data

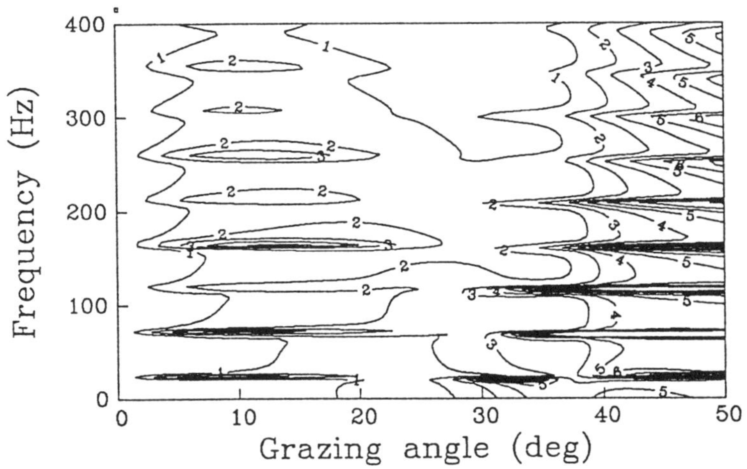


Figure 7. Contoured reflection loss in decibels versus frequency and grazing angle for a layered elastic bottom.

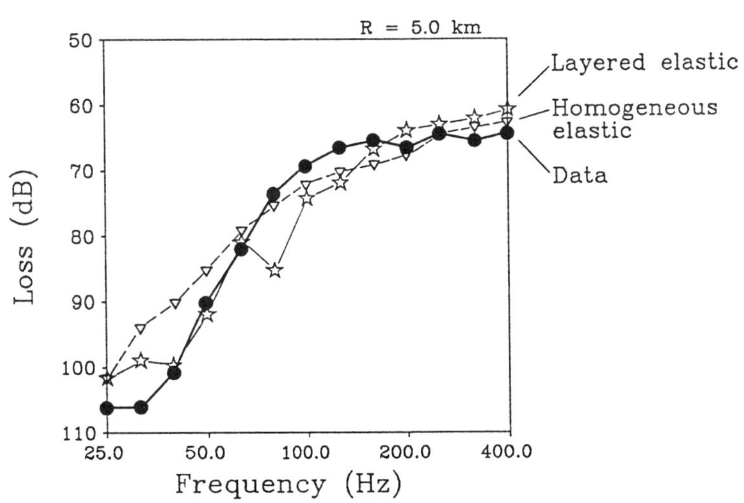


Figure 8. Comparison of measured propagation losses in the Barents Sea at a range of 5 km with model predictions for a homogeneous elastic bottom (Table 2) and a layered elastic bottom (Table 3).

collected at a range of 5 km. Hence, we have constructed two geoacoustic models which are both plausible and which agree equally well with the experimental data over the frequency range considered. The question of which model is most similar to reality cannot be answered with the current data set. The two elastic models (homogeneous and layered) would provide quite different answers for frequencies below 25 Hz, but no data are available in this frequency range.

The point is, however, that both of these models have high shear speeds in the upper 5 m of the bottom, and the high shear speeds are required in order to explain the measured low-frequency losses in the Barents Sea. Therefore, to return to the question posed in the title of this paper, the answer must be affirmative, i.e. the measured excess attenuation in this area is indeed a shear wave effect. Of course, this conclusion does not exclude the presence of other loss mechanisms in the seafloor. Our modeling results are based on the assumption that sediment materials are homogeneous and isotropic. In reality, sediment materials are often heterogeneous and anisotropic, which could give rise to additional coupling between different wave types in the sediment and also to additional energy dissipation. However, so far there is no evidence (theoretical or experimental) that these effects can account for the excess losses observed in the Barents Sea, and, consequently, shear effects remain the most likely cause for the loss of low-frequency acoustic energy to the seabed.

## 4. Summary and Conclusions

It is clear from this model study that the observed low-frequency excess attenuation in certain shallow-water cases cannot be explained within the framework of a fluid sediment model. It is also evident that the low shear speeds associated with unconsolidated sediments ($\sim$ 100 m/s) have negligible effect on acoustic propagation in the water column. Only the assumption of a high shear speed right at the water/bottom interface (consolidated sediments) or the presence of a solid substrate covered by a thin sediment layer, can account for the measured low-frequency losses. In both cases the principal loss mechanism is the coupling of acoustic energy into shear waves in the seabed.

## References

[1] Schmidt, H. (1988) "SAFARI: Seismo-acoustic fast-field algorithm for range independent environments," Rep. SR-113, SACLANT Undersea Research Centre, La Spezia.

[2] Hamilton, E.L. (1987) "Acoustic properties of sediments," in A. Lara-Sáenz, C. Ranz-Guerra and C. Carbó-Fité (eds.), *Acoustics and Ocean Bottom* (C.S.I.C., Madrid, Spain), pp. 3–58.

[3] Stoll, R.D. (1989) *Sediment Acoustics* (Springer-Verlag, Berlin, Germany).

[4] Hughes, S.J., Ellis, D.D., Chapman, D.M.F. and Staal, P. (1990) "Low-frequency acoustic propagation loss in shallow water over hard-rock seabeds covered by a thin layer of elastic–solid sediment," J. Acoust. Soc. Amer. **88**, 283–297.

[5] Faleide, J.I. (1984) "Geoacoustic modeling of the sea floor — with examples from the Barents Sea, North Sea, Skagerak, Kattegat and Oslofjord," Rep. 84/2007, Norwegian Defence Research Establishment, Kjeller, Norway.

# SENSITIVITY OF BOTTOM LOSS TO ATTENUATION AND SHEAR CONVERSION.

ÅGE KRISTENSEN and JENS M. HOVEM
*SACLANT Undersea Research Centre*
*19026 La Spezia*
*Italy*

ABSTRACT. Acoustic wave propagation in shallow water is highly dependent on the properties of the bottom. For grazing angles lower than the critical angle, which corresponds to propagating modes, the reflection loss will depend on absorpsion and conversion to shear waves in the bottom. This paper reports some results of a sensitivity study aimed at obtaining a better quantitative understanding of how the bottom parameters influence the propagation loss. First we assume that the bottom can be modelled as an unconsolidated homogeneous halfspace. The study shows that propagation loss is governed by the p-wave parameters and that the s-wave parameters are of minor importance. Conversion to shear waves will be important for a hard bottom. Second we consider the case where the bottom has a thin sedimentary layer over a hard elastic substrate. The results show that abnormally high losses may occur over a relatively wide angle-frequency band. A minimum requirement for these abnormal losses to occur is that the subbottom has a high shear speed and that there is attenuation either in the layer or the substrate, but it is not neccesary that the sediment layer supports shear waves.

## 1. Introduction

Reflection loss at the ocean bottom is an important part of the total loss for long range sound propagation in shallow water. For grazing angles lower than critical the bottom loss depends on absorption and conversion of energy into shear waves in the bottom . To give accurate transmission loss (TL) predictions the bottom parameters, i.e. compressional and shear wave velocity and attenuation, must be known.

At SACLANTCEN work is in progress to develop instruments and techniques that can determine these parameters directly and remotely. Before methods to measure the parameters are devised, we have carried out a parameter study aimed at determining quantitatively how the plane wave reflection coefficient depends on the properties of the bottom.

In order to establish the acceptable errors in predicting the bottom loss it is assumed that the total transmission loss is to be predicted with an accuracy of 3 dB at 20 km range in isovelocity water with constant depth of 100 m. For this idealized case, it is assumed that the bottom loss contribution to the total transmission loss can be calculated as



*J. M. Hovem et al. (eds.), Shear Waves in Marine Sediments*, 431–438.

$$TL = BL\frac{R}{2D}\tan\theta_0 \quad (1)$$

where BL is the bottom loss in dB per bounce, R is the range, D is the water depth and $\theta_0$ is the grazing angle with the bottom corresponding to a particular ray or mode. As demonstrated later in this paper comparison with a numerical code such as SNAP[1] shows that this simple expression is sufficiently accurate for a sensitivity analysis.

In the environmental model we will consider this corresponds, for the lowest order mode, to a grazing angle of approximately $7^0$ at 50 Hz, this translates into a required accuracy of about 0.2 dB for the prediction of the reflection loss for a single bounce.

The first part of this study considers a homogenous bottom and we discuss the effect of shear wave conversion and absorption on the bottom reflectivity. We also briefly discuss the case where the bottom is poroelastic by application of the Biot theory. In the second part we study a simple layered bottom with a sediment layer over a solid substrate and we look at the effect of different medium parameters. Since our prime interest is long range propagation we will in all cases consider only grazing angles smaller than critical, i.e. the propagating modes.

## 2. Homogeneous Bottom

For the homogeneous bottom the density $\rho = 2000$ kg/m$^3$ and compressional wave speed $v_p = 1700$ m/s are kept constant. The compressional attenuation $\alpha_p$, the shear speed $v_s$ and the shear attenuation $\alpha_s$ are varied one at the time in order to study the effect of the individual parameters.

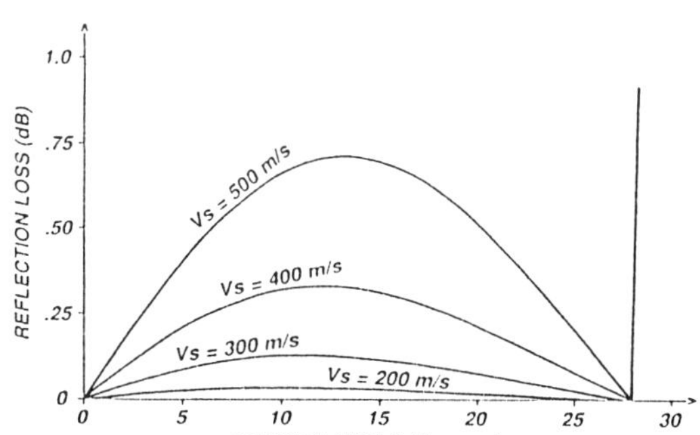


Figure 1. Reflection loss versus grazing angle for various shear speeds for a thick sedimentary bottom.

### 2.1. SHEAR WAVE CONVERSION

The effect of shear wave conversion (no compressional loss) on the bottom loss is illustrated in Fig. 1, showing bottom loss per bounce versus grazing angle for different values of the shear velocity. Notice that quite high shear speeds are required to produce a significant bottom loss. For instance, a shear speed higher than 400 m/s is needed to account for a loss per bounce higher than 0.2 dB for the lowes order mode in the above example. In the literature[2] shear wave speeds higher than 400 m/s are hardly ever reported for sedimentary bottoms.

In order to gain further insight in the loss due to shear wave conversion at the interface, a first order perturbation approach shows that the reflection intensity at angles lower than critical can be expressed as

$$|R|^2 = 1 - 4\epsilon^4 |\sin^2\theta_1 T_{01} T_{10}|^2 \tag{2}$$

where $\epsilon$ is a perturbation parameter given by $\epsilon = v_s/v_p$ and $T_{ij}$ are the normal transmission coefficients at fluid-fluid interfaces and $\theta_1$ is the refracted angle in the second medium. A similar expression has earlier been derived by Vidmar[3]. The accuracy of this expression is found to be quite good at least up to $\epsilon = 0.1$ corresponding to $v_s/v_p = 0.3$ or up to $v_s$=500 m/s for $v_p$=1700 m/s. Equation 2 shows that reflection loss due to shear conversion is a secondary effect since $|R|^2$ is approximately proportional to $(v_s/v_p)^4$. Consequently its effect on the bottom loss will be small for small $\epsilon$ but increasing rapidly for higher $\epsilon$-values.

These results show that for thick sedimentary bottoms shear wave conversion is of minor interest for TL-predictions. However for consolidated bottoms, where the shear speed may be well over 400 m/s, shear wave conversion is an important loss mechanism.

In conclusion we find that the importance of the bottom loss due to shear wave attenuation is rather modest and with the given accuracy requirement of no interest for realistic values of shear speed and attenuation.

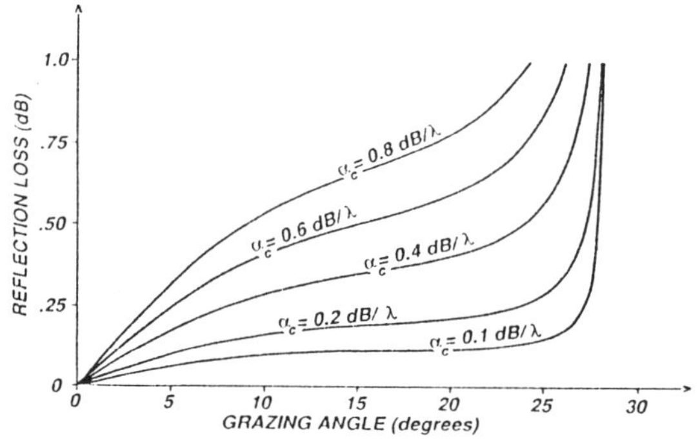


Figure 2. Reflection loss versus grazing angle for various compressional attenuation factors for a liquid bottom.

## 2.2. COMPRESSIONAL WAVE ATTENUATION

The effect of compressional wave attenuation on the bottom loss is illustrated in

Fig. 2 showing bottom loss per bounce versus grazing angle for different values of the compressional attenuation, $\alpha_p$. In this case the bottom is considered as a fluid. We notice that a compressional attenuation of 0.4 dB/$\lambda$ will produce a loss per bounce of 0.2 dB at $7^0$.

For the reflection loss due to compressional wave attenuation the following approximate expression can be derived:

$$|R|^2 = 1 - \frac{\delta}{2\pi|\cos\theta_1|^2} T_{01} T_{10} \tag{3}$$

where $\delta$ is the log decrement in Neper/$\lambda$. The bottom loss increases linearly with the compressional wave attenuation and this expression has been found to be accurate up to a value $\delta = 0.25$ or to about 2.0 dB/$\lambda$.

Since the bottom loss increases linearly with the compressional attenuation it is necessary for this particular environment and at 50 Hz to know the attenuation factor to an accuracy of 0.4 dB/$\lambda$ in order to reach our TL-prediction accuracy requirement of 3.0 dB at 20 km range.

*2.2.1. Comparison With SNAP.* The simple model of Equation 1 and the numerical model SNAP has been compared. The results are shown in Table 1 where the bottom loss part for the two lowest order modes of the transmission loss is given at 20 km range and 50 Hz. M1 and M2 are loss at 20 km range for mode 1 and mode 2. The results show that the simplified model gives very good agreement with the SNAP model.

Table 1: Bottom loss at 20 km range

| $V_s$ (m/s) | $\alpha_p$ (dB/$\lambda$) | Refl.Coef. M1 | M2 | SNAP M1 | M2 |
|---|---|---|---|---|---|
| 200 | 0.0 | 0.4 | 0.8 | 0.4 | 0.7 |
| 300 | 0.0 | 1.3 | 3.0 | 1.3 | 2.8 |
| 400 | 0.0 | 3.2 | 8.0 | 3.0 | 7.4 |
| 200 | 0.1 | 0.8 | 2.4 | 0.9 | 2.5 |
| 200 | 0.2 | 1.5 | 4.4 | 1.6 | 4.4 |
| 200 | 0.4 | 2.3 | 8.6 | 2.2 | 8.2 |

## 2.3. POROELASTIC EFFECTS

In this section we shall model the bottom as a porous solid using the Biot theory and the appropriate boundary conditions. The problem of reflection and transmission at an interface with a porous medium has been studied before and in particular the work of Stoll and Kan[4] is relevant for the present study. We will use their notation for the many parameters of the Biot theory.

Table 2: Parameters for the porous solid model

| Parameter | Unit | Value |
|---|---|---|
| Bulk modulus of grains, $K_r$ | (Pa) | $3.60*10^{10}$ |
| Bulk modulus of fluid, $K_f$ | (Pa) | $2.25*10^{9}$ |
| Specific gravity of grains | | 2.65 |
| Specific gravity of fluid | | 1.0 |
| Shear modulus of frame, $\mu$ | (Pa) | 0 |
| Bulk modulus of frame,$K_b$ | (Pa) | $3.50*10^{8}$ |
| Log decrement of frame | (shear) | 0.15 |
| Log decrement of frame | (dilitation) | 0.15 |
| Porosity | | 0.37 |
| Permeability, k | ($cm^2$) | $1.30*10^{-7}$ |
| Viscosity of fluid, $\eta$ | (Poise) | 0.01 |
| Poresize parameter, a | (cm) | $1.37*10^{-3}$ |
| Structure constant, c | | 1.25 |

Anticipating that the most significant effect of the porous interface would be with a highly permeable medium, we study the case of the bottom being composed of medium sand with compressional and shear velocities of 1700 m/s and 200 m/s and attenuations being 0.87 dB/$\lambda$ for both waves.

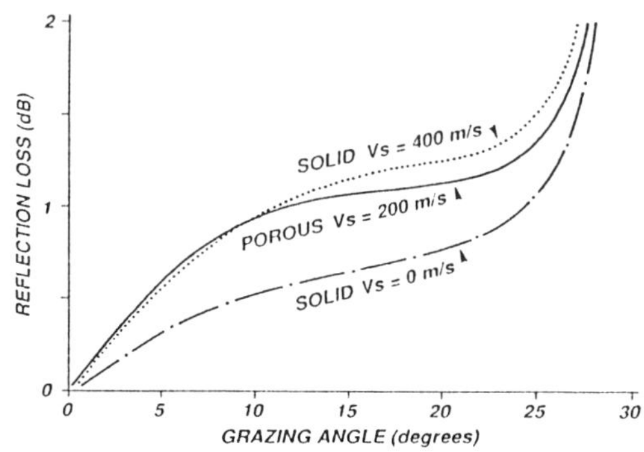


Figure 3. Comparison of reflection loss versus grazing angle between a elastic bottom and a poroelastic bottom.

The values of the parameters in the Biot theory are listed in Table 2. Figure 3 compares the reflection loss calculated by the two models and shows that in this case the porous effect is marginally significant adding about 0.3 to 0.35 dB to the reflection loss at $7^0$. With the solid model a shear wave velocity as high as 400 m/s or a compressional wave attenuation of about 1.3 dB/$\lambda$ is required to model the same reflection loss. However this was considering a very permeable sand; when applying the porous-solid boundary conditions to all other bottom types we find that the porous effect is rather insignificant.

Table 3: Layered elastic bottom

| Layer | | Thickness (m) | Sound speed (m/s) | Attenuation (dB/$\lambda$) | Density (kg/m$^3$) |
|---|---|---|---|---|---|
| Water | | 100 | 1500 | 0.0 | 1000 |
| Sediment | $p$ | 10 | 1700 | 0.8 | 2000 |
| | $s$ | | varying | 0.0 | |
| Substrate | $p$ | $\infty$ | 4700 | 0.30 | 2500 |
| | $s$ | | 2300 | 0.15 | |

## 3. Layered Bottom

The layered elastic model has a sediment layer over a solid infinite substrate. The layer thickness of 10 m has a density $\rho = 2000$ kg/m$^3$ and constant compressional wave parameters, $v_p = 1700$ m/s and $\alpha_p = 0.8$ dB/$\lambda$. The sediment is either considered as a fluid, $v_s = 0.0$ m/s, or an elastic material with $v_s = 200$ m/s and $\alpha_s = 0.0$ dB/$\lambda$. The model parameters are summarized in Table 3.

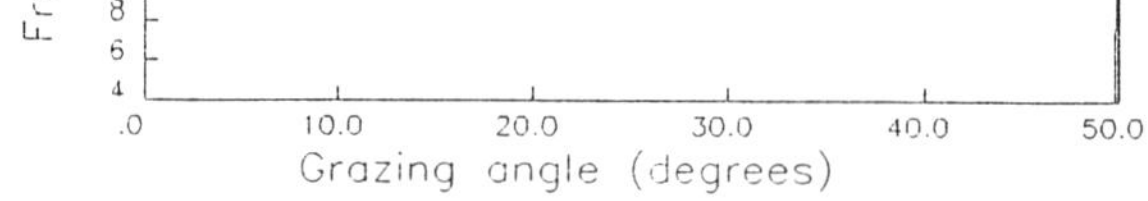


Figure 4. Reflection loss versus grazing angle and frequency for a liquid sediment layer over a hard elastic substrate.

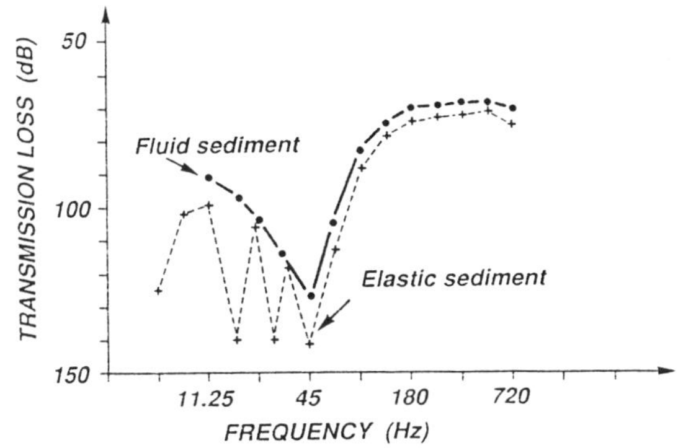


Figure 5. Transmission loss versus frequency at 30 km range both for a liquid and a elastic sediment layer over a hard elastic substrate.

## 3.1. FLUID SEDIMENT LAYER

We will first study the influence of a thin fluid sediment layer over a hard elastic bottom by calculating the plane wave reflection coefficient. The result is shown in Fig. 4 where the plane wave reflection coefficient is plotted versus grazing angle and frequency. The contours represent loss per bounce in decibels. Fig. 4 shows that even with a fluid sediment layer we can have very high losses over a fairly broad frequency range which in this case is from 20 to 150 Hz. This high loss region occurs at grazing angles smaller than the critical angle. In our environmental model the two lowest order modes propagates at angles which are in the middle of this region.

We have used SAFARI[5] to calculate TL versus frequency at 30 km range. The result is shown in Fig. 5. The beforementioned frequency range shows high losses.

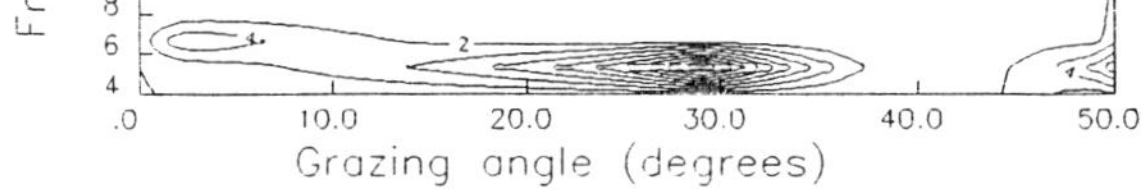


Figure 6. Reflection loss versus grazing angle and frequency for a elastic sediment layer over a hard elastic substrate.

### 3.2 ELASTIC SEDIMENT LAYER

We have also calculated the plane wave reflection coefficient versus grazing angle and frequency for an elastic sediment layer. The result is shown in Fig. 6. A similar high loss region as we saw in the fluid sediment case is apparent. In addition there are resonance-like responses at lower frequencies. This resonance-like phenomenon has also been reported by other investigators[6].

SAFARI is used to calculate the TL versus frequency. The results in Fig. 5 shows that at high frequencies the results for the two cases are very similar. The relatively wide frequency band with high losses occurs for both environmental models while we also have the resonance-like phenomenon at low frequencies for the elastic sediment case.

## 4. Summary and Conclusions

The parameter study shows that for a homogeneous bottom the loss due to shear conversion is only important when the shear speed is higher than 400 m/s. In practice this means that for homogeneous bottoms shear conversion is only important for consolidated bottoms. For unconsolidated bottoms the loss caused by compressional wave attenuation is important and for the environmental model we have used attenuation coefficient should be known to an accuracy of 0.4 dB/$\lambda$. Poroelastic effects may be marginally important for very permeable bottom materials but are otherwise of no importance. The bottom loss of a layered structure is very complicated showing a number of angle-frequency domains where exceptionally high losses may occur. A minimum requirement for these abnormal losses to occur is that the substrate supports shear waves and that there is attenuation in at least one of the layers, but it is not necessary that the sediment layer supports shear waves. Further research is needed to understand the cause of these high losses and to predict their occurrence.

## References


[1] Jensen, F.B. and Ferla, M. C. (1979) "SNAP: The SACLANTCEN normal-mode acoustic propagation model," SM-121, SACLANT Undersea Research Centre, La Spezia.

[2] Hamilton, E.L. (1987) "Acoustic properties of sediments," in A. Lara-Sáenz, C. Ranz-Guerra and C. Carbó-Fité (eds.), *Acoustics and Ocean Bottom* (C.S.I.C., Madrid, Spain), pp. 3–58.

[3] Vidmar,P.J (1980) "Ray path analysis of shear wave effects on bottom reflection loss," J. Acoust. Soc. Am. **68**, 639–648.

[4] Stoll, R. D and Kan, T. K. (1981) "Reflection of acoustic waves at a water-sediment interface," J. Acoust. Soc. Am. **70**, 149–156.

[5] Schmidt, H. (1988) "SAFARI: Seismo-acoustic fast-field algorithm for range independent environments," SR-113, SACLANT Undersea Research Centre, La Spezia, Italy.

# THE EFFECT OF SHEAR WAVE ATTENUATION ON ACOUSTIC BOTTOM LOSS RESONANCE IN MARINE SEDIMENTS

STEVEN J. HUGHES and DAVID M.F. CHAPMAN
*Defence Research Establishment Atlantic*
*Box 1012, Dartmouth, Nova Scotia*
*Canada, B2Y 3Z7*

N. ROSS CHAPMAN
*Defence Research Establishment Pacific*
*Forces Mail Office, Victoria, British Columbia*
*Canada, V0S 1B0*

ABSTRACT. Acoustic propagation experiments in shallow water over thinly-sedimented bedrock sometimes show high propagation loss, with resonant-like structure, in the 10 to 100 Hz band. This can be interpreted with the aid of plane-wave reflection models that associate high bottom reflection loss at certain combinations of frequency and grazing angle with shear-wave resonances in the sediment layer. In this paper, the seabed is modelled as a thin layer of elastic-solid sediment overlying a hard rock substrate, with plane interfaces at layer boundaries. The bottom reflection loss resonances are explained by both full wave-theoretical reflection loss calculations and a simplified coherent ray reflection loss model. For the ray model, two simplifying assumptions are applied: (1) there is no conversion of energy between compressional waves and shear waves at the water-sediment boundary, and (2) sediment shear waves are attenuated so strongly that they traverse the sediment layer once up and down only. The paper investigates the effect of sediment shear-wave attenuation on bottom reflection loss. We find that the resonance frequencies are controlled by the shear speed and sediment layer thickness, and that the amount of bottom reflection loss is sensitive to the shear-wave attenuation.

## 1. Introduction

Some acoustic propagation measurements over partially exposed hard-rock seabeds show high loss with resonant-like structure in the frequency band below about 100 Hz (MacPherson and Fothergill (1962), Worley and Walker (1982), Beebe and Holland (1986), Staal *et al* (1987)). For example, Figure 1 shows measured propagation loss versus frequency over a rough, partially exposed granite seabed (Staal *et al* (1987)). A resonant-like structure is evident in the measured propagation loss at frequencies of about 25 and 70 Hz. Treating the seabed as a single-layer, solid substrate, the SAFARI fast-field program (Schmidt, 1988) can be used to calculate theoretical propagation loss, also shown in Figure 1. The measured loss exceeds the calculation by nearly 60 dB at 25 Hz.

Vidmar (1980a) predicts that shear-wave resonances in a thin layer of solid sediment overlying a hard substrate could produce 25 dB peaks in the bottom reflection loss. For sediments with low shear speed, these resonant peaks occur at frequencies for which odd multiples of quarter-wavelengths of the shear wave are equal to the thickness of the sediment layer. Beebe and Holland (1986) successfully model the high propagation loss they measured by including a thin layer of solid sediment in their geoacoustic model of the seabed. Hughes *et al* (1990) show that the measured propagation loss in Figure 1 can also be modelled by including a thin layer of solid



*J. M. Hovem et al. (eds.), Shear Waves in Marine Sediments*, 439–446.

sediment in the seabed model. They show that the high propagation loss is due to the excitation and attenuation of sediment shear-waves at the frequencies predicted by Vidmar (1980a).

This paper investigates the effect of sediment shear-wave attenuation on the bottom reflection loss resonances. We use a simple two-layer seabed model, with plane boundaries and iso-speed layers, to calculate bottom reflection loss. We find that the amount of bottom reflection loss at resonance is sensitive to the attenuation of shear-waves in the seabed sediment layer, that the resonance frequencies are controlled by the shear speed and sediment layer thickness, and that the calculated propagation loss is controlled in large part by the bottom reflection loss.

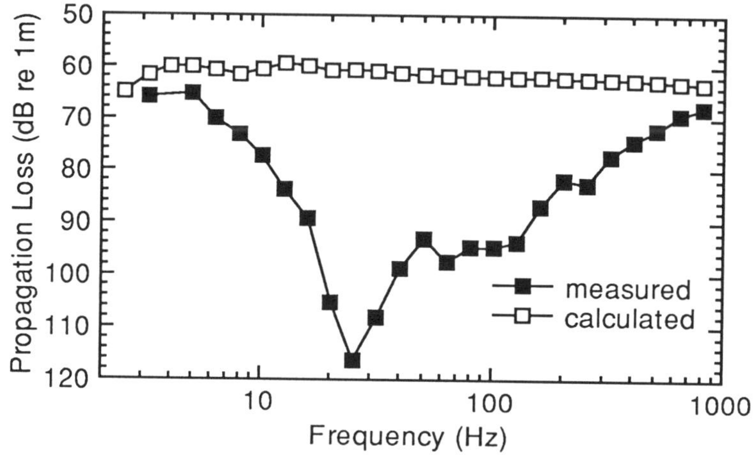


*Figure 1.* *Acoustic propagation loss versus frequency measured over a rough, partially exposed granite seabed (filled squares), and theoretical propagation loss calculated for a single-layer, solid substrate (open squares).*

## 2. Bottom Reflection Loss Models

We use a simple two-layer model of the seabed for the calculation of bottom reflection loss: a thin layer of solid sediment (sand) overlying a solid half-space substrate (granite). All layers are iso-speed with plane boundaries. The geoacoustic parameters for the model are from Hughes *et al* (1990), and are given in Table 1.

*Table 1.* *Geoacoustic model for the calculation of bottom reflection loss. Sound speed and attenuation in the sediment layer and hard-rock substrate are given for both compressional (p) and shear (s) waves. Note that attenuation coefficients are given as decibels per wavelength.*

| Layer | Thickness | Sound Speed | | Attenuation | | Density |
|---|---|---|---|---|---|---|
| | (m) | (m/s) | | (dB/λ) | | (g/cm$^3$) |
| Water | ∞ | *p* 1460 | | 0.0 | | 1.0 |
| Sediment | 1.75 | *p* 1780 | *s* 170 | *p* 1.25 | *s* 2.2 | 2.0 |
| Substrate | ∞ | *p* 5500 | *s* 2400 | *p* 0.55 | *s* 0.14 | 2.6 |

We have three computer software models with which we are able to calculate bottom reflection loss. The first model, SAFARI (Schmidt, 1988), calculates plane-wave reflection coefficients as a function of frequency and grazing angle for horizontally stratified fluid/solid media. The second model, described by Hughes *et al* (1990), is an analytic solution for plane-wave reflection from a thin, solid plate sandwiched between a fluid half-space and a solid half-space. The third is a simplified coherent-ray reflection loss model described below.

In principle, ray theory can be used to calculate the reflection loss from a multi-layered seabed by keeping track of all the possible rays that are reflected, transmitted, and generated at each interface. In the water, the compressional wave incident upon the water-sediment boundary gives rise to a reflected compressional wave in the water, and transmitted shear and compressional waves in the sediment. Each of these waves reflects at the sediment-substrate boundary and converts to a wave of the other type, resulting in four upward-travelling rays in the sediment. These rays multiply again at the sediment-water boundary, and so on. For a multiple-ray calculation to give the same result as a wave-theory calculation, an infinite number of rays must be combined coherently. Hastrup (1980) uses this technique with compressional waves for a mud and sand seabed. Vidmar (1980b) includes shear waves in his analysis of a sediment and rock seabed, and demonstrates how complex the calculations can become. With the rapid proliferation of both compressional and shear wave rays, some simplifications must be made so that the ray theory can help us understand reflection from the two-layer sediment and rock seabed.

Depending upon the acoustic characteristics of the layers, we can avoid the population explosion of shear and compressional waves by eliminating entire unimportant families of rays. We then have a partial sum that is an approximation of the wave-theory reflection coefficient. Specifically, in our analysis of reflection from the two-layer sediment and granite seabed, we make two simplifying assumptions: (a) there is no conversion of energy between compressional waves and shear waves at the water-sediment boundary, and (b) shear waves in the sediment are attenuated so strongly that they transit the sediment layer once up and down only. The reflection coefficient is then the sum of terms from only two ray families: (1) a family of compressional waves that travel up and down through the sediment layer, and (2) a related family in which one of the compressional wave reflections at the sediment-substrate boundary is replaced by: (a) a compressional-to-shear wave conversion, (b) an up-and-down transit by the shear wave, and (c) a shear-to-compressional wave conversion. A member of the latter family is illustrated in Figure 2. For both ray families, a compressional wave is re-radiated into the water when an upward-travelling compressional wave reflects from the sediment-water boundary. Each ray family gives rise to a geometric series in the expression for the reflection coefficient. These series have closed-form sums, resulting in an analytic approximation to the reflection coefficient, including most of the features of the full wave-theory, as will be shown.

## 3. Bottom Reflection Loss Calculations

Bottom reflection loss as a function of frequency and grazing angle for the simple two-layer seabed model (Table 1), calculated using the plane-wave model of Hughes *et al* (1990), is shown in Figure 3. The reflection loss data are displayed as grey levels, with white corresponding to less than 1 dB loss and black corresponding to greater than 16 dB loss. Resonances of the sediment shear waves result in high bottom loss at frequencies harmonically related to the thickness of the sediment layer (Vidmar (1980a), Hughes *et al* (1990)). This effect can be seen in the Figure as dark (i.e. high loss) horizontal bands at 25 Hz, 70 Hz, 120 Hz, and so on.

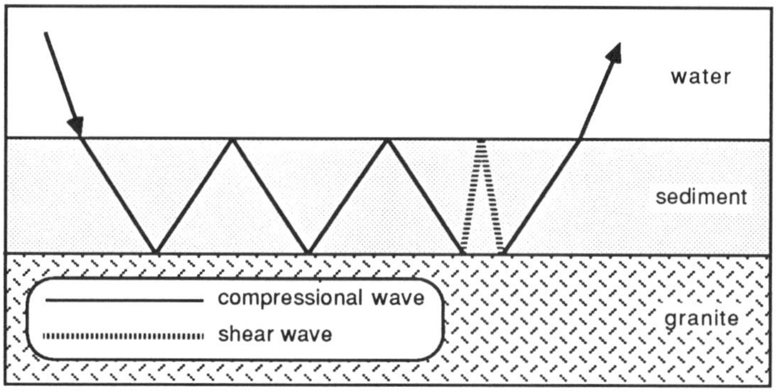


*Figure 2.* *A typical ray-path associated with a term in the ray-theory calculation of reflection loss from a two-layer sediment and granite seabed. At each reflection or transmission, there are other rays (not shown) associated with other terms.*

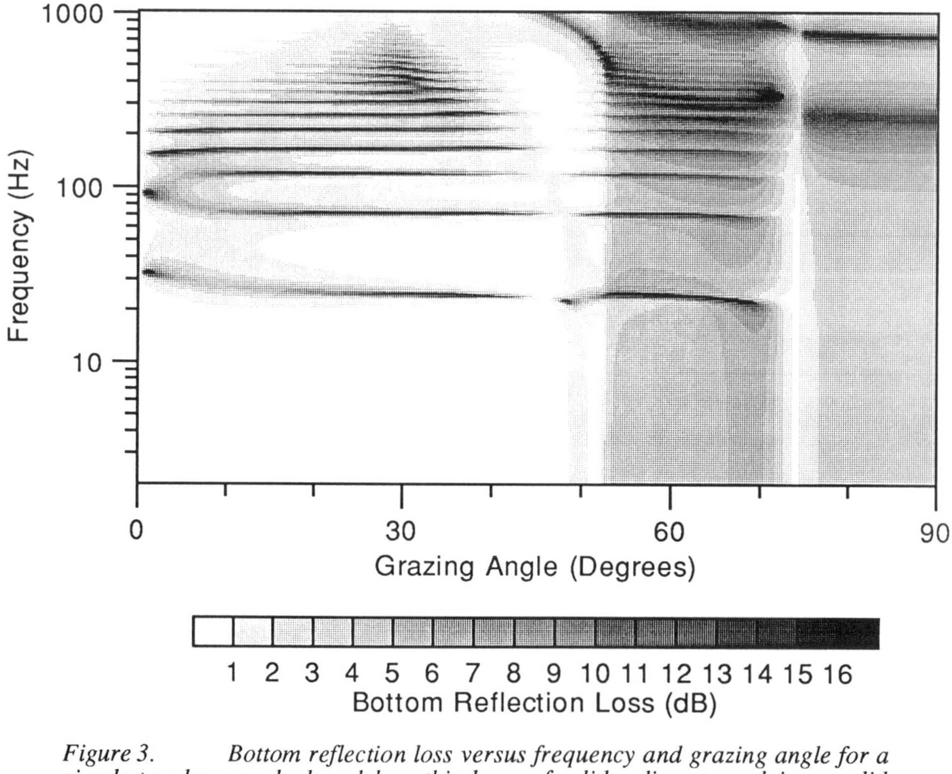


*Figure 3.* *Bottom reflection loss versus frequency and grazing angle for a simple two-layer seabed model: a thin layer of solid sediment overlying a solid half-space substrate. White corresponds to less than 1 dB bottom loss, while black is greater than 16 dB loss.*

To investigate the effect of sediment shear-wave attenuation on the bottom loss resonances, we have calculated bottom reflection loss at 30° grazing angle using three different computer software models: SAFARI (Schmidt, 1988), the analytic plane-wave reflection model of Hughes *et al* (1990), and the simplified coherent-ray model described here. In Figure 4, we show bottom reflection loss versus frequency at 30° grazing angle for each model. For all models, the bottom loss resonances predicted by Vidmar (1980a) are evident as indicated by peaks in the reflection loss at frequencies of 25 Hz, 70 Hz, 120 Hz, and so on. The width in frequency of the fundamental resonance is broadened considerably by the coherent-ray model. This model also predicts negative reflection loss at frequencies below the fundamental resonance and between the fundamental and first harmonic. Although the three models do not, in general, predict the same magnitude at each resonant peak, they do agree well on the frequencies and the overall shape of the resonance structure.

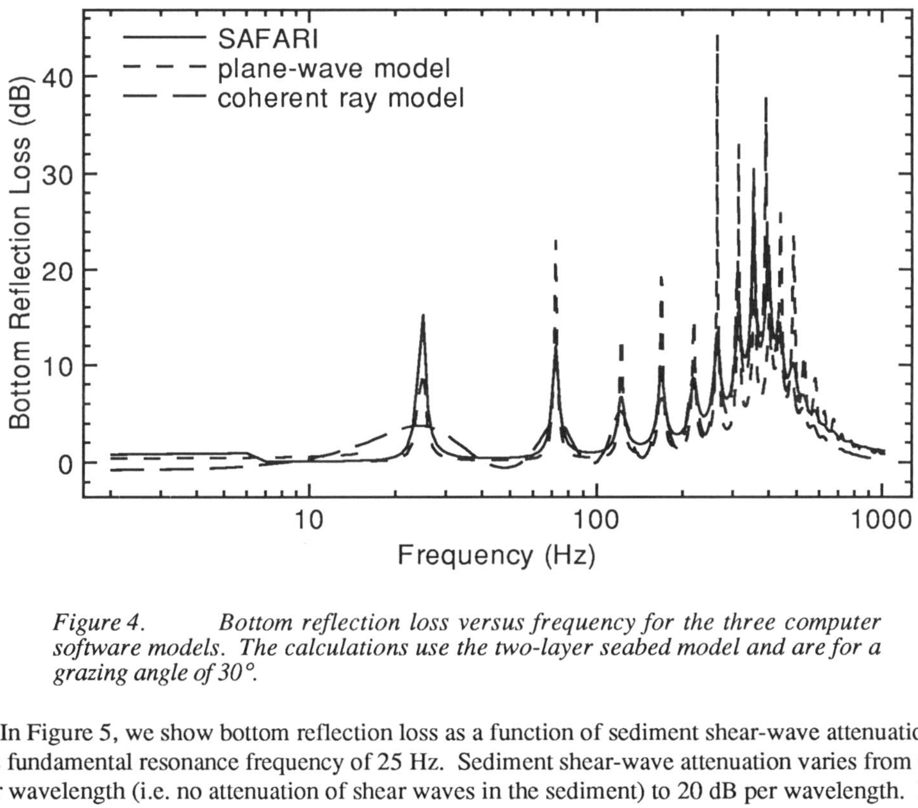


*Figure 4. Bottom reflection loss versus frequency for the three computer software models. The calculations use the two-layer seabed model and are for a grazing angle of 30°.*

In Figure 5, we show bottom reflection loss as a function of sediment shear-wave attenuation at the fundamental resonance frequency of 25 Hz. Sediment shear-wave attenuation varies from 0 dB per wavelength (i.e. no attenuation of shear waves in the sediment) to 20 dB per wavelength. Both SAFARI and the plane-wave reflection model predict 0 dB bottom loss with no shear wave attenuation, and both tend toward 0 dB bottom loss for large shear-wave attenuation. Between these two extremes is a maximum value of bottom reflection loss: 15 dB bottom loss at 2.2 dB per wavelength attenuation for SAFARI, and nearly 14 dB bottom loss at 5 dB per wavelength attenuation for the plane-wave model. The coherent ray model agrees well with SAFARI for values of shear-wave attenuation greater than about 10 dB per wavelength, but fails for small attenuations. This is to be expected, as the model assumes that the sediment shear waves are attenuated so

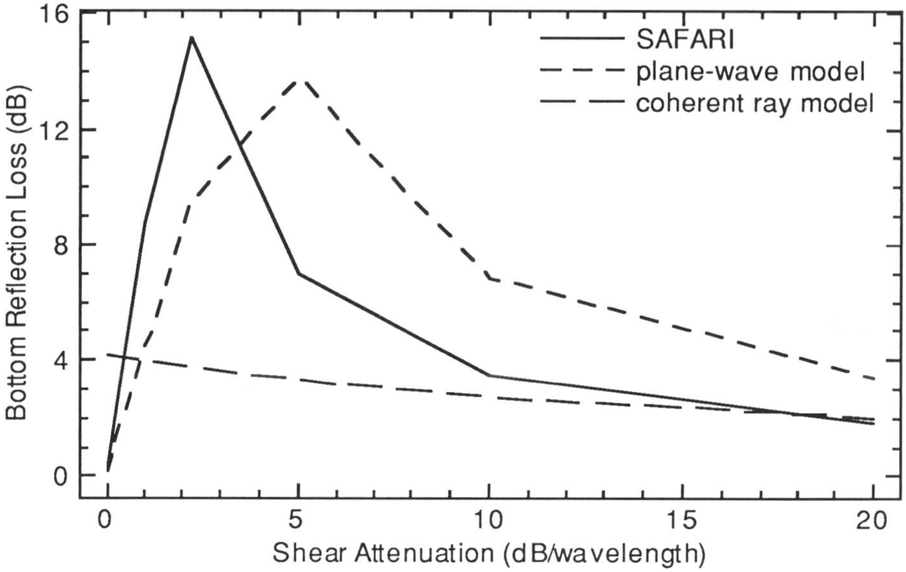


*Figure 5. Bottom reflection loss versus shear-wave attenuation for the three computer software models. The calculations use the two-layer seabed model and are for 25 Hz at a grazing angle of 30°.*

strongly that they traverse the sediment layer once up and down only. This assumption is not valid for the case of very small shear-wave attenuation.

In Figure 6 we show how the frequency of the fundamental bottom loss resonance changes with the thickness of the sediment layer and with the sediment shear speed. For each curve in the Figure, one parameter (layer thickness or shear speed) is variable while the other is held constant, with the value specified in Table 1. From Figure 6, we see that the frequency of the fundamental bottom loss resonance is inversely proportional to the thickness of the sediment layer and is proportional to the sediment shear speed. The frequency dependence of the bottom loss resonance is determined by the thickness of the sediment layer in terms of acoustic wavelength, which is determined only by the sound speed in the layer and the thickness of the layer. Increasing or decreasing the sediment layer thickness has the effect of shifting the bottom loss resonance structure shown in Figure 1 down or up, respectively, in frequency. Changes in shear speed not only alter the resonance frequency proportionally, but also change the overall shape of the resonance structure. At shear speeds above about 500 m/s, the well defined structure seen in Figure 1 begins to collapse, and at even higher frequencies disappears altogether.

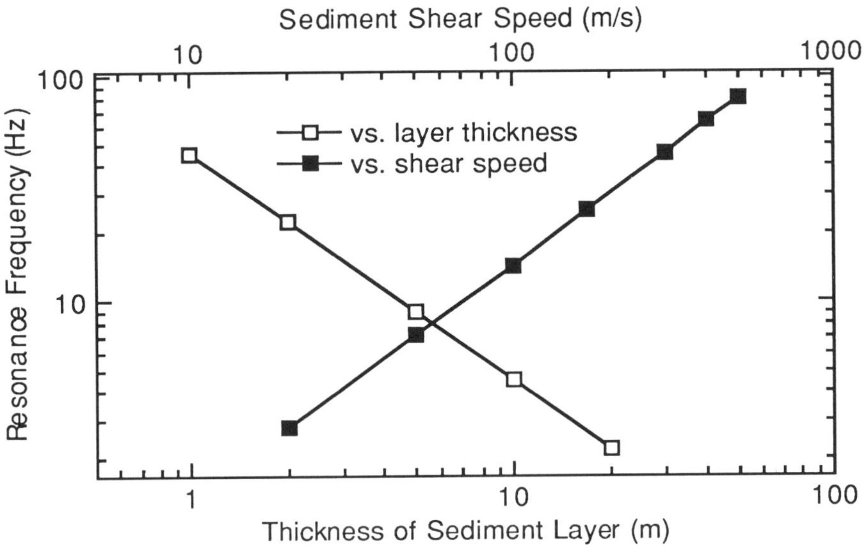


*Figure 6.* *Frequency of the fundamental bottom loss resonance as a function of the thickness of the sediment layer (open squares) and also as a function of the shear speed (filled squares).*

## 4. Conclusions

We have used a simple two-layer sediment-substrate seabed model with plane boundaries and iso-speed layers to calculate bottom reflection loss. We have shown that the amount of bottom reflection loss is sensitive to the shear-wave attenuation, and that the bottom reflection loss exhibits resonant structure. The magnitude of the bottom loss resonance goes to zero both for zero sediment shear-wave attenuation and for large shear attenuation. Between these limits, there is a maximum bottom loss of about 15 decibels at a sediment shear-wave attenuation of 2 to 5 dB per wavelength. We have also shown that the bottom loss resonance frequencies are proportional to the shear speed and inversely proportional to the sediment layer thickness.

We have shown calculations of bottom reflection loss versus frequency for a grazing angle of 30° using three different computer software models: SAFARI (Schmidt, 1988), the plane-wave model of Hughes *et al* (1990), and a simplified coherent ray model described here. All three models show resonances in the bottom reflection loss at frequencies for which odd multiples of quarter-wavelengths of the shear wave are equal to the thickness of the sediment layer, as predicted by Vidmar (1980a). We have shown that the resonance frequencies are controlled by the shear speed and by the thickness of the sediment layer.

The simplified coherent ray model broadens the fundamental bottom reflection loss resonance in frequency, but otherwise agrees well with SAFARI and the plane-wave model. The coherent ray model shows that most of the important features of the bottom loss resonance can be understood by ignoring shear wave conversion at the sediment-water boundary, and by allowing shear waves generated at the sediment-substrate boundary to transit the sediment layer up and down only once.

## References


Beebe, J.H. and Holland, C.W. (1986). 'Shallow-water propagation effects over a complex, high-velocity bottom', J. Acoust. Soc. Am. 80, 244-250.

Hastrup, O.F. (1980) 'Some bottom-reflection loss anomalies near grazing and their effect on propagation in shallow water', W.A. Kuperman and F.B. Jensen (eds.), *Bottom-Interacting Ocean Acoustics*, Plenum Press, pp. 135-152.

Hughes, S.J., Ellis, D.D., Chapman, D.M.F., and Staal, P.R. (1990) 'Low-frequency acoustic propagation loss in shallow water over hard-rock seabeds covered by a thin layer of elastic-solid sediment', J. Acoust. Soc. Am. 88, 283-297.

MacPherson, J.D., and Fothergill, N.O. (1962). 'Study of low-frequency sound propagation in the Hartlen Point region of the Scotian Shelf', J. Acoust. Soc. Am. 34, 967-971.

Schmidt, H. (1988) 'SAFARI: Seismo-Acoustic Fast field Algorithm for Range-Independent environments - User's Guide', SACLANTCEN Report SR-113.

Staal, P.R., Chapman, D.M.F., and Zakarauskas, P. (1987) 'The effect of variable roughness of a granite seabed on low-frequency shallow-water acoustic propagation', H.M. Merklinger (ed.), *Progress in Underwater Acoustics, Plenum Press*, pp. 485-492.

Vidmar, P.J. (1980a) 'The effect of sediment rigidity on bottom reflection loss in a typical deep sea sediment', J. Acoust. Soc. Am. 68, 634-638œ.

Vidmar, P.J. (1980b) 'Ray path analysis of sediment shear wave effects on bottom reflection loss', J. Acoust. Soc. Am. 68, 639-648.

Worley, R.D. and Walker, R.A. (1982). 'Low-frequency ambient ocean noise and sound transmission over a thinly sedimented rock bottom', J. Acoust. Soc. Am. 71, 863-870.

THE INFLUENCE OF SEDIMENT RIGIDITY ON THE PLANE-WAVE REFLECTION COEFFICIENT

M A AINSLIE
*YARD Ltd*
*233 High Holborn*
*London WC1V 7DJ*
*UK*

**ABSTRACT.** A simple method for calculating bottom loss for a two-layer solid sea-bed is compared with results from SAFARI, and a criterion discussed for neglecting sediment rigidity. A sensitivity analysis is performed, revealing that at low frequency and 20° grazing angle, parameters requiring highest measurement accuracy are the substrate shear speed ($v_3$) and the sediment compression and shear speeds ($c_2$, $v_2$).

## 1. Introduction

Historically, for the purpose of calculating a reflection coefficient, the ocean bottom has often been treated as a fluid [1, 2, 3, 4]. This is partly because for sufficiently thick unconsolidated sediments such as silt or clay this is a good approximation [5], and partly because of the extra complications involved when shear effects are included, often requiring the introduction of approximations [6] or numerical models [7, 8]. Nevertheless, the influence of sediment elasticity can be important for more rigid or thin sediments [9, 10] and in this paper we present a method which can be used to calculate the reflection coefficient for a two-layer (sediment and substrate) solid sea-bed, either coherently or incoherently. The example chosen for illustration is that defined by Vidmar [6] and the emphasis is placed on incoherent calculation in order to prevent interference effects masking slower but more significant changes in the background level.

The total reflection coefficient can be written in the form [11]

$$E = R_{12}^{pp} + PT_{12}^{pp} + ST_{12}^{ps} \tag{1.1a}$$

where the two transmitted waves are weighted by the coefficients P and S,

$$P = \frac{\tau^{sp}\rho^{ps} - \tau^{pp}(\rho^{ss} - 1)}{(\rho^{pp} - 1)(\rho^{ss} - 1) - \rho^{sp}\rho^{ps}} \qquad S = \frac{\tau^{pp}\rho^{sp} - \tau^{sp}(\rho^{pp} - 1)}{(\rho^{pp} - 1)(\rho^{ss} - 1) - \rho^{sp}\rho^{ps}} \tag{1.1b}$$

and $\rho^{\alpha\beta}$, $\tau^{\alpha\beta}$ are generalised reflection, transmission coefficients [11]

$$\rho^{\alpha\beta} = R_{23}^{\alpha p} R_{21}^{p\beta} + R_{23}^{\alpha s} R_{21}^{s\beta} \qquad \tau^{\alpha\beta} = R_{23}^{\alpha p} T_{21}^{p\beta} + R_{23}^{\alpha s} T_{21}^{s\beta}. \tag{1.1c}$$

The reflection (transmission) coefficient $R_{12}$ ($T_{12}$) is for a plane wave incident on the sediment (medium 2) from the water (medium 1). The superscripts pp and ps indicate p-wave reflection/transmission and p-s conversion respectively. Other coefficients are similarly defined, with the substrate layer as medium 3. Eq (1.1) is most useful when the sediment layer is



*J. M. Hovem et al. (eds.), Shear Waves in Marine Sediments*, 447–454.

sufficiently thin for shear waves to cross it with little attenuation. In practice, the decay rate is so large [12] that this constrains the sediment thickness to at most a few shear wavelengths $\lambda_s$. If the sediment thickness is large compared to $\lambda_s$, little energy can cross it in the form of shear waves and in this limit eq (1.1a) reduces to

$$E = R_{12}^{pp} + T_{12}^{pp} R_{23}^{pp} T_{21}^{pp} / (1 - R_{21}^{pp} R_{23}^{pp}) . \qquad (1.2)$$

Both eqs (1.1) and (1.2) can in principle be evaluated either coherently or incoherently as described in Reference 11. In practice, if the sediment is thin enough to require the use of eq (1.1), interference effects are likely to be important, especially for odd multiples of $\lambda_s/4$ [13] and therefore coherent addition is necessary. Conversely, eq (1.2) implies a thick sediment for which incoherent addition can be justified [3] and we now proceed to use eq (1.2) to explain some of the observations made by Vidmar [5].

## 2. Loss vs Angle

The above formulae are applicable to an inhomogeneous transition layer (sediment) in which the sound speed and attenuation (p and s) vary with depth, and an example of this for a fluid sediment is given in Reference 3. For simplicity, an homogeneous sediment layer is adopted here because what is of interest is the sediment-substrate reflection. Specifically we consider the same geoacoustic parameters (See Table 1) as used by Vidmar [5] in his sensitivity study and attempt to explain some of his results. The notation has also been borrowed from Vidmar's paper, with subscripts 1, 2, 3 used to denote each of the three layers as indicated in Table 1.

Table 1. Geoacoustic parameters used in the sensitivity analysis for thick sediments.

| Layer (subscript) | Density $\rho$ (g/cm$^3$) | Sound Speed c(m/s) | Compressional Attenuation $k_p$ (dB/m/kHz) | Shear Speed v (m/s) | Shear Attenuation $k_s$ (dB/m/kHz) |
|---|---|---|---|---|---|
| Water (1) | 1.053 | 1540 | 0.0 | 0 | 0.0 |
| Sediment (2) | 1.250 | 1525 | 0.100 | 100 | 10.0 |
| Substrate (3) | 2.800 | 5700 | 0.035 | 2900 | 0.07 |

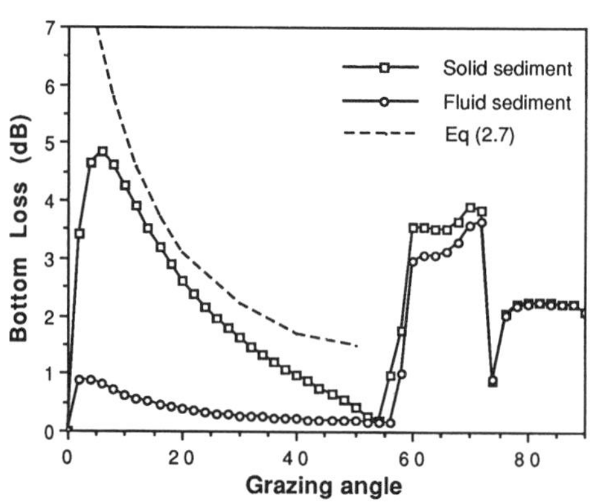


Figure 1 - Bottom loss vs grazing angle using eq (1.2) (incoherent) with the parameters of Table 1

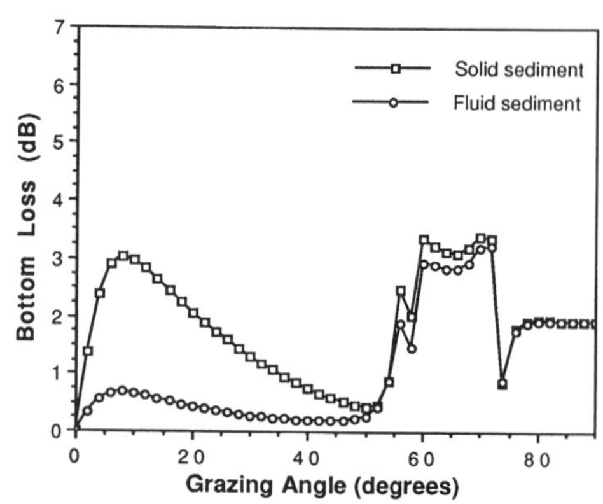


Figure 2 - Bottom loss vs grazing angle using SAFARI with the parameters of Table 1.

For a sediment depth H = 50m and frequency f = 14Hz, we obtain the bottom loss vs angle (BL = -10 $\log_{10}$E) curves shown in Figure 1 (eq (1.2)) and Figure 2 (SAFARI [7]).

The remainder of this section is devoted to explaining the shape of these curves for angles less than about 55°. (The high loss region for steeper angles is due to the generation of p and s waves in the substrate which are not considered further). Considering the fluid sediment ($v_2 = 0$) results first and assuming $R_{12} << 1$, we note that eq (1.2) simplifies further to

$$E^{(0)} \approx a_p^2 R_{23}^{(0)} + R_{12}^{(0)}\left[1 - a_p^2 R_{23}^{(0)}\right]^2 \tag{2.1}$$

where the pp superscripts have been replaced with zeros to denote evaluation for a fluid sediment. Note that the 2-way attenuation factor $a_p^2$ implicit in $R_{23}^{pp}$ [11] has now been included explicity, i.e.

$$R_{23}^{pp} \longrightarrow a_p^2 R_{23}^{(0)} \tag{2.2}$$

where, if $\theta_2$ is the sediment grazing angle

$$-10\log_{10} a_p = k_{p2} f H \operatorname{cosec}\theta_2 . \tag{2.3}$$

Concentrating in particular on angles up to the critical angle for substrate shear waves (58°), rays reaching the substrate are totally internally reflected. Noting also that for the low frequency example chosen, $a_p^2 \approx 1$ and therefore

$$E^{(0)} \approx a_p^2 . \tag{2.4}$$

For the solid sediment, the equivalent result is

$$E \approx R_{23}^{pp} = a_p^2 R_{23} \tag{2.5}$$

assuming that the solid-solid reflection coefficient $R_{23}$ is also close to one.

The measure we use here to judge the effect of sediment rigidity is

$$\delta = \left(E^{(0)} - E\right)/E^{(0)} \approx 1 - R_{23} . \tag{2.6}$$

It is shown in the Appendix that for small sediment grazing angle $\theta_2$ and a shear critical angle $\theta_s$ greater than about 55°, $\delta$ is given approximately by

$$\delta \sim \begin{cases} 4(v_2/c_2)\operatorname{cosec}^2\theta_0 \sin\theta_2 & \theta_2 < \theta_0 \quad (2.7a) \\ 4(v_2/c_2)\operatorname{cosec}\theta_2 & \theta_2 > \theta_0 \quad (2.7b) \end{cases}$$

where $\theta_0$ is the angle which approximately minimises $R_{23}$, given by

$$\sin\theta_0 = \frac{\cot^4\theta_s}{4(\rho_3/\rho_2)(\operatorname{cosec}\theta_s - \operatorname{cosec}\theta_p)} \tag{2.8}$$

$$\theta_s = \cos^{-1}(c_2/v_3) \tag{2.9}$$

$$\theta_p = \cos^{-1}(c_2/c_3) . \tag{2.10}$$

The dashed line in Figure 1 shows the quantity (10 $\log_{10}$e) $\delta$, an approximation to the bottom loss expected for the solid sediment, plotted vs grazing angle, with $\delta$ from eq (2.7). This approximation is a good one for $\theta_2 << \theta_s$ and $R_{12} << 1$, and where the dashed line differs significantly from eq (1.2) for the solid sediment (squares), at small angles $\theta_1 \to 0$ this is because $R_{12} \to 1$ and at steep angles $\theta_1 \sim 55°$, $R_{23} \to 1$. An explanation of this last point is given in Section 3.

The parameters chosen by Vidmar ($\theta_1 = 20°$, $\theta_2 = 21.5°$, $\theta_0 = 6.8°$) are such that equation (2.7b) applies and we expect $v_2$ and $c_2$ to be the most important parameters. Vidmar does indeed pick out $v_2$ and $c_2$ in his numerical study (Table IV), but in addition notes that $v_3$ and $c_3$ also influence the value of $\delta A_0(\varepsilon)/A_{0c}(\varepsilon)$ (a parameter closely related to the above $\delta$), albeit only for the lower values of these parameters. A clue as to why this should be so can be found in the derivation of eq (2.7) (Appendix ), which requires the inequality

$$\alpha \equiv (\rho_3/\rho_2) \sin\theta_0 \operatorname{cosec}\theta_p \ll 1 \qquad (2.11)$$

to be satisfied. An increase in the value of $c_3$ or $v_3$ from its central value has the effect of reducing this ratio still further, so the inequality, and hence eq (2.7) still holds. By contrast, a reduction in either one to the values considered by Vidmar results in $\alpha$ increasing from its reference value of 0.27 (for the parameters of Table 1) to 3.61 and 0.71 respectively for changes in $v_3$ (to 1700 m/s) and $c_3$ (to 3400 m/s). It should come as no surprise therefore that a significant change in the reflection loss occurs as a new regime is entered.

## 3. Loss vs Substrate Shear Speed

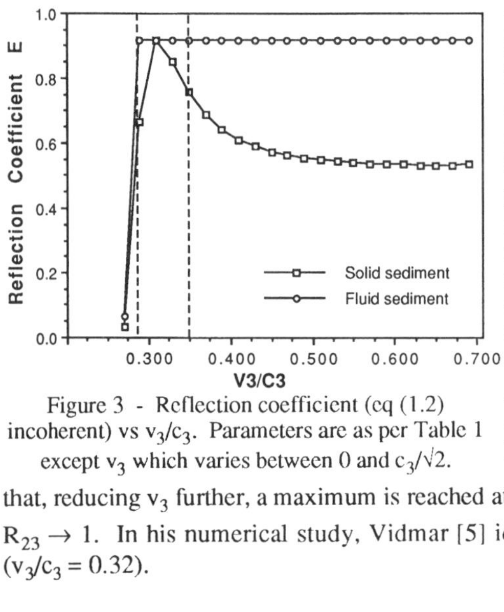


Figure 3 - Reflection coefficient (eq (1.2) incoherent) vs $v_3/c_3$. Parameters are as per Table 1 except $v_3$ which varies between 0 and $c_3/\sqrt{2}$.

It is apparent from the above discussion that the substrate shear speed $v_3$ plays an important part in determining the form of the reflection coefficient via the critical angle $\theta_s$. To examine this dependence, E is plotted as a function of the ratio $v_3/c_3$ in Figure 3, with $c_3$ held fixed at 5700 m/s. The region of interest here is $\theta_2 < \theta_s$ corresponding to $v_3/c_3 > 0.29$ to the right of the first vertical dashed line, for which the reflection coefficient is close to one for the fluid sediment and virtually independent of $v_3$. But for the solid sediment ($v_2 = 100$ m/s), E changes rapidly with $v_3$, especially for $v_3/c_3 < 1/2$. From eq (2.7) we expect the rapid change in the vicinity of $\alpha = 1$ ($v_3/c_3 = 0.35$, to the left of the second dashed line) and it can be shown (Appendix, eq (A.10)) that, reducing $v_3$ further, a maximum is reached at $\cos^2\theta_s \approx (9/10)\cos^2\theta_2$ ($v_3/c_3 = 0.30$) for which $R_{23} \to 1$. In his numerical study, Vidmar [5] identified a similiar maximum at $v_3 = 1800$ m/s ($v_3/c_3 = 0.32$).

## 4. A Criterion for Neglecting Sediment Rigidity

It is clear from Figure 3 that the highest losses induced by the sediment rigidity occur in the region $\alpha < 1$, for which $\delta \approx 4\,(v_2/c_2)\operatorname{cosec}\theta_2$ for moderate $\theta_2$. Putting values of $\theta_2 = 21.5°$, $c_2 = 1525$ m/s and insisting that $\delta \ll 1$ gives

$$v_2 \ll c_2 \sin\theta_2/4 = 140 \text{ m/s}. \qquad (4.1)$$

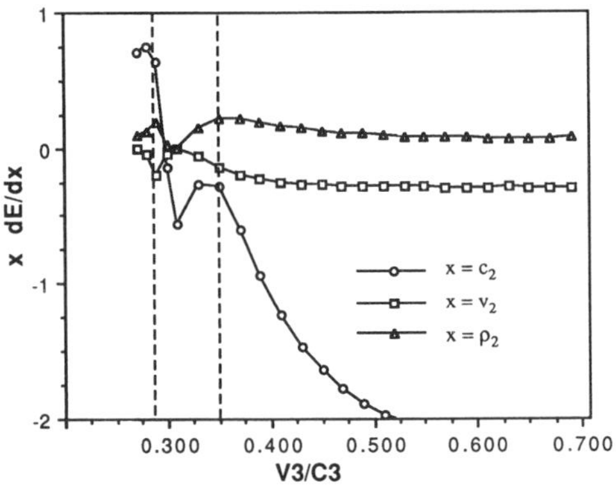


Figure 4 - Sensitivity of reflection coefficient E to sediment parameters. The parameters are as per Table 1 except $v_3$ which varies between 0 and $c_3/\sqrt{2}$.

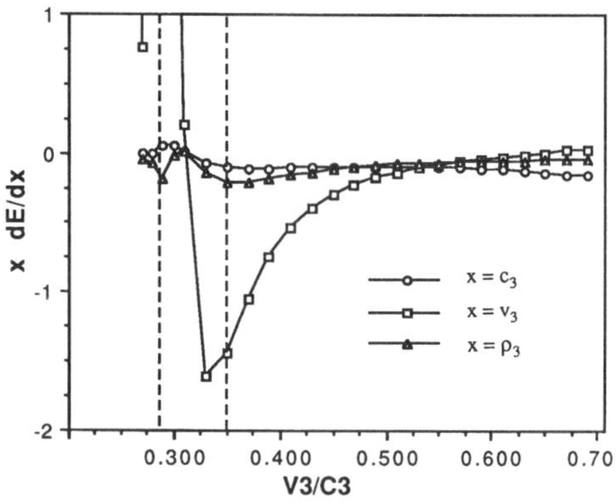


Figure 5 - Sensitivity of reflection coefficient E to substrate parameters. The parameters are as per Table 1 except $v_3$ which varies between 0 and $c_3/\sqrt{2}$.

A review by Hamilton [14] shows that for both sand and silt/clay sediments, almost all measurements of $v_2$ exceed 50 m/s, and it could be that inequality (4.1) is never satisfied. In other words, the only circumstances for which shear wave generation can be safely ignored is if virtually no acoustic energy (p or s) reaches the substrate boundary, i.e.

$$k_{p2} f H > 20 \text{ dB}. \qquad (4.2)$$

## 5. Sensitivity vs Substrate Shear Speed

Here we address the question of sensitivity of the reflection coefficient E to changes in the input parameters. Sensitivity is important because it determines the accuracy to which parameters must be measured for making bottom loss predictions. The normalised rate of change of the reflection coefficient xdE/dx is plotted vs $v_3/c_3$ for the six parameters which determine $R_{23}$ in Figures 4 ($x = c_2, v_2, \rho_2$) and 5 ($c_3, v_3, \rho_3$). The vertical dashed lines mark $\theta_2 = \theta_s$ and $\alpha=1$ as in Figure 3. Close to the reference value ($v_3/c_3 \approx 0.5$) the importance of $c_2$ and $v_2$ is confirmed. Other points worth noting are:

- For values of $\alpha$ of order one or greater, $v_3$ takes over from $c_2$ as the most important parameter.
- The increased sensitivity to $c_3$ noted by Vidmar [5] for the smaller values of this parameter is not reproduced here. The reason for this is simply that the value used by Vidmar ($c_3$ = 3400 m/s) corresponds to an unphysical negative value of Lamé's constant $\lambda$ [15] and here we have only allowed $\lambda > 0$ ($v_3/c_3 < 1/\sqrt{2}$).
- When $\cos^2\theta_s \approx (9/10)\cos^2\theta_2$ ($v_3/c_3 \approx 0.3$), $\delta$ and its first derivative are both zero. Hence the convergence of all curves (except $c_2$) on the axis at this point.
- Sensitivity to $\rho_2$ and $\rho_3$ are mirror images of each other. This is because the densities appear in $\delta$ only through the ratio $\rho_3/\rho_2$.

To put these findings into context, note that for a fluid sediment (i.e. $v_2 = 0$, $v_3 > c_2$), only the sensitivity to $c_2$, $v_2$ and $\rho_2$ would have been non-zero because $R_{23} \to 1$.

## 6. Discussion and Conclusions

We set out to assess the effect that sediment rigidity has on bottom loss, using the incoherent plane-wave reflection coefficient as a tool and comparing our results with those of a numerical study by Vidmar [5]. The main conclusions are

(i) At low frequency (small p attenuation) one of the most important controlling parameters is $\alpha = \cot^4\theta_s/4(\sin\theta_p \operatorname{cosec}\theta_s - 1)$. If $\alpha$ is small compared to unity (large $v_3/c_2$), the reflection coefficient is determined primarily by the sediment properties $v_2$ and $c_2$ as can be seen from eq (2.7) and Figure 4. For larger $\alpha$ (smaller $v_3/c_2$), $v_3$ replaces $v_2$ in this role as $\delta \propto \alpha^{-2} \sim \tan^8\theta_s \operatorname{cosec}^2\theta_s$ (Figure 5).

(ii) Contrary to Vidmar's observation, we find that $c_3$ is never of primary importance. It is probably sufficient to estimate it by multiplying $v_3$ by the appropriate ratio from Reference 16.

(iii) For realistic sediment shear speeds, the departure of $R_{23}$ from unity (as measured by eq (2.7)) is never insignificant for all grazing angles, and sediment rigidity should be taken into account unless the p-wave attenuation is so great that no energy reaches the substrate.

We deliberately concentrated on the interactions at the sediment-substrate interface by adopting a low frequency and $c_2 < c_1$, thus ensuring that the influence of frequency, sediment depth, attenuation and the layer 1 parameters $\rho_1$, $c_1$ were negligible. Higher frequencies and/or $c_2 > c_1$ would have meant a greater proportion of energy reflected at the water-sediment boundary, and hence greater sensitivity to these parameters, detracting from the p-s interaction of interest here. The circumstances for which f, H, $k_p$, $k_s$ can be important are described by Vidmar [5].

It should be emphasised that the analysis (in particular the Appendix) is equally applicable to a fast sediment (with $c_2 > c_1$, characteristic of continental shelf areas), but only for sediment grazing angles $\theta_2$ in the interval $[0, \theta_s]$. Smaller grazing angles in the water are totally internally reflected (at least in the absence of shear) but the evanescent p-wave can still result in significant shear conversion and high loss [17, 18]. Eq (1.1) is completely general and can be applied to this problem if required. In fact it is particularly suitable for a ray path analysis of the kind described by Hughes et al [17] because of the ease with which individual ray paths or ray families can be systematically excluded simply by setting the relevant coefficients to zero in eqs (1.1).

## Acknowledgement

The work has been carried out with the support of the Procurement Executive, Ministry of Defence (UK).

## APPENDIX - CALCULATION OF $R_{23}$ TO FIRST ORDER IN $v_2/c_2$

### A.1 FIRST ORDER EXPANSION

The solid-solid amplitude reflection coefficient $r_{23}$ is given by (see for example Reference 6)

$$r_{23} = r_p - 2(v_2/c_2)\cos\theta_2 \cot\theta_2 \Gamma_p^2 + O(v_2/c_2)^2 \qquad \text{(A.1)}$$

where $\Gamma_p = [1 + r_p - t_p - \tan\phi_3\, t_s]/2$ and the three coefficients $r_p$, $t_p$, $t_s$ are the reflection and transmission coefficients for a fluid-solid boundary, given by [19]. Here we define the impedance ratios

$$\zeta = \zeta_p \cos^2 2\phi_3 + \zeta_s \sin^2 2\phi_3 \qquad \text{(A.2)}$$

$$\zeta_p = \rho_3 c_3 \sin\theta_2 / (\rho_2 c_2 \sin\theta_3) \qquad \text{(A.3)}$$

$$\zeta_s = \rho_3 v_3 \sin\theta_2 / (\rho_2 c_2 \sin\phi_3) \qquad \text{(A.4)}$$

and the grazing angles $\theta$, $\phi$ are given by Snell's law $\cos\theta_i / c_i = \cos\phi_i / v_i$ = constant. By assumption (remember that we are dealing with the regime $0 < \theta_2 < \theta_s$) $\cos\phi_3$, $\cos\theta_3$, $\cos\theta_2$, $\sin\theta_2$ are all real, and similarly $\sin\phi_3$, $\sin\theta_3$ are imaginary. After some algebra, the power reflection $R_{23}$ is then

$$R_{23} = |r_{23}|^2 \approx 1 - 4(v_2/c_2)\cos^2\theta_2\, N^2 / \left(1 + |\zeta|^{-2}\right) \qquad \text{(A.5)}$$

where

$$N = 1 + (\zeta_p \cos 2\phi_3 + \zeta_s \sin 2\phi_3 \tan\phi_3)/\zeta. \qquad \text{(A.6)}$$

## A.2 LARGE $\theta_s$ LIMIT

Introducing the notation $s = (\cot^2\theta_s)/2$ and making the approximations $s \ll 1$, $s^2 \ll (\sin\theta_p/\sin\theta_s - 1)$, $\theta_2 \ll \theta_p = \cos^{-1}(c_2/c_3)$ and $\theta_2 \ll \theta_s = \cos^{-1}(c_2/v_3)$ valid for a sufficiently large shear critical angle, we obtain

$$N \rightarrow 1 \qquad \text{(A.7)}$$

$$\zeta \rightarrow -i(\rho_3/\rho_2)\sin\theta_2 \left(\operatorname{cosec}\theta_p - \operatorname{cosec}\theta_s\right)/s^2 \qquad \text{(A.8)}$$

and eqs (2.7a), (2.7b) follow by substituting the limits $|\zeta| \ll 1$ and $|\zeta| \gg 1$ respectively in eq (A.5).

## A.3 SMALL $\theta_s$ LIMIT

Eq (A.6) can be written

$$\frac{\zeta}{2\zeta_p} N = \cos 2\phi_3 \cos^2\phi_3 + \frac{\cos\theta_p}{\cos\theta_s} \sin\phi_3 \sin\theta_3 \left(3 - 2\sin^2\phi_3\right). \qquad \text{(A.9)}$$

For small $\theta_s$ ($\approx\theta_2$) we can put $\cos\phi_3 \approx 1$. If in addition we have $\cos\theta_p \ll \cos\theta_2$, then so long as $\rho_3/\rho_2$ is of order one, eq (A.5) simplifies to

$$R_{23} \approx 1 - 16\,(v_2/c_2)\,(\rho_3/\rho_2)^2 \sin^2\theta_2 \left[1 - 3\left(\frac{\cos^2\theta_2}{\cos^2\theta_s} - 1\right)^{\frac{1}{2}}\right]^2. \qquad \text{(A.10)}$$

**References**

[1] Mitchell S K and Lemmon J J (1979) 'A ray theory model of acoustic interaction with the ocean bottom', J. Acoust. Soc. Am. 66 (3), 855-861.
[2] Chapman N R, Bannister R W and Falconer R K H (1990) 'The effect of an elastic basement in the interpretation of bottom loss data', J. Acoust. Soc. Am. 87, (5), 2044-2051.
[3] Ainslie M A and Harrison C H (1989) 'Reflection of sound from deep water inhomogeneous fluid sediments', Proc. Annual Meeting of the Canadian Acoustical Association, A J Cohen (ed.), Halifax, pp. 12-17.
[4] Robins A J (1991) 'Reflection of a plane wave from a fluid layer with continuously varying density and sound speed', J. Acoust. Soc. Am. 89.
[5] Vidmar P J (1980) 'The dependence of bottom reflection loss on the geoacoustic parameters of deep sea (solid) sediments', J. Acoust. Soc. Am. 68 (5), 1442-1453.
[6] Vidmar P J (1980) 'Ray path analysis of sediment shear wave effects on bottom reflection loss', J. Acoust. Soc. Am. 68 (2), 639-648.
[7] Schmidt H (1988) 'Seismo-Acoustic Fast field Algorithm for Range Independent environments - User's guide', SACLANT Report SR-113.
[8] Vidmar P J and Foreman TL (1979) 'A plane-wave reflection loss model including sediment rigidity', J. Acoust. Soc. Am. 66 (6), 1830-1835.
[9] Kuperman WA and Jensen FB (eds 1980) Bottom-Interacting Ocean Acoustics, Plenum Press, New York.
[10] Akal A and Berkson J M (eds 1986) Ocean Seismo-Acoustics, Plenum Press, New York.
[11] Ainslie M A (1990) 'Reflection of plane waves from inhomogeneous solid sediments', Proc I.O.A 12 (1), 191-198.
[12] Hamilton E L (1976) 'Attenuation of shear waves in marine sediments', J. Acoust. Soc. Am. 60, 334-338.
[13] Hughes S J, Ellis D D, Chapman D M F and Staal P R (1990) 'Low-frequency acoustic propagation loss in shallow water over hard-rock seabeds covered by a thin layer of elastic-solid sediment', J. Acoust. Soc. Am. 88 (1), 283-297.
[14] Hamilton E L (1976) 'Shear-wave velocity versus depth in marine sediments: A review', Geophysics 41 (5), 985-996.
[15] Auld B A (1973) Acoustic Fields and Waves in Solids Volume I, John Wiley and Sons, New York.
[16] Hamilton E L (1979) '$V_p/V_s$ and Poisson's ratios in marine sediments and rocks', J. Acoust. Soc. Am. 66 (4), 1093-1101.
[17] Hughes S J, Chapman D M F and Chapman N R (1991) 'The effect of shear wave attenuation on acoustic bottom loss resonance in marine sediments', in this volume.
[18] Kristensen A and Hovem J M (1991) ' Sensitivity of bottom loss to attenuation and shear conversion', in this volume.
[19] Ewing W M, Jardetzky W S and Press F (1957) Elastic Waves in Layered Media, McGraw-Hill, New York.

# SHEAR WAVE CONVERSION DUE TO SUBBOTTOM FACETS

PETER GERSTOFT
*Ødegaard and Danneskiold-Samsøe*
*Kroghsgade 1*
*2100 Copenhagen, Denmark*
*and Massachusetts Institute of Technology*

HENRIK SCHMIDT
*Massachusetts Institute of Technology*
*Cambridge*
*MA 02139 USA*

OLE VILMANN
*Ødegaard and Danneskiold-Samsøe*
*Kroghsgade 1*
*2100 Copenhagen, Denmark*

ABSTRACT. A hybrid propagation model for propagation in stratified media has recently been developed, which combines the boundary element method with the direct global matrix method. This model allows for simulation of scattering and reverberation from elastic facets in a stratified seismo-acoustic ocean environment. The Green's functions used in the boundary elements method can be determined very efficiently by means of a modified SAFARI code. This allows for modelling of very general stratifications, including any combination of fluid, isotropic elastic and transversily isotropic elastic layers. The use of the boundary element propagation code also allows calculation of the whole spectrum of output features from the direct global matrix method. This enables interpretation of the results in terms of basic wave physics and the separation of the total and the scattered field. Since only the facet boundary needs to be discretized, the hybrid method is computationally efficient compared with other numerical methods solving the two-way wave equation (i.e. finite difference and finite element models).

## 1. Introduction

*Facet* reverberation in the ocean environment may be the result of local, abrupt lateral changes in the medium properties. Discrete methods such as the finite difference approach have been applied to such reverberation problems as scattering from the bottom or Arctic ice keels [1, 2]. These discrete methods are very general in terms of geometries that can be treated, but the computational requirements are substantial because the entire environment must be discretized with a grid size smaller than the wavelengths involved. These methods are therefore limited to short range 2-dimensional reverberation problems. Other problems with discrete methods are related to the radiation condition at the boundary of the computational mesh and numerical dispersion introduced by the discretization.

The facet reverberation problem can also be solved in a self-consistent manner using Green's theorem for the surface of this facet, which requires a finite discretization of the surface for numerical implementation. The boundary element method (BEM) combines



*J. M. Hovem et al. (eds.), Shear Waves in Marine Sediments*, 455–462.

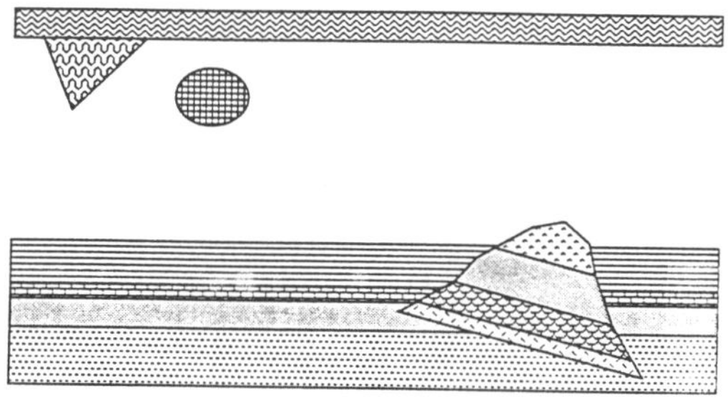

Figure 1: A horizontally stratified ocean seismo-acoustic environment is interrupted by three examples of facets, an ice keel, a cylinder and a layered sea mount.

both an integral representation of the wave field within a volume and a point representation of stresses and displacements on the boundary between the domains. The need for a dense mesh is limited to the domain boundary alone, eliminating the problem of discretely representing the wave field throughout the volume. This feature makes the BEM useful for representing scattering and radiation problems. Several boundary element formulations based on wavenumber integration for the Green's functions have been presented, e.g. [3, 4]. but, they have all been formulated for either a purely fluid or a purely elastic media. None of the formulations are directly applicable to the ocean seismo-acoustic reverberation problem.

The present paper is an extension of the boundary element approach in [5, 6], in which, the approach of Kawase [7] was generalized to incorporate linear variation of the field parameters over the elements. A new boundary element representation was formulated which allows stratification of both the exterior and interior domains with any combination of acoustic, elastic and transversely isotropic layers. This was achieved by combining the boundary element method with the direct global matrix approach [8, 9] for propagation in a stratified ocean seismo–acoustic environment.

## 2. The Boundary Element Method

The 2–dimensional problem under consideration is generalized in Fig. 1. A horizontally stratified ocean, bounded above by air or ice and below by an elastic stratified bottom, is assumed to be interrupted by a facet (i.e. a local deviation from the ideal, horizontal stratification). Examples of facets include ice keels, grooves in an arctic ice cover, sea mounts or diapirs in the ocean bottom. We assume a discrete facet of local extend surrounded by a stratified environment. The facet is also assumed to be stratified but it could be of any composition, provided a mathematical/numerical model for its dynamic behavior is available. Thus, the interior could be modelled by a finite element method, for example.

The boundary element method is based on Green's theorem for an elastic media. By assuming both the interior and exterior domains are plane stratified, the wave field can be expressed in terms of a boundary integral along the surface of the facet, provided the Green's functions are chosen to be the ones satisfying all boundary conditions in the stratifications.

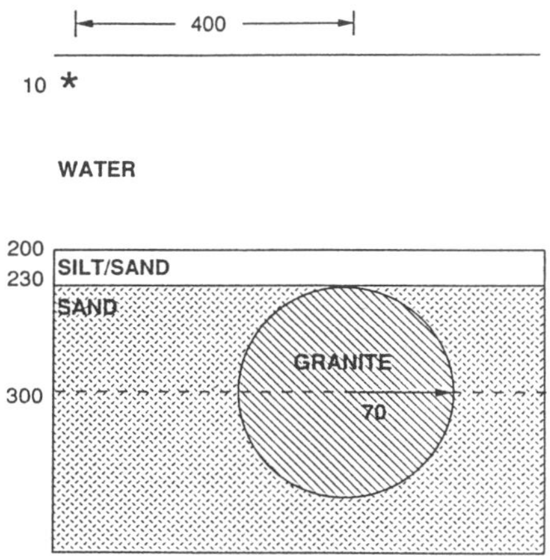


| Medium | Compression Speed m/s | Compression Atten dB/$\lambda$ | Shear Speed m/s | Shear Atten dB/$\lambda$ | Density g/cm$^3$ |
|---|---|---|---|---|---|
| Water | 1500 | 0 | - | - | 1.0 |
| Silt/sand | 1600 | 0.2 | 400 | 0.5 | 1.6 |
| Sand | 1800 | 0.1 | 600 | 0.2 | 1.8 |
| Granite | 5000 | 0.01 | 2500 | 0.02 | 2.4 |

Figure 2: A diagram of a cylindric granite inclusion in a stratified elastic sea bed. A transient source (*) is placed at 10 m depth and 400 m from the center of the inclusion. The water depth is 200 m and the center of the inclusion is located at a depth of 300 m. The pressure field is determined for a 2–dimensional range–depth window 500 m in depth and between 100 and 700 m in range from the source.

We then discretize the integral around the facet to obtain a system of linear equations, which describes the distribution of displacements and tractions around the boundary of the facet. Each element of the equations is a double integral over both geometric elements and wavenumbers. By interchanging those integrals and assuming a linear variation of the physical variables, an analytical integration along each element is possible. This reduces the dimension of the integral by one. Solving the resulting system of equations yields the set of displacements and tractions along the boundary, which represent the reflections from the facet, due to the real source. The total wave solution is now found by adding the real and virtual source contribution. The theory is described in detail by Gerstoft and Schmidt [6].

## 3 Numerical Examples

In the first example the hybrid approach simulates short range, time domain reverberation from a subbottom elastic facet. The example, shown in Fig 2, is an extension of the one given by Gerstoft and Schmidt [6]. A cylindric granite inclusion is embedded in the subbottom of a stratified shallow water environment. The bottom consists of a silt/sand sediment layer overlaying a sand subbottom. A transient acoustic source the facet, is transmitting a second order 3-loop Blackman-Harris pressure pulse, with a center frequency of 20 Hz. The acoustic field in the water column and sediment was computed over a fine grid (60x60 points) in the range window 100-700 m relative to the source, which covers both the back- and forward scattering regions from the facet.

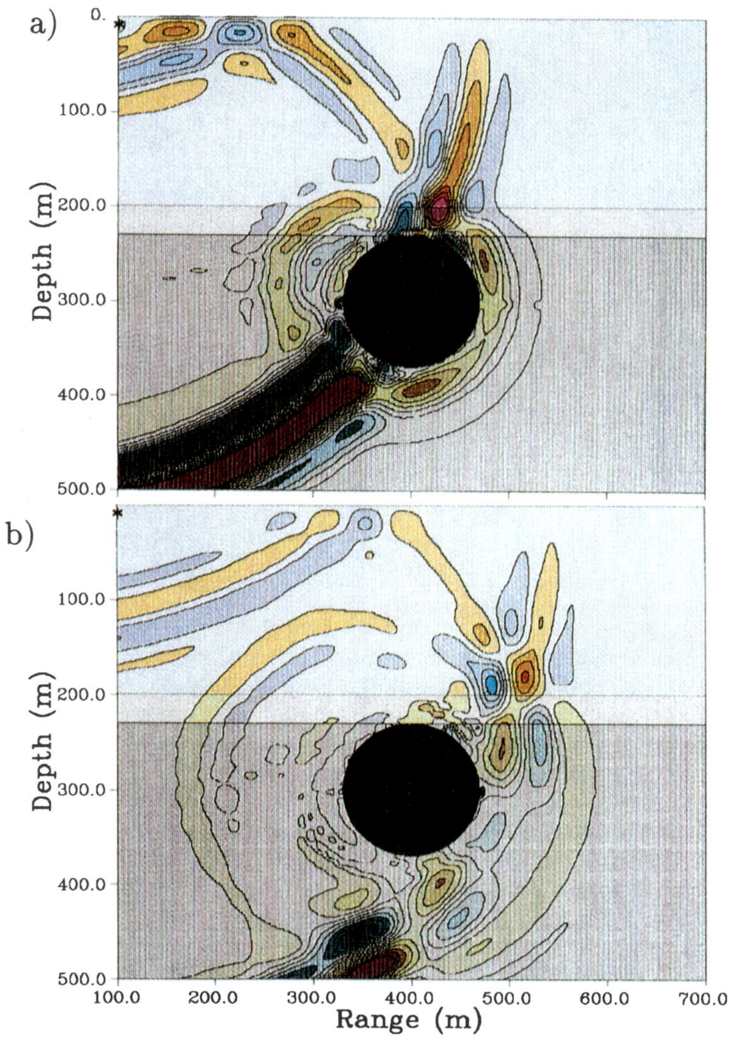


**Figure 3: Snap-shot contours of the full acoustic field at times a) 0.35** s **and** b) 0.40 s. Positive stress (negative pressure) amplitude is indicated by red colors, negative by blue colors.

Fig. 3a shows the field contours at $t$ = 0.35 s, shortly after, the pressure wave first interacts with the facet. The incident and bottom bounced wavelets are clearly identifiable, together with the secondary surface bounce. In addition, the facet reverberated wavelet is visible in the bottom left half of Fig. 3a. To the left of the figure it is seen that the incident wave comes faster trough the inclusion due to the higher velocities in the granite. At $t$ = 0.40 s, where the major part of the incident scattered and reverberated wavelets are still visible, but very difficult to interpret in detail (Fig. 3b). Note the concentration of energy close to the seabed.

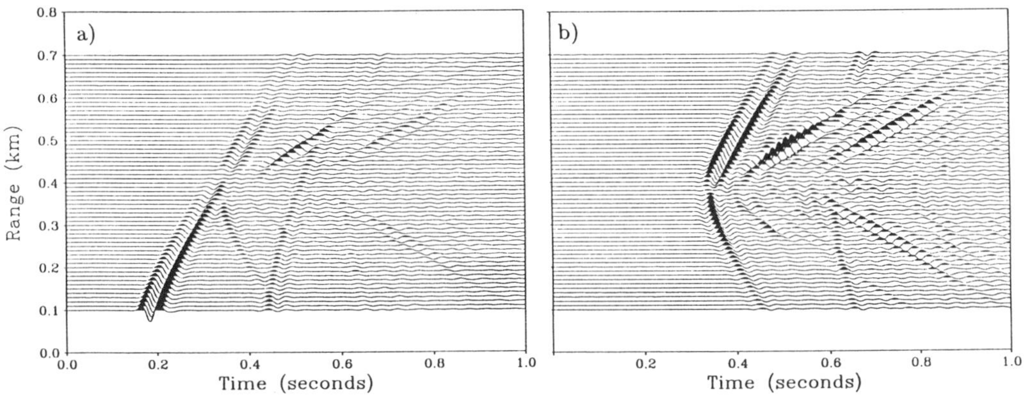


Figure 4: Stacked, synthetic time series for geophones recording vertical velocities on the seabed 100 to 700 m in range from the source. a) The total field and b) the scattered field.

Stacked time series of vertical velocities received by a geophone array placed on the seabed at depth 200 meter is presented in Fig. 4. The 100 to 700 meter range, covers both the forward- and backscattering regions. Multiple scattering from the granite inclusion is evident, with the earliest arrival appearing at slightly shorter range than the center of the facet. An advantage of the BEM approach is that we can directly exclude the source term and analyze the scattered wave field alone, by simply excluding the real source term. From the scattered wave field, (Fig. 4b), it is seen that the P-wave is scattered in both the forward and backwards directions, whereas the slower converted shear and interface waves are propagated primarily in the forward direction. The importance of including shear in the simulation of bottom reverberation is evident from the amplitude of the converted shear waves relative to the amplitude of the scattered P-waves.

The seabed often exhibits a significant amount of anisotropy as a result of layering of marine sediments. As the stratified model can efficiently handle transverse isotropy by the method of Schoenberg[10, 11], where the elastic constants of an transversely isotropic layer can be computed from a combination of infinitely fine layers. The first subbottom layer was replaced with a finely layered medium alternating between ($c_p = 2000m/s, c_s = 600m/s$) and ($c_p = 1200m/s, c_s = 200m/s$). In order to obtain the same main arrival times as for the isovelocity sediment the mean velocities are identical. A computation of the reverberated field for this scenario shows some change in the scattered wavefield, the interpretation of which is obstructed by the complexity of the field. However, we can take advantage of the inherent spatial and temporal, spectral decomposition in the current approach. The wavenumber kernel of the scattered field for the vertical particle velocity at 10 Hz is presented for both the isovelocity and the transversely isotropic cases (Fig 5). Positive wavenumbers correspond to waves propagating in the positive direction and

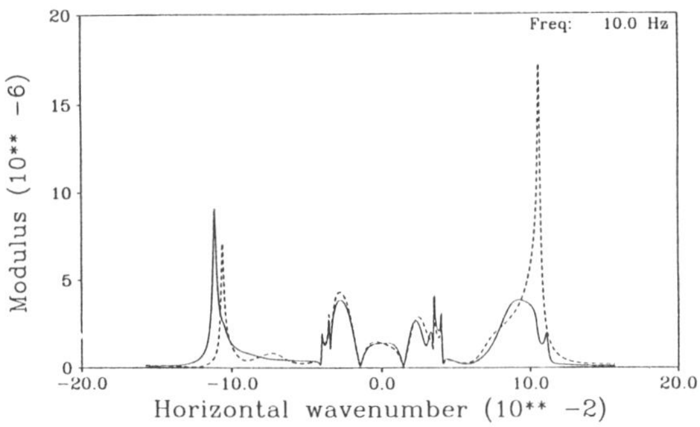


Figure 5: Wavenumber kernel for scattered vertical velocities at the seabed. Solid line: isotropic seabed; dashed line: transversely isotropic seabed.

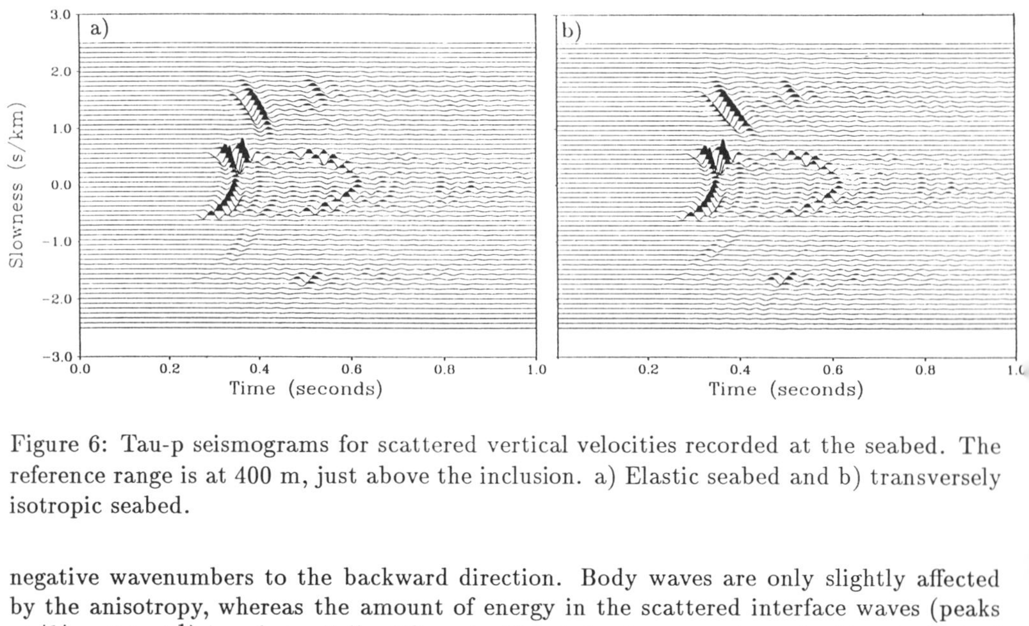


Figure 6: Tau-p seismograms for scattered vertical velocities recorded at the seabed. The reference range is at 400 m, just above the inclusion. a) Elastic seabed and b) transversely isotropic seabed.

negative wavenumbers to the backward direction. Body waves are only slightly affected by the anisotropy, whereas the amount of energy in the scattered interface waves (peaks at $|k| \approx 0.1m^{-1}$) is substantially different. This behavior suggests a significant effect of layering of sediments on the excitation and propagation of seismic interface waves.

In addition to the calculation of wavenumber kernels, the present approach can be used to compute the $\tau - p$ diagrams which are an important interpretation tool used in seismology. Mapping into the $\tau - p$ domain, which is essentially a stacking of plane wave seismograms, is a well established process for the interpretation of seismic records, [12, 13]. The intercept time ($\tau$) and the ray parameter ($p$) is directly related to the time, ($t$), and to the wavenumber $k$:

$$\tau = t - p(x - x^*) \; , \quad p = k/\omega \; , \tag{1}$$

where $x$ is the distance from the origin and $x^*$ is the pivoting or reference range. The

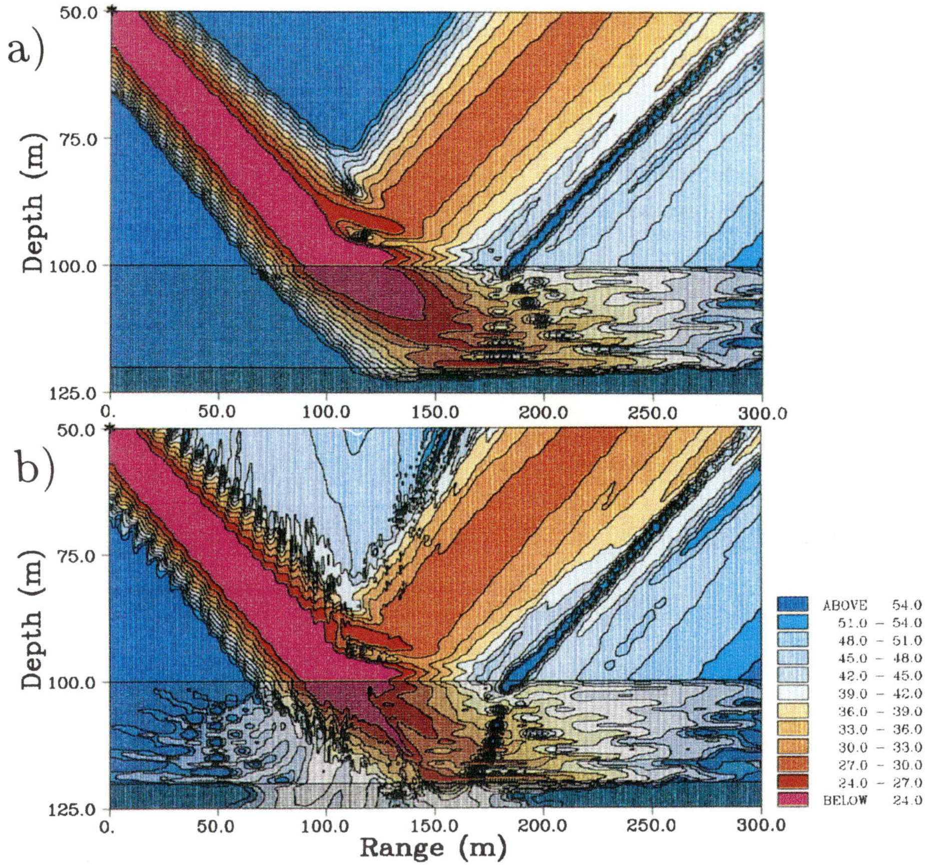


Figure 7: Gaussian beam penetrating into the subbottom. The Pressure transmission loss is displayed in dB. A hard reflecting cylinder of radius 1 meter is placed 1 meter below the seabed. (a) Cylinder at range 70 meter and (b) cylinder at range 120 meters.

synthetic $\tau - p$ diagram can be computed directly from the wavenumber–time response, and can be used as a test result for signal processing algorithms. Since only a few wavenumbers are necessary to produce a $\tau - p$ diagram the computation is much faster than for a complete seismogram. The $\tau - p$ diagram for the scattered vertical velocities at the seabed for a reference range of 400 meters (i.e., at the center of the bottom inclusion) is presented in Fig. 6. The waves corresponding to an absolute slowness less than $0.7s/km$ represent body waves in the water column; whereas, large slowness values correspond to the shear and interface waves which are evanescent in the water. Waves propagating in the forward and backward directions correspond to positive and negative slowness, respectively. It is clearly seen, in accordance with the previous figures, that the shear and interface waves are dominating the scattering and propagating mainly in the forward direction, whereas the scattered body waves in the water propagate more or less equally in both directions.

Finally we apply the hybrid propagation model to scattering from a small object due to an incident beam. In this case a linear vertical array generates a 1000 Hz Gaussian beam which impings on the bottom at a nominal grazing angle of 25°, [14]. The ocean is semi–infinite and the seabed at 100 meters consists of a soft sediment ($c_p = 1600m/s, c_s = 400m/s$) overlaying a higher velocity layer ($c_p = 1800m/s, c_s = 600m/s$) . The inclusion has a radius of 1 meter and is modelled as a hard reflector at ranges of either 70 (Fig. 7a) or 120 meters (Fig. 7b). The beam–generated acoustic field is displayed as transmission loss in dB. At 70 meters no scattering is evident because the inclusion is outside the beam; whereas, at 120 meters significant scattering occurs.

# References


[1] M.E. Dougherty and R.A. Stephen (1987) 'Geoacoustic scattering from sea-floor features in ROSE area', J. Acoust. Soc. Am., 82:238–256.

[2] J.R. Fricke, R.A. Stephen, and A.B. Baggeroer (1988) 'Numerical modeling of the scattered acoustic field from elastic ice', J. Acoust. Soc. Am., 83(S1):S37.

[3] G.T. Schuster and L.C. Smith (1985) 'Modeling scatterers embedded in a plane-layered media by a hybrid Haskell-Thomson and boundary integral equation method', J. Acoust. Soc. Am., 78:1387–1394.

[4] T.W. Dawson and J.A. Fawcett (1990) 'A boundary integral equation method for acoustic scattering in a waveguide with nonplanar surfaces', J. Acoust. Soc. Am., 87:1110–1125.

[5] H. Schmidt and P. Gerstoft (1990) 'Boundary element simulation of long range facet reverberation', J. Acoust. Soc. Am., 87(S1):S83.

[6] P. Gerstoft and H. Schmidt (1991) 'A boundary element approach to ocean seismo acoustic facet reverberation', J. Acoust. Soc. Am., April 1991.

[7] H. Kawase (1988) 'Time-domain response of a semi-circular canyon for incident SV, P and Rayleigh waves calculated by the discrete wavenumber boundary element method', Bull. Seismol. Soc. Am., 78:1415–1437.

[8] H. Schmidt and G. Tango (1986) 'Efficient global matrix approach to the computation of synthetic seismograms', Geophys. J. R. Astr. Soc. 84:331–359.

[9] H. Schmidt and F.B. Jensen (1985) 'A full wave solution for propagation in multilayered viscoelastic media with application to Gaussian beam reflection at fluid-solid interfaces', J. Acoust. Soc. Am. 77:813–825.

[10] Schoenberg M (1983) 'Reflection of elastic waves from periodically stratified media with interfacial slip', Geophysicical Prospecting, 31:265–292.

[11] J.S. Kim (1989) Radiation from directional seismic sources in laterally stratified media with application to Arctic ice cracking noise, PhD thesis, Massachusetts Institute of Technology, May 1989.

[12] R.A. Phinney, K. R. Chowdhury, and L. N. Frazer (1981) 'Transformations and analysis of record sections', J. Geophys. Res., 86:359–377.

[13] S. Treitel, P. R. Gutowski, and D. E. Wagner (1982) 'Plane-wave decomposition of seismograms'. Geophysics, 47:1375–1401.

[14] H. Schmidt (1987) SAFARI: Seismo-acoustic fast field algorithm for range independent environments. user's guide, SR 113, SACLANT ASW Research Centre, La Spezia, Italy.

# A SEISMO-ACOUSTIC FINITE ELEMENT MODEL FOR UNDERWATER ACOUSTIC PROPAGATION

Joseph E. Murphy
University of New Orleans
Department of Physics
New Orleans, LA 70148
USA

Stanley A. Chin-Bing
Naval Oceanographic and
Atmospheric Research Laboratory
Stennis Space Center, MS 39529
USA

ABSTRACT. The finite element technique can provide an accurate description of the physics of wave propagation in inhomogeneous media. The authors have developed a finite element acoustic model which handles range and depth dependence in both sound speed and density, as well as almost arbitrary variations in bottom topography. We discuss modifications to this model which will allow us to treat the ocean bottom as an elastic medium with range and depth dependence in shear speed and shear attenuation.

## 1. INTRODUCTION

The applicability of the finite element (FE) technique to the solution of the Helmholtz equation and related underwater acoustic propagation problems has been examined by a number of investigators[1-3]. Recently, the FE method has been successfully applied by Murphy and Chin-Bing[4-7] to develop computer models that directly calculate the complex pressure field resulting from acoustic propagation and scattering in a inhomogeneous ocean with range-varying boundaries (sea surface and ocean bottom). These computer models are computationally intensive but are capable of giving benchmark accurate predictions of the underwater complex pressure field[4,5]. In this paper we present modifications to our FE ocean acoustic model, FOAM, that will allow it to include the ocean bottom as an anelastic medium, having range and depth dependence in shear speed and in shear attenuation.

The theory, development, and implementation of the FOAM have been discussed previously[4,5] and only the salient points will be repeated to the extent that the development of FOAM may be contrasted with that of the present work.

The time-harmonic complex pressure resulting from a point source of unit strength and angular frequency $\omega$ at position $\mathbf{r}_s$ in an inhomogeneous ocean is determined by the wave equation

$$\rho\nabla \cdot (\rho^{-1} \nabla P) + k^2 P = - \delta(\mathbf{r} - \mathbf{r}_s) \tag{1}$$

and appropriate boundary conditions. The common assumption is made of a three-dimensional (3-D) ocean environment axially symmetric about a vertical line passing through the source, so that in the above equation (written in cylindrical coordinates), $\rho = \rho(z, r)$ is the density, $k = k(z, r) = \omega/c(z, r)$ is the wavenumber and $c(z, r)$ is the sound speed. Attenuation is included as an imaginary part of the complex wavenumber.



*J. M. Hovem et al. (eds.), Shear Waves in Marine Sediments*, 463–470.

FOAM simulates a typical ocean acoustic scenario in the following way: An acoustic source is placed somewhere in the water column (or fluid ocean bottom) and the sound speed, both in the water and in the fluid ocean bottom, may vary with both depth and range. The depth of the water-bottom interface may vary with range. The acoustic pressure is taken to vanish at the air-water interface (pressure release surface). The ocean bottom is assumed to be homogeneous beyond some maximum depth, and the sound speed profiles are assumed not to vary beyond some maximum range. Thus, along the right and bottom sides of the domain to be modeled, FOAM imposes an approximate boundary condition (radiation condition) which does not reflect any energy back into the interior.

In applying the FE technique to solving the ocean acoustic wave equation, the differential equation and its unknown solution are replaced by a system of algebraic equations in terms of the parameters defining an approximate solution. The domain of the problem is partitioned into non-overlapping *elements* and assumes a simple form for the approximate solution within each element. These local representations are then joined together by using the appropriate physical continuity conditions to provide a global solution. The ocean acoustic wave equation and the associated boundary conditions give a system of matrix equations which are sparse, banded and symmetric.

The FE technique gives a set of complex pressure values at the nodes of the FE grid. FOAM uses triangular elements in forming its FE grid. One of the key features of the method is the ability to use a grid of elements adapted to the specific problem. For example, if the density has a discontinuous jump in going across the water-bottom interface, the mesh would be formed so that the interface coincides with the boundaries of the appropriate interior elements.

## 2. THE SEISMO-ACOUSTIC FINITE-ELEMENT MODEL: SAFE

The FE techniques developed in creating the underwater acoustic (fluid) model, FOAM, have been employed to create a Seismo-Acoustic Finite Element model, SAFE, that includes the coupled effects of shear waves and compressional waves, together with shear and compressional attenuations. The SAFE model departs from FOAM in three major ways: (1) Rectangular elements are used (Fig. 1) rather than triangular elements; (2) Displacements are calculated by the SAFE model and the complex pressure field determined from derivatives of these displacements whereas FOAM solves for the complex pressures directly; and, (3) SAFE allows for an inhomogeneous elastic ocean bottom with attenuations whereas FOAM assumes an inhomogeneous fluid ocean bottom.

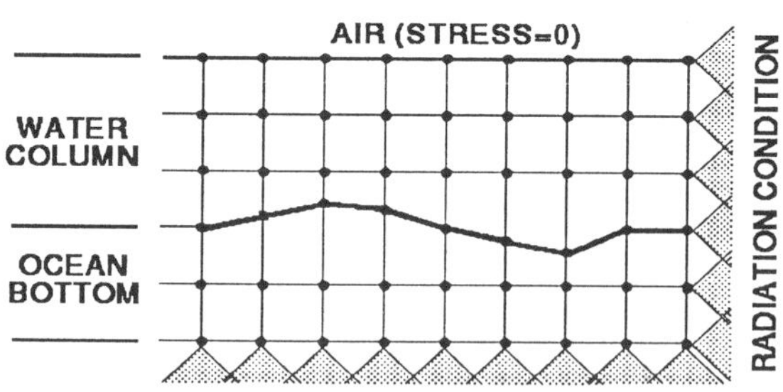


Figure 1. The finite element grid with boundary conditions. Element nodes are adjusted to lie along the water-bottom interface.

2.1 2-D ELASTIC EQUATIONS OF MOTION. The equations of motion for the 2-D elastic medium are

$$- \omega^2 \rho\, u - (\partial/\partial x)\,(c_{11}\, \partial u/\partial x + c_{12}\, \partial v/\partial y) - c_{33}\,(\partial/\partial y)\,(\partial u/\partial y + \partial v/\partial x) = f_x\ , \quad (2)$$

and

$$- \omega^2 \rho\, v - c_{33}\,(\partial/\partial x)\,(\partial u/\partial y + \partial v/\partial x) - (\partial/\partial y)\,(c_{12}\, \partial u/\partial x + c_{22}\, \partial v/\partial y) = f_y\ , \quad (3)$$

where Cartesian coordinates have been used to simplify the form. (In the SAFE computer model, axial symmetry is assumed and the equations are written appropriately.)

On the boundary of each element, the tractions are

$$t_x = (c_{11}\, \partial u/\partial x + c_{12}\, \partial v/\partial y)\, n_x + c_{33}\,(\partial u/\partial y + \partial v/\partial x)\, n_y\ , \quad (4)$$

and

$$t_y = c_{33}\,(\partial u/\partial y + \partial v/\partial x)\, n_x + (c_{12}\, \partial u/\partial x + c_{22}\, \partial v/\partial y)\, n_y\ , \quad (5)$$

where for plane strain,

$$c_{11} = c_{22} = E(1-\nu)\,/\,(1+\nu)(1-2\nu), \quad c_{12} = E\nu\,/\,(1+\nu)(1-2\nu), \quad c_{33} = E\nu\,/\,2(1+\nu)\ , \quad (6)$$

E is the modulus of elasticity, and $\nu$ is Poisson's ratio.

Within each element, the displacements are interpolated as

$$u = \sum_{i=1}^{4} u_i\, \Psi_i\ , \qquad v = \sum_{i=1}^{4} v_i\, \Psi_i\ , \quad (7)$$

$$\Psi_i\,(\xi_i, \eta_j) = \delta_{ij} \qquad \sum_{i=1} \Psi_i\,(\xi, \eta) = 1\ . \quad (8)$$

with $\Psi_i$ being linear interpolation functions and $u_i$, $v_i$ being the node displacements.

Insertion of Eq. (7) into the equations of motion, Eqs. (2) and (3), then gives

$$[K^{11}]\,\{u\} + [K^{12}]\,\{v\} = \{F^1\}\ , \qquad [K^{21}]\,\{u\} + [K^{22}]\,\{v\} = \{F^2\}\ , \quad (9)$$

where

$$K_{ij}^{11} = \int_\Omega (c_{11}\, \partial\Psi_i/\partial x\, \partial\Psi_j/\partial x + c_{33}\, \partial\Psi_i/\partial y\ \ \partial\Psi_j/\partial y - \omega^2 \rho\, \Psi_i\, \Psi_j)\, dx\, dy\ , \quad (10)$$

$$K_{ij}^{12} = K_{ji}^{21} = \int_\Omega (c_{12}\, \partial\Psi_i/\partial x\, \partial\Psi_j/\partial y + c_{33}\, \partial\Psi_i/\partial y\ \ \partial\Psi_j/\partial x)\, dx\, dy\ , \quad (11)$$

$$K_{ij}^{22} = \int_\Omega (c_{33}\, \partial\Psi_i/\partial x\, \partial\Psi_j/\partial x + c_{22}\, \partial\Psi_i/\partial y\ \ \partial\Psi_j/\partial y - \omega^2 \rho\, \Psi_i\, \Psi_j)\, dx\, dy\ , \quad (12)$$

$$F_i^1 = \int_\Omega \Psi_i\, f_x\, dx\, dy + \int_\Gamma \Psi_i\, t_x\, dS\ , \quad (13)$$

$$F_i^2 = \int_\Omega \Psi_i\, f_y\, dx\, dy + \int_\Gamma \Psi_i\, t_y\, dS\ . \quad (14)$$

At present, we allow different material properties in each rectangular element. The material constants, $c_{ij}$, and the density are kept constant within an element, thus allowing all of the integrations required for the K-matrices to be done analytically. Later we will include the ability to deform the rectangular elements so that the nodes can follow the range-dependent ocean bottom bathymetry, or any other surfaces across which there are sharp discontinuities in properties. Numerical integration can be performed for the contributions that these quadrilateral elements make to the FE system of equations.

2.2 THE BOUNDARY INTEGRAL. The FE rectangular elements are influenced by neighboring elements and boundaries depending upon whether the elements are interior elements or elements with one or two of its sides on a boundary.

Elements within the interior of the FE mesh are exactly cancelled by similar terms from a neighboring element. This is due to the condition that the stresses must be continuous. The FE rectangular elements can also have one or possibly two of their sides on the actual boundary of the computer model. When that is the case, two different boundary conditions are presently allowed: (a) Stress-Free: $t_x = t_y = 0$. This is treated exactly and the boundary term is zero. (b) Radiation Boundary Condition: Waves should pass without reflection through the boundaries that terminate the model. This boundary condition is treated approximately.

2.3 RADIATING BOUNDARY CONDITIONS. In FOAM, which is a scalar CW model for pressure waves, $\partial P/\partial n$ appears in boundary integrals computed along the bottom and the right side of the computational grid. The approximate radiation condition first used was

$$\partial P/\partial n = i\,k\,P = i\,(\omega/c)\,P \tag{15}$$

which gives a small effective reflection from the boundary.

In the seismo-acoustic model, SAFE, we are incorporating the approximate boundary conditions discussed by Kalinowski[2]:

$$\sigma_n = \rho\, c_p\, v_n \quad \text{(normal traction)}\,, \qquad \sigma_t = \rho\, c_s\, v_t \quad \text{(tangential traction)}\,, \tag{16}$$

where $\sigma_n$ , $\sigma_t$ are the normal and tangential stress, respectively; $c_p$ , $c_s$ are the compressional and the shear wave speeds, respectively; and, $v_n$ , $v_t$ are the normal and the tangential components of particle velocity on the boundary, respectively.

These conditions for both the scalar wave model and the seismo-acoustic model, are exact for plane waves propagating normal to the boundary. In that case, they give "zero" reflection. For waves slightly off normal, the numerical reflections are small and can be extinguished by a combination of the radiation boundary condition and increasing the attenuation near the boundary.

2.4 ATTENUATION. Attenuation for CW propagation can be treated in the standard way by allowing a complex wavenumber, $k=(\omega/c)(1+i\eta\beta)$, where $\beta$ is the attenuation in dB/wavelength, and $\eta$ is a numerical constant given by $\eta=(40\pi\log_{10}e)^{-1}$. Equivalently, SAFE uses a complex wave speed for both compressional waves and shear waves, i.e. $1/c_p \longrightarrow 1/c_{p,eff} = (1/c_p)(1+i\,\eta\,\beta_p)$, $1/c_s \longrightarrow 1/c_{s,eff} = (1/c_s)(1+i\,\eta\,\beta_s)$.

## 3. NUMERICAL EXAMPLES

For purposes of numerical comparisons the results given below make reference to a pressure which we are computing as $P=-B(\partial u/\partial x+\partial v/\partial y)$, where P is the pressure and B is the bulk modulus.

3.1 SANITY CHECKS. In the process of developing any new computational model, a number of basic, often trivial, computations are made to ensure that the model is performing correctly. These "sanity checks" are important and often take place before any comparisons are made with analytic and/or benchmark numerical solutions. We report below two such sanity checks.

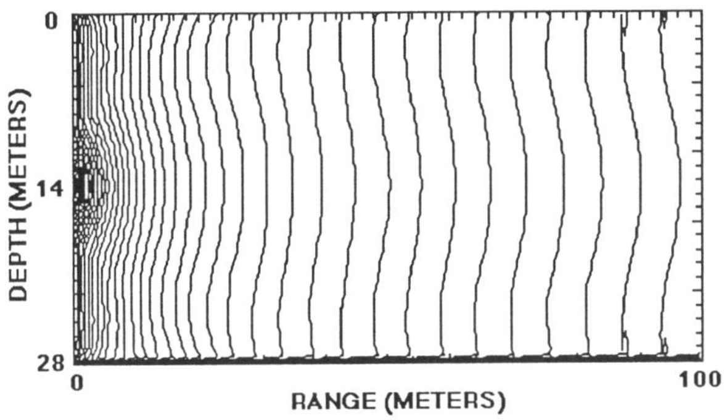


Figure 2. Sanity check on the radiation boundary condition by simulating a point source in a homogeneous unbounded media. The pressure contours are nearly circular.

3.1.1 *Symmetric Contours-Radiation Boundary Conditions.* A check of how well the radiation boundary condition actually "extinguishes" unwanted reflections can be obtained by simulating a point source in free space. Figure 2 shows this sanity check where the upper, lower, and right boundaries of the SAFE model have the radiation boundary condition imposed. Pressure contours emanating from the source should be spherical. Figure 2 shows this to be the case in the center with small deviations away from spherical at the upper and lower boundaries. This is to be expected since the radiation boundary conditions are not perfect. In practice, a small amount of gradually increasing attenuation precedes each boundary on which the radiation condition is applied.

3.1.2 *Lloyd's Mirror.* Another useful sanity check is to create the Lloyd's mirror interference pattern and examine the smoothness of the pressure contours. This is shown in Fig. 3 for the case of a shear speed of 500 m/s and for zero shear speed. The compressional wave speed is 1500 m/s in each case. The pressure contours are fairly smooth with some distortion evident and due to small unwanted reflections. In practice, these unwanted reflections would be eliminated by a small amount of attenuation near the boundaries (as discussed above).

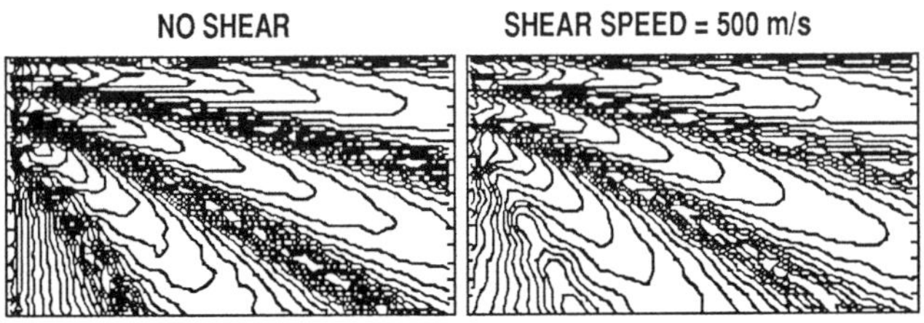


Figure 3. Pressure contours from the SAFE model showing the Lloyd's mirror interference for a medium with no shear and for a medium with shear.

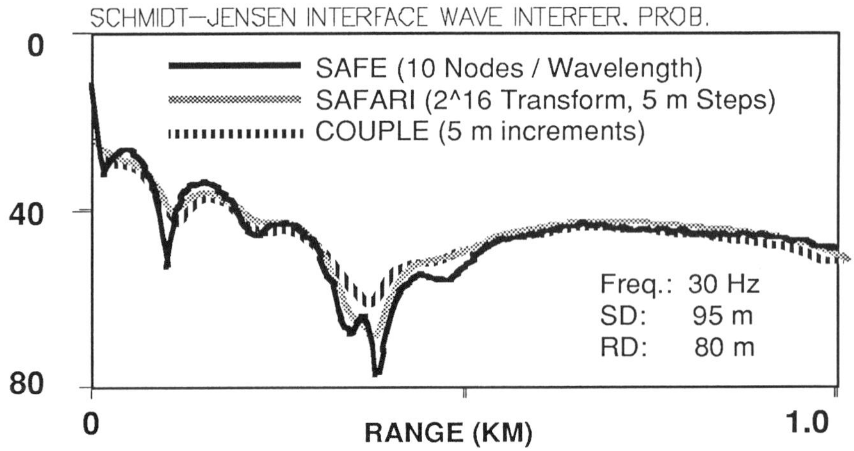


Figure 4. A comparison between the SAFE and SAFARI models, each without shear in the bottom. The non-shear model, COUPLE, is also shown for comparisons.

3.2 SCHMIDT-JENSEN INTERFACE WAVE INTERFERENCE EXAMPLE. Figures 4 and 5 present a test of the SAFE model against a benchmark numerical solution. For the benchmark problem, we have chosen an example given by Schmidt and Jensen[8]. A 100 m deep homogeneous water column (1500 m/s) overlies a solid homogeneous bottom (compression speed of 1800 m/s, shear speed of 600 m/s, compressional attenuation of 0.1 dB/λ, shear attenuation of 0.2dB/λ, and a 2.0 density ratio between bottom and water.

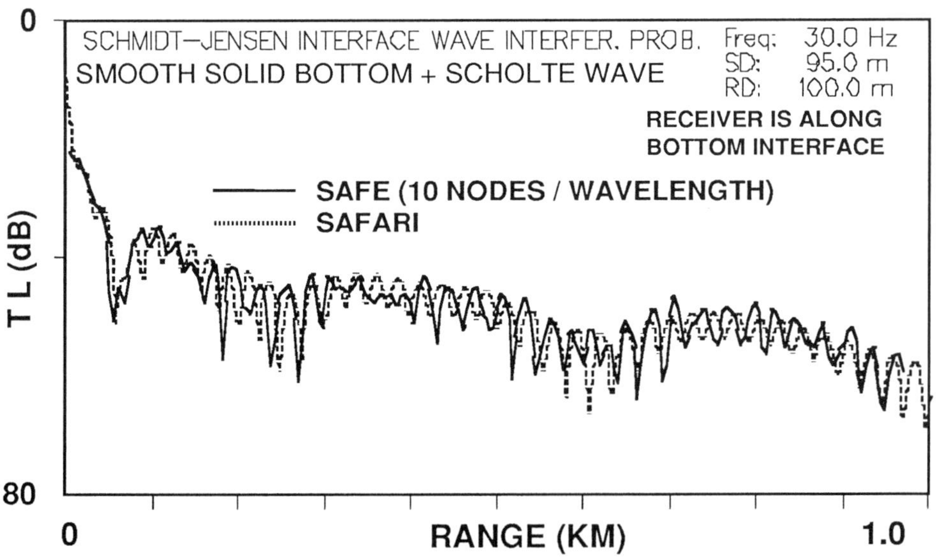


Figure 5. A comparison of predictions made by the SAFE model and the SAFARI model.

The 30 Hz. acoustic point source is placed at a depth of 95 m so as to excite the Scholte interface wave that travels along the water-solid interface. There are two trapped modes in the water column. An interference pattern between the interface (Scholte) wave and the two trapped modes should be evident in the water near the water-solid boundary.

An comparison between SAFE and two very accurate underwater acoustic models is shown in Fig. 4 for the fluid bottom case. The receiver is placed at a depth of 95 m (5 m above the interface) and the receiver is at a depth of 80 m. The results of the SAFE and SAFARI[8] models *without* shear, and Evan's COUPLE model[9] (which does not contain shear) are shown.

Figure 5 compares predictions from the SAFE and SAFARI models *with* shear in the bottom. The source is again placed at a depth of 95 m but the receiver is along the fluid-solid interface (at a depth of 100 m). The interference pattern between the two trapped modes and the interface (Scholte) wave is quite noticeable in each model's transmission loss (TL) predictions (as rapid oscillations). Less dramatic but also quite noticable are the longer oscillations that are caused by the two mode interference.

The SAFE model that produced the curves given in Figs. 4 and 5 had a grid density of 10 nodes per wavelength in both depth and range. From our experience with the FOAM model, a node density of 10 nodes per wavelength is at the "edge" of that needed to produce highly accurate results. A node density of 15 to 20 nodes per wavelength would be expected to produce better results, but of course would require more computer memory and longer computation time.

Figure 6 shows pressure contour plots of SAFE with and without shear for the Schmidt-Jensen example. The noteworthy feature is the contribution from the interface wave that appears as small semi-circles about the interface in the SAFE *with* shear pressure-contour plot while the SAFE *without* shear plot shows no such effect.

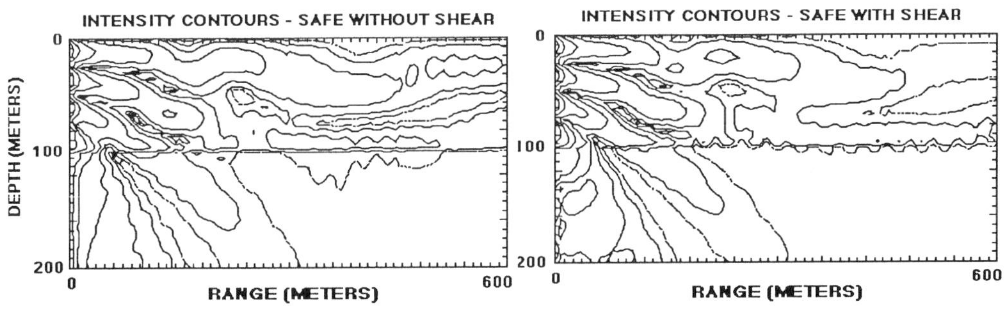


Figure 6. Pressure contours from the SAFE model. The figure on the left is from the SAFE model without shear. The figure on the right is from the SAFE model with shear.

## 4. SUMMARY

The FE fluid model, FOAM, has been modified to allow both compressional and shear speeds and compressional and shear attenuations in the bottom. This new computational model with a lossy elastic bottom (now called the SAFE model) will allow the study of the effects of compressional-shear wave conversions on the water-borne complex pressure field. The theory and validations of the SAFE model have been presented and the results appear favorable and encouraging.

## 5. ACKNOWLEDGEMENTS

This work was supported by the Office of Naval Research (ONR), the Naval Oceanographic and Atmospheric Research Laboratory, and the ONR Acoustic Reverberation Special Research Program.

## 6. REFERENCES

[1] Goldstein, C. I. (1986), "Finite element methods applied to nearly one-way wave propagation," J. Comput. Phys., 64, 56-81. *See also*, Bayliss, A., Goldstein, C. I. and Turkel, E. T. (1983), "An iterative method for the Helmholtz equation," J. Comput. Phys., 49, 443-457; Bayliss, A., Goldstein, C. I. and Turkel, E. T. (1985), "The numerical solution of the Helmholtz equation for wave propagation problems in underwater acoustics," Comp. & Maths. with Appls., 11, 655-665.

[2] DiNapoli, F. R. and Deavenport, D. L. (1979), "Numerical Models of Underwater Acoustic Propagation," in J. A. DeSanto, editor, *Ocean Acoustics*, Springer-Verlag, Berlin. Chapter 3 contains a synopsis by A. J. Kalinowski on the applicability of finite element techniques to acoustic propagation problems, pp. 142-154.

[3] Kuo, J. T. (1988), "The influence of the elasticity of the ocean bottom on wave propagation," in *Computational Acoustics: Wave Propagation*, Lee, D., Sternberg, R. L., and Schultz, M. H. (eds.), Elsevier Science Publishers B. V. (North-Holland), pp.199-223. *See also,* Teng, Y. C. and Kuo, J. T. (1988), "A finite element algorithm for solving the transient problem in a fluid-solid coupled medium," in *Computational Acoustics: Wave Propagation*, Lee, D., Sternberg, R. L., and Schultz, M. H. (eds.), Elsevier Science Publishers B. V. (North-Holland), pp.239-257; Kuo, J. T., Teng, Y. C., Pecholcs, P., and Blair, J. (1985), "A note on the influence of the elasticity on wave propagation in an acoustic/elastic coupled medium," Comp. & Maths. with Appls., 11, 887-896.

[4] Murphy, J. E. and Chin-Bing, S. A. (1989), "A finite element model for ocean acoustic propagation and scattering," J. Acoust. Soc. Am. 86, 1478-1483.

[5] Murphy, J. E. and Chin-Bing, S. A. (1988), "A finite element model for ocean acoustic propagation," Mathematical and Computer Modeling, 11, 70-74.

[6] Chin-Bing, S. A. and Murphy, J. E. (1988), "Application of the finite element model to low-frequency extreme short-range range-dependent ocean acoustic propagation," in R. Vichnevetsky, P. Borne, and J. Vignes (eds.), *Proceedings of the IMACS 12th World Congress on Scientific Computation*, Vol. 2, Gerfidn Cite Scientifique, Villeneuve d'Ascq, France, pp. 250-252.

[7] Chin-Bing, S. A. and Murphy, J. E. (1990), "Acoustic wavefield distortion by sea-mounts: A finite element analysis," in D. Lee, A. Cakmak, R. Vichnevetsky (eds.), *Computational Acoustics*, Vol. 1, Elsevier Publishers, North-Holland, pp. 37-52.

[8] Schmidt, H. and Jensen, F. B. (1985), "Efficient numerical solution technique for wave propagation in horizontally stratified environments," in *Computational Ocean Acoustics*, Schultz, M. H. and Lee, D. (eds.), Pergamon, New York, pp. 699-715.

[9] Evans, R. B. (1983), "A coupled mode solution for acoustic propagation in a waveguide with stepwise depth variations of a penetrable bottom," J. Acoust. Soc. Am., 74, 188-194 .

# FINITE DIFFERENCE MODELLING OF SHEAR WAVES

R.A. STEPHEN [1]
*SACLANT Undersea Research Centre*
*19026 La Spezia*
*Italy*

ABSTRACT. The finite difference synthetic seismogram method provides the capability to study compressional and shear wave interaction in range dependent marine models. A study has been carried out of the pressure, vertical displacement, and horizontal displacement of the sound field at the seafloor for the P-SV problem. Compressional, shear and interface waves can be clearly identified in both snapshot and time series formats of the sound field for soft sediments (shear wave velocity = 450m/s). Key features have been identified. 1)The coupled shear head wave in the bottom can drain energy away from the compressional head wave. 2)The evanescently generated shear wave can be quite energetic and will create precursors to the Stoneley wave when there is sub-bottom structure. 3)Backscattered, secondary Stoneley waves are generated when the direct water wave hits a change in slope at the seafloor. 4)Weak backscattered water waves are generated from slope discontinuities in laterally heterogeneous waveguides. 5)At the seafloor, Stoneley waves are most evident on vertical displacement sensors. 6)Below the seafloor, at non-normal offsets, body wave energy is largest on horizontal component sensors.

## 1. Introduction

The finite difference synthetic seismogram method provides the capability to study compressional-to-shear wave conversion at the seafloor in range dependent environments [1,2]. In this paper, the method is used to study shear wave conversion and propagation in range dependent, shallow and deep water environments with soft sedimentary bottoms.

The finite difference approach is particularly useful for pulse sources, for strong backscattering, and for studies of the geoacoustic field at and below the seafloor. Since calculations are carried out in the time domain, solutions for a given source pulse can be obtained directly. The method can also obtain a CW solution by using a CW source and running the computations until steady state is reached [2]. The finite difference solution has been applied to the two-way elastic wave equation. Solutions are obtained which include both forward scattering and back scattering as well as conversions to shear and interface waves. Multiple interactions between

[1]Currently on leave from Woods Hole Oceanographic Institution, Woods Hole, MA 02543, USA.



*J. M. Hovem et al. (eds.), Shear Waves in Marine Sediments*, 471–478.

Table 1: Velocity and Density Functions

| Material | Compressional Velocity (k/s) | Shear Velocity (k/s) | Density (gm/cc) |
|---|---|---|---|
| Water | $\alpha_1 = 1.50$ | $\beta_1 = 0.00$ | $\rho_1 = 1.00$ |
| Soft Sediment | $\alpha_2 = 1.70$ | $\beta_2 = 0.45$ | $\rho_2 = 1.42$ |

scatterers are also completely included. Since the finite difference method treats the whole model as a discrete grid, vertical and horizontal displacements (or velocities or accelerations) at the seafloor and within the bottom are obtained at the same time as the pressure field in the water. Insight into multi-pathing and bottom and sub-bottom scattering is obtained directly. Given enough computer memory, the whole wavefield as a function of time can be stored (it is always computed) and displayed as a video animation to obtain even further insight into complex propagation problems.

In this paper propagation is considered in sediment with a shear velocity of 450m/s. The wave types and mode conversions that occur for the examples presented would be the same for lower and higher shear velocity sediments but the travel time relationships and amplitudes could differ considerably. The key requirement is that shear velocity be less than water velocity (a soft bottom). If the shear velocity is greater than water velocity (a hard bottom) the wave types and mode conversions can be quite different. Propagation over hard bottoms will be treated in a separate paper.

## 2. Results

### 2.1. DEEP WATER WITH A SOFT BOTTOM

The first finite difference model considers the case of a thick (600m) homogeneous water layer overlying a homogeneous sedimentary bottom (Velocity and density values are summarized in Table 1). A compressional point source is located at 50m depth and its time function is a Gaussian pulse with a dominant frequency of 10Hz [3]. The water depth is sufficiently large, at this frequency, that distinct individual wavefronts can be observed.

Snapshots of compressional and shear energy density at 2.5sec after the shot instant and the time series for a line of receivers (vertical displacement, horizontal displacement, and pressure) at the seafloor (600m depth) are presented in Figure 1. Wave types for this problem are summarized by Brekhovskikh[4]. The furthest propagating energy corresponds to a ground wave packet which consists of the compressional head wave, transmitted compressional wave, and coupled shear head wave. The water wave packet, which is considerably more energetic, comes next and it consists of the direct wave, reflected wave, converted shear wave and the compressional component of the direct wave root. After this packet come the first and second water column multiples and their associated transmitted waves.

The evanescent waves, including the Stoneley wave, are not observable in this case

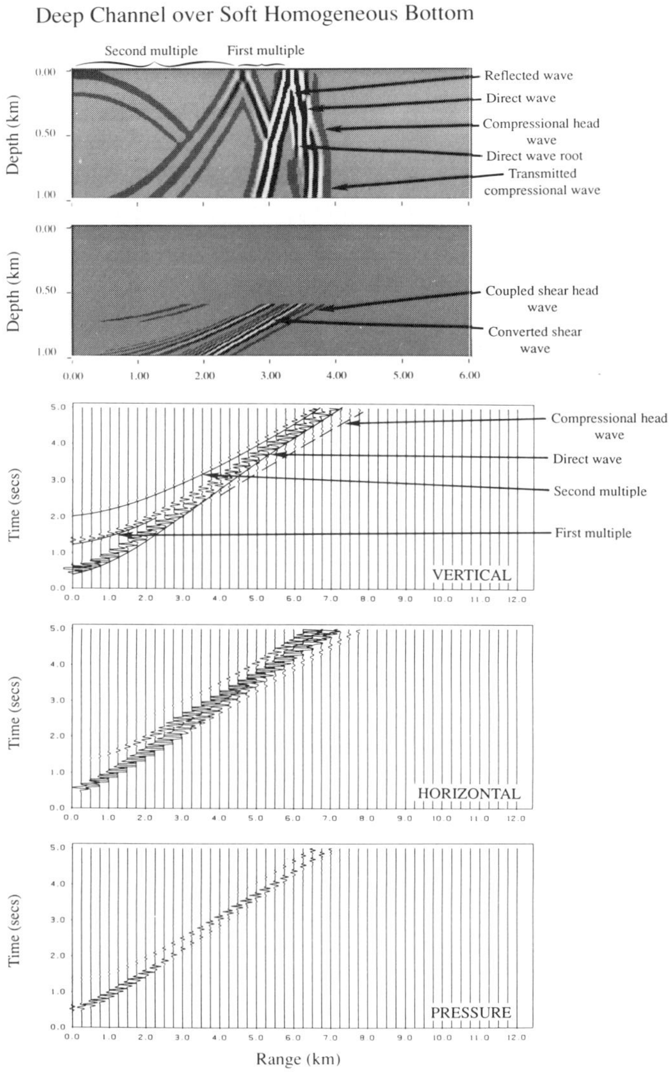


Figure 1: Finite difference synthetic seismogram snapshots and time series are shown for a deep water, laterally homogeneous model. The top and bottom snapshots are compressional and shear energy density, respectively, at 2.5sec after the shot instant. From the top, the time series represent vertical displacement, horizontal displacement, and pressure for a horizontal line of receivers at the seafloor (600m depth). In deep water energy packets can be identified with individual ray paths as discussed in the text. In the time series there is little difference in the relative energy on the three receivers.

because the source is too far from the interface. (The wavelength at the dominant frequency is 150m and the source is 550m away from the seafloor.)

## 2.2. SHALLOW WATER WITH A SOFT BOTTOM

The second finite difference model considers the case of a thin (100m) homogeneous water layer overlying a homogeneous, sedimentary bottom. The source is the same as the previous model and at the same depth (50m), so it is midway in the channel. Snapshots of the compressional and shear energy density at 2.5sec after the shot instant and the time series at the seafloor are shown in Figure 2. The compressional wave packet in the water column has two parts: the ground wave, consisting of the compressional head wave and its multiples, and the water wave, consisting of the direct, internally reflected compressional wave and its multiples.

The leading edge of the shear wave group in the bottom, corresponding to the ground wave, is the coupled shear head wave ($P_1P_2S_2$). The leading edge of the water wave packet in the bottom is the converted shear wave. Because the source is near to the seafloor, there is a third packet of energy consisting of the Stoneley wave and the evanescently generated shear wave. The evanescently generated shear wave corresponds to evanescent energy (complex angles) from the point source.

## 2.3. DOWNSLOPE PROPAGATION WITH A SOFT BOTTOM

The third finite difference model considers the case of downslope propagation. The source is located in a shallow channel as in the previous example. However at 1.25km down range the bottom changes slope from 0° to 3.8° (a slope of 1:15). Snapshots for the wavefield at 2.5sec and time series for a line of receivers at 100m depth are shown in Figure 3. The horizontal phase velocity of the head wave changes to adjust for the slope of the bottom (1640m/s). This head wave is much weaker than the head wave for the laterally homogeneous waveguide. The front edge of the water wave is similar to the laterally homogeneous case. Some near sub-critical energy in the channel becomes super-critical at the shelf break. These correspond to weak, higher order modes that are trapped by 'down slope enhancement'. Shear continues to leak out of these higher order modes blurring the distinction between the converted shear wave branches. Within the channel are backscattered water waves and Stoneley waves which can be more easily identified on the time series. In this snapshot the original Stoneley wave has not reached the shelf break.

The time series plots (Figure 3) show clearly backscattering from the shelf break. The water wave generates both a backscattered water wave and a backscattered Stoneley wave. The backscattered Stoneley wave is strongest on the vertical component and it dominates all other scattered waves.

## 2.4. UPSLOPE PROPAGATION WITH A SOFT BOTTOM

The fourth finite difference model considers the case of upslope propagation. The source is located in deep water as in Model 1. However the bottom gets shallower at an angle of 3.8° (a slope of 1:15) beyond 1.25km. For a line of receivers at 600m depth, receivers beyond 1.25km are buried beneath the seafloor. The snapshots at 2.5sec (Figure 4) are very similar to the laterally homogeneous deep water example.

Shallow Channel over Soft Homogeneous Bottom

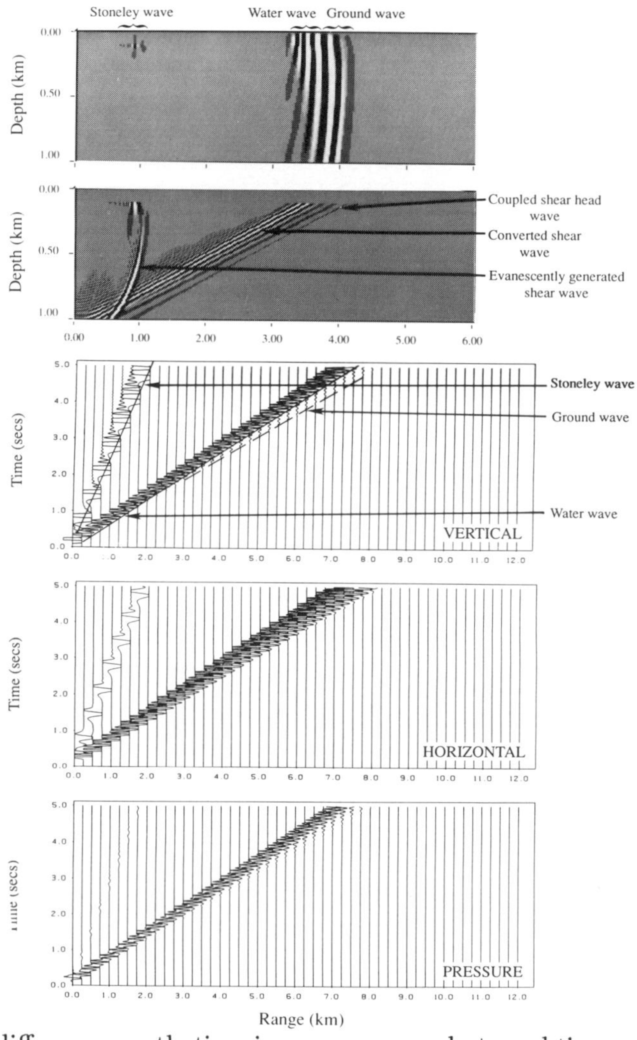


Figure 2: Finite difference synthetic seismogram snapshots and time series are shown for a shallow water, laterally homogeneous model. The format and amplitude scales are the same as Figure 1. The horizontal line of receivers is at the seafloor (100m depth). In shallow water, energy packets can be identified primarily to modes of propagation which correspond to the interference of ray paths. The energy packets are described in the text. The Stoneley wave is much stronger, relative to other arrivals, in vertical displacement and it is much weaker in pressure. The evanescently generated shear wave, seen in the lower snapshot, is important because in media with sub-bottom structure it can be focussed back to the water column generating precursors to the Stoneley wave.

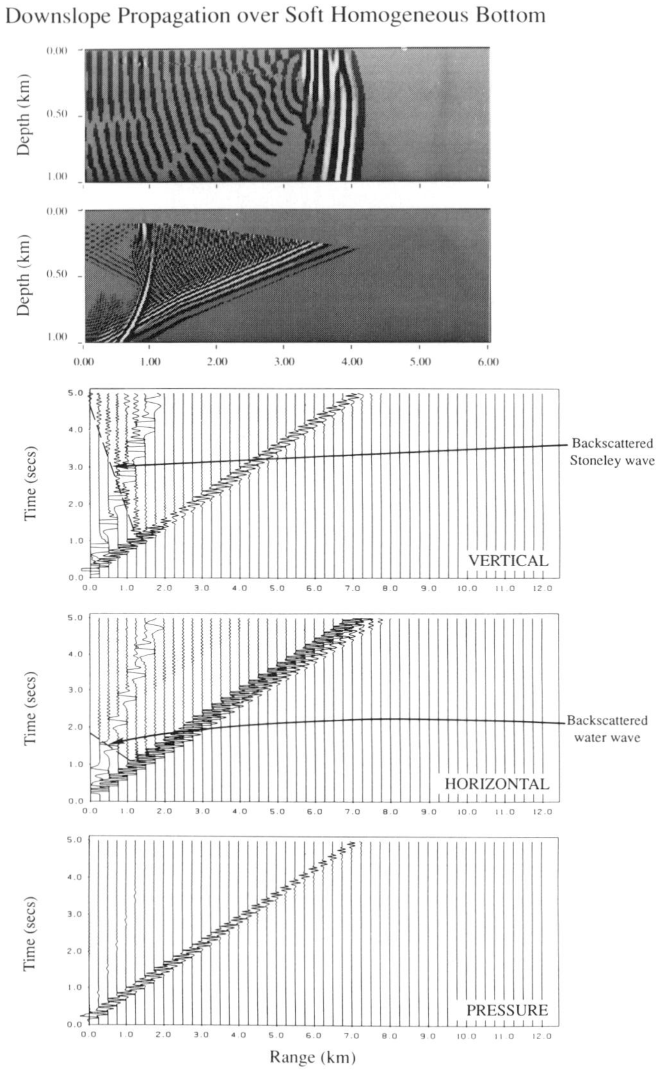


Figure 3: Finite difference synthetic seismogram snapshots and time series are shown for a downslope propagation model. The format and amplitude scales are the same as Figure 1. The horizontal line of receivers is at 100m depth. (It does not follow the seafloor. At ranges beyond 1.25km the receivers are progressively farther from the seafloor.) The most interesting feature is the backscattered Stoneley wave which was converted from the direct wave on incidence with the shelf break. There is also a weak backscattered water wave created by the impedance contrast in the waveguide. The amplitudes of the principle water and ground waves decay beyond the shelf break, compared with the laterally homogeneous case in Figure 2.

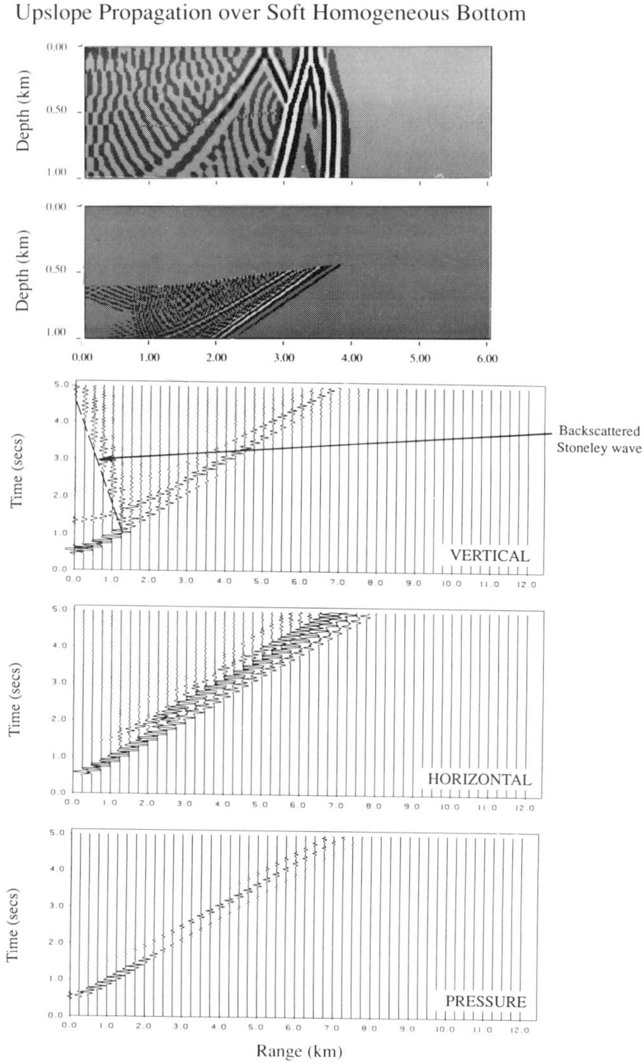


Figure 4: Finite difference synthetic seismogram snapshots and time series are shown for an upslope propagation model. The format and amplitudes are the same as Figure 1. The horizontal line of receivers is at 600m depth. As for the downslope case, there is a backscattered Stoneley wave generated when the direct wave hits the change in slope. This example shows the phenomenon occuring in deep water. The secondary Stoneley wave is most evident on the vertical displacement. At ranges beyond 1.0km, body wave amplitudes are much larger on the horizontal component than on the vertical component.

There is some weak body wave scattering from the change in slope. However the most dramatic feature is a strong back-scattered Stoneley wave generated when the water waves hit the change in slope. This secondary Stoneley wave is most readily observed on the vertical component time series. Note that in the bottom (beyond 1.25km range) the horizontal displacement is much larger in amplitude than the vertical dispacement.

## 3. Summary and Conclusions

The finite difference synthetic seismogram technique is a useful method for studying wave propagation in laterally homogeneous media and in media with changes in bottom slope. In this paper, we considered propagation in models with soft, homogeneous bottoms. Some key features have been identified. 1) The coupled shear head wave in the bottom can drain energy away from the compressional head wave. In shallow water this corresponds to leaking energy from the ground wave packet. 2) The evanescently generated shear wave can be quite energetic and will create precursors to the Stoneley wave when there is sub-bottom structure. 3) Backscattered, secondary Stoneley waves are generated when the direct water wave hits a change in slope at the seafloor. These secondary Stoneley waves are generated even in deep water and can be an important loss mechanism for bottom interacting energy. 4) Weak backscattered water waves are generated from slope discontinuities in laterally heterogeneous waveguides. 5) At the seafloor, Stoneley waves are most evident on vertical displacement sensors. 6) Below the seafloor, at non-normal offsets, body wave energy is largest on horizontal component sensors.

## 4. Acknowledgements

The work described in this paper was supported by the Office of Naval Research under Contract Number N00014-90-J-1541 to Woods Hole Oceanographic Institution(WHOI). The computer work and the preparation of figures were carried out at WHOI. The paper was written while the author was on sabbatical leave at the SACLANT Undersea Research Centre. Woods Hole Oceanographic Institution Contribution Number 7631.

## 5. References

[1] Stephen, R.A. (1988) "A review of finite difference methods for seismo-acoustic problems at the sea floor", Reviews of Geophysics **26**, 445-458.

[2] Stephen, R.A. (1990) "Solutions to range-dependent benchmark problems by the finite difference method", J. Acoust. Soc. Am. **87**, 1527-1534.

[3] Stephen, R.A., Pardo-Casas, F. and Cheng, C.H. (1985) "Finite difference synthetic acoustic logs", Geophysics **50**, 1588-1609.

[4] Brekhovskikh, L.M. (1960) "Waves in layered media", Academic Press, New York.

# A STABLE HIGHER-ORDER ELASTIC PARABOLIC EQUATION WITH APPLICATION TO SCHOLTE WAVE PROPAGATION

M.D. Collins and W.A. Kuperman
*Naval Research Laboratory*
*Washington, DC 20375*
*USA*

ABSTRACT. An accurate and stable elastic parabolic equation (PE) is derived from a higher-order Padé approximation. Accuracy is achieved by constraining some of the derivatives of the Padé approximation near the point corresponding to the reference wavenumber. Stability is achieved by requiring that the Padé approximation map part of the lower-left quadrant of the complex plane into the upper half of the complex plane. The stable higher-order elastic PE accurately handles all wave types including interface waves, very wide propagation angles, and large depth variations and gradual range variations in the seismo-acoustic parameters. The rotated PE, which offers improved accuracy for some problems involving sloping ocean bottoms, is generalized to elastic media. The elastic PE is applied to illustrate the variation of phase velocity as a Scholte wave propagates up a sloping ocean bottom and evolves into a Rayleigh wave and to illustrate how low shear speed acts as a loss mechanism for a problem involving a broadband source.

## 1. INTRODUCTION

The parabolic equation (PE) underwater sound propagation model [1] has recently been extended to handle problems involving elastic ocean bottoms [2,3]. However, the first successful elastic PE models are not suitable for general use even for range-independent problems. The original higher-order elastic PE [2], which is a generalization of the first rigorous factorization of an elastic wave equation [4], is very accurate but weakly unstable. The instability can be suppressed for some problems with the inefficient approach of truncating the computational domain at a large depth. A completely stable elastic PE was obtained [3] by replacing the real coefficients of the Claerbout approximation [5] with complex coefficients. Although this approximation allows smaller truncation depths, it is not sufficiently accurate for problems involving wide propagation angles, large variations in the seismo-acoustic parameters, or thin elastic layers.

In this paper, the approaches used to derive the original elastic PE models are combined and generalized to obtain a higher-order elastic PE that is very accurate and completely stable. The basis for this elastic PE is a Padé approximation consisting of a sum of rational-linear terms with complex coefficients. Accuracy is achieved by requiring that some of the derivatives of the Padé approximation be correct at the point corresponding to the reference wavenumber. Stability is achieved by requiring that the Padé approximation map part of the lower-left quadrant of the complex plane into the upper half of the complex



*J. M. Hovem et al. (eds.), Shear Waves in Marine Sediments*, 479–486.

plane. Complex Padé coefficients that satisfy these constraints are obtained by solving a system of nonlinear equations using Newton's method. The stable higher-order elastic PE is accurate for problems that involve propagation of all wave types including interface waves, very wide propagation angles, and large depth and gradual range variations in the seismo-acoustic parameters. The rotated PE [6], which achieves improved accuracy for some range-dependent problems by marching parallel to a sloping ocean bottom, is generalized to elastic media. Examples are presented to illustrate accuracy and stability, the variation of phase speed as a Scholte wave propagates up a sloping ocean bottom and evolves into a Rayleigh wave, and how low shear speed in the ocean bottom acts as a loss mechanism for a problem involving a broadband source.

## 2. THE ELASTIC PARABOLIC EQUATION

We work in cylindrical coordinates with the depth $z$ being the distance below the ocean surface and the range $r$ being the horizontal distance from a point source at $z = z_0$. The solution is assumed to be independent of azimuth due to symmetry. Cylindrical spreading is accounted for by factoring the vertical displacement $W$ as $w = \sqrt{r}W$ and the dilatation $D$ as $\Delta = \sqrt{r}D$. For range-independent problems and for $r$ much larger than representative wavelengths, the following equations of motion can be derived from the formulation derived in [7]:

$$\mu \frac{\partial^2 w}{\partial r^2} + \mu \frac{\partial^2 w}{\partial z^2} + \rho\omega^2 w + (\lambda + \mu)\frac{\partial \Delta}{\partial z} + 2\frac{\partial \mu}{\partial z}\frac{\partial w}{\partial z} + \frac{\partial \lambda}{\partial z}\Delta = 0, \tag{1}$$

$$(\lambda + 2\mu)\frac{\partial^2 \Delta}{\partial r^2} + (\lambda + 2\mu)\frac{\partial^2 \Delta}{\partial z^2} + \rho\omega^2\Delta + 2\frac{\partial \mu}{\partial z}\frac{\partial^2 w}{\partial r^2} + \omega^2\frac{\partial \rho}{\partial z}w + \left(\frac{\partial \lambda}{\partial z} + 2\frac{\partial \mu}{\partial z}\right)\frac{\partial \Delta}{\partial z} + \frac{\partial}{\partial z}\left(\frac{\partial \lambda}{\partial z}\Delta\right) + 2\frac{\partial}{\partial z}\left(\frac{\partial \mu}{\partial z}\frac{\partial w}{\partial z}\right) = 0, \tag{2}$$

where $\omega$ is the circular frequency of the source. Loss is handled by using the complex compressional and shear speeds $C_p = c_p / (1 + i\eta\beta_p)$ and $C_s = c_s / (1 + i\eta\beta_s)$, where $c_p$ and $c_s$ are the real wave speeds, $\beta_p$ and $\beta_s$ are the compressional and shear attenuations in decibels per wavelength, and $\eta = (40\pi \log_{10} e)^{-1}$. The Lamé constants $\lambda$ and $\mu$ are related to the wave speeds and the density $\rho$ by $\lambda = \rho\left(C_p^2 - 2C_s^2\right)$ and $\mu = \rho C_s^2$.

Equations 1 and 2 provide a coupled system of the form

$$\mathrm{L}\frac{\partial^2}{\partial r^2}\begin{pmatrix}\Delta \\ w\end{pmatrix} + \mathrm{M}\begin{pmatrix}\Delta \\ w\end{pmatrix} = 0, \tag{3}$$

where the matrices L and M contain depth operators. Since Eq. 3 does not contain a term involving $\partial/\partial r$ and the depth operators in L and M commute with $\partial/\partial r$, we may factor Eq. 3 to obtain the outgoing elastic wave equation

$$\frac{\partial}{\partial r}\begin{pmatrix}\Delta\\ w\end{pmatrix}=ik_0\sqrt{1+\frac{\mathrm{L}^{-1}\left(\mathrm{M}-k_0^2\mathrm{L}\right)}{k_0^2}}\begin{pmatrix}\Delta\\ w\end{pmatrix}, \tag{4}$$

where $k_0$ is defined in terms of the representative sound speed $c_0$ as $k_0 = \omega/c_0$. Removing the plane-wave factor $\exp(ik_0 r)$ from $\Delta$ and $w$, we obtain

$$\frac{\partial}{\partial r}\begin{pmatrix}\Delta\\ w\end{pmatrix}=ik_0\left[1-\sqrt{1+\frac{\mathrm{L}^{-1}\left(\mathrm{M}-k_0^2\mathrm{L}\right)}{k_0^2}}\right]\begin{pmatrix}\Delta\\ w\end{pmatrix}, \tag{5}$$

To derive an elastic PE from Eq. 5, an approximation for the function

$$f(\zeta)=-1+\sqrt{1+\zeta} \tag{6}$$

is used to evaluate the depth operator appearing in Eq. 5 as a differential operator, where the square-root function is defined in terms of the branch lying in the upper half of the complex plane. The following approximation is a generalization of the Padé series of [8,9]:

$$f(\zeta)\cong g(\zeta)=\sum_{j=1}^{n}\frac{a_{2j-1}\zeta}{1+a_{2j}\zeta}, \tag{7}$$

where the coefficients depend on $n$ and may be complex. Applying the Padé series to Eq. 5, we obtain the higher-order elastic PE:

$$\frac{\partial}{\partial r}\begin{pmatrix}\Delta\\ w\end{pmatrix}=ik_0\sum_{j=1}^{n}\frac{a_{2j-1}\mathrm{L}^{-1}\left(\mathrm{M}-k_0^2\mathrm{L}\right)}{k_0^2+a_{2j}\mathrm{L}^{-1}\left(\mathrm{M}-k_0^2\mathrm{L}\right)}\begin{pmatrix}\Delta\\ w\end{pmatrix}. \tag{8}$$

Equation 8 can be solved with the numerical method described in [2] and may be applied to obtain leading-order solutions for weakly range-dependent problems by allowing the parameters appearing in L and M to depend on $r$. This approach is often accurate for the acoustic PE; however, it can lead to large errors if the range dependence is not sufficiently gradual. For some acoustics problems involving upward-sloping ocean bottoms, improved accuracy can be achieved by working in a rotated coordinate system in which $r$ is replaced by the ocean-bottom tangent $\tau$ and $z$ is replaced by the normal $\sigma$. With this approach, which is also valid for elastic media, the preferred direction is taken to be $\tau$.

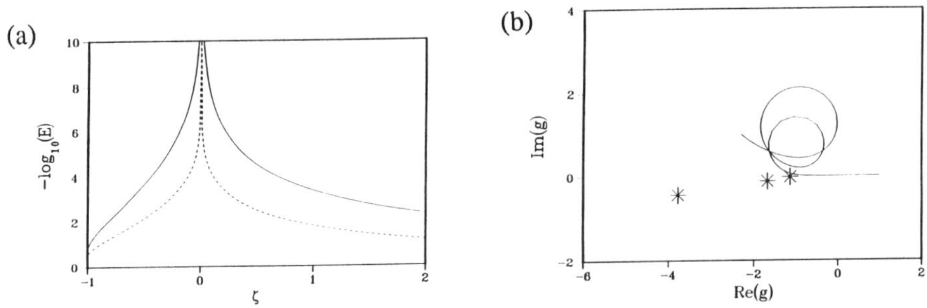


Figure 1. Accuracy and stability. (a) The error in the $n = 1$ (dashed curve) and the $n = 4$ (solid curve) Padé approximations. (b) The image of the interval [-3,3] of the real line under the $n = 4$ mapping. The asterisks mark the locations of the poles of the mapping.

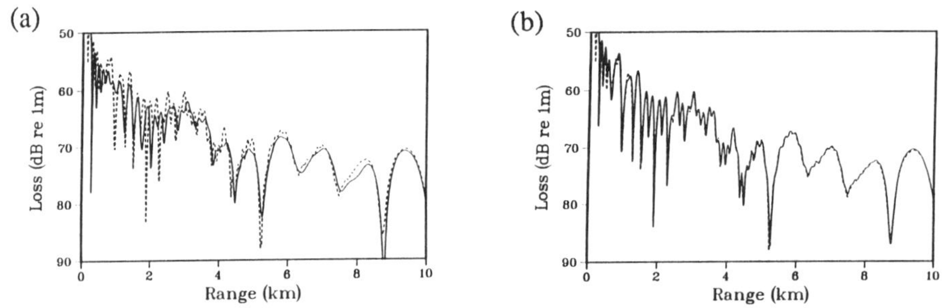


Figure 2. Results at $z = 25$ m for Example A. The dashed curves are the spectral solution. The solid curves are (a) the $n = 1$ solution and (b) the $n = 4$ solution.

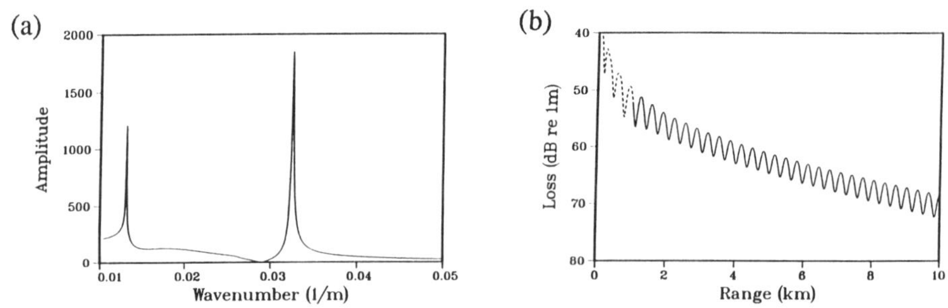


Figure 3. Results at $z = 85$ m for Example B. (a) The wavenumber spectrum. (b) The spectral solution (dashed curve) and the $n = 4$ solution (solid curve).

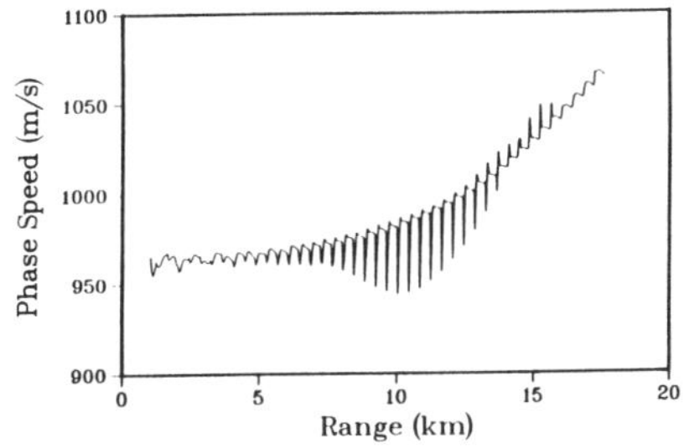


Figure 4. Results of Example C. Numerical estimate of the horizontal phase velocity of an interface wave.

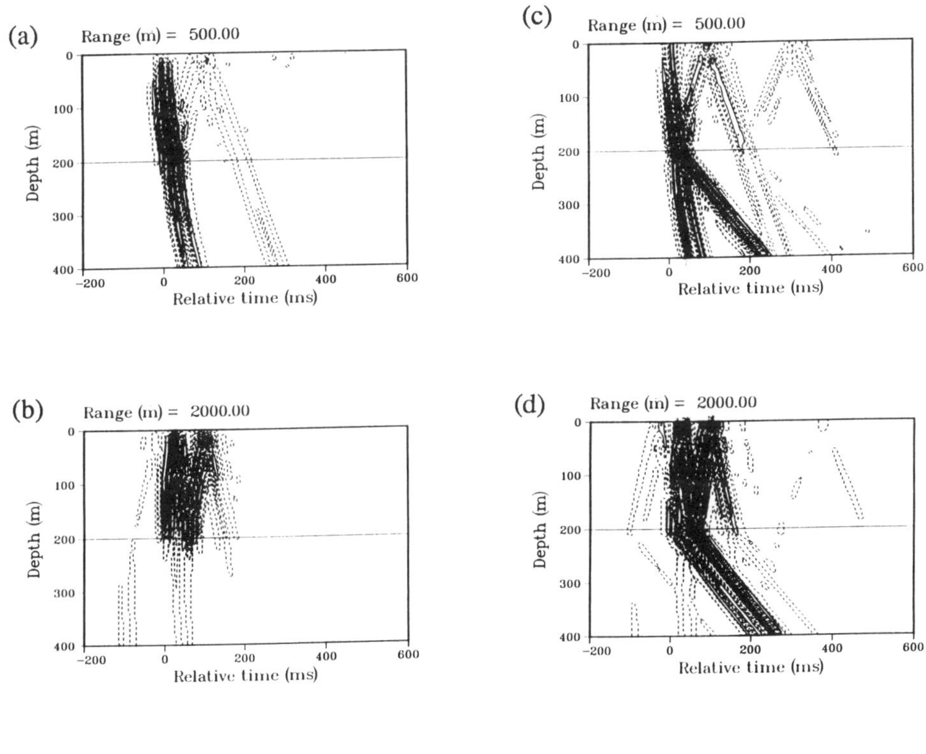


Figure 5. Time-domain solutions for Example D. The dilatation (a) in the nearfield and (b) in the farfield. The vertical displacement (c) in the nearfield and (d) in the farfield.

The rotated elastic PE is of the same form as Eq. 8 with $r$ replaced by $\tau$ and $z$ replaced by $\sigma$.

The Padé coefficients are defined by placing $2n$ constraints on $g(\zeta)$. A high level of accuracy is achieved by requiring that the first $2n$-$m$ derivatives of $g(\zeta)$ and $f(\zeta)$ agree at $\zeta = 0$. The remaining $m$ constraints are devoted to stability, which can be achieved by ensuring that all of the eigenvalues in the lower-left quadrant of the complex plane [10] are mapped into the upper half of the complex plane [3]. For problems that do not involve thin elastic layers such as ice cover, these eigenvalues are very close to the real line; and it suffices to take $m = 1$ and require that $g(\zeta_0) > 0$, where $\mathrm{Re}(\zeta_0) \leq -1$ and $\mathrm{Im}(\zeta_0) \leq 0$. Coefficients that satisfy these constraints can be determined using Newton's method. The accuracy and stability of the $n = 1$ and $n = 4$ Padé approximations are illustrated in Figure 1. The error $E(\zeta)$ is defined to be $|f(\zeta)$-$g(\zeta)|$. The $n = 4$ approximation is accurate over a much wider region. The image of the interval [-3,3] lies in the upper half of the complex plane for $n = 4$. Thus part of the lower-left quadrant of the complex plane is mapped into the upper half of the complex plane. For problems involving thin elastic layers, it may be necessary to take $m > 1$ because the eigenvalues in the lower half of the complex plane are farther from the real line.

## 3. EXAMPLES

For each of the examples, the sound speed is 1500 m/s in the homogeneous water column, and the seismo-acoustic parameters are constant in the ocean bottom. Reference solutions are generated for the range-independent examples by solving Eq. 3 using a finite-difference numerical approach for the spectral solution that is based on the Hankel transform [11]. Once the matrices described in [2] are constructed, this is a relatively simple solution to implement. Example A involves a 20-Hz source at $z = 25$ m in an ocean of depth 600 m with $c_p = 3400$ m/s, $c_s = 1700$ m/s, $\rho = 2$ g/cc, and $\beta_p = \beta_s = 0.5$ in the bottom. Results for Example A appear in Figure 2. The $n = 1$ solution exhibits phase errors due to the very wide propagation angles supported by the hard elastic bottom. The $n = 4$ solution is in excellent agreement with the spectral solution.

Example B involves a 5-Hz source at $z = 85$ m in an ocean of depth 100 m with $c_p = 2400$ m/s, $c_s = 1200$ m/s, $\rho = 1.5$ g/cc, $\beta_p = 0.1$, and $\beta_s = 0.2$ in the bottom. Results for Example B appear in Figure 3. The peaks in the wavenumber spectrum correspond to a trapped mode near cutoff (horizontal phase velocity $\cong$ 2388 m/s) and a Scholte wave (horizontal phase velocity $\cong$ 965 m/s). The spectral solution is used to initialize the elastic PE at $r = 1$ km. The $n = 4$ elastic PE solution is in excellent agreement with the spectral solution.

Example C involves a gradually-sloping ocean bottom and the same ocean-bottom parameters as Example B. The ocean depth decreases linearly from 100 m to zero over $0 < r < 20$ km. We take $n = 2$ and $c_0 = 1000$ m/s and initialize the rotated elastic PE with the component of the spectral solution corresponding to the Scholte wave. The interface wave phase velocity estimated from the rotated elastic PE solution near the ocean bottom appears in Figure 4. The trend of the estimated phase velocity increases from 965 m/s and approaches 1100 m/s near $r = 20$ km, which is approximately the correct phase velocity for a Rayleigh wave. The irregular behavior of the curve is apparently due to the approximate method used to estimate the phase velocity.

Example D involves a broadband point source with frequency content centered at 40 Hz. Since the Padé coefficients are complex, we obtain pulse solutions with Fourier synthesis rather than with the direct time-domain approach that has been used for the acoustic PE [9,12]. The source is at $z = 25$ m in an ocean of depth 200 m. In the ocean bottom, $c_p = 1700$ m/s, $c_s = 800$ m/s, $\rho = 1.5$ g/cc, and $\beta_p = \beta_s = 0.5$. Results for Example D appear in Figure 5. At $r = 500$ m, the vertical displacement shows a compressional wavefront and a shear wavefront being transmitted into the ocean bottom by the leading wavefront in the water column; the dilatation, of course, only shows the compressional wavefront. At $r = 2$ km, energy is no longer being transmitted into a compressional wave in the bottom; however, a shear wavefront is still being transmitted into the bottom. This illustrates how low shear speed acts as a loss mechanism for underwater sound propagation.

## 4. DISCUSSION

An accurate and stable elastic PE has been derived from a Padé series with complex coefficients. At this stage in its development, the elastic PE can handle all waves types including interface waves, very wide propagation angles, and large depth variations and gradual range variations in the seismo-acoustic parameters. The elastic PE was applied to illustrate the convergence of the phase velocity of a Scholte wave to the phase velocity of a Rayleigh wave as an interface wave propagates up a sloping ocean bottom and to illustrate in the time domain how low shear speed in the ocean bottom acts as a loss mechanism. The rotated elastic PE should offer improved accuracy for some upslope propagation problems. Improving the accuracy of the elastic PE for range-dependent problems is an important topic for future research.

[1] Tappert, F.D. (1977), "The Parabolic Approximation Method," in *Wave Propagation and Underwater Acoustics*, edited by J.B. Keller and J.S. Papadakis, Lecture Notes in Physics, Volume 70 (Springer, New York).

[2] Collins, M.D. (1989), "A Higher-Order Parabolic Equation for Wave Propagation in an Ocean Overlying an Elastic Bottom," J. Acoust. Soc. Am. 86, 1459-1464.

[3] Wetton, B.T.R. and Brooke, G.H. (1990), "One-Way Wave Equations for Seismoacoustic Propagation in Elastic Waveguides," J. Acoust. Soc. Am. 87, 624-632.

[4] Greene, R.R. (1985), "A High-Angle One-Way Wave Equation for Seismic Wave Propagation along Rough and Sloping Interfaces," J. Acoust. Soc. Am. 77, 1991-1998.

[5] Claerbout, J.F. (1976), *Fundamentals of Geophysical Data Processing* (McGraw-Hill, New York), pp. 206-207.

[6] Collins, M.D. (1990), "The Rotated Parabolic Equation and Sloping Ocean Bottoms," J. Acoust. Soc. Am. 87, 1035-1037.

[7] Kolsky, H. (1963), *Stress Waves in Solids* (Dover, New York), p.11.

[8] Bamberger, A., Engquist, B., Halpern, L., and Joly, P. (1988), "Higher Order Paraxial Wave Equation Approximations in Heterogeneous Media," SIAM J. Appl. Math. 48, 129-154.

[9] Collins, M.D. (1989), "Applications and Time-Domain Solution of Higher-Order Parabolic Equations in Underwater Acoustics," J. Acoust. Soc. Am. 86, 1097-1102.

[10] Miklowitz, J. (1984), *The Theory of Elastic Waveguides* (North-Holland, Amsterdam), p.198.

[11] Schmidt, H. and Jensen, F.B. (1985)," A Full Wave Solution for Propagation in Multilayered Viscoelastic Media with Application to Gaussian Beam Reflection at Fluid-Solid Interfaces," J. Acoust. Soc. Am. 77, 813-825.

[12] McDonald, B.E. and Kuperman, W.A. (1987), "Time Domain Formulation for Pulse Propagation Including Nonlinear Behavior at a Caustic," J. Acoust. Soc. Am. 81, 1406-1417.

# ELASTIC_PE: A PARABOLIC APPROXIMATION FOR PROPAGATION MODELLING OF SHEAR WAVE EFFECTS IN SEDIMENT LAYERS

G. H. BROOKE
*DREP*
*FMO Victoria, BC Canada*
*V0S 1B0*

ABSTRACT. **ELASTIC_PE** is a finite-difference parabolic approximation adapted to propagation problems involving a water layer over an elastic bottom. It is well suited to modelling low-speed shear wave effects associated with the uppermost sediment layers of the ocean bottom where shear and compressional speed gradients and range dependence are important parameters. **ELASTIC_PE** allows for linear sound speed gradients within layers and treats range dependence by updating the material parameters at each range step. This paper outlines briefly the **ELASTIC_PE** method, including recent developments concerning implementation of a wide-angle square-root approximation and a field-compensation technique for treating range dependent bathymetry. Standard numerical examples are used to demonstrate the application of the method to typical underwater acoustic problems involving the interaction of primarily water-borne acoustic energy with low speed shear waves in the ocean bottom. Sources in the water column, well away from the bottom, couple significant energy to shear waves when the shear speed in the uppermost bottom layer is greater than ~0.4 km/s. Sources near the ocean bottom couple additional energy to interface waves. Interference effects involving the interface waves are only partially accounted for by the current version of the **ELASTIC_PE** algorithm.

## 1. Introduction

One-way approximations for seismo-acoustic propagation in elastic waveguides have been the subject of renewed interest. Greene's [1] SHAPE algorithm and the perturbational approach of Wales [2] have stimulated much of the recent work. Collins [3], for example, extended his FEPE algorithm to treat Greene's one-way matrix equation associated with elastic waveguides by directly applying the wide-angle Padé series square root approximation with real coefficients. Wetton and Brooke [4], working with the same matrix equation, introduced the bilinear square root approximation, a rational linear form with complex coefficients, in their **ELASTIC_PE** algorithm. The bilinear form impacts on the numerical stablility of one-way approximations for waveguides involving elastic media. After a brief summary of the **ELASTIC_PE** method, this paper describes two recent extensions. First, a second-order Padé series square root approximation with complex coefficients has been implemented giving the method a wider angle capability. Second, a consistent approach to the problem of a sloping interface has been developed whereby the interface is represented closely by a staircase approximation and at each of the resulting vertical interfaces a correction is applied to the **ELASTIC_PE** wavefield to ensure continuity of normal stress. Brief descriptions of the underlying concepts associated with these two extensions of **ELASTIC_PE** are presented here along with numerical results demonstrating their implementation. Finally, numerical examples are used to illustrate



*J. M. Hovem et al. (eds.), Shear Waves in Marine Sediments*, 487–494.

**ELASTIC_PE** calculations of the interaction of low-speed shear waves in the ocean sediment layers and acoustic energy in the water column. Both range-independent and range-dependent waveguide configurations are used for this purpose. It is shown that the interface wave, when considered as part of the total field, represents a challenging problem for the one-way approximation.

## 2. ELASTIC_PE

### 2.1. THE ALGORITHM

**ELASTIC_PE** is a finite-difference solution technique implemented from the following first order matrix equation in the dilatation, Δ, and the vertical displacement, W,

$$( \partial_x I - ik_o L^{1/2} ) \begin{bmatrix} \Delta(x,z) \\ W(x,z) \end{bmatrix} = 0 \qquad (1)$$

where x is the range coordinate, **I** is the identity matrix, $k_o = \omega / c_o$ is a reference wavenumber whose value is specified by the particular choice, $c_o$, for the reference sound speed, $\omega$ is the radian frequency and **L** is a matrix operator involving partial derivatives and refractive indices which are functions of the z or depth coordinate only. For specific details concerning the operator **L**, the reader is directed to reference [4]. Solution of equation (1) involves the calculation of the operator $\mathbf{L}^{1/2}$, a formidable task which is accomplished only approximately in most applications. In this work, a series of bilinear transformations is used. This approximation, denoted by $F_B$, has the form

$$F_B(L) = I + \sum_{i=1}^{N} ( I + b_i L )^{-1} a_i ( I - L ) \qquad (2)$$

where the coefficients { $a_i, b_i$ } are complex. With phase factor, $e^{ik_o x}$, removed, equations (1) and (2) combine to yield

$$( \partial_x I - ik_o (F_B(L) - I ) ) \begin{bmatrix} \Delta(x,z) \\ W(x,z) \end{bmatrix} = 0. \qquad (3)$$

In **ELASTIC_PE**, the numerical solution of equation (3) is obtained by a straightforward application of the Crank-Nicholson algorithm to discretize in range ( Δr ) combined with a layered waveguide scheme [1],[4] for discretizing in depth ( Δz ). The discretization grid is terminated in depth by a rigid surface preceeded by an absorbing layer to prevent significant reflection. This scheme couples a total of seven field variables on any three consecutive depth-grid points. For wide-angle applications (i.e. higher order square root approximations), the method of alternating directions is used to apply sequentially the individual operator terms in $F_B(\mathbf{L})$. As a result, the algorithm involves the solution of N heptadiagonal systems of equations at each range step. It should also be mentioned that, for numerical purposes, **ELASTIC_PE** treats all layers as if they were elastic layers; fluids are modelled by choosing the shear speed to be a low value equal to 10m/sec. Finally, **ELASTIC_PE** can be started using one of three different initial fields. They are, (i) Tappert's gaussian source [5], (ii) Greene's wide-angle source [6] and (iii) a normal mode summation. The first two source fields are the standard initial fields used in acoustic PE codes. The normal-mode starter has been incorporated into **ELASTIC_PE** to provide starting fields which satisfy the boundary conditions at the water-sediment interface. The normal modes are derived from a waveguide consisting of a fluid layer over an elastic solid bounded below by a rigid surface.

## 2.2. A WIDE-ANGLE SQUARE ROOT APPROXIMATION

The bilinear square root approximation is designed to allow the complex-valued spectral components associated with elastic waveguides to be treated in a numerically stable manner by the one-way approximation. The coefficients of $F_B(\mathbf{L})$ in equation (2) are chosen by extending the approach used in reference [4]. To facilitate the discussion, consider the approximation of $F_B(x)$ to $x^{1/2}$ over real values of x in the interval $0 < x < 1$. Define a mapping circle, of radius R, centred at (1,-R) in the complex x-plane. With these definitions, the coefficients $\{ a_i, b_i \}$ are chosen such that (i) the mapping circle is mapped onto the real line in the $F_B(x)$-plane, (ii) the point $x = 1$ is mapped onto the point $F_B(1) = 1$ and (iii) the approximation minimizes the difference between the first 2N terms in the Taylor's expansions of $x^{1/2}$ and of $F_B(x)$ about the value x=1. In this context, $x^{1/2}$ represents a horizontal wavenumber normalized to the reference wavenumber, $k_0$. The value x=1 corresponds to energy propagating with phase speed equal to the reference sound speed. Essentially, the mapping is designed to yield an accurate square-root approximation for wavenumbers close in value to $k_0$ while satisfying the stability constraints imposed by the complex-valued spectral components of the elastic waveguide.

To date, coefficients for the cases N=1 and N=2 have been derived and implemented in the **ELASTIC_PE** algorithm. Higher order coefficients ( N>2 ) require the solution of a system of nonlinear equations. Not surprisingly, the real part of the complex coefficients so derived are similar in value to the Padé coefficients used by Collins [3] - by definition, Padé coefficients are chosen to match coefficients of the respective Taylor's expansions. Furthermore, like their real counterparts, the complex coefficients result in a good approximation which extends to values of x much greater than one. TABLE 1 summarizes the performance of the square root approximations for N=1,2 in the interval $0 < x < 2$ in comparison with the exact values. The most obvious property of the bilinear approximation is that it introduces a small imaginary part to the value of the square root. In terms of horizontal wavenumber, the imaginary part equates to an artificial attenuation introduced into the propagating field. This attenuation is mimimized by making R, the mapping radius of the bilinear approximation, large. For thick sediment layers with low values of the shear speed, R can usually be chosen to make the artificial attenuation negligible. Conversely, as the shear speed increases or the thickness of the elastic layer decreases, the value of R must be reduced in order to maintain stable one-way propagation. As might be expected, accurate wide-angle prediction using the **ELASTIC_PE** algorithm is more restricted in such cases. It is also apparent from Table 1 that the case N=2 provides a better approximation over a wider range of values of x than the case N=1. It is worth noting, however, that for the same value of R, the artificial attenuation is slightly higher for N=2. In applications of the **ELASTIC_PE** algorithm, second order approximations improve the field prediction most noticeably close to the source where high angle energy is more important.

TABLE 1. Square root approximations, $F_B(x)$: N=1 and N=2, R=60

| x | $x^{1/2}$ | $F_B$:N=1 | $F_B$:N=2 |
|---|---|---|---|
| 2.0 | 1.4142 | 1.4000 +i 0.0027 | 1.4139 +i 0.0029 |
| 1.5 | 1.2247 | 1.2222 +i 0.0008 | 1.2247 +i 0.0009 |
| 1.0 | 1.0000 | 1.0000 +i 0.0000 | 1.0000 +i 0.0000 |
| 0.8 | 0.8944 | 0.8947 +i 0.0002 | 0.8944 +i 0.0002 |
| 0.6 | 0.7746 | 0.7778 +i 0.0008 | 0.7747 +i 0.0009 |
| 0.4 | 0.6325 | 0.6471 +i 0.0021 | 0.6334 +i 0.0023 |
| 0.2 | 0.4472 | 0.5000 +i 0.0042 | 0.4556 +i 0.0055 |

### 2.3. THE SLOPING INTERFACE

**ELASTIC_PE**, derived under the assumption of range-independence, is applied to problems involving a sloping interface by closely approximating the interface with a staircase and by compensating the fields at the resulting vertical interfaces in order to satisfy boundary conditions. That is, the range-dependent problem is replaced by an associated problem made up of range-independent sections. Within each section, the assumption of range-independence is upheld and the parabolic approximation is used to march the elastic wavefield forward by one range step. It is a relatively simple programming matter to sense when a vertical interface is encountered. In such instances, the parabolic wavefield is compensated to approximate the condition of continuity of normal stress ( $(\lambda+2\mu)\Delta-2\mu\partial W/\partial z$ ) across the interface where $\lambda$ and $\mu$ are the Lamé parameters associated with the elastic media. This correction involves only a few field values in the immediate vicinity of the interface and requires minimal addition to the computer code.

## 3. Numerical results.

### 3.1. RANGE-INDEPENDENT WAVEGUIDES

Consider seismo-acoustic propagation at a frequency of 30Hz in the range-independent waveguide consisting of an isospeed fluid layer over an elastic bottom as shown in Fig. 1. This

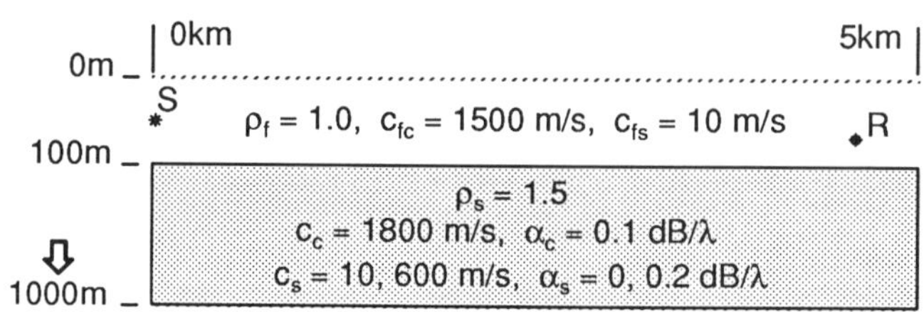


Figure 1. Range-independent waveguide configuration.

configuration [7] is an interesting one because it combines two different aspects of the problem concerning shear wave interactions in marine sediments. First, for sources in the fluid, well off the bottom, the acoustic energy in the water column senses the presence of shear rigidity in the sediments through a modification of the reflection coefficient at the fluid/elastic interface. Consequently, a large proportion of the energy at low grazing angles is coupled into shear wave energy in the bottom and radiated away. Second, for sources near the bottom, energy is partitioned further into the interface wave which has a strong shear field component. The phase speed of the interface wave is low, nearly 530m/s in the current example. Modelling the interaction of this wave with the rest of the acoustic energy travelling at phase velocities greater than 1500m/s, requires that **ELASTIC_PE** accurately represent the field over a wide range of wavenumbers. In order to establish a basis for comparison, consider the case when the bottom is a fluid. Propagation loss for the pressure ( $\rho c^2 \Delta$ ) as a function of range for a source and receiver both located at a depth of 50m is shown in Fig. 2. Results calculated using **ELASTIC_PE** ( $\Delta r$=5m , $\Delta z$=1m, N=1 and R=60 ) are shown as the dashed curve and agree very well with corresponding results, shown as the solid triangles, generated using **SAFARI**. The regular interference pattern exhibited by these results is caused by interference between the two discrete modes in the waveguide. Good agreement is also obtained between **ELASTIC_PE** and **SAFARI** when the bottom is an elastic medium ( $c_S$=600m/s and $\alpha_S$=0.5m/s ). This is demonstrated by the solid curve in Fig.2 which was generated using **ELASTIC_PE** ( $\Delta r$=5m, $\Delta z$=1m, and N=1). The corresponding circles are the **SAFARI** results[8]. Once again, there is a distinct interference pattern in the propagation loss, however, in this case it is caused by two resonant horizontal wavenumbers associated with the continuous spectrum. The effect of shear rigidity in the bottom sediments is quite dramatic in this shallow water case, resulting in additional loss to the acoustic energy in the water column of nearly 10dB at a range of 5km.

In the next example, the source is located at a depth of 95m, very near the bottom. Propagation loss results in this case are shown in Fig. 3. Five sets of data are presented in the figure. First, the dashed curve ( **ELASTIC_PE** ) and the corresponding solid triangles ( **SAFARI** ) are the results

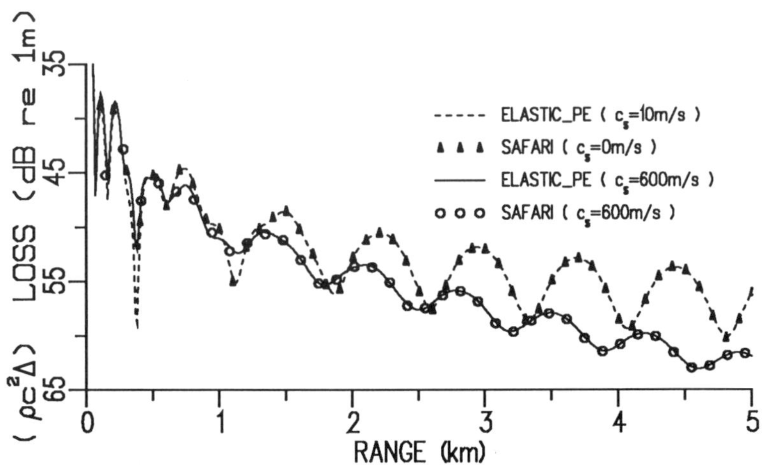


Figure 2. Propagation loss versus range. Range-independent waveguide with source depth at 50m and receiver depth at 50m. **ELASTIC_PE** results obtained with $c_0$=1550m/s.

when the bottom is a fluid. As was the case previously, **ELASTIC_PE** is capable of accurate field predictions when media are fluids. The uppermost solid curve in this figure corresponds to the loss calculated by **SAFARI** when the bottom is an elastic medium. Once again, there is substantial loss of acoustic energy in the water column to the shear waves in the sediments. Of particular interest, however, are the rapid oscillations in this loss curve extending out to ranges near 3km. This fine structure represents an interference between the interface wave and the acoustic energy in the water column. It fluctuates rapidly with range because the phase speeds of the two interfering wave types differ by ~1000m/s. The interface wave is rapidly attenuated by the lossy bottom leaving only the characteristic interference pattern described earlier at longer ranges. It is precisely this problem which presents a major challenge for the **ELASTIC_PE** algorithm - the ability to predict accurately and simultaneously the fields due to two or more wave types with vastly different phase speeds. Propagation loss generated using **ELASTIC_PE** is shown in the lower two solid curves in Fig. 3. The middle solid curve corresponds to relatively low-angle parameter values given by N=1 and R=60; the lowermost curve corresponds to wide-angle parameter values given by N=2 and R=600. The loss in each case has been artificially increased by 10dB and 20dB, respectively, to make the curves distinguishable. Clearly, with a choice of $c_0$=1550m/s, the low-angle **ELASTIC_PE** does not predict the fine structure in the propagation loss and, hence, does not represent well the interface wave propagation with phase speed equal to 530m/s. Had the reference sound speed been chosen close in value to the phase speed of the interface wave, it would have been propagated accurately but the acoustic energy in the water column would not. The high-angle prediction is much better; it shows a substantial component due to the interface wave as well as the familiar acoustic component in the water column. The agreement with **SAFARI**, while improved, is still not exact, indicating that the second-order square-root approximation cannot quite cover the wide range of phase speeds present in this shallow water example.

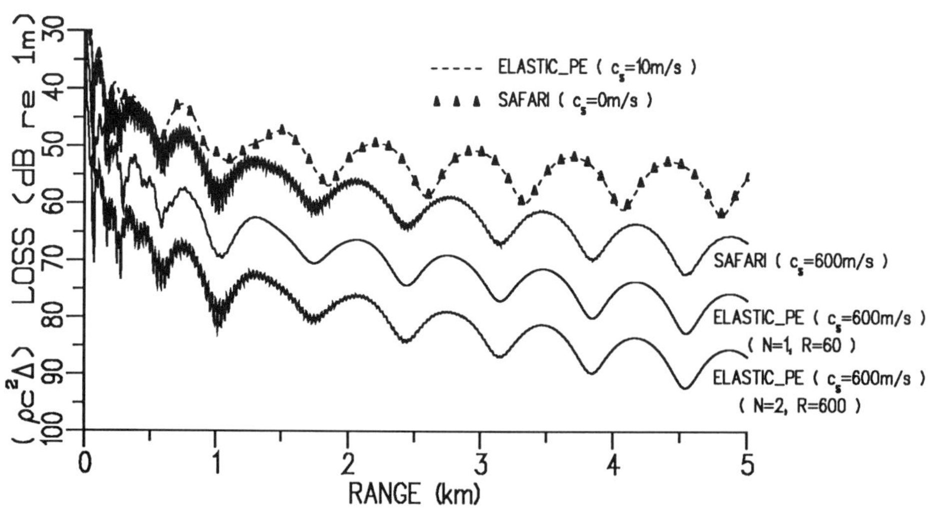


Figure 3. Propagation loss versus range. Range-independent waveguide with source depth at 95m and receiver depth at 100m. **ELASTIC_PE** results obtained with $c_0$=1550m/s. Note that the two lowermost curves have been displaced by 10dB and 20dB, respectively.

3.2. RANGE-DEPENDENT WAVEGUIDES

Consider the benchmark wedge, Problem 2A, described in a recent JASA publication [8] and illustrated in Fig. 4. Source (S) and receiver (R) depths are 100m and 30m, respectively, the source frequency is 25 Hz and the media are assummed to be lossless fluids ( i.e. $c_S$=10m/s, $\alpha_S$=0 and $\alpha_C$=0 ). Propagation loss calculated using the **ELASTIC_PE** algorithm ( N=1, $\Delta$r=5m and $\Delta$z=1m ) as a function of range is shown as the solid curve in Fig. 5. For comparison, the one-way coupled mode results [8] are shown as the dotted curve. The field compensation technique employed in the **ELASTIC_PE** algorithm clearly performs very well in this acoustic case and, based on results not shown here, in all of the other variations of this problem considered in the benchmark issue. Equivalent results for a lossy elastic bottom ( $c_S$=800m/s, $\alpha_S$=0.5dB/$\lambda$ and $\alpha_C$=0.5 dB/$\lambda$ )) are shown as the dashed curve in Fig. 5. These results compare favourably with those presented by Collins [3]. Because the dilatation is not proportional to stress ( pressure ) in the elastic medium, stress normal to the sloping interface has been plotted when the receiver depth passes into the bottom. Although continuity is not perfect, the results are encouraging.

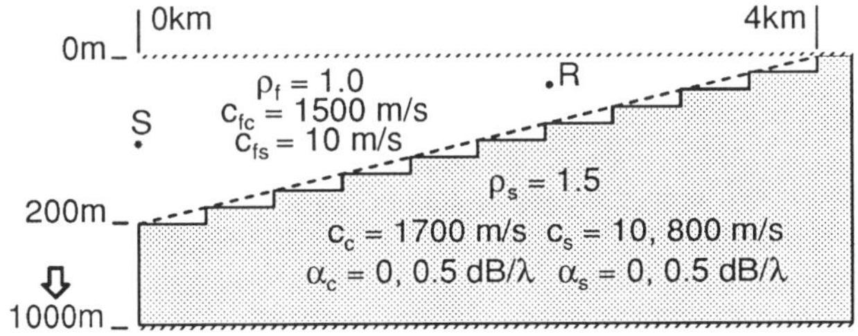


Figure 4. Range-dependent waveguide configuration.

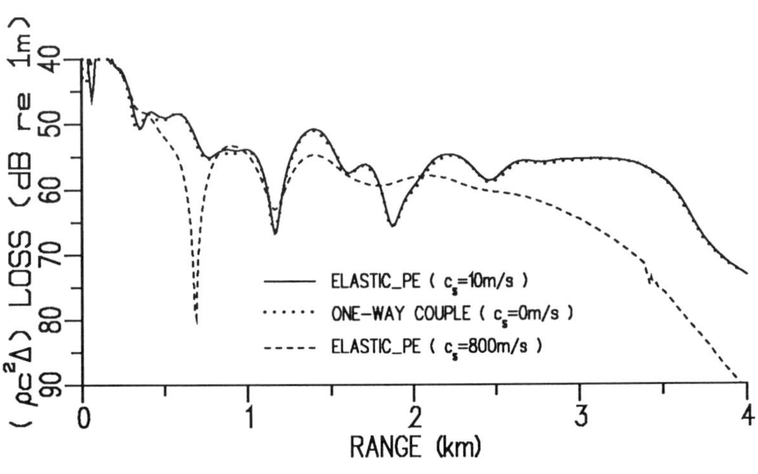


Figure 5. Propagation loss versus range. Benchmark wedge Problem 2A and an elastic wedge.

The oscillations in the curve at the longer ranges may be due in part to inadequate step sizes or inacurracies in the treatment given the sloping interface. Greene's acoustic source was used to generate all of the **ELASTIC_PE** results shown in Fig. 5.

## 4. Summary and conclusions

This paper represents a brief outline of the current version of **ELASTIC_PE**, a one-way propagation algorithm adapted to elastic media. The algorithm uses a series of bilinear transforms ( up to second order ) to approximate the square-root matrix operator inherent for elastic media. The result is a substantial wide-angle capability, although improvement is required in this area if full field predictions for propagation problems involving both interface waves and acoustic energy in the water column are to be obtained. The range of horizontal wavenumbers in such problems exceeds that over which the square-root approximation maintains its accuracy. Work continues on this aspect of the algorithm. For shear wave interactions which do not involve the interface wave, **ELASTIC_PE** performs very well in the range-independent case and should allow effects associated with shear speed and attenuation gradients in marine sediments to be investigated in an efficient and thorough manner. Range dependence presents its own set of restrictions on **ELASTIC_PE.** Currently, staircase approximations are used to represent sloping interfaces and require that a field compensation technique be employed to maintain continuity of normal stress across the associated vertical interfaces. Agreement with the one-way coupled mode results for several benchmark wedge problems involving only acoustic media, verifies that the technique leads to results consistent with other one-way algorithms. Verification of the results for range-dependent elastic media is limited at this point by the scarsity of solutions for such problems, obtained by other techniques.

## 5. References

[1]. Greene, R.R. (1985) 'A high-angle one-way wave equation for seismic wave propagation along rough and sloping interfaces', J. Acoust. Soc. Am. 77, 1991-1998.
[2]. Wales, S.C. (1986) 'A vector parabolic equation model for elastic propagation', in T. Akal and J. M. Berkson (eds.), Ocean Seismo-Acoustics, Plenum, New York, 57-66.
[3]. Collins, M.D. (1989) 'A higher-order parabolic equation for wave propagation in an ocean overlying an elastic bottom', J. Acoust. Soc. Am. 86, 1459-1463.
[4]. Wetton, B.T.R. and Brooke, G.H. (1990) 'One-way wave equations for seismoacoustic propagation in elastic waveguides', J. Acoust. Soc. Am. 87, 624-632.
[5]. Tappert, F.D. (1977) 'The parabolic approximation method', in J.B. Keller and J.S. Papadakis (eds.), Wave Propagation and Underwater Acoustics, Springer, New York, 224-287.
[6]. Greene, R.R. (1984) 'The rational approximation to the acoustic wave equation with bottom interaction', J. Acoust. Soc. Am. 76, 1764-1773.
[7]. Schmidt, H. and Jensen, F. (1985) 'Efficient numerical solution technique for wave propagation in horizontally stratified environments', in M. Schultz and D. Lee (eds.), Computational Ocean Acoustics, Pergamon, New York, 699-715
[8]. Jensen, F. and Ferla, C.M. (1990) 'Numerical solutions of range-dependent benchmark problems in ocean acoustics', J. Acoust. Soc. Am. 87, 1499-1510.

# COMPUTATION OF SHEAR WAVES BY INTEGRAL EQUATIONS METHODS IN STRATIFIED MEDIA

Jacques MEUNIER
*IFREMER DITI/SOM*
*BP 70*
*29280 Plouzané*
*France*

Claude GUENNOU
*UBO - Faculté des Sciences*
*6 Avenue Victor Le Gorgeu*
*29287 Brest Cedex*
*France*

ABSTRACT : This paper presents the SISMOS code developed by IFREMER. The principle of SISMOS code lies on integral equations applied to the Kennett's formalism. Sediments are caracterized by homogeneous horizontal stratified layers. Several examples of synthetic seismograms on which shear waves appear are presented in homogeneous and stratified media.

## 1 Introduction

This paper takes back the Kennett's formalisation, which, after a Fourier transform over time and a Hankel transform with respect to the horizontal cylindrical coordinates, enables to calculate the response of the medium through reflection-transmission operators for every values of the angular frequency $\omega$ and the slowness $p$ (where $p$ is defined as the ratio of horizontal wave number $k$ over angular frequency). The restitution of the displacements in every point of the medium is performed with the help of a fast Hankel transform (FHT) ; a fast Fourier transform (FFT) over frequency gives the time response of the stratification when the frequency content of the source is available. This approch is known as the spectral method.

The paper briefly describes the method and a program developed by IFREMER applying this method. It gives some examples of synthetic seismograms dealing with shear waves.

## 2 Kennett's formalisation

The wave equation in an elastic homogeneous medium is the well known Navier's equation:

$$(\lambda + 2\mu)\nabla\nabla.\vec{u}(r,\theta,z;t) - \nabla\wedge\nabla\wedge\vec{u}(r,\theta,z;t) + \rho\vec{f}(r,\theta,z;t) = \rho\partial_{tt}\vec{u}(r,\theta,z;t) \qquad (1)$$

where $\lambda$ and $\mu$ are the Lamé Moduli, $\vec{u}$ the displacement vector and $\vec{f}$ a body force per unit mass $\rho$. The hypothesis of small displacements is required and a downward oriented set of cylindrical coordinates is considered.



*J. M. Hovem et al. (eds.), Shear Waves in Marine Sediments*, 495–502.

Kennett (1981) proposed a change of variables with settling the new functions $u_V$, $u_H$, $\sigma_{Vz}$, $\sigma_{Hz}$ :

$$
\begin{aligned}
u_V &= \partial_r u_r + r^{-1}u_r + r^{-1}\partial_\theta u_\theta = \nabla_H.\vec{u}, \\
u_H &= \partial_r u_\theta + r^{-1}u_\theta - r^{-1}\partial_\theta u_r = \vec{e_z}.(\nabla\wedge\vec{u}), \\
\sigma_{Vz} &= \partial_r \sigma_{rz} + r^{-1}\sigma_{rz} + r^{-1}\partial_\theta \sigma_{\theta z} = \nabla_H.\vec{t_z}, \\
\sigma_{Hz} &= \partial_r \sigma_{\theta z} + r^{-1}\sigma_{\theta z} - r^{-1}\partial_\theta \sigma_{rz} = \vec{e_z}.(\nabla\wedge\vec{t_z}),
\end{aligned} \tag{2}
$$

where $\nabla_H$ is the horizontal divergency operator, and $\vec{t_z}$ the stress vector acting on an horizontal facet.

Using these new functions, the second order derivatives equation (1) may be transformed into the two decoupled sets (3) and (4), involving only first order derivatives with respect to the deph coordinate $z$ :

$$
\partial_z \begin{bmatrix} u_z \\ u_V \\ \sigma_{zz} \\ \sigma_{Vz} \end{bmatrix} = \begin{bmatrix} 0 & 2C_S^2/C_P^2 - 1 & (\rho C_P^2)^{-1} & 0 \\ -\Delta_H & 0 & 0 & (\rho C_S^2)^{-1} \\ \rho\partial_{tt} & 0 & 0 & -1 \\ 0 & \rho[\partial_{tt} - 4C_S^2(1 - C_S^2/C_P^2)\Delta_H] & 2C_S^2/C_P^2 - 1 & 0 \end{bmatrix} \begin{bmatrix} u_z \\ u_V \\ \sigma_{zz} \\ \sigma_{Vz} \end{bmatrix} - \rho \begin{bmatrix} 0 \\ 0 \\ f_z \\ f_V \end{bmatrix} \tag{3}
$$

$$
\partial_z \begin{bmatrix} u_H \\ \sigma_{Hz} \end{bmatrix} = \begin{bmatrix} 0 & (\rho C_S^2)^{-1} \\ \rho\partial_{tt} - \rho C_S^2 \Delta_H & 0 \end{bmatrix} \begin{bmatrix} u_H \\ \sigma_{Hz} \end{bmatrix} - \rho \begin{bmatrix} 0 \\ f_H \end{bmatrix} \tag{4}
$$

where the functions $f_H$ and $f_V$ are defined from $\vec{f}$ after using the transformations settled in relation (2) and $C_P$ and $C_S$ are the compressional and shear waves velocities.

The three-dimensional problem is separated into two problems: a two-dimensional solution couples compressional and shear waves propagation in the vertical plane (PSV propagation, equation (3)) and a one-dimensional solution with shear wave motion confined to a horizontal plane (SH propagation, equation (4)).

It is easier to handle the two decoupled wave equations after a Hankel transform with respect to the horizontal coordinates $(r, \theta)$. Sets of ordinary differential equations, with regard to the depth, are obtained after a Fourier transform over time. Kennett defines the $m^{th}$ order Fourier-Hankel transform $H_m$ as :

$$
H_m(f(r,\theta,z;t)) = \hat{f}(k,m;\omega) = \frac{1}{2\pi}\int_{-\infty}^{+\infty} e^{i\omega t}dt \int_0^{+\infty} J_m(kr)rdr \int_{-\pi}^{+\pi} e^{-im\theta} f(r,\theta;t)d\theta \tag{5}
$$

where $k$ is the horizontal wave number, $\omega$ the angular frequency (the horizontal slowness $p = k/\omega$ is often used in Kennett's formalisation). For an isotropic medium, the $H_m$ transform applied to the two decoupled sets (PSV propagation and SH propagation), easily relies the displacements and the stresses at two different levels of the medium by mean of a propagator matrix (independent of the order of the Fourier-Hankel transform).

Kennett introduced a transformation which connects the stress displacement vector $\vec{B}$ to a new wave vector $\vec{V}$ :

$$\vec{B} = \mathbf{D}\,\vec{V} \tag{6}$$

where he chose the matrix $\mathbf{D}$ to give a simple form for the evolution of $\vec{V}$ with $z$. The matrix $\mathbf{D}$ is only dependent on the elastic parameters of the medium and the horizontal slowness $p$. The components of the wave vector $\vec{V}$ may be directly connected with a system of up-going ($V_U$) and down-going ($V_D$) waves (Kennett, 1983) :

$$\vec{V}(z) = [V_U(z), V_D(z)]^t \tag{7}$$

where ($V_U$) and ($V_D$) are both vectors for PSV propagation and scalars for SH propagation.

The reflection-transmission operators are defined as propagators of up-going and down-going waves. In the region $z_1 \leq z \leq z_2$, we may write :

$$\begin{pmatrix} V_U(z_1) \\ V_D(z_2) \end{pmatrix} = \begin{pmatrix} R_D^{12} & T_U^{12} \\ T_D^{12} & R_U^{12} \end{pmatrix} \begin{pmatrix} V_D(z_1) \\ V_U(z_2) \end{pmatrix} \tag{8}$$

where $R_D^{12}, R_U^{12}, T_D^{12}, T_U^{12}$ are the reflection-transmission operators.

At an interface, the reflection-transmission operators are obtained by writing the continuity of the stress-displacement vector $\vec{B}$. The problem is entirely resolved by adding the following boundary conditions : stress components of the vector $\vec{B}$ are null at a free surface, and there is no up-going wave in the lower uniform half-space.

Kennett introduced the source as a discontinuity of the stress-displacement vector across the source plane. From the expression of the discontinuity of the vector $\vec{B}$, we may calculate the discontinuity of the wave vector $\vec{V}$ across the source plane (depth $z_s$). The reflection-transmission operators, enable us to calculate the wave vector $\vec{V}$ at the receptor plane (depth $z_r$). Finally, the displacements at any point of the receptor plane can be recovered from the displacements which appear as elements of the vector $\vec{B}(z_r)$ by inverting the Fourier-Hankel transform.

## 3 SISMOS code

Boelle (1987) applied the Kennett's formalisation to a program called SISMOS code. The main characteristics of this code are the following ones :

The medium is defined by homogeneous solid or liquid horizontal stratified layers bounded by an upper free surface (i.e. air).

Each layer is defined by the thickness (the thickness of the lower layer is infinite), the density, the velocity and damping coefficient of compressional waves and, for solid layers, by the velocity and damping coefficient of shear waves.

SISMOS code calculates the impulse response from a seismic source placed at any point in the medium. Two kinds of seismic sources can be considered : a) a punctual vertical, horizontal or lenead force either in a solid layer or at a liquid-solid interface, or b) a spherical impact in the liquid layer.

The receivers simulate hydrophones when they are placed in a liquid layer. They simulate three-dimensionnal geophones when they are placed either in a solid layer or at a liquid-solid interface and measure displacements as a function of the time.

## 4 Results

Synthetic sismograms were calculated for four sediments models in the conditions of marine experiments with the aim to visualize shear waves :

1. a rather soft homogeneous half-space (Fig 3),
2. a stratified media: a 5 meters thick soft layer lying over a rigid half-space (Fig 4),
3. a stratified media: a 1 meter thick soft layer lying over a rigid half-space (Fig 5),
4. a rather rigid homogeneous half-space (Fig 6),

In every case, a 20 meters thick water layer overwhelms the soil. Characteristics of water are: $\rho = 1000\ kg/m^3$ and $C_P = 1500\ m/s$.

For the visualisation of shear wave in marine sediments, a punctual horizontal force $\vec{F_x}$ is chosen. The source is placed at the sediment-water interface. The analytic expression of the punctual horizontal force $\vec{F_x}$ is given by :

$$\vec{F_x}(t) = f_{xmax} 16 \Big[\frac{t}{t_{max}}(1 - \frac{t}{t_{max}})\Big]^2 \qquad (9)$$

where $0 \leq t \leq t_{max}$. In the examples presented in this paper, $t_{max}$= 2 ms. (Fig. 1)

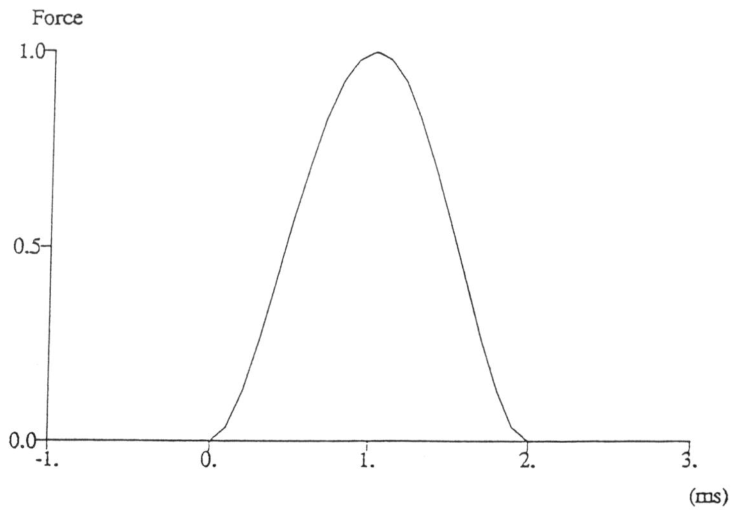


Fig 1. Source signal. Horizontal force applied at point source.

The receivers represent three-dimensional geophones. They are placed at the same interface than the source and lined with it. The spacing between the receivers is 2 meters as well as the spacing between the source and the first receiver. The receiver line is perpendicular to the applied force $\vec{F_x}$ (Fig. 2).

In this configuration there are neither vertical displacement nor radial displacement, but only horizontal displacement parallel to the force $\vec{F_x}$. The displacements due to compressional and interface waves are eliminated. Displacement due to shear waves are only visible.

Synthetic seismograms were drawn with SITHERE, a seismic data processing program (Lericolais, 1990)

In each example, the horizontal force $\vec{F_x}$ is presented in channel 1 and the calculated normelized displacements for each of the ten receivers are presented in channels 2 to 11. Signal duration is 128 ms. The displacement relative to the shear wave propagation, is horizontal and parallel to the applied force $\vec{F_x}$. The vertical and radial components of the displacement are null. Each channel is normalized ; the maximum value is reduced to 1. This form of representation is purposely chosen to obtain easy reading signals (the amplitude of displacements is highly decreasing with the distance from the source). The noise, which sometimes appears on the last channels, is in fact insignificant.

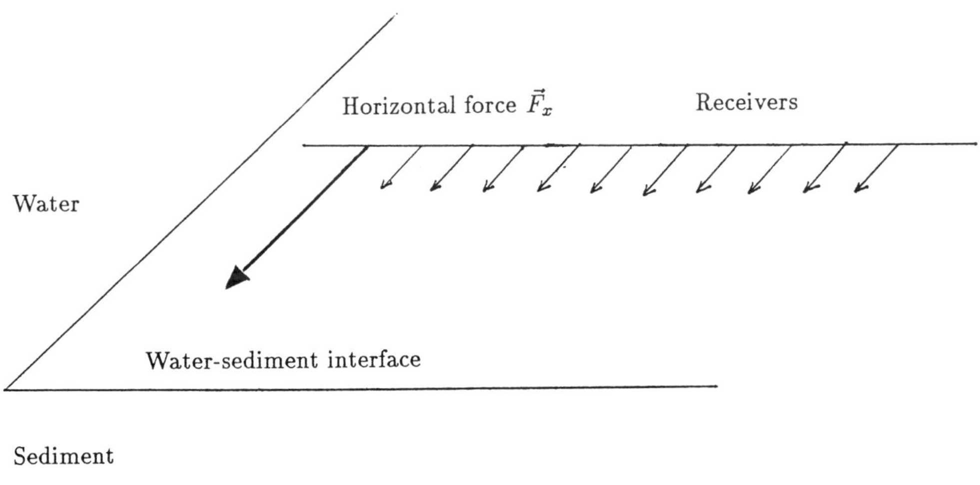


Fig 2. Scheme of implantation of source and receivers.

In case of an uniform half-space, the direct shear wave in the sediment is clearly visible in the synthetic seismograms. Since the geometry of the implantation of source and receivers is well known, the velocity of the direct shear wave can be calculated from the synthetic seismogram and compared to input values (Fig 3 and 6).

In case of a stratified medium several kinds of shear waves are visible (Fig 4 and 5). Shear waves propagating in the upper sediment layer were evident as in figures 3 and 6. Shear waves reflected on the soil interface appear after the direct shear wave and may be mingled with it when the thickness of the layer is small ; after a reflection over the soil-water interface, this wave generates a twice reflected shear wave which is also visible. Shear wave refracted in the lower layer are visible when the upper sediment layer is thin (Fig 5).

## 5 Conclusions

SISMOS code correctly computes direct, reflected and refracted seismic waves.

In this paper, only four simple examples where shear waves are clearly visible are presented. But SISMOS code can also be used to calculate the response of seismic sources (horizontal, vertical and leaned force in a solid layer, spherical impact in a liquid layer) in every kind of horizontaly stratified medium. Receivers plane can be placed anywhere in the medium (receivers simulate geophones when they are placed in sediment layers, and hydrophones in water column).

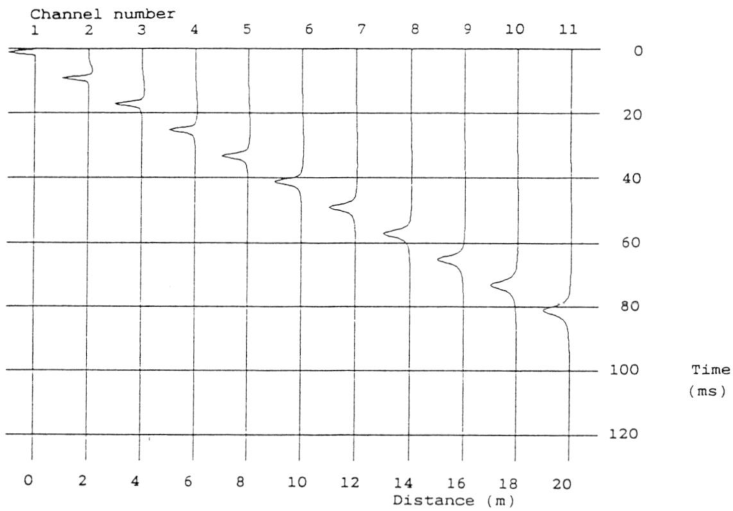


Fig 3. Normalized synthetic seismograms calculated from SISMOS code for the case of a 20 m thick layer of water ($\rho = 1000\ kg/m^3$, $C_P = 1500\ m/s$) overlying a soft sediment ($\rho = 1500\ kg/m^3$, $C_P = 1550\ m/s$ and $C_S = 250\ m/s$) of infinite thickness. Only direct shear wave displacements are shown.

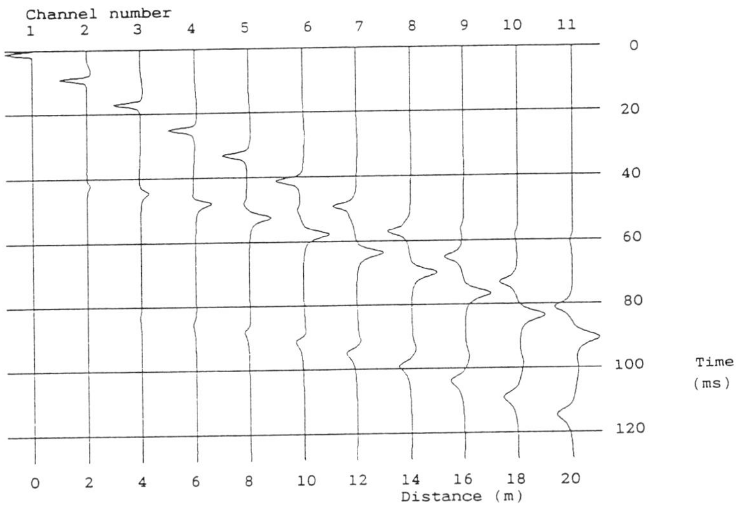


Fig 4. Normalized synthetic seismograms calculated from SISMOS code for the case of a 20 m thick layer of water ($\rho = 1000\ kg/m^3$, $C_P = 1500\ m/s$) overlying a statified medium : an upper 5 m thick layer of soft sediment ($\rho = 1500\ kg/m^3$, $C_P = 1550\ m/s$ and $C_S = 250\ m/s$) and a hard sediment ($\rho = 1800\ kg/m^3$, $C_P = 1600\ m/s$ and $C_S = 1000\ m/s$) of infinite thickness. Direct and reflected shear wave displacements are shown.

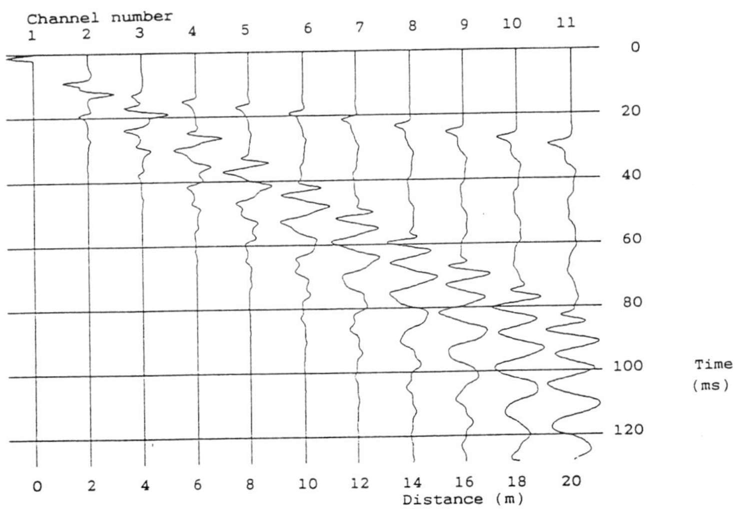


Fig 5. Normalized synthetic seismograms calculated from SISMOS code for the case of a 20 m thick layer of water ($\rho = 1000\ kg/m^3$, $C_P = 1500\ m/s$) overlying a statified medium : an upper 1 m thick layer of soft sediment ($\rho = 1500\ kg/m^3$, $C_P = 1550\ m/s$ and $C_S = 250\ m/s$) and a hard sediment ($\rho = 1800\ kg/m^3$, $C_P = 1600\ m/s$ and $C_S = 1000\ m/s$) of infinite thickness. Direct, reflected and refracted shear wave displacements are shown.

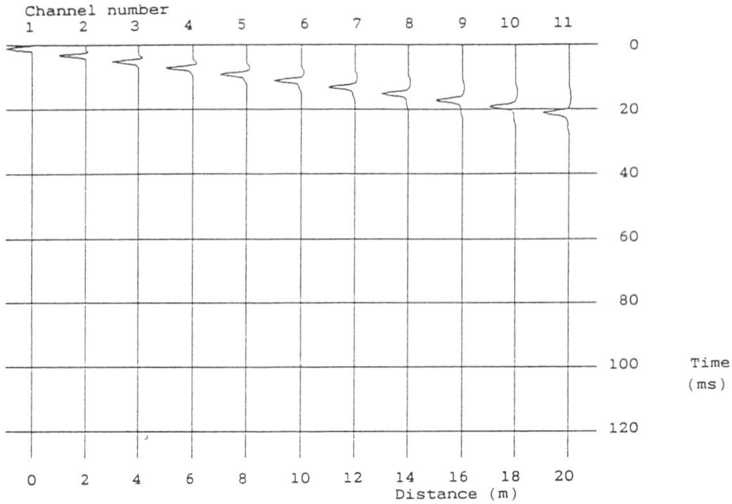


Fig 6. Normalized synthetic seismograms calculated from SISMOS code for the case of a 20 m thick layer of water ($\rho = 1000\ kg/m^3$, $C_P = 1500\ m/s$) overlying a hard sediment ($\rho = 1800\ kg/m^3$, $C_P = 160\ m/s$ and $C_S = 1000\ m/s$) of infinite thickness. Only direct shear wave displacements are shown.

So, SISMOS code is able to predict the wave propagation in stratified media and to help the interpretation of seismic records in terms of characterization of sediment properties.

SISMOS code is easy to use on SUN 4 workstation ; results are obtained within few mimutes.

## 6 Acknowledgments

The writers would like to thank J.L. BOELLE, author of SISMOS code at IFREMER.

## 7 References

Boelle, J.L. (1985) 'Les méthodes sismiques au service de la géotechnique',DIT/GO S/R43 (IFREMER)

Boelle, J.L. (1987) 'Calcul de sismogrammes synthétiques en milieux stratifiés plans',DIT /GO S/R26 (IFREMER)

Kennett, B.L.N. (1981) 'Elastic wave propagation in stratified media', Advances in applied mechanics 21, 79-165

Kennett, B.L.N. (1983) Seismic wave propagation in stratified media, Cambridge University Press.

Lericolais, G., Allenou, J.P., Berné, S., Morvan, P., (1990) 'A new system for acquisition and processing of very high-resolution seismic reflection data', Geophysics 55, 1036-1046

# NORMAL MODES IDENTIFICATION IN SHALLOW WATER USING SPECTRAL ANALYSIS : THEORY AND EXPERIMENTS

P. CRISTINI, J.L. GARNIER and C. GAZANHES
*Laboratoire de mécanique et d'acoustique*
*Equipe Ultrasons*
*31, Chemin Joseph Aiguier*
*13009 MARSEILLE*
*FRANCE*

ABSTRACT. A model of shallow water propagation in an isovelocity ocean is investigated making only the assumption of horizontal stratification. The Hankel transform of the sound field pressure in the water layer shows peaks which correspond to normal modes propagating in that layer. The spectrum obtained is characteristic of the bottom nature. We carry out experiments on a reduced scale model of shallow water propagation and analyse, using an FFT algorithm, the signal obtained, by moving at a constant speed, a receiver in the sound field. We compare the results given by the application of the FFT algorithm and those given theoretically by the Hankel transform. We then provide results of the FFT process on such signals for differents types of the ocean bottom. Semi-infinite fluid and solid bottom cases are first considered, then our interest focuses on a geo-acoustical model of the sea bed consisting of a fluid layer (sedimentary layer) overlying a semi-infinite solid.

## 1. Introduction :

Shallow water propagation, because of its particular nature, is widely considered as an important subject. The mechanisms that control propagation are in a close relationship with nature of the bottom so the knowledge of the sound pressure can be used to investigate the seabed. That is the reason why we use a method valid for any laterally homogeneous bottom for calculating the sound field pressure. By using this method, based on the Hankel transform [1,2], we can see that the reflection coefficient of the seabed is the parameter that controls propagation.

The purpose of this article is the normal modes identification based on the following considerations. In shallow water the acoustic field appears to be the sum of cylindrically propagating modes. Therefore the application of an FFT algorithm to process the signal of a receiver moving at a constant speed, in a zone where the cylindrical radiation has a little influence, will show peaks corresponding to the modes whose knowledge will lead to mode identification.

## 2. Acoustic model of shallow water propagation :

Our study deals with shallow water propagation implying that the water depth is less than ten wavelengths. The shallow water model consists of an isovelocity water layer laying on the seabed in which we place a harmonic point source (figure 1).



*J. M. Hovem et al. (eds.), Shear Waves in Marine Sediments*, 503–509.

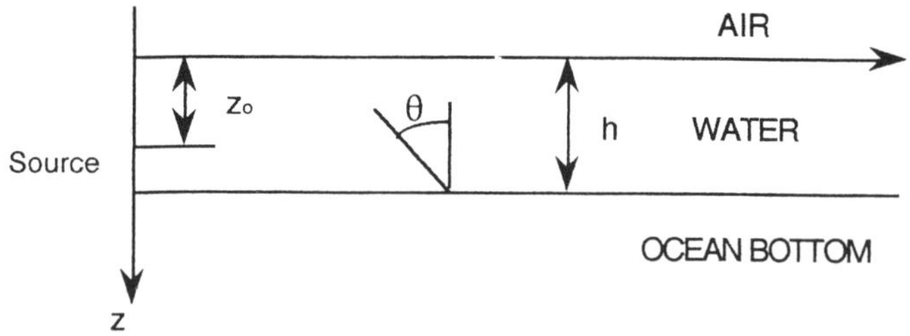


figure 1. Shallow water model

## 3. Representation of the acoustic field in the water layer :

The acoustic field generated in the water layer can be expressed as the inverse Hankel transform of the depth-dependent Green's function [1] :

$$(1) \qquad p(r,z,z_o) = \int_0^{+\infty} P(\xi,z,z_o)\, J_o(\xi r)\, \xi \, d\xi$$

where $\xi$ is the horizontal wavenumber and $P(\xi,z,z_o)$ the depth-dependent Green 's function whose expression is :

$$(2) \qquad P(\xi,z,z_o) = - \frac{e^{iK|z-z_o|} - e^{iK(z+z_o)} + V(\xi)\, e^{2iKh}\left( e^{-iK(z+z_o)} - e^{-iK|z-z_o|} \right)}{iK\left[ 1 + V(\xi)\, e^{2iKh} \right]}$$

where K is the vertical wavenumber and V the reflection coefficient of the bottom.

The inverse Hankel transform can be performed by means of the residues theorem for large values of the distance between the source and the receiver [3,4]. The acoustic field then appears as the sum of propagating normal modes corresponding to the residues :

$$(3) \qquad p(r,z,z_o) = 2\pi \sum_{n=1}^{N} \frac{2 \sin K_n z \, \sin K_n z_o}{2ih - \dfrac{K_n \frac{d}{d\xi}\left[V\right]_{\xi=\xi_n}}{\xi_n V(\xi_n)}} \; H_o^{(1)}(\xi_n r)$$

with $\xi_n$ the real solutions of the *characteristic equation* :

$$(4) \qquad 1 + V(\xi)\, e^{2iKh} = 0$$

## 4. Spectral analysis of the acoustic field :

We can use the first order approximation of the Hankel function to simplify the expression of equation (3) [5] :

$$(5) \qquad p(r,z,z_o) = P_o \sum_{n=1}^{N} \frac{G_n(z,z_o)}{\sqrt{r}} \exp[i\ \xi_n r]$$

with $G_n(z,z_o) = A_n \sin(K_n z) \sin(K_n z_o)$

For larges values of r the function $1/\sqrt{r}$, for a limited zone of exploration, can be considered as constant. So if we use an FFT algorithm on the signal obtained by moving, at constant speed, the receiver in the sound field, we must see peaks corresponding to the modes propagating in the water layer. These peaks will be the $\xi_n v$ because of the variable change introduced by the receiver displacement. The knowledge of these peaks positions will lead to mode identification. The results given by the FFT algorithm are similar to those given by the depth-dependent Green's function because its poles correspond to the modes. Therefore we will use this function for the comparison between experimental and theoretical results and for the calculation of the theoretical spatial frequencies of the modes.

## 5. Experimental results :

We have in our laboratory a five meter long tank which is a reduced scale model of shallow water propagation. The equipment of this experimental tank allows changes in the bottom configuration. Therefore it has been possible to realize experiments for semi-infinite sand and concrete bottoms and also for a stratified configuration consisting of a sand layer overlying a concrete bottom.The semi-infinite bottoms were in fact 20 cm thick for the sand bottom and 6 cm thick for the concrete bottom.

The sound pressure signal received by a hydrophone moving along the waveguide is cross-multiplied with the source signal and sent to a spectral analyser (figure 2). We start acquiring data at a distance large enough to ensure the validity of equation (5) for the frequencies we used.The spatial exploration extends from 0,5 m to 2,5 m. We have also put absorbers along the tank walls in order to avoid reflexions. For each type of bottom, we give as a comparison the plot of the depth-dependent Green's function .

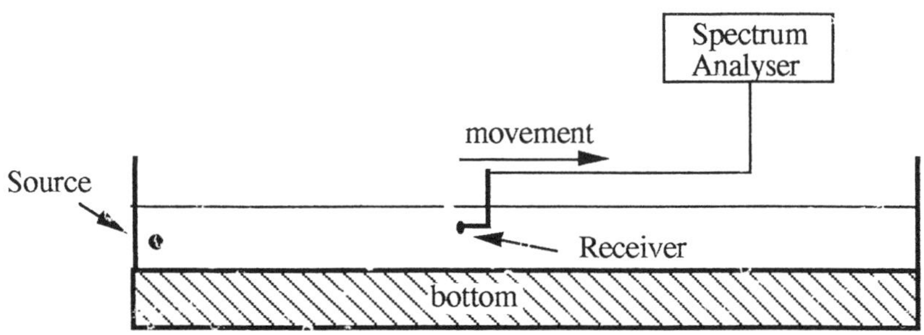


Figure 2. Principle of the experimental process

### 5.1. SAND BOTTOM :

The sand thickness is large enough (20 cm) to assume that this medium is semi-infinite. The results given here were obtained for two frequencies of the source (75 kHz and 60 kHz) with a water depth of 82 mm. The receiver speed is 0,00754 m/s and the theoreticals parameters of the sand bottom used for plotting the depth-dependent Green's function are : $\rho = 2000$ kg m$^{-3}$, c = 1750 m/s.

TABLE 1. Comparison between experimental and theoretical spatial frequencies for a sand bottom

| | SAND BOTTOM AT 75 kHz | | SAND BOTTOM AT 60 kHz | |
|---|---|---|---|---|
| Mode | Theoretical | Experimental | Theoretical | Experimental |
| 1 | 0,3800 | 0,384 | 0,3030 | 0,304 |
| 2 | 0,3737 | 0,376 | 0,2951 | 0,296 |
| 3 | 0,3621 | 0,364 | 0,2807 | 0,281 |
| 4 | 0,3440 | 0,344 | 0,2593 | 0,260 |
| 5 | 0,3226 | 0,324 | -- | -- |

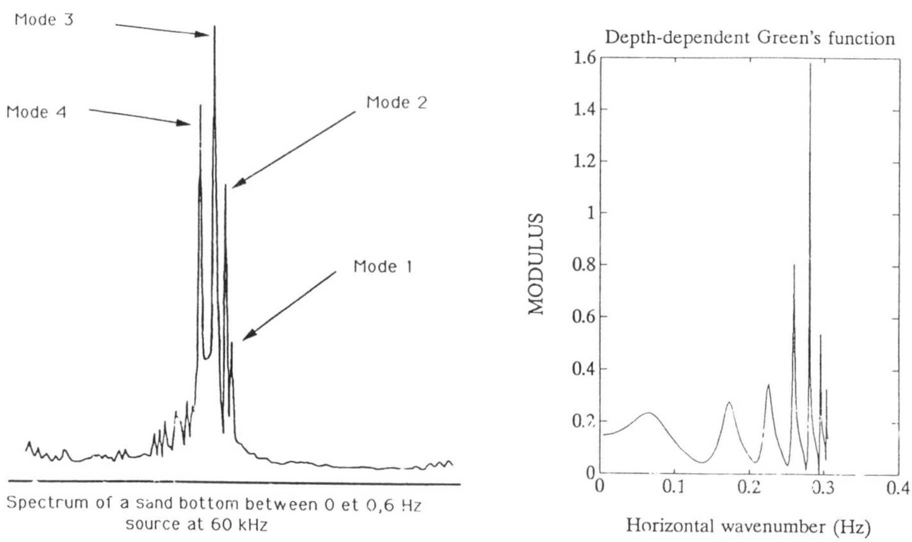


Figure 3. Results of the FFT process and plot of the depth-dependent Green's function for a sand bottom at 60 kHz.

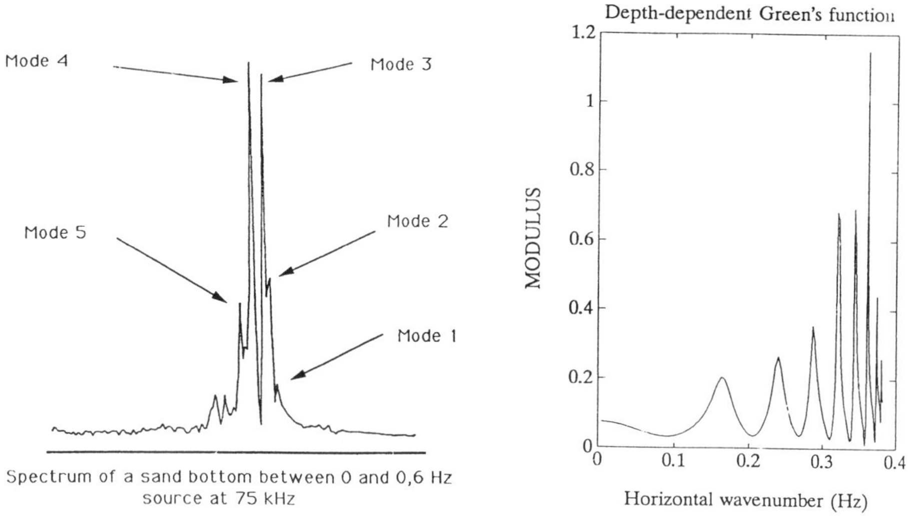


Figure 4. Results of the FFT process and plot of the depth-dependent Green's function for a sand bottom at 75 kHz.

5.2. CONCRETE BOTTOM :

The results given here were obtained for a source frequency of 75 kHz with a water depth of 82 mm. The receiver speed is 0,00754 m/s and the theoretical parameters of the concrete bottom used for plotting the depth-dependent Green's function are : $\rho = 2350$ kg m$^{-3}$, $c_l = 4500$ m/s and $c_t = 2500$ m/s

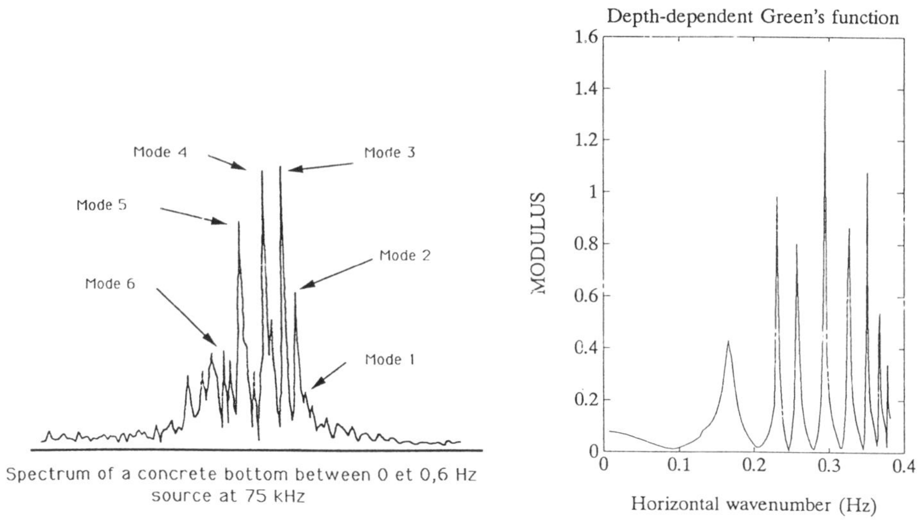


Figure 5. Results of the FFT process and plot of the depth-dependent Green's function for a concrete bottom at 75 kHz.

TABLE 2. Comparison between experimental and theoretical spatial frequencies for a concrete bottom at 75 kHz

| Mode | Theoretical | Experimental |
|---|---|---|
| 1 | 0,3780 | 0,380 |
| 2 | 0,3670 | 0,368 |
| 3 | 0,3503 | 0,352 |
| 4 | 0,3253 | 0,328 |
| 5 | 0,2939 | 0,296 |
| 6 | 0,2566 | 0,256 |

## 5.3. STRATIFIED BOTTOM :

The configuration of the seabed consists of a 23 mm sand layer lying on a concrete bottom.The results given here were obtained for a source frequency of 80 kHz with a water depth of 60 mm. The receiver speed is 0,004717 m/s and the theoretical values for sand and concrete are identical to those given previously.

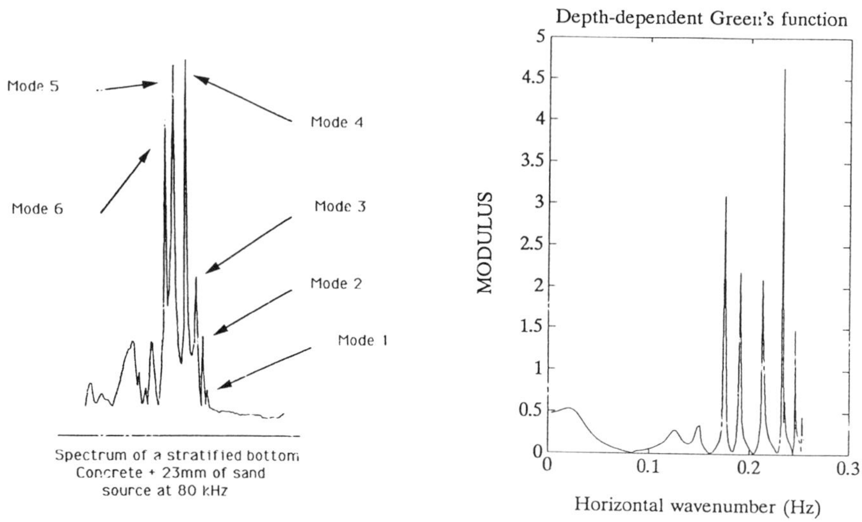


Figure 6. Results of the FFT process and plot of the depth-dependent Green's function for a stratified bottom at 80 kHz.

TABLE 3. Comparison between experimental and theoretical spatial frequencies for a stratified bottom at 80 kHz

| Mode | Theoretical | Experimental |
|---|---|---|
| 1 | 0,2527 | 0,253 |
| 2 | 0,2449 | 0,245 |
| 3 | 0,2318 | 0,233 |
| 4 | 0,2120 | 0,212 |
| 5 | 0,1910 | 0,188 |
| 6 | 0,1735 | 0,173 |

## 6. Analysis of the previous results :

The experimental results show that the separation of the modes propagating in the water layer is realized. For the sand and stratified bottoms the modes identification is done with good agreement with the theory. But for the concrete bottom, the seventh mode theoretically foreseen is not observed and one mode observed between the fifth and sixth mode is not expected. First the behaviour of the concrete bottom may not be like a semi-infinite bottom because of its thickness not large enough. But another and more general explanation of the phenomenons encountered can be found in the fact that we do not know accurately the theoretical parameters of the bottoms. This error is more important for the high order modes than for the low order modes because what we identify is in fact the sine of the modal angle. For low order modes, i.e. for angles near 90°, the slope of the sine function is near zero which is not the case for the high order ones, so the spatial frequencies of low order modes will not be so affected by a change in the bottom parameters as the frequencies of high order modes.

## 7. Conclusion :

We have shown in this article that the use of an FFT algorithm on the signal given by a moving receiver along the waveguide is efficient for the separation and identification of the modes propagating in shallow water. We have also shown that the spatial frequencies of higher level modes are more sensitive to the bottom nature. Their knowledge with a higher frequency resolution will be useful for the recovery of the bottom parameters.

## REFERENCES

[1] FRISK, G.V.and LYNCH, J.K. (1984) 'Shallow water waveguide characterization using the Hankel transform', J. Acoust. Soc. Am. 76, 205-216.

[2] MERCER, J.A. (1988) 'Geoacoustic paramaters and depth dependent Green's function for a silty sediment at 100 Hz in shallow water', J. Acoust. Soc. Am. 83, 611-616.

[3] OFFICER, C.B. (1958) *Introduction to the theory of sound transmission,* Mc Graw Hill .

[4] BREKHOVSKIKH, L.M. (1980) *Waves in layered media,* Second ed. Academic Press Inc.

[5] ABRAMOVITZ, M. and STEGUN, I.A. (1969) Handbook of Mathematical functions, New York : Dover .

[6] GORSKAYA, N.V. NIKOLAEV, G.N. RYCHOVA, T.A. and SALIN, B.M. (1981) 'Spectral analysis in the investigation of fields of harmonic sources in an acoustic waveguides', Sov. Phys. Acoust. 27 (2) 110-112 .

# ESTIMATION OF GEOACOUSTIC PROPERTIES BY INVERSION OF ACOUSTIC FIELD DATA

N. R. CHAPMAN
Defence Research Establishment Pacific
FMO Victoria, B.C. V0S 1B0
CANADA

ABSTRACT. The reconstruction of a geoacoustic model of the ocean bottom from acoustic field data is an important inverse problem in ocean acoustics. This paper presents a review of techniques which have been developed for inverting acoustic field measurements in deep and shallow water to infer geoacoustic model parameters. These include inversion of total field data by analysis of the propagation loss or the depth-dependent Green's function, inversion of Scholte wave dispersion properties, and inversion of bottom reflected field data by analysis of bottom loss or the transient waveform. An example is also given for the use of matched field processing for the estimation of bottom properties. The results derived from the various methods are compared with estimates from non-acoustic methods.

## INTRODUCTION

The interaction of low frequency sound with the ocean bottom is an important consideration in a wide variety of applications in ocean acoustics ranging from propagation modelling to the evaluation of sonar system performance. The traditional approach in describing the interaction has been to develop a geoacoustic model, assuming that the bottom is layered and the bottom materials are fluids. It is apparent, however, that for many applications in both deep and shallow water, the shear rigidity of the bottom materials has a significant effect on sound propagation in the water through the conversion of acoustic energy into shear waves. Consequently it is appropriate to consider a more realistic approach and model the ocean bottom as a visco-elastic solid described by the depth dependence of the compressional (P) and shear (S) wave speeds and attenuations, and the density.

This paper presents a review of methods used in ocean acoustics to estimate geoacoustic model parameters by inversion of acoustic field data. The inversion process can be described as a method for obtaining a model of the ocean bottom which satifies a set of acoustic observations. The procedure involves various aspects of acoustic data analysis, including modelling, processing and interpretation. Four approaches have been used for obtaining inverse solutions, including exact inversion, perturbative inversion, inversion by iteration of forward models, and inversion based on specific features of the acoustic field [1]. The most widely used approach in ocean acoustics is the iteration of forward models; this method utilizes all or most of the available data, and is very effective in estimating information from real data whereas exact methods are often unstable. Specific feature inversion is also in common use; this technique is relatively easy to implement, however, it does not make use of all available data. We shall describe several inversion schemes that utilize these two approaches. These include inversion of total field data, inversion of



*J. M. Hovem et al. (eds.), Shear Waves in Marine Sediments*, 511–520.

Scholte wave dispersion properties, inversion of bottom reflected field data, and inversion using the matched field processing concept. For each example, we first describe the experiment and then discuss the inversion method and the results that were obtained. Where possible, comparisons are made with parameter values derived from non-acoustic methods.

Although it is comparatively straightforward to invert the density and the compressional wave properties, the shear wave properties of marine sediments are more difficult to estimate due to the high attenuation of these waves. In addition to reviewing the various inversion techniques, we will discuss the application of each example to the problem of inverting shear wave properties, and indicate those methods that have been successful in obtaining realistic parameter values.

## INVERSION OF TOTAL FIELD DATA

Techniques for inverting total acoustic field data have been applied to measurements made in deep and shallow water. The first technique that we will discuss has been developed by Frisk and co-workers [2-4] to obtain measurements of the spatial part of the steady state acoustic field. The experiment consists of deploying hydrophones near the sea floor and recording a pulsed cw signal as the sound source is towed away at constant depth from the receivers (Fig. 1). Data are collected over the range 0-10 km at intervals of between 0.5 to 1.0 wavelengths, depending on the drift rate. The signal amplitude and phase versus range, obtained from the measured quadrature components of the signal, are used in the inversion.

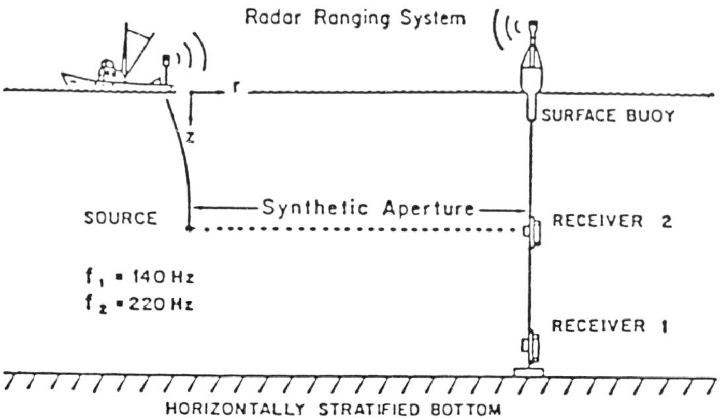


Figure 1. Experiment geometry for the Frisk technique.

The deep water experiment was carried out at sites where previous seismic refraction surveys had provided independent information about the bottom. The data shown in Fig. 2a were inverted using the method of iteration of forward models to obtain estimates of the parameters of the geoacoustic model shown in Fig. 2b [1]. The oscillations in the data are due to interference between the direct, bottom reflected, and subbottom paths; at short ranges the subbottom energy reflects from the basement interface whereas at longer ranges the energy is refracted by the gradient layer. The parameters of primary importance in the model are the sediment layer thickness, and the sound speed gradient and discontinuity at the sea floor; other parameters such as density, attenuation and basement sound speed were less significant. The inverse solution was obtained by comparing the measured field with calculations of the field, based on approximate solutions of the wave equation, while systematically varying the geoacoustic profile parameters. The parabolic equation was used to obtain estimates of the primary parameters, and a wave model based on a numerical Hankel transform of the ocean bottom reflection coefficient was used to refine these values and estimate the other parameters. The estimated values (Fig. 2b) were consistent with values reported by Hamilton [5] for similar bottom types.

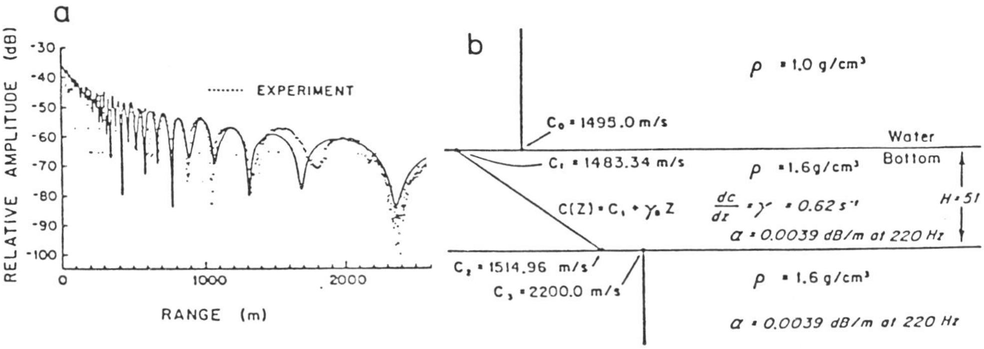


Figure 2. (a) Deep water data at 220 Hz and the modelled result. (b) The geoacoustic model derived from the data.

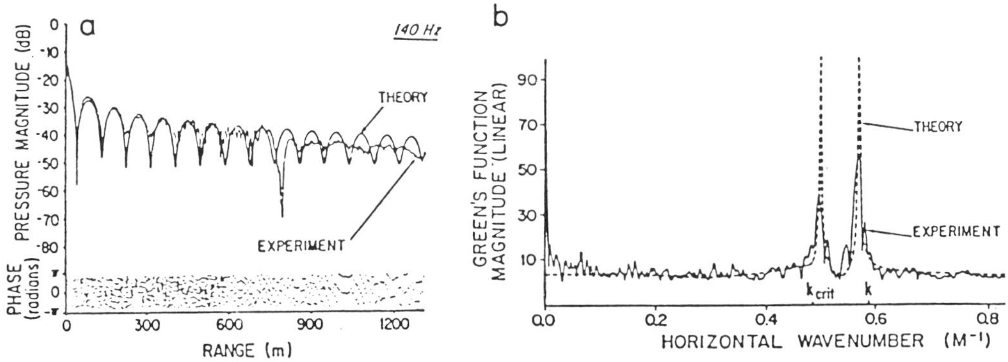


Figure 3. (a) Shallow water data at 140 Hz and the modelled result. (b) The horizontal wavenumber spectrum obtained from the data.

The same experimental technique has also been used to determine geoacoustic properties in shallow water waveguides. In this case, a numerical Hankel transform method [6] was used to obtain the Green's function versus horizontal wavenumber from the pressure field versus range data. In order to avoid aliasing errors in taking the Hankel transform, the field was sampled at least every half wavelength in range, starting from zero range. A variation of the method involves an approximate Hankel transform in which a horizontal line array is used to beamform the tow ship noise [7]; the method of Frisk et al.[3] is analogous to beamforming with a synthetic aperture array, and the two approaches are essentially equivalent. The Green's function contains information about the discrete and continuous modal spectra of the waveguide; specifically, the prominent peaks in the Green's function occur at wavenumbers corresponding to the eigenvalues for the trapped or virtual modes in the waveguide. The inversion makes use of the sensitivity of the wavenumber spectrum to the acoustic properties of the bottom; the method proceeds by assuming a geoacoustic model and uses the iteration of forward models technique to compare the measured Green's function with wavenumber spectra calculated for specfic values of the bottom parameters. The solution is determined as the best fit, using a least squares criterion, over the complete data set, including two frequencies and the two receivers. An example of the pressure field data and the Green's function are shown in Fig. 3 for 140 Hz; there are two dominant modes giving rise to the observed interference pattern. The geoacoustic model inferred from the data included a surficial sediment layer of 15 m with a strong sound speed gradient of 14 $s^{-1}$, a density of 1.7 $gm/cm^3$ and an attenuation of 0.05 dB/m/kHz. These parameters, particularly the large near-surface sound speed gradient, are consistent with the results of Hamilton [5] for the sandy bottoms that were typical of the site. The analysis has also been extended to invert data in a range dependent environment, assuming that the variation in the geoacoustic properties is adiabatic [3]. For this case, the waveguide was separated into sections where the geoacoustic properties were constant, and the method was applied to each part.

Although the shear wave properties were not estimated in the analysis, it is possible to do so by including the effects of sediment rigidity in the models. However, for the slow S wave speeds typical of marine sediments, the effect is not significant; only a small shift in the wavenumber spectral peaks is observed. For larger values of S wave speed, such as in a chalk bottom, shear waves have an important effect, as shown by Ellis and Chapman [8]. In their experiment, broadband propagation loss measurements were made using explosive charges at a shallow water site where a thin sediment layer covered a chalk basement. The loss was determined from the received energy by taking the difference (in dB) between the charge source level and shot energy level at each range. The data were characterized by an increase in loss of about 20 dB below 300 Hz which could not be accounted for on the basis of a fluid sediment model. The iteration of forward models technique was used to invert the bottom sound speed and attenuation, assuming a single homogeneous solid layer for the geoacoustic model. The values obtained were 2.4 and 1.0 km/s for the P and S wave speeds, and 0.1 and 1.0 dB/m/kHz for the attenuations of the respective waves; these values are consistent with the acoustic properties of chalk [5].

## INVERSION OF SCHOLTE WAVE DISPERSION DATA

A practical technique used by several groups for inferring shear wave properties in shallow water sediments is the inversion of Scholte wave dispersion properties. This wave is a guided seismic interface wave that propagates along the boundary between the water and the sediment; the particle motion is elliptical in the range/depth plane, and the amplitude decays exponentially from the interface. For homogeneous media in contact, the Scholte wave speed and attenuation are related in a simple way to the shear wave properties. For a layered sediment environment, the Scholte wave is dispersive with no low frequency cutoff. The phase speed decreases from the Rayleigh wave speed at low frequency to the Scholte wave speed at high frequency. If the S wave speed is greater than the sound speed in water, the Scholte wave is the only trapped mode; the other modes are leaky and this leads to large propagation loss.

The experiment consists of measuring the signal from explosive charges at hydrophones or geophones on the sea floor. The Scholte wave appears as a strongly dispersed low frequency arrival following the water wave, as shown by the geophone data described by Jensen and Schmidt [9] in Fig. 4a. The modal dispersion is measured by filtering the time series data in a narrow band about the frequency of maximum amplitude of the interface wave and examining the filtered output for zero crossings. The group speed of the wavelet is determined from the experimental geometry and the measured travel time. The measured dispersion is shown in Fig. 4b; there are three discrete modes of which the slowest is the Scholte mode with its energy peaked at 2 Hz. The other modes are shear modes.

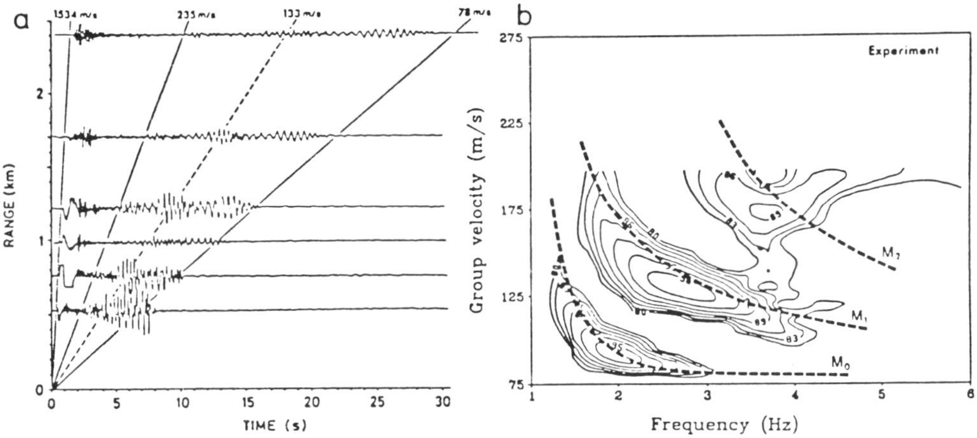


Figure 4. (a) Horizontal component geophone data showing the Scholte wave. (b) Modal dispersion curves derived from the data.

These data were inverted using the iteration of forward models technique to determine a geoacoustic profile that would reproduce the measured dispersion [9]. Since the Scholte wave dispersion is relatively insensitive to the densitiy and the P wave properties the values of these quantities were fixed using 1.8 gm/cm$^3$, 1.6 km/s and 0.1 dB/$\lambda$ for the density, sound speed and attenuation respectively and only the S wave properties were varied in the inversion. The inferred profile consisted of an 80-m sediment layer in which the shear speed varied from 85 - 300 m/s with a gradient of 6 s$^{-1}$, and the attenuation decreased from 0.5 to 0.1 dB/$\lambda$. The shear speed profile agrees well with Hamilton's values for sand-silt sediments [5].

## INVERSION OF BOTTOM REFLECTED FIELD DATA

In the deep ocean it is possible to measure the bottom reflected component of the field separately, since, for impulsive sources, the travel time differences of the various multipaths are much longer than the pulse duration. The most common and straightforward technique is the bottom loss experiment, in which explosive charges are used to measure the propagation loss of the bottom reflected paths. The experiment geometry is similar to that in Fig. 1 except that both the source and receiver are usually at shallow depths far from the bottom. The charges are deployed as the shooting platform, a ship or an aircraft, opens range from the receiver to a distance of at least one convergence zone (about 50 km). The bottom loss versus grazing angle is determined by subtracting (in dB) the modelled geometric spreading loss of the specular ray from the measured loss,

$$BL = (H_m - H_c + 6)/n \qquad (1)$$

where $H_m$ is the measured loss for the nth bottom bounce, $H_c$ is the calculated loss and the 6 dB correction accounts for the surface reflected multipath energy. Although the bottom loss concept is simple, the interpretation of the data is not a straightforward process, and some care must be taken in the inversion. We will describe here an iteration of forward models technique based on a ray theory bottom loss simulator to invert the geoacoustic profile [10]. The simulator consists of a 3-layer geoacoustic model for which all ray quantities such as range, path length, spreading loss and travel time can be expressed as analytic functions of the grazing angle. The eigenrays for the sea floor reflected and sediment interacting paths are determined for each range, and the intensity is computed coherently. Subsequently, the bottom loss is determined using equation 1.

For the case of moderately thin sediments, the ray paths are shown in Fig. 5a, and the measurements for a 1/3 octave band at 63 Hz in Fig. 5b. These data were obtained in a turbidite sediment basin where seismic records indicated the presence of a strong reflector about 200 - 400 m beneath the sea floor. The inversion was most sensitive to the sound speed gradient (0.6 $s^{-1}$), thickness (400 m) and attenuation (0.02 dB/m/kHz) of the sediment, and the P and S wave speeds (2.2 and 0.7 km/s) of the basement. The values obtained for these parameters were consistent with the results of Hamilton for turbidite sediments [5] and with the seismic data for the site. For this example, the shear wave properties of the sea floor sediments were less significant than those of the materials at the basement interface.

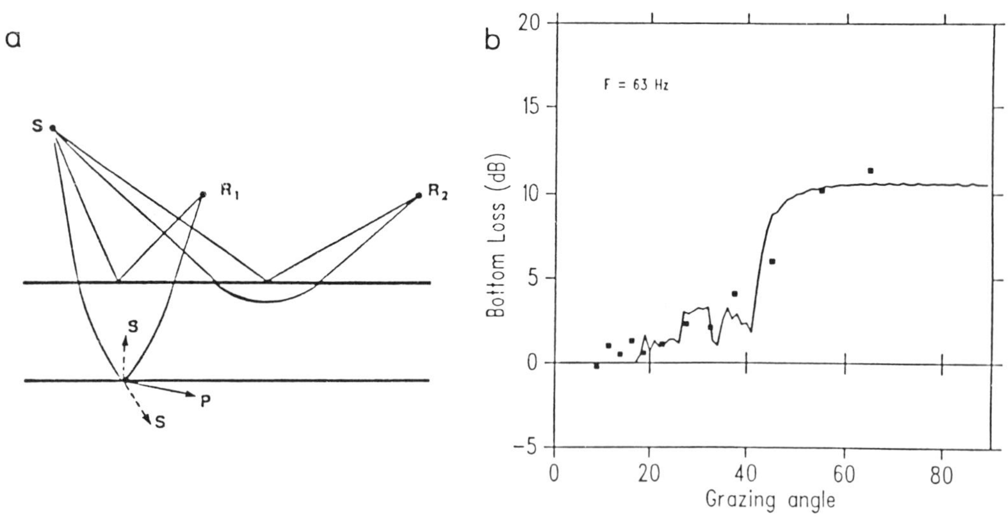


Figure 5. (a) Ray paths for short and long ranges for moderately thin sediment. (b) Measured and modelled bottom loss for 63 Hz.

A similar technique to the bottom loss method is the inversion of reflection amplitude versus offset or angle information [11]. An example of this method is provided by the experiment of Diachok et al. at a deep ocean site with a thin sediment layer overlying very young basalt [12]. In this case, the bottom reflection coefficient was determined versus grazing angle from the ratio of the scaled bottom reflected and direct path energies. The data were inverted using a numerical fast field calculation to model the reflectivity for a geoacoustic model consisting of a constant sound speed gradient layer for the basalt. The inferred values of 2.8 and 0.8 km/s for the interfacial P and S wave speeds respectively were consistent with recent results for young basalt obtained from seismic methods. The reflectivity of older basalt was measured in the same manner during the PACIFIC ECHO experiment, a joint exercise between the Naval Research Laboratory and the Defence Research Establishment Pacific. The ECHO data shown in Fig. 6 were inverted using the iteration of forward models technique based on plane wave reflection from an elastic solid. The results indicated larger values of the P and S wave speeds, 3.4 and 1.6 km/s, consistent with the hypothesis that the basalt sound speeds are age dependent [12].

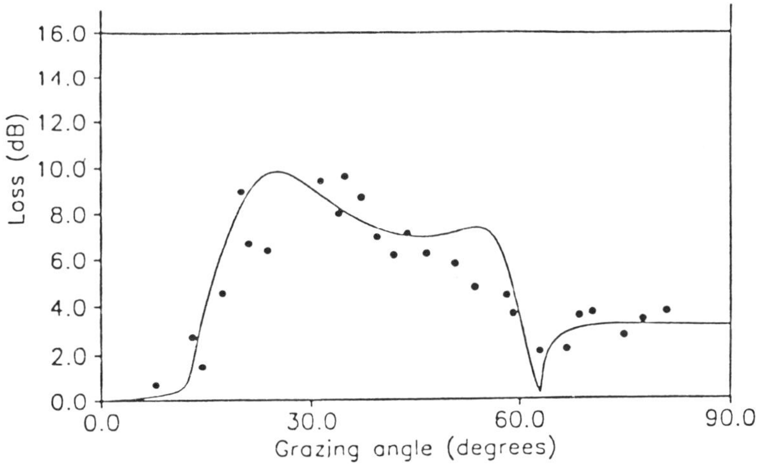


Figure 6. Bottom reflection loss for the basalt bottom. The S and P wave critical angles are at $15^0$ and $65^0$, respectively.

Shear wave properties of the surficial sediments were inverted by Sen et al. using the reflectivity method for modelling the bottom reflected waveform [13]. This technique, which involves the calculation of a synthetic seismogram for an assumed bottom model, has been used extensively in the analysis of marine seismic data. The data shown in Fig. 7a were obtained using shallow explosive charges in an experiment carried out at a turbidite basin site. The inversion proceeds by calculating a Green's function for the model that includes contributions from all orders of multiple bottom bounce paths; the waveform is constructed from this information and an estimate of the charge wavelet. A series of synthetics for a variety of models were computed by varying the sediment parameters; the modelled waveforms shown in Fig. 7a (based on the geoacoustic profile in Fig. 7b) have the correct moveout for the branch SS'. The S wave speed in the top 40 m of sediment was estimated as 50 m/s, with an increase to 200 m/s beneath this layer. The shear attenuations were 18.1 and 1.4 dB/m/kHz respectively for the two layers

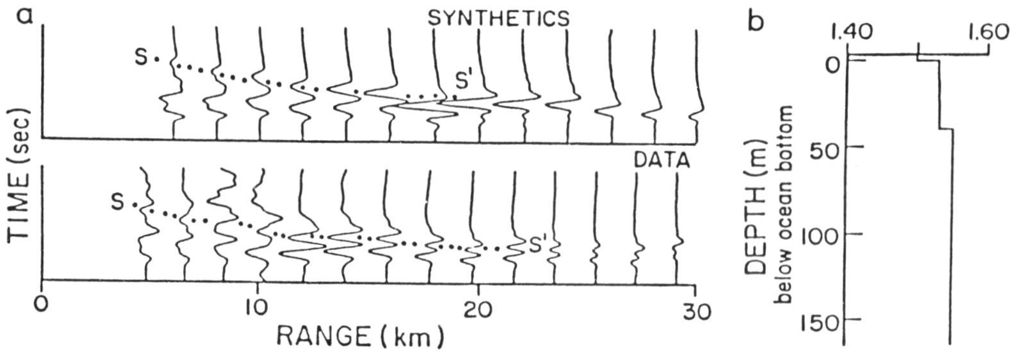


Figure 7. (a) Synthetics and measured waveforms for the turbidite sediment bottom. (b) The geoacoustic model estimated from the data.

The final example in this section illustrates an inversion method based on a specific feature of the acoustic field. In this case, travel time versus angle information is inverted for bottom interacting arrivals in a thick sediment turbidite bottom. The data in Fig. 8 represent the deconvolved impulse response obtained for the first bottom bounce arrival; the signals are stacked to align the sea floor reflection. The prominent bipolar signal following the first arrival is interpreted as a sediment refracted arrival; the phase shift of $\pi/2$ is due to the interaction with the caustic formed by the refracted rays. The inversion is based on the effect of the caustic; another approach using the amplitude of the caustic signal has been described by Frisk [14] and Spofford [15]. Here we assume a pseudo-linear sound speed profile and determine the travel time difference, $\delta t$, between the refracted and sea floor reflected arrivals,

$$\delta t = 2/3p \left[ 1 - (c_s/c_w)^2 \cos^2\theta \right]^{3/2}, \quad \text{with} \quad c(z) = c_s \left( 1 - 2pz/c_s \right)^{-1/2}; \tag{2}$$

p is the gradient and $c_s/c_w$ is the ratio of the sediment and water sound speeds at the bottom. The gradient and sound speed ratio were inverted from the travel time data by plotting $\delta t$ versus $\cos^2\theta$. The values obtained were $p = 0.6\ s^{-1}$, and $c_s/c_w = 1.02$; again,these results compare favourably with those of Hamilton for turbidites [5], and also with the results of seismic experiments carried out in the region [13].

## INVERSION BY MATCHED FIELD PROCESSING

A technique of great potential for inferring geoacoustic profiles is inversion using the matched field processing concept [16]. In a typical experiment the amplitude and phase of the total field is measured with a vertical array of hydrophones having a large enough aperture to span a significant portion of the water column. The method is conventionally used to invert an unknown source position by matching the data with modelled replicas of the field that are generated by a propagation model for specified positions of the source. The best match is determined using traditional estimators such as the Bartlett (conventional) or the minimum variance method. Conversely, for a known experiment geometry, the bottom properties can be inferred by a similar procedure; in this case replica fields can be calculated for specific values of an assumed geoacoustic profile. This method has the advantages that data from a large number of receivers are used in the inversion, and the search over the parameter space can be carried out efficiently using fully elastic normal mode methods [17].

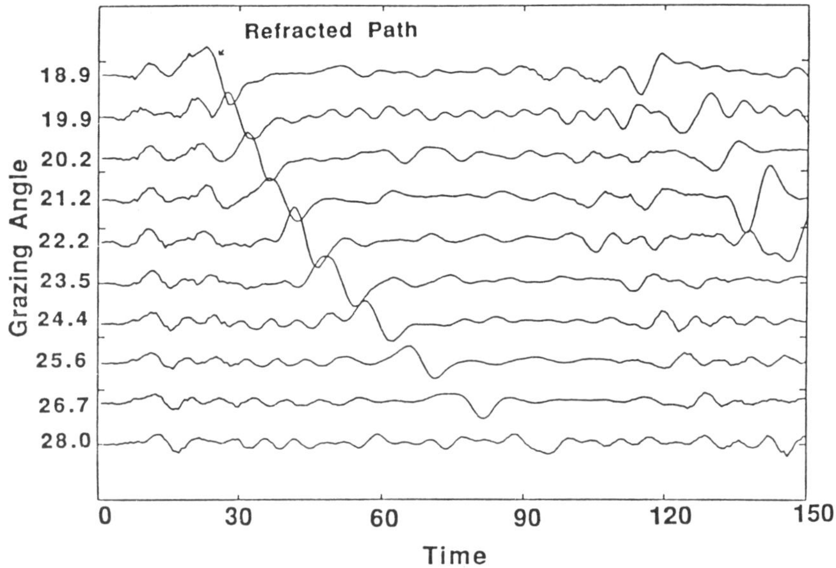


Figure 8. Deconvolved waveforms showing the refracted arrival for a thick sediment site.

## SUMMARY

This paper has reviewed techniques for inverting measurements of the acoustic field in order to infer geoacoustic models of the ocean bottom. In most of the examples, the inversion is carried out using an iteration of forward models approach; although this is not an exact inversion, the method is easy to apply to real data, and gives reasonable values for the model parameters. Depending on the design of the experiment it is possible to infer shear wave properties of the bottom materials from these data. Methods based on inversion using matched field processing are potentially very useful and are likely to make an important contribution to this field of research in the immediate future.

## REFERENCES

1. Frisk, G.V. (1989) 'Inverse methods in ocean bottom acoustics', in *Geophysical Tomography*, Y. Desaubies, A. Tarantola and J. Zinn-Justin (eds), Elsevier Science Publishers B.V., pp. 3-32.
2. Frisk, G.V., Doutt, J.A., and Hays, E.A. (1986) 'Geoacoustic models for the Icelandic Basin', J. Acoust. Soc. Am. **80**, 591-600.

3. Frisk, G.V., Lynch, J.F., and Rajan, S.D. (1989) 'Determination of compressional wave speed profiles using modal inverse techniques in a range dependent environment in Nantuckett Sound', J. Acoust. Soc. Am. **86**, 1928-1939.
4. Lawrence, M.L. (1988) 'Determination of geoacoustic properties by a two-frequency synthetic aperture method', proceedings of *Conference on Seismic and Underwater Acoustics,* University of New South Wales, Sydney, Australia.
5. Hamilton, E.L. (1980) 'Geoacoustic modeling of the sea floor', J. Acoust. Soc. Amer. **68**, 1313-1340.
6. Frisk, G.V. and Lynch, J.F. (1984) 'Shallow water waveguide characterization using the Hankel transform', J. Acoust. Soc. Am. **76**, 205-216.
7. Kuperman, W.A., Werby, M.F., Gilbert, K.E. and Tango, G.J. (1985) 'Beamforming on bottom-interacting tow-ship noise', IEEE J. Oceanic Eng. **OE-10**, 290-298.
8. Ellis, D.D. and Chapman, D.M.F. (1985) 'A simple shallow water model including shear wave effects', J. Acoust. Soc. Am. **78**, 2087-2096.
9. Jensen, F.B. and Schmidt, H. (1986) 'Shear Properties of ocean sediments determined from numerical modelling of Scholte wave data', in *Ocean Seismo-acoustics*, T. Akal and J.M. Berkson, (eds.), Plenum Press, New York, pp 683-692.
10. Chapman, N.R., Bannister, R.W. and Falconer, R. (1990) 'The effect of an elastic basement on the interpretation of bottom loss data', J. Acoust. Soc. Am. **87**, 2044-2051.
11. Chapman, N.R., Stinson, K., Levy, S., Cabrera, J., and Oldenburg, D.W. (1988) 'Estimation of the elastic properties of sea floor sediments by inversion of precritical reflection data', IEEE J. Oceanic Eng. **OE-13**, 215-221.
12. Diachok, O.I., Dicus, R.L. and Wales, S.C. (1984) 'Elements of a geoacoustic model of the upper crust', J. Acoust. Soc. Am. **75**, 324-334.
13. Sen, M.K., Frazer, L.N., Chapman, N.R., and Mallick, S. (1988) 'Analysis of multipath sound propagation in the ocean near 49° N 128° W', J. Acoust. Soc. Amer.,**83**, 588-597.
14. Frisk, G.V. (1980) 'Determination of sediment sound speed profiles using caustic range information', in *Bottom-interacting Ocean Acoustics,* W. Kuperman and F. Jensen, eds., Plenum Press, New York, pp 153-158.
15. Spofford, C.W., (1980) 'Inference of geoacoustic parameters from bottom loss data', in *Bottom-interacting Ocean Acoustics*, W. Kuperman and F. Jensen, eds., Plenum Press, New York, pp 159-171.
16. Fizell, R.G. (1987) 'Application of high-resolution processing to range and depth estimation using ambiguity function methods', J. Acoust. Soc. Am. **82**, 606-613.
17. Porter, M.B. and Reiss, E.L. (1985) 'A numerical method for bottom interacting ocean acoustic normal modes', J. Acoust. Soc. Am. **76**, 244-252.

# A FAST SIMULATED ANNEALING ALGORITHM FOR THE INVERSION OF MARINE SEDIMENT SEISMO-ACOUSTIC PARAMETERS

W.A. Kuperman and M.D. Collins
*Naval Research Laboratory*
*Washington, DC 20375*
*USA*

H. Schmidt
*Massachusetts Institute of Technology*
*Cambridge, Massachusetts 02139*
*USA*

ABSTRACT. Nonlinear seismo-acoustic inversion is performed using a fast simulated annealing algorithm. A high-resolution cost function and the matched-field approach are used to estimate the seismo-acoustic parameters. The parabolic equation method, which has recently been extended to handle elastic ocean bottoms, is used for range-dependent matched-field inversion with a vertical array. For locally range-independent problems, the inversion is performed in wavenumber space. This approach offers efficiency gains of a few orders of magnitude because the number of wavenumber samples required for inversion is much smaller than the number of wavenumber samples required for computing the acoustic field. Simulated annealing is used to search the large parameter space of ocean bottoms for the parameter set corresponding to the best replica field.

## 1. INTRODUCTION

Although several methods have been proposed for estimating the seismo-acoustic parameters of the ocean bottom by remote sensing [1-5], no existing method is capable of accurately solving this nonlinear optimization problem in a reasonable amount of time. If given infinite computing power and unlimited acoustic data, accurate parameter estimates can be obtained by performing an exhaustive search for the set of ocean-bottom parameters corresponding to the best replica field. This matched-field approach has been applied for point source localization [6,7], a problem for which the parameter search space is two or three dimensional, and to the inverse problem of estimating the speed of sound in the ocean [8]. To apply the matched-field approach to the seismo-acoustic inversion problem, the large parameter space of ocean bottoms must be searched. This brute-force approach requires very intensive calculations because the wave equation must be solved for each test ocean bottom. However, matched-field inversion has become practical due to the development of efficient nonlinear optimization methods such as simulated annealing [9,10], which has been applied to remote-sensing problems [11,12].



*J. M. Hovem et al. (eds.), Shear Waves in Marine Sediments*, 521–528.

Simulated annealing is more efficient than other methods for some optimization problems involving large numbers of parameters. This Monte Carlo method is based on the analogy of slowly cooling a pure liquid to form a perfect crystal. Simulated annealing consists of a Markov process in which each step involves randomly perturbing the parameters, evaluating an energy function, and testing if the energy is lowered. If the energy is lowered, the perturbation is accepted; if the energy is raised, the perturbation is accepted according to a Boltzmann probability distribution to allow escape from local minima. The most general simulated annealing algorithm requires that the temperature (an artificial control parameter) decrease inverse logarithmically with the number of iterations [13]. A fast simulated annealing algorithm allows an inverse-linear cooling schedule by occasionally allowing relatively large perturbations [14]. Simulated annealing for mixtures is based on the analogy of a mixture of liquids with different freezing temperatures that correspond to different parameter types [15]. This analogy is appropriate for the seismo-acoustic inverse problem; for example, the array data might not be equally sensitive to the two wave speeds or might be more sensitive to parameters near the ocean-bottom interface than to parameters deep within the ocean bottom.

In this paper, simulated annealing is applied to matched-field seismo-acoustic inverse simulations. The parabolic equation (PE) method [16], which is useful for range-dependent problems and has recently been extended to accurately handle elastic ocean bottoms [17-19], is applied for matching complex pressure on a vertical array. For problems that are locally range independent, matched-field inversion can be speeded up by a few orders of magnitude by using a horizontal array and working in wavenumber space using an efficient spectral model [20,21]. This approach has been used previously for estimating reflection coefficients [22]. Seismo-acoustic inverse simulations are presented for problems involving unknown wave speeds, attenuations, and density including a range-dependent problem.

## 2. MATCHED-FIELD INVERSION

In this section, we describe a simulated annealing algorithm for matched-field inversion. Simulated annealing algorithms consist of a sequence of iterations involving random perturbations of the unknown parameters. The perturbations are controlled by the temperature $T$, which decreases slightly after each iteration. All perturbations that produce decreases in the energy $E$ are accepted (i.e., the test values of the parameters are modified) at all values of $T$. Perturbations that produce increases in $E$ are accepted with a probability that decreases as $T$ decreases (a random number generator is used to determine whether or not to accept the perturbations). Since $E$ must be computed many times, it is important to have a means for computing $E$ efficiently.

The initial value of the control parameter $T$ is $T_0$ and $T = T_0/j$ at the $j$-th iteration. The $M$ test ocean-bottom parameters $q_{m,j}$ are initialized with an estimate (not necessarily accurate) of the true parameters for $j = 1$. For the complex pressure matching algorithm, simulated annealing is used to search the space of test parameters for the ocean bottom that minimizes the energy

$$E = \sum_{n=1}^{N} [P_n - Q_n]^2, \tag{1}$$

where the data complex pressures $P_n$ are measured on an $N$-hydrophone vertical array and the replica complex pressures $Q_n$ are computed using the test parameters. The random perturbations are determined by the the random numbers $\chi_{m,j}$, which are selected uniformly from the interval (-1,1), and the random numbers $\psi_{m,j}$, which are selected uniformly from (0,1).

At the $j$-th iteration, the parameters are perturbed one at a time. The perturbed test parameters are given in terms of the perturbation constants $\Delta q_m$ by

$$q'_{m,j} = q_{m,j} + \chi^3_{m,j}\Delta q_m . \tag{2}$$

The random numbers are cubed so that large perturbations are possible but small perturbations are more likely. With this approach, the algorithm is useful for the inverse-linear cooling schedule. At the $m$-th step of the $j$-th iteration, the energy change is

$$\Delta E_{m,j} = E\left(q'_{m,j}\right) - E\left(q_{m,j}\right). \tag{3}$$

The new test parameter is

$$q_{m,j+1} = \begin{cases} q'_{m,j} & \text{if } \psi_{m,j} < \exp(-\Delta E_{m,j}/T) \\ q_{m,j} & \text{if } \psi_{m,j} > \exp(-\Delta E_{m,j}/T), \end{cases} \tag{4}$$

where $\exp(-\Delta E_{m,j}/T)$ is the acceptance probability. In particular, the perturbation is accepted if $\Delta E_{m,j} < 0$ (i.e., lower energy states are always accepted). It is possible to escape from local minima because the acceptance probability is positive for $\Delta E_{m,j} > 0$. However, escape from local minima becomes more difficult as $T$ decreases.

An approximate expression for $\Delta E_{m,j}$ is

$$\Delta E_{m,j} \cong \chi^3_{m,j}\Delta q_m \, \partial E/\partial q_{m,j} . \tag{5}$$

We apply the chain rule, and Eq. 5 becomes

$$\Delta E_{m,j} \cong \chi^3_{m,j}\Delta q_m \vec{A}\cdot\vec{B}_{m,j}, \tag{6}$$

$$\vec{A} = \left(\partial E/\partial Q_1, \cdots, \partial E/\partial Q_N\right), \tag{7}$$

$$\vec{B}_{m,j} = \left(\partial Q_1/\partial q_{m,j}, \cdots, \partial Q_N/\partial q_{m,j}\right). \tag{8}$$

For the parameters to behave like the molecules of a pure substance, the expected value of $|\Delta E_{m,j}|$ must be independent of $m$. Since the expected values of $\chi_{m,j}$ and $\vec{A}$ are independent of $m$, this requirement can be approximately achieved near the minimum-energy state by taking

$$\Delta q_m = \alpha / \left\| \vec{B}_{m,j} \right\|, \tag{9}$$

where $\alpha$ is a constant and $\left\| \vec{B}_{m,j} \right\|$ is evaluated at an *a priori* estimate of the true parameter set.

For problems that can be approximated as locally range independent, the seismo-acoustic parameters can be inverted very efficiently in wavenumber space. A horizontal array is used for this type of problem. Since the array length must be finite, the measured data can not be transformed directly into wavenumber space without introducing a truncation error that might be significant. If the array extends out to the farfield region that is dominated by normal-mode propagation, this difficulty can be overcome by applying simulated annealing to estimate the $L$ amplitudes $A_l$ and eigenvalues $k_l$ that minimize

$$E = \sum_{n=n_0}^{N} \left[ P(r_n) - \sum_{l=1}^{L} A_l \exp(ik_l r_n) \right]^2, \tag{10}$$

where the $r_n$ are the ranges of the farfield hydrophones $n_0, \cdots, N$. The following estimate is obtained for the acoustic field in the region beyond the array:

$$P(r) \cong \sum_{l=1}^{L} A_l \exp(ik_l r). \tag{11}$$

Since it is often necessary to compute the wavenumber spectrum for a few thousand wavenumber samples to accurately construct the complex pressure, it might be possible to speed up the inversion process by a few orders of magnitude by working in wavenumber space and performing matched-field inversion using approximately ten to twenty wavenumber samples. Simulated annealing is applied to search for the minimum of:

$$E = \sum_{i=1}^{I} \left[ \hat{P}_i - \hat{Q}_i \right]^2, \tag{12}$$

$$\hat{P}_i = \int_0^\infty r \, J_0(rh_i) P(r) dr, \tag{13}$$

where the $I$ wavenumber samples $h_i$ need not be selected with equal spacing. The replica values $\hat{Q}_i$ are computed using the test parameters.

## 3. EXAMPLES

In this section, we consider an example for each of the matched-field inversion methods. The sound speed in the homogeneous water column is 1500 m/s for both of the examples. For Example A, we consider a range-dependent problem for which the ocean depth is known to be 550 m at the source location and 500 m at the array location 2 km

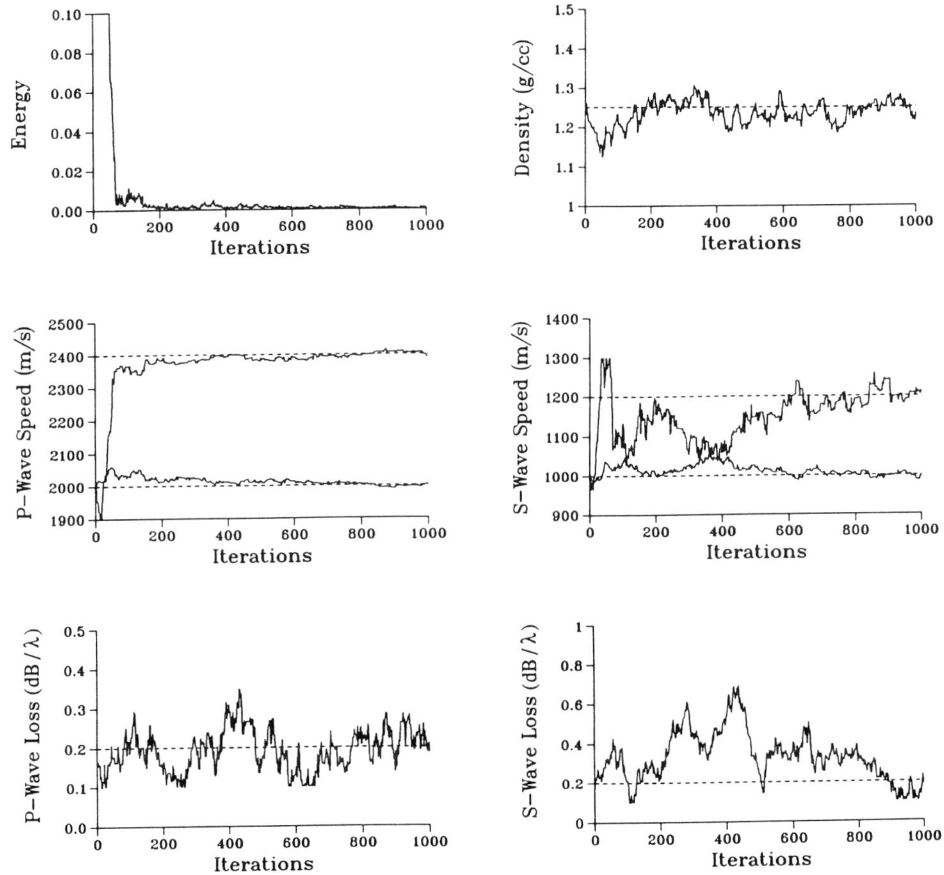


Figure 1. The Markov process for Example A.

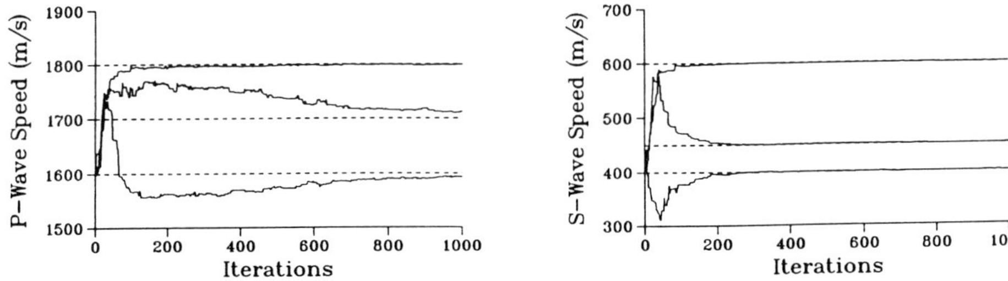


Figure 2. The Markov process for Example B.

away with linear variation between. Twenty hydrophones in a vertical array are spaced 20 m apart with the top hydrophone at depth $z = 20$ m below the ocean surface. The frequency of the point source at $z = 25$ m is 10 Hz. Within the ocean bottom, the seismo-acoustic parameters depend only on $z$. The wave speeds are linear for 500 m $< z <$ 700 m and constant for $z >$ 700 m. The compressional speed $c_p$ is 2000 m/s at $z = 500$ m and 2400 m/s at $z = 700$ m. The shear speed $c_s$ is 1000 m/s at $z = 500$ m and 1200 m/s at $z = 700$ m. The other seismo-acoustic parameters are constant throughout the ocean bottom. The compressional and shear attenuations $\beta_p$ and $\beta_s$ are both 0.5 dB/$\lambda$ and the density $\rho$ is 1.25 g/cc. The depths of the parameter nodes are assumed to be known.

The perturbation constants are as follows: 10 m/s at $z = 500$ m and 80 m/s at $z = 700$ m for both $\Delta c_p$ and $\Delta c_s$; 0.025 g/cc for $\Delta\rho$; and 0.05 dB/$\lambda$ for both $\Delta\beta_p$ and $\Delta\beta_s$. The Markov process appears in Figure 1. The energy drops sharply as the compressional speeds converge quickly to the correct values. The array data is less sensitive to the other parameters, which do not begin to converge until long after the compressional speeds have locked in. However, all of the parameters are recovered fairly accurately. Example A required 40 min of computation time on a Cray X-MP/24 computer. Since run times increase approximately linearly with the number of unknown parameters, efficient parameterization and use of *a priori* information regarding the seismo-acoustic parameters are important factors for reducing computation time. We expect that further algorithm improvements will lead to significant improvements in efficiency.

We illustrate matched-field inversion in wavenumber space with Example B. A 5-Hz source is placed at $z = 95$ m in a 100 m deep ocean with a horizontal array lying on the ocean bottom, which consists of two 10 m thick homogeneous layers overlying a homogeneous half space. In the respective layers, $c_p$ values are 1600 m/s, 1700 m/s, and 1800 m/s; $c_s$ values are 400 m/s, 450 m/s, and 600 m/s; $\rho$ values are 1.8 g/cc, 1.9 g/cc, and 2 g/cc; $\beta_p$ values are 0.2 dB/$\lambda$, 0.15 dB/$\lambda$, and 0.1 dB/$\lambda$; and $\beta_s$ values are 0.5 dB/$\lambda$, 0.4 dB/$\lambda$, and 0.2 dB/$\lambda$. All parameters except the wave speeds are assumed to be known. The perturbation constants $\Delta c_p = 50$ m/s and $\Delta c_s = 30$ m/s are used in each layer. The sixteen equally-spaced wavenumber samples correspond to phase speeds between 300 m/s and 3000 m/s. The Markov process is illustrated in Figure 2. The shear speeds converge much faster than the compressional speeds due to an interface wave. Example B required just 3 min of computation time on an Alliant FX-40 computer with two processors.

## 4. CONCLUSION

Matched-field seismo-acoustic inversion with simulated annealing can handle remote-sensing problems involving complicated depth dependence and gradual range dependence. This optimization approach is based on a high-resolution cost function that can accurately estimate all of the seismo-acoustic parameters. For range-independent problems, matched-field inversion can be performed very efficiently in wavenumber space. The appropriate analogy for adapting simulated annealing to the inverse problem is that of a mixture of substances with different freezing temperatures slowly cooling to form a perfect crystal.

[1] Stickler, D.C. (1978), 'Inverse Scattering in a Stratified Medium,' J. Acoust. Soc. Am. 74, 994-1005.

[2] Coen, S. (1982), 'Velocity and Density Profiles of a Layered Acoustic Medium from Common Source-Point Data,' Geophysics 47, 898-905.

[3] Shaw, P.R. and Orcutt, J.A. (1985), 'Waveform Inversion of Seismic Refraction Data and Applications to Young Pacific Crust,' Geophys. J. R. Astr. Soc. 82, 375-414.

[4] Kuperman, W.A., Werby, M.F., Gilbert, K.E., and Tango, G.J. (1985), 'Beam Forming on Bottom-Interacting Tow-Ship Noise,' IEEE J. Ocean. Eng. 10, 290-298.

[5] Rajan, S.D., Lynch, J.F., and Frisk, G.V. (1987), 'Perturbative Inversion Methods for Obtaining Bottom Geoacoustic Parameters in Shallow Water,' J. Acoust. Soc. Am. 82, 998-1017.

[6] Bucker, H.P. (1976), 'Use of Calculated Sound Fields and Matched-Field Detection to Locate Sound Sources in Shallow Water,' J. Acoust. Soc. Am. 59, 368-373.

[7] Baggeroer, A.B., Kuperman, W.A., and Schmidt, H. (1988), 'Matched Field Processing: Source Localization in Correlated Noise as an Optimum Parameter Estimation Problem,' J. Acoust. Soc. Am. 83, 571-587.

[8] Tolstoy, A. (1989), 'Low-Frequency Acoustic Tomography Using Matched Field Processing,' J. Acoust. Soc. Am. Suppl. 1 86, S7.

[9] Metropolis, N., Rosenbluth, A.W., Rosenbluth, M.N., Teller, A.H., and Teller, E. (1953), 'Equations of State Calculations by Fast Computing Machines,' J. Chem. Phys. 21, 1087-1091.

[10] Kirkpatrick, S., Gellatt, C.D., and Vecchi, M.P. (1982), 'Optimization by Simulated Annealing,' (IBM Thomas J. Watson Research Center, Yorktown Heights, NY).

[11] Rothman, D.H. (1985), 'Nonlinear Inversion, Statistical Mechanics, and Residual Statics Estimation,' Geophysics 50, 2784-2796.

[12] Basu, A. and Frazer, L.N. (1989), 'Inversion Using Simulated Annealing,' J. Acoust. Soc. Am. Suppl. 1 86, S52.

[13] Geman, S. and Geman, D. (1984), 'Stochastic Relaxation, Gibbs Distributions, and the Bayesian Restoration of Images,' IEEE Trans. Patt. Anal. Mach. Intel. 6, 721-741.

[14] Szu, H. and Hartley, R. (1987), 'Fast Simulated Annealing,' Phys. Let. 122, 157-162.

[15] Kuperman, W.A., Collins, M.D., Perkins, J.S., and Davis, N.R. (1990), 'Optimal Time-Domain Beamforming with Simulated Annealing Including Application of *A Priori* Information,' J. Acoust. Soc. Am. 88, 1802-1810.

[16] Tappert, F.D. (1977), 'The Parabolic Approximation Method,' in *Wave Propagation and Underwater Acoustics*, edited by J.B. Keller and J.S. Papadakis, Lecture Notes in Physics, Vol. 70 (Springer, New York).

[17] Collins, M.D. (1989), 'A Higher-Order Parabolic Equation for Wave Propagation in an Ocean Overlying an Elastic Bottom,' J. Acoust. Soc. Am. 86, 1459-1464.

[18] Wetton, B.T.R. and Brooke, G.H. (1990), 'One-Way Wave Equations for Seismoacoustic Propagation in Elastic Waveguides,' J. Acoust. Soc. Am. 87, 624-632.

[19] Collins, M.D. and Kuperman, W.A., 'A Stable Higher-Order Elastic Parabolic Equations with Application to Scholte Wave Propagation,' this proceedings.

[20] Schmidt, H. and Jensen, F.B. (1985),' A Full Wave Solution for Propagation in Multilayered Viscoelastic Media with Application to Gaussian Beam Reflection at Fluid-Solid Interfaces,' J. Acoust. Soc. Am. 77, 813-825.

[21] Schmidt, H. (1988), 'SAFARI User's Guide,' SACLANTCEN SR-113, SACLANT Undersea Research Centre, La Spezia, Italy.

[22] Frisk, G.V. and Lynch, J.F. (1984), 'Shallow Water Waveguide Characterization Using the Hankel Transform' J. Acoust. Soc. Am. 76, 205-216.

# DETERMINATION OF THE SEDIMENT SHEAR SPEED PROFILES FROM PHASE AND GROUP VELOCITY DISPERSION DATA OF SH WAVE

SUBRAMANIAM D. RAJAN and CINTHIA S. HOWITT
*Woods Hole Oceanographic Institution*
*Woods Hole, MA 02543*

ABSTRACT The geoacoustic properties of the sediments are important parameters for modeling acoustic propagation in the ocean. The geoacoustic parameters are the compressional and shear wave speed profiles, their attenuation and density. An inverse method for extracting the density and shear speed profiles in the sediment material from the phase velocity and group velocity dispersion relationships of SH waves is presented. The performance of this approach is studied using noise free and noisy synthetic data. The method is then applied to the group velocity dispersion relationship obtained from field data.

## 1. Introduction

Propagation of acoustic energy in the ocean is dependent on the geoacoustic properties of the sea bed. The geoacoustic properties are the compressional and shear wave speeds, their attenuation and the density of the sediment material and these are functions of frequency and spatial position. However, for a narrow band of frequencies we can assume these properties to be independent of frequency and if we assume horizontal stratification, the properties are a function of depth only.

Methods for determining the geoacoutic parameters of the sediments in shallow water are areas of active research. While experimental techniques like VSP and crosshole tomography can be used to determine the elastic properties of sediment material, these approaches do not lend themselves to adaptation as a survey tool. However, experimental methods like the one performed by Ewing[1] and Stoll et al.[2] can be used in a survey mode and the shear speed profile in the sediment layers estimated. In the experimental set up of Ewing[1] a broadband source, designed to generate SH waves at the ocean floor, is employed. The source and receiver are deployed on the ocean floor and measurements are made with the source to detector offsets of up to 200 m. This measured field contains information on the shear properties of the sediments. We will present methods of obtaining the shear speed profile of the sediment material from such field measurement of SH wave data.

## 2. Theory

The propagation of SH waves in the sediment is governed by the differential Eqn. [3]

$$-\omega^2\rho(z)l(\omega,z) = \frac{d}{dz}\left(\mu(z)\frac{dl(\omega,z)}{dz}\right) - k^2(\omega)\mu(z). \tag{1}$$



*J. M. Hovem et al. (eds.), Shear Waves in Marine Sediments*, 529–536.

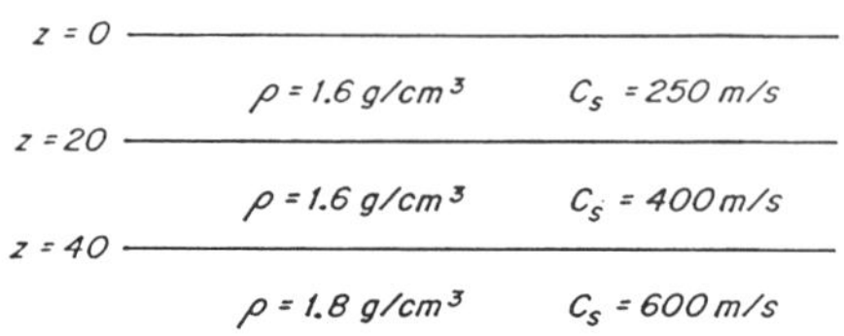


Figure 1: Ocean bottom model. $Z$ - Depth in meters from ocean bottom

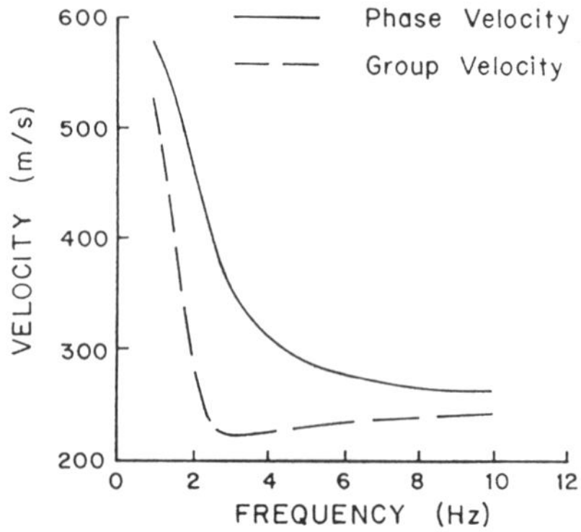


Figure 2: Phase and group velocity dispersion curves

In the above expression $l(\omega, z)$ is the particle displacement, $\mu(z)$ the shear modulus, $\rho(z)$ the density, $\omega$ the radian frequency and $k(\omega)$ the wave number. The equation must satisfy the boundary conditions at $z = 0$ and $z = \infty$. At $z = 0$ the free surface condition is to be met and at $z = \infty$ the amplitude must vanish since no energy is supplied at infinity. Under such boundary conditions nontrivial solutions to the equation exist only when the wave number $k(\omega)$ takes on special values $k_n(\omega)$. These correspond to various modes propagating in the medium. Figure 1 depicts a two layer ocean bottom model. In Figure 2 the relationship between the phase velocity $c_n(\omega) = \frac{\omega}{k_n(\omega)}$ and frequency for the fundemental mode of this model is shown. Similar dispersion relationships exist for higher order modes as well. These modes when excited travel at different speeds (the group velocities). It can be shown that the group velocity $\nu(\omega)$ is given by the expression $\nu_n(\omega) = \frac{d\omega}{dk_n}$. In Figure 2 the group velocity dispersion relationship of the fundemental mode is also presented.

Methods for estimating the sediment shear parameters consist of estimating a quantity like the phase velocity of one or several modes at different frequencies and using this as data in an inverse method to infer the unknown parameters. The relationship between the measured quantity $c_m(\omega_n)$ and the unknown parameters can be written as

$$c_m(\omega_n) = f_m(\omega_n, \boldsymbol{p}) \tag{2}$$

where $c_m(\omega_n)$ is the phase velocity of the mth mode and $\boldsymbol{p}$ is a vector containing the unknown parameters namely the shear wave speed $c_s(z)$, and density $\rho(z)$. In compact notation, the observations are related to the unknown parameters by

$$\boldsymbol{c} = \boldsymbol{f}(\boldsymbol{p}) \tag{3}$$

where $\boldsymbol{c}$ is a vector containing the data ( i.e. phase or group velocities of different modes at various frequencies) and $\boldsymbol{f}(\boldsymbol{p})$ relates the data to the unknown parameters. If we model the sediment as a stack of homogeneous layers, the vector $\boldsymbol{p}$ contains the shear speed and density in the layers. The estimation problem then reduces to the determination of the set of parameters $\boldsymbol{p}$ which satisfy Eqn. (3) in the least squared sense i.e. the set that minimizes the cost function $\phi(\boldsymbol{p}) = \sum_n \mid c_n - f_n(\boldsymbol{p}) \mid^2$.

The maximum likelihood estimate of the unknown parameters is obtained by the minimization of the cost function[4]

$$\phi(\boldsymbol{p}) = (\boldsymbol{c} - \boldsymbol{f}(\boldsymbol{p}))^T \boldsymbol{C}_d^{-1} (\boldsymbol{c} - \boldsymbol{f}(\boldsymbol{p})) + (\boldsymbol{p} - \boldsymbol{p}_\circ)^T \boldsymbol{C}_p^{-1} (\boldsymbol{p} - \boldsymbol{p}_\circ) \tag{4}$$

where $\boldsymbol{C}_d$ and $\boldsymbol{C}_p$ are covariances of the data and model parameters. A number of algorithms exist for solving this least squares minimization problem. We have used the variable metric method due to Fletcher and Powell[5].

Having obtained the solution by this method, the posterior covariance of the estimated parameters $\boldsymbol{C}_p'$ is obtained from[4]

$$\boldsymbol{C}_p' = (\boldsymbol{G}_\infty^T \boldsymbol{C}_d^{-1} \boldsymbol{G}_\infty + \boldsymbol{C}_p^{-1})^{-1} \tag{5}$$

where $\boldsymbol{G}_\infty$ contains the gradients and $\infty$ stands for the maximum likelihood point obtained by an iterative algorithm.

## 3. Performance study with synthetic data

We now present examples of inversion for shear wave speed by two methods: (a) from phase velocity dispersion data and (b) group velocity dispersion data. The input data for inversion algorithm was obtained by sampling the dispersion curve of the fundemental mode in Figure 2 at discrete points in the frequency space in the range 1 Hz to 10 Hz. A total of 14 data points were used in the inversion. In performing the inversion it was assumed that the densities were known and only the shear speeds were estimated. The covariance of the shear speed in the layers was assumed to be infinite corresponding to no apriori knowledge. The elements of the data vector were assumed to be uncorrelated with unit variance. With these assumptions the cost function in Eqn. (4) reduces to $\phi(\boldsymbol{p}) = (\boldsymbol{c} - \boldsymbol{f}(\boldsymbol{p}))^T (\boldsymbol{c} - \boldsymbol{f}(\boldsymbol{p}))$.

The estimated shear speed profile, the starting profile and the true profile for the two methods are shown in Figures 3 and 4 for the cases of noise free and noisy data. The noisy data was obtained by adding a Gaussian random variable with a standard deviation of 10 m/s to the noise free data. A smooth curve was fitted to the data thus obtained. This smoothed curve was sampled as before and used as input data to the inversion algorithm.

From the results obtained the following observations are made:

(i) The reconstructed profiles are in good agreement with the true profiles except in the region adjoining the interface at 20 m depth where the reconstructed profiles do not follow the sharp discontinuity at the interface. However, as shown in Figure 5, the phase velocities predicted by the reconstructed profile in Figure 3 agree with the values for the true profile. Similar agreement is also noted in the case of the group velocities.

(ii) The reconstruction is good even when the data is corrupted with noise.

The ability of the reconstruction algorithms to estimate the shear speeds in the layers depends on the sensitivity of the phase and group velocity to changes in the layer shear speed i.e. to the partial derivative of the phase and group speed to the layer parameters. For the

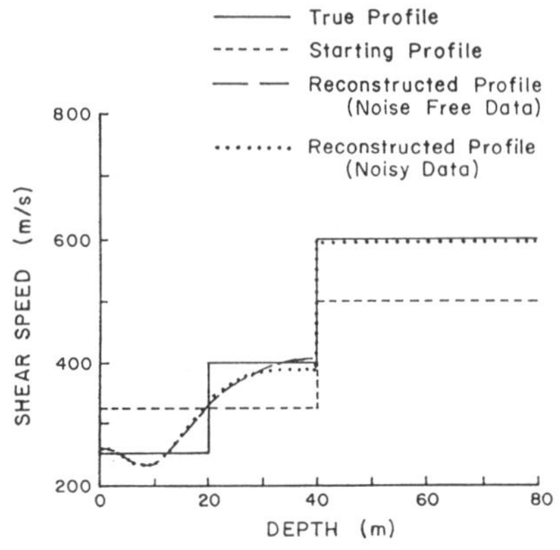


Figure 3: Reconstructed shear speed profile from phase velocity data.

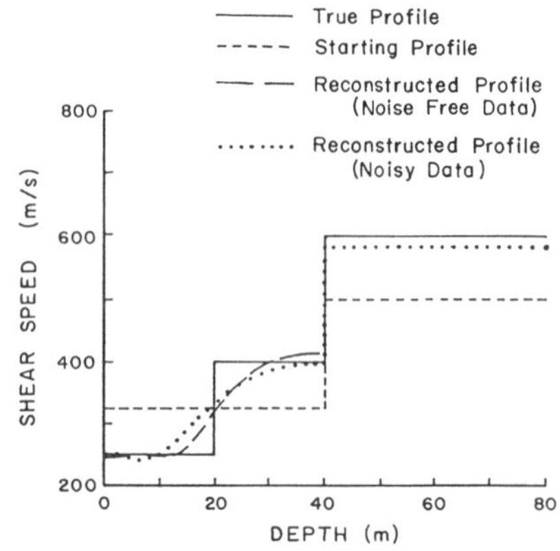


Figure 4: Reconstructed shear speed profile from group velocity data.

starting model, the partial derivatives of the phase and group velocities with respect to the shear speed in the layers vary slowly as shown in Figure 6. This results in the reconstructed profile being smoothed over in regions adjoining sharp discontinuities. Further iteration will improve the results obtained but such improvement is slow.

## 4. Inversion of field data

Ewing et al. [1] conducted experiments at various sites in the Atlantic Ocean with the objective of determining the geoacoustic properties of the sediment layers. These sites were chosen because of the wealth of geophysical and geotechnical information available for this area from previous experiments. This would provide an opportunity to compare the results determined from inversion of seismic data with the "ground truth" properties. In this paper we restrict our analysis to data obtained at the AMCOR site 6011 (34 °43'30"N,73°56'36"W).

The source used in the experiment was a specially designed sled fitted with port and starboard mounted air powered cannons. The receivers (geophones and hydrophones) were mounted on another sled of similar design. The design of the sleds was such as to ensure their upright attitude on the sea floor. At the start of the experiment the two sleds with proper orientation were positioned on the ocean floor with an offset of approximately 200m. In recording the profile, the receiver sled was pulled towards the source sled in increments of approximately 4 to 8 m. One or more shots were fired at each range. Alternate firing of the port and starboard guns produced alternately polarized SH excitation. The source also produced P and SV waves. Subtraction of sequential shot data produced primarily SH data while addition gave P and SV data. This approach, in contrast to other experiments where SV data was obtained[2], was taken because vertically polarized waves interact with compressional wave energy whereas horizontally polarized waves do not; thus simplifying the modeling problem.

There are two approaches to the analysis of this data. The multiple filter technique[6] can be applied to the data and the group velocity dispersion data estimated. Alternatively

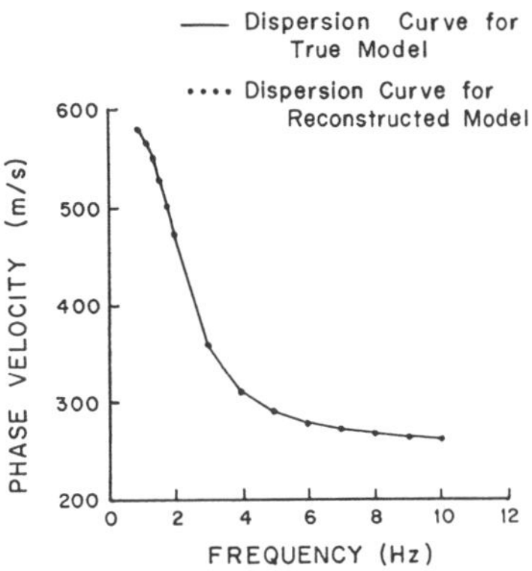


Figure 5: Phase velocity dispersion data of true model and reconstructed model.

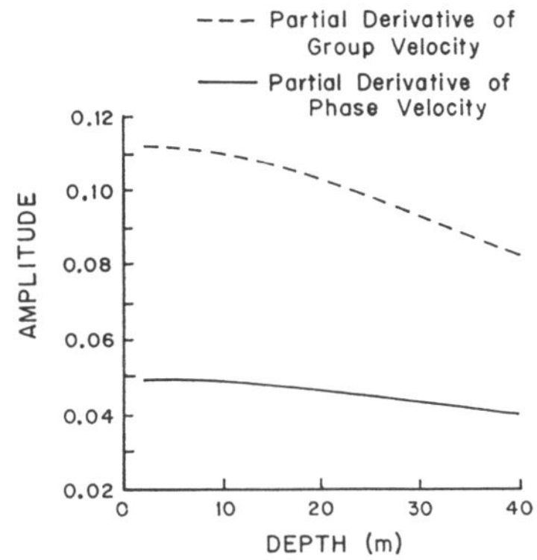


Figure 6: Partial derivative of phase and group velocity with respect to shear speed at various depths. Computed for starting model in Figure 3

by performing 2D Fourier transform of the measured field, the phase velocity dispersion data can be obtained[7]. However, the data recorded by Ewing[1] were not adequately sampled in space to perform the 2D Fourier transform. We, therefore, restrict our analysis to the group velocity dispersion.

To perform the multiple filter analysis, the data were low pass filtered in the band 0-25 Hz, a closely spaced Gaussian window applied to the filtered data and in-phase and quadrature components of the signal (individual shot data) in the frequency domain obtained. The amplitude of the resulting complex signal in the time domain was plotted as a function of frequency and group velocity. The contour plot for one of the shots is shown in Figure 7. The group velocity dispersion curves are along the crests of the contour diagram and the curve for the fundemental mode is shown in the figure. From this curve the group velocity at 0.5 Hz intervals in the range 4Hz to 10 Hz was determined and used as input data in the inversion algorithm described earlier. In performing the inversion, a constant density of 1.6 g/cm$^3$ (representative of the sediment material in the area) was assumed and only the shear speed $C_s$ in the layers estimated. The reconstructed profile is shown in Figure 8.

The variance of the estimates obtained by inversion is dependent on the data and model covariance matrices as indicated in Eqn. (5). To compute the data covariance matrix, we assume that the errors are uncorrelated and have equal variance. The covariance matrix, then, has the form $\sigma_d^2$I where I is the identity matrix and $\sigma_d^2$ is the variance of the data error. The error in the data is related to the error in range measurement and arrival time. The range measurement error is assumed to be $\pm 1$ m (Ref. 1). The error in arrival time is dependent on the bandwidth of the source and signal to noise ratio. For a bandwidth of 16 Hz and a signal to noise ratio of 20 dB ( a conservative estimate ), the devaition $\sigma_d$ is 6.5 m/s. The model parameters are assumed to be correlated with a correlation length of 12 m which corresponds to the thickness of three layers. The deviation of model parameter error is assumed to be 30 m/s. This is based on the values of shear wave speed for sediments and their error bars published by Hamilton [8].

Using the data and model covariance matrices computed as described above the posterior

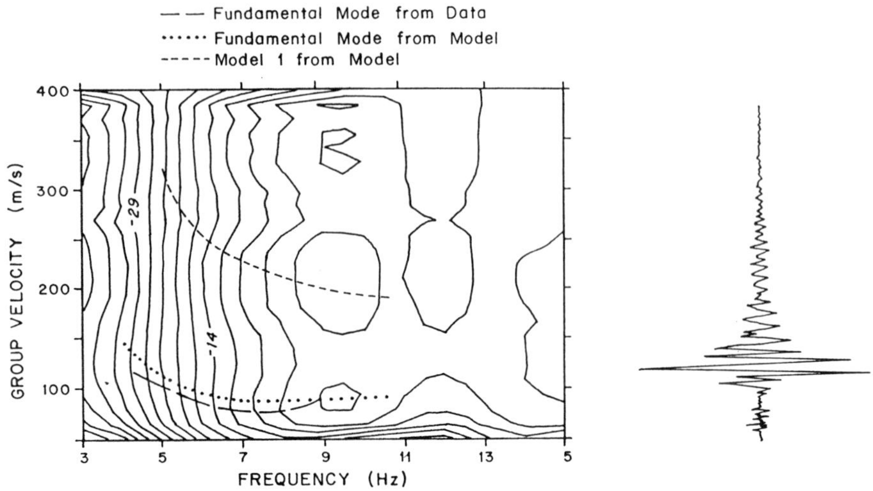


Figure 7: Contour plot from multiple filter analysis. Signal at range 61m from source. Signal shown on the right

| Layer depth | Shear Speed (m/s) | Deviation (m/s) |
|---|---|---|
| 0 - 4 m | 101.14 | 1.29 |
| 4 - 8 m | 226.79 | 6.93 |
| 8 - 12 m | 232.83 | 7.26 |
| 12 - 16 m | 274.99 | 12.60 |
| 16 - 20 m | 312.49 | 16.25 |
| 20 - 24 m | 326.41 | 17.69 |
| 24 - 40 m | 330.00 | 18.30 |
| above 40 m | 404.90 | 20.66 |

Table 1: Estimted shear wave speed in the layers and deviation of the estimate.

variance $\sigma_s^2$ of the shear wave speed estimate was determined. The upper and lower bounds of the shear wave speed in the layer were obtained by plotting $C_s \pm \sigma_s$ as shown in Figure 8. The values of $C_s$ and $\sigma_s$ for the various layers are tabulated in Table 1.

The sediment strata at site AMCOR 6011 consists of a layer of silty clay about 10 m thick overlying a 40 m layer of sand. Stoll et al[2] estimated the geoacoustic properties of the sediment layers at this site based on the physical properties of the sediments measured in earlier experiments and a Biot model. The shear speed profile obtained by this method is also shown in Figure 8.

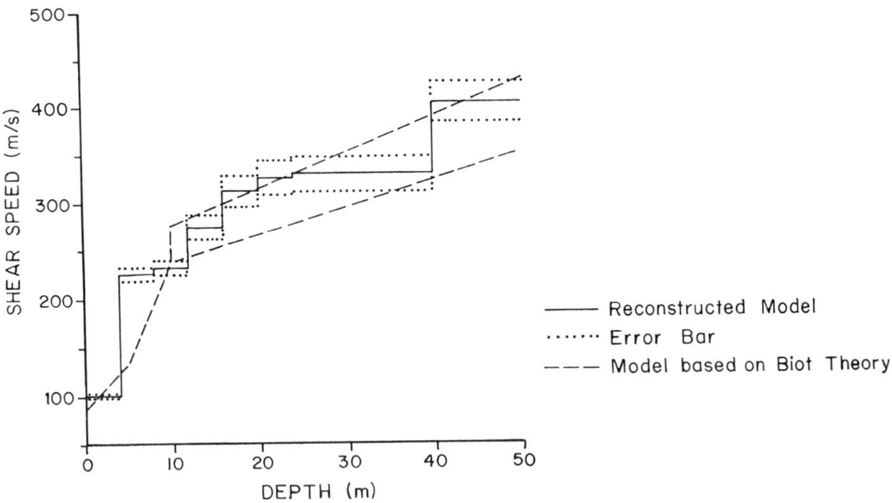


Figure 8: Shear speed model from data with error bars and model based on physical properties of the sediment and Biot theory from Reference 3.

We observe that:

(a) the inversion was able to estimate the profile only in the top 20 m of the sediment layers and beyond this depth the starting profile remains virtually unaffected,

(b) in the depth range of 0 to 16 m the posterior error in model parameter is much less than the value assumed for computing the model covariance matrix indicating the reliability of the estimates for these layers, and

(c) in the top 20 m the profile compares favourably with the profile predicted by the Biot model.

The depth limitation in reconstruction arises because the group velocity dispersion data for the fundemental mode alone was used in inversion. Since the mode function for the fundemental mode has little energy at depths greater than 20 m for the frequencies considered, the inversion algorithm cannot reconstruct the shear wave speeds at these depths and the solution merges with the starting model.

The group velocity dispersion data for the fundemental and the first mode were computed using the inferred bottom model. These are indicated in Figure 7. It is observed that the dispersion curves of the modes fit reasonably well with the dispersion curves in the data. It is interesting that though the shear speed profile was obtained using fundemental mode data, the dispersion curve of the first mode predicted by the model agrees well with the first mode dispersion curve in the data. This is comforting and suggests that the inferred model is a good approximation to the true bottom model.

## 5. Conclusion

We have presented methods for obtaining the shear speed profiles in sediments from dispersion data of SH waves. Performance studies conducted with noise free and noisy synthetic data show that both methods perform equally well. In terms of computational time, phase velocity inversion is faster.

The method was then applied to field data obtained by Ewing[1]. The spectral content of the signal was in the band 4Hz to 20 Hz. This limited the depth to which the shear speed profile could be obtained from the dispersion curve of the fundemental mode to approximately 20 m. For estimation of sediment properties in deeper layers, the source bandwidth has to be modified to contain sufficient energy at frequencies well below 4 Hz. The model obtained by inversion agrees reasonably well with the model predicted using physical properties of the sediment and Biot theory.

## 6. Acknowledgement

The authors acknowledge the help and suggestions given by their colleagues J. I. Ewing, G. V. Frisk and J. F. Lynch at various stages of their work. The authors also thank J. I. Ewing for providing the data from his experiment. We also acknowledge the programming help given by A. Newhall in the early stages of this work. The authors gratefully acknowledge the support of ONR contract N00014-89-k-0055. This paper is WHOI contribution No. 7452.

## References

1. Personal communication from J. I. Ewing.

2. Stoll, R. D., Bryan, G. M., Flood, R., Chayes, D., and Manley, P. (1988) ' Shallow seismic experiments using shear waves, ' J. Acoust. Soc. Am., 83, 93 - 102.

3. Aki, K., and Richard, P. G. (1980) ' Quantitative seismology: Theory and methods,' Freeman Press, San Francisco.

4. Tarantola, A. (1987) ' Inverse problem theory: Methods for data fitting and model parameter estimation,' Elsevier Science Publishers, New York.

5. Fletcher, R. and Powell, M. J. D. (1963) ' A rapidly convergent descent method for minimization,' Computer J., 6, 163 - 168.

6. Dziewonkski, A., Block, S., and Landisman, M. (1969) 'A technique for the analysis of transient seismic signals,' Bull. Seismol. Soc. Am., 59, 427 - 444.

7. Gabriels, P., Snieder, R., and Nolet, G. (1987) ' In situ measurement of shear wave velocity in sediments with higher-mode Rayleigh waves,' Geophysical Prospecting, 35,187 -196.

8. Hamilton, E. L. (1980) ' Geoacoustic modeling of the sea floor,' J. Acoust.Soc. Am., 68, 1313 -1340.

# MEASUREMENT OF THE ELASTIC PROPERTIES OF THE OCEAN BOTTOM BY INVERSION OF REFLECTION AMPLITUDE DATA

N. R. CHAPMAN
Defence Research Establishment Pacific
FMO Victoria, B.C. V0S 1B0 CANADA

K. ROHR
Geological Survey of Canada
Pacific Geoscience Centre
Sidney, B.C. V8L 4B2 CANADA

ABSTRACT. A method is described for estimating the elastic properties of ocean bottom materials by inversion of acoustic reflection amplitude data. The amplitude of the reflected signal depends on the angle of incidence and the contrasts in the elastic properties of the media across the sea floor, and we assume that this relationship can be described by the plane wave reflection coefficient. Ocean bottom reflections are analyzed using a nonlinear parametric inversion to infer the density and the compressional and shear wave speeds in the surficial layer of sea floor material. Results are presented for inversion of data from experiments in sedimented and basaltic regions off the west coast of Canada. The data were obtained using either a fixed receiver and a moving source, or a towed source and reciever. For the latter case the reflections were first processed as common midpoint data. The estimates of the elastic properties are in good agreement with ground truth data from sediment cores and with seismic refraction profiles obtained in the same region.

## INTRODUCTION

The measurement of the elastic properties of sea floor sediments is an important consideration for various aspects of underwater acoustics research. A number of methods have been developed for estimating ocean bottom properties based on inversion of acoustic field data. These include the inversion of travel time data from seismic reflection and refraction surveys to obtain sound speed profiles in the bottom [1]. Direct methods have been described by Thomson [2] and Bregman et al. [3] for reconstructing the impedance profile of a layered medium by inversion of its plane-wave reflection response. In addition, Dicus [4], Berge et al. [5] and Sen et al. [6] have used other methods to determine sediment properties by acoustic modeling of bottom reflection data.

In this paper, we present a technique for estimating the elastic properties of the topmost layer of ocean bottom sediments. The method is based on the inversion of the amplitude versus angle information contained in the signal reflected from the sea floor, and is capable of estimating both the compressional (P) and the shear (S) wave speeds. A similar approach has been considered in the analysis of marine seismic data [7]. The amplitude of the reflected signal depends on the source-receiver separation and the physical properties of the sea floor material. A non-linear parametric inversion based on the model



*J. M. Hovem et al. (eds.), Shear Waves in Marine Sediments*, 537–544.

of plane-wave reflection at a fluid/elastic interface is used to obtain estimates of the density and the P and S wave sound speeds. This work extends a previous development [8] to include both precritical and postcritical reflection amplitude information. In the remainder of this paper the design of the experiment and the inversion method are described, and the sensitivity of the inversion to the bottom properties is discussed. The method is then applied to ocean bottom reflection data obtained off the west coast of Vancouver Island.

## EXPERIMENT DESIGN

The geometry of the experiment is shown in Fig. 1. A point source is excited at depth $z_s$ in the water and the signals are received by a series of hydrophones, such as a towed line array, at depth $z_r$. Both the source and the receivers are assumed to be omnidirectional. In practice, the experiment may also be carried out by fixing the receiver position and moving the source out in range. The data consist of a set of seismograms from which the absolute strength of the sea floor refected arrival is to be determined at each range.

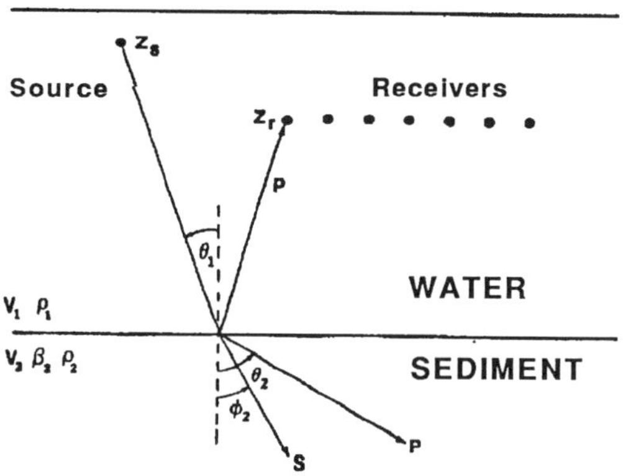


Figure 1. Experiment geometry and ray paths of plane waves interacting at a fluid/solid interface.

## INVERSION PROCEDURE

The inversion assumes a one dimensional model of the environment in which the properties of the water layer are known, and the density and sound speeds in the bottom layer are constant. The water-bottom interface is assumed to be flat, and the source/receiver geometry and the water depth are assumed to be known. The bottom is assumed to be sufficiently far away from the source so that, for the frequency band of interest (10-200 Hz in this case), a plane-wave approximation is adequate for the incident sound wave. Under these conditions, ray theory can be used to trace the propagation paths and determine the angles of incidence at each range. The effect of attenuation in the water is assumed to be negligible. Finally, it is assumed that the direct arrival (after correction for geometrical spreading) is a reasonable approximation of the source waveform. The reflection coefficient can then be determined from the ratio of the peak amplitudes of the reflected and direct path signals. The method can, in principle, be extended to obtain estimates of the properties of deeper sediment layers by deconvolving the impulse response of the bottom to determine the amplitudes of subbottom reflectors.

The bottom layer is modelled as an elastic solid with constant density, $\rho_2$, and constant P and S wave speeds, $v_2$ and $\beta_2$ respectively. The plane-wave reflection coefficient for the compressional wave with ray parameter p (equal to $\sin\theta_1/v_1$) is [9]

$$R(p) = (mx + my - z)/(mx + my + z) \tag{1}$$

where $m = \rho_2/\rho_1$

$$x = v_2 \cos^2 2\phi_2 / v_1 \cos\theta_2$$

$$y = \beta_2 \sin^2 2\phi_2 / v_1 \cos\phi_2$$

and $z = 1/\cos\theta_1$.

Given a set of reflection amplitudes, $R_k$, observed at N angles of incidence, the inversion seeks a solution for the elastic properties ($v_2$, $\beta_2$, $\rho_2$) of the sediment. The values are those which minimize the rms error, $\varepsilon$, between the observed reflection amplitudes and those computed from (1)

$$\varepsilon = \{ 1/N \sum_{k=1}^{N} (R_k^o - R_k^c)^2 \}^{1/2} \tag{2}$$

where the superscripts o and c denote observed and calculated values respectively. Since there are three parameters to determine it is convenient to find the optimum values by assuming trial values for the P and S wave speeds in the sediment, and then solve for the density. If (1) is rearranged in the form

$$m(1 - R_k)(x_k + y_k) = (1 + R_k)z_k \tag{3}$$

the new expression involves only a single unknown, m, if x and y are specified. The set of equations (3) are solved to obtain a solution for m and an associated pair of values $v_2$ and $\beta_2$ which best fits the data in a least squares sense. The algorithm takes into account that the data can be precritical or postcritical, depending on the value of $v_2$. To find the optimum solution, a systematic search of possible solutions is carried out over a selected range of trial values for $v_2$ and $\beta_2$. This procedure generates an error matrix of values of $\varepsilon$ which is in turn searched to determine the global minimum. The values of $v_2$, $\beta_2$, and $\rho_2$ corresponding to this global minimum constitute the best estimate of the elastic properties.

## SENSITIVITY OF THE METHOD

The viability of the inversion technique has been demonstrated previously using synthetic data [8]. In this section, the sensitivity of the inversion is described using the model defined in (1) to show the effect of sediment density and the P and S wave speeds on the reflectivity. The parameter values were taken from the recent work of Hamilton [10]. In Fig. 2, the reflection coefficient is shown for three values of P wave speed corresponding to silt (1.58 km/s), sand-silt (1.65 km/s) and coarse sand (2.0 km/s) sediments. The

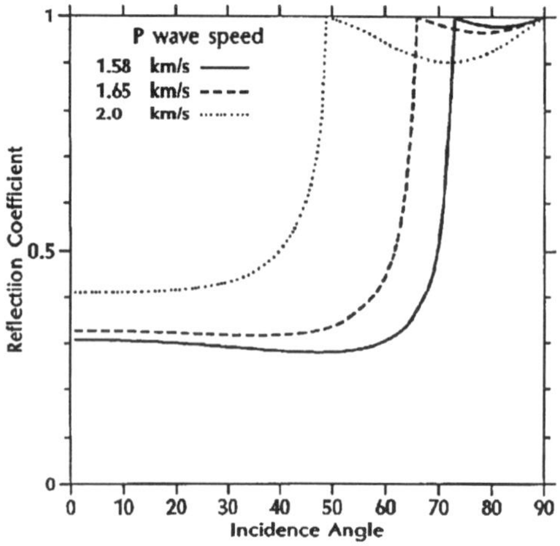


Figure 2. Plane wave reflection coefficient calculated for silt, sand-silt and coarse sand sediments.

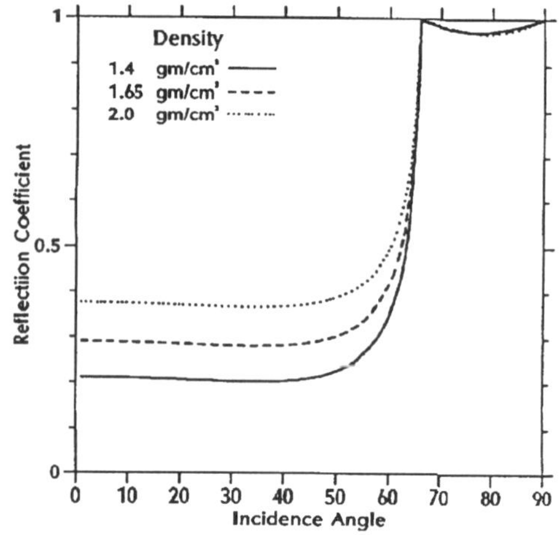


Figure 3. Dependence of the plane wave reflection coefficient on sediment density.

density and the ratio $\beta_2/v_2$ were held constant at 1.8 gm/cm$^3$ and 0.25 respectively. There is strong sensitivity to the P wave speed, provided that the reflectivity is sampled at least to the critical angle. The density dependence is shown in Fig. 3 where the reflection coefficient is calculated for sediment densities of 1.4, 1.65, and 2.0 gm/cm$^3$, with the P and S wave speeds held constant at 1.65 and 0.4 km/s respectively. The variation is significant only for precritical angles, and it is clear that the inversion of density is sensitive to the absolute strength of the relection coefficient. The dependence on S wave speed is shown in Fig. 4 where the reflection coefficient is calculated for speeds of 0.2, 0.4, and 0.6 km/s, with constant density and P wave speed of 1.8 gm/cm$^3$ and 1.65 km/s respectively. Surficial sea floor sediments are only weakly capable of supporting shear wave propagation, and it is common practice to model the sediment as a fluid in most underwater acoustics applications. However, the reflection coefficient is sensitive to the S wave speed, with the greatest variation in amplitude for angles near the P wave critical angle and at postcritical angles. The solid curve is typical of the fluid-like behaviour for low speeds, but for increasing S wave speeds, the reflectivity decreases significantly near the critical angle. The effect of P and S wave attenuation is negligible; there is no significant change in the reflection coefficient amplitude for typical values of these parameters (0.5 dB/$\lambda$ for P waves and 2.5 dB/$\lambda$ for S waves). It is true, however, that sediment attenuation is a significant energy loss for waves that are refracted in the subbottom layers.

The case of consolidated bottom materials is considered in Fig. 5 where we show the sensitivity of the reflection coefficient to S wave speeds of 0.8, 1.0 and 1.2 km/s in relatively young basalt with P wave speed of 2.5 km/s and density 2.2 gm/cm$^3$. The relatively low P wave speed is based on refraction results from mid-ocean ridges by Purdy [11], Cudrak [12] and Harding et al [13]. It is clear that the amplitude is strongly sensitive to S wave speed, particularly at angles greater than the P wave critical angle.

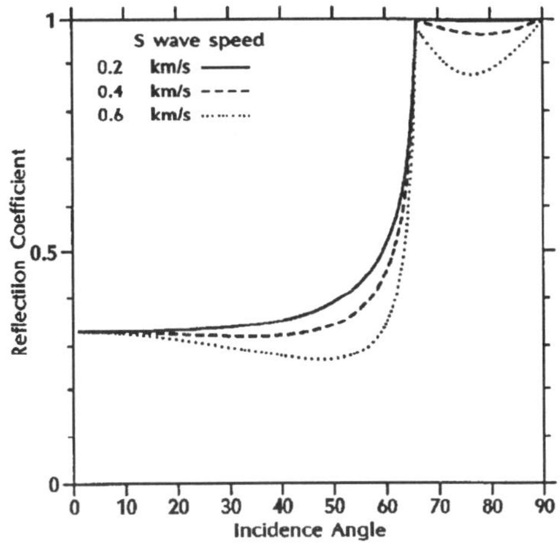


Figure 4. Dependence of the plane wave reflection coefficient on sediment S wave speed.

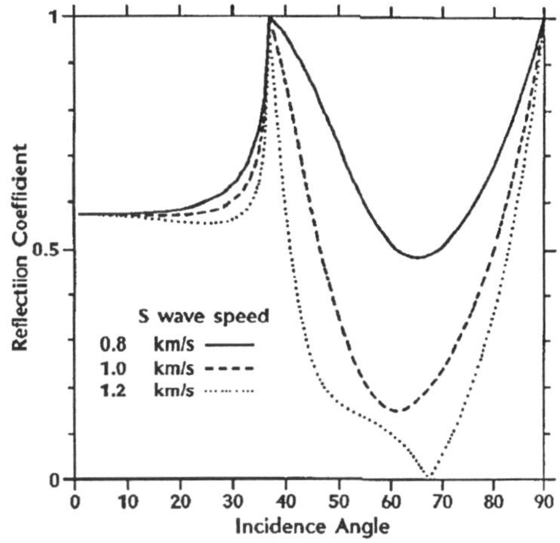


Figure 5. Dependence of the plane wave reflection coefficient on basalt S wave speed.

## ANALYSIS OF DATA

The inversion was first applied to data obtained in an experiment carried out in the Cascadia Basin off the west coast of British Columbia. The water depth was 2.5 km, and the sediment column at the site is at least 1 km of turbidite sand, silt-sand and silt-clay layers. The data were recorded with a vertical array of hydrophones at a depth of 430 m. Small explosive charges (0.8 kg) were deployed at nominal depths of 180 m out to a range of 20 km to sample the reflectivity into the postcritical region.

In order to estimate the amplitudes of pulses from explosive charges, the sampling rate of the data must be sufficiently high to include the full bandwidth of the shot waveform. Although the spectral peak of the charge was at about 50 Hz, there was appreciable energy (within 10-15 dB of the peak level) generated at frequencies out to 10 times this value. Consequently the original sampling rate of 1500 samples/s was increased by a factor of 4. The data were low-pass filtered at 120 Hz, and the amplitudes of the direct path and sea-floor reflected signals were determined at each range. Uncertainty in the measurements was reduced by averaging over three hydrophones at the bottom of the array. It is common practice to use spherical spreading in correcting the amplitudes for geometrical spreading, however, this is not a good assumption for the data near the critical angle, since the direct path is in a ray shadow at this range. In this work, a modified ray theory [14] was used with a measured sound speed profile to correct the signal amplitudes for propagation in the water, and also to estimate the angles of incidence. The results of the inversion are given in Table 1. The error bounds were determined from the error surface generated by the search over the trial values of the model parameters. The values of these parameters spanned a range between 1.50-1.75 km/s for $v_2$, and from 0.0-0.6 for the ratio $\beta_2/v_2$, consistent with the values expected for turbidite sediment near the continental rise [10]. The fit with the data is shown in Fig. 6, where the solid curve is calculated using the estimated values.

TABLE 1. Estimated values of the elastic properties of the bottom materials

| | Density (gm/cm$^3$) | P wave speed (km/s) | S wave speed (km/s) |
|---|---|---|---|
| Cascadia Basin | 1.41 + 0.1 | 1.58 + 0.02 | 0.21 + 0.1 |
| Juan de Fuca Ridge | 1.8 + 0.1 | 2.8 + 0.05 | 1.12 + 0.1 |

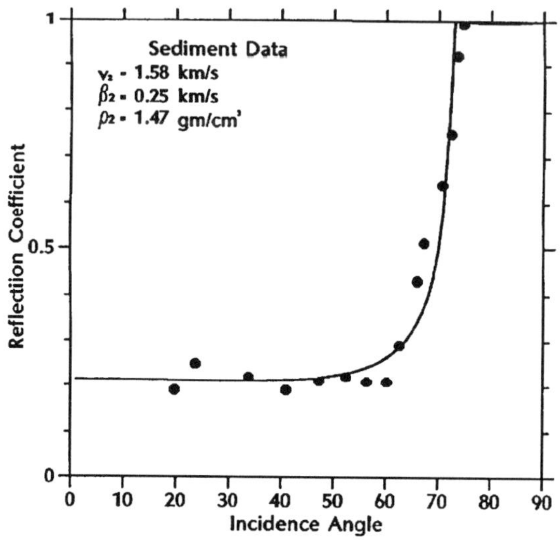


Figure 6. The measured data (•) and the model calculation using the estimated parameters for the sediment bottom.

The estimates obtained from the amplitude inversion compare very well with results from other experiments in this region. The P wave profile has been estimated by inversion of seismic travel time data [8] and by waveform modelling of bottom reflections [6], with values between 1.52-1.60 km/s within the first 15 m of the sediment column. A ground truth measure of the density was estimated from the analysis of a sediment core taken near the experiment site [15]. The composition of the first 2 m of sediment was primarily clay (65%) and silt (34%), for which the density is predicted in the range from 1.35-1.45 gm/cm$^3$ [10]. For marine sediments of this type the surficial S wave speed is 0.1-0.2 km/s, with an increase to 0.4-0.6 km/s within 200-300 m of the sea floor. The estimate reported here likely represents an average over a few wavelengths (around 30-50 m), so the value is somewhat higher than that obtained by Sen et al. [6] for the surficial sediments.

The second example is for data from a seismic reflection line obtained with an airgun and a streamer in 2.5 km water over young basalt on the Juan de Fuca Ridge. These data were first processed as common midpoint data, and then Warner's method [16] was used to calibrate the data acquisition system and determine the reflection coefficients. This involved first measuring the ratio of the sea floor reflection and the first water column multiple at a sedimented portion at the end of the line in order to determine the calibration constant, c. Then the reflection coefficient was estimated for the basalt data using the relationship

$$V = A_p c/d \tag{4}$$

where $A_p$ is the reflection amplitude and d is the geometric spreading correction. The results of the inversion are listed in Table 1, and the comparison of the model with the data is shown in Figure 7. Since there is information only to $32^0$, these estimates are strongly biased by the limited range of the data, so the value for P wave speed should be considered as an upper limit, and that for the S wave speed as a lower limit. Nevertheless, the estimates are very reasonable values for 0.5 ma basalt [11-13].

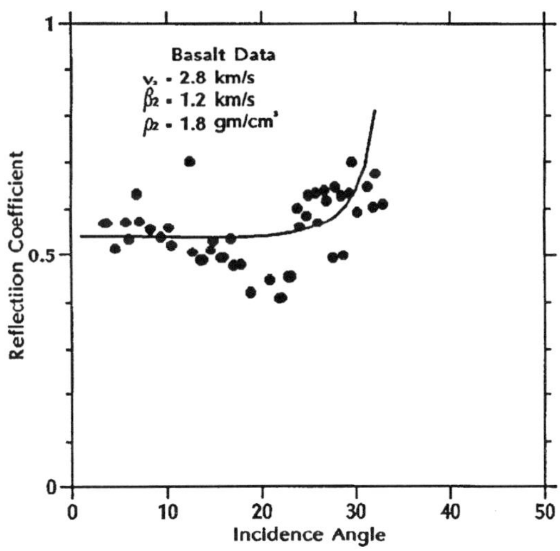


Figure 7. The measured data ( • ) and the model calculation using the estimated parameters for the basalt bottom.

## SUMMARY

A method has been described for estimating the elastic properties of surficial sea floor sediments, based on the inversion of reflection amplitude data. The technique is simple and direct, and can be used with data obtained in conventional seismic reflection experiments in deep or shallow water. The inversion was applied to bottom reflectivity measurements from a deep water basin and a mid-ocean ridge, and the results are in good agreement with archival seismic and geological information from the same region.

## REFERENCES

1. Aki, K., and Richards, P. (1980) *Quantitative Seismology*, W.H. Freeman and Co, San Francisco.

2. Thomson, D.J. (1987) 'Inversion of ocean subbottom reflection data', in *Progress in Underwater Acoustics,* H.M. Merklinger,(ed.), Plenum Press, New York, pp 271-278.
3. Bregman , N. D., Chapman, C. H. and Bailey, R.C. (1985) 'A noniterative procedure for inverting plane-wave reflection data at several angles of incidence using the Riccati equation', Geophys. Prospect. **33** 185-200.
4. Dicus, R.L. (1976) 'Preliminary investigation of the ocean bottom impulse response at low frequencies', Naval Oceanographic Office, Tech. Note 6130-4-76.
5. Berge, A.M., Drivenes, G., Kanestrom, R., and Beskow, B. (1986) "Acoustic modeling of the sea floor", Geophys. Prospect. **34**, 11-29.
6. Sen, M.K., Frazer, L.N., Chapman, N.R., and Mallick, S. (1988) 'Analysis of multipath sound propagation in the ocean near 49° N 128° W', J. Acoust. Soc.Amer.,**83**, 588-597.
7. Poley, D.F., Lawton, D.C. and Blasco, S.M. (1989) 'Amplitude-offset relationships over shallow velocity inversions', Geophysics **54**, 1114-1122.
8. Chapman, N.R., Stinson, K., Levy, S., Cabrera, J., and Oldenburg, D.W. (1988) 'Estimation of the elastic properties of sea floor sediments by inversion of precritical reflection data', IEEE J. Oceanic Eng. **OE-13**, 215-221.
9. Brekhovskikh, L.M. (1980)*Waves in Layered Media*, Academic Press, New York.
10. Hamilton, E.L. (1980) 'Geoacoustic modeling of the sea floor', J. Acoust. Soc. Amer. **68**, 1313-1340.
11. Purdy, G.M. (1987) 'New observations of the shallow seismic structure of young oceanic crust', J. Geophys. Res., **92**, 9351-9362.
12. Cudrak, C.F. (1988) 'Shallow crustal structure of the Endeavour Ridge segment from a detailed seismic refraction survey', Masters thesis, University of British Columbia.
13. Harding, A.J., Orcutt, J.A., Kappus, M.E., Vera, E.E., Mutter, J.C., Buhl, P., Detrick, R.S. and Brocher, T.M. (1989) 'Structure of young oceanic crust at 13°N on the East Pacific Rise from expanding spread profiles', J. Geophys Res., **94**, 12163-12196.
14. Weinberg, H. (1985) 'Generic sonar model', Naval Underwater Systems Centre, NUSC Technical Document 5971D.
15. Bornhold, B. (1989) Pacific Geoscience Centre, Sidney, B.C.
16. Warner, M. (1990) 'Absolute reflection coefficients from deep seismic reflections', Technophysics **173**,15-23.

# SURFACE WAVE INVERSION FOR SHEAR WAVE VELOCITY

R. B. HERRMANN and G. I. AL-EQABI
*Department of Earth and Atmospheric Sciences*
*Saint Louis University*
*3507 Laclede Avenue*
*St. Louis, Missouri 63103*
*U. S. A.*

ABSTRACT: Land based seismic exploration has typically viewed the surface wave or ground roll as undesirable noise while earthquake seismology has been able to use the surface wave to invert for lateral variations in crust-upper mantle shear-wave velocity. Although surface-wave inversion techniques for shear-wave velocity and shear-wave Q are easy to apply once the phase velocity, group velocity and anelastic attenuation dispersion are known, the manner in which these parameters are obtained efficiently from large exploration data sets has not been fully explored. This paper will address the processing steps required and apply them to surface-waves recorded in a refraction experiment in the State of Maine in the U. S.

## 1. Introduction

In earthquake seismology, surface waves have long been used to infer shear wave velocity models of the crust and upper mantle of the earth (Ewing *et al.*, 1957). Analysis techniques have been developed to invert for shear velocity as a function of depth, and recently progress has been made in using these data to invert for lateral variations in earth structure (Nolet, 1987; Nolet, 1990). On the other hand, less attention has been paid to surface waves in seismic exploration, where the surface wave is viewed as undesirable noise that affects the imaging of subsurface features.

The objective of the paper is to demonstrate the use of surface wave data acquired as part of a refraction survey for the determination of shallow shear velocity and Q. Even though surface-wave data are available in the distance range of 0 - 40 km, which is much larger than in typical seismic exploration data spreads, the need to process multichannel data is a shared feature. Analysis techniques are reviewed, and are actual data sets are processed.

## 2. Data Processing for Dispersion

Given multichannel data, dispersion information can be obtained by single channel as well as multichannel processing. Single channel processing can provide group velocity dispersion, while multichannel processing is required to obtain phase velocities without making assumptions about source phase or the number of wavelengths



*J. M. Hovem et al. (eds.), Shear Waves in Marine Sediments*, 545–556.

between the source and single receiver. The techniques of multiple filter analysis and phase velocity stacking are used to define the dispersion parameters.

### 2.1. MULTIPLE FILTER ANALYSIS

Multiple filter analysis was introduced by Dziewonski *et al.* (1969) for the determination of group velocity from a dispersed waveform. Herrmann (1973) studied the use of multiple filter analysis to estimate spectral amplitudes of the various modes. Let the Fourier transform of dispersed propagating wave at distance $r$ be $A(\omega,r)e^{-j(k(\omega)r-\phi)}$. Define a Gaussian filter as

$$H(\omega) = \begin{cases} e^{(-\alpha(\omega-\omega_c)^2/\omega_0^2)} & |\omega-\omega_0| \le \omega_c \\ 0 & |\omega-\omega_0| > \omega_c \end{cases} \tag{1}$$

where

$$\omega_c = \omega_0(\pi/\alpha)^{1/2}$$

The result of applying this filter to a dispersed signal is a filtered time signal whose envelope reaches a maximum at the group velocity arrival times. The envelope of the filtered time history for a single mode is approximately

$$g(t,r) = \frac{A(\omega_0,r)\omega_0}{2\pi}\left(\frac{\pi}{\alpha}\right)^{1/2} e^{(-\omega_0^2(t-r/U_0)^2/4\alpha)} \tag{2}$$

Computationally, the time series is filtered using (1), a Hilbert transform of the filtered time series is taken, and the two are combined to form the envelope. The envelope is searched for the relative maxima, from which the group velocity is computed from the time of the peak amplitude and the spectral amplitude is estimated from (2) by the relation $A(\omega_0,r) = g_{max}(2\pi/\omega_0)(\alpha/\pi)^{1/2}$.

For ease of analysis each two plots are made for each trace analyzed. The first presents the spectral amplitude versus frequency, annotating each with a symbol related to the value of the relative amplitude. The second plot provides contours of the envelope as a function of group velocity and frequency, with the group velocities of the envelope peaks indicated by the same symbol as the first plot. Examination of both plots assists in the identification of modes.

For a given shot, individual traces are processed using multiple filter analysis, and a table containing the frequency, group velocity and spectral amplitude is produced. Assuming a laterally homogeneous waveguide, the individual group velocity estimates of a given mode will be identical, and the decay of spectral amplitude with distance, $r$, will have a functional form

$$A(\omega,r) = A_0 e^{-\gamma r}\, r^{-1/2}. \tag{3}$$

An interactive program, which works on non-graphic computer displays, is used first to identify the modes on the basis of a group velocity histogram, and then to define the anelastic attenuation coefficient, $\gamma$, through a least squares fit of observed spectral amplitude of this mode according to the model (3). Provisions are made for editing extreme outliers. The result of this interactive program is a tabulation of group velocities and anelastic attenuation values together with their standard errors, for use with an inversion program.

### 2.2. GROUP VELOCITY STACKING

The above technique for estimating the average group velocity and anelastic attenuation over the profile requires running multiple filter analysis for each trace, consolidation of output into a large file, and then interactively editing the data set.

Barker (1988) introduced the idea of determining the group velocity by stacking the envelopes of different traces for a given frequency. This is possible if the envelopes are transformed from a function of time to a function of group velocity. The group velocity of the spread is obtained by searching for the maximum of the stack. If a program automatically selects the maximum, it can then return to the envelope of each trace to search for the maximum in the neighborhood of the velocities from the automatic selection so that error estimates of the group velocity can be made. In addition, spectral amplitudes can be used to estimate anelastic attenuation coefficients.

### 2.3. PHASE VELOCITY STACKING

McMechan and Yedlin (1981) introduced a technique for the determination of phase velocity by performing a $p-\tau$ stack followed by a transformation into the $p-\omega$ domain. Mokhtar *et al* (1988) described a procedure for accomplishing this by first Fourier transforming the time series, and then performing the stacking operation in the frequency domain. Let the Fourier spectrum of a seismic signal of the $i$'th station at distance $r_i$ be

$$A(\omega,r_i)e^{j\psi(\omega)_i}, \tag{4}$$

where the phase term $\psi(\omega)$ includes a distant dependent component for propagation. The $p$–$\omega$ stack of $N$ traces is defined by the relation

$$U(p,\omega)=\sum_{i=1}^{N} C(\omega)^{-1}A(\omega,r_i)e^{j\psi_i}e^{j\omega p r_i}. \tag{5}$$

where

$$C(\omega) = A(\omega,r_1)e^{j\psi_1}.$$

Division by $C(\omega)$ is a simple artifice to normalize the spectrum and remove the source phase if the signal is that of a single mode. The observed spectrum is assumed to be the superposition of $M$ surface-wave modes such that

$$A(\omega,r_i)e^{j\phi_i}=\sum_{k=1}^{M} S_k(\omega,r_i)e^{j[\phi_{0k}(\omega)-\omega p_{0k}(\omega)r_i]}, \tag{6}$$

where $S_k(\omega,r_i)$ is the amplitude spectrum of the $k$'th mode and the phase is separated into distance-independent, $\phi_{0k}(\omega)$, and distance-dependent, $\omega p_{0k}(\omega)r_i$ components. $p_{0k}$ is the inverse of the phase velocity of the $k$'th mode at frequency $\omega$.

If the signal consists primarily of a single noise free surface-wave mode at a given frequency, then the quantity $U(p,\omega)$ will have a maximum when $p = p_{0k}$. Alternatively it is convenient to search for the maximum of the function

$$|\, U(p,\omega)\, | \tag{7}$$

to find the dispersion curves since it is a real function. Since there are $N$ distances, the maximum value of the quantity $| U(p,\omega) |$ should be equal to $N$.

If this value is less than $N$, then we may attribute this to an error in the ray parameter between the stations and thus estimate the error in the corresponding phase velocity. If we assume that in the neighborhood of a stack maximum the quantity $\Delta p = p - p_{0k}$ has a normal distribution with zero mean and a variance $\sigma^2$, then the expected value of any term in (4) is

$$E[e^{-j2\pi f \Delta p \, r_i}] = e^{-2\sigma^2\pi^2 f^2 r_i^2} \tag{8}$$

where we used the definition $\omega = 2\pi f$.

Since each term in (7) is always positive, the expected value of the stack (5) of a single mode is just

$$\left| \sum_{i=1}^{N} e^{-2\sigma^2\pi^2 f^2 (r_i - r_1)^2} \right| \tag{9}$$

Given the stack value (7), a Newton-Raphson technique is used to find the value of $\sigma$ from (9) that corresponds to this value. The error in phase velocity is obtained using the definition $c = 1/p$ and the relation

$$\Delta c = \sigma c^2. \tag{10}$$

This relation was tested by numerically modeling a stacking operation in which the ray parameter error had the assumed distribution.

In the more realistic case of multimode surface waves, (7) will not yield a maximum independent of the amplitude spectrum of the other modes. Thus the stack value at a given frequency will typically be largest for one mode and smaller for others. The simplistic error analysis will yield larger, perhaps unwarranted, errors for the other modes. This is an inherent problem with this technique, that can only be resolved if a phase matched filter technique is first applied to each input spectra to isolate a single mode before the phase velocity stack is performed.

### 2.4. EVALUATION OF TECHNIQUES

All these techniques suffer from the need for manual intervention to correctly define the dispersion data, especially identification of the mode. The only real identification of the mode number is to count the zero crossings of the surface-wave eigenfunctions, with the fundamental mode Love and vertical component Rayleigh eigenfunctions having zero zero-crossings with depth. Eigenfunction-depth data are not available with surface sensors. An alternative technique uses the fact that the fundamental mode has the lowest phase, and usually group, velocities at any frequency. For surface sources, the fundamental mode is often the one with largest amplitudes.

The fact that the spectral amplitude level of the surface wave modes varies with frequency, the source excitation forces, and with distance due to anelastic attenuation, means that not necessarily the same mode is dominant at a given frequency. Because of this, the phase velocity stacking and multiple filter techniques will find it difficult to follow the dispersion curves of a low amplitude mode in the presence of a larger amplitude.

The group velocity stacking technique works reasonably well for estimating the group velocity value, even though the determination of the average group velocity should be weighted as a function of distance, since the group velocity resolution is less at shorter distances. The automatic determination of the anelastic attenuation coefficient, $\gamma$, also requires such a weighting. In addition, since this depends upon spectral amplitude estimates, the interference of modes at short distances will give erroneously spectral amplitudes, and thus overestimate the $\gamma$ values. In such a case, it is better to run the multiple filter analysis on each individual trace, and then interactively determine the values.

## 3. Surface-wave Inversion

Surface wave inversion is relatively straight forward once the dispersion is correctly defined. This requires a linearization of the problem and the subsequent iterative inversion of the observations.

### 3.1. LOVE-WAVE OR SH FORMULATION.

Following Keilis-Borok *et al.* (1989), the equation for the SH-wave eigenfunction, $V^{(3)}$, in a cylindrical coordinate system is

$$\frac{d}{dz}\begin{bmatrix} V^{(3)} \\ p_{\phi z} \end{bmatrix} = \begin{bmatrix} 0 & 1/\mu \\ \mu k^2 - \rho\omega^2 & 0 \end{bmatrix} \begin{bmatrix} V^{(3)} \\ p_{\phi z} \end{bmatrix}, \tag{11}$$

where $k$ is the wavenumber, $\omega$ is the angular frequency, $\mu$ is the rigidity and $\rho$ is density. The boundary conditions for surface waves are that $p_{\phi z}(z=0)=0$, and $V^{(3)}(z)\to 0$, as $z\to\infty$.

A discrete solution to the boundary conditions satisfies the equation

$$\begin{aligned} L &= \omega^2 I_0 - k^2 I_1 - I_2 \qquad (12)\\ &= \omega^2 \int_0^\infty \rho (V^{(3)})^2 dz - k^2 \int_0^\infty \mu (V^{(3)})^2 dz - \int_0^\infty \mu \left(\frac{dV^{(3)}}{dz}\right)^2 dz\,, \end{aligned}$$

where $L$ is the Lagrangian. The group velocity is determined from the relation

$$U = \frac{d\omega}{dk} = kI_1/\omega I_0\,. \tag{13}$$

If the medium is split into constant velocity-density layers, of thickness $d_m$, density $\rho_m$, and shear-wave velocity $\beta_m$, and the layer is bounded by $z=z_m-d_m$ and $z=z_m$, then the partials of the phase velocity with shear velocity for fixed frequency and density, and with density for fixed frequency and shear-wave velocity are

$$\left(\frac{\partial c}{\partial \beta}\right)_m = \frac{\beta_m \rho_m}{UI_0} \int_{z=z_m-d_m}^{z_m} \left[ (V^{(3)})^2 + \left(\frac{1}{k}\frac{dV^{(3)}}{dz}\right)^2 \right] dz \tag{14}$$

and

$$\left(\frac{\partial c}{\partial \rho}\right)_m = \frac{\beta_m}{2\rho_m}\left(\frac{\partial c}{\partial \beta}\right)_m - \frac{c^2}{2UI_0}\int_{z=z_m-d_m}^{z_m} (V^{(3)})^2 dz \ . \tag{15}$$

3.2. RAYLEIGH WAVE OR P-SV FORMULATION

In a cylindrical coordinate system, the equation of motion governing the eigenfunctions is

$$\frac{d}{dz}\begin{bmatrix} V^{(1)} \\ V^{(2)} \\ p_{zz} \\ p_{rz} \end{bmatrix} = \begin{bmatrix} 0 & k\lambda/(\lambda+2\mu) & 1/(\lambda+2\mu) & 0 \\ -k & 0 & 0 & 1/\mu \\ -\rho\omega^2 & 0 & 0 & k \\ 0 & -\rho\omega^2+4k^2\mu(\lambda+\mu)/(\lambda+2\mu) & -k\lambda/(\lambda+2\mu) & 0 \end{bmatrix}\begin{bmatrix} V^{(1)} \\ V^{(2)} \\ p_{zz} \\ p_{rz} \end{bmatrix}, \tag{16}$$

where the vertical component eigenfunction is $V^{(1)}$, and the radial component eigenfunction is $V^{(2)}$. The ellipticity at the free surface is $\epsilon = V^{(2)}(0)/V^{(1)}(0)$. The boundary conditions for surface waves are that $p_{rz}(z=0) = 0$, $p_{zz}(z=0) = 0$, $V^{(1)}(z) \to 0$, as $z \to \infty$. and $V^{(2)}(z) \to 0$, as $z \to \infty$.

A discrete solution to the boundary conditions satisfies the equation

$$L = \omega^2 I_0 - k^2 I_1 - 2kI_2 - I_3 \ , \tag{17}$$

where $L$ is the Lagrangian, and

$$I_0 = \int_0^\infty \rho \left[(V^{(1)})^2 + (V^{(2)})^2\right] dz \tag{18}$$

$$I_1 = \int_0^\infty \left[\mu (V^{(1)})^2 + (\lambda+2\mu)(V^{(2)})^2\right] dz \tag{19}$$

$$I_2 = \int_0^\infty \left[\mu V^{(1)} \frac{dV^{(2)}}{dz} - \lambda V^{(2)} \frac{dV^{(1)}}{dz}\right] dz \tag{20}$$

$$I_3 = \int_0^\infty \left[(\lambda+2\mu)\left(\frac{dV^{(1)}}{dz}\right)^2 + (\mu)\left(\frac{dV^{(2)}}{dz}\right)^2\right] dz \ . \tag{21}$$

The group velocity is determined from the relation

$$U = \frac{d\omega}{dk} = (kI_1 + I_2)/\omega I_0 \ . \tag{22}$$

If the medium is split into constant velocity-density layers, of thickness $d_m$, density $\rho_m$, compressional-wave velocity $\alpha_m$, and shear-wave velocity $\beta_m$, and the layer is bounded by $z = z_m - d_m$ and $z = z_m$, then the partials of the phase velocity with shear velocity for fixed frequency, compressional velocity and density, with compressional velocity for fixed frequency, shear velocity and density, and with density for fixed frequency and compressional and shear velocity are

$$\left(\frac{\partial c}{\partial \alpha}\right)_m = \left(\frac{\alpha_m \rho_m}{UI_0}\right) \int_{z=z_m-d_m}^{z_m} \left[V^{(2)} - \frac{1}{k}\frac{dV^{(1)}}{dz}\right]^2 dz \tag{23}$$

$$\left(\frac{\partial c}{\partial \beta}\right)_m = \left(\frac{\beta_m \rho_m}{UI_0}\right) \int_{z=z_m-d_m}^{z_m} \left[\left(V^{(1)} + \frac{1}{k}\frac{dV^{(2)}}{dz}\right)^2 + \frac{4}{k}V^{(2)}\frac{dV^{(1)}}{dz}\right] dz \tag{24}$$

$$\left(\frac{\partial c}{\partial \rho}\right)_m = \frac{1}{2\rho}\left[\alpha\left(\frac{\partial c}{\partial \alpha}\right)_m + \beta\left(\frac{\partial c}{\partial \beta}\right)_m\right] - \frac{c^2}{2UI_0}\int_{z=z_m-d_m}^{z_m}\left[(V^{(1)})^2 + (V^{(2)})^2\right]dz \; . \tag{25}$$

If the material properties within each layer are constant, propagator techniques can be used to determine the dispersion and eigenfunctions and the integrals can be analytically evaluated (Harkrider, 1979; Wang, 1981).

### 3.3. NUMERICAL PARTIAL DERIVATIVE COMPUTATION AND CAUSALITY

The partial of group velocity with respect to a layer parameter $v$ is

$$\frac{\partial U_o}{\partial v} = \frac{U_o}{c_o}\frac{\partial c_o}{\partial v} + \frac{U_o^2}{c_o^2}\left(\frac{T}{c}\frac{dc_o}{dt}\frac{\partial c_o}{\partial v} - T\frac{\partial}{\partial T}\frac{\partial c_o}{\partial v}\right) \tag{26}$$

The phase velocity partial with respect to period is *numerically* computed using the relation

$$\frac{\partial}{\partial T}\frac{\partial c_o}{\partial v} = T\left[\left(\frac{\partial c_o}{\partial v}\right)_{T+hT} - \left(\frac{\partial c_o}{\partial v}\right)_{T-hT} / 2hT\right]$$

In these expressions, the parameter $v$ can take on the values $\alpha$, $\beta$ or $\rho$, and the subscript $o$ represents the value in the parameter in the purely elastic model.

If a causal Q is introduced, a Futterman causality tied to a reference angular frequency $\omega_r$, the elastic parameters will equal the anelastic values when the angular frequency $\omega$ equals $\omega_r$. The causal phase velocity $c$ is given by

$$c = c_o + \frac{1}{\pi}\ln\left(\frac{\omega}{\omega_r}\right)\sum\left(\frac{\partial c_o}{\partial \beta}\beta Q_\beta^{-1} + \frac{\partial c_o}{\partial \alpha}\alpha Q_\alpha^{-1}\right) \tag{27}$$

and the value of the spatial anelastic attenuation factor, $\gamma$, is

$$\gamma = \frac{\omega}{2c_o^2}\sum\left(\frac{\partial c_o}{\partial \beta}\beta Q_\beta^{-1} + \frac{\partial c_o}{\partial \alpha}\alpha Q_\alpha^{-1}\right) \tag{28}$$

In order to obtain the correct partial derivatives of causal phase velocity and gamma with respect to the model parameters, the following are used. A partial or a parameter without the subscript $o$ represents the causal value. Note that it is assumed that $Q_\beta/Q_\alpha$ is a fixed ratio, hence a partial with respect to $Q_\beta^{-1}$ may involve a partial of $c_o$ with respect to $\alpha$.

$$\frac{\partial c}{\partial v} = \frac{\partial c_o}{\partial v}\left(1 + \frac{1}{\pi Q_v}\ln\left(\frac{\omega}{\omega_r}\right)\right) \tag{29}$$

$$\frac{\partial c}{\partial Q_\beta^{-1}} = \frac{1}{\pi}\ln(\frac{\omega}{\omega_r})\left(\frac{\partial c_o}{\partial \beta}\beta + \frac{\partial c_o}{\partial \alpha}\alpha\frac{Q_\beta}{Q_\alpha}\right) \tag{30}$$

$$\frac{\partial \gamma}{\partial v} = \frac{\omega}{2c_o^2}\frac{\partial c_o}{\partial v}Q_\beta^{-1} - 2\frac{\gamma}{c_o}\frac{\partial c_o}{\partial v} \tag{31}$$

$$\frac{\partial \gamma}{\partial Q_\beta^{-1}} = \frac{\omega}{2c_o^2}\left(\frac{\partial c_o}{\partial \beta}\beta + \frac{\partial c_o}{\partial \alpha}\alpha\frac{Q_\beta}{Q_\alpha}\right) \tag{32}$$

$$U = U_o\left[1 + (2-\frac{U_o}{c_o})\left(\frac{c-c_o}{c_o}\right) + \frac{2\gamma U_o}{\pi\omega}\right]. \tag{33}$$

where it is assumed in the derivation that $\left(\frac{c-c_o}{c_o}\right)^2$ is negligible.

$$\frac{\partial U}{\partial v} = \frac{\partial U_o}{\partial v}\left[\frac{U}{U_o} - \frac{U_o}{c_o}\left(\frac{c-c_o}{c_o}\right) + \frac{2\gamma U_o}{\pi\omega}\right] \tag{34}$$

$$+ \frac{\partial c_o}{\partial v}\frac{U_o}{c_o}( -2\frac{2\gamma U_o}{\pi\omega} + \frac{U_o}{c_o}\frac{1}{\pi Q_v}$$

$$+ (2-\frac{U_o}{c_o})(\frac{1}{\pi Q_v}\ln(\frac{\omega}{\omega_r}) - \left(\frac{c-c_o}{c_o}\right)))$$

$$+ \frac{U_o^2}{c_o^2}\left(\frac{c-c_o}{c_o}\right)\frac{\partial c_o}{\partial v}$$

$$\frac{\partial U}{\partial Q_\beta^{-1}} = \frac{U_o}{c_o}(2-\frac{U_o}{c_o})\frac{\partial c}{\partial Q_\beta^{-1}} \tag{35}$$

$$+ \frac{1}{\pi}\frac{U_o^2}{c_o^2}(\frac{\partial c_o}{\partial \beta}\beta + \frac{\partial c_o}{\partial \alpha}\alpha\frac{Q_\beta}{Q_\alpha})$$

An iterative linear inversion is performed because of the non-linear nature of the problem. At any stage there is a current model which is used to predict the observations and there is also a lack of fit. For simplicity, the compressional wave velocity $\alpha$ is not directly inverted, rather it is tied to the shear velocity by an *a priori* assumption, such as a constraint that constant Poisson's ratio be constant. We may thus express the difference between observed and predicted values as a linear function of changes in shear velocity model and changes in the inverse Q model. Since the effect of compressional wave $Q_\alpha$ may not be negligible, the compressional wave $Q_\alpha$ is related to the shear wave $Q_\beta$ by a ratio $Q_\alpha/Q_\beta$ for the layers. Given these assumptions, the inversion can take on two appearances, non-causal and causal, depending upon which partial derivatives are used. To keep the notation general, the partial derivatives with respect to layer velocity in the next three equations can either be causal or non-causal.

The differences in observed and predicted phase velocities are modeled as

$$c_{obs} - c_{pred} = \frac{\partial c}{\partial \beta_1}\Delta\beta_1 + \cdots + \frac{\partial c}{\partial \beta_n}\Delta\beta_n + \frac{\partial c}{\partial Q_{\beta_1}^{-1}}\Delta Q_{\beta_1}^{-1} + \cdots + \frac{\partial c}{\partial Q_{\beta_n}^{-1}}\Delta Q_{\beta_n}^{-1} , \qquad (36)$$

the difference between observed and predicted group velocities is given by

$$U_{obs} - U_{pred} = \frac{\partial U}{\partial \beta_1}\Delta\beta_1 + \cdots + \frac{\partial U}{\partial \beta_n}\Delta\beta_n + \frac{\partial U}{\partial Q_{\beta_1}^{-1}}\Delta Q_{\beta_1}^{-1} + \cdots + \frac{\partial U}{\partial Q_{\beta_n}^{-1}}\Delta Q_{\beta_n}^{-1} , \qquad (37)$$

and the difference between observed and gamma values predicted by the current model is

$$\gamma_{obs} - \gamma_{pred} = \frac{\partial \gamma}{\partial \beta_1}\Delta\beta_1 + \cdots + \frac{\partial \gamma}{\partial \beta_n}\Delta\beta_n + \frac{\partial \gamma}{\partial Q_{\beta_1}^{-1}}\Delta Q_{\beta_1}^{-1} + \cdots + \frac{\partial \gamma}{\partial Q_{\beta_n}^{-1}}\Delta Q_{\beta_n}^{-1} . \qquad (38)$$

Once set up, the system of linear equations is solved using singular value decomposition (Russell, 1987). Provisions are made to damp the solutions as a mean of controlling convergence. Inversion can be performed in an unweighted manner, or can be perform using *a priori* weights available from the error estimates from the dispersion analysis. In addition, a differential inverse can be use to solve for the change in inversion parameter across a layer boundary, rather than for the parameter itself. This constraint guarantees a smooth model and avoids unrealistic alternating low velocity layers in the resulting model.

## 4. Maine Refraction Experiment

In 1984, the U. S. Geological Survey ran a series of refraction experiments in the state of Maine (Murphy and Luetgert, 1986; Murphy and Luetgert, 1987). The refraction profiles went both along (NE) and cross (NW) the strike of the Appalachian Mountains. Surface geology varied significantly in lithology and geologic age. Because of the excellent data set, both in spatial sampling and calibrated instruments, this data set was analyzed to learn about shallow shear-wave velocity and Q.

Figure 1 shows the data from one shot point, SP1 with distance increasing in a southeast direction. Fundamental mode phase velocity and spatial attenuation values were inverted, with the resulting earth model given in Table 1. To avoid any bias of the initial model upon the inverted model, the inversion started with a twelve layered model with velocities and densities of each layer equal to that of the halfspace. The inferred velocities in the lower part of the model are not well resolved. The high values are not unexpected given the highly metamorphosed geology sampled by the surface waves.

Figure 2 presents the fundamental mode synthetic seismograms for an explosion source in this model. The arrival times and waveforms compare well to the observed data of Figure 1. As a check on the correctness of the inversion, both the synthetic and observed data were filtered through the same narrow band band-pass filters, and the peak amplitudes plotted as a function of distance. This exercise showed that the derived $Q_\beta$ model explained the observed decrease of amplitude with distance.

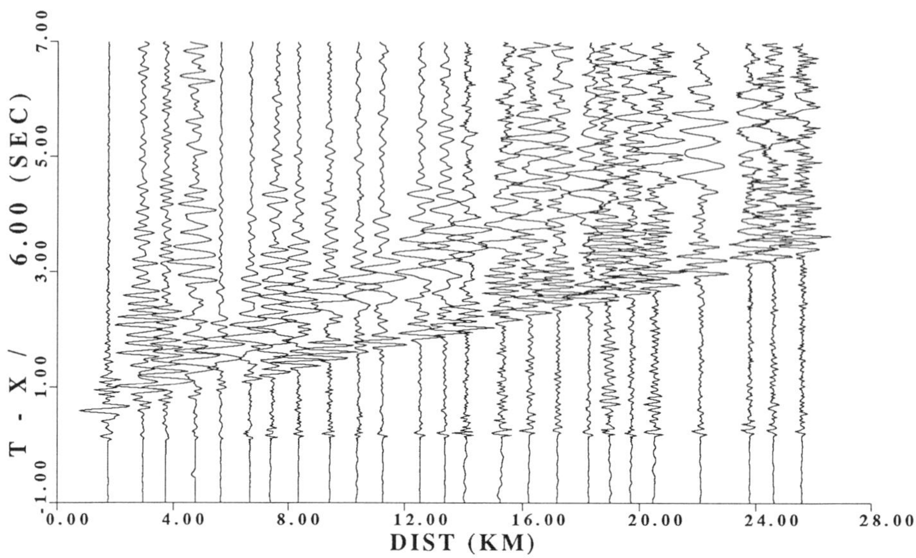


Figure 1. Data for SP1, plotted using a reduction velocity of 6.0 km/s. The data have been low pass filtered between 0.5 and 5.0 Hz.

TABLE 1. Velocity Model for SP1

| $H$ (km) | $\alpha$ (km/s) | $\beta$ (km/s) | $\rho$ (g/cm$^3$) | $Q_\alpha$ | $Q_\beta$ |
|---|---|---|---|---|---|
| 0.25 | 4.45 | 2.57 | 2.37 | 64 | 32 |
| 0.25 | 5.61 | 3.24 | 2.62 | 64 | 32 |
| 0.25 | 5.85 | 3.38 | 2.67 | 64 | 32 |
| 0.25 | 6.15 | 3.55 | 2.74 | 70 | 35 |
| 0.25 | 6.30 | 3.64 | 2.79 | 74 | 37 |
| 0.25 | 6.40 | 3.70 | 2.82 | 84 | 42 |
| 0.25 | 6.41 | 3.70 | 2.82 | 96 | 48 |
| 0.25 | 6.44 | 3.72 | 2.83 | 114 | 57 |
| 0.50 | 6.45 | 3.72 | 2.83 | 142 | 71 |
| 0.50 | 6.49 | 3.75 | 2.85 | 190 | 95 |
| 0.50 | 6.50 | 3.76 | 2.85 | 284 | 142 |
| | 6.65 | 3.84 | 2.89 | 568 | 284 |

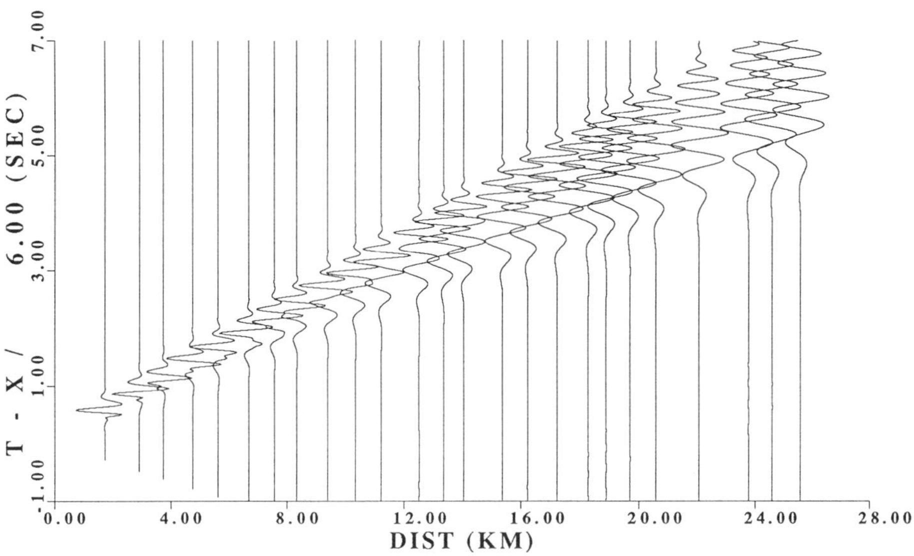


Figure 2. Synthetic traces for SP1, plotted using a reduction velocity of 6.0 km/s. The data have been low pass filtered between 0.5 and 5.0 Hz and also passed through a 2 Hz seismometer for proper comparison with Figure 1.

## 5. Future Directions

Experience with limited data sets indicate certain rules for estimating the depth limits of surface-wave inversion. Since longer periods or longer wavelength surface waves penetrate deeper, and since longer periods are better resolved at larger distances, two such rules are 1) the surface wave provides resolvable information to depths on the order of 1/3 – 1 $\lambda$, where $\lambda$ is the longest observed wavelength, and 2) the maximum depth is on the order of 1/20 – 1/10 $r_{max}$, where $r_{max}$ is the of the maximum distance of observation. This relatively limited depth resolution, compared to refraction analysis, is mitigated by the ability to resolve changes in the surface-wave waveguide within a given shot spread.

The data sets analyzed and the inversion programs developed both considered land based data. Introduction of a water layer is simple, and is already encompassed in the theory presented. There is no difficulty in using surface-wave inversion techniques for data acquired in shallow marine sediments.

## 6. Acknowledgments

The assistance of W. Kohler and W. Mooney of the USGS in providing the data set is appreciated. This work was sponsored in part by the U. S. Geological Survey under Grant 14-08-0001-G1922, and by the Air Force Geophysics Lab under Contract F1962090K0040.

## 7. References

Barker, T. (1988). Array processing of Rayleigh waves for shallow shear wave velocity structure (abs), *Seismological Research Letters* **59**, 12.

Dziewonski, A. M., S. Bloch, and M. Landisman (1969). A technique for the analysis of transient seismic signals, *Bull. Seism. Soc. Am.* **59**, 427-444.

Ewing, W. M., W. S. Jardetzky, and F. Press (1957). *Elastic Waves in Layered Media*, McGraw Hill, Inc., New York.

Harkrider, D. G. (1979). Eigenfunction normalization for energy calculations in surface wave codes (abs), Earthquake Notes **50**, No. 4, 28.

Herrmann, R. B. (1973). Some aspects of band-pass filtering of surface waves, *Bull. Seism. Soc. Am.* **63**, 663-671.

Keilis-Borok, V. I., A. L. Levshin, T. B. Yanovskaya, A. V. Lander, B. G. Bukchin, M. P. Barmin, L. I. Ratnikova, and E. N. Its (1989). *Seismic surface waves in a laterally inhomogeneous earth*, Kluwer Academic Publishers, Dordrecht.

McMechan, G. A. and M. J. Yedlin (1981). Analysis of dispersive waves by wave field transformation, *Geophysics* **46**, 869-874.

Mokhtar, T. A., R. B. Herrmann and D. R. Russell (1988). Seismic velocity and Q model for the shallow structure of the Arabian shield from short-period Rayleigh waves, *Geophysics* **53**, 1379-1387.

Murphy, J. M. and J. H. Luetgert (1986). Data report for the Maine-Quebec cross-strike seismic-refraction profile, *U. S. Geological Survey Open-File Report 86-47*, Menlo Park, CA.

Murphy, J. M. and J. H. Luetgert (1987). Data report for the Maine along-strike seismic-refraction profiles, *U. S. Geological Survey Open-File Report 87-133*, Menlo Park, CA.

Nolet, G. (1987). Waveform Tomography, in *Seismic Tomography*, edited by G. Nolet, pp 301-322, D. Reidel, Hingham, Mass.

Nolet, G. (1990). Partitioned waveform inversion and two-dimensional structure under the network of autonomously recording seismographs, *J. Geophys. Res.* **95**, 8499-8512.

Russell, D. R. (1987). Multichannel processing of dispersed surface waves, *Ph. D. Dissertation*, Saint Louis University.

Wang, C. Y. (1981). Wave theory for seismogram synthesis, *Ph. D. Dissertation*, Saint Louis University, St. Louis, Missouri.

# DETERMINATION OF SHEAR VELOCITY PROFILES BY INVERSION OF INTERFACE WAVE DATA

A. CAITI, T. AKAL and R.D. STOLL(§)
*SACLANT Undersea Research Centre*
*19026 La Spezia, Italy*
*(§) Lamont-Doherty Geological Observatory of Columbia University*
*Palisades, New York 10964, USA*

ABSTRACT. An inversion scheme for estimation of marine sediment shear velocity and attenuation from interface wave dispersion curves is presented. A model of shear velocity versus depth in the upper part of the sediment that gives a best fit to phase and group velocity dispersion curves in a least square sense is obtained through an automatic inversion procedure. This procedure is based on a Thomson-Haskell integration without taking into account compressional and shear attenuation. The inversion is regularized with smoothness constraints on the velocity versus depth curve to avoid numerical instability. The shear wave velocity profile is used as input to SAFARI, a full wave field forward model, to interactively estimate shear attenuation and to obtain synthetic seismograms for data-model comparison

## 1. Introduction

The measurement of shear wave velocity and attenuation in seafloor sediment has always been considered a difficult and challenging task, and relatively few data have been collected [1]. Direct measurements, both *in situ* and in laboratory, can provide data only locally and for a limited depth in the sediment [2, 3]. The limitation of these techniques, for studying large areas or making extensive surveys, suggests the importance of utilizing seismic data inverted in an efficient manner to obtain seismo-acoustic properties of the bottom. In this paper we present an inversion procedure applied to data taken during a joint experiment by Lamont-Doherty Geological Observatory (LDGO) of Columbia University and Saclant Undersea Research Centre (Saclantcen) at one of the Saclantcen test sites in the Ligurian Sea. In this experiment the propagation of seismic interface waves at the water sediment interface was studied and a model of the shear wave velocity in the sediment was determined that gives a best fit to phase and group velocity dispersion curves. This model was subsequently "tuned" using the SAFARI full wave field forward model interactively to estimate shear wave attenuation and to compare measured and simulated time responses. In this paper the experimental techniques are described and the inversion method is discussed in detail.



*J. M. Hovem et al. (eds.), Shear Waves in Marine Sediments*, 557–565.

## 2. Experiment

The joint experiments took place in October 1989 at two Saclantcen test sites in the Gulf of La Spezia, Italy. These sites were selected so that the results of the experiments and the inversion could be critically evaluated by comparison to direct *in situ* shear wave velocity measurements.

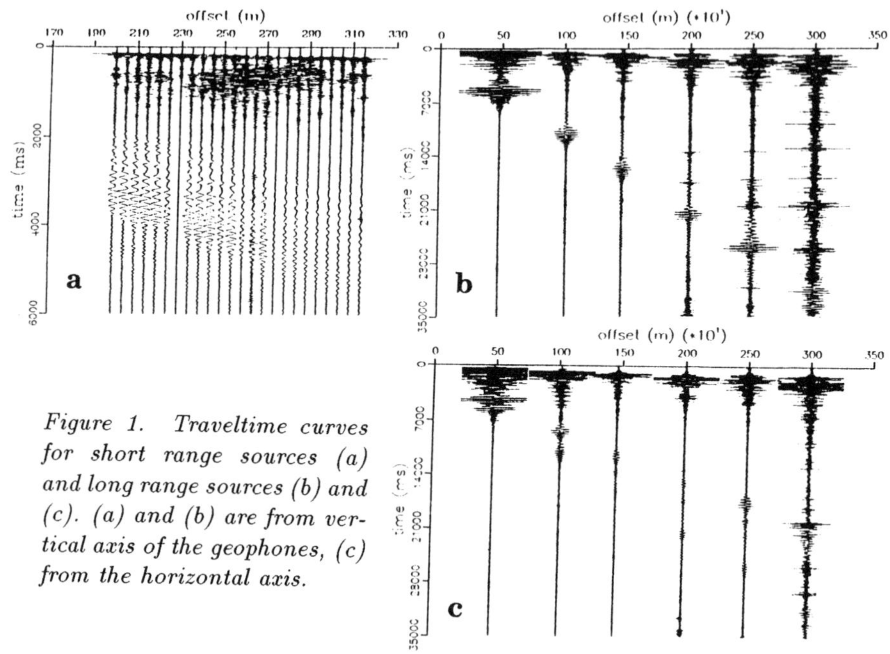


*Figure 1. Traveltime curves for short range sources (a) and long range sources (b) and (c). (a) and (b) are from vertical axis of the geophones, (c) from the horizontal axis.*

Data were acquired using two different techniques. One configuration consisted of a 24-geophone array, with gimballed geophones spaced 5m apart. The geophone array was lowered from the ship's stern and dragged a short distance to insure as straight a line as possible. After each shot the cable was pulled in approximately 100m to obtain a complete range stacked display of traveltime curves covering about 350m of distance. The source used in these experiments was an eletrically detonated seismic shell loaded into a specially designed gun that was placed on the seafloor. In the second technique three triaxial geophones were clamped to the sides of a 6 meters deep borehole. The source was an electrically detonated TNT charge fired at or close to the bottom at ranges up to 4000m. A more detailed description of the experimental approaches is given in this volume [4]. Using these techniques, it was possible to obtain the traveltime curves shown in Fig. 1. The presence of strong interface waves is clearly evident in all three examples. Signals from both the down hole and the bottom array geophones were recorded simultaneously in order to observe variation in signals with both depth and range.

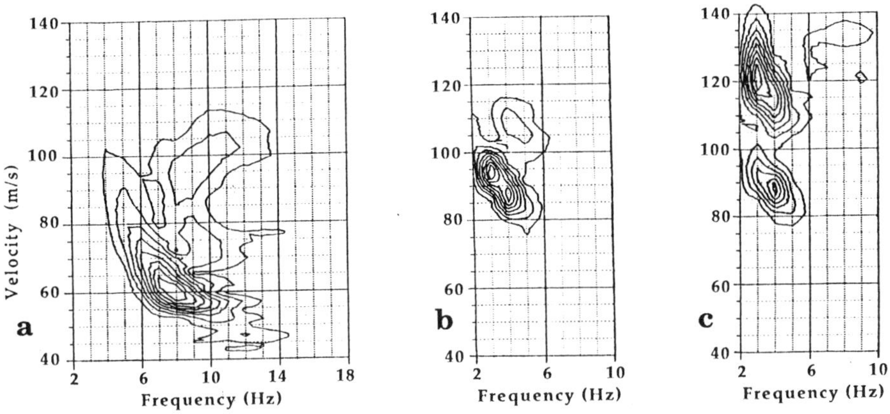


*Figure 2. Dispersion analysis of interface waves presented as Gabor contours. (a): vertical component and short range source; (b): vertical component and long range source; (c): horizontal component and long range source; (d): combined Gabor matrix including data from (a), (b) and (c).*

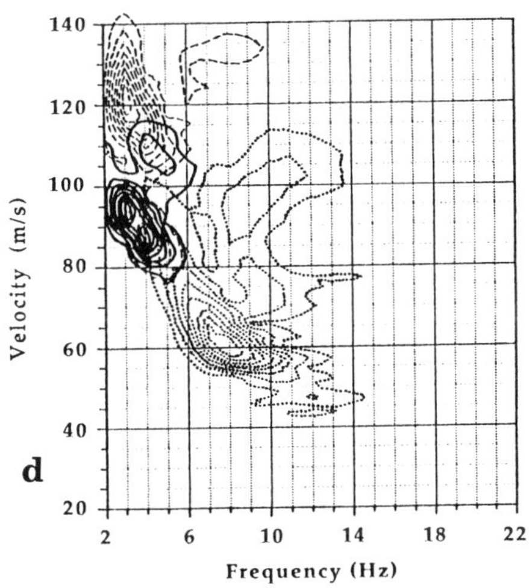


Multiple filter analysis was used to obtain a Gabor diagram showing group velocity dispersion [5]. Group velocity dispersion curves were obtained by locating the maximum values (ridge crest) in the resulting contour diagrams (Fig. 2). The long range signals have a significative frequency content in the range of 2 to 8 Hz, whereas in the short range signals the frequency range is about 5-15 Hz. Simultaneous use of both data sets allows therefore analysis over the wider frequency range of 2 to 15 Hz. The fundamental mode is more evident in the vertical axis Gabor diagrams (Figs. 2a, 2b); whereas the first shearing mode is more evident in the data collected from the horizontal axis of the long range source. By superimposing the dispersion diagrams from both the long range and short range data we are able to construct a composite group curve (Fig. 2d), which permits both better resolution in the near bottom portion of the model and deeper overall penetration.

### 3. Data inversion

A method for estimating shear velocity as a function of depth in the sediment has to specify: a seismo-acoustic model of the seafloor; a measure by which we can critically compare our data and the estimates produced by the model; some *a priori* information about the solution of the problem. We assume sediment can be modelled by $n$ horizontally stratified layers each characterized by a layer thickness, $(h_i)$, p-wave velocity, $(\alpha_i)$, s-wave velocity, $(\beta_i)$, and density, $(\rho_i)$, where the first layer is the water column. From this geoacoustic model we compute theoretical dispersion curves for interface waves propagating at the water-sediment boundary with the Thomson-Haskell recursive method [6, 7]. In sediments where the p-wave velocity is much higher than the s-wave velocity, dispersion curves for interface waves are fairly insensitive to density and p-wave velocity parameters. These geoacoustic values may be treated as "passive" parameters [8] which are assumed to be constant during inversion. As a result, we have a relation between the group velocity of the interface wave, $(c)$, and the shearing properties of the subbottom model in the form of:

$$c(k, \omega, \underline{\beta}) = f(k, \omega, \beta_1, \cdots, \beta_i, \cdots, \beta_n) \tag{1}$$

where $k$ is the mode number, $k = 0, 1, \cdots$, and $\omega$ is the angular frequency. The mode number can be considered fixed since the inversion is carried out one mode at a time.

In order to compare the measured dispersion with the estimate produced by the model, we define a cost function, $(J)$, as:

$$J = \| d(k, \omega) - c(k, \omega, \underline{\beta}) \|^2 \tag{2}$$

where $d(k, \omega)$ is the measured dispersion and the norm is the $l_2$ norm. We need to determine the vector $\underline{\beta}$ that minimizes $J$ (i.e. the best fit of the dispersion data in a least square sense); however, the inverse problem posed in this way is unstable [9], because "small" variations in the measured data cause "large" changes in the solution. Following Thikonov's regularization theory, we modify the cost function $J$ by inserting an additional term that takes into account *a priori* information on the solution. The modified cost function $J^*$ and the solution $\underline{\beta}^*$ are then defined as:

$$\underline{\beta}^* = min_{\underline{\beta}} J^* = min_{\underline{\beta}}(\| d(k, \omega) - c(k, \omega, \underline{\beta} \|^2 + \lambda \| H\underline{\beta} \|^2) \tag{3}$$

where $H$ is an operator (possibly linear or affine) through wich we can constrain the solution. $H$ must be such that $H^T H$ is positive definite, so that there always exists an unique solution of the minimum problem for an appropriate choice of $\lambda$. Lambda $(\lambda)$ is the Lagrange multiplier that allows a trade-off between pure data fitting and complete *a priori* solution. The user can specify different values of $\lambda$ for each layer shear velocity, in order to vary the strenght of the constraint from one layer to another. The constraint operators we have chosen are first, second and third derivatives of the shear velocity vs. depth curve. This kind of constraint has the physical meaning of avoiding rapid variations and unrealistic oscillations of the shear velocity behaviour with depth.

The solution of the minimum problem is found with the classical iterative Newton method. First we discretize the frequency space on a set of $m + 1$ samples, where $m$ is chosen by the user, so that:

$$\begin{aligned} \omega_j &= \omega_0 + \frac{j}{m}(\omega_m - \omega_0) \\ c_j &= c(\omega_j) \quad j = 0, 1, \cdots, m \end{aligned} \tag{4}$$

We linearize Eq. (1) by writing:

$$\Delta c_k = \sum_{i=1}^{n} \frac{\partial c(\omega)}{\partial \beta_i} |_{c=c_k} \Delta \beta_i \tag{5}$$

which in matrix form becomes:

$$T \Delta \underline{\beta} = \Delta \underline{c} \tag{6}$$

where $T$ is the jacobian of the function in Eq. (1) with respect to the vector $\underline{\beta}$. The solution of the linearized minimum problem is:

$$\Delta \underline{\beta} = (T^T T + \lambda H^T H)^{-1} T^T \Delta \underline{c} \tag{7}$$

Starting from an initial estimate for the shear wave velocity vector $(\underline{\beta})$, we iteratively solve Eq. (7) until the distance between the simulated and the measured dispersion is below some specified threshold value or until the difference between two consecutive updates of $\underline{\beta}$ is negligible for all components of $\underline{\beta}$. This method clearly suffers from the limit of deterministic non-linear minimization, because of the possibility of reaching a local minimum of the cost function. The inversion must therefore be repeated with different initial estimates of $\underline{\beta}$ to determine if the solution is a true global minimum.

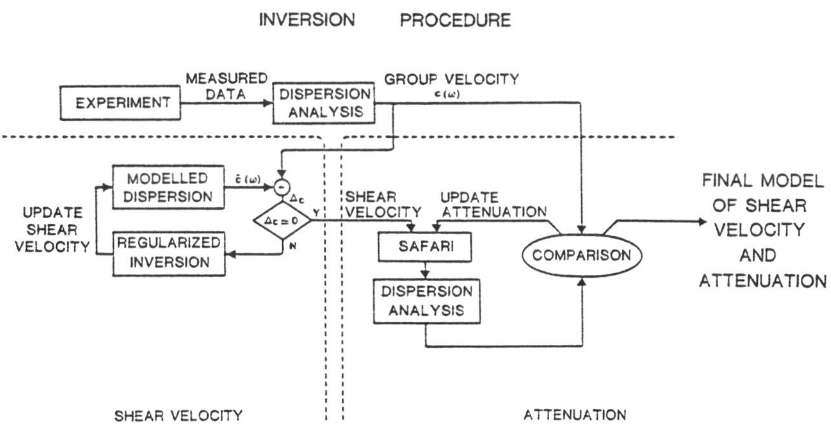


*Figure 3. Block diagram of the inversion process.*

The shear velocity versus depth profiles are used to calculate synthetic seismograms using SAFARI [10]. The synthetic seismograms are compared directly in time with the measured data and are used to calculate dispersion curves that can be compared with the dispersion curves obtained from experimental data. This provides a check to insure that local minima have not been reached, and allows "fine tuning" of the model to determine best fit to data. Moreover, since shear wave attenuation can be included as a parameter in SAFARI, an estimate of the shear wave attenuation with depth can be made by adjusting the model to produce the proper amplitude in the synthetic signals. The overall inversion process is summarized in Fig. 3.

The sensitivity of the solution is defined as the percentage variation $\| \Delta\underline{\beta} \| / \| \underline{\beta} \|$ required to obtain a prescribed variation $\| \Delta\underline{c} \| / \| \underline{c} \|$ in the simulated data. Dealing with a non linear minimization, it is not possible to give a general estimate of sensitivity. For the cases studied in this work, we have found empirically that variations up to 8-10% in $\underline{\beta}$ lead to variations of 2-3% in the dispersion curves.

## 4. Inversion results

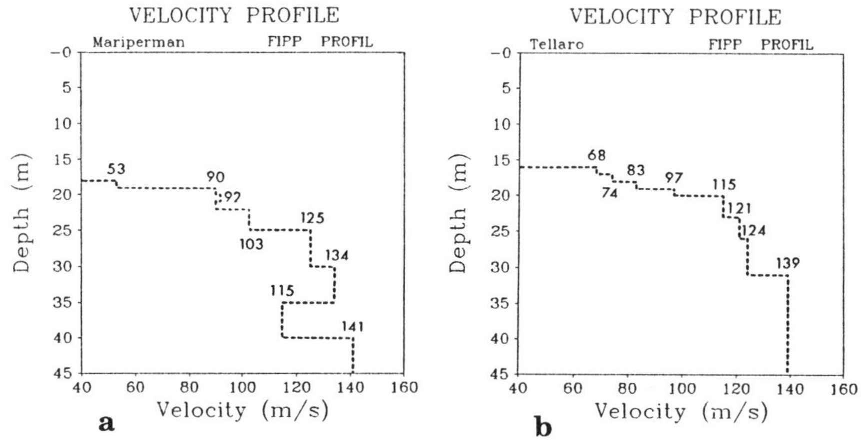


*Figure 4. Shear velocity vs. depth profile for the sites "Mariperman" (a) and "Tellaro" (b).*

Results of the inversion procedure for two test-sites (Fig. 4) were used to calculate the dispersion curves presented in Fig. 5. Modelled data are in good agreement with measured data for the fundamental mode, but there is some discrepancy in the low frequency part of the first mode. Long range synthetic seismograms calculated with SAFARI are similar to corresponding measured signals for both vertical and radial particle motion (Fig. 6). These synthetic seismograms were calculated using the attenuation versus depth profile presented in Fig. 7.

Direct measurements of shear wave velocity have been reported for the Mariperman and Tellaro sites by Richardson [3] and for the Mariperman site by Muir [4]. Richardson, who utilized direct measurements, reported shear velocity gradients for the upper 2 meters of sediment at both sites. Muir measured cross-hole shear velocity gradients to depth of 5 meters below the seafloor. Comparison of their data with the results of our inversion shows reasonable agreement among data obtained with three very diverse methods of measurements (Fig. 8 and 9).

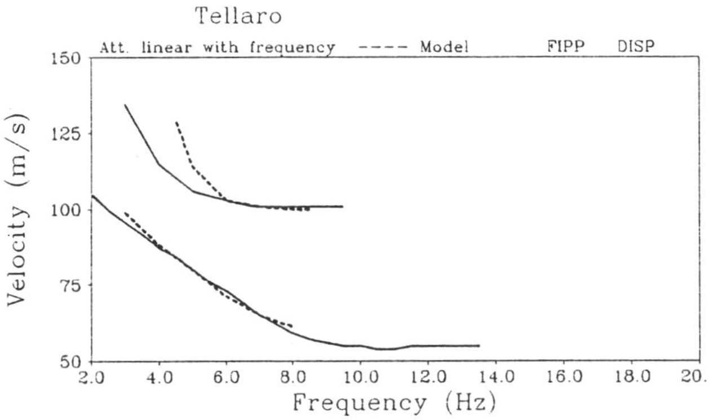


*Figure 5. Measured and modelled group velocity for fundamental and first mode; dotted line is the group velocity obtained from the dispersion analysis of SAFARI time response with the environmental model of fig. 4.*

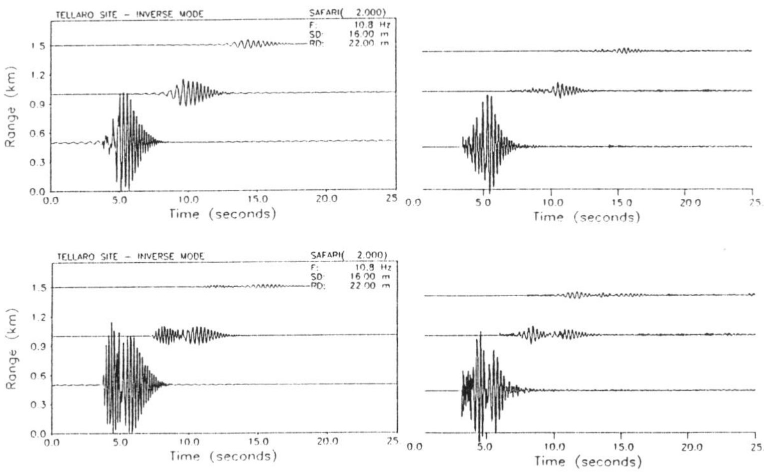


**a**

**b**

*Figure 6. Synthetic seismograms from SAFARI (left) compared with the measured data (right). (a): vertical particle velocity. (b): horizontal particle velocity.*

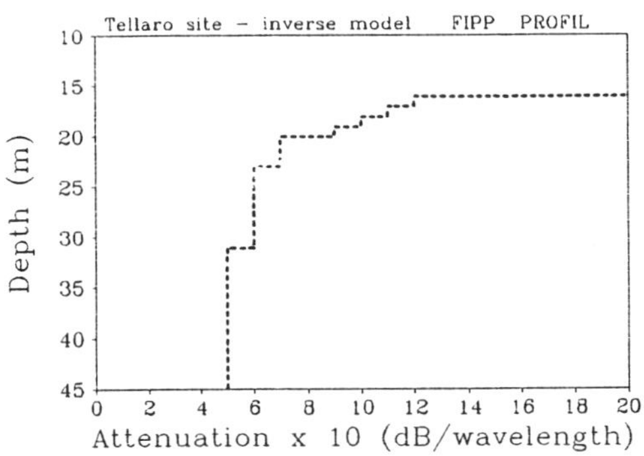


*Figure 7. Attenuation estimate from SAFARI for the site "Tellaro".*

## 5. Conclusions

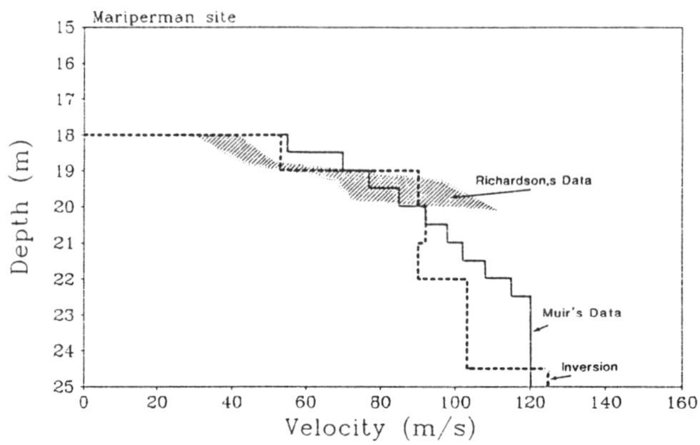


*Figure 8. Site "Mariperman"; comparison of the inverse model (dotted line) with the direct measurements of Muir (continuous line) and Richardson (shaded area).*

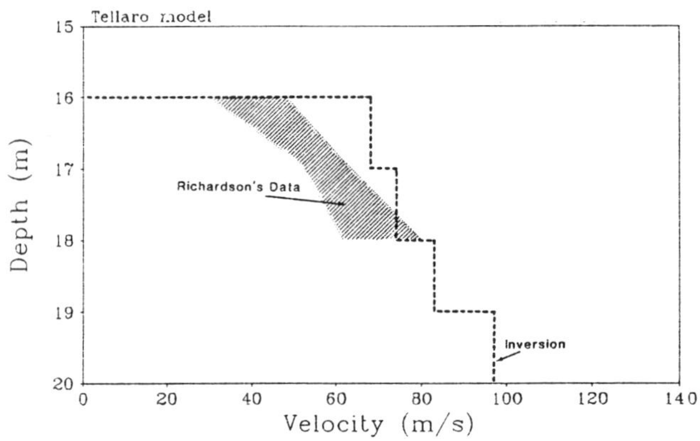


*Figure 9. Site "Tellaro"; comparison of the inverse model (dotted line) with the direct measurements of Richardson (shaded area).*

In this paper we have shown that interface wave dispersion can be used to derive a geoacoustic model of the seafloor for the upper several meters of sediments. We have developed software which allows the derivation of a model in an iterative, semi-automatic, manner by using techniques pionereed by earthquake seismologists for studying surface waves produced by earthquakes. This method replaces tedious trial-and-error procedures which were formerly used to solve the inverse problem. At the present time we are able to obtain satisfactory models with little computation time, so that experiments may be modified in the field to optimize data collection. By combining both short and long range data we have been able to derive models with improved resolution and to obtain information at deeper depths. In the future we hope to improve further the near bottom resolution by adding closely spaced phones to exploit the information contained in the higher frequency portion of the source spectrum.

## Acknowledgements

We gratefully thank the referees and the editors for their helpful comments


## References


[1] Hamilton, E.L. (1987) "Acoustic properties of sediments," in A. Lara-Sáenz, C. Ranz-Guerra and C. Carbó-Fité (eds.), *Acoustics and Ocean Bottom* (C.S.I.C., Madrid, Spain), pp. 3–58.

[2] Stoll, R.D. (1989) "Sediment acoustics," Springer & Verlag, New York.

[3] Richardson, M.D., Muzi, E., Miaschi, B., Turgutcan, F. (1990) "Shear velocity gradients in near surface marine sediments," in this volume.

[4] Muir, T.G., Akal, T., Richardson, M.D., Stoll, R.D., Caiti, A., Hovem, J.M. (1990) "Comparison of different techniques for shear wave velocity and attenuation measurements," in this volume.

[5] Dziewonsky, A., Bloch, S., Landisman, M. (1969) "A technique for the analysis of transient seismic signals" Bull. Seism. Soc. Am., **59**, 427–444

[6] Grant, F.S., West, G.F. (1965) "Interpretation theory in applied geophysics," McGraw-Hill.

[7] Haskell, N.A. (1953) "Dispersion of surface waves on multilayered media," Bull. Seism. Soc. Am., **43**, 17–34

[8] Dorman, J., Ewing, M. (1962) "Numerical inversion of seismic surface wave dispersion data and crust-mantle structure in the New York-Pennsylvania area" Journ. Geoph. Res. **67**, 13, 5227–5241

[9] Thikonov, A.N., Arsenin, V.Y. (1977), "Solution of ill-posed problems", J. Wiley & Sons, New York.

[10] Schmidt, H. (1988) "SAFARI: Seismo-acoustic fast-field algorithm for range independent environments," SR-113, SACLANT Undersea Research Centre, La Spezia, Italy.

# USING SYNTHETIC SEISMOGRAMS TO DETERMINE SHEAR WAVE VELOCITIES IN SEDIMENTS FROM OCEAN BOTTOM SEISMOMETER REFRACTION OBSERVATIONS

Gregory T. Leger and Hugh A. MacPherson
Oceanroutes Canada, Inc.
271 Brownlow Ave.
Dartmouth, Nova Scotia
Canada, B3B 1W6

ABSTRACT. The synthetic seismogram method is applied to observations of singly converted shear waves in a marine fan sedimentary environment. The method is used to identify the shear wave phases and to calculate the shear wave velocity in the sediments. Observations indicate that the shear waves are trapped in the sediment layer by strong reflections at the seafloor and the basement. Essentially, the sediments are acting as a shear wave guide.

## INTRODUCTION

Our aim in this paper is to use the WKBJ synthetic seismogram method to analyze observations obtained from ocean bottom seismometers (OBS) of shear waves in marine sediments. The observations were made over the distal Bengal Fan during a seismic refraction experiment performed from the R.V. Conrad during the RC2706 cruise in July 1986 (Leger and Louden,1990). The strong sediment shear waves observed in that data set were reported by Leger (1989).

Acoustic propagation loss experiments indicate that there are areas of the oceans especially on the continental shelves which exhibit large losses at frequencies less than 100 Hz. It is generally accepted that the losses are largely due to excitation of shear waves through mode conversion from compressional (termed P or acoustic waves) to shear (S or elastic waves) in the seabed.

In acoustic propagation experiments such as those described by Hughes et al (1990), propagation loss is measured as a function of frequency at a fixed shot receiver spacing. A forward modelling process with computerized propagation loss algorithms, is used to analyze these data. Various combinations of seabed acoustic and elastic parameters are used until an adequate match between observations and calculations is found. Essentially, the



J. M. Hovem et al. (eds.), Shear Waves in Marine Sediments, 567–574.

analyst infers a seabed geo-acoustic model from an observed absence of acoustic energy in certain frequency bands. In contrast, a seismic refraction experiment is a direct observation of bottom propagating acoustic and elastic energy. Travel time and amplitude characteristics of the series of reflections and refractions from geophysical horizons in the sediments and crust are used to determine the geo-acoustic parameters of the seabed.

## SHEAR WAVE OBSERVATIONS

There is a fundamental physical difference in the interaction of P-waves and S-waves at the seafloor. P-wave reflections and refractions from sub-seafloor boundaries return into the water layer attenuated by various scattering and absorption processes. On the other hand, mode converted shear waves returning to the seafloor from horizons deeper in the seabed are almost totally reflected downward due to the high shear impedance contrast at the interface. An analogous situation occurs at the sea surface for P-waves. The process of mode conversion therefore results in additional losses of acoustic energy from the water layer.

Acoustic basement is another strong shear wave reflector. S-waves are reflected upwards to propagate in a wave guide formed by the sediment layer with strong reflectors at the top and bottom boundaries. These S-waves are observed as long codas on geophone profiles and can be used to determine the elastic parameters of the seabed.

Shear waves which are mode converted at the seafloor as well as four types of crustal seismic waves are observed on the record sections produced by OBS in marine refraction experiments. To identify these phases, we use the terminology of Lewis and McClain (1977), consisting of three letters designating the type of ray (P or S) in the sediments on the down going leg of the ray path, through the crust, and in the sediments on the upward leg of the ray path.

The crustal P-wave arrivals are referred to as the PPP branch since they travel through both the sediments and crust as P-waves. Doubly converted crustal shear waves generated by mode conversion at the sediment - basement interface are referred to as the PSP branch since they travel to and from the base of the sediments as P-waves, and through the crust as S-waves. Both PPP and PSP are P-waves at the seafloor and are commonly observed on sonobuoy and ocean bottom hydrophone records (figure 1) and are well understood (White and Stephen, 1980).

Two types of singly converted crustal shear waves which undergo just one mode conversion at the basement (PPS and PSS) can be identified on OBS geophone profiles. In the sediments, the singly converted phases are P-waves on the downward leg but S-waves on the upward leg. Because they

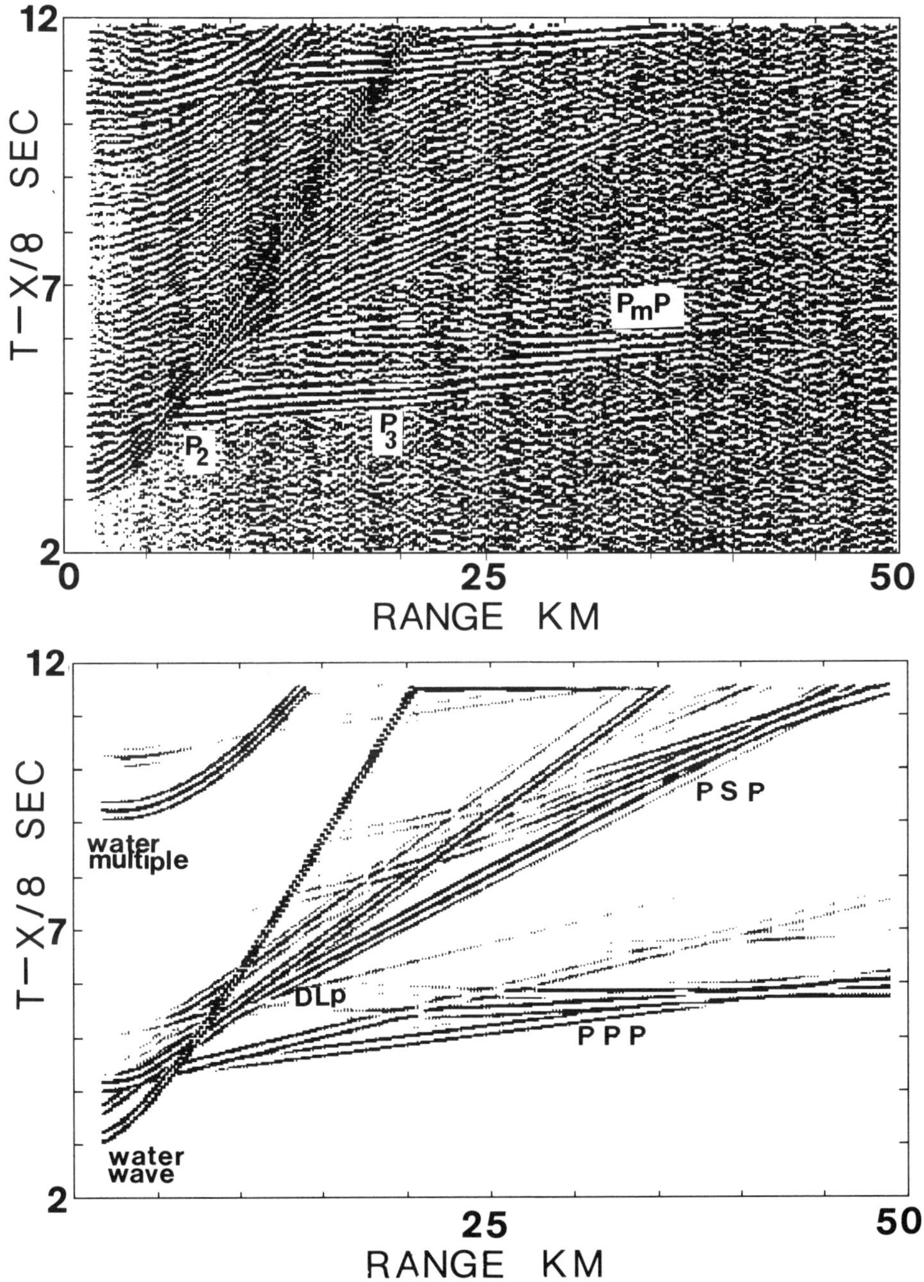


Figure 1. (above) Record section recorded on the hydrophone channel of an ocean bottom seismometer located at -4.2593N, 79.8372E. Certain PPP and PSP refraction and wide angle reflection phases are identified. The coherent background noise is due to cross talk from the time code channel. (below) WKBJ synthetic seismogram with the PPP and PSP branches used to interpret the observed record section.

arrive at the ocean bottom receiver as S-waves, they are observed on geophone profiles but are absent from hydrophone profiles separated from the seafloor.

The singly converted crustal shear waves can be distinguished by several characteristics (Duennebier et al. 1987). The PPS branch corresponds to PPP rays which are mode converted at the basement on the up going leg of their ray path. On the geophone profiles they are characterized by phase velocities identical to the PPP branch but offset from the PPP branch by a constant delay in time (figure 2).

The PSS branch is mode converted at the basement on the down going leg of its ray path and travels as an S-wave thereafter. This branch is characterized by the phase velocity of the PSP branch and is offset from the PSP branch by an identical delay as observed between the PPP and PPS branches. This delay time is a characteristic of the ratio of the S-wave to P-wave velocity in the sediments.

The seabed converted shear wave phases are observed on the horizontal geophone record as distinctly separated reflections just below the direct water wave arrivals. They have normal moveouts characteristic of sediment shear wave velocities, and are labelled S on figures 2 and 3.

## METHOD

The method of obtaining the sediment shear wave velocity used here, exploits the OBS response to shear waves in three steps.

**Step 1. Determine the P-wave velocity structure.**

The total thickness and P-wave velocity structure of the sediments are found by analyzing the P-wave phases on the hydrophone records in the normal fashion (Dix, 1955). Deeper crustal P and S velocity structure can be determined by a number of methods (eg. White, 1979; Spudich and Orcutt, 1980; Purdy, 1983). We then use a synthetic seismogram method to compute a representation of the PPP and PSP branches. The result of this calculation is shown in figure 1. The direct water wave, its multiple, the PPP and PSP branches were calculated and are labelled on the diagram. The P-wave sediment dogleg multiple (bouncing between the seafloor and the basement) of the PPP branch is labelled DLp.

For the purpose of studying the sediment shear waves, it is important only to match travel times and the major amplitude features of the P-wave phases. The effects of small dips can be compensated for by adjusting the velocity - depth crustal model without being overly concerned whether an accurate estimate of true crustal structure is being made. Essentially, we are preparing an artificial shear wave source at the base of the sediments. We used the WKBJ2 program (Chapman et al.,1988).

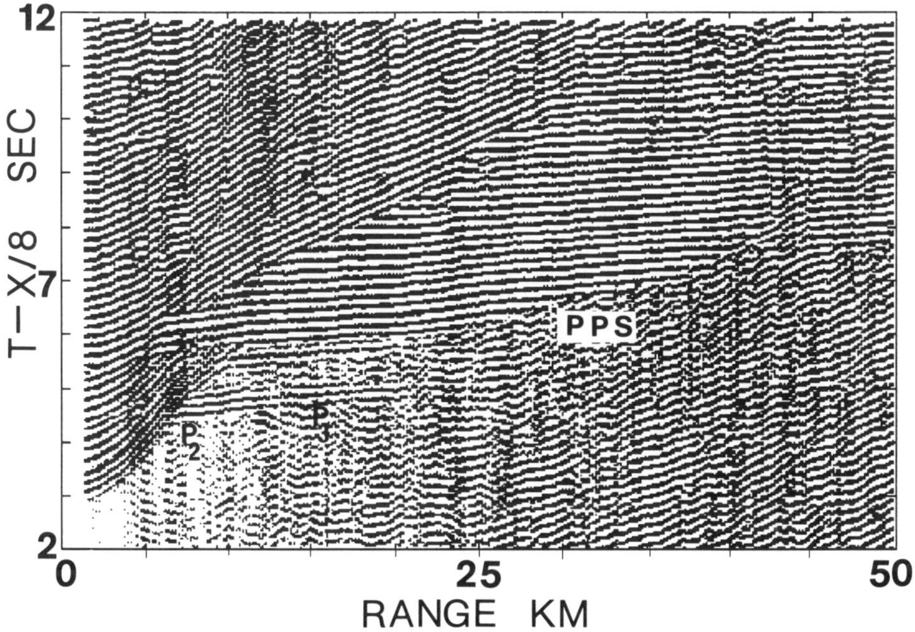


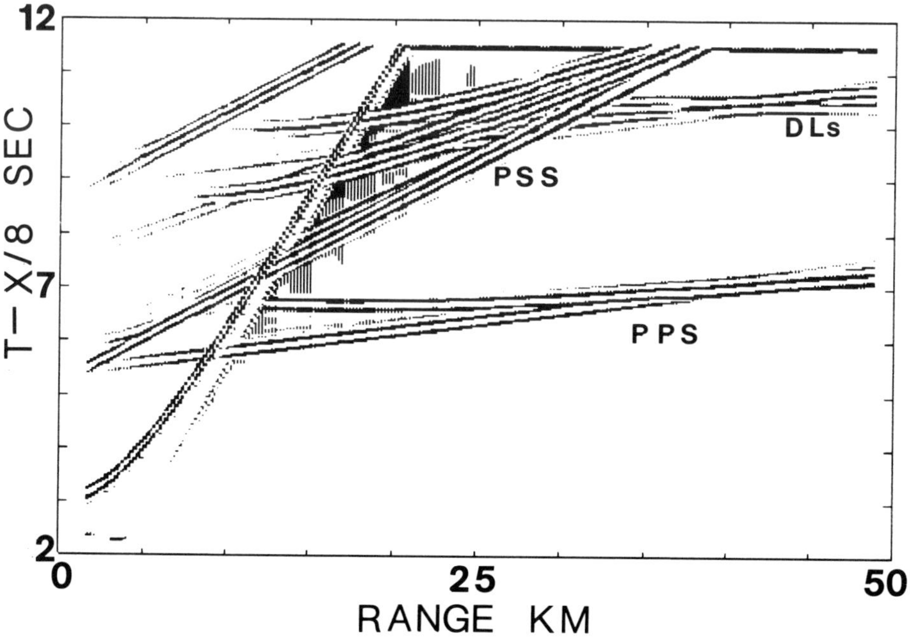


Figure 2. (above) Record section recorded on the horizontal geophone channel of an ocean bottom seismometer located at -4.2593N, 79.8372E. The distinct differences between the hydrophone and geophone channels are due to the sensitivity of the geophone channel to shear waves. The mode converted shear wave phases are identified on the diagram.
(below) WKBJ synthetic seismogram used to find the bulk shear velocity in the sediments by matching the travel time positions of the PPS and PSS phase.

**Step 2. Find the bulk sediment S-wave velocity.**

The average S-wave velocity in the sediments can be estimated from the observed travel time delay between the arrival of the P-wave phases and the S-wave phases (figure 2). An approximation of the shear wave velocity can be found by assuming that the travel time delay is equivalent to the difference in the vertical incidence one way travel time between a P and an S wave. This sediment shear velocity is then used to produce a synthetic seismogram in which the PPS and PSS branches appear. The S-wave velocity of the input file to the WKBJ program is then adjusted until an adequate match is found between calculations and observations. The result of this calculation is shown in figure 2. The branch labelled DLs is the S-wave sediment dog leg multiple of the PPS branch.

**Step 3. Determine a more detailed sediment S-wave velocity structure.**

The detailed S-wave velocity structure in the sediments is found by computing synthetic seismograms to match the travel time paths of the seabed mode converted S-wave reflections and selected sediment multiples of the PPS and PSS branches. These are specified in the WKBJ input parameter file. The physical properties and numeric ray codes specified in the input parameter file of the seismogram program, serve as a measure of both the shear velocity as well as the modes of propagation of the shear waves.

Figure 3 shows one such calculation. Five S-wave dogleg reflections of the PPS branch (labelled DL1 through DL5) and of the PSS branches are shown on this diagram along with the seabed converted S-wave reflections (labelled S). The input earth model of the sediments for this calculation consisted of five sediment layers ranging in S-wave velocity between 0.42 km/sec at the seafloor and 0.89 km/sec at the basement.

## DISCUSSION

Using these methods, the determination of the average bulk shear wave velocity is relatively simple and estimates of this quantity can be made directly by comparisons between hydrophone and geophone profiles. A more precise determination of the sediment shear wave velocity structure is found through a more detailed analysis of the seabed mode converted S-wave reflections and the reverberations which are caused by multiple S-wave dogleg reflections within the sediment layers. The accuracy of the method will improve when digital ocean bottom instruments with better recording fidelity, dynamic range and frequency response become available.

The strong reverberations of the shear waves observed on the geophone record indicate that the shear waves are trapped in the sediment layer. There are strong arrivals

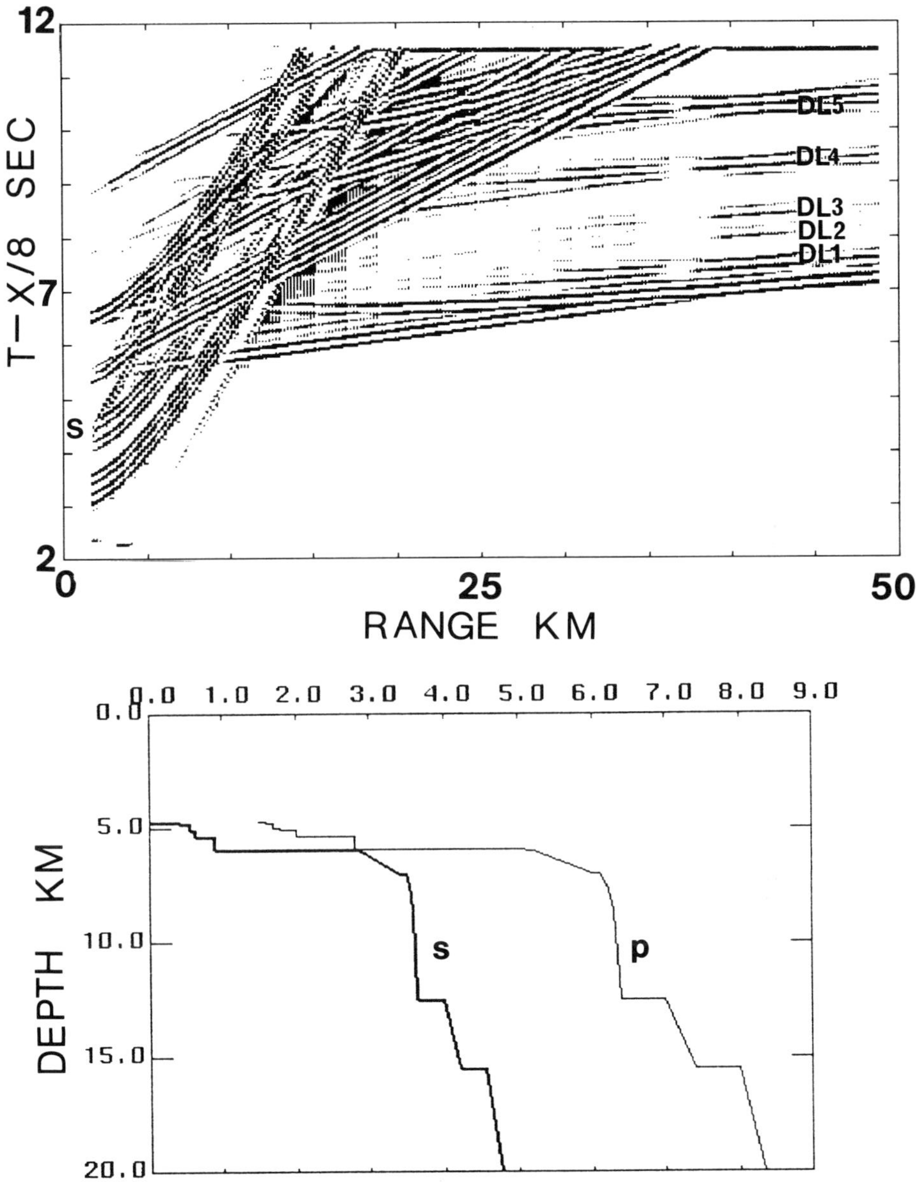


Figure 3. (above) WKBJ synthetic seismogram used to find a more detailed shear velocity - depth profile in the sediments by matching the seabed converted shear wave reflectors and the internal sediment S-wave reflections of the PPS and PSS phases.
(below) The velocity - depth profile used for the calculation.

with crustal S-wave phase velocities occurring later than 60 seconds after the shot initiation. These wrap around the record section plot and appear as arrivals coming in before the first crustal arrivals (figure 2). In this environment, the sediments are acting as a waveguide to shear waves. This property of shear wave propagation in sediments can be exploited to overcome low frequency propagation losses by placing geophone sensors at the seafloor sensitive to the shear energy. In other words the energy loss from the water layer can be treated as an energy gain in the seabed.

In geographic areas where basement P to S mode conversions are significant, doubly converted PSP phases will be observed in refraction experiments using both sonobuoys and ocean bottom seismometers. The presence or absence of doubly converted phases on sonobuoy profiles indicates where OBS measurements can be used to find the shear wave velocities in the sediments. These will also be the areas which are likely to exhibit anomalously large low frequency acoustic propagation losses in the water layer.

## ACKNOWLEDGMENTS

This work was funded through a joint research program between Oceanroutes Canada and the Department of National Defence DIRP (22070EA01). The RC2706 experiment was planned and conducted by Donald Forsyth (Brown University), Jeffery Weissel (LDGO) and Keith Louden (Dalhousie University) who generously provided us with the data. Ian Fergusson (Oceanroutes Canada) provided computer expertise. The RC2706 experiment was funded by grants from the Natural Sciences and Engineering Research Council of Canada (OGP0008459) and the U.S. National Science Foundation (OCE8511980).

## REFERENCES

Chapman, C.H., Chu Jen-Yi and D.G. Lyness (1988), 'The WKBJ Seismogram Algorithm' in Doornbos, D.J.(ed) 'Seismological Algorithms, Computational Methods and Computer Programs', Academic Press, Toronto.

Dix, C.H. (1955), 'Seismic velocities from surface measurements', Geophysics 20, 68-86.

Duennebier, F.K., B. Lienert, R. Cessaro, P. Anderson and S. Mallick (1987), 'Controlled-source seismic experiment at Hole 581C', in Duennebier, F.K., Stephen,R., Gettrist, J.F. et al., 1987, Initial Reports of the Deep Sea Drilling Project, Volume LXXXVIII; Washington.

Hughes, S.J., Ellis, D.D., Chapman, D.M.F., Staal, P.R. (1990), 'Low-frequency acoustic propagation loss in shallow water over hard-rock seabeds covered by a thin layer of elastic-solid sediment', Journal of the Acoustical Society of America, Vol. 88 No.1.

Leger, G.T. (1989), 'Was the oceanic crust in the central Indian Basin deformed by lithospheric boudinage? Evidence from seismic refraction [M.S. thesis]. Dalhousie University, Halifax.

Leger, G.T. and Louden K.E. (1990), 'Seismic refraction measurements in the central Indian Basin: Evidence for crustal thickening related to intraplate deformation', in Cochran, J.R. and Stow, D.A.V., et al., 1990, Proceedings of the Ocean Drilling Program, Scientific Results, Vol 116: College Station TX.

Lewis, B.T.R., and McClain, J. (1977), 'Converted shear waves as seen by ocean bottom seismometers and surface buoys', Bulletin of the Seismological Society of America, Vol. 67, No. 5, 1291-1302.

Purdy, G.M. (1983), 'The seismic structure of 140 Myr old crust in the western central Atlantic Ocean', Geophysical Journal of the Royal Astronomical Society, 72, 115-137.

Spudich, P., and Orcutt, J. (1980), 'A new look at the seismic velocity structure of the oceanic crust', Reviews of Geophysics and Space Physics, 18, 627-645.

White, R.S. (1979), 'Oceanic upper crustal structure from variable angle seismic reflection - refraction profiles', Geophysical Journal of the Royal Astronomical Society, 57, 683-726.

White, R.S. and Stephen R.A. (1980), 'Compressional to shear wave conversion in oceanic crust', Geophysical Journal of the Royal Astronomical Society, 60, 547-565.

# ANALYTICAL INVESTIGATION OF SEISMIC SURFACE WAVES IN THE SEAFLOOR

J.M. Roesset, S.G. Wright, and M. Sedighi-Manesh
*Department of Civil Engineering*
*ECJ 4.604*
*The University of Texas*
*Austin, Texas 78712-1076 USA*

ABSTRACT. Analytical studies are being conducted to assess the applicability of the Spectral Analysis of Surface Waves (SASW) method to the determination of the properties of the soil in the sea floor, and to understand the dispersive characteristics of the waves generated by a source applied on the surface of a soil deposit overlaid by water. A classical analytical model assuming two dimensional plane interface waves and determining the roots of the characteristic equation has been implemented in a computer program. In addition a three dimensional solution has been developed in order to reproduce more accurately the actual conditions of the test. In this case a dynamic load is applied over a small area at the soil water interface and the resulting displacements at various points along this interface (locations of the receivers) are computed. The apparent velocity of propagation is then obtained from the phase differences between receivers as a function of frequency or wavelength.

## 1. Introduction

The Spectral Analysis of Surface Waves (SASW) method has been successfully used on land to backfigure the elastic moduli of soils (or pavement systems) and their variation with depth. In this method a source and two receivers are placed directly on the surface of the soil deposit. A dynamic, impulsive, load is applied by the source and the time histories of the displacements (velocities or accelerations) are recorded at each one of the receivers. From a spectral analysis of the records the phase differences are obtained as a function of frequency and the apparent velocities of propagation are then computed at each frequency in order to obtain experimental dispersion curves. An inversion procedure is finally used to backfigure the soil properties (shear wave velocities) and their variation with depth. The possibility of extending this method to determine the properties of the soil in the seafloor and to detect the possible existence of gas hydrates is now being investigated. A combined analytical and experimental program has been initiated under this research. In this paper the analytical work is presented while the experimental results are the subject of a companion paper by Stokoe, Gauer and Bay.

Although the amount of work reported in the literature on the characteristics of interface waves at the sea bottom is relatively limited there are a number of significant papers which have dealt with this topic. As early as 1956 Strick and Ginzbarg [9] addressed the problem of waves along a fluid-solid interface referring to them as Stoneley waves. In 1959 three papers by Roever, Vining and Strick [7] presented the results of a series of experiments using two solid media with shear wave velocities smaller and larger than the wave velocity in the fluid, a theoretical formulation, and an interpretation of the waveforms and arrival times in terms of a critically refracted P-wave, direct and reflected waves and a Stoneley type of interface wave. In addition the existence of a pseudo



*J. M. Hovem et al. (eds.), Shear Waves in Marine Sediments*, 575–582.

Rayleigh wave which travels with the velocity of the Rayleigh wave in the solid without any fluid, in the case of a hard bottom (with shear wave velocity larger than the wave propagation velocity in the fluid), was discussed.

Davies [1] reported in 1965 the results of a series of seismic tests in the Indian Ocean bottom. The shear wave velocities of the top 16 m of the seabed were determined matching the experimental dispersion curves with analytical predictions. Davies pointed out that the main contribution to the results was a Stoneley type wave with a velocity of propagation of approximately 0.88 times the shear wave velocity of the soil.

Hamilton et al [2] presented in 1970 the results of tests conducted in the seafloor off San Diego, California from a research submersible computing the shear wave velocity of the bottom sediments (assumed half-space) from the measured velocities and attenuation of compressional waves and the velocities of Stoneley waves. The studies of Davies and Hamilton et al dealt with soft sediments (shear wave velocity much smaller than the acoustic wave velocity of water).

Two papers by Rauch [4,5] in 1980 and 1986 have clarified the differences between the case of a soft and a hard bottom and the types of waves which propagate along a water-solid interface. Rauch refers to the interface waves as Scholte waves. For a soft seabed ($c_S < c_W$ where $c_S$ is the shear wave velocity of the soil and $c_W$ the acoustic wave velocity in water) Rauch indicated that Scholte waves propagate with speeds of the order of 0.88 $c_S$ as suggested by Davies. For a hard bottom ($c_S > c_W$), on the other hand, Rauch suggested that Scholte wave speeds are of the order of 1250 to 1500 m/sec, always slightly smaller than $c_W$. Although Rauch stated that on an almost homogeneous rock basement only the fundamental mode, corresponding to the Scholte wave, can be expected, he also pointed out that for short wavelengths there is a large number of modes and that even in deep water the free surface may be sensed in the case of a hard bottom. There is therefore an apparent difference of opinion and unresolved issue in relation to the role of the Scholte and pseudo Rayleigh waves in the case of deep water over a stiff bottom.

Although soft sediments are more likely to be encountered at the seafloor than rock, pockets of gas hydrates can have a very large stiffness. It was decided therefore to explore both cases in the present series of studies.

## 2. Formulation for Water

Considering the water as an inviscid compressible fluid with irrotational motion the general equation for the potential $\phi$ in two dimensions can be written as

$$\nabla^2\phi = \frac{\partial^2\phi}{\partial x^2} + \frac{\partial^2\phi}{\partial z^2} = -\frac{\omega^2}{c_W^2}\phi \tag{1}$$

where $\omega$ is the frequency, $c_W$ is the compression wave velocity for the water and $\nabla^2$ is the Laplace operator.

The general solution of equation (1) is of the form

$$\phi = (a_1 e^{\alpha z} + a_2 e^{-\alpha z})\, e^{i(\omega t - kx)} \tag{2}$$

where k is the wave number, c is the phase velocity and

$$\alpha^2 = k^2 - \frac{\omega^2}{c_W^2} = \frac{\omega^2}{c^2}\left(1 - \frac{c^2}{c_W^2}\right) \tag{3}$$

The vertical fluid particle velocity and displacements are

$$v = \frac{\partial \phi}{\partial z} = \alpha(a_1 e^{\alpha z} - a_2 e^{-\alpha z}) e^{i(\omega t - kx)} \qquad d = \frac{v}{i\omega} \tag{4}$$

whereas the pressure is

$$p = -i\omega\rho\phi \tag{5}$$

Writing $v = \bar{v}\, e^{i(\omega t - kx)}$ and $p = \bar{p}\, e^{i(\omega t - kx)}$, at the free surface $z = 0$

$$\bar{v}_1 = \alpha\,(a_1 - a_2) \qquad \bar{d}_1 = \frac{-i\,\bar{v}_1}{\omega} \qquad \bar{p}_1 = -i\omega\rho\alpha(a_1 + a_2) \tag{6}$$

and at the bottom $z = h$

$$\bar{v}_2 = \alpha(a_1 e^{\alpha h} - a_2 e^{-\alpha h}) \qquad \bar{d}_2 = \frac{-i\bar{v}_2}{\omega} \qquad \bar{p}_2 = i\omega\rho\alpha\,(a_1 e^{\alpha h} + a_2 e^{-\alpha h}) \tag{7}$$

Obtaining the coefficients $a_1$ and $a_2$ in terms of $\bar{d}_1$ $\bar{d}_2$ and substituting into the expressions for $\bar{p}_1$ $\bar{p}_2$, the tractions at the top and bottom are

$$\begin{aligned} \bar{t}_1 = \bar{p}_1 &= -\frac{\omega^2\rho}{\alpha}\,\frac{e^{\alpha h} + e^{-\alpha h}}{e^{\alpha h} - e^{-\alpha h}}\,\bar{d}_1 + \frac{\omega^2\rho}{\alpha}\,\frac{2}{-e^{\alpha h} - e^{-\alpha h}}\,\bar{d}_2 \\ \bar{t}_2 = -\bar{p}_2 &= \frac{\omega^2\rho}{\alpha}\,\frac{2}{e^{\alpha h} - e^{-\alpha h}}\,\bar{d}_1 - \frac{\omega^2\rho}{\alpha}\,\frac{e^{\alpha h} + e^{-\alpha h}}{e^{\alpha h} - e^{-\alpha h}}\,\bar{d}_2 \end{aligned} \tag{8}$$

Applying the linear free surface boundary condition at $z = 0$ $\bar{t}_1 = -\rho g \bar{d}_1$ and eliminating $\bar{d}_1$

$$\bar{t}_2 = -\frac{\omega^2\rho}{\alpha}\;\frac{(e^{\alpha h} - e^{-\alpha h}) - \frac{\alpha g}{\omega^2}(e^{\alpha h} + e^{-\alpha h})}{(e^{\alpha h} + e^{-\alpha h}) - \frac{\alpha g}{\omega 2}(e^{\alpha h} - e^{-\alpha h})}\;\bar{d}_2 = k_w\ \bar{d}_2 \tag{9}$$

where $\rho$ is the mass density of the water, g is the acceleration of gravity and h is the water depth.

When the propagation velocity c is larger than $c_w$ the variable $\alpha$ becomes imaginary. Writing then $\alpha = i\,\beta$ so that $\beta = \frac{\omega}{c}\sqrt{\frac{c^2}{c_w{}^2} - 1}$

$$k_w = -\frac{\omega^2\rho}{\beta}\frac{\sin\beta h - \frac{\beta g}{\omega^2}\cos\beta h}{\cos\beta h + \frac{\beta g}{\omega^2}\sin\beta h} \tag{10}$$

It should be noticed that as long as $\alpha$ is real ($c < c_w$) as the water depth increases $e^{-\alpha h}$ becomes negligible and $k_w$ tends to $-\frac{\omega^2\rho}{\alpha}$. In this case the solution will be independent of the water depth. On the other hand if $c > c_w$, $\sin\beta h$ and $\cos\beta h$ remain finite for any water depth and the presence of the free surface will always be felt. $k_w$ represents then the relation between the traction (vertical force per unit of length) and the vertical displacement at the sea bottom.

### 3. Formulation for Soil

Consider a horizontal soil layer with thickness $h_i$, shear wave velocity $c_{si}$, compressional wave velocity $c_{pi}$ and mass density $\rho_i$. Using the formulation suggested by Kausel and Roesset [3] the tractions and displacements at the top and bottom surfaces of the soil layer can be related through a dynamic stiffness matrix $K^i$. $K^i$ will be a 4x4 matrix, which can be partitioned into 4 submatrices $K^i_{11}$ $K^i_{12}$ $K^i_{21}$ $K^i_{22}$ each 2x2. The expressions for these matrices are given in [3].

For a half space opening downwards the stiffness matrix is 2x2 relating tractions and displacements at the top only. Calling it $K^h_{11}$

$$K^h_{11} = 2kG\left\{\frac{1 - s^2}{2(1 - rs)}\begin{bmatrix} r & -1 \\ -1 & s \end{bmatrix} + \begin{bmatrix} 0 & 1 \\ 1 & 0 \end{bmatrix}\right\} \tag{11}$$

where $\quad G = \rho\, c_s^2 \qquad r = \sqrt{1 - (\frac{\omega}{kc_p})^2} \qquad s = \sqrt{1 - (\frac{\omega}{kc_s})^2} \qquad (12)$

The dynamic stiffness matrix for a complete soil deposit can then be obtained assembling the stiffness matrices of the individual layers and the underlying half space. The resulting system of equations for a given frequency $\omega$ and wave number k is then of the form

$$\begin{bmatrix} K^1_{11} & K^1_{12} & & & \\ K^1_{21} & K^1_{22}+K^2_{11} & K^2_{12} & & \\ & K^2_{21} & K^2_{22} + K^3_{11} & K^3_{12} & \\ & & & & \\ & & & & K^n_{22} + K^h_{11} \end{bmatrix}\begin{Bmatrix} \bar{U}_1 \\ \bar{U}_2 \\ \bar{U}_3 \\ \cdot \\ \bar{U}_n \end{Bmatrix} = \begin{Bmatrix} \bar{P}_1 \\ \bar{P}_2 \\ \bar{P}_3 \\ \cdot \\ \bar{P}_n \end{Bmatrix} \tag{13}$$

where $\bar{U}_j$ are the displacements $-i\ \bar{u}_j$, $\bar{w}_j$ (horizontal and vertical) of the jth interface and $\bar{P}_j$ are the net tractions $-i\ \bar{p}_{xj}$, $\bar{p}_{zj}$ (horizontal and vertical) applied at that interface.

When dealing with a layer of water over the soil deposit compatibility of vertical displacements at the soil water interface requires that $\bar{d}_2 = \bar{w}_1$ using the previous notation. From equilibrium $\bar{p}_{z1} + \bar{t}_2$ must be the net force per unit of length applied at the soil water interface. It is thus sufficient to add to the (2,2) term of the matrix $K^1_{11}$ (or $K^h_{11}$ when dealing with a soil half space directly under the water) the term $k_w$ computed in the previous section.

To determine the wave numbers and mode shapes of the waves propagating through the water-soil system one must determine the values of k that make zero the determinant of the modified stiffness matrix K. In this formulation the wave number k corresponding to the Airy wave along the free surface of the water, given by $\alpha \tanh(\alpha h) = \omega^2/g$ and $k^2 = \alpha^2 + \omega^2/c_w{}^2$ has been eliminated because it is of no practical interest for the present problem (it should be noticed that for this wave number the term $k_w$ would become infinity).

For the case of a soil half space the characteristic equation becomes

$$4rs - (1 + s^2)^2 + r(1 - s^2)\ \frac{k_w}{kG} = 0 \qquad (14)$$

## 4. Three Dimensional Formulation

When dealing with a load distributed over a small circular area at the soil-water interface it is convenient to express the solution in cylindrical coordinates. For an actual determination of the time histories of the motions recorded by the different receivers it would be necessary to transform first the time history of the load from the time to the frequency domain and perform at the end an inverse Fourier transformation of the results. For the purpose of the present studies, however, this transformation is not necessary, since the variation with frequency of the phases is desired. Considering then a harmonic force at a frequency $\omega$ uniformly distributed over a circle of radius R, and calling $\bar{w}_1(k)$ the solution of equation (13) with the addition of the $k_w$ term to the stiffness matrix and a load vector with all components zero except $\bar{p}_{z1} = 1$

$$w(r) = R\int_0^\infty \bar{w}_1\ (k)\ J_0(kr)\ J_1(kR)\ dk \qquad (15)$$

For relatively simple cases, where the poles of the integrand can be easily obtained the integral in expression (15) can be performed using contour integration. This procedure becomes extremely laborious however when dealing with a more complex problem. In the present application the integral is evaluated numerically assuming a small amount of internal damping, of a hysteretic nature, in the soil. The damping is accounted for through the use of a complex shear modulus G(1 + 2iD) where D is the damping ratio. The same amount of damping was assumed for shear and compressional waves (shear and constrained moduli).

Once the displacements $\omega$ are computed for each frequency at the distances of the receivers the apparent propagation velocity (phase velocity) is computed from the phase differences.

## 5. Parametric Studies

To validate the two dimensional (plane waves) program the soil profile described by Rauch [4] in his 1980 paper was used. The resulting dispersion curve is compared to Rauch's results in Figure 1. It can be seen that the agreement is excellent.

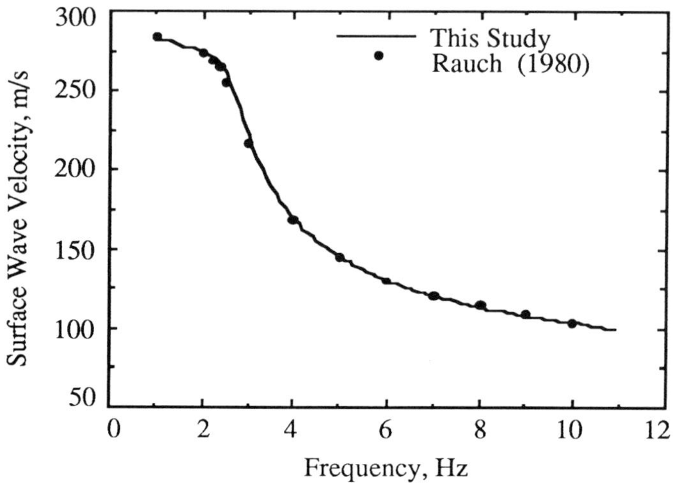


Fig. 1 Comparison of theoretical dispersion curves from two independent models for a given shear wave velocity profile of a layered half-space overlaid by water

A number of parametric studies were then conducted for a soft bottom half space ($c_s < c_w$) From these studies it was found that the phase velocity of the first mode, which will be referred to as the Scholte wave following Rauch, is a function of Poisson's ratio of the soil (or the ratio of the P wave velocity $c_p$ to the shear wave velocity $c_s$ of the soil), the mass density of the soil, and the ratio of the shear wave velocity of the soil $c_s$ to the compression wave velocity of the water $c_w$. The ratio of the velocity of propagation of the Scholte wave to the shear wave velocity of the soil increases as Poisson's ratio of the soil and its mass density increase. It decreases, on the other hand, as the shear wave velocity of the soil, or the ratio $c_s/c_w$ increases. For ratios $c_s/c_w$ from 0.04 to 0.4 the velocity ratio of the Scholte wave to the shear wave in the soil ranges from 0.8 ($c_s/c_w$ = 0.4, $\rho_s = 1.5\ \rho_w$, $\nu = 0.2$) to 0.91 ($c_s/c_w = 0.4$, $\rho_s = 2.25\ \rho_w$, and $\nu$ close to 0.5), where $\rho_s$ and $\rho_w$ are the mass densities of the soil and the water respectively and $\nu$ is the soil's Poisson's ratio. The value of 0.88 suggested by Davis and Rauch is an appropriate one for typical soft soil deposits and it can provide a reasonable first estimate for many practical cases. It is important to realize, however, that it can change depending on the actual soil properties, if a more accurate determination of these properties is desired. It should be noticed also that even for the case of a soft bottom there are a number of modes with propagation velocities of the order of $c_w$ and larger. These modes, however, are not important. The results from the three dimensional analyses for soft soil ($c_s < c_w$) give apparent velocities of propagation computed from the phase differences which are almost identical to the velocity of the Scholte wave. Figure 2 shows the theoretical dispersion curves from the two dimensional model for a saturated soil with a shear wave velocity of 0.04 $c_w$ and different water depths. It can be seen that for wavelengths smaller than half the water depth the propagation velocity is essentially independent of the water depth.

When dealing with a stiff bottom ($c_s > c_w$) the dispersion curve for the first mode starts with a wave velocity slightly smaller than $c_w$ for short wavelengths in relation to the water depth and reaches approximately the Rayleigh wave velocity of the soil at long wavelengths. For wavelengths smaller than about five times the water depth there are however several modes with velocities of propagation ranging from $c_w$ to $c_s$. At very small wavelengths the number of modes is extremely large. These results are consistent with those reported by Rauch. Figure 3 shows the ratio of the propagation velocity to the shear wave velocity of the soil $c_S$ as a function of $c_S/c_W$ for the first mode and a wavelength equal to one half of the water depth. For $c_S < c_W$ it was assumed that the soil was saturated with a compressional wave velocity $c_P = c_W$. As a result for $c_S < 0.577\ c_W$ the Poisson's ratio of the soil is variable increasing from 0.25 to nearly 0.5 as the value of $c_S$ decreases, while for $c_S > 0.577\ c_W$ the soil has a constant Poisson's ratio of 0.25. This is the reason for the kink in the curve.

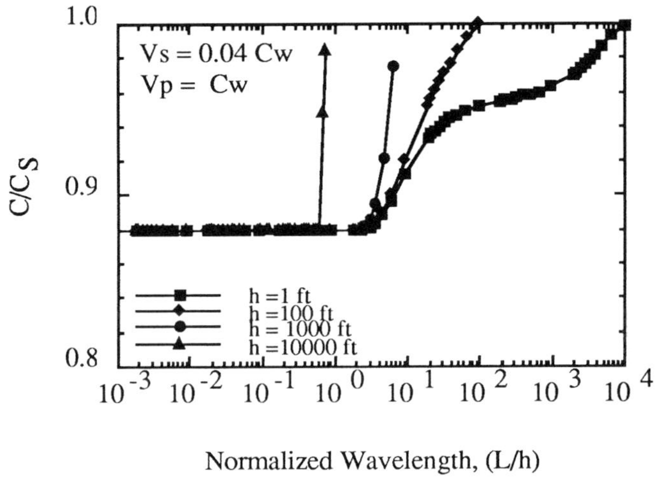


Fig. 2 Theoretical dispersion curves in a saturated half-space overlaid by varying depths of water

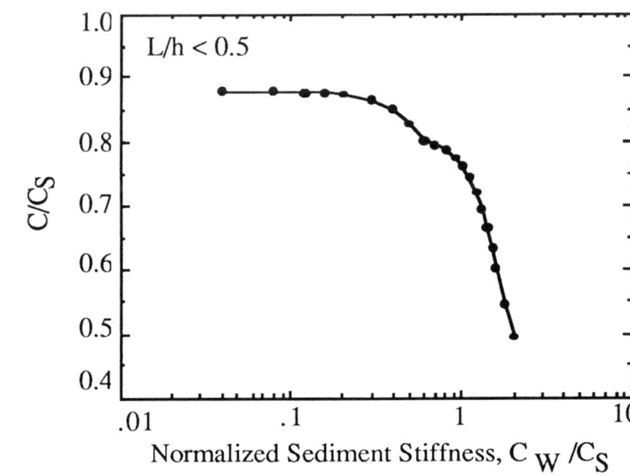


Fig. 3 Variation of normalized surface wave velocity with changes in stiffness for a saturated uniform sediment in deep water ($L/h < 0.5$, $\nu = 0.25$)

The question of which mode is the one excited by a dynamic load applied at the soil water interface still remains. The results of the three dimensional program for a limited number of cases studied to date show propagation velocities with large fluctuations between receivers (depending on the distance between receivers). The average propagation velocity appears to be, however, substantially larger than $c_w$ (of the order of 0.8 to 0.9 $c_s$). Because of the large number of modes, however, the accuracy of the integration procedure must be evaluated before reaching a definitive conclusion. Experimental results on a concrete slab under water showed in one case propagation velocities of the order of $c_w$ but in other cases values close to the Rayleigh wave velocity of the concrete.

## 6. Acknowledgement

The work reported in this paper is part of a research project conducted at The University of Texas at Austin under the sponsorship of the Offshore Technology Research Center created by the National Science Foundation at The University of Texas at Austin and Texas A&M University.

## 7. References

Davies, D., (1965), "Dispersed Stoneley Waves on the Ocean Bottom," Bull. Seism. Soc. Am. Vol. 55, pp. 903-918.

Hamilton, E.L., H.P. Bucker, D.L. Keir, and J.A. Whitney, (1970), "Velocities of Compressional and Shear Waves in Marine Sediments Determined In Situ from a Research Submersible," Journal of Geophysical Research, vol. 75, No. 20, July, pp. 4039-4049.

Kausel, E. and J.M. Roesset, (1981), "Stiffness Matrices for layered Soils," Bull. Seism. Soc. Am. Vol 71, No. 6, December, pp. 1743-1761.

Rauch, D., (1980) "Experimental and Theoretical Studies of Seismic Interface Waves in Coastal Waters," In: Kuperman, W.A and Jensen, F.B., Eds., Bottom-Interacting Ocean

Rauch, D., (1986), "On the Role of Bottom Interface Waves in Ocean Seismo-Acoustics: A Review," Ocean Seismo-Acoustics: Low-Frequency Underwater Acoustics conference, La Spezia, Italy, June, 1985.

Roesset, J.M., K.H. Stokoe, II, D.W. Chang, and M. Aouad, (1990), "Modulus and Thickness of the Pavement Surface Layer from SASW Tests," Transportation Research Board Annual Meeting, Washington, D.C

Roever, W.L., T.F. Vining, and E. Strick, (1959), "Propagation of Elastic Wave Motion from an Impulsive Source Along a Fluid/solid Interface: I. Experimental Pressure response, II Theoretical Pressure Response, III The Psuedo-Rayleigh Wave", Publication No. 173, Shell Development Company, Exploration and Production Research Division, Philosophical Transactions of the Royal Society of London, Vol. 251a, pp. 455-523.

Stokoe, K.H. II, S.G.Wright, J.M. Roesset, R. Gauer, and M. Sedighi-Manesh, (1990) "In Situ Measurement of Stiffness Profiles in Ocean Bottom Materials Using the SASW Method", 1990 Offshore Technology Conference, OTC 6234, pp299-305.

Strick, E. and A.S. Ginzbarg (1956), "Stoneley Wave Velocities for a Fluid-Solid Interface", Bull. Seis. Soc. Amer, Vol. 46, pp. 281-292.

## LIST OF PARTICIPANTS

M.A. Ainslie
YARD Ltd.
233 High Holborn
London WC1V 7DJ
UK

Hassan B. Ali
NOARL Code 245
Stennis Space Center
MS 39529-5004
USA

James F. Andrews
ONR Europe
231, Old Marylebone Road
London NW1 5TH
UK

Kenneth C. Baldwin
Ocean Engineering Center
University of New Hampshire
Kingsbury Hall
Durham, New Hampshire 03824
USA

Alessandro Barbagelata
SACLANT Undersea Research Centre
Viale San Bartolomeo, 400
La Spezia 19026
Italy

J.D. Bennell
School of Ocean Sciences
University College of North Wales
Bangor LL59 5EY
UK

Patricia A. Berge
University of Hawaii at Manoa
2525 Correa Rd.
Honolulu HI 96822
USA

Leonart Berghult
Swedish Defence Research Establishment
102 54 Stockholm, Linnegatan 89
Stockholm
Sweden

David Berry
The Open University
Walton Hall
Milton Keynes, MK7 6AA
UK

L. Dale Bibee
NOARL Code 362
Stennis Space Center
MS 39529-5004
USA

Samantha K. Breeding
MAR, Inc.
101 N.W. Interchange
Bay St. Louis, MS 39520
USA

Kevin B. Briggs
NOARL Code 361
Stennis Space Center
MS 39529-5004
USA

Peter D. Bromirski
University of Hawaii at Manoa
2525 Correa Road
Honolulu, HI 96822
USA

Gary H. Brooke
Defence Research Establishment Pacific
FMO, Victoria, British Columbia
VOS 1BO
Canada

Burlie A. Brunson
Planning Systems Inc.
7925 Westpark Drive
McLean, Virginia 22102
USA

Paul J. Bucca
NOARL Code 222
Stennis Space Center
MS 39529-5004
USA

Andrea Caiti
SACLANT Undersea Research Centre
Viale San Bartolomeo, 400
La Spezia 19026
Italy

Robert Carter Jr.
US Naval Oceanographic Office
Stennis Space Center
MS 39522-5001
USA

Jacques R. Chamuel
Sonoquest Adv. Ultrasonics Research
P.O. Box 153
Wellesley Hills, MA 02181-5339
USA

N.R. Chapman
Defence Research Establishment Pacific
FMO, Victoria, British Columbia
VOS 1BO
Canada

David M.F. Chapman
Defence Research Establishment Atlantic
9 Grove Street, P.O. Box 1012
Dartmouth, Nova Scotia B2Y 3Z7
Canada

C.H. Cheng
M.I.T./ERL, E34-440
42 Carleton St.
Cambridge, MA 02139
USA

Stanley A. Chin-Bing
NOARL Code 221
Stennis Space Center
MS 39529-5004
USA

Michael D. Collins
Naval Research Laboratory
4555 Overlook Avenue
Washington DC 20375-5000
USA

P. Cristini
Laboratoire de Mécanique et
d'Acoustique
31, Chemin Joseph Aiguier
13009 Marseille
France

Angela M. Davis
School of Ocean Sciences
University College of North Wales
Wales
UK

Ger de Lange
Fugro-McClelland Engineers B.V.
P.O. Box 250
2260 AG Leidschendam
The Netherlands

O. Diachok
Naval Research Laboratory, Code 5120
4555 Overlook Avenue
Washington DC 20375-5000
USA

Leroy M. Dorman
Scripps Institution of Oceanography
Geological Research Division
La Jolla CA 92093-0215
USA

Peter Gerstoft
Ødegaard & Danneskiold-Samsøe ApS
Kroghsgade 1
2100 Copenhagen
Denmark

Joseph F. Gettrust
NOARL Code 362
Stennis Space Center
MS 39529-5004
USA

Stewart A.L. Glegg
Florida Atlantic University
P.O. Box 3091
Boca Raton, FL 33431-0991
USA

D. Goldberg
Lamont-Doherty Geological Observatory
Palisades, New York 10964
USA

Louis A. Hargenrader
Syscon Corporation
1000 Thomas Jefferson Street
Washington DC 20007
USA

Robert B. Herrmann
Saint Louis University
3507 Laclede Avenue
St. Louis, Missouri 63103
USA

Charles W. Holland
Planning Systems Inc.
7925 Westpark Drive
McLean, Virginia 22102
USA

Rune M. Holt
IKU Sintef Group Research Institute
N-7034 Trondheim
Norway

Jens M. Hovem
SACLANT Undersea Reseach Centre
Viale San Bartolomeo, 400
La Spezia 19026
Italy

Steven J. Hughes
Defence Research Establishment Atlantic
P.O. Box 1012
Dartmouth, Nova Scotia B2Y 3Z7
Canada

Michael T. Hurley
BP Research, Sunbury Research Centre
Chertsey Road
Sunbury-on-Thames,
Middlesex TW16 7LN
UK

D.G. Huws
School of Ocean Sciences
Menai Bridge
Wales LL57 5EY
UK

Finn B. Jensen
SACLANT Undersea Research Centre
Viale San Bartolomeo, 400
La Spezia 19026
Italy

Neil Jones
Department of Ocean Sciences
University of Cambridge
Cambridge CB3 OEZ
UK

S.E. Jones
School of Ocean Sciences
Menai Bridge
Gwynedd LL59 5EY
UK

Reidar Kanestrøm
Norsk Hydro Bergen Research Centre
P.O. Box 4313
N-5028 Bergen
Norway

Ilkka Karafalo
F.O.A., Box 27322
10254 Stockholm
Sweden

Age Kristensen
SACLANT Undersea Research Centre
Viale San Bartolomeo, 400
La Spezia 19026
Italy

William A. Kuperman
Naval Research Laboratory
Code 5166
Washington DC 20375
USA

Dawn L. Lavoie
NOARL Code 361
Stennis Space Center
MS 39529-5004
USA

Gregory T. Leger
Ocean routes Canada, Inc.
271 Brownlow Avenue
Dartmouth, Nova Scotia B3B 1W6
Canada

Klaus C. Leurer
Institut für Geophysik
Universität Kiel
Kiel
Germany

B.T.R. Lewis
School of Oceanography
University of Washington, WB-10
Seattle, WA, 98195
USA

Colin MacBeth
Edinburgh Anisotropy Project
British Geological Survey
Edinburgh EH9 3LA
UK

Murli H. Manghnani
School of Ocean and Earth Science
and Technology, University of Hawaii
Honolulu, HI 96822
USA

Rudi G. Markl
US Naval Oceanographic Office
Code OAG
Stennis Space Center
MS 39522-5001
USA

Betty Mayfield
Avenue de Catalunya 120
Cerveira
Lleida
Spain

Ian R. McDermott
School of Ocean Sciences
Menai Bridge
Gwynedd LL57 5EY
UK

T.M. McGee
Dept. of Exploration Geophysics
University of Utrecht
P.O. Box 80.021
3508 TA Utrecht
The Netherlands

Rolf Meissner
Institut für Geophysik, Univ. Kiel
Olshausenstr. 40
2300 Kiel
Germany

Roger Merrifield
NAVOCEANO
Stennis Space Center
MS 39529 5004
USA

Jacques Meunier
IFREMER DITI/SOM
B.P. 70, 29280 Plouzané
Brest
France

Halcyon E. Morris
NOARL Code 114
Stennis Space Center
MS 39529-5004
USA

Gerald Morris
NOARL
Stennis Space Center
MS 39529-5004
USA

Thomas G. Muir
Applied Research Laboratories
P.O. Box 8029
Austin, TX 78713
USA

Siv Jorum Nakken
STATOIL
Postuttak
7004 Trondheim
Norway

Amos M. Nur
Stanford University
Department of Geophysics
Stanford, California 94305
USA

John S. Papadakis
Inst. of Applied & Computat. Mathematics
P.O. Box 1527 711 10
Heraklio, Crete
Greece

A.M. Raaen
IKU, N-7034
Trondheim
Norway

G. Rabau
Laboratoire de Mécanique et d'Acoustique
31, Chemin Joseph Aiguier
13009 Marseille Cedex
France

Subramaniam D. Rajan
Woods Hole Oceanographic Institution
Blgd. 302
Woods Hole, MA 02543
USA

P.N.J. Rasolofosaon
Institut Français du Pétrole
1-4 Avenue des Bois-Préau, BP 311
92506 Rueil-Malmaison Cedex
France

Dieter Rauch
USEA S.p.A.
Via delle Pianazze 74
19027 Termo di La Spezia
Italy

Michael D. Richardson
NOARL Code 361
Stennis Space Center
MS 39529-5004
USA

Jose M. Roesset
Department of Civil Engineering
The University of Texas
Austin, Texas 78712-1076
USA

Henrik Schmidt
Department of Ocean Engineering
Massachusetts Institute of Technology
Cambridge, MA 02139
USA

Anthony E. Schreiner
Scripps Institution of Oceanography 0215
UCSD
La Jolla, CA 92093-0215
USA

Peter J. Schultheiss
SCHULTHEISS Geotek
Fern Cottage, Marley Lane
Haselmere, Surrey GU27 3RF
UK

Michael Snoek
Clonganny House
Killmuckridge/Co. Wexford
Eire
Ireland

Robert A. Spence
R.A. Spence Engineering Inc.
7787 Hudson Street
Vancouver B.C. V6P 4L7
Canada

R.A. Stephen
SACLANT Undersea Research Centre
Viale San Bartolomeo, 400
La Spezia 19026
Italy

Kenneth H. Stokoe
Department of Civil Engineering
ECJ 9227
The University of Texas
Austin, Texas 78712-1076
USA

Robert D. Stoll
Lamont-Doherty Geological Observatory
of Columbia University
Palisades, New York 10964
USA

M. Taroudakis
Inst. of Applied & Comput.
Mathematics
P.O. Box 1527 711 10
Heraklio, Crete
Greece

Denzil Taylor Smith
School of Sciences
University College of North Wales
Bangor LL59 5EY
UK

F. Theilen
Institut für Geophysik, Univ. Kiel
Olshausenstr. 40
D-2300 Kiel
Germany

Altan Turgut
RSMAS, University of Miami
4600 Rickenbacker Causeway
Miami, FL 33149
USA

J.H.N. van den Berg
Ministry of Defence
P.O. Box 20701
NL-2500 ES The Hague
The Netherlands

R.B. Whitmarsh
Institute of Oceanographic Sciences
Brook Road, Wormley
Godalming, Surrey GU8 5UB
UK

Richard D. Woods
2630 G.G. Brown Laboratory
University of Michigan
Ann Arbor, Michigan 48109-2125
USA

Tokuo Yamamoto
RSMAS, University of Miami
4600 Rickenbacker Causeway
Miami, Florida 33149
USA

## AUTHOR INDEX

## SUBJECT INDEX

**r**



**s**



**t**



**u**



**v**



**w**